Quantum and Statistical Field Theory

Quantum and Statistical Field Theory

MICHEL LE BELLAC
Department of Physics
University of Nice

Translated by
G. Barton

Clarendon Press • Oxford
1991

Oxford University Press, Walton Street, Oxford OX2 6DP
Oxford New York Toronto
Delhi Bombay Calcutta Madras Karachi
Petaling Jaya Singapore Hong Kong Tokyo
Nairobi Dar es Salaam Cape Town
Melbourne Auckland
and associated companies in
Berlin Ibadan

Oxford is a trade mark of Oxford University Press

Published in the United States
by Oxford University Press, New York

A catalogue record for this book is available from the British Library

Library of Congress Cataloging-in-Publication Data
Le Bellac, Michel.
[*Phenomenes critiques aux champs de jauge. English*]
Quantum and statistical field theory/Michel Le Bellac;
translated by G. Barton.
Translation of: Des phenomenes critiques aux champs de jauge.
Includes bibliographical references and index.
1. Quantum field theory. 2. Statistical mechanics. I. Title.
QC174.45.L4 1991 530.1'4—dc20 91-12653
ISBN 0-19-853929-0
ISBN 0-19-853964-9 (pbk.)

Typeset by Macmillan India Ltd
Printed in Great Britain by Bookcraft Ltd, Midsomer Norton

Aidé par le Ministère français chargé de la culture

Preface

In 1965, at the end of their classic text which became the bible of a whole generation of particle physicists, Bjorken and Drell (1965, Chapter 19) warned pessimistically 'Therefore, conclusions based on the renormalization group arguments . . . are dangerous and must be viewed with due caution. So is it with all conclusions from local relativistic field theories'. In the same year, Dyson (1965) one of the founding fathers of quantum electrodynamics, wrote 'It is easy to imagine that in a few years the concepts of field theory will drop totally out of the vocabulary of day-to-day work in high-energy physics'. We know that in fact the evolution of physics did not bear out these predictions; instead, modern elementary-particle physics is inextricably bound up with non-Abelian gauge theories, which are direct generalizations of the quantum electrodynamics developed in the aftermath of the Second World War by Schwinger, Feynman, Dyson, and Tomonaga.

Since quantum field theory had after all been invented to describe the creation and annihilation of particles, its reconquest of this domain after an eclipse lasting roughly a decade was not altogether a surprise. Much more astonishing was the impact of field theory, in the early seventies, on our understanding of critical phenomena. Of course, field-theory methods (Green's functions, Feynman diagrams, etc.) had been borrowed by solid-state and by nuclear physicists in the context of the 'many-body problem' long before; but this was largely a matter of technicalities rather than of basically new ideas (except for superconductivity and superfluidity, where the fundamental notion of broken symmetry was introduced for the first time). By contrast, the irruption into statistical physics of the concept of renormalization was truly revolutionary, because there had been no premonition that the removal of infinities, the original bane of field theory, would have a part to play in a domain without infinities to begin with. It was Wilson's achievement to realize that, at large distances, the correlation functions of a system near a critical point were described by a renormalized theory. Moreover, this fundamental insight turned out to shed new light on renormalization itself.

It seemed useful to write an introductory text that would put the developments over the last twenty years within the reach of students embarking on research, and also of physicists in other areas wishing to be initiated into field theory. The overall plan differs significantly from the traditional pattern, which is of books wholly devoted either to statistical or to high-energy physics. The approach we adopt takes one straight to the heart of field theory, namely to renormalization and to the renormalization group in the context of the simplest possible model; this is the φ^4 or Ginzburg–Landau model, which in spite of its simplicity has basic applications to the physics of critical phenomena.

The book is divided into four parts. The first is an introduction to critical phenomena and to Wilson's version of the renormalization group. Applications like the ε expansion and the XY model are treated in detail. The second part discusses the perturbation expansion, still in the context of statistical physics. Chapter 5 introduces the basic techniques (generating functionals and Feynman diagrams). This chapter is rather technical and somewhat tedious, but is essential to the two following chapters which tackle the fundamental concepts of the theory, namely renormalization and the renormalization group in the version using the Callan–Symanzik equations.

The third part abandons the Euclidean space of statistical physics for the Minkowski space of relativistic quantum theory. The essentials of the theory are already in place, because the Green's functions of the quantum theory are nothing but analytic continuations of the correlation functions of the Euclidean theory. This is explained in Chapter 8 for the case of ordinary quantum mechanics. While functional integrals are used systematically up to this point, Chapter 9 digresses briefly to describe canonical quantization, and the following chapter collects the irreducible minimum of results for application to particle physics, in a hypothetical world where all particles have spin zero. Finally, part four introduces gauge theories. Chapter 11 describes the quantization of the Dirac and of the electromagnetic fields. Chapter 12 studies quantum electrodynamics in some detail, and Chapter 13 is an introduction to non-Abelian gauge fields. The book ends with a brief look at lattice gauge theories, which provide the occasion for a synthesis of the central concepts introduced in the course of the earlier discussions. Readers interested only in quantum field theory could start with Chapter 5, skip Section 3 of Chapter 7, and likewise skip any later passages devoted specifically to critical phenomena.

Since the book is meant to be introductory rather than exhaustive, complicated general proofs are occasionally replaced by illustrative examples. Similarly, most calculations are carried through in some detail; it is hoped that this will not obscure the underlying physics. Finally, several important topics have been omitted altogether in order to keep the book to a reasonable length; amongst these are the exact solution of the two-dimensional Ising model; the scaling laws for the equation of state; invariance properties (under the Lorentz or under discrete symmetry groups); operator-product expansions; anomalies; geometrical approaches to gauge theories; and so on. The reader can find excellent discussions of these topics in the books and papers listed in the references. As a general rule priority is given to *methods*, with relatively little detail about the *physical systems* used to illustrate them.

To tackle the first part one needs only the basic notions of statistical mechanics. In fact, disregarding a few passages of fairly heavy algebra, the gist of the first three chapters could be taught to final-year undergraduates to supplement a course on that subject. Part II uses no advanced ideas at all, though it does require some quite lengthy calculations. But a knowledge of quantum mechanics and special relativity at advanced undergraduate/introductory postgraduate level is indispensable for Parts III and IV. Here one would also benefit from an elementary acquaintance with functions of a complex variable, with group theory, and with elementary-particle physics. Although the book is meant as an introduction to the field, readers may find its different parts of unequal difficulty. Chapters 1, 2, 3, 8, and 9 are probably rather easy, while Chapters 7 and 13, and some sections of Chapters 10 and 12, are more advanced.

The book comes with 130 problems of varying degrees of difficulty. Some are simple applications of material in the text; others can serve to open the way to small research topics. There has been an attempt (not invariably successful) to avoid the pattern 'derive equation (36)', through more detailed statements, through hints, and by giving partial solutions. In some cases a reference will enable the reader to check or to complete the solution.

The general references indicated with an asterisk in the list at the end of this work are a selection of books and review articles. They include those that have proved most useful in preparing the present text, plus those that the reader is likely to find most accessible. A complete bibliography is of course out of the question; in time-hallowed words, I address my apologies to colleagues whose work has not been properly cited.

This book stems from a *cours de troisième cycle*, taught in various forms in the DEA de Physique Théorique (Marseille–Nice), the DEA de Physique de la Matière Condensée (Nice), and in the Magistère of Constantine. I am most grateful to the students, whose comments have been invaluable. Victor Alessandrini has read the entire manuscript and has made me the beneficiary of many very pertinent observations, especially about the general organization of the book. I have profited also from criticism and suggestions by E. Brezin, J. P. Provost, J. L. Meunier, and F. Guérin. P. de Giovanni has commented most usefully both on the text and on the problems. I am very grateful to Michèle Leduc, who has done much to bring the book into being. I have sorely tried the patience of Chantal Djankoff, who has typed the many versions of the manuscript with her customary competence and efficiency, and I thank her warmly. Finally, Joanna has shared with me my doubts and uncertainties while this book was taking shape; without her constant support, it could not have been written.

Nice M. Le Bellac
December 1986

Preface to the English edition

There have been a number of modifications to the French text. I have reorganized and largely rewritten Chapters 7 and 10, and entirely rewritten the end of Chapter 3 from Section 3.5 on. A new appendix on Ward identities has been added, some problems have been modified or expanded and a few have been added.

After publication of the French edition, there have appeared three remarkable books, written at a somewhat more advanced level than the present one: the first by Parisi (1988); the second by Itzykson and Drouffe (1989); and the last by Zinn-Justin (1989). They have been very useful in the preparation of this edition, and I would like to thank Claude Itzykson and Jean Zinn-Justin, who provided me with their manuscripts prior to publication.

I am grateful to Dr Gabriel Barton for the high quality of his translation; this English version owes much to his style. I would like also to thank him for his useful remarks on the text and for his patience in agreeing to include my numerous modifications in his work. I am indebted to Jean-Pierre Provost, who suggested the new approach in Section 3.6.1, and to Pierre Méry and Michel Perrottet, who pointed out some misprints in the French version. Finally I am grateful to Jocelyne Bettini for her excellent work on the manuscript.

Nice M. Le Bellac
July 1991

Notation and conventions

As a rule we respect standard notation, which inevitably leads to some clashes; for instance S represents both the action and the S-matrix, β both the inverse temperature and the Callan–Symanzik function, and so on. The context should prevent confusion. Paragraphs in small print are either digressions or technicalities, and can be skipped at a first reading.

Throughout the book we observe the convention that repeated suffixes are summed over:

$$a_i b_i = \sum_i a_i b_i.$$

The notation in Parts III and IV is generally that of Itzykson and Zuber (1980). In particular we use the Minkowski metric

$$x^2 = x_0^2 - \boldsymbol{x}^2, \quad \text{or} \quad g^{\mu\nu} = \text{diag}(1, \ -1, \ -1, \ -1).$$

(Seeing how often we cross between Euclidean and Minkowski space, it might have been better to choose the metric $x^2 = \boldsymbol{x}^2 - x_0^2$, but the habits of twenty years are not so easily broken.) The Dirac matrices are defined by

$$\{\gamma^\mu, \gamma^\nu\} = \gamma^\mu\gamma^\nu + \gamma^\nu\gamma^\mu = 2g^{\mu\nu}.$$

We do depart from the notation of Itzykson and Zuber in the following respects:

- the propagators (Δ_{F}, S_{F}, . . .) differ by a factor i;
- the Dirac spinors are normalized through $\bar{u}u = 2m$;
- the generators of Lie algebras are chosen to be Hermitean.

Contents

Part I Critical phenomena

1 Introduction to critical phenomena 3

1.1 The ferromagnetic transition 3
1.2 The Ising model 5
1.3 The mean field 12
1.4 Correlation functions 19
1.5 Qualitative description of critical phenomena 27
Problems 31
Further reading 38

2 Landau theory 39

2.1 The Ginzburg–Landau Hamiltonian and the Landau approximation 40
2.2 The Landau theory of phase transitions 48
2.3 Correlation functions 50
2.4 Critique of the Landau approximation, and the Ginzburg criterion 52
Problems 59
Further reading 66

3 The renormalization group 67

3.1 Basic concepts: blocks of spins, critical surface, and fixed points 69
3.2 Behaviour near a fixed point. Critical exponents 77
3.3 The Ising model on a triangular lattice, and the approximation by cumulants 86
3.4 The Gaussian model 90
3.5 Calculation of the critical exponents to order ε 100
3.6 Marginal fields, the function $\beta(g)$, and logarithmic corrections in $4D$ 109
Problems 116
Further reading 125

4 Two-dimensional models 126

4.1 The XY model: qualitative study 127
4.2 Renormalization-group analysis 133
4.3 Nonlinear σ-models 140
Problems 145
Further reading 147

Part II Perturbation theory and renormalization: the Euclidean scalar field

5 The perturbation expansion and Feynman diagrams 151

5.1 Wick's theorem and the generating functional 151
5.2 The perturbation expansion of $G^{(2)}$ and of $G^{(4)}$: Feynman diagrams 157
5.3 Connected correlation functions and proper vertices 171
5.4 The effective potential: the loop expansion 179
5.5 The evaluation of Feynman diagrams 182
5.6 Power-counting: ultraviolet and infrared divergences 190
Problems 196
Further reading 201

6 Renormalization 202

6.1 Introduction 203
6.2 The renormalization of mass and coupling constant 206
6.3 Field renormalization: counterterms 209
6.4 The general case 214
6.5 Composite operators and their renormalization 220
6.6 The minimal subtraction scheme (MS) 225
Problems 229
Further reading 235

7 The Callan–Symanzik equations 236

7.1 The Callan–Symanzik equations at the critical temperature 238
7.2 The Callan–Symanzik equations for $T > T_c$ 250
7.3 Renormalization and the renormalization group 256
7.4 The renormalization group in $D = 4$ dimensions 264

7.5 The renormalization group in dimensions $D < 4$ 267
Problems 274
Further reading 279

Part III The quantum theory of scalar fields

8 Path integrals in quantum and in statistical mechanics 283
8.1 Quantum spin and Ising model 286
8.2 Particle in a potential 291
8.3 Euclidean continuation, and comments 303
Problems 308
Further reading 312

9 Quantization of the Klein–Gordon field 313
9.1 The quantization of elastic vibrations 315
9.2 Quantization of the Klein–Gordon field 323
9.3 Coupling to a classical source, and Wick's theorem 331
Problems 340
Further reading 345

10 Green's functions and the S matrix 346
10.1 Perturbation expansion of the Green's functions 347
10.2 Path integrals and the Euclidean theory 355
10.3 Cross-sections and the S-matrix 362
10.4 The unitarity of the S-matrix 377
10.5 Generalizations 387
Problems 394
Further reading 397

Part IV Gauge theories

11 Quantization of the Dirac field and of the electromagnetic field 401
11.1 Quantization of the Dirac field 402
11.2 Wick's theorem for fermions 414
11.3 The Lagrangean formalism for the classical electromagnetic field 419

11.4 Quantization of the electromagnetic field 423
Problems 431
Further reading 435

12 Quantum electrodynamics 436

12.1 The Feynman rules for quantum electrodynamics 437
12.2 Applications 445
12.3 One-loop diagrams in electrodynamics 451
12.4 Ward identities, unitarity, and renormalization 469
Problems 483
Further reading 493

13 Non-Abelian gauge theories 494

13.1 Non-Abelian gauge fields: the classical theory 495
13.2 The quantization of non-Abelian gauge theories 506
13.3 The Glashow–Salam–Weinberg model of electroweak interactions 514
13.4 Quantum chromodynamics 530
13.5 Lattice gauge theories 544
Problems 554
Further reading 562

Appendices

A Fourier transforms and Gaussian integration 565

A.1 Fourier transforms 565
A.2 Gaussian integrals 567
A.3 Integrals in *D* dimensions 569

B Feynman integrals with dimensional regularization (Euclidean case) 571

C Noether's theorem, conserved currents and Ward identities 573

C.1 Noether's theorem and conserved currents 573
C.2 Energy–momentum tensor 574
C.3 Ward identities 575

D Formulary 577

D.1 The Lorentz group 577
D.2 Dirac matrices 578
D.3 Cross-sections 580
D.4 Feynman rules 581

References 585

Index 589

I Critical Phenomena

1 Introduction to critical phenomena

This first chapter is an elementary introduction to the models and methods that will be used later. The ferromagnetic (Curie) transition serves as our prototype for all second-order phase transitions, and serves also to introduce the notions of broken symmetry and of order parameter. The basic model of ferromagnetism, namely the Ising model, is described in Section 1.2, and solved for an elementary case. The reason why this model is so important is that it has an exact solution in two dimensions, which exhibits spontaneous magnetization. Section 1.3 is devoted to a first approach to mean-field methods, which remain amongst the most frequently exploited approximations even today. They will lead us to define a first set of critical exponents; the listing of such exponents will then be completed in Section 1.4, once correlation functions have been defined. Finally, in Section 1.5 a qualitative description of critical phenomena will enable us, in a first approximation, to define the concept of scale invariance and its relation to the behaviour of the theory under dilatations.

1.1 The ferromagnetic transition

The first part of this book aims to study second-order phase transitions ('transitions of the second kind'), also called critical phenomena for reasons to be explained in Section 1.5. To be specific, we shall restrict ourselves to the ferromagnetic–paramagnetic transition; this is the most familiar example of a second-order phase transition, and its general characteristics can easily be transposed to all the others. Ferromagnetism is a very complex phenomenon and we shall settle for just a sketchy description, referring for the 'solid-state physics' aspects to the book by Kittel (1986). Certain materials (iron, nickel, cobalt, etc.) can be magnetized at room temperature. Microscopically speaking this means that electrons in an incomplete inner shell have their spins effectively aligned in one and the same direction. Since there is a magnetic moment associated with every spin, such alignment implies that all these magnetic moments add, thus producing a macroscopic magnet.

When a ferromagnet is heated above a certain temperature T_c of the order of 10^3 K, called the Curie temperature, its magnetization disappears, and the material becomes paramagnetic. It is easy to imagine, intuitively, that the tendency of the spins to align is due to an interaction between them that favours such a configuration. At high temperatures thermal agitation tends to destroy this configuration, thus causing the magnetization to vanish. This explanation does contain some but by no means all of the truth; indeed it has been one of the basic problems of statistical physics to prove that a phase transition really does occur. To be precise, the discussion which follows applies not to an arbitrary ferromagnetic sample, but to a single 'domain' (of linear dimensions $\sim 10^{-2}$ mm) in such a sample. (Again we refer to Kittel for an explanation of just what a domain is.) At temperatures above T_c the domain is not magnetized (nor is the sample); when the temperature drops below T_c the domain is magnetized, while the total magnetization of the sample, which consists of many different domains, may very well be zero. The magnetization $\mathscr{M}$ of the domain grows as the temperature falls, whence it peaks at $T = 0$. All the spins are then aligned in the same direction. For $0 < T < T_c$ the spins still have a tendency to align in the same direction, but thermal agitation allows only a partial alignment (Fig. 1.1).

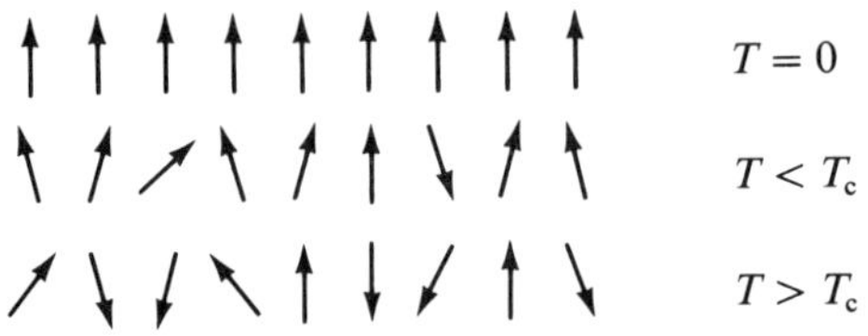

Fig. 1.1

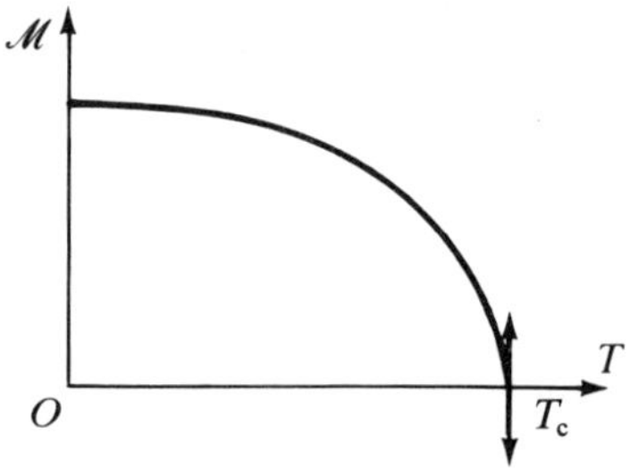

Fig. 1.2 Qualitative behaviour of the magnetization

The curve in Fig. 1.2 sketches $\mathscr{M}$ as a function of temperature, in the absence of an externally applied magnetic field: $\mathscr{M}$ is called the *spontaneous magnetization*, since it is not due to any applied field. Note the vertical tangent at $T = T_c$.

That spontaneous magnetization should exist is, a priori, a remarkable phenomenon: the spin Hamiltonian is rotation-invariant, and there are no preferred directions in space. For $T > T_c$ one can identify no preferred spatial direction in the ferromagnetic sample either. By contrast, for $T < T_c$ there is a preferred direction, that of the magnetization. The state of a domain is no longer invariant under all rotations, but only under rotations around axes parallel to the magnetization direction. In other words the symmetry group of the low-temperature phase is (only) a subgroup of that of the high-temperature phase; it is called the isotropy group (or little group) of the privileged direction. This phenomenon is known as *spontaneous symmetry breaking*. The magnetization, zero in the high-temperature phase and non-zero in the low-temperature phase, is called the *order parameter* of the transition. The phenomenon of symmetry breaking and the notion of an order parameter are encountered in almost all second-order phase transitions.

One might well be surprised that a preferred direction should exist even though there is nothing to identify it a priori. But in fact the smallest inhomogeneity or the least residual magnetic field, or *B*-field for short, allow one to define a direction. (Think of a vertical cylindrical rod subject to a vertical compressive force $\boldsymbol{F}$ along its axis (Fig. 1.3): under large-enough $\boldsymbol{F}$ the rod eventually buckles, even though the situation is, a priori, perfectly invariant under rotations around the axis. Here too one is dealing with a broken symmetry.)

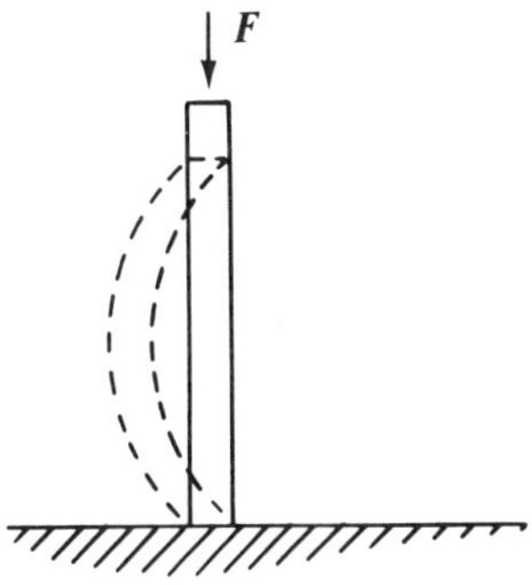

Fig. 1.3 Broken symmetry

1.2 The Ising model

1.2.1 Description of the model

Following a strategy familiar to physicists, we will try to construct a *model* for ferromagnetism, simplifying the actual situation while attempting nevertheless to preserve those of its features that seem to be fundamental. Constructing a model amounts to looking for the best compromise between two contradictory requirements, namely

(a) to find equations simple enough to be solved, analytically if possible or failing that on a computer;

(b) not to lose in this process of simplification any of the essential features of the physics one wishes to study.

What seems essential to ferromagnetism is the interaction between spins, tending to align them. It seems reasonable to assume that one may replace each atom of the ferromagnet by an electron to be held responsible for the ferromagnetism. Accordingly, the first step of our model-building describes the ferromagnet by electrons placed at the sites of the underlying crystal lattice. Though it is not essential, we shall generally simplify the discussion by considering a cubic lattice.

The spin–spin interaction has short range; between two spins separated by ten lattice spacings it may be neglected*. The second step consists in writing down an interaction between nearest-neighbour spins only; on a two-dimensional (i.e. plane, $D = 2$) lattice every spin has four nearest neighbours, while on a three-dimensional ($D = 3$) lattice it has six. (See Fig. 1.4; in the rest of this book D will always stand for the dimensionality of space. A plane has dimension $D = 2$; ordinary space has $D = 3$.) Again we may hope that nothing essential has been lost by this approximation.

The simplest Hamiltonian with a tendency to align the spins is

$$H = -J \sum_{\langle i,j \rangle} \boldsymbol{\sigma}_i \cdot \boldsymbol{\sigma}_j, \tag{1.2.1}$$

where J, the *coupling constant*, is a positive constant, and the $\boldsymbol{\sigma}_i$ are Pauli

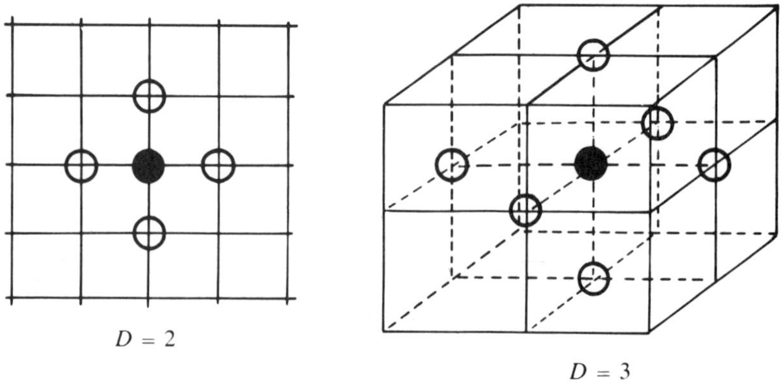

Fig. 1.4 ○ = Nearest neighbours of spin ●

* Remember that this spin–spin interaction has its origin in exchange forces, a combination of Coulomb repulsion and Pauli principle. The magnetic dipole–dipole interaction is much too weak to explain ferromagnetism, although it plays a role in the formation of domains.

matrices; the notation $\sum_{\langle i,j\rangle}$ indicates summation over nearest neighbours. The Hamiltonian (1.2.1) defines the *quantum Heisenberg model.* One can show however that quantum effects are unimportant in the immediate neighbourhood of T_c, unless $T_c = 0$: in the neighbourhood of the critical point, quantum fluctuations are overwhelmed by statistical fluctuations. Hence we may replace the Pauli matrices $\boldsymbol{\sigma}_i$ by classical vectors $\boldsymbol{S}_i$ of length 1; this defines the *classical Heisenberg model.*

The resultant model is still too complicated: no analytic solutions are known (except for $D = 1$; Chapter 4 will investigate it for $D = 2$ by the renormalization-group method), and computer studies demand much nicety. That is why one introduces another, supplementary, approximation (the last!): the vectors $\boldsymbol{S}_i$ are replaced by scalars S_i that can take only two values:

$$S_i = +1 \quad \text{or} \quad S_i = -1.$$

Thus the spins are always parallel to a fixed axis, and along this axis there are two possible orientations (spin up: $S_i = +1$, and spin down: $S_i = -1$).

When Lenz suggested this model to his student Ising as a thesis subject (in 1920), there existed absolutely no methods for assessing its approximations. Today we know that *qualitatively* the Ising model is a good model of ferromagnetism, but that *quantitatively* some of its predictions are poor. The point is that in the Heisenberg model the magnetization is a vector, and one needs three numbers to specify it (e.g. its magnitude plus two angles): one says that *the dimensionality n of the order parameter* is three: $n = 3$. But in the Ising model the direction of magnetization is fixed, and a single (algebraic) number suffices to specify it: the dimensionality of the order parameter is $n = 1$*.

Now the critical exponents (defined in Section 1.3) depend on n, which entails quantitative differences between the two models. Worse, in two dimensions ($D = 2$) the difference becomes qualitative: the Ising model exhibits spontaneous magnetization, while the Heisenberg model does not (Chapter 4).

The discussion above allows us to write the Hamiltonian H of the Ising model as

$$H = -J \sum_{\langle i,j\rangle} S_i S_j, \quad S_i = \pm 1, \tag{1.2.2}$$

and thence (in terms of the temperature T and of Boltzmann's constant k) the partition function Z as

$$Z = \sum_{[S_i]} e^{\frac{J}{kT}\sum_{\langle i,j\rangle} S_i S_j}. \tag{1.2.3}$$

* For simplicity we restrict attention to scalar or vectorial order parameters. For a more general definition see Mermin (1979).

Here the first sum runs over all configurations, so that

$$\sum_{[S_i]} = \sum_{S_1=\pm 1} \sum_{S_2=\pm 1} \cdots \sum_{S_N=\pm 1} \tag{1.2.4}$$

if there are N sites in the lattice. The number of terms in the partition function is 2^N.

1.2.2 The Ising model in one dimension

In order to familiarize oneself with the model it is worth considering the very simple case of a linear lattice, i.e. N spins on a straight line. This is just the case $D = 1$ (Fig. 1.5). Let us evaluate the partition function Z from the Hamiltonian H:

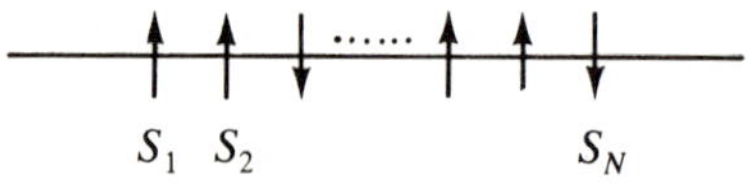

Fig. 1.5

$$H = -J \sum_{l=1}^{N-1} S_l S_{l+1},$$

$$Z = \sum_{[S_l]} e^{-H/kT} = \sum_{[S_l]} e^{\frac{J}{kT}\sum_{l=1}^{N-1} S_l S_{l+1}} = \sum_{[S_l]} \prod_{l=1}^{N-1} e^{KS_l S_{l+1}},$$

where $K = J/kT$.

By virtue of the identity

$$e^{KS_l S_{l+1}} = \cosh K + S_l S_{l+1} \sinh K$$

we can rewrite Z as

$$Z = (\cosh K)^{N-1} \sum_{[S_l]} \prod_{l=1}^{N-1} (1 + S_l S_{l+1} \tanh K). \tag{1.2.5}$$

One method very common in statistical mechanics is the use of a *high-temperature expansion*: as $T \to \infty$ one tries to expand the partition function as a series in powers of some parameter $\kappa(T)$ such that $\kappa(T) \to 0$ as $T \to \infty$. In our present case $\tanh K$ is such a parameter: $\tanh K = \tanh(J/kT) \to 0$ as $T \to \infty$.

Let us therefore try to expand in powers of $\tanh K$:

$$\prod_l (1 + S_l S_{l+1} \tanh K) = 1 + \tanh K \sum (SS) + (\tanh K)^2 \sum (SSSS) + \cdots. \tag{1.2.6}$$

With every term of this expansion one can associate a graph; to the term

$$(\tanh K)^6 (S_2 S_3)(S_4 S_5)(S_5 S_6)$$

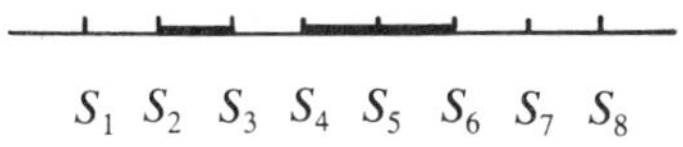

Fig. 1.6

for instance there corresponds the graph in Fig. 1.6, where heavy lines link the nearest-neighbour pairs featuring in the expansion. Within the sum over configurations let us isolate say the sum over S_4:

$$\sum_{[S_i]} = \sum_{S_4 = \pm 1} \left[\sum \cdots \sum (\tanh K)^6 (\ )(\ )(\ ) \right].$$

When S_4 changes sign, the term within square brackets changes sign, so that all the terms in the sum over configurations cancel in pairs. The sum fails to vanish only if an even number (0 or 2) of links originate from each spin. Only the first term in (1.2.6) gives a non-zero result (i.e. the high-temperature expansion reduces to just one term), whence

$$Z = 2^N (\cosh K)^{N-1}. \tag{1.2.7}$$

The partition function yields access to all the thermodynamic functions; in particular it allows one to determine whether or not there is a phase transition. However, one must note that for a *finite* system the partition function is a finite sum of analytic functions of the temperature (provided $T \neq 0$); moreover, all the terms in the sum are positive, so that the free energy $-kT \ln Z$ is also analytic. On the other hand a phase transition corresponds to a singularity of the thermodynamic functions, and before deciding whether there is such a transition one must take the *thermodynamic limit* $N \to \infty$. In effect it is impossible, *mathematically*, to observe a phase transition in a finite system.

As an example we evaluate the free energy per spin, $\hat{F}$, in the thermodynamic limit:

$$\left.\begin{aligned} \hat{F} &= \lim_{N \to \infty} \frac{1}{N} F = \lim_{N \to \infty} \left(-\frac{kT}{N} \ln Z \right), \\ \hat{F} &= -kT \ln \left(2 \cosh \frac{J}{kT} \right). \end{aligned}\right\} \tag{1.2.8}$$

$\hat{F}$ is an analytic function of T (except at $T = 0$), whence *the Ising model has no phase transition in one dimension.* This result has been generalized by Peierls: *in the absence of long-range interactions, no one-dimensional system can display a phase transition.*

1.2.3 The correlation function of the one-dimensional Ising model

We proceed to calculate the *correlation function* of two spins S_i and S_j; it is defined as the average value $\langle S_i S_j \rangle$ of the product $S_i S_j$*. The correlation function is a measure of the influence exerted by a given spin, say S_i, whose direction is fixed; it is easy to show that the conditional probability for having $S_j = +1$, given that $S_i = +1$, is $\frac{1}{2}(1 + \langle S_i S_j \rangle)$. Since the interaction favours the alignment of spins, a nearby spin S_j will tend to assume the same orientation as S_i; however, thermal agitation counteracts this tendency and exerts a de-correlating effect. Qualitatively speaking one expects some correlation that weakens as the distance between S_i and S_j increases; at a fixed distance apart the correlation will be stronger when the temperature is lower. (Only if there is no phase transition does this argument apply at all temperatures. Why?)

The value of $\langle S_i S_j \rangle$ is calculated as a standard statistical average:

$$\langle S_i S_j \rangle = \frac{1}{Z} \sum_{[S_l]} S_i S_j \, \mathrm{e}^{-H/kT}$$
$$= \frac{1}{Z} (\cosh K)^{N-1} \sum_{[S_l]} S_i S_j \prod_{l=1}^{N-1} (1 + S_l S_{l+1} \tanh K).$$

With the factors S_i and S_j outside the product we associate one additional linkage, and once again the only non-zero term is that in which an even number of links originate from every spin (Fig. 1.7). The end-result for $\langle S_i S_j \rangle$ reads

$$\left.\begin{aligned} \langle S_i S_j \rangle &= \frac{1}{Z} (\cosh K)^{N-1} 2^N (\tanh K)^{|i-j|} = (\tanh K)^{|i-j|}, \\ \langle S_i S_j \rangle &= \mathrm{e}^{-|i-j|\,|\ln \tanh K|} = \mathrm{e}^{-|i-j|\,|\ln \tanh (J/kT)|}. \end{aligned}\right\} \qquad (1.2.9)$$

The correlation function decreases exponentially with the distance $|i - j|$ (Fig. 1.8). With a the lattice spacing, the distance between spins S_i and S_j in cm is $a|i - j| = r_{ij}$; the *correlation length* ξ is defined by

$$\langle S_i S_j \rangle = \mathrm{e}^{-r_{ij}/\xi}, \qquad (1.2.10)$$

and for the one-dimensional Ising model equation (1.2.9) yields

$$\xi = \frac{a}{|\ln \tanh J/kT|}. \qquad (1.2.11)$$

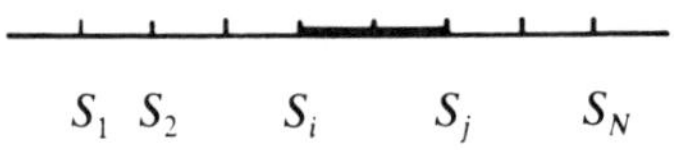

Fig. 1.7

* In the general case one must subtract $\langle S_i \rangle \langle S_j \rangle$: see Section 1.4.

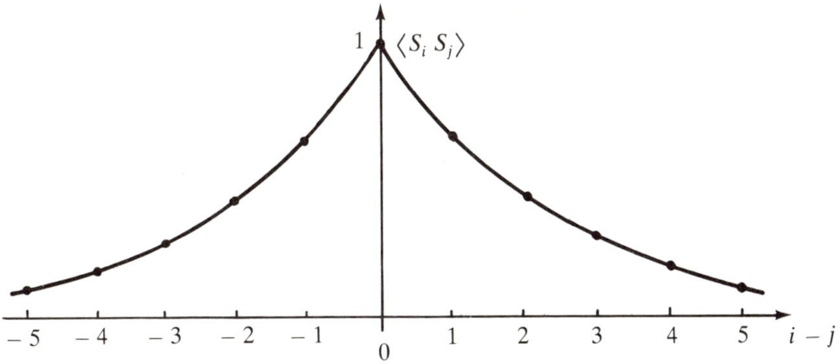

Fig. 1.8 $\langle S_i S_j \rangle$ as a function of $(i - j)$

This expression shows that the correlation length decreases with rising temperature; it tends to zero as $T \to \infty$, and to infinity as $T \to 0$, thus confirming the intuitive argument given earlier. Note also that (1.2.9) confirms the absence of spontaneous magnetization:

$$\lim_{|i-j| \to \infty} \langle S_i S_j \rangle = \langle S_i \rangle \langle S_j \rangle = \langle S \rangle^2 = 0.$$

1.2.4 The two-dimensional Ising model

For two dimensions ($D = 2$) Onsager in 1944 calculated the partition function and its thermodynamic limit exactly. This 'solution of the Ising model' is a veritable mathematical *tour de force*, and even with certain simplifications introduced later the calculation is too long to reproduce here. In presence of an applied B-field there is no exact solution of the two-dimensional Ising model to date; nor is there in three dimensions, with or without an applied field.

Onsager's solution proves the existence of a ferromagnetic transition in two dimensions. The transition temperature T_c is given by

$$\sinh 2K_c = \sinh(2J/kT_c) = 1,$$

or in other words by

$$kT_c = \frac{2J}{\ln(1 + \sqrt{2})} \simeq 2.27\,J. \tag{1.2.12}$$

Near the transition temperature the specific heat diverges like $\ln|T - T_c|$, and the mean value $M_0 = \langle S \rangle$, defined for one spin as

$$M_0 = \lim_{B \to 0^+} \lim_{N \to \infty} \left[\frac{1}{N} \sum_i \langle S_i \rangle \right], \tag{1.2.13}$$

is given by

$$M_0 = [1 - \sinh(2J/kT)^{-4}]^{1/8}.$$

The crucial point is that, near T_c, M_0 (and therefore $\mathscr{M}$) behaves like $(T_c - T)^{1/8}$:

$$M_0 \sim (T_c - T)^{1/8}. \tag{1.2.14}$$

One should note the order of the limits in (1.2.13). In zero external field, a spontaneous magnetization can exist only *after* the thermodynamic (or infinite volume) limit has been taken.

1.3 The mean field

1.3.1 The mean-field equation

Since the solution of the Ising model is complicated or out of reach, one needs approximation methods. The basic one, proposed by Weiss in 1907, is the mean-field (or molecular-field) method. It is not at all peculiar to the Ising model, and continues to be employed in current work; when studying a model for a phase transition, very often one's first reaction is to try the mean-field approximation. We give below the simplest version of mean-field theory. A more sophisticated and more powerful approach is described in Problem 1.5.

The approximation is based on the following idea: consider one particular spin S_i, and assume that its energy E_i can be calculated by replacing all the other spins by their average value $\langle S_j \rangle$. This leads one back to a classic problem of paramagnetism. It proves convenient to subject the spin system to an externally applied B-field. A classical spin is a vector $\boldsymbol{S}$ with an associated magnetic moment $\boldsymbol{\mu} = \mu \boldsymbol{S}$. In the field $\boldsymbol{B}$ the energy of such a spin is $-\boldsymbol{\mu}\cdot\boldsymbol{B} = -\mu\boldsymbol{S}\cdot\boldsymbol{B}$. In the Ising model $\boldsymbol{B}$ has a fixed direction (that of the spins), and the energy can be written simply as $-\mu SB$. In a B-field the Hamiltonian becomes

$$H = -J \sum_{\langle i,j \rangle} S_i S_j - \mu B \sum_i S_i,$$

and in the mean-field approximation the energy of the spin S_i is given by

$$E_i = -JS_i \sum_j \langle S_j \rangle - \mu B S_i.$$

Depending on whether the spin is up $(+1)$ or down (-1), we find the energy levels E_{i+} and E_{i-},

$$E_{i+} = -J \sum_j \langle S_j \rangle - \mu B = -qJM - \mu B,$$

$$E_{i-} = J \sum_j \langle S_j \rangle + \mu B = qJM + \mu B.$$

Here M is the average value of S_j: $M = \langle S_j \rangle$; and q is the number of nearest neighbours. The classical calculation for paramagnetism yields the mean value of S_i:

$$\langle S_i \rangle = \tanh\left(\frac{qJM + \mu B}{kT}\right).$$

Since all spins are on the same footing, $\langle S_j \rangle$ too must equal M. This is just the condition for the approximation to be self-consistent; it yields the equation

$$M = \tanh\left(\frac{qJM + \mu B}{kT}\right),$$

which we shall find convenient to rewrite as

$$\tanh^{-1} M\left(= \frac{1}{2}\ln\frac{1+M}{1-M}\right) = \frac{qJ}{kT}M + \frac{\mu B}{kT}. \qquad (1.3.1)$$

This is the *basic equation of the mean-field approximation*, and the rest of the present section consists in deriving its consequences.

1.3.2 The ferromagnetic transition in mean-field theory

Equation (1.3.1) is transcendental and must be solved numerically. We can form a qualitative idea of the solutions by proceeding graphically: the solutions (there may be one or several) are given by the intersections of the straight line $(qJ/kT)M + \mu B/kT$ with the curve $\tanh^{-1} M$ (Fig. 1.9). It is useful to recall that the curve $\tanh^{-1} M$ has two vertical asymptotes, at $M = \pm 1$, and that its tangent at the origin has unit slope.

The graphs in Fig. 1.9 show that for $B > 0$ there can be three solutions; but solutions having $M < 0$ are metastable or unstable (Problem 2.2). The physically acceptable solution corresponds to a magnetization in the same direction as the field.

Since the slope of the tangent of $\tanh^{-1} M$ at the origin is 1, we see that as $B \to 0^+$ the solution tends to a finite positive value $M_0 \neq 0$ if qJ/kT is greater than 1, and to $M_0 = 0$ if qJ/kT is smaller than 1. (Starting with a negative field $B < 0$ we would, given $qJ/kT > 1$, obtain $-M_0$.) Accordingly, the mean-field approximation predicts a spontaneous magnetization $\pm M_0$. In passing we note the broken-symmetry phenomenon: the two spin orientations are equivalent, but at low temperatures the spontaneous magnetization chooses one or the other. An infinitesimally small change in the B-field suffices to produce $-M_0$ instead of $+M_0$. Under the addition of an infinitesimal B-field the solution $M_0 = 0$ is unstable, provided only that $qJ/kT > 1$.

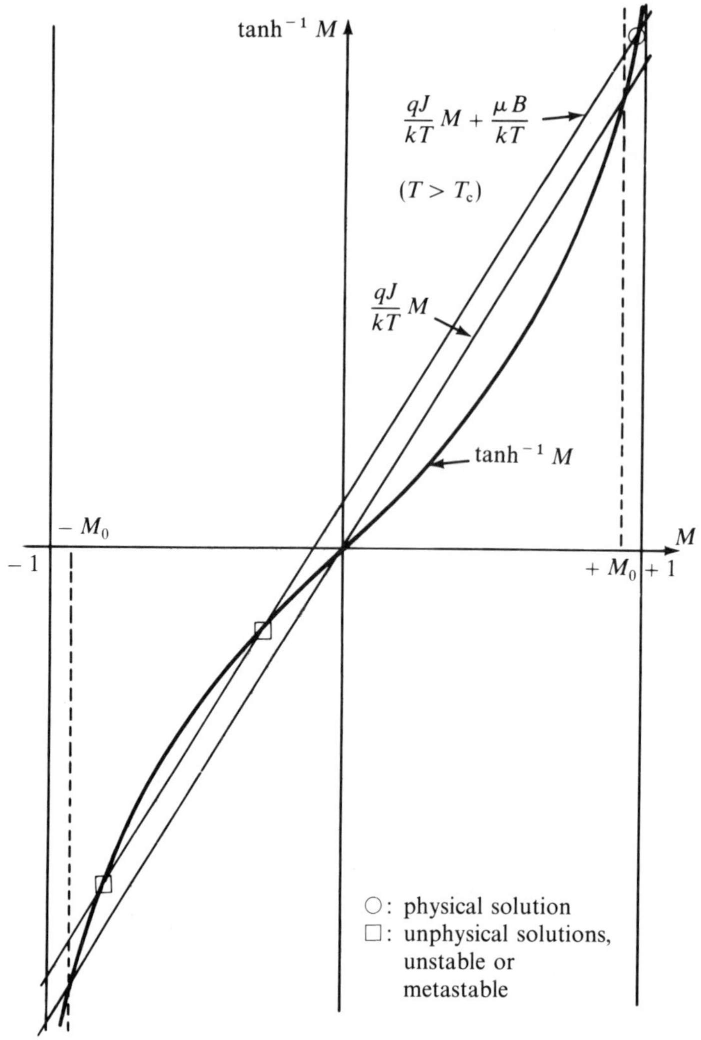

Fig. 1.9 Graphical solution of equation (1.3.1)

We can summarize by saying that in zero B-field the mean-field approximation predicts

(a) a non-zero spontaneous magnetization if $T < T_c = qJ/k$;

(b) zero spontaneous magnetization if $T > T_c$.

Thus the transition temperature is

$$T_c = qJ/k. \tag{1.3.2}$$

1.3.3 Behaviour close to the phase transition

Later in this section we shall solve equation (1.3.1), approximately, near T_c for small B. Under these conditions the magnetization is weak ($M \ll 1$) and we can use the series expansion

$$\tanh^{-1} M = M + \frac{1}{3} M^3 + O(M^5). \tag{1.3.3}$$

The magnetization in zero field

Define the 'reduced temperature' t,

$$t = \frac{T - T_c}{T_c}; \quad \frac{T}{qJ} = \frac{T}{T_c} = 1 + t.$$

The mean-field equation becomes

$$M \simeq (1 + t)\left(M + \frac{1}{3} M^3\right),$$

whence

$$M_0 \simeq \sqrt{-3t}.$$

Near T_c the spontaneous magnetization therefore varies like $(T_c - T)^{1/2}$:

$$M_0 \sim (T_c - T)^{1/2}. \tag{1.3.4}$$

The graph giving M_0 for arbitrary T can be constructed by solving (1.3.1) numerically (Fig. 1.10).

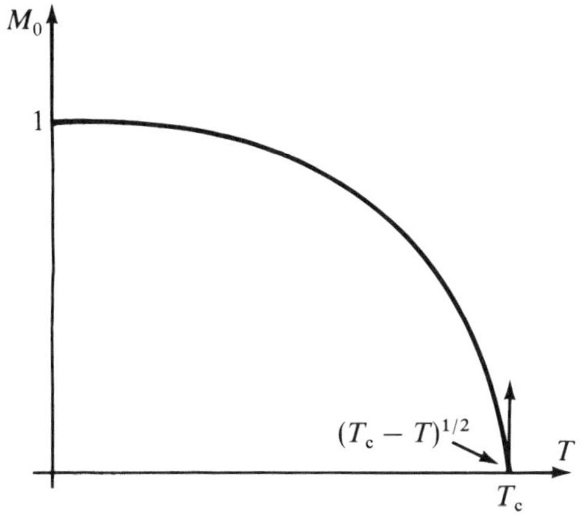

Fig. 1.10 The magnetization as a function of T

The susceptibility in zero field

(a) $T > T_c$

For T close but not equal to T_c and for $B \to 0$, the term with M^3 in (1.3.3) can be neglected:

$$M \simeq (1 + t)M - \frac{\mu B}{kT_c},$$

whence

$$M \simeq \frac{\mu B}{k(T - T_c)}.$$

The total magnetization $\mathscr{M}$ equals $N\mu M = \mu^2 NB/(T - T_c)$, and the magnetic susceptibility in zero field is

$$\chi = \left.\frac{\partial \mathscr{M}}{\partial B}\right|_{B=0} = \frac{\mu^2 N}{k(T - T_c)}.$$

Thus the susceptibility varies like $(T - T_c)^{-1}$:

$$\chi \sim (T - T_c)^{-1}. \tag{1.3.5a}$$

(b) $T < T_c$

In this case one must take account of the spontaneous magnetization M_0, with $M_0^2 = -3t$. We write

$$M = M_0 + \varepsilon.$$

With the higher-order terms neglected, equation (1.3.1) becomes

$$M_0 + \varepsilon = (1 + t)(M_0 + \varepsilon) + \frac{1}{3}(M_0 + \varepsilon)^3 - \frac{\mu B}{kT_c}.$$

Since ε is of order B (as $B \to 0$), terms in ε^2 and ε^3 are negligible, and we find

$$\varepsilon = \frac{-\mu B}{2k(T - T_c)}, \quad \chi = \frac{N\mu^2}{2k(T_c - T)};$$

therefore once again

$$\chi \sim (T_c - T)^{-1}. \tag{1.3.5b}$$

We have the same power law in (1.3.5a) and (1.3.5b), but the numerical coefficients differ by a factor of 2.

The critical isotherm

Set $T = T_c$ and calculate B as a function of $\mathscr{M}$:

$$M = M + \frac{1}{3}M^3 - \frac{\mu B}{kT_c} \Rightarrow B = \frac{kT_c}{3\mu}M^3,$$

whence

$$B = \frac{kT_c}{3\mu}(\mathscr{M}/N\mu)^3.$$

In this case the power law we find reads

$$B \sim \mathscr{M}^3. \tag{1.3.6}$$

The specific heat in zero field

In the mean-field approximation, the internal energy (i.e. the expectation value of the Hamiltonian) for zero field can be found at once:

$$E = -\frac{1}{2}qJNM_0^2 \qquad T < T_c,$$

$$E = 0 \qquad T > T_c.$$

One need merely note that every spin is replaced by its average value M_0, and that there are $qN/2$ pairs of spins. For $T < T_c$ but still close to T_c, we exploit the value already found for M_0:

$$E = -\frac{1}{2}qJN\frac{3(T_c - T)}{T_c} = \frac{3}{2}kN(T - T_c).$$

The specific heat C in zero field is given by the derivative of E with respect to T,

$$C = \left.\frac{\mathrm{d}E}{\mathrm{d}T}\right|_{B=0} = \frac{3}{2}kN. \tag{1.3.7}$$

Since C evidently vanishes for $T > T_c$, *the specific heat is discontinuous*, with a discontinuity $3kN/2$ at $T = T_c$. C too can be calculated for arbitrary T by solving equation (1.3.1) numerically (Fig. 1.11).

1.3.4 The critical exponents α, β, γ, δ

It is found experimentally that the spontaneous magnetization, the susceptibility, the critical isotherm, and the specific heat all obey power laws near $T = T_c$, and one defines the *critical exponents** α, β, γ, δ (also called *critical indices*)

$$C \sim |T - T_c|^{-\alpha} \tag{1.3.8a}$$

$$M_0 \sim (T_c - T)^{\beta} \quad (T < T_c) \tag{1.3.8b}$$

* On occasion people have introduced exponents α', γ' for $T < T_c$; for instance

$$\chi \sim (T - T_c)^{-\gamma}: T > T_c; \quad \chi \sim (T_c - T)^{-\gamma'}: T < T_c.$$

But it seems (and theory confirms) that one always has $\alpha = \alpha'$, $\gamma = \gamma'$, and these primed exponents have fallen into disuse.

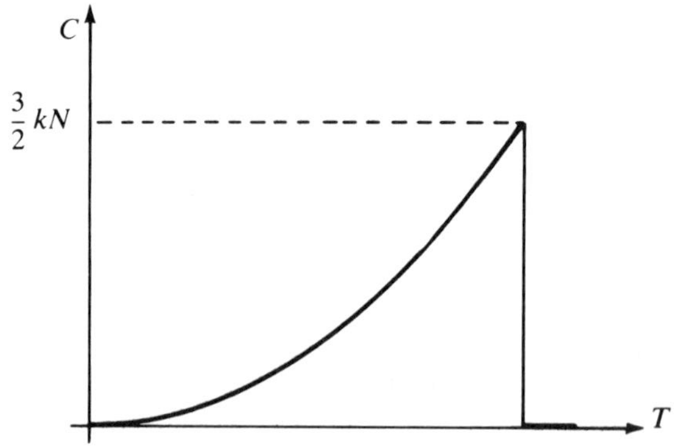

Fig. 1.11 The specific heat as a function of T

$$\chi \sim |T - T_c|^{-\gamma} \tag{1.3.8c}$$

$$B \sim \mathscr{M}^{\delta} \qquad (T = T_c). \tag{1.3.8d}$$

The values predicted by the mean-field approximation are $\alpha = 0$ (for the discontinuity in C); $\beta = \frac{1}{2}$; $\gamma = 1$; $\delta = 3$.

Let us now consider the mean-field approximation more closely. It disregards all fluctuations of the spins, since every spin is replaced by its average. Given nothing more than the formulation explained above, we have no means of assessing this approximation in advance; all we can do is to assess it a posteriori by comparing its output with the exact results ($D = 1$ and $D = 2$), or with numerical calculations ($D = 3$).

For $D = 1$ the mean-field approximation is an unmitigated disaster: it predicts a transition in a system where we have just proved that there is none. (It is worth noting that *in the mean-field approximation the critical exponents are independent of D*).

The situation is somewhat better for $D = 2$ where we know that a transition does exist. However, the mean-field approximation gives the transition temperature

$$kT_c = 4J \quad (q = 4),$$

while Onsager's solution gives

$$kT_c = 2.27\,J.$$

Generally speaking, the mean-field approximation favours the occurrence of a phase transition. For $D = 1$ it predicts one where there is none; for $D = 2$ it predicts a transition temperature higher than the true value. The reason is that the fluctuations (which the mean-field approximation neglects) tend to inhibit the transition.

It is interesting to compare the mean-field exponents with those of the analytic calculation ($D = 2$) or those obtained from a numerical computation using a high temperature expansion ($D = 3$):

		$D = 2$	$D = 3$
α	discont.	$\ln\|T - T_c\|$	0.110 ± 0.005
β	0.5	0.125	0.312 ± 0.003
γ	1	1.75	1.238 ± 0.002
δ	3	15*	$5.0 \quad \pm 0.05$

One can see from this table that the mean-field results improve as the dimensionality D rises. Intuitively it seems plausible that to disregard fluctuations is less of a risk when the number of nearest neighbours is high ($q = 6$ for $D = 3$, while $q = 2$ for $D = 1$). In fact, however, this reasoning is not altogether correct, because it is the dimensionality of the space that matters: we shall see later that merely increasing the number of interactions leaves the critical exponents unchanged. Indeed the mean-field approximation becomes exact as $D \to \infty$ (the critical exponents are given correctly as soon as $D > 4$: see Chapter 3). It becomes exact also in the case of very long-range interactions: one can show for instance that the exact solution coincides with the mean-field result if every spin couples to all the other spins on the lattice with a coupling constant J/N (see Problem 1.3). The number of spins coupled to a given spin is then so large that it does become legitimate to replace S_i by its average value M.

1.4 Correlation functions

1.4.1 Definition and generating function

The correlation function for two spins has been introduced already, in Section 1.2. Since this construct will play a crucial role through all that follows, we proceed to list several definitions and useful properties. The present section uses an elementary example to introduce some techniques that will be developed more systematically in Chapter 5.

Section 1.2 defined the correlation function G_{ij} of two spins as the expectation value $\langle S_i S_j \rangle$. This definition is satisfactory when $\langle S_i \rangle = 0$, or in other words when $T > T_c$ and $B = 0$. When $\langle S_i \rangle \neq 0$, the assertion that two spins are uncorrelated means that $\langle S_i S_j \rangle = \langle S_i \rangle \langle S_j \rangle = M^2$. Thus it is logical to formulate the definition of G_{ij} in the general case as

$$G_{ij} = \langle S_i S_j \rangle - \langle S_i \rangle \langle S_j \rangle. \tag{1.4.1}$$

* Actually, the result $\delta = 15$ is very plausible (in view of the scaling laws), but it has not been established analytically. Numerical calculations yield $\delta = 15.04 \pm 0.07$.

If we assume that G_{ij} falls exponentially ($G_{ij} \sim \exp(-r_{ij}/\xi)$), then we find the behaviour shown qualitatively in Fig. 1.12 for the two cases $T > T_c$ and $T < T_c$ ($\langle S_i^2 \rangle = \langle 1 \rangle = 1$) respectively.

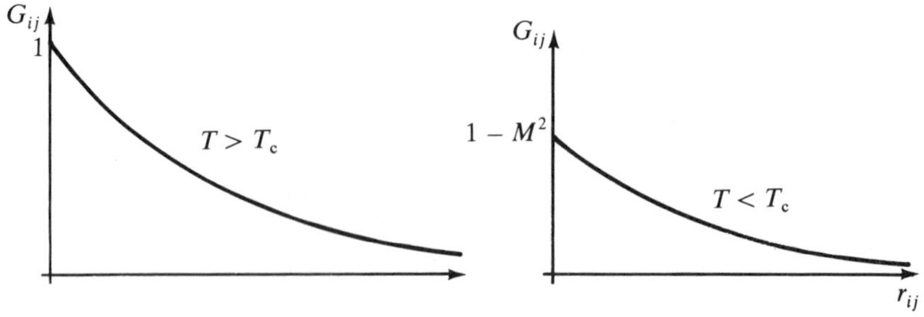

Fig. 1.12

It proves useful to relate G_{ij} to a second derivative of the partition function $Z[B_i]$ in a site-dependent, nonuniform field B_i,

$$Z[B_i] = \sum_{[S_k]} \exp\left(-\beta\left(H_0 - \mu \sum_k B_k S_k\right)\right), \tag{1.4.2}$$

where $\beta = 1/kT$* and H_0 is the Hamiltonian (1.2.2). The expectation value $\langle S_i \rangle$ is given by

$$\langle S_i \rangle = \frac{1}{Z} \sum_{[S_k]} S_i \exp\left(-\beta\left(H_0 - \mu \sum_k B_k S_k\right)\right),$$

whence

$$\langle S_i \rangle = \frac{1}{\beta\mu Z} \frac{\partial Z}{\partial B_i} = \frac{1}{\beta\mu} \frac{\partial \ln Z}{\partial B_i}. \tag{1.4.3}$$

Continue by differentiating once more, with respect to B_j:

$$\langle S_i S_j \rangle = \frac{1}{Z} \sum_{[S_k]} S_i S_j \exp\left(-\beta\left[H_0 - \mu \sum_k B_k S_k\right]\right),$$

whence

$$\langle S_i S_j \rangle = \frac{1}{(\beta\mu)^2} \frac{1}{Z} \frac{\partial^2 Z}{\partial B_i \partial B_j}.$$

* Not to be confused with the critical exponent β. One cannot respect all the traditional symbols without some (small) risk of ambiguities. This risk has been judged a lesser evil than redesigning the notation.

This entails

$$G_{ij} = \frac{1}{(\beta\mu)^2 Z}\frac{\partial^2 Z}{\partial B_i \partial B_j} - \left(\frac{1}{\beta\mu Z}\frac{\partial Z}{\partial B_i}\right)\left(\frac{1}{\beta\mu Z}\frac{\partial Z}{\partial B_j}\right),$$

whence

$$G_{ij} = \frac{1}{(\beta\mu)^2}\frac{\partial^2 \ln Z}{\partial B_i \partial B_j}. \tag{1.4.4}$$

The mean value $\langle S_i \rangle$ and the correlation G_{ij} are found by differentiating the partition function; the process could be extended to correlation functions for several spins (as we shall do in Chapter 5). Because of this, $Z[B_i]$ is called the *generating function* (*or generator*) *of the correlation functions*. Note further that differentiation of Z yields $\langle S_i S_j \rangle$, while differentiation of $\ln Z$ yields G_{ij} directly; the latter is called the *connected correlation function*. More generally (see Chapter 5) the logarithm $\ln Z$ of Z is the *generator of the connected correlation functions*. Moreover, even if there is no physical applied B-field, it can prove useful to introduce a fictitious one in order to calculate the correlation functions from (1.4.4); we merely set $B = 0$ in the end-result:

$$G_{ij}|_{B=0} = \frac{1}{(\beta\mu)^2}\frac{\partial^2 \ln Z}{\partial B_i \partial B_j}\bigg|_{B=0}.$$

One very important property of the correlation function is that (up to a factor $\beta\mu$) it represents *the response of the spin S_i to a variation of the B-field at the site j.* In order to show this, we need merely evaluate $\partial\langle S_i \rangle/\partial B_j$ from (1.4.3):

$$\frac{\partial \langle S_i \rangle}{\partial B_j} = \frac{1}{\beta\mu}\frac{\partial^2 \ln Z}{\partial B_i \partial B_j} = \beta\mu G_{ij}. \tag{1.4.5}$$

Notice that it is indeed the *connected* correlation function that is featured in (1.4.5). This leads us to make an important observation on the state at $T < T_c$. On physical grounds, we expect that the response of a spin S_i to the variation of the B-field at site j should vanish at large values of $\|\boldsymbol{r}_i - \boldsymbol{r}_j\|$; this will indeed be the case if the *connected* correlation function G_{ij} tends to zero when $\|\boldsymbol{r}_i - \boldsymbol{r}_j\| \to \infty$. One then says that the corresponding state obeys the *clustering property*. We have obtained the state at $T < T_c$ in the thermodynamic limit in the presence of an infinitesimal, spatially homogeneous, magnetic field: see (1.2.13); the magnetization takes then one of two possible values $+M_0$ or $-M_0$ (for simplicity we restrict our discussion to an order parameter with dimension $n = 1$). Such states are called *pure states*. With other external conditions, we could obtain a state whose magnetization would not be equal to $+M_0$ or $-M_0$: for example, in zero external field the magnetization would be zero, from symmetry arguments. Such states are called *mixed states*. There is a rigorous theorem which states that pure states are in one-to-one correspondence with clustering states; thus, in a pure state, the connected correlation function vanishes at large distances, while this is not the case for mixed states.

One can also obtain a pure state by taking a very large lattice and assigning a well-defined orientation to the spins on the boundary. If for example all spins on the boundary are directed upward, one can show that the magnetization M_0 is obtained in the limit of an infinite lattice.

1.4.2 The fluctuation-response theorem

Let $\mathcal{M}$ denote the magnetization (i.e. the total magnetic moment):

$$\mathcal{M} = \mu \sum_i \langle S_i \rangle = \mu \langle \mathcal{S} \rangle,$$

where $\mathcal{S} = \sum_i S_i$ is the total spin. The derivative $\partial \mathcal{M} / \partial B_j$ reads

$$\frac{\partial \mathcal{M}}{\partial B_j} = \mu \sum_i \frac{\partial \langle S_i \rangle}{\partial B_j} = \beta \mu^2 \sum_i G_{ij}.$$

For a macroscopic system ($N \to \infty$), or under periodic boundary conditions, G_{ij} in a uniform B-field is translation-invariant (Fig. 1.13). Since all the B_j are equal to B, $\partial \mathcal{M} / \partial B$ is given by

$$\frac{\partial \mathcal{M}}{\partial B} = \sum_j \frac{\partial \mathcal{M}}{\partial B_j} \frac{\partial B_j}{\partial B} = \beta \mu^2 \sum_{i,j} G_{ij}.$$

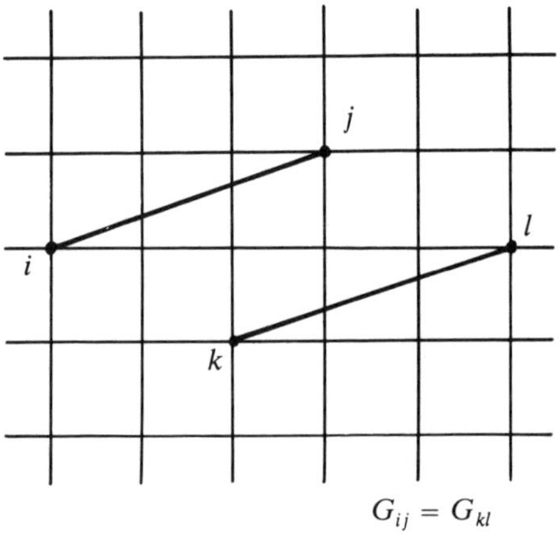

Fig. 1.13

One establishes a 'fluctuation-response theorem' by relating the susceptibility χ (which measures the response to an external field) to the correlation function (which measures the spin fluctuations):

$$\chi = \beta \mu^2 \sum_{ij} G_{ij} = \beta \mu^2 (\langle \mathcal{S}^2 \rangle - \langle \mathcal{S} \rangle^2),$$

where the second equality shows that for a spin system χ is always positive*.

When the correlation length is much smaller than the size of the system, translation invariance allows us to write

$$\chi \approx N\beta\mu^2 \sum_j G_{ij}, \tag{1.4.6}$$

where the sum over j is in fact independent of i.

1.4.3 Measurement of the correlation function

The correlation function is an interesting theoretical tool, but most of its importance stems from the fact that it can be measured experimentally. The experiment consists in scattering slow neutrons from a ferromagnet. The magnetic moments of the neutrons interact with those of the electrons, and the scattering amplitude of a neutron by an electron at the site i is proportional to the spin S_i. Let A be the amplitude for scattering by an electron at the origin; the rules of quantum mechanics (see for instance Messiah 1961, Chapter XIX) then entail that the scattering amplitude A_i from a spin having the same orientation but situated at $\boldsymbol{r}_i$ on site i is

$$A_i = A\mathrm{e}^{\mathrm{i}(\boldsymbol{k}-\boldsymbol{k}')\cdot\boldsymbol{r}_i} = A\mathrm{e}^{\mathrm{i}\boldsymbol{q}\cdot\boldsymbol{r}_i},$$

where $\boldsymbol{k}$ is the incident and $\boldsymbol{k}'$ the scattered wave-vector. Moreover, since the scattering amplitude is proportional to the spin,

$$A_i \sim S_i\mathrm{e}^{\mathrm{i}\boldsymbol{q}\boldsymbol{r}_i},$$

statistical averaging makes the cross-section proportional to

$$\sigma = \left\langle \left| \sum_i S_i \mathrm{e}^{\mathrm{i}\boldsymbol{q}\cdot\boldsymbol{r}_i} \right|^2 \right\rangle = \sum_{i,j} \langle S_i S_j \rangle \mathrm{e}^{\mathrm{i}\boldsymbol{q}\cdot(\boldsymbol{r}_i-\boldsymbol{r}_j)}.$$

By appeal to translation-invariance, one notes that the term $\langle S_i\rangle\langle S_j\rangle$ of G_{ij} contributes only for $\boldsymbol{q}=0$, and one finds (for $\boldsymbol{q}\neq 0$) that the cross-section is proportional to the Fourier transform $\tilde{G}(\boldsymbol{q})$ of the correlation function:

$$\sigma \sim N\sum_j G_{ij}\mathrm{e}^{\mathrm{i}\boldsymbol{q}\cdot(\boldsymbol{r}_i-\boldsymbol{r}_j)} = N\tilde{G}(\boldsymbol{q}). \tag{1.4.7}$$

(The sum over j in (1.4.7) is independent of the site i.) Notice also that the fluctuation-response theorem may be written

$$\chi \sim N\tilde{G}(\boldsymbol{q}=0). \tag{1.4.8}$$

* Of course there exist also materials having negative susceptibility (namely diamagnetics), but there the mechanism responsible for the magnetization is completely different.

1.4.4 The critical exponents η and ν

In the neighbourhood of $T = T_c$ experiment shows that for $q \ll 1/a$ (a = lattice spacing) the correlation function is well represented by an expression of the type

$$\tilde{G}(\boldsymbol{q}) = \frac{1}{q^{2-\eta}} f(q\xi) \quad (q \ll 1/a). \tag{1.4.9}$$

The quantity ξ here has the dimensions of length, and is called the correlation length. It will play a crucial role in the following chapters, because it relates the length scale to the (reduced) temperature. It diverges according to a power law as $T \to T_c$:

$$\xi \sim |T - T_c|^{-\nu} \sim |t|^{-\nu}. \tag{1.4.10}$$

The form (1.4.9) of the correlation function will be made plausible in Section 1.5, and established in Chapter 3.

Equations (1.4.9) and (1.4.10) define the critical exponents η and ν. The function $f(x)$ tends to a finite limit as $x \to \infty$, so that at $T = T_c$ one has

$$\tilde{G}(\boldsymbol{q}) \sim q^{-2+\eta} \quad (T = T_c). \tag{1.4.11}$$

At this point we can return to ordinary space through an inverse Fourier transformation (see Appendix A):

$$G(\boldsymbol{r}) = \int \frac{d^D q}{(2\pi)^D} e^{-i\boldsymbol{q}\cdot\boldsymbol{r}} \frac{f(q\xi)}{q^{2-\eta}}.$$

A change of variable $\boldsymbol{q} = \boldsymbol{u}/\xi$ yields

$$G(\boldsymbol{r}) = \xi^{-D} \int \frac{d^D u}{(2\pi)^D} \frac{e^{-i\boldsymbol{u}\cdot(\boldsymbol{r}/\xi)} f(u)}{(q\xi)^{2-\eta} \xi^{\eta-2}}$$

$$= \frac{h(r/\xi)}{\xi^{D+\eta-2}} = \frac{h(r/\xi)}{r^{D+\eta-2}(\xi/r)^{D+\eta-2}} = \frac{g(r/\xi)}{r^{D+\eta-2}}.$$

Since the form of $\tilde{G}(\boldsymbol{q})$ has been assumed to be valid for $q \ll 1/a$, the result for $G(\boldsymbol{r})$ will be valid if $r \gg a$:

$$G(\boldsymbol{r}) = \frac{g(r/\xi)}{r^{D+\eta-2}} \quad (r \gg a). \tag{1.4.12}$$

As $r \to \infty$, the function $g(r/\xi)$ varies exponentially: $g(r/\xi) \sim \exp(-r/\xi)$.

At $T = T_c$ the correlation function decreases as a power of r. For $\eta < 2$ (which is the case in practice since one always finds that η is close to zero), the integral of $G(\boldsymbol{r})$ over $\boldsymbol{r}$ (i.e. its Fourier transform of argument $\boldsymbol{q} = 0$) diverges at $T = T_c$. In the light of (1.4.8) this entails that at $T = T_c$ the susceptibility diverges as well. The behaviour of the functions $G(r)$ and $\tilde{G}(q)$ is sketched in Fig. 1.14.

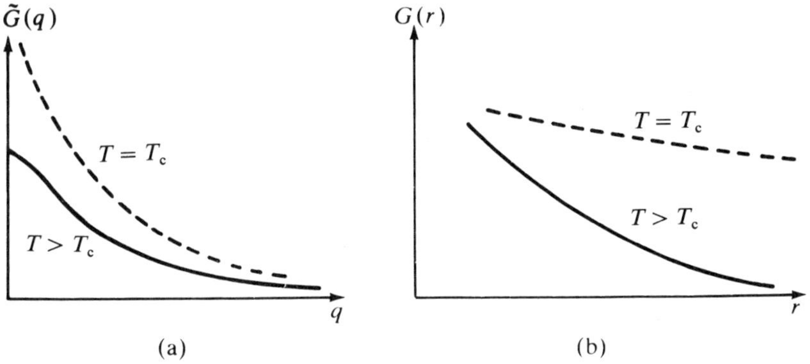

Fig. 1.14 Qualitative behaviour of the correlation function in configuration space (b), and in Fourier space (a).

A scaling law

When $T \neq T_c$, $\tilde{G}(0)$ is finite. In order to compensate for the divergent factor $q^{-2+\eta}$ it must therefore be the case that $f(q\xi) \sim (q\xi)^{2-\eta}$ as $q \to 0$. This entails

$$\tilde{G}(0) \sim \xi^{2-\eta} \sim |T - T_c|^{-\nu(2-\eta)}.$$

But $\tilde{G}(0) \sim \chi$ (equation (1.4.8)), while the exponent γ is defined by

$$\chi \sim |T - T_c|^{-\gamma}$$

(see (1.3.8c)). Comparison yields a relation between the critical exponents, also called a *scaling law*:

$$\gamma = \nu(2 - \eta). \tag{1.4.13}$$

Summary of critical exponents

It is useful to summarize the six critical exponents α, β, γ, δ, η, and ν, defined by equations (1.3.8), (1.4.9), and (1.4.10):

specific heat:	$C \sim \|T - T_c\|^{-\alpha}$
order parameter:	$M \sim (T_c - T)^{\beta} \quad (T < T_c)$
susceptibility:	$\xi \sim \|T - T_c\|^{-\gamma}$
critical isotherm:	$B \sim \mathcal{M}^{\delta} \quad (T = T_c)$
correlation function at $T = T_c$:	$\tilde{G}(\boldsymbol{q}) \sim q^{-2+\eta} \quad (T = T_c)$
correlation length:	$\xi \sim \|T - T_c\|^{-\nu}$.

1.4.5 Legendre transformations

We introduce one final tool, to be used in the next chapter for studying correlation functions by generalizing an idea familiar from thermodynamics, namely Legendre transformations (see Problem 1.6). In classical thermodynamics, the differential $\mathrm{d}F$ of the (Helmholtz) free energy is given by

$$\mathrm{d}F = -S\,\mathrm{d}T - \mathcal{M}\,\mathrm{d}B; \quad \mathcal{M} = -\left.\frac{\partial F}{\partial B}\right|_T.$$

The Gibbs free energy (thermodynamic potential) Γ is the Legendre transform of F, defined as

$$\Gamma = F + \mathcal{M}B,$$

whence

$$\mathrm{d}\Gamma = -S\,\mathrm{d}T + B\,\mathrm{d}\mathcal{M}, \quad B = \partial\Gamma/\partial\mathcal{M}|_T.$$

Since $-F$ is a convex function of B (from the fluctuation-response theorem), it is indeed possible to define Γ, which is in turn a convex function of $\mathcal{M}$ (Problem 1.6). We shall generalize this transformation to non-uniform fields B_i; to ease the notation we set $\beta = \mu = 1$, and ignore the T-dependence, which plays no role in the argument. Then the relation (1.4.4) becomes

$$G_{ij} = \frac{\partial^2 \ln Z}{\partial B_i \partial B_j} = \frac{\partial^2 W}{\partial B_i \partial B_j},$$

while the Gibbs free energy is defined by

$$\Gamma = \sum_i M_i B_i - W \quad \text{with} \quad M_i = \langle S_i \rangle = \left.\frac{\partial W}{\partial B_i}\right|_T.$$

Under these conditions

$$\mathrm{d}\Gamma = \sum_i B_i\,\mathrm{d}M_i,$$

which implies

$$B_i = \left.\frac{\partial \Gamma}{\partial M_i}\right|_T.$$

Now consider

$$\frac{\partial B_i}{\partial M_j} = \frac{\partial^2 \Gamma}{\partial M_i \partial M_j}.$$

This quantity is nothing but the (matrix) inverse of G_{ij}, which exists since G_{ij} is a positive matrix: a simple generalization of the fluctuation-response theorem shows that $\sum_{i,j} \lambda_i \lambda_j G_{ij} \geqslant 0$. Thus

$$\sum_j \frac{\partial^2 \Gamma}{\partial M_i \partial M_j} G_{jk} = \sum_j \frac{\partial B_i}{\partial M_j}\frac{\partial M_j}{\partial B_k} = \delta_{ik},$$

whence the end-result reads

$$G_{ij}^{-1} = \frac{\partial^2 \Gamma}{\partial M_i \partial M_j}. \tag{1.4.14}$$

1.5 Qualitative description of critical phenomena

Consider a ferromagnetic material in zero external field at a temperature $T > T_c$. One observes clusters of up-spins and clusters of down-spins, the linear dimensions of these clusters being of the order of the correlation length ξ. (This is true on average: the spin system fluctuates, and obviously one will find some clusters with dimensions greater and some with dimensions smaller than ξ.) Moreover, there may be islands of down-spins surrounded by up-spins, and so on. Very instructive numerical simulations can be found in Wilson (1979).

Let the temperature now decrease towards T_c. The average size of the clusters (and the correlation length) increases. At $T = T_c$ one finds clusters of all possible sizes. In the sea of up-spins one finds islands of down-spins, on which there are lakes of up-spins, and so on. At the transition temperature, *fluctuations extend over regions of all possible dimensions: there is no longer any scale of length*, a fact expressed by saying that at the critical point the physics is invariant under *scale transformations*. In geometrical terms, one says that a physical system at criticality displays a *fractal structure* (Mandelbrot 1982). The relation between the fractal dimension and the critical exponents has been worked out by Coniglio and collaborators (see e.g. Coniglio 1989).

The phase transition realizes a most remarkable scenario: having started with short-range interactions (between nearest neighbours), we encounter correlations of long range ($\sim \xi$), and at $T = T_c$ even of infinite range. These correlations invalidate all the classic perturbative expansions, since the latter apply only when the range of correlations is short (a few ångströms). It can happen for instance that, in implementing perturbation theory, there appear integrals like

$$\int_a^\xi \frac{d\lambda}{\lambda} \tag{1.5.1}$$

(see Chapter 5). Here λ is a fluctuation wavelength, obviously bounded below by the lattice spacing a (the Brillouin condition). This result shows that for the study of critical phenomena all wavelengths between a and ξ are equally important: the contribution to the integral from fluctuations with wavelengths between $\lambda = a$ and $\lambda = 2a$ is $\ln 2$; the contribution from wavelengths between $\lambda = 10^3 a$ and $\lambda = 2 \times 10^3 a$ is also $\ln 2$. In view of the fact that $\xi/a \to \infty$ as $T \to T_c$, perturbation theory is inapplicable, because the integral (1.5.1) diverges logarithmically at large wavelengths.

On the other hand, such complexity has its positive side. Since a phase transition is a large-scale cooperative phenomenon, one can hope that some of its properties will depend only on very general features (like the dimensionality D of space, the dimensionality n of the order parameter, and the symmetries of the local couplings), but not on details of the interactions. This aspect is called *universality**. For example, the Ising models on a square and on a triangular lattice have the same critical exponents, these being universal quantities. By contrast, the transition temperature depends on the details of the interaction, and is not universal.

Another property of the transition point is worth mentioning: the two-dimensional Ising model on a square lattice is evidently not invariant under rotations around an axis perpendicular to the plane of the lattice. Indeed if one measures the correlation length in a given direction $\hat{\boldsymbol{n}}$, specified by an angle α (Fig. 1.15), then it is found to depend on α; e.g.

$$\frac{\xi(\alpha = 0)}{\xi(\alpha = \pi/4)} = \sqrt{2}\,\frac{\ln\left(\dfrac{2v}{1 - v^2}\right)}{\ln\left(v\,\dfrac{1 + v}{1 - v}\right)},$$

where $v = \tanh(J/kT)$. At high temperatures this ratio is very different from 1. But as we approach T_c the ratio does tend to 1: the correlation length is then the same in all directions, and the system does become rotation-invariant.

v	$\xi(\alpha = 0)/\xi(\alpha = \pi/4)$	ξ along the axes (in units of a)
0.05	1.12	0.35
0.1	1.08	0.48
0.2	1.03	0.83
0.3	1.01	1.71
$\sqrt{2} - 1 = 0.414$	1.00	∞

As soon as the correlation length has grown to $2a$, departures from rotation-invariance shrink to less than 1%. This suggests that near T_c the role of the lattice is not fundamental, a fact that will allow us to use a continuum model later on.

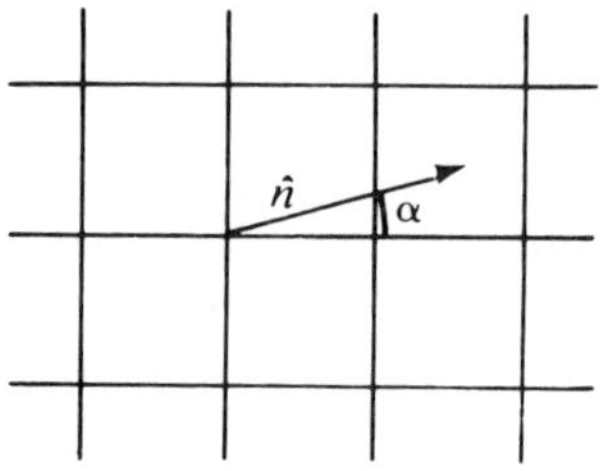

Fig. 1.15

* One exception is Baxter's '8-vertex model', whose critical exponents vary continuously.

The remarks above enable one to hazard a guess at the form of the correlation function. Isotropy near the critical temperature allows us to write

$$G(\mathbf{r}) = h(r, a),$$

where a stands for the set of microscopic parameters specifying the interaction (lattice spacing and coupling constants), written so that they have the dimensions of length*. For the ratio of two correlation functions with arguments r_1 and r_2 ($r_1, r_2 \gg a$) we find

$$\frac{G(r_2)}{G(r_1)} = \varphi\left(\frac{r_1}{r_2}, \frac{r_1}{a}\right), \tag{1.5.2}$$

where we have used the fact that φ is dimensionless. Scale-invariance at the critical point implies that, at $T = T_c$, φ cannot depend on a:

$$\frac{G(r_2)}{G(r_1)} = \varphi\left(\frac{r_1}{r_2}\right).$$

(Later chapters will need to define the ratio (1.5.2) in the limit $a \to 0$). This equation expresses scale-invariance at the critical point, and can be rewritten as

$$G(r/s) = \varphi(s)G(r). \tag{1.5.3}$$

It shows how the correlation function behaves under dilatations: s is the *dilatation factor*. On the other hand, the group composition law reads

$$G\left(\frac{r}{s_1 s_2}\right) = \varphi(s_1 s_2)G(r) = \varphi(s_1)\varphi(s_2)G(r).$$

This entails that $\varphi(s)$ is a power: $\varphi(s) = s^\lambda$, whence $G(r) = cr^{-\lambda}$. Writing $\lambda = D - 2 + \eta$, we can justify the power-law behaviour at the critical point:

$$G(r) \sim \frac{1}{r^{D-2+\eta}}.$$

When T is close but not equal to T_c, we assume that there exists just one characteristic length, namely the correlation length ξ. If so, then $G(r)$ must have the form

$$G(r) = \frac{g(r/\xi)}{r^{D-2+\eta}} \tag{1.5.4}$$

(see equation (1.4.12)); and $g(x) \to c$ as $x \to 0$ ($\xi \to \infty$).
Accordingly, if $a \ll r \ll \xi$, we get characteristically critical behaviour; which is what we want, because in this regime everything happens as if the correlation length were infinite.

Finally some comments on the term 'critical phenomena'. Standard media (like argon) have phase diagrams of the classic form shown in Fig. 1.16. At the

* If a parameter b has dimension L^α, we need merely adopt $b^{1/\alpha}$ instead.

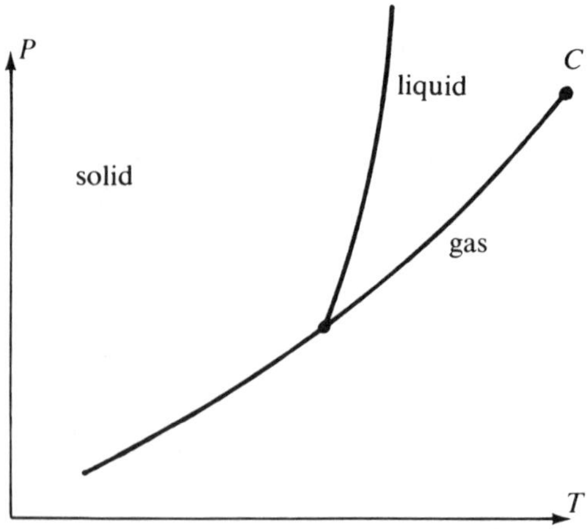

Fig. 1.16 Phase diagram

critical point C the difference between liquid and gas vanishes. (It is merely *quantitative*, while the difference between solid and liquid (or gas) is *qualitative*, since a solid possesses long-range order. Correspondingly there is no critical point on the curve separating solid from liquid.) As one approaches the critical point, density fluctuations become larger, and their correlation length tends to infinity. The scattering of light by these fluctuations becomes very marked when their linear dimensions become comparable with the wavelength of the light. The fluid then assumes a milky appearance (critical opalescence). This phenomenon is closely analogous to what happens at the ferromagnetic transition, with the analogies given by

fluctuating magnetization $\rightarrow$ fluctuating density,

order parameter M $\rightarrow$ $\rho_L - \rho_G$ ($\rho_{L(G)}$ = density of the liquid (gas)),

neutron scattering $\rightarrow$ light scattering.

Because of this analogy (which goes deep), second-order phase transitions are often discussed in the phraseology of critical phenomena; in particular the transition temperature T_c is often called the *critical temperature*. There is even a mean-field approximation to the properties of the critical point: it is no other than the Van der Waals equation (see for instance Reif 1965, Chapter 10).

Problems

1.1 Broken symmetry in mechanics

Consider a bead of mass m sliding without friction on a vertical ring. The bead is joined to the top of the ring by a spring of natural length $l_e = a$ and of stiffness constant C (Fig. 1.17).

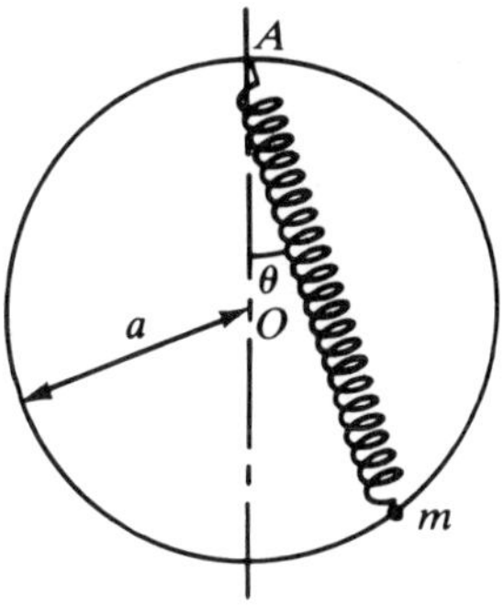

Fig. 1.17

(a) If $Ca < 2mg$ (g is the acceleration of gravity) show that the position of stable equilibrium is $\theta \doteq 0$.

(b) If $Ca > 2mg$, show that there are *two* positions of stable equilibrium.

(c) Sketch the potential energy $U(\theta)$ of the bead for $Ca = mg$ and for $Ca = 4mg$, and discuss the forms of these two curves.

1.2 Further results on the $D = 1$ Ising model

In this problem we work with periodic boundary conditions: the spin numbered $N + 1$ is identified with the spin numbered 1.

(a) Show that the partition function may be written

$$Z_N = \sum_{[S_i]} \mathbb{V}_{S_1 S_2} \mathbb{V}_{S_2 S_3} \cdots \mathbb{V}_{S_N S_1} = \mathrm{Tr}(\mathbb{V})^N,$$

where $\mathbb{V}$ is the matrix

$$\mathbb{V} = \begin{matrix} & S = 1 & S = -1 \\ S = 1 & \mathrm{e}^{K} & \mathrm{e}^{-K} \\ S = -1 & \mathrm{e}^{-K} & \mathrm{e}^{K} \end{matrix}.$$

Hence prove that, as $N \to \infty$, $\lim Z_N^{1/N} = \lambda_1 = 2\cosh K$, where λ_1 is the largest eigenvalue of $\mathbb{V}$.

(b) Show that in the presence of a uniform B-field

$$\lim_{N \to \infty} Z_N^{1/N} = \mathrm{e}^K \cosh L + [\mathrm{e}^{2K}(\sinh L)^2 + \mathrm{e}^{-2K}]^{1/2}, \quad L = \frac{\mu B}{kT}.$$

(c) Calculate the internal energy and the specific heat in zero field.

(d) Still in zero field, calculate the correlation function $\langle S_i S_j \rangle$ using the same method as in (a). Assume that $N \to \infty$. From this result, compute the susceptibility χ in zero magnetic field and check with the result of (c). Compute the Fourier transform of the correlation function

$$\tilde{G}(q) = \sum_{n=-\infty}^{+\infty} \mathrm{e}^{iqn} \langle S_0 S_n \rangle.$$

Examine the $T \to 0$ limit and show that $\eta = 1$. Could you infer this result in x-space? Hint: the variable $t = \exp(-2\beta J)$ may be useful.

(e) Compute the 4-point correlation function in zero magnetic field, namely

$$\langle S_i S_j S_k S_l \rangle, \quad i < j < k < l,$$

and its connected part $\langle S_i S_j S_k S_l \rangle_c$.

(f) Let us introduce, for arbitrary D, the quantity

$$E_i = \frac{1}{2} \sum_{\langle j,\, i \rangle} S_i S_j, \quad i \text{ fixed}.$$

Show that the specific heat per spin, $\hat{C}$, in zero magnetic field, can be related to the connected correlation function $\langle E_i E_j \rangle_c$ ('energy–energy correlation')

$$\hat{C} = k\beta^2 \sum_j \langle E_i E_j \rangle_c = \beta^2 k \sum_j \langle (E_i - \langle E_i \rangle)(E_j - \langle E_j \rangle) \rangle.$$

(g) Compute $\langle E_i E_j \rangle_c$ in the $D = 1$ case, and check with the result found in (c). Hint: examine separately the cases $j = i$, $j = i + 1$ and $j = i + p$, $p \geqslant 2$.

1.3 Mean-field theory and long-range interactions

We aim to show that mean-field theory becomes exact for the Ising model, *provided the same interaction acts between all pairs of spins.* Let the number of spins be N ($N \gg 1$). Then there are $N^2/2$ pairs of spins, and the interaction must be proportional to $1/N$ in order for the energy to be proportional to N.

(a) The Hamiltonian of the model is

$$H = -\frac{J}{N}\sum_{i,j} S_i S_j; \quad S_i = \pm 1; \quad S_i^2 = 1,$$

where the sum runs over *all* pairs (i, j) (and not merely over nearest neighbours), and where J is a constant. Show that one may equally well write

$$H = -\frac{J}{2N}\left[\left(\sum_{i=1}^{N} S_i\right)^2 - N\right].$$

(b) Let $M = \langle S \rangle$ be the magnetization per spin: $-1 \leqslant M \leqslant +1$. Show that the degeneracy of the state with magnetization M is

$$W(M) = \frac{N!}{\left(\frac{N}{2}(1 + M)\right)!\left(\frac{N}{2}(1 - M)\right)!}.$$

(c) Show that the partition function may be written

$$Z = \sum_M W(M) \mathrm{e}^{-H(M)/kT}, \tag{1}$$

where $H(M)$ is the value of the Hamiltonian when the magnetization is M. Evaluate Z by using Stirling's approximation $n! \approx n^n \mathrm{e}^{-n}$, and by showing that in (1) the sum over M may be replaced by its largest term. Show that this term is given by the maximum of the function

$$-F = JM^2 - kT[(1 + M)\ln(1 + M) + (1 - M)\ln(1 - M)]. \tag{2}$$

What is the physical interpretation of F?

(d) By looking for the maximum of $-F$, prove that there is a phase transition, and calculate the critical temperature T_c. In recovering the equation characteristic of mean-field theory it may prove useful to notice that $\tanh^{-1} x = \frac{1}{2}\ln[(1 + x)/(1 - x)]$. What is the sign of $F''(M_0)$, where M_0 is the magnetization?

(e) For $M \to 0$, show that F assumes the form anticipated by Landau theory (see Chapter 2), namely

$$F(T) = F_0 + \frac{1}{2}B(T)M^2 + \frac{1}{4}C(T)M^4,$$

where $B(T_c) = 0$. Evaluate $B(T)$ and $C(T)$.

(f) Generalize to the case of a uniform external B-field.

1.4 The specific heat in mean-field approximation

We aim to determine not only the discontinuity in the specific heat, but also the slope of the curve at $T = T_c$.

(a) By expanding $\tanh^{-1} M$ to order M^5, show that the magnetization M_0 in zero field is

$$M_0 \approx \sqrt{-3t}(1 + \alpha t),$$

and determine α.

(b) Hence show that the slope of the specific heat at T_c (see Fig. 1.18) is

$$B = \lim_{T \to T_c -} \frac{dC}{dT}.$$

(c) How does the specific heat behave as $T \to 0$?

(d) From (b) and (c) derive a qualitative graphical representation of $C(T)$ in mean-field theory.

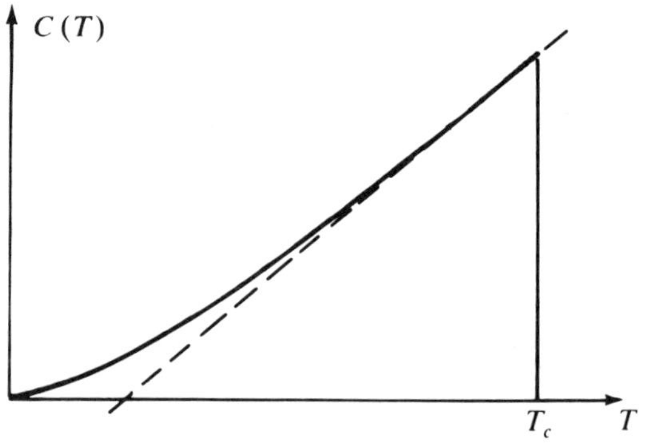

Fig. 1.18

1.5 A variational approach to mean-field theory

(a) Let $P(x)$ be a probability distribution ($P(x) \geq 0$, $\int_{-\infty}^{\infty} P(x)dx = 1$), and

$$g(\lambda) = \ln\left[\int_{-\infty}^{\infty} dx\, e^{\lambda x} P(x)\right].$$

Show that $g''(\lambda) \geq 0$ (the function $g(\lambda)$ is *convex*); from $g(\lambda) \geq g(0) + \lambda g'(0)$ deduce

$$\langle e^x \rangle \geq e^{\langle x \rangle}.$$

What is the geometrical interpretation of this inequality? Can you generalize to any convex function?

(b) Lower bound for the free energy. Let us consider a system of spins and its Hamiltonian $H[S]$. Then the Boltzmann distribution ρ_B reads

$$\rho_B[S] = [Z(\rho_B)]^{-1}\mathrm{e}^{-\beta H[S]}.$$

Let $\rho[S] = [Z(\rho)]^{-1}\mathrm{e}^{-\beta H_\rho[S]}$ be another probability distribution. By writing

$$\mathrm{e}^{-\beta F(\rho_B)} = \sum_{[S]} \mathrm{e}^{-\beta(H - H_\rho)} \frac{\mathrm{e}^{-\beta H_\rho}}{\mathrm{e}^{-\beta F(\rho)}} \mathrm{e}^{-\beta F\rho}$$

($F = -\beta^{-1}\ln Z$ = free energy), and by using the identity proved in (a), show that

$$F(\rho_B) \leqslant F(\rho) + \langle H - H_\rho \rangle_\rho \equiv \Psi(\rho),$$

where $\langle A \rangle_\rho$ stands for the expectation value of A, calculated with the probability distribution $\rho[S]$.

(c) Let H and H_ρ be defined by

$$H = -J \sum_{\langle i,j \rangle} S_i S_j - \sum_i B_i S_i,$$

$$H_\rho = -\sum_i g_i S_i.$$

Compute $Z(\rho)$. One now uses the bound derived in (b) and minimizes $\Psi(\rho)$ with respect to the g_i's. Compute $\Psi(\rho)$, setting $M_i = \langle S_i \rangle_\rho$.

(d) Let us consider the case of a uniform magnetic field: $B_i = B$. Minimize $\Psi(\rho)$ with respect to g and recover the fundamental equation (1.3.1) of mean-field theory. What is the physical interpretation of g?

(e) Let us now go back to the case of a variable magnetic field B_i. What then is the basic equation of mean-field theory? Compute $\partial M_i/\partial B_j$, and the connected correlation function $\langle S_i S_j \rangle_c$ in zero magnetic field. Why is it possible to find a non-trivial result while $\langle S_i S_j \rangle_\rho = \langle S_i \rangle_\rho \langle S_j \rangle_\rho$? *Hint*: show that $\langle S_i S_j \rangle_c$ obeys the equation

$$\left(\frac{\delta_{ij}}{\beta(1 - M^2)} - J_{ij}\right)\langle S_j S_k \rangle_c = \delta_{ik},$$

where $J_{ij} = 1$ if i and j are nearest neighbours, and $J_{ij} = 0$ otherwise; M is the (uniform) magnetization. Solve this equation by Fourier transformation.

1.6 The Legendre transform

Let $f(x)$ be a *convex* function of x. Consider the function $F(p, x) = px - f(x)$, and let $x(p)$ be a solution of $F'_x(p, x) = 0$. Show that $x(p)$ is unique. The Legendre transform $g(p)$ of $f(x)$ is defined by

$$g(p) = \text{Max}|_x F(p, x).$$

Show that $g'(p) = x(p)$ and that $g(p)$ is a convex function of p. Prove that the Legendre transform is involutive (i.e. the Legendre transform of $g(p)$ is $f(x)$).

1.7 Scaling laws

We assume that (as will be proved in Chapter 3) the correlation function satisfies the 'scaling law'

$$G(t, B, r) = s^{-(D+\eta-2)} G\left(s^{1/\nu} t, s^{y_B} B, \frac{r}{s} \right).$$

Here $t = (T - T_c)/T_c$, and s is a positive real number.

(a) Show that this scaling law may be written

$$G(t, B, r) = r^{-(D+\eta-2)} f\left(\pm 1, \frac{B}{|t|^{\Delta}}, \frac{r}{|t|^{-\nu}} \right),$$

$(+1\colon t > 0;\ -1\colon t < 0;\ \Delta = \nu y_B)$.

(b) Using the fluctuation-response theorem, show that the susceptibility χ obeys

$$\chi = |t|^{-\nu(2-\eta)} g\left(\pm 1, \frac{B}{|t|^{\Delta}} \right).$$

From this, recover the scaling law $\gamma = \nu(2 - \eta)$.

(c) By noting that $\chi = -(\partial^2 F/\partial B^2)_T$, show that the free energy F behaves according to

$$F \sim |t|^{2\Delta-\gamma} h\left(\pm 1, \frac{B}{|t|^{\Delta}} \right).$$

(d) Hence show, by eliminating Δ, that

$$\alpha + 2\beta + \gamma = 2; \quad \gamma = \beta(\delta - 1).$$

(e) Show that the equation of state may be written in the form

$$\frac{M}{B^{1/\delta}} = f\left(\frac{t}{B^{1/\beta\delta}} \right).$$

(f) Chapter 3 will show that $y_B = \frac{1}{2}(D + 2 - \eta)$. Assuming this, derive the scaling law $2 - \alpha = \nu D$.

(g) Does the $D = 2$ Ising model obey the four scaling laws

(1) $\gamma = \nu(2 - \eta)$; (2) $2 - \alpha = \nu D$; (3) $\alpha + 2\beta + \gamma = 2$; (4) $\gamma = \beta(\delta - 1)$?

Further reading

For a general introduction to the physics of ferromagnetism, see Kittel (1986, Chapter 15). The solution of the Ising model in 2 dimensions is given for instance by Landau and Lifshitz (1980, Chapter 14) or by Itzykson and Drouffe (1989, Chapter II). A very clear discussion of mean-field theory can be found in Reif (1965, Chapter 10), and an introduction to correlations in Ma (1976, Chapter 1). Pure states, mixed states, and the role of the thermodynamic limit are examined by Parisi (1988, Chapter 2). Particularly recommended are Wilson (1979), and Brush (1967).

2 Landau Theory

The mean-field approximation studied in the last chapter is not always very reliable, and one would like to be able to calculate the effects of fluctuations, which the approximation neglects. The formulation presented in Chapter 1 has the virtue of simplicity, but it also has the disadvantage that it is ill-adapted to dealing with fluctuations. The present chapter aims at a reformulation that does admit such calculations, which will lead us to the theory of Ginzburg and Landau. This theory adopts a 'Hamiltonian' $H_{\mathrm{GL}}[\varphi_i]$ depending on certain random variables φ_i defined on the sites i of a lattice, the probability of a configuration $[\varphi_i]$ being proportional to $\exp(-H_{\mathrm{GL}}[\varphi_i])$. The random variable φ_i is called *a field variable* or simply *a field*. It varies continuously in the interval $]-\infty, +\infty[$, and plays the same part as did the Ising spin in Chapter 1, though the latter was restricted to the two values $S_i = \pm 1$. The order parameter remains one-dimensional: the generalization to n-dimensional order parameters will be studied later, in Section 3.5.3.

It is far from clear at the outset that there is any connection at all between Ginzburg–Landau theory and the Ising model; for the moment we merely assert that both theories belong to the same universality class, leaving the justification of this statement to the next chapter. (One can also recover the Ising model in some limit: see Problem 2.5.) In fact the probability $\exp(-H_{\mathrm{GL}}[\varphi_i])$ describes a spin system only in the vicinity of a phase transition. One should question, too, the relation between the parameters of the Hamiltonian H_{GL} and the (microscopic) parameters of the underlying spin model. This connection is fairly obscure, though Problem 2.1 supplies one clue towards an answer. However, the 'proof' in that problem contains a weak link, which is why we do not rely on it in the text. Instead, we shall arrive at H_{GL} heuristically, by trying to guess what properties it must have, and by requiring that it reproduce the equations of mean-field theory at a certain level of approximation, the so-called *Landau approximation*. The corrections to this approximation then allow for the effects of fluctuations, and allow one to decide on its validity (through the *Ginzburg criterion*).

Section 2.1 presents a heuristic derivation of the Ginzburg–Landau Hamiltonian; mean-field theory is then recovered in Section 2.2. As a byproduct one obtains an instructive comparison between phase transitions of first and of second order. The correlation function is calculated in Section 2.3, leading to the determination of the critical exponents η and ν. Finally, Section 2.4 formulates the Ginzburg criterion, which allows one to assess the validity of the Landau approximation and thence of mean-field theory.

2.1 The Ginzburg–Landau Hamiltonian and the Landau approximation

2.1.1 The case with just one site

The mean-field approximation consists in pretending that a spin S_i on site i is subject only to the averaged influence of all the other spins; one can then start by considering the chosen spin S_i by itself. The mean value M of S_i is given by (1.3.1). In order to calculate the critical exponents one need merely expand $\tanh^{-1} M$ to order M^3; on recalling that the critical temperature T_c is given by $T_c = qJ/k$, equation (1.3.1) can then be written

$$\frac{\mu B}{kT} = M\left(\frac{T - T_c}{T}\right) + \frac{1}{3}M^3. \tag{2.1.1}$$

The results for the critical exponents depend on only two features of equation (2.1.1):

(i) the coefficient of M vanishes linearly at $T = T_c$; it is > 0 for $T > T_c$ and < 0 for $T < T_c$;

(ii) the coefficient of M^3 is positive.

Let us try to reproduce equation (2.1.1) by means of a Hamiltonian $H(\varphi)$ depending on a continuous random variable φ $(-\infty < \varphi < \infty)$, such that $\langle\varphi\rangle = M$. By analogy to the invariance of the Ising model under the transformation $S_i \to -S_i$ in the absence of an external field, we now adopt invariance under the parity operation, $H(\varphi) = H(-\varphi)$; clearly we are trying to preserve the symmetries of the Ising model. Further, the right-hand side of (2.1.1) is a polynomial in M depending on two coefficients, and we can try a similar polynomial for H as well:

$$H(\varphi) = \frac{1}{2!}r_0\varphi^2 + \frac{1}{4!}u_0\varphi^4, \quad u_0 > 0. \tag{2.1.2}$$

The notations r_0 and u_0 are conventional in statistical physics, and the factors $1/2!$ and $1/4!$ will prove useful in Chapter 5. The form of H in (2.1.2) is not chosen at random. On the contrary, for $r_0 < 0$, H has *two* minima, suggesting the possibility of a broken symmetry (Fig. 2.1). By analogy with the Ising model one adds a coupling $-B\varphi$ to an externally applied B-field (setting $\mu = 1$ in order to ease the notation), and one writes the partition function as

$$Z = \int \mathrm{d}\varphi\, \mathrm{e}^{-H(\varphi) + B\varphi}. \tag{2.1.3}$$

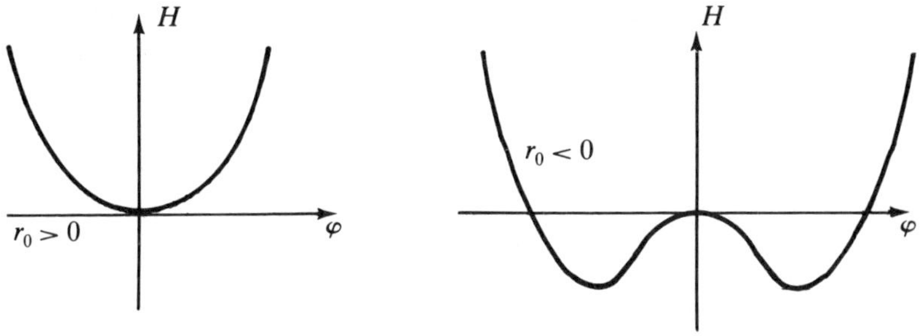

Fig. 2.1

Note that this definition of Z omits the customary factor $1/kT$: we are trying to describe a physical system only in the near-vicinity of a critical point, where the slowly-varying factor $1/kT$ can be subsumed into the definition of H (just as the factors $1/T$ in (2.1.1) could be replaced by $1/T_c$). The coefficient u_0 in (2.1.2) must be positive if the integral in (2.1.3) is to converge, since $\lim_{\varphi \to \pm\infty} H(\varphi) = +\infty$; but the sign of r_0 is arbitrary.

The combination $H_1(\varphi) = H(\varphi) - B\varphi$ has an absolute minimum at the point $\varphi = \varphi_0$ satisfying the condition

$$H'(\varphi_0) = B, \tag{2.1.4}$$

and the exponential in (2.1.3) has a maximum at $\varphi = \varphi_0$ (see Fig. 2.2). The *Landau approximation* consists in replacing the integral (2.1.3) by the value of its integrand at $\varphi = \varphi_0$:

$$Z \approx e^{-H_1(\varphi_0)} = e^{-H(\varphi_0) + B\varphi_0}. \tag{2.1.5}$$

This approximation might seem crude, which indeed it is for just one site: in particular, if B is small and $r_0 < 0$, then the contribution from the other minimum at $\varphi = \varphi_0'$ is almost equally important. This objection will be met

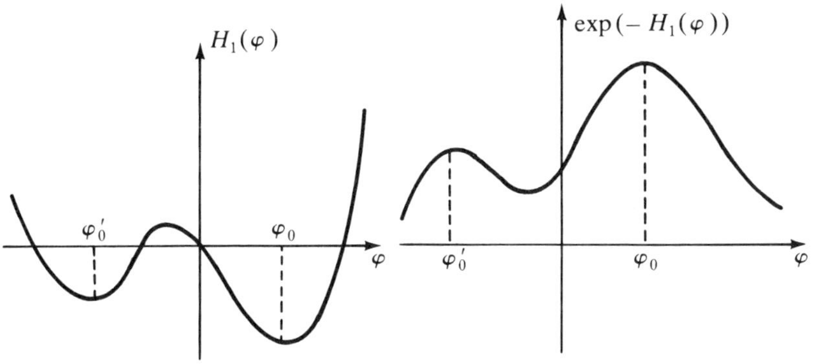

Fig. 2.2

a little later; let us meanwhile disregard such niceties and continue with the calculation (which also serves as a pedagogic introduction to what will follow presently). From (2.1.5) we obtain, up to a factor, the Helmholtz free energy

$$W = \ln Z = -H(\varphi_0) + B\varphi_0. \tag{2.1.6}$$

Since φ_0 is determined by the stationarity condition (2.1.4), the magnetization $M = \langle \varphi \rangle$ equals φ_0:

$$M = \frac{\partial W}{\partial B} = -H'(\varphi_0)\frac{\partial \varphi_0}{\partial B} + \varphi_0 + B\frac{\partial \varphi_0}{\partial B} = \varphi_0. \tag{2.1.7}$$

Equations (2.1.6) and (2.1.7) then yield the Gibbs free energy

$$\Gamma = MB - W = \varphi_0 B + H(\varphi_0) - B\varphi_0 = H(\varphi_0) = H(M); \tag{2.1.8}$$

in view of (2.1.2) this becomes

$$\Gamma(M) = \frac{1}{2}r_0 M^2 + \frac{1}{4!}u_0 M^4. \tag{2.1.9}$$

It is important to notice that φ_0 is an increasing function of B, so that W is a convex function of B ($\partial^2 W/\partial B^2 \geqslant 0$). Then Γ can be defined as a convex function of M (see Problem 1.6).

If we now calculate the magnetic field

$$B = \frac{\partial \Gamma}{\partial M} = r_0 M + \frac{1}{3!}u_0 M^3, \tag{2.1.10}$$

then we notice that (2.1.10) has all the desired properties, provided r_0 vanishes linearly at some temperature T_0:

$$r_0 = \bar{r}_0(T - T_0). \tag{2.1.11}$$

In that case (2.1.10) is a mean-field equation describing a second-order phase transition at a transition temperature $T_c = T_0$. Notice the analogy between the approximation (2.1.5) and that of mean-field theory: (2.1.5) ignores the fluctuations of φ around its average value $\varphi_0 = \langle \varphi \rangle$; indeed, the random variable φ has been replaced outright by its average. We must recall however that $\langle \varphi \rangle$ and φ_0 coincide only at the level of the Landau approximation (see Section 2.4).

2.1.2 Generalization to *N* sites

We must of course remember that our real problem involves N sites rather than just one. The generalization to N sites is nontrivial, because it allows one to introduce interactions (couplings) between different sites. A random variable $\varphi(\boldsymbol{x}_i)$ (or simply φ_i) is assigned to each site $\boldsymbol{x}_i$ (or i), varying over the domain

$$-\infty < \varphi_i < \infty.$$

It proves convenient to impose periodic boundary conditions; in one dimension for instance, these demand

$$\varphi(x_i + aN) = \varphi(x_i).$$

By analogy with the Ising model we introduce interactions between nearest neighbours, e.g. between $\varphi(\boldsymbol{x}_i)$ and $\varphi(\boldsymbol{x}_i + \boldsymbol{\mu})$, where the vector $\boldsymbol{\mu}$ links site $\boldsymbol{x}_i$ to one of its nearest neighbours. More in detail, starting with a site $\boldsymbol{x}_i$ we construct a set of D orthogonal vectors of length a

$$\{\boldsymbol{\mu}\} = \{\boldsymbol{e}_1, \boldsymbol{e}_2, \ldots, \boldsymbol{e}_D\}$$

linking $\boldsymbol{x}_i$ to half its nearest neighbours (see Fig. 2.3); and we define a 'discretized gradient' of $\varphi(\boldsymbol{x}_i)$, having the components

$$\partial_\mu \varphi(\boldsymbol{x}_i) = \frac{1}{a}[\varphi(\boldsymbol{x}_i + \boldsymbol{\mu}) - \varphi(\boldsymbol{x}_i)], \tag{2.1.12}$$

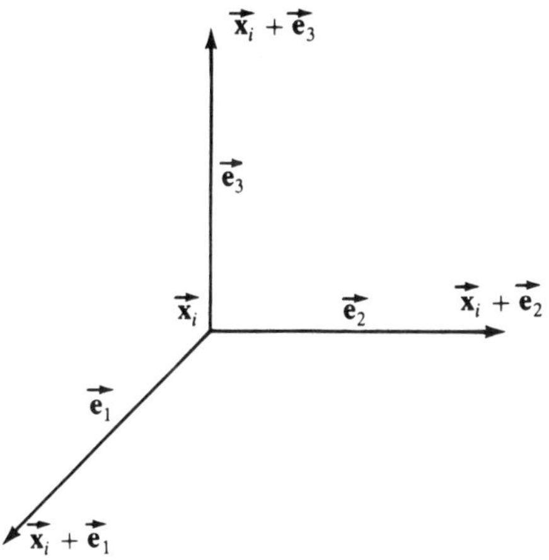

Fig. 2.3

where a is the lattice spacing. Interactions between nearest neighbours can then be expressed by means of this gradient,

$$\sum_{i,\mu} \frac{1}{a^2}[\varphi(\boldsymbol{x}_i + \boldsymbol{\mu}) - \varphi(\boldsymbol{x}_i)]^2 = \sum_i (\nabla\varphi(\boldsymbol{x}_i))^2, \tag{2.1.13}$$

and we shall postulate the Ginzburg–Landau Hamiltonian in the form

$$H_{\mathrm{GL}}[\varphi_i] = a^D \sum_{i=1}^{N} \left[\frac{1}{2}(\nabla\varphi_i)^2 + \frac{1}{2}r_0(T)\varphi_i^2 + \frac{1}{4!}u_0\varphi_i^4\right]. \tag{2.1.14}$$

The interaction (2.1.13) appears to differ from an interaction between nearest neighbours $\sum_{\mu,i}\varphi(\boldsymbol{x}_i+\boldsymbol{\mu})\varphi(\boldsymbol{x}_i)$, but the terms involving $\sum_i(\varphi(\boldsymbol{x}_i))^2$ amount simply to a redefinition of $r_0(T)$ in (2.1.14) (Problem 2.5(b)). The factor a^D will serve, later, to facilitate the passage to the continuum limit. We stress that H_{GL} is intended to describe the physical system *only in the vicinity of the critical point*. That the relation between H_{GL} and Z is not the standard one is manifested by the expression for the partition function, namely by

$$Z=\int\prod_{i=1}^{N}\mathrm{d}\varphi_i\exp\left(-a^D\sum_{i=1}^{N}\left[\frac{1}{2}(\nabla\varphi_i)^2+\frac{1}{2}r_0(T)\varphi_i^2+\frac{1}{4!}u_0\varphi_i^4\right]\right); \qquad (2.1.15)$$

evidently H_{GL} depends on the temperature through $r_0(T)$, this being the only factor to vary rapidly with T near the transition. Strictly speaking 'Hamiltonian' here is a misnomer; all that one is asserting is that $\exp(-H_{\mathrm{GL}}[\varphi_i])$ is proportional to the probability of observing the configuration $[\varphi_i]$ (see also Problem 2.5).

The calculation of Section 2.1.1 can now be generalized to N variables without more ado. H_{GL} is augmented by the effects of an external B-field (and the notation is eased by writing H instead of H_{GL}):

$$Z=\int\prod_{i=1}^{N}\mathrm{d}\varphi_i\exp\left(-H[\varphi_i]+\sum_i B_i\varphi_i\right); \qquad (2.1.16)$$

and one determines the maximum of the integrand through the condition

$$B_i=\left.\frac{\partial H}{\partial\varphi_i}\right|_{\varphi_i=\varphi_{i0}}, \qquad (2.1.17)$$

which defines a set of values $\{\varphi_{i0}\}=\{\varphi_{i0},\ldots,\varphi_{N0}\}$ of the random variables φ_i. The Landau approximation yields the free energy W in the form

$$W=\ln Z\approx -H[\varphi_{i0}]+\sum_i B_i\varphi_{i0};$$

it also yields the magnetization

$$M_i=\frac{\partial W}{\partial B_i}=-\sum_j\left.\frac{\partial H}{\partial\varphi_j}\right|_{\varphi_{j0}}\frac{\partial\varphi_{j0}}{\partial B_i}+\varphi_{i0}+\sum_j B_j\frac{\partial\varphi_{j0}}{\partial B_i}=\varphi_{i0},$$

and as before we find the very simple result

$$\Gamma(M_i)=H(M_i). \qquad (2.1.18)$$

Evidently this relation applies only in the Landau approximation (see Section 2.4), and one must not confuse the magnetization M_i, which is an average value, with φ_i, which is a random variable.

The passage to N sites allows us to dispose of the problem of secondary maxima that was raised earlier: in a *uniform* B-field ($B_i=B$), one has $\varphi_{i0}=\varphi_0$, where φ_0 is the value determined in the single-site problem; hence

$$W \approx -N[H(\varphi_0) - B\varphi_0].$$

As $N \to \infty$, the would-be contribution to W from the secondary maximum of $\exp(-H)$ is suppressed by a factor $\exp(-cN)$, whence, for any nonzero B, the value of φ_0 and thereby the value of M are unique. Evidently φ_0 changes sign with B. The mechanism of spontaneous symmetry-breaking is at work in plain view: the magnetization M_0 ($M_0 > 0$) in zero field must be defined as in (1.2.13) by

$$M_0 = \lim_{B \to 0^+} \lim_{N \to \infty} \langle \varphi_i \rangle,$$

and the role of the thermodynamic limit is evidently essential.

Finally let us remark that, from Fig. 2.1, $\Gamma(M)$ does not appear to be convex for $r_0 < 0$. In fact one must discard the part of the curve between the two minima, and replace it with a straight line; in other words, one must take the convex envelope of Γ (see also Problem 2.2).

2.1.3 Continuum formulation

As a rule one does not work with the form (2.1.14) of the Ginzburg–Landau Hamiltonian, since the calculations it entails can become rather laborious. One prefers instead to adopt a continuum formulation, where $\boldsymbol{x}_i$ varies continuously over all of the space occupied by the physical system, instead of being confined to the sites of some lattice; then

$$\boldsymbol{x}_i \to \boldsymbol{x},$$

and the field variable $\varphi(\boldsymbol{x}_i)$ becomes a function of the point $\boldsymbol{x}$:

$$\varphi(\boldsymbol{x}_i) \to \varphi(\boldsymbol{x}).$$

This can be viewed as a 'coarse-graining' of the original Ising model over a volume b^D, where $b \sim$ a few lattice spacings, provided one can choose $b \ll \xi$ (Parisi 1988, Chapter 5).

Since we are interested in the critical region, where the important fluctuations have wavelengths $\gg a$, the continuum version should be equivalent to the lattice version. Under such conditions the 'discrete gradient' (2.1.12) may be replaced by the ordinary gradient, a good approximation for fluctuations having wavelengths $\gg a$. In the continuum limit the Ginzburg–Landau Hamiltonian (2.1.14) becomes

$$H_{\mathrm{GL}} = \int \mathrm{d}^D x \left[\frac{1}{2}(\nabla\varphi)^2 + \frac{1}{2} r_0(T)\varphi^2 + \frac{1}{4!} u_0 \varphi^4 \right], \tag{2.1.19}$$

since the factor a^D in (2.1.14) allows an immediate passage from Riemann sum to integral. However, there does remain one legacy from the initial definition on a lattice, where the wavevector $\boldsymbol{k}$ was subject to the Brillouin condition, e.g. to

$$-\frac{\pi}{a} \leqslant k_x \leqslant \frac{\pi}{a}. \tag{2.1.20}$$

Accordingly, (2.1.19) must be supplemented by the following condition: with $\tilde{\varphi}(\boldsymbol{k})$ the Fourier transform of $\varphi(\boldsymbol{x})$, the magnitude of the wavevector $\boldsymbol{k}$ is restricted by a cutoff Λ:

$$\|\boldsymbol{k}\| \leqslant \Lambda \sim \frac{\pi}{a}. \tag{2.1.21}$$

By contrast, in problems from quantum field theory, where space is truly continuous, there is no limit on $\|\boldsymbol{k}\|$: $\Lambda \to \infty$ (see Part III).

The Hamiltonian (2.1.19) is a *functional* of the field $\varphi(\boldsymbol{x})$, and in order to manipulate it with ease we must generalize the notions of ordinary differentiation and integration so that they extend to functionals. Start with derivatives. Let $I(\varphi)$ be a functional of the field φ, obtained by taking the continuum limit of a function of N variables φ_i. In order to simplify the notation we suppose at first that space has only one dimension ($D = 1$). Then the functional derivative $\delta I/\delta\varphi(x)$ is defined by

$$\frac{\delta I}{\delta\varphi(x)} = \lim_{a\to 0} \frac{1}{a}\frac{\partial I}{\partial\varphi_i}. \tag{2.1.22}$$

Examples:

$$\text{(a)}\ I = \int \mathrm{d}y\, f(y)\varphi^p(y) = \lim_{a\to 0} a\sum_k f_k\varphi_k^p,$$

$$\frac{\partial I}{\partial\varphi_i} = apf_i\varphi_i^{p-1}, \quad \frac{\delta I}{\delta\varphi(x)} = pf(x)\varphi^{p-1}(x);$$

$$\text{(b)}\ I = \int \mathrm{d}y\, V(\varphi(y)), \quad \frac{\delta I}{\delta\varphi(x)} = V'(\varphi(x)); \tag{2.1.23}$$

$$\text{(c)}\ I = \int \mathrm{d}y\left(\frac{\partial\varphi}{\partial y}\right)^2 = \lim_{a\to 0} a\sum_k \frac{1}{a^2}(\varphi_{k+1} - \varphi_k)^2,$$

$$\frac{\partial I}{\partial\varphi_i} = \frac{2}{a}(2\varphi_i - \varphi_{i+1} - \varphi_{i-1}), \quad \frac{\delta I}{\delta\varphi(x)} = -2\frac{\partial^2\varphi}{\partial x^2};$$

$$\text{(d)}\ \frac{\delta\varphi(x)}{\delta\varphi(y)} = \delta(x-y), \quad \text{because} \quad \frac{\partial\varphi_i}{\partial\varphi_k} = \delta_{ik}; \tag{2.1.24}$$

$$\text{(e)}\ \frac{\delta I}{\delta\psi(x)} = \int \mathrm{d}y \frac{\delta I}{\delta\varphi(y)}\frac{\delta\varphi(y)}{\delta\psi(x)}, \quad \text{because} \quad \frac{\partial I}{\partial\psi_i} = \sum_k \frac{\partial I}{\partial\varphi_k}\frac{\partial\varphi_k}{\partial\psi_i}. \tag{2.1.25}$$

The generalization to an arbitrary number D of dimensions is trivial; example (c) for instance becomes

$$\text{(f)}\ I = \int \mathrm{d}^D y (\nabla\varphi(y))^2; \quad \frac{\delta I}{\delta\varphi(x)} = -2\nabla^2\varphi(x). \tag{2.1.26}$$

Here we have used periodic boundary conditions; more generally one finds surface terms, which in turn lead to boundary conditions on the fields. An alternative and more usual way of introducing functional derivatives is described in Problem 2.8.

In order to write down the partition function one must integrate over all configurations of the field $\varphi(x)$, i.e. one must evaluate a functional integral. The integration measure $\mathscr{D}\varphi(x)$ is defined by

$$\mathscr{D}\varphi(x) = \lim_{a\to 0} \mathscr{N}(a) \prod_{i=1}^{N} \mathrm{d}\varphi_i, \tag{2.1.27}$$

where $\mathscr{N}(a)$ is a factor chosen so as to ensure that the limit $a \to 0$ exists at least in simple cases (see Section 8.2.1). In fact multiplicative constants (independent of B) are irrelevant in calculating correlation functions, since they cancel between numerator and denominator (see Section 1.4.1); the constant $\mathscr{N}(a)$ is unimportant, and there is no need to define it precisely. On the other hand it is clear that at this stage equation (2.1.27) is purely formal, and that the existence of such an integration measure deserves much deeper mathematical study. However, the theory of functional integration is complicated, and we shall make do with the intuitive definition given above, appealing to the lattice formulation if we get into difficulties. We merely mention that (2.1.28) can be given a rigorous meaning for $D = 2$ and $D = 3$ (see Glimm and Jaffe 1987). In this spirit our end-result for the partition function in an external field $B(x)$ can be written as

$$Z = \int \mathscr{D}\varphi(x) \exp\left(-\int \mathrm{d}^D x \left[\frac{1}{2}(\nabla\varphi)^2 + \frac{1}{2}r_0\varphi^2 + \frac{1}{4!}u_0\varphi^4 - B\varphi\right]\right). \tag{2.1.28}$$

In order to become familiar with the continuum formulation, the reader is invited to recover the Landau approximation by starting from (2.1.28), and to show that in this approximation the Gibbs free energy is a functional $\Gamma(M)$ of the magnetization $M(x)$, given by

$$\Gamma(M) = \int \mathrm{d}^D x \left(\frac{1}{2}(\nabla M)^2 + \frac{1}{2}r_0 M^2 + \frac{1}{4!}u_0 M^4\right). \tag{2.1.29}$$

One can of course obtain equation (2.1.29) immediately as the continuum limit of the relation $\Gamma(M_i) = H(M_i)$. The two following sections exploit the Landau approximation (2.1.29) in order to discuss the nature of phase transitions, and then to discuss correlation functions.

Before going to these applications of Landau theory, it is worth noticing that the line of arguments we have been following is quite different from that originally used by Landau. He assumed that he could write a functional of $M(\boldsymbol{x})$ and $B(\boldsymbol{x})$, call it the Landau functional $\mathscr{L}(M,B)$, using the following arguments.

(i) From symmetry considerations, in the case of an order parameter with dimension $n = 1$, one must have $\mathscr{L}(M,B) = \mathscr{L}(-M,-B)$. If the order parameter has a more complicated structure, group theoretical arguments must be used to generalize this property.

(ii) Since the magnetization is small at the critical point, one can try to expand $\mathscr{L}(M,B)$ by powers of M: Landau kept terms of order M^2 and M^4.

(iii) Since one is interested in long-wavelength fluctuations, $M(\boldsymbol{x})$ is slowly varying and one keeps only the simplest gradient term, namely $[\nabla M(\boldsymbol{x})]^2$.

One thus writes

$$\mathscr{L}(M,B) = \int \mathrm{d}^D x \left[\frac{1}{2}(\nabla M(\boldsymbol{x})^2) + \frac{1}{2} r_0 M^2(\boldsymbol{x}) + \frac{u_0}{4!} M^4(\boldsymbol{x}) - M(\boldsymbol{x}) B(\boldsymbol{x})\right],$$

where r_0 vanishes at $T = T_0$, r_0 and u_0 being *analytic* functions of T near $T = T_0$. The magnetization $M(\boldsymbol{x})$ is obtained by minimizing $\mathscr{L}$,

$$\frac{\delta \mathscr{L}}{\delta M(\boldsymbol{x})} = 0,$$

and taking the *absolute* minimum in the case of a constant magnetization. The equations derived in this way are of course completely equivalent to those written previously, although one should not confuse the Landau functional $\mathscr{L}(M,B)$ with the Gibbs potential $\Gamma(M)$. The incorrect assumption is that r_0 and u_0 are analytic functions of the temperature. Indeed, if b is the distance over which one performs the 'coarse graining', r_0 and u_0 must depend on b and this introduces non-analyticity in the temperature.

2.2 The Landau theory of phase transitions

2.2.1 Second-order transitions

Suppose the field B is uniform. Then the magnetization M is likewise uniform, and $\Gamma(M)$ can be written

$$\Gamma(M) = V\left(\frac{1}{2} r_0(T) M^2 + \frac{1}{4!} u_0 M^4\right). \tag{2.2.1}$$

(In the continuum version, the site magnetization M_i becomes magnetization per unit volume.) The B-field is given by

$$B = \frac{\partial \Gamma}{\partial \mathcal{M}} = \frac{1}{V}\frac{\partial \Gamma}{\partial M} = r_0(T)M + \frac{1}{6}u_0 M^3, \tag{2.2.2}$$

which redelivers all the results of mean-field theory.

It is easy to show that the specific heat is discontinuous at $T = T_0$, with a discontinuity

$$\Delta C = 3\bar{r}_0^2 T_0^2 / u_0$$

(see Problem 2.7).

We have now reached the goal outlined at the start of this chapter: to construct a partition function which, evaluated in a certain approximation equivalent to that of mean-field theory, faithfully reproduces the equations of the latter. In fact equation (2.1.28) will allow us to go further, i.e. to calculate the effects of fluctuations, which were neglected in deriving (2.1.29).

2.2.2 First-order transitions

A simple modification allows one to describe first-order transitions. Suppose that $\Gamma(M)$ is given by

$$\frac{1}{V}\Gamma(M) = \frac{1}{2}r_0(T)M^2 + \frac{1}{4!}u_0 M^4 + \frac{1}{6!}v_0 M^6, \tag{2.2.3}$$

where we now take $u_0 < 0$ and $v_0 > 0$. When $T = T_0$, the function $\Gamma(M)$ at $M = 0$ is concave downwards, whence $\Gamma(M)$ has two negative minima. By starting from a temperature $\gg T_0$, it is easy to convince oneself that, at some temperature $T_c > T_0$, $\Gamma(M)$ must have three minima at points where $\Gamma(M) = 0$; one of these minima is at $M = 0$, and the other two at $\pm \Delta M$. In other words, at $T = T_c$ there are two phases in equilibrium with each other, one with $M = 0$ and the other with $M = \Delta M$ (or $-\Delta M$): see Fig. 2.4(a). The phase with $M = 0$ is stable when $T > T_c$, and metastable when $T_0 < T < T_c$, conformably with the fact that for $M = 0$ the second derivative of the Gibbs free energy is positive at these temperatures. *The spontaneous magnetization, i.e. the order parameter, is discontinuous at the transition*: hence we are dealing with a first-order transition, where two phases can coexist, one with zero and the other with nonzero magnetization. *Accordingly, one can observe metastability*: for instance, even at $T < T_c$ the system can persist for a certain time in the phase with $M = 0$. Examples of transitions of this kind are discussed by Kittel (1986). Note that there exist first-order transitions for which no order parameter can be defined; one such example is the liquid-to-gas transition.

For second-order transitions the order parameter vanishes continuously at $T = T_c$ ($= T_0$ in the Landau approximation). In such cases there can never be

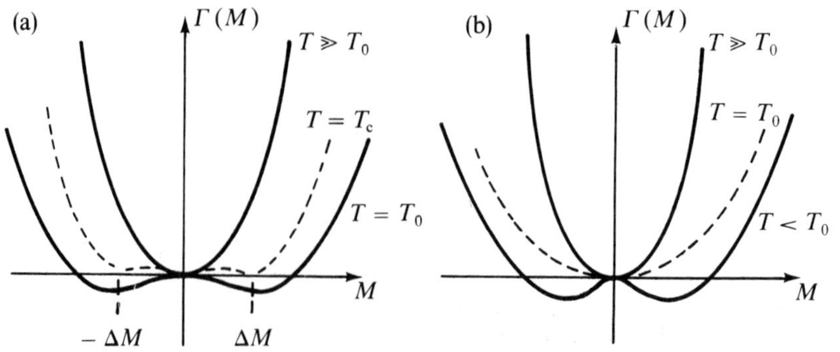

Fig. 2.4 (a) First-order phase transition; (b) second-order phase transition

coexistent phases, nor metastability. Indeed the phase with $M = 0$ is unstable rather than metastable: $\Gamma''(0) < 0$ for $T < T_c$ (Fig. 2.4b).

2.3 Correlation functions

In a field $B(\boldsymbol{x})$, the relation between Γ and B is given by

$$B(\boldsymbol{x}) = \frac{\delta \Gamma}{\delta M(\boldsymbol{x})} = -\nabla^2 M(\boldsymbol{x}) + r_0(T)M(\boldsymbol{x}) + \frac{u_0}{6}M^3(\boldsymbol{x}) \tag{2.3.1}$$

if one uses the continuum formulation. Differentiating with respect to $B(\boldsymbol{y})$ one obtains

$$\left[-\nabla_x^2 + r_0(T) + \frac{1}{2}u_0 M^2(\boldsymbol{x})\right] G(\boldsymbol{x}, \boldsymbol{y}) = \delta(\boldsymbol{x} - \boldsymbol{y}), \tag{2.3.2}$$

since the correlation function is defined by $G(\boldsymbol{x}, \boldsymbol{y}) = \delta M(\boldsymbol{x})/\delta B(\boldsymbol{y})$.

If the B-field is uniform, translation invariance entails $G(\boldsymbol{x}, \boldsymbol{y}) = G(\boldsymbol{x} - \boldsymbol{y})$; since M is independent of $\boldsymbol{x}$, the Fourier transform of equation (2.3.2) takes the very simple form

$$\left(\boldsymbol{q}^2 + r_0(T) + \frac{1}{2}u_0 M^2\right)\tilde{G}(\boldsymbol{q}) = 1,$$

or

$$\tilde{G}(\boldsymbol{q}) = \frac{1}{\boldsymbol{q}^2 + r_0(T) + \frac{1}{2}u_0 M^2}. \tag{2.3.3}$$

In $\boldsymbol{x}$-space this yields

$$G(\boldsymbol{x}) = \int \frac{\mathrm{d}^D q}{(2\pi)^D} \frac{\mathrm{e}^{-i\boldsymbol{q}\cdot\boldsymbol{x}}}{\boldsymbol{q}^2 + r_0(T) + \frac{1}{2}u_0 M^2}. \tag{2.3.4}$$

It is instructive to rederive these results from the discrete version (Problem 2.5).

Consider zero field ($B = 0$). One must distinguish between two cases:

(a) $T > T_0$: then $M = 0$, and

$$\tilde{G}(\boldsymbol{q}) = \frac{1}{\boldsymbol{q}^2 + \bar{r}_0(T - T_0)} = \frac{1}{\boldsymbol{q}^2\left(1 + \dfrac{\bar{r}_0(T - T_0)}{\boldsymbol{q}^2}\right)}. \tag{2.3.5a}$$

From the definitions (1.4.9, 1.4.10) of η, ν, and ξ, one sees that

$$\begin{cases} \eta = 0 \\ \xi = [\bar{r}_0(T - T_0)]^{-1/2} \\ \nu = 1/2. \end{cases}$$

(b) $T < T_0$: then $M^2 = -6r_0/u_0$, and

$$\tilde{G}(\boldsymbol{q}) = \frac{1}{\boldsymbol{q}^2 + 2\bar{r}_0(T_0 - T)}. \tag{2.3.5b}$$

In this case $\xi = [2\bar{r}_0(T_0 - T)]^{-1/2}$, but the critical exponents remain unchanged: $\eta = 0$ and $\nu = \frac{1}{2}$.

As in the case of the exponents $\alpha \ldots \delta$, it is interesting to compare the above with the exact results ($D = 2$) or with those calculated numerically from a high-temperature expansion ($D = 3$):

Exponent	Landau	$D = 2$	$D = 3$
η	0	0.25	0.0375 ± 0.0025
ν	1/2	1	0.6305 ± 0.0015

Again one notes that the Landau approximation improves as the dimensionality of the space rises.

The values $\alpha = 0$, $\beta = \frac{1}{2}$, $\gamma = 1$, $\delta = 3$, $\eta = 0$, $\nu = \frac{1}{2}$ are called the *classical values* of the critical exponents: these are the values given by mean-field theory, or by the Landau approximation which is its generalization.

Let us end by calculating, in the Landau approximation, the correlation function in $\boldsymbol{x}$-space, for $D = 3$ (for the general case see Problem 2.6(e)). We shall extend the integration over q from zero to infinity, rather than from zero to Λ; the result is correct provided $r = \|\boldsymbol{x}\| \gtrsim 1/\Lambda$. One obtains

$$G(\boldsymbol{x}) = \int \frac{d^3q}{(2\pi)^3} \frac{e^{-i\boldsymbol{q}\cdot\boldsymbol{x}}}{q^2 + \xi^{-2}}$$
$$= \frac{1}{(2\pi)^2} \int_0^\infty \frac{q^2\,dq}{q^2 + \xi^{-2}} \int_{-1}^{1} e^{-iqr\cos\theta}\, d(\cos\theta)$$
$$= \frac{1}{(2\pi^2 r)} \int_0^\infty \frac{q\,dq \sin qr}{q^2 + \xi^{-2}}$$
$$= \frac{1}{4\pi^2 r} \operatorname{Im} \int_{-\infty}^{\infty} \frac{q\,dq\, e^{iqr}}{q^2 + \xi^{-2}}.$$

The last integral is easily found by the calculus of residues with the contour shown in Fig. 2.5. The contour encloses a pole at $q = i\xi^{-1}$, whence

$$G(\boldsymbol{x}) = \frac{1}{4\pi r} e^{-r/\xi}. \qquad (2.3.6)$$

As it must, the form (2.3.6) of the correlation function agrees with (1.4.12), provided one takes $\eta = 0$ and $g(r/\xi) = \exp(-r/\xi)$.

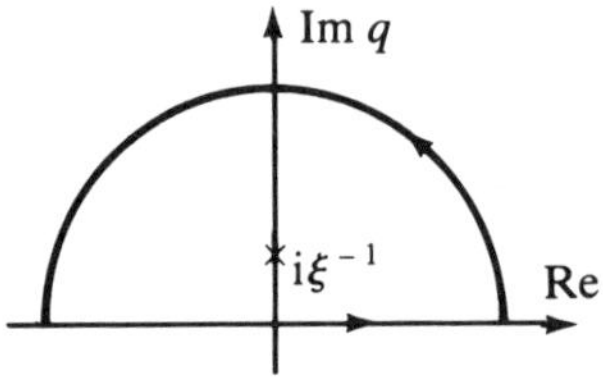

Fig. 2.5 Contour for calculating $G(\boldsymbol{x})$

2.4 Critique of the Landau approximation: the Ginzburg criterion

The Landau like the mean-field approximation neglects the fluctuations of φ. Hence it fails when these fluctuations are large compared to $\langle\varphi\rangle$. We proceed to a first derivation of a criterion for the validity of Landau theory, called the Ginzburg criterion.

2.4.1 The Ginzburg criterion: first derivation

Consider the average magnetization $\mathcal{M}$ over a volume V whose linear dimensions are of order ξ. For $T < T_0$ one has

$$\mathcal{M}^2 = \frac{6\bar{r}_0}{u_0}(T_0 - T)V^2.$$

Now compare $\mathcal{M}^2$ with the mean-square fluctuation $(\Delta\mathcal{M})^2$ over the same volume:

$$(\Delta\mathcal{M})^2 = \int \mathrm{d}^D x\, \mathrm{d}^D y [\langle\varphi(x)\varphi(y)\rangle - \langle\varphi(x)\rangle\langle\varphi(y)\rangle] = V\int \mathrm{d}^D x\, G(x).$$

Since the linear dimensions of V are of order ξ, we have $V \sim \xi^D$, and straightforward dimensional analysis yields

$$\int_V \mathrm{d}^D x\, G(x) \sim \int \frac{r^{D-1}\,\mathrm{d}r}{r^{D-2}} \mathrm{e}^{-r/\xi} \sim \xi^2 \sim \frac{1}{\bar{r}_0(T_0 - T)}.$$

The ratio $(\Delta\mathcal{M})^2/\mathcal{M}^2$ is given by

$$\frac{(\Delta\mathcal{M})^2}{\mathcal{M}^2} = \frac{u_0/6}{\bar{r}_0^2 V(T_0 - T)^2} = \left(\frac{u_0}{6}\bar{r}_0^{D/2-2}\right)(T_0 - T)^{D/2-2}. \tag{2.4.1}$$

For $D > 4$, $(T_0 - T)^{D/2-2}$ vanishes as $T \to T_0^-$, and $(\Delta\mathcal{M})^2/\mathcal{M}^2 \ll 1$. Under these conditions, Landau theory seems plausible; it is, at least, internally consistent.

For $D < 4$ the situation is different: in this case $(T_0 - T)^{D/2-2}$ diverges as $T \to T_0^-$, and Landau theory certainly fails. Nevertheless it may still be possible to find a temperature range ΔT around T_0 such that $(\Delta\mathcal{M})^2/\mathcal{M}^2 \ll 1$ when $T < T_0 - \Delta T$; then Landau theory remains applicable for such T even if $D < 4$. It can even happen (as in superconductors) that the interval ΔT is so narrow that the region where Landau theory fails remains invisible in practice. By contrast, for the superfluid transition in helium-4 (even though it is in many respects closely analogous to the superconducting transition) the interval ΔT is $\gg T_0$, and one can observe no region where Landau theory applies. For $T > T_0$ these arguments lead to no conclusion. However, another argument from the study of the specific heat yields a result analogous to (2.4.1) (see Problem 2.7).

We proceed to a more detailed version of the Ginzburg criterion, and preface it by calculating the first correction to Landau theory.

2.4.2 Corrections to the Landau theory

As we did for the Landau approximation, we start with a single variable. Notice that the Landau approximation improves as the maximum of the exponential becomes sharper. Let us introduce a parameter $\hbar$* such that the smaller $\hbar$, the better the approximation:

$$Z = \int \mathrm{d}\varphi\, \mathrm{e}^{-\frac{1}{\hbar}[H(\varphi) - B\varphi]}.$$

We shall show that Z can be written as a power series in $\hbar$ (of course, at the end of the calculation one sets $\hbar = 1$). As in Section 2.1, we define φ_0 by

$$B = H'(\varphi_0).$$

Now expand $H(\varphi) - B\varphi$ in the vicinity of $\varphi = \varphi_0$ by setting $\psi = \varphi - \varphi_0$. Since

* Of course, in the present problem the parameter $\hbar$ has no connection with Planck's constant. Nevertheless the notation is chosen deliberately, because the same constant also enters an analogous problem from quantum field theory (see Chapter 8). The Landau approximation in statistical physics is the analogue of the classical approximation in quantum field theory.

the derivative vanishes at $\varphi = \varphi_0$ by construction, the expansion has no term linear in ψ, and one finds

$$H(\varphi) - B\varphi = H(\varphi_0) - B\varphi_0 + \frac{1}{2}D(\varphi_0)\psi^2 + \frac{1}{3!}u_0\varphi_0\psi^3 + \frac{1}{4!}u_0\psi^4,$$

where $D(\varphi_0)$ is defined by

$$D(\varphi_0) = H''(\varphi_0) = \frac{\partial B}{\partial \varphi_0} = r_0 + \frac{1}{2}u_0\varphi_0^2. \tag{2.4.2}$$

Substitution into Z yields

$$Z = \mathrm{e}^{-\frac{1}{\hbar}(H(\varphi_0) - B\varphi_0)} \int \mathrm{d}\psi \exp\left(-\frac{1}{2\hbar}D(\varphi_0)\psi^2 - \frac{1}{3!\,\hbar}u_0\varphi_0\psi^3 - \frac{1}{4!}\frac{u_0}{\hbar}\psi^4 \right),$$

where we now change the integration variable to $\psi' = \psi/\sqrt{\hbar}$. The Jacobian, being a constant, may be dropped (see the comment following equation (2.1.27)), and the expression for Z becomes

$$Z = \mathrm{e}^{-\frac{1}{\hbar}(H(\varphi_0) - B\varphi_0)} \int \mathrm{d}\psi' \exp\left(-\frac{1}{2}D(\varphi_0)\psi'^2 \right) \times \exp\left(-\left(\frac{u_0\hbar^{1/2}}{3!}\varphi_0\psi'^3 + \frac{u_0\hbar}{4!}\psi'^4 \right) \right). \tag{2.4.3}$$

As $\hbar \to 0$, the second exponential in (2.4.3) may be expanded in powers of $\hbar$; every term in this expansion yields a Gaussian integral. Assigning to W a factor $\hbar$ to ensure that $M = \partial W/\partial B$, we obtain for $W = \hbar \ln Z$ the result

$$W = \hbar \ln Z = -H(\varphi_0) + B\varphi_0 - \frac{\hbar}{2}\ln D(\varphi_0) + O(\hbar^2). \tag{2.4.4}$$

We evaluate the magnetization

$$M = \frac{\partial W}{\partial B} = \varphi_0 - \frac{\hbar}{2}\frac{D'(\varphi_0)}{D(\varphi_0)}\frac{\partial \varphi_0}{\partial B} = \varphi_0 - \frac{\hbar}{2}\frac{D'(\varphi_0)}{D^2(\varphi_0)},$$

and, with $H_1 = H - B\varphi$, note that

$$H_1(\varphi_0) = H_1(M) + (H_1(\varphi_0) - H_1(M)).$$

Since $H_1'(\varphi_0) = 0$, the second term in this equation is of order $\hbar^2$, and for the Gibbs free energy to order $\hbar$ one finds

$$\Gamma(M) = H(M) + \frac{\hbar}{2}\ln D(M) + O(\hbar^2). \tag{2.4.5}$$

Proceed now to the general case, using the continuum formulation; then $\varphi_0(\boldsymbol{x})$ is governed by the equation

$$B(\boldsymbol{x}) = \left.\frac{\delta H}{\delta \varphi(\boldsymbol{x})}\right|_{\varphi(\boldsymbol{x}) = \varphi_0(\boldsymbol{x})},$$

and the Taylor expansion around $\varphi_0(\boldsymbol{x})$ is

$$H_1 \simeq H(\varphi_0(\boldsymbol{x})) - \int \mathrm{d}^D x\, B(\boldsymbol{x})\varphi_0(\boldsymbol{x}) + \frac{1}{2}\int \mathrm{d}^D x\, \mathrm{d}^D y\, \psi(\boldsymbol{x}) D(\boldsymbol{x}, \boldsymbol{y}) \psi(\boldsymbol{y}).$$

Here, $\psi(\boldsymbol{x}) = \varphi(\boldsymbol{x}) - \varphi_0(\boldsymbol{x})$, and

$$D(\boldsymbol{x}, \boldsymbol{y}) = \left.\frac{\delta^2 H}{\delta\varphi(\boldsymbol{x})\delta\varphi(\boldsymbol{y})}\right|_{\varphi = \varphi_0}.$$

The integral over $\psi(\boldsymbol{x})$ is Gaussian (see Appendix A); ignoring multiplicative constants, we find

$$\int \mathscr{D}\psi \exp\left[-\frac{1}{2}\int \mathrm{d}^D x\, \mathrm{d}^D y\, \psi(\boldsymbol{x}) D(\boldsymbol{x}, \boldsymbol{y}) \psi(\boldsymbol{y})\right] \sim [\det D]^{-1/2}$$

$$= \exp\left(-\frac{1}{2}\operatorname{Tr}\ln D\right).$$

It remains to interpret this result, which at this stage is evidently somewhat formal. Start by evaluating $D(\boldsymbol{x}, \boldsymbol{y})$:

$$D(\boldsymbol{x}, \boldsymbol{y}) = \left(-\nabla_x^2 + r_0(T) + \frac{1}{2}u_0\varphi_0^2\right)\delta^{(D)}(\boldsymbol{x} - \boldsymbol{y}),$$

where, in our present approximation, $\varphi_0 \approx M$.

$D(\boldsymbol{x}, \boldsymbol{y})$ is the limit of a matrix D_{ij} diagonalizable through a Fourier transformation, provided that, as we now assume, φ_0 is independent of $\boldsymbol{x}$ (i.e. provided the system is translation-invariant). By equation (A.1.7),

$$\operatorname{Tr} D(\boldsymbol{x}, \boldsymbol{y}) = N a^D \int \frac{\mathrm{d}^D q}{(2\pi)^D}\left(\boldsymbol{q}^2 + r_0(T) + \frac{1}{2}u_0 M^2\right),$$

whence

$$\operatorname{Tr}\ln D = V \int \frac{\mathrm{d}^D q}{(2\pi)^D}\ln\left(\boldsymbol{q}^2 + r_0(T) + \frac{1}{2}u_0 M^2\right).$$

The generalization of (2.4.5) is

$$\Gamma = \int \mathrm{d}^D x\left[H(M) + \frac{\hbar}{2}\int \frac{\mathrm{d}^D q}{(2\pi)^D} \times \ln\left(\boldsymbol{q}^2 + r_0(T) + \frac{u_0}{2}M^2\right) + O(\hbar^2)\right]. \tag{2.4.6}$$

The first term in equation (2.4.6) is nothing but the Landau approximation (2.1.29). Accordingly, equation (2.4.6) gives the first two terms in the *loop expansion*, which will be investigated more thoroughly in Chapter 5, Section 4.

Important comment: the Gaussian model

If one sets $u_0 = 0$ in the Ginzburg–Landau Hamiltonian (2.1.19), then the integral over φ is Gaussian, and the result (2.4.6) is exact to order $\hbar$, since the Hamiltonian is then quadratic in φ. This case of $u_0 = 0$ is called the *Gaussian model*; in fact it is defined only for $r_0(T) > 0$. While M in (2.4.6) is uniform by assumption, the Gaussian model can dispense with this assumption, because $D(x, y)$ is independent of M; the Gibbs free energy of the model is *exactly*

$$\Gamma(M) = \int \mathrm{d}^D x \left[\frac{1}{2} (\nabla M)^2 + \frac{1}{2} r_0(T) M^2 + \frac{1}{2} \int \frac{\mathrm{d}^D q}{(2\pi)^D} \ln(r_0 + \boldsymbol{q}^2) \right].$$

The correlation function, which is the inverse of $\delta^2 \Gamma / \delta M(\boldsymbol{x}) \delta M(\boldsymbol{y})$, is given by the inverse Fourier transform of $(\boldsymbol{q}^2 + r_0(T))^{-1}$:

$$\text{Gaussian model:} \quad \tilde{G}(\boldsymbol{q}) = \frac{1}{\boldsymbol{q}^2 + r_0(T)}. \tag{2.4.7}$$

An alternative derivation of this result can be found in Problem 2.5(c).

2.4.3 The Ginzburg criterion: second derivation

In view of the correction given by equation (2.4.6), and setting $\hbar = 1$, we obtain the B-field (uniform by assumption)

$$B = r_0 M + \frac{u_0}{6} M^3 + \frac{1}{2} u_0 M \int \frac{\mathrm{d}^D q}{(2\pi)^D} \frac{1}{\boldsymbol{q}^2 + r_0(T) + \frac{1}{2} u_0 M^2},$$

and the susceptibility in the limit $B \to 0$,

$$\chi^{-1} = \left. \frac{\partial B}{\partial M} \right|_{B=0}$$

$$= r_0 + \frac{u_0}{2} M^2 + \frac{u_0}{2} \int \frac{\mathrm{d}^D q}{(2\pi)^D} \frac{1}{\boldsymbol{q}^2 + r_0(T) + \frac{u_0}{2} M^2} + \cdots. \tag{2.4.8}$$

The terms that have been dropped vanish for $T > T_c$, because at such temperatures $M = 0$. In order to simplify the discussion we restrict it to $T > T_c$. The standard notation for χ^{-1} is $\chi^{-1} = r$ (not to be confused with $\|\boldsymbol{x}\|$) and, in the Landau approximation,

$$\chi_0^{-1} = r_0 = \bar{r}_0 (T - T_0).$$

Equation (2.4.8) (with $M = 0$) becomes

$$r = r_0 + \frac{1}{2} u_0 \int \frac{\mathrm{d}^D q}{(2\pi)^D} \frac{1}{\boldsymbol{q}^2 + r_0}. \tag{2.4.9}$$

Equation (2.4.9) shows that the critical temperature T_c differs from T_0; in fact, $T_c < T_0$: we check that fluctuations decrease the transition temperature with respect to that of mean field theory. The critical temperature is, effectively, defined by $r(T_c) = 0$ (infinite susceptibility), while for $T = T_0$ one has $r > 0$. It seems therefore that we must enter the region $r_0 < 0$, which is problematic because the integral in (2.4.9) is no longer defined. To sidestep the difficulty, we note that actually (2.4.9) represents the start of an expansion in powers of u_0:

$$r = r_0 + a_1 u_0 + a_2 u_0^2 + \cdots .$$

To show this, we assign dimensions to $\hbar$, say those of action; since $H/\hbar$ must be dimensionless, φ has the dimensions $\hbar^{1/2} l^{1-D/2}$, where l is length. The dimensions of $\mathrm{d}^D x$ and of $\boldsymbol{\nabla}$ are l^D and l^{-1} respectively. It follows at once that r_0 has dimensions l^{-2}, while u_0 has dimensions $\hbar^{-1} l^{D-4}$. Dimensional analysis of the equation for r implies that the term in $\hbar^L$ is proportional to u_0^L. This implies in turn that under the integral in (2.4.9) r_0 may be replaced by r; the resulting error is of order u_0^2, and we have neglected terms of this order already. Thus

$$r = r_0 + \frac{1}{2} u_0 \int \frac{\mathrm{d}^D q}{(2\pi)^D} \frac{1}{\boldsymbol{q}^2 + r}. \tag{2.4.10}$$

The critical temperature is given by

$$0 = \bar{r}_0 (T_c - T_0) + \frac{1}{2} u_0 \int \frac{\mathrm{d}^D q}{(2\pi)^D} \frac{1}{\boldsymbol{q}^2}. \tag{2.4.11}$$

Therefore

$$\bar{r}_0 (T_c - T_0) = -\frac{u_0}{2(2\pi)^D} S_D \int_0^{\Lambda} q^{D-3} \mathrm{d}q = \frac{-u_0 S_D \Lambda^{D-2}}{2(2\pi)^D (D-2)},$$

where S_D is the surface area of the unit sphere in D dimensions (Appendix A). It follows that the theory makes no sense at all unless $D > 2$.

Let us subtract equation (2.4.11) from (2.4.10):

$$r = \bar{r}_0 (T - T_c) - \frac{1}{2} r u_0 \int \frac{\mathrm{d}^D q}{(2\pi)^D} \frac{1}{\boldsymbol{q}^2 (\boldsymbol{q}^2 + r)}. \tag{2.4.12}$$

It proves convenient to distinguish between the two cases $D > 4$ and $D < 4$.

$D > 4$: The integral in (2.4.12) converges at $q^2 = 0$ even if $r = 0$. Hence one finds

$$r = \bar{r}_0 (T - T_c) - Cr, \quad r = \frac{\bar{r}_0 (T - T_c)}{1 + C}, \tag{2.4.13}$$

where C is a finite constant. We see that *the order-$\hbar$ correction to Landau theory does not alter the critical exponent γ; $\gamma = 1$ for $D > 4$*. This is equally true for the corrections of order $\hbar^N$ (see Section 3.5.1). For $D > 4$, the Landau (or mean-field) approximation gives the correct critical indices. By contrast, the critical temperature, the coefficient of $(T - T_c)^{-1}$, etc. are modified; but these are quantities that are not universal.

$D < 4$: If $r = 0$, the integral in (2.4.12) diverges as $q \to 0$: one says that it is *infrared-divergent*. In fact it behaves like $\int dq/q^{5-D}$. It is the long wavelengths that are important for critical behaviour, because it is they that make the integrals diverge. With $\boldsymbol{q} = \boldsymbol{k}\sqrt{r}$, the integral becomes

$$I = \int \frac{d^D k}{(2\pi)^D} \frac{(\sqrt{r})^D}{r^2 \boldsymbol{k}^2(\boldsymbol{k}^2 + 1)} = (\sqrt{r})^{D-4} \frac{S_D}{(2\pi)^D} \int_0^\infty \frac{k^{D-1} dk}{k^2(k^2 + 1)}.$$

Since it converges at infinity, we have taken the limit $\Lambda \to \infty$. Accordingly, we can rewrite (2.4.12) as

$$r = \bar{r}_0(T - T_c) - u_0 C r(r^{-\varepsilon/2}), \tag{2.4.14}$$

where we have set

$$\varepsilon = 4 - D \tag{2.4.15}$$

and

$$C = \frac{S_D}{2(2\pi)^D} \int_0^\infty \frac{k^{D-1} dk}{k^2(k^2 + 1)}.$$

Equation (2.4.14) is incompatible with $r \sim (T - T_c)$, and *for $D < 4$ the critical exponents are not given correctly by the Landau theory*. Equation (2.4.14) also makes it clear that the true expansion parameter is not u_0 but rather $u_0 r^{-\varepsilon/2}$, which diverges at the critical point. One could have guessed this result from purely dimensional arguments.

Again one can look for a temperature range where the Landau theory applies even if $D < 4$; this demands

$$u_0 C r^{-\varepsilon/2} \ll 1,$$

or

$$u_0 C [\bar{r}_0(T - T_c)]^{D/2-2} \ll 1,$$

compatibly with equation (2.4.1). However, our second derivation as just given is somewhat more informative, in that it allows us to determine the coefficient C.

Problems

2.1 'Derivation' of the Ginzburg–Landau Hamiltonian (Brézin *et al.* 1976, Section IV; Amit 1984, Chapter 2)

We aim to derive the Hamiltonian (2.1.19) from the Ising model. One starts from the following expression for $Z(B)$:

$$Z(B) = \int \prod_i \mathrm{d}S_i \delta(S_i^2 - 1) \exp\left(\sum_{i,j} S_i V_{ij} S_j + \sum_i B_i S_i \right),$$

$$\begin{aligned} V_{ij} &= J/2kT \quad (i \text{ and } j \text{ nearest neighbours}) \\ &= 0 \qquad\quad (\text{otherwise}). \end{aligned}$$

Generalization: $\delta(S_i^2 - 1) \to \rho(S_i)$, where $\rho \geqslant 0$ and V_{ij} depends on $\boldsymbol{x}_i - \boldsymbol{x}_j$, decreasing with $\| \boldsymbol{x}_i - \boldsymbol{x}_j \|$. Note that the factor μ/kT has been set equal to 1.

(a) Using Gaussian integration (Appendix A) show that

$$Z(B) = \int \prod_i \mathrm{d}\varphi_i \mathrm{d}S_i \rho(S_i) \exp\left(-\frac{1}{4}(\varphi - B)^{\mathrm{T}} V^{-1} (\varphi - B) + \varphi^{\mathrm{T}} S \right).$$

The notation is that of Appendix A. Set

$$\int \mathrm{d}S\, \rho(S) \mathrm{e}^{\varphi S} = \mathrm{e}^{A(\varphi)}.$$

Show that $A''(\varphi) \geqslant 0$, and that

$$Z(B) = \int \prod_i \mathrm{d}\varphi_i \exp\left(-\frac{1}{4}(\varphi - B)^{\mathrm{T}} V^{-1} (\varphi - B) + \sum_i A(\varphi_i) \right). \tag{1}$$

(b) The Landau approximation: find the maximum of the exponent in (1), and thence determine the Gibbs free energy

$$\Gamma(M) = -\sum_{i,j} M_i V_{ij} M_j + \sum_i C(M_i),$$

where $C(M)$ is the Legendre transform of $A(\varphi)$:

$$C(M) = \sum_i M_i \varphi_{i0} - \sum_i A(\varphi_{i0}).$$

(c) Show that, for the Ising model, $A(\varphi) = \ln \cosh \varphi$, and for a uniform applied field recover the mean-field equation (1.3.1). Show that in the general case $C''(M) > 0$, and that one can have either a first-order or a second-order transition, depending on the sign of the term featuring M^4.

(d) What is the relation between the original correlation function $\langle S_i S_j \rangle$ and the correlation function $\langle \varphi_i \varphi_j \rangle$ in zero B-field? Compare their

Fourier transforms, and show that they are proportional to each other as $q \to 0$. You may wish to introduce the Fourier transform of V,

$$\tilde{V}(\boldsymbol{q}) = \sum_j V_{ij} e^{i\boldsymbol{q}(x_i - x_j)}.$$

(e) Introduce the Fourier transform of φ_i,

$$\tilde{\varphi}(\boldsymbol{q}) = \frac{1}{\sqrt{N}} \sum_i e^{i\boldsymbol{q}\cdot\boldsymbol{x}} \varphi_i.$$

Show that, for $q \to 0$,

$$\tilde{V}(\boldsymbol{q}) \simeq V_0(1 - \rho \boldsymbol{q}^2), \quad \rho > 0.$$

(You may restrict yourself to the Ising model on a cubic lattice.) Derive

$$\sum_{i,j} \varphi_i V_{ij}^{-1} \varphi_j \simeq \frac{1}{V_0} \sum_q \tilde{\varphi}(\boldsymbol{q}) \tilde{\varphi}(-\boldsymbol{q}) + \frac{\rho}{V_0} \sum_q \boldsymbol{q}^2 \tilde{\varphi}(\boldsymbol{q}) \tilde{\varphi}(-\boldsymbol{q}),$$

and show that in $\boldsymbol{x}$-space the second term becomes

$$\frac{\rho}{V_0} \int d^D x (\nabla \varphi)^2.$$

Rescaling $\varphi \to \lambda \varphi$ then allows one to recover (2.1.19).

(This 'derivation' is due to Berlin and Kac and to Hubbard and Stratonovich; it has the weak point that the matrix V_{ij}^{-1} (which enters the calculation only at an intermediate stage) is not positive definite (which would make its eigenvalues strictly positive). In the case of the Ising model, it is easy to show that V_{ij}^{-1} has negative eigenvalues.)

2.2 Metastability for $T < T_c$

(a) Show for $T < T_c$, and in a certain domain

$$-B_0 \leqslant B \leqslant B_0$$

of B (to be determined), that the equation

$$B = r_0 M + (u_0/6) M^3, \quad (u_0 > 0)$$

has three roots. Thence determine the general shape of the graph of M as a function of B (see Fig. 2.6).

(b) Show that solutions on the segments (AB) and (A′B′) are metastable, while those on (BB′) are unstable. Draw the correct shape of $\Gamma(M)$.

(c) Show that, as B varies from a negative to a positive value at fixed $T < T_c$, one observes a phase transition of the first kind.

(d) Find a mechanical analogue (see for instance Problem 1.1).

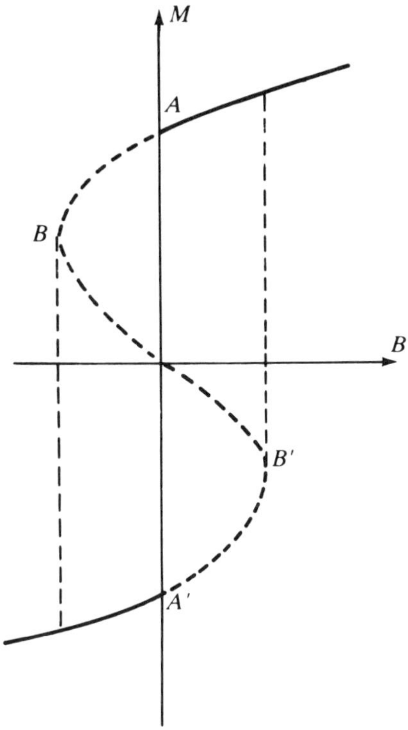

Fig. 2.6

2.3 A first-order transition

Assume that, in zero field,

$$\Gamma(M) = \frac{1}{2}\bar{r}_0(T - T_0)M^2 - \frac{1}{4}M^4 + \frac{1}{6}M^6.$$

What is the temperature T_c of the (first-order) phase transition? What is the discontinuity in the magnetization?

Determine the behaviour of the susceptibility χ in zero field for $T > T_c$. What is the discontinuity in χ at $T = T_c$?

(At a first-order transition, the fluctuations are not fully developed; the fluctuation-dominated regime is aborted before they can make the susceptibility diverge.)

2.4 Goldstone bosons

Suppose that the order parameter has dimension $n = 2$ (the generalization to arbitrary n is trivial).

In the Landau approximation, the Gibbs free energy is written

$$\Gamma(M) = \int \mathrm{d}^D x \left[\frac{1}{2} \sum_{i=1}^{2} (\nabla M_i)^2 + \frac{1}{2} r_0(T) \vec{M}^2 + \frac{1}{4!} u_0 (\vec{M}^2)^2 \right]$$

$$\vec{M} = (M_1, M_2); \quad \vec{M}^2 = M_1^2 + M_2^2.$$

(a) Assume that in zero field at $T < T_0$ the components of the magnetization $\vec{M}$ are $M_1 = M$, $M_2 = 0$. Determine the value of M.

(b) Let $G_{ij}(\boldsymbol{x}, \boldsymbol{y})$ be the Landau approximation to the correlation function,

$$G_{ij}^{-1}(\boldsymbol{x}, \boldsymbol{y}) = \frac{\delta^2 \Gamma}{\delta M_i(\boldsymbol{x}) \delta M_j(\boldsymbol{y})}.$$

Calculate the Fourier transforms $\tilde{G}_{11}(\boldsymbol{q})$, $\tilde{G}_{12}(\boldsymbol{q})$, and $\tilde{G}_{22}(\boldsymbol{q})$ under the conditions of part (a). Show that $\tilde{G}_{22}(\boldsymbol{q})$ diverges as $q \to 0$. What are the physical consequences of this?

2.5 The Hamiltonian and the correlation function on a lattice

(a) Let us start from a Gaussian Hamiltonian on a cubic lattice with unit lattice spacing ($a = 1$),

$$H = -J \sum_{\langle i,j \rangle} \varphi_i \varphi_j,$$

where $\langle i, j \rangle$ denotes nearest neighbours, and let the probability for the configuration $[\varphi_i]$ be

$$P[\varphi_i] = \frac{1}{Z} \exp\left[-\beta H + \frac{1}{2} \sum_i \varphi_i^2 \right].$$

By going to Fourier space, show that the correlation function $\langle \varphi_k \varphi_l \rangle$ is given by

$$\langle \varphi_k \varphi_l \rangle = \int \frac{\mathrm{d}^D q}{(2\pi)^D} \frac{\mathrm{e}^{\mathrm{i}\boldsymbol{q} \cdot (\boldsymbol{r}_k - \boldsymbol{r}_l)}}{1 - 2\beta J \sum_{\mu=1}^{D} \cos q_\mu},$$

where q_μ is the μ-component of $\boldsymbol{q}$ ($-\pi \leqslant q_\mu \leqslant \pi$). Show that the integral is defined only if $\beta < \beta_c = (2JD)^{-1}$.

(b) Show that one can rewrite the exponent in $P[\varphi_i]$ as

$$\frac{1}{2} \beta J \sum_{i,\mu} (\partial_\mu \varphi_i)^2 + \frac{1}{2} (1 - 2\beta DJ) \sum_i \varphi_i^2.$$

Add to this expression a term

$$-g\sum_i(\varphi_i^2-1)^2,$$

and show that $P[\varphi_i]$ is equivalent to that obtained from the GL Hamiltonian (2.1.14). Show that the Ising model is recovered in the limit $g\to\infty$.

(c) Now revert to the GL Hamiltonian (2.1.14) with lattice spacing a, and define the Fourier transforms (see Appendix A)

$$\tilde{G}(\boldsymbol{q})=a^D\sum_{\boldsymbol{y}}G(\boldsymbol{x}-\boldsymbol{y})e^{i\boldsymbol{q}\cdot(\boldsymbol{x}-\boldsymbol{y})}.$$

Show that in the Landau approximation the correlation function $\tilde{G}(\boldsymbol{q})$ in a uniform field is given by

$$\tilde{G}(\boldsymbol{q})=\frac{1}{(2/a^2)\sum_1^D(1-\cos(aq_\mu))+r_0+\frac{1}{2}u_0M^2},$$

where $q_\mu=(\boldsymbol{q}\cdot\boldsymbol{\mu})/a$.

What is now the correlation function of the Gaussian model? Show that, if $qa\ll 1$, then this correlation function is approximately invariant under rotations. Show that the term in q^4 breaks the invariance.

(d) Determine the matrix D_{ij} which is the lattice equivalent of the function $D(\boldsymbol{x},\boldsymbol{y})$ featured in Section 2.4.2. Show that in the limit $a\to 0$ one indeed recovers the result (2.4.5).

2.6 Correlation functions in D dimensions for $T>T_0$

(a) For $T>T_0$ the correlation function in the Landau approximation (or in the Gaussian model) satisfies

$$(-\nabla^2+r_0)G(\boldsymbol{x})=\delta(\boldsymbol{x}).$$

In view of the fact that $G(\boldsymbol{x})$ depends only on $r=\|\boldsymbol{x}\|$, show that $G(r)$ satisfies

$$-\frac{d^2G}{dr^2}-\frac{D-1}{r}\frac{dG}{dr}+r_0G=\delta(\boldsymbol{x}).$$

(b) Show that for $D=2$ and $r_0=0$ (i.e. exactly at the critical point) the equation $-\nabla^2G(\boldsymbol{x})=\delta(\boldsymbol{x})$ is solved by $-\ln r/2\pi$. (*Hint*: think of the electrostatic potential of an infinitely extended line charge.)

(c) Show that in D dimensions

$$G(r)=\frac{1}{(2\pi)^{D/2}r^{D/2-1}}\int_0^\Lambda\frac{q^{D/2}dq}{q^2+r_0}J_{D/2-1}(qr)$$

where J_ν is the Bessel function of order ν. In D-dimensional space the integration measure is

$$\mathrm{d}^D x = r^{D-1}\mathrm{d}r \sin^{D-2}\theta_{D-1}\,\mathrm{d}\theta_{D-1}\sin^{D-3}\theta_{D-2}\mathrm{d}\theta_{D-2}\ldots\mathrm{d}\theta_1$$

$$0 \leqslant \theta_k \leqslant \pi, \quad k \neq 1; \quad 0 \leqslant \theta_1 \leqslant 2\pi.$$

(d) Thence, for $D = 4$ and $r_0 = 0$, deduce that

$$G(r) = \frac{1}{4\pi^2 r^2}(1 - J_0(\Lambda r)).$$

(e) Write

$$G(r) = \int_0^\infty \mathrm{d}t \int \frac{\mathrm{d}^D q}{(2\pi)^D} \exp[-\mathrm{i}\boldsymbol{q}\cdot\boldsymbol{r} - t(q^2 + r_0)].$$

Integrate over q and estimate the integral using a saddle point method. Check the large-r behaviour (1.4.12) for $\eta = 0$.

2.7 The specific heat and the Ginzburg criterion

(a) Show that in zero field the specific heat per unit volume is

$$C = -\frac{1}{V}T\frac{\mathrm{d}^2(\Gamma T)}{\mathrm{d}T^2}.$$

In parts (b) and (c) we shall use the fact (which must itself be proved) that the discontinuous part of the specific heat is

$$C_\mathrm{s} = -\frac{1}{V}T^2\frac{\mathrm{d}^2\Gamma}{\mathrm{d}T^2}.$$

(b) Show that in the Landau approximation

$$C = 0, \quad T > T_0,$$

$$C \approx T^3\frac{3\bar{r}_0^2}{u_0}, \quad T < T_0.$$

Hence the specific heat is discontinuous, with a discontinuity

$$\Delta C = 3\bar{r}_0^2 T_0^2/u_0.$$

(c) Show that, if the leading correction to the Landau theory is taken into account, the specific heat does not vanish for $T > T_0$. What then is the expression for C?

From here on we restrict ourselves to $T > T_0$.

(d) Show that for $D > 4$ the correction merely augments the specific heat by a term whose variation with T is slow. Show that, by contrast, for

$D < 4$ the specific heat behaves like

$$(T - T_c)^{-\alpha}.$$

What is the value of α?

(e) For $D < 4$ one can define the domain of validity of the Landau approximation by demanding that the specific heat calculated in (c) be smaller than the discontinuity calculated in (b):

$$C \lesssim \Delta C.$$

Use this argument to recover the Ginzburg criterion.

(f) Show that the average energy $\langle E \rangle$ per unit volume is proportional to the correlation function $G(\boldsymbol{x})$ evaluated at $\boldsymbol{x} = 0$: $\langle E \rangle \sim G(\boldsymbol{x} = 0)$. Exploit the formula (2.3.5a) to reproduce the conclusion of part (c).

2.8 Functional derivatives

The usual definition of a functional derivative is as follows: let $I(f)$ be a functional of f, and let $\varepsilon h(x)$ be a small variation around f. Write

$$I(f + \varepsilon h) = I(f) + \varepsilon \int \mathrm{d}x \frac{\delta I}{\delta f(x)} h(x).$$

This equation *defines* the functional derivative $\delta I/\delta f(x)$. From this definition, derive equations (2.1.23) to (2.1.26). In addition find the functional derivative of $I(f) = \operatorname{Max} f(x)$. Show that it is defined only if the maximum is unique.

2.9 Behaviour of the magnetization at a boundary surface

Consider a ferromagnet for $T < T_c$ with a boundary surface: $M \to M_0$ when $z \to \infty$, $M \to -M_0$ when $z \to \infty$; the system is assumed to be invariant for any translation parallel to the (x, y) plane. It is also assumed to be described by Landau's theory (see e.g. equation (2.1.29)), with $M_0 = (6|r_0|/u_0)^{1/2}$.

(a) Find the differential equation which is obeyed by the function $f(z) = M(z)/M_0$. Integrate this differential equation, taking into account the boundary conditions at $z = \pm\infty$, and assuming that $f(0) = 0$. What happens if instead $f(z_0) = 0$?

(b) Show that the surface energy σ per unit area is given by

$$\sigma = \frac{3r_0^2}{2u_0} \int_{-\infty}^{\infty} \mathrm{d}z (1 - \tanh^2(z/2\xi)),$$

where ξ is the correlation length.

Further reading

The Ginzburg–Landau Hamiltonian and the Landau approximation are discussed by Pfeuty and Toulouse (1977, Chapter 2), and by Ma (1976, Chapter III). See also Shenker (1982, Section 2). Mathematical details on functional integration can be found in Glimm and Jaffe (1987, Chapter 3). The Landau theory of phase transitions is described by Landau and Lifshitz (1980, Section 138), and by Kittel (1986, Chapter XIII); correlation functions are considered by Ma (1976, Chapter III). Pfeuty and Toulouse (1977, Chapter 2) give a very complete discussion of the Ginzburg criterion. At a more advanced level, see Amit (1984, Chapters 2 and 6), and Brézin *et al.* (1976, Section IV).

3 The renormalization group

Consider a physical system near a critical point. The number of degrees of freedom effectively interacting with each other is $\sim \xi^D$, where ξ is the correlation length, and at the critical point this number diverges since $\xi \to \infty$. Traditional perturbative methods fail completely in this kind of problem, because they are designed for cases where only a few degrees of freedom interact. The renormalization-group method, invented by Wilson, consists in systematically reducing the number of degrees of freedom, by integrating over short-wavelength fluctuations. Suppose we start with a spin system on a lattice, with lattice spacing a; then the minimum wavelength for fluctuations is of order a. Now integrate over fluctuations having wavelengths $a \lesssim \lambda \lesssim sa$, where $s > 1$ is called the dilatation factor. This makes no difference to the behaviour of correlation functions for $r > sa$: *integration over short-wavelength fluctuations assigns to the original physical system* another corresponding physical system having the same behaviour at long distances.* The transformation from one to the other is called a renormalization-group transformation (RGT); it can be iterated by integrating over fluctuations having wavelengths $sa \lesssim \lambda \lesssim s^2 a$ etc., thus establishing a whole sequence of corresponding physical systems all with the same long-distance behaviour.

At this point it is essential to be explicit about what one means by the long-distance behaviour being 'the same'. If we use the same unit of length to describe both systems, then this expression is quite correct. However, each system possesses a natural unit of length, its effective lattice spacing: the effective lattice spacing of the (transformed) system after the first transformation is sa, and the correlation length of the transformed system measured in the appropriate natural units is ξ/s. Accordingly, it is logical to associate with every RGT *an increase (dilatation) of the unit of length by a factor s*; this will allow us to compare the two systems on a lattice common to both. After sufficiently many transformations we can make the initial system correspond to another, whose correlation length is of the order of the lattice spacing, and whose behaviour we can therefore hope to determine by perturbation methods.

The dilatation of the unit of length by a factor s, transforming a length r into r/s, is evidently reminiscent of a scaling transformation. A RGT does indeed link two correlation functions of arguments r and r/s respectively; but we are not

* Here, 'physical system' means 'a model for a phase transition'. The model need not describe any actual physical system, not even approximately.

dealing with simple dimensional analysis, because the transformation affects the system parameters as well: the integration over short-wavelength fluctuations is taken into account by changing (renormalizing) the system parameters. An iterated RGT yields no useful information without further input. The requisite input is the existence of a *fixed point*: near a fixed point we can restrict attention to a *finite* number of (so-called relevant) parameters. In that case one finds that the correlation functions exhibit simple behaviour under scale transformations $r \to r/s$, which in turn allows one to determine the critical exponents.

Intuitively, one can visualize an RGT by imagining the physical system observed through two microscopes with different resolutions. The first uses a wavelength λ, permitting the observation of details having dimensions $\sim \lambda$; the second uses a wavelength $s\lambda$, whence its resolving power is less ($s > 1$). The change from the first microscope to the second integrates out details having dimensions between λ and $s\lambda$. To represent the dilatation of the unit of length, one need merely suppose that the magnification of the second microscope is less than that of the first by a factor $1/s$.

Through the second microscope, positively magnetized clusters (say) will appear s times smaller than through the first, provided the correlation length is finite (recall that these clusters are of size $\sim \xi$) (Fig. 3.1).

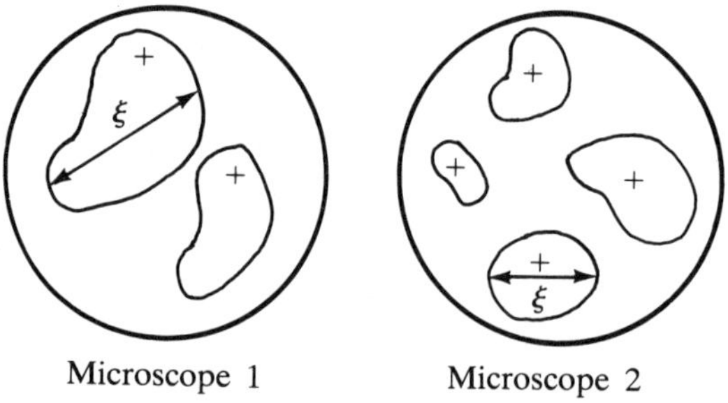

Fig. 3.1 Intuitive view of an RGT

At the critical point ($T = T_c$) there are fluctuations of all possible sizes, and the images seen through the two microscopes are analogous if they are averaged over time (at any one time of course a cluster of given size covers a space s times smaller in the field of view of the second microscope as compared with the first).

Before embarking on our study of the renormalization group, an important preliminary remark is in order: the renormalization group is a recent invention, and it is not backed by rigorous theorems except in certain special cases. Hence we shall need to introduce assumptions whose domain of validity is not precisely defined, and which at the end of the day are justified by their successes in practice.

Section 3.1 introduces the basic concepts: the definition of blocks of spins, critical surfaces, and fixed points. Section 3.2 shows how to obtain critical exponents by studying the vicinity of a fixed point. These two initial sections give the general results, and are followed by others devoted to illustrating them in special cases. Section 3.3 gives an example of RGTs on a lattice, with an approximate solution of the equations of the renormalization group (RG). Section 3.4 introduces the RG in Fourier space, with the Gaussian model as an example. Critical exponents are calculated to order $\varepsilon = (4 - D)$ in Section 3.5. Section 3.6 considers marginal variables, and allows us to make the connection with the classic renormalization methods of field theory, which will be described in Chapters 6 and 7.

3.1 Basic concepts: blocks of spins, critical surface, and fixed points

One possible strategy for integrating over short-wavelength fluctuations is to form blocks of spins. This is not the only possible strategy: we shall meet another in Section 3.4. Moreover there are two strategies for blocking spins: one linear, and another nonlinear. We begin with the latter because, initially, it seems to be the simpler. But difficulties soon appear, and eventually the linear strategy emerges as the more efficient, at least as regards explicit calculations.

3.1.1 Blocks of spins and nonlinear transformations

For definiteness we consider the Ising model in two dimensions ($D = 2$), on a square lattice of lattice spacing a. We shall group the spins by fours (= 1 block of spins), and to each block we shall attribute a spin determined by the following rules:

(a) the spin of the block is $+1$ (-1) if the sum of the spins in the block is positive (negative);

(b) if the sum of the spins in the block is zero, then we attribute, to the block, spin $+1$ or -1 by tossing a coin; alternatively we assign, by any arbitrary convention, the value $+1$ to three such configurations, and the value -1 to the other three. Blocks of nine spins would of course eliminate this problem, but the figures would take longer to draw.

The blocking of the spins is the first step of the renormalization-group transformation. The second step consists in reverting to the initial lattice, by doubling the unit of length; thus the dilatation factor is 2. This procedure allows us to compare two different physical systems (the initial spins and the blocks) on one and the same lattice (Fig. 3.2).

Let us write the initial spins as S_i and those of the blocks as S'_α, the latter being given by

$$S'_\alpha = f(S_i), \quad i \in \text{block } \alpha. \tag{3.1.1}$$

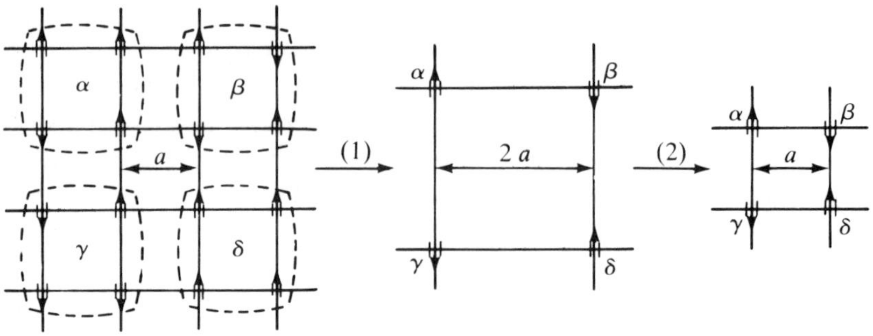

Fig. 3.2 The formation of blocks of spins

One example of such a function $f(S_i)$ has been given just above; more generally we could take any function $f(S_i)$ that reflects the tendency of the spins in the block to adopt a common orientation.

The probability of observing a given configuration $[S'_\alpha]$ of the blocks is uniquely determined once we know the interaction Hamiltonian of the spins. Consequently there must exist a Hamiltonian $H'[S'_\alpha]$ such that the probability of observing the configuration $[S'_\alpha]$ is proportional to $\exp(-H'[S'_\alpha])$. It is not hard to write down a formal expression for H',

$$e^{-H'[S'_\alpha]} = \sum_{[S_i]} \prod_{\alpha} \delta(S'_\alpha - f(S_i)|_{i\in\alpha}) e^{-H[S_i]} \tag{3.1.2}$$

($H'[S'_\alpha]$ contains a part G independent of S'_α, which has not been segregated explicitly for the moment: see Section 3.2, equation (3.2.12).) Given a configuration $[S_i]$ there exists just one configuration $[S'_\alpha]$ for which the Kronecker delta is nonzero; hence

$$\sum_{[S'_\alpha]} \prod_{\alpha} \delta(S'_\alpha - f(S_i)|_{i\in\alpha}) = 1, \tag{3.1.3}$$

which immediately entails

$$Z = \sum_{[S_i]} e^{-H[S_i]} = \sum_{[S'_\alpha]} e^{-H[S'_\alpha]} = Z'. \tag{3.1.4}$$

The partition function of the transformed system equals that of the initial system. But there is no reason why the Hamiltonian $H'[S'_\alpha]$ should have the Ising form (1.2.2). One is led to generalize the Hamiltonian (1.2.2) by writing

$$-H = K_1 \sum_{\langle i,j\rangle} S_i S_j + K_2 \sum_{\langle\langle i,j\rangle\rangle} S_i S_j + K_3 \sum_{\langle ijkl\rangle} S_i S_j S_k S_l + \cdots \tag{3.1.5}$$

(see Problem 3.1), where the first term features nearest neighbours, the second next-nearest neighbours, the third 'plaquettes', etc. (see Fig. 3.3). Note that the factor $1/kT$ is subsumed into the definition of the coefficients K_i. The only constraint is the symmetry property $H(S_i) = H(-S_i)$.

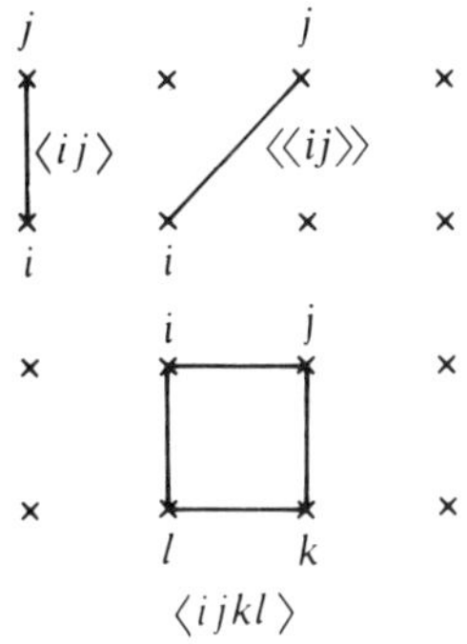

Fig. 3.3

The coefficients $K_1, K_2, \ldots, K_n \ldots$ are also called *coupling constants*, and are written collectively as μ; they define a *parameter space. A physical system at a given temperature is represented by a point in parameter space.* Accordingly, the RGT which establishes the correspondence between the system of spins and the system of blocks, call it R_2, is *a transformation acting on parameter space*: it maps a point μ of this space into another such point μ'. One could write

$$\mu = \{K_1, K_2, \ldots, K_n, \ldots\},$$

$$\mu' = \{K'_1, K'_2, \ldots, K'_n, \ldots\},$$

and

$$\mu' = R_2 \mu. \tag{3.1.6}$$

At this point we must introduce two assumptions that will be needed later:

A1. The coefficients $(K_1, K_2, \ldots)$ are analytic functions of the temperature even at $T = T_c$, and the K'_α are analytic functions of the K_α.

A2. The RGTs do not introduce long-range interactions, i.e. they do not introduce any coefficients that mediate strong couplings between distant spins: the interaction between any two spins must fall faster than any inverse power of their separation.

These assumptions are reasonable, since the transformation (3.1.6) is local in space (to each block it assigns only a finite number of spins); but it is not known under what general conditions they apply. Certain transformations (see Section 3.4) do introduce long-range interactions, but these are just artefacts of particular methods. Such interactions should have no physical consequences, though they can pose technical problems for the unwary.

Two generalizations are possible immediately.

(a) Work in a D-dimensional space: each block then contains 2^D spins.

(b) Form blocks containing s^D spins instead of 2^D, and multiply the unit of length by s. The corresponding RGT is written R_s:

$$\mu' = R_s \mu. \tag{3.1.7}$$

For a spin system, $s = \sqrt{2}, 2, \sqrt{3}$, etc.; in Section 3.4 we shall meet a technique that admits continuously variable s.

As explained at the start of this chapter, the renormalization-group strategy consists in iterating the RGT R_s many times: we *define* R_{s^n} as the nth iterate of R_s: $R_{s^n} = R_s \ldots R_s$*. At this point it is convenient to answer the following question: why must one iterate R_s, instead of proceeding directly to R_{s^n}? The fact is that one cannot carry out the integration for $s \gg 1$, except in trivially solvable models (where the RG is not needed in the first place); by contrast, the calculation for $s \sim 1$ is feasible at least approximately, because it involves only fluctuations having wavelengths $\sim sa$ ($s^n a$ after n iterations), corresponding to only a limited number of degrees of freedom. Whenever an argument features a dilatation factor $s \gg 1$, this factor will have to be interpreted as the result of many iterations of the RGT.

3.1.2 Linear transformations

In forming blocks as described above, the function $f(S_i)$ is nonlinear, whence the reference to 'nonlinear RGTs'. One method of defining the spin of a block that *is* linear constructs the average of the spins in the block, and writes

$$S'_\alpha = \frac{\lambda(s)}{s^D} \sum_{i \in \alpha} S_i. \tag{3.1.8}$$

Here, $\lambda(s)$ is a function of s whose role will be spelled out later. Needless to say the block spins no longer equal ± 1; after a few iterations of the RGT the value of the spin becomes practically continuous, and thus analogous to the variable φ_i (or $\varphi(\boldsymbol{x})$) used in the Ginzburg–Landau theory.

The form (3.1.5) of the Hamiltonian is no longer suited to defining parameter space. The Hamiltonian to which one is led after a few iterations is analogous to that of Ginzburg and Landau; it is augmented however by an infinite number of further terms of the kind φ^6, φ^8, $\varphi^2(\nabla\varphi)^2$, etc., over and above the standard components $(\nabla\varphi)^2$, φ^2, and φ^4. *Parameter space now becomes the space spanned by the coefficients of all these various terms*; in other words once again it becomes the space of the coupling constants.

Although at first sight this linear strategy seems more complicated than the nonlinear one, it has the very remarkable feature that the correlation length ξ' of

* Note that R_{s^2} is *not* equivalent to the formation of blocks of s^{2D} spins, for a reason well-known from indirect elections: what constitutes a majority of the 4 blocks does not always constitute a majority of the 16 spins!

the transformed system equals that of the initial system divided by s:

$$\xi' = \xi/s. \tag{3.1.9}$$

To see this, consider two blocks far enough apart, and calculate $\langle S'_\alpha S'_\beta \rangle$:

$$\begin{aligned}
\langle S'_\alpha S'_\beta \rangle &= \frac{1}{Z} \sum_{[S']} S'_\alpha S'_\beta \, e^{-H'[S']} \\
&= \frac{1}{Z} \sum_{[S']} \sum_{[S]} S'_\alpha S'_\beta \prod_\gamma \delta\left(S'_\gamma - \frac{\lambda(s)}{s^D} \sum_{i \in \gamma} S_i \right) e^{-H[S]} \\
&= \frac{\lambda^2(s)}{s^{2D}} \frac{1}{Z} \sum_{[S]} \left(\sum_{i \in \alpha} S_i \right) \left(\sum_{j \in \beta} S_j \right) e^{-H[S]} \\
&= \frac{\lambda^2(s)}{s^{2D}} \sum_{i \in \alpha} \sum_{j \in \beta} G_{ij} \simeq \lambda^2(s) G_{ij}.
\end{aligned}$$

Here we have used equations (3.1.2) and (3.1.8), and have then interchanged the order of the summations over $[S]$ and $[S']$; in the last step we have assumed $\xi \gg sa$, in which case G_{ij} varies little as i, j range over the blocks α, β respectively. If $G_{ij} \sim \exp(-r_{ij}/\xi)$ for sufficiently distant i and j, and in view of $r_{\alpha\beta} \approx r_{ij}/s$ (recall that in fact $r_{\alpha\beta} = r_{ij}$, but that these two distances are measured in different units), we have

$$G_{\alpha\beta} \sim \exp(-s r_{\alpha\beta}/\xi),$$

which proves (3.1.9). For $r \gg a$ we can write the general relation

$$G\left(\frac{r}{s}, \mu'\right) \simeq \lambda^2(s) G(r, \mu). \tag{3.1.10}$$

The dilatation of the unit of length ($r \to r/s$) and the dilatation of the spins ($\lambda(s)$) must not be allowed to obscure the fact that the systems parametrized by μ and μ' have basically the same long-distance behaviour if expressed in the same units. Note moreover that one can interpret equation (3.1.10) as the transformation law for the correlation function under a scale transformation (see Section 1.5). However, this transformation law is complicated, because the parameters of the Hamiltonian are modified: $\mu \to \mu'$. Indeed we still have some way to go before we are in a position to exploit this relation.

Strictly speaking the relations (3.1.9) and (3.1.10) have been derived only for linear RGTs. But in order to simplify the argument we shall assume them to hold, at least approximately, even in the nonlinear case.

Under linear RGTs, the formation of blocks of $(s_1 s_2)^D$ spins yields the same transformation as does the product of two RGTs corresponding to the formation of blocks of s_1^D and of s_2^D spins respectively; this is true provided $\lambda(s)$ satisfies

$$\lambda(s_1 s_2) = \lambda(s_1)\lambda(s_2).$$

If this equation is obeyed, then $\lambda(s)$ is of the form

$$\lambda(s) = s^{d_\varphi}, \tag{3.1.11}$$

where d_φ is a constant called the *anomalous dimension* of the field. This nomenclature will be explained later (Section 3.4). In actual fact the argument just given is not quite correct, because $\lambda(s)$ can be a function also of the parameters of H; equation (3.1.11) applies locally in parameter space, and in principle one should write $d_\varphi(\mu)$ (see Problem 3.4(c)).

3.1.3 Critical surface and fixed points

Though the argument applies generally, let us for definiteness revert to the Ising model in two dimensions, and to nonlinear RGTs. At a given temperature the model is represented by a point in parameter space (see (3.1.5)),

$$K_1 \neq 0; \quad K_2 = K_3 = K_4 = \cdots = 0.$$

The T-dependence enters through K_1; as T varies, the point representing the Ising model describes a line in parameter space, called the *physical line of the Ising model.*

At a certain value

$$K_1 = K_{1c} \approx 0.44,$$

corresponding to $T = T_c = 2.27J/k$ (see Section 1.2.4), the Ising model displays a second-order phase transition, and its correlation length diverges.

Let us start from a point $(K_{1c}, 0, 0, \ldots)$ of parameter space, and apply an RGT. The correlation length ξ' remains infinite ($\xi' = \xi/s$), and the transformed system is critical as before. The locus of points in parameter space that correspond to systems at the critical point ($\xi = \infty$) is called *the critical surface* (*or critical manifold*) S_∞. An RGT applied to a point on S_∞ transforms it into another point likewise on S_∞. Since the correlation length is infinite at the transition, we are restricting ourselves to second-order transitions (or, possibly, to transitions of higher order).

If we start from a point $Q \notin S_\infty$, then the successive RGTs $Q \to Q' \to Q'' \ldots$ move the representative point further and further away from the critical surface, since each operation divides the correlation length by s, leading successively to systems that are more and more remote from the critical region. Under iterations of the RGTs the representative points describe a system of trajectories called the *renormalization flow.* If we disregard the dilatation of the units of length and of spin, then *all the points on a given trajectory correspond to one and the same long-range behaviour* (Fig. 3.4).

The behaviour of the points $P', P'', \ldots, P^{(n)}$ arising from $P \in S_\infty$ by successive iterations of the RGT could, a priori, be completely arbitrary: there might be double points, limit cycles, etc. The physically interesting case is that where the

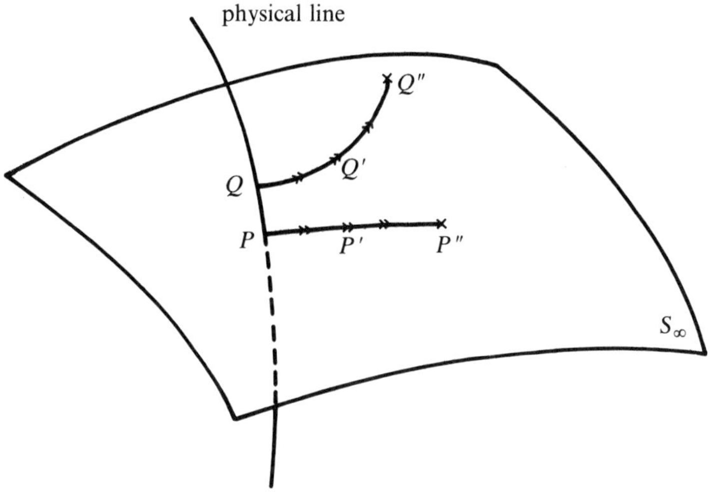

Fig. 3.4 Evolution under an RGT

sequence $P \to P' \to \ldots \to P^{(n)} \to \ldots$ converges to a fixed point P^*:

$$\lim_{n\to\infty} P^{(n)} = P^*,$$

with the fixed point characterized by a set μ^* of parameters such that

$$R_s\mu^* = \mu^*. \tag{3.1.12}$$

Speaking more precisely, there exists on the critical surface a catchment area ('basin of of attraction') $\mathscr{D}(P^*)$ of the fixed point, such that $\lim_{n\to\infty} P^{(n)} = P^*$ if $P \in \mathscr{D}(P^*)$. We shall ignore cases where $\mathscr{D}(P^*)$ is of lower dimensionality than the critical surface. They lead to so-called 'cross-over' behaviour (see Pfeuty and Toulouse (1977, Chapter 8), and Problem 3.9).

It is possible for there to be several fixed points, each with its own basin of attraction. It is also possible for a fixed point to recede to infinity. At present, no general conditions are known for the existence of fixed points, nor for their basins of attraction: in any particular case one must proceed by explicit calculation. Accordingly we adopt the following assumption:

A3. If $P \in S_\infty$, and if one carries out a large number of RGTs, then $P^{(n)}$ converges to a fixed point $P^* \in S_\infty$ which obeys $R_s\mu^* = \mu^*$. There may be several fixed points, each with its own basin of attraction.

We end this section with four comments.

(i) In general the position of the fixed point depends on the particular form chosen for the RGTs. On the other hand, physical results (critical exponents) must not depend on the particular form of the RGTs: see Problem 3.8.

(ii) The role of $\lambda(s) = s^{d_\varphi}$. According to (3.1.10) one finds that at the fixed point

$$G(r, \mu^*) = s^{-2d_\varphi} G\left(\frac{r}{s}, \mu^*\right).$$

(Strictly speaking, one should specify that d_φ be evaluated for the fixed-point parameters: $d_\varphi = d_\varphi(\mu^*)$.)

The choice $s = r/a$ in the preceding equation shows that at the fixed point the correlation function behaves according to

$$G(r, \mu^*) = \left(\frac{a}{r}\right)^{2d_\varphi} G(a, \mu^*).$$

Since $P^* \in S_\infty$, we know from the definition of the critical exponent η that $G(r)$ must behave like $(r)^{-D+2-\eta}$. Hence we may identify

$$d_\varphi = \frac{1}{2}(D - 2 + \eta). \tag{3.1.13}$$

The result (3.1.13) shows that one must introduce a dilatation factor $\lambda(s) = s^{d_\varphi}$ into the definition of the RGTs: it is impossible to arrive at a fixed point unless this dilatation factor has been chosen appropriately, and, in particular, unless it is related to η according to (3.1.13). Study of the Gaussian model (Section 3.4.2) will allow us to illustrate the necessity of the factor $\lambda(s)$.

(iii) An isolated fixed point leads to critical exponents that are unique. However one can also meet lines (or surfaces) of fixed points: see Chapter 4. In that case critical exponents depend continuously on parameters like the temperature.

(iv) Why call it the 'renormalization group'?

The idea of the 'renormalization group' was formulated by Stueckelberg and Petermann (1953), and independently by Gell-Mann and Low (1954), in the context of the divergences of quantum field theory (see Chapter 6). 'Renormalization' is a procedure for making the theory finite; such procedures are not unique, and, originally, the 'renormalization group' expressed the invariance of the physics under changes in the procedure. This version of the RG may be considered as a special case of the version introduced by Wilson in 1971, which is the one we have been describing. The links between the two versions will be examined in Chapter 7. Note meanwhile that the spin dilatation in (3.1.8) by the factor $\lambda(s)$ is often called 'spin renormalization', by analogy to a similar operation in quantum field theory.

3.2 Behaviour near a fixed point. Critical exponents

The behaviour of RGTs in the vicinity of a fixed point will enable us to calculate the critical exponents by linearizing the RG equations near that point. In order to produce an elementary explanation, we start by pretending that parameter space has only one dimension, i.e. that one parameter suffices to describe the sequence of RGTs. Obviously this assumption is oversimplified, and we shall show later how it can be relaxed.

3.2.1 Elementary treatment

Thanks to the analyticity assumption (A1) (Section 3.1.1), the equation

$$K' = R_s K$$

can be linearized in the vicinity of the fixed point. If K is near K^*, a Taylor expansion around K^* yields

$$K' = K^* + (K - K^*) \left.\frac{\mathrm{d}R_s}{\mathrm{d}K}\right|_{K^*} + O(K - K^*)^2,$$

where the terms of order $(K - K^*)^2$ have been neglected; since $R_{s_1} R_{s_2} = R_{s_1 s_2}$, one can write $\mathrm{d}R_s/\mathrm{d}K|_{K^*} = s^y$:

$$K' - K^* = s^y(K - K^*).$$

By virtue of assumption (A1), K is an analytic function of T, and barring a coincidence it must vanish linearly at T_c where it changes sign:

$$K - K^* \sim T - T_c.$$

(Recall that K^* lies on the critical surface which here shrinks to a point.)

We choose $|K - K^*| \to 0$, and $|K' - K^*|$ finite but small enough to validate the linear approximation; the point K' in parameter space represents a system far from criticality. Henceforth we write such a condition as $|K' - K^*| \sim 1$. Then the correlation length $\xi(K')$ is finite and of order a; measuring the correlation length in units of a one has $\xi(K') \sim 1$*. One can now evaluate $\xi(K)$ in view of

$$\xi(K) = s\xi(K') = \left(\frac{K' - K^*}{K - K^*}\right)^{1/y} \xi(K') \sim |T - T_c|^{-1/y},$$

* One can be be somewhat more precise by taking $|K' - K^*| \sim k_0$ and $\xi(K') = \xi_0$, where k_0 and ξ_0^{-1} may be small (10^{-2} ?) but *finite*, in order to validate the Taylor expansion. It is easy to show that this makes no difference to the result.

whence

$$\nu = \frac{1}{y}. \tag{3.2.1}$$

Thus the critical exponent ν relates to the derivative of the RGT at the fixed point. Note that the result applies equally for $T > T_c$ and for $T < T_c$; in other words $\nu = \nu'$ (see the footnote on p. 17). The conclusion will be extended to parameter spaces of an arbitrary number of dimensions.

3.2.2 Linearization near a fixed point

Let $\mu = \{K_\alpha\}$ be a point in parameter space near to $\mu^* = \{K_\alpha^*\}$:

$$K_\alpha = K_\alpha^* + \delta K_\alpha;$$

and let $\mu' = \{K'_\alpha\}$ be its transform under an RGT: $\mu' = R_s\mu$.

The relation between $\delta K'_\alpha$ and δK_α is approximately linear if μ and μ^* are close enough:

$$\delta K'_\alpha \simeq \sum_\beta T_{\alpha\beta}(s)\delta K_\beta, \tag{3.2.2}$$

where

$$T_{\alpha\beta}(s) = \left.\frac{\partial K'_\alpha}{\partial K_\beta}\right|_{\mu^*}. \tag{3.2.3}$$

From the relation $T(s_1 s_2) = T(s_1)T(s_2)$ written for $s_1 = 1 + \delta$, $s_2 = s$, and for infinitesimal δ, one can show that

$$T(s) = \exp(\mathscr{T} \ln s),$$

where $\mathscr{T} = \mathrm{d}T/\mathrm{d}\ln s|_{s=1}$. Let $e^{(i)}$ be an eigenvector of $\mathscr{T}_{\alpha\beta}$, corresponding to an eigenvalue y_i:

$$\sum_\beta \mathscr{T}_{\alpha\beta}\, e_\beta^{(i)} = y_i e_\alpha^{(i)}.$$

Accordingly this vector satisfies

$$\sum_\beta T_{\alpha\beta}(s) e_\beta^{(i)} = s^{y_i} e_\alpha^{(i)}.$$

In general the matrix $\mathscr{T}_{\alpha\beta}$ is not symmetric, and there is no guarantee whatever that its eigenvalues are real nor that its eigenvectors form a complete set. We shall assume nevertheless that, in the event, they do (somewhat weaker assumptions would also serve). If so, then any point of parameter space can be referred to the basis $\{e^{(i)}\}$,

$$\delta K_\beta = \sum_i t_i e_\beta^{(i)};$$

in the linear approximation this becomes

$$\delta K'_\alpha = \sum_i t_i s^{y_i} e_\alpha^{(i)}. \tag{3.2.4}$$

The coefficient t_i is called a *scaling field.* An RGT multiplies the scaling field by s^{y_i}. Equation (3.2.4) shows that we must distinguish between three cases.

(i) $y_i > 0$: the scaling field increases under iterations of the RGTs; t_i is called a *relevant field.*

(ii) $y_i = 0$: in the linear approximation the scaling field remains constant. To determine its actual behaviour one must go beyond this approximation. Such fields are called *marginal.*

(iii) $y_i < 0$: the scaling field decreases under iterations of the RGTs. Such fields are *irrelevant.* It should be stressed that the relevance, irrelevance, or marginality of fields are defined *with respect to a given fixed point.*

In the vicinity of the fixed point the Hamiltonian may be written

$$H = H^* + \sum_i t_i O_i,$$

where the coefficients O_i of the scaling fields are the *scaling operators* conjugate to the fields. If the field t_i is, respectively, relevant, irrelevant, or marginal, then the operator O_i will likewise be called relevant, irrelevant, or marginal.

Suppose for the moment that there are no marginal fields; cases where there are will be studied in Section 3.6. The vectors $e^{(i)}$ span a vector space whose origin is the fixed point ($t_i = 0\ \forall i$). If there are N relevant fields, then, in order to be on the critical surface, one must in the linear approximation fix N parameters $t_1 = t_2 = \cdots = t_N = 0$. To see this, note that if t_1 say is nonzero, then iteration of the RGT will move the representative point away from the fixed point, whence the starting point cannot have been on the critical surface. (This assumes tacitly that there is no cross-over behaviour.) Conversely, if $t_1 = t_2 = \cdots = t_N = 0$, then in the linear approximation the representative point converges to the fixed point, and the starting point must have been on the critical surface. *Accordingly, the hyperplane* $t_1 = t_2 = \cdots = t_N = 0$ *is the tangent plane of the critical surface at the fixed point.*

The most important cases are those where one can reach the critical surface by varying just one parameter (namely the temperature). They correspond to second-order transitions; conversely, such transitions correspond to situations where, at the critical point in question, there exists just one relevant field. (For clarity, we work with zero B-field. Some authors prefer to include external fields and to define a second-order phase transition by $t = B = 0$; hence the terminology 'tricritical point' in what follows).

Tricritical points are obtained by fixing two parameters (say temperature and pressure); these cases arise when there are two relevant fields. The generalization to polycritical points of order N is obvious: one needs N relevant fields. Henceforth we confine ourselves to second-order transitions, and accordingly to just one relevant field t_1.

Figure 3.5 shows, schematically, the vicinity of the fixed point in the vector space for the case of three parameters. The double-headed arrows indicate the direction in which the representative point moves under an RGT. Notice the one divergent axis ($e^{(1)}$) and the two convergent axes ($e^{(2)}$ and $e^{(3)}$): *the fixed point has an instability of order 1.*

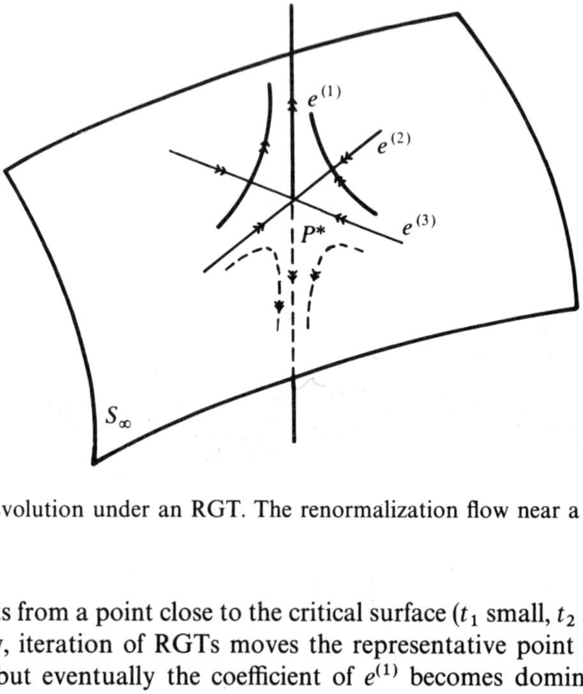

Fig. 3.5 Evolution under an RGT. The renormalization flow near a fixed point

If one starts from a point close to the critical surface (t_1 small, t_2 and t_3 finite), then, initially, iteration of RGTs moves the representative point closer to the fixed point; but eventually the coefficient of $e^{(1)}$ becomes dominant, and the point then moves away from P^*. This observation allows one to sketch the *renormalization flow* at least qualitatively (Fig. 3.5).

The linear approximation fails if the parameters t_i are large; but even then the general pattern of the renormalization flow remains the same, though its technical description may become more complex. Note in particular that all trajectories which do not start from a point located on the critical surface converge either to the $T = \infty$ fixed point, or to the $T = 0$ fixed point: indeed $\xi = 0$ and $\xi = \infty$ are the only solutions to $\xi = \xi/s$.

3.2.3 The correlation function in zero field

Before coming to grips with the equations it is useful to give a heuristic justification of the role played by the fixed point in determining critical exponents. Consider therefore a fixed point with corresponding scaling fields $t_1, t_2, t_3, \ldots, t_i, \ldots$, where

$$y_1 > 0; \quad \cdots < y_i < \cdots < y_3 < y_2 < 0.$$

Start from a physical system where t_1 is very small (10^{-6}) while $t_2, t_3, \ldots$ are small but finite (10^{-1}), these numbers being, obviously, somewhat arbitrary. Under iterations of the RGTs, the representative point rapidly approaches the fixed point, and then spends a long time in its vicinity before finally diverging along the $e^{(1)}$ axis (see Fig. 3.5). A wide interval of s will be dominated by the neighbourhood of the fixed point, and it is this interval that is responsible for the critical behaviour.

Let us now make the argument quantitative. As in our simplified discussion, the critical indices will be related to the eigenvalues s^{y_i} of the matrix $T_{\alpha\beta}(s)$. The correlation function $G(r)$ is a function of the scaling fields t_1, t_2, $t_3, \ldots$; and, barring accidents, the field t_1 must vanish linearly on the critical surface:

$$t_1 \sim t = \frac{T - T_c}{T_c}.$$

Next, we appeal to the relation (3.1.10) and to the transformation law for the scaling fields, and adopt the convention that all lengths (ξ, r) are to be measured in units of the lattice parameter a:

$$G(r; t_1, t_2, \ldots) = s^{-2d_\varphi} G\left(\frac{r}{s}; t_1', t_2', \ldots\right). \tag{3.2.5}$$

Start on the critical surface, with $t_1 = 0$. Equation (3.2.5) then becomes

$$G(r; 0, t_2, \ldots) = s^{-2d_\varphi} G\left(\frac{r}{s}; 0, s^{y_2} t_2, \ldots\right).$$

Let us now choose $s = r$; this choice corresponds to integrating over all fluctuations having wavelengths between a and r. One finds

$$G(r; 0, t_2, \ldots) = r^{-2d_\varphi} G(1; 0, r^{y_2} t_2, \ldots).$$

If $r^{y_2} t_2 \ll 1$, then this equation shows that at the critical point the correlation function indeed obeys a power law, the critical exponent η being given as a function of d_φ by equation (3.1.13). Thus we have verified, for $r/a \gg 1$, the property of scale-invariance at the critical point that was considered in Section 1.5. *The width of the critical region*, i.e. of the region displaying scale-invariance, is governed by

$$\text{constant} \times t_2 (r/a)^{y_2} \ll 1.$$

The constant depends on the microscopic details of the model; as y_2 becomes more negative, scale invariance applies for lower values of r. Note that the length-scale a has dropped out of sight completely, but only provided $r/a \gg 1$. This is the feature that distinguishes scale-invariance at the critical point from *naïve scale-invariance* of purely dimensional origin (see Section 3.4.2); the latter would apply for all values of r.

Suppose now that we are not on the critical surface ($t_1 \neq 0$), and that the RGTs have been iterated a sufficient number of times for the correlation function on the right of (3.2.5) to be evaluated far from the critical region:

$$s^{y_1} t_1 \sim \pm 1.$$

Let ξ denote the quantity $|t_1|^{-1/y_1} \sim |t|^{-1/y_1}$ (ξ will obviously be identified with the correlation length); then equation (3.2.5) yields

$$G(r) = s^{-2d_\varphi} G\left(\frac{r}{s};\ \pm\left(\frac{s}{\xi}\right)^{y_1}, s^{y_2} t_2, \ldots\right).$$

Here the $+$ sign corresponds to $t > 0$ ($T > T_c$), and the $-$ sign to $t < 0$ ($T < T_c$). We have $s \sim \xi$ by construction, and find the desired result

$$G(r) = \xi^{-2d_\varphi} G\left(\frac{r}{\xi};\ \pm 1, \xi^{y_2} t_2, \ldots\right).$$

Physically, the choice $s = \xi$ corresponds to integrating over all fluctuations with wavelengths $a \lesssim \lambda \lesssim \xi$; after iteration of the RGTs the correlation length is of order 1, and the correlation function is nonsingular with respect to $(T - T_c)$.

According to the last equation the behaviour of $G(r)$ is simple provided $\xi^{y_2} t_2 \ll 1$; and once again we define the critical region as the region where this condition is obeyed, which it will be if $|T - T_c|$ is small enough. Inside the critical region we therefore find for the correlation function the form (1.4.9), which allows us to identify ξ with the correlation length:

$$G(r) = r^{-2d_\varphi} f_\pm\left(\frac{r}{\xi}\right). \tag{3.2.6}$$

Here, a priori, the functions f_+ ($T > T_c$) and f_- ($T < T_c$) are different. Equation (3.2.6) identifies the critical exponents and

$$d_\varphi = \frac{1}{2}(D - 2 + \eta), \quad \nu = \frac{1}{y_1}, \tag{3.2.7}$$

where we note again that the exponents ν and ν' are equal (see the footnote on p. 17).

The proof just given presupposes that the starting point is close to the fixed point, validating the linear approximation. If the starting point is far from P^*

but still belongs to its basin of attraction, one can reach points right next to the critical surface. By continuity, the trajectory of the representative point will arrive near the fixed point. Let s_0 be the parameter of the RGT (or of the successive RGTs) at which the representative point has reached the linear region:

$$G(r, \mu) = s_0^{-2d_\varphi} G\left(\frac{r}{s_0}, \mu'\right).$$

We need merely apply the preceding argument to $G(r/s_0, \mu')$, and find

$$G\left(\frac{r}{s_0}, \mu'\right) = \left(\frac{r}{s_0}\right)^{-2d_\varphi} f_\pm\left(\frac{r}{s_0 \xi}\right).$$

This equation shows that $G(r, \mu)$ has the form (3.2.6), with the critical exponents given in (3.2.7). *The same exponents are found whatever point we start from in the basin of attraction of the fixed point.* Since different points correspond to different Hamiltonians, this entails the *universality property of critical exponents*: they do not depend on details of the Hamiltonian, but only on certain very general properties.

The width of the critical region is governed by the condition

$$\xi^{y_2} t_2 \sim |T - T_c|^{-\nu y_2} t_2 \ll 1.$$

It depends on microscopic details through t_2, and on the fixed point through y_2. If t_2 is large or y_2 small (in absolute value), then the critical behaviour may be difficult to observe. Moreover, when y_2 is small, the evolution along the $e^{(2)}$ axis is slow, which can obscure the role played by the fixed point. The extreme case where $y_2 = 0$ (marginal field) will be studied in Section 3.6.

3.2.4 The correlation function with $B \neq 0$

Let us now apply a uniform external B-field. To the Hamiltonian this adds the term

$$B \sum_i S_i = B \frac{s^D}{\lambda(s)} \sum_\alpha S'_\alpha = B s^{D - d_\varphi} \sum_\alpha S'_\alpha.$$

Hence the transformation law for the B-field reads

$$B \to B' = s^{D - d_\varphi} B = s^{y_B} B; \quad y_B = \frac{1}{2}(D + 2 - \eta). \tag{3.2.8}$$

If we assume that η is small (as it is in practice: $\eta \approx 0$ to 0.1), then y_B is positive, and under an RGT B increases.

By the same method as before one finds

$$G(r, t, B) = s^{-2d_\varphi} G\left(\frac{r}{s}, s^{y_1} t, s^{y_B} B\right),$$

where we have identified t_1 with t, and have stopped indicating the irrelevant fields $t_2, t_3, \ldots$. This equation was written down without proof in Problem 1.5. It enables one to determine the exponents β, γ, δ as functions of η and ν; but one can proceed more directly by studying M, the mean magnetization per spin. The magnetization $M = \langle S \rangle$ transforms according to

$$\begin{aligned} M(t, B) &= s^{-d_\varphi} M(s^{y_1} t, s^{y_B} B) \\ &= s^{-d_\varphi} M\left(\pm \left(\frac{s}{\xi}\right)^{1/\nu}, s^{y_B} B \right). \end{aligned} \tag{3.2.9}$$

For $T = T_c$, $B \neq 0$, one takes $s = B^{-1/y_B}$, so that

$$M(0, B) = B^{d_\varphi / y_B} M(0, 1),$$

(where, effectively, $\xi \to \infty$). This equation yields the exponent δ,

$$\delta = \frac{y_B}{d_\varphi} = \frac{D + 2 - \eta}{D - 2 + \eta}. \tag{3.2.10}$$

For $T < T_c$ and $B = 0$, one sets $s = \xi$:

$$M(t, 0) = \xi^{-d_\varphi} M(-1, 0) \sim |t|^{-\nu d_\varphi}.$$

This equation yields the exponent β,

$$\beta = \nu d_\varphi = \frac{1}{2} \nu (D - 2 + \eta). \tag{3.2.11}$$

It remains to determine the critical exponent α; to this end, one exploits the free energy in zero field.

3.2.5 The free energy

First we must explicate equation (3.1.2), which links the Hamiltonians $H'[S'_\alpha]$ and $H[S_i]$. In fact, in the integration over short wavelengths there appears a constant term G, independent of the spins S'_α. Equation (3.1.2) must be amended to read

$$\mathrm{e}^{-G - H'[S'_\alpha]} = \sum_{[S_i]} \prod_\alpha \delta(S'_\alpha - f(S_i)|_{i \in \alpha}) \mathrm{e}^{-H[S_i]}. \tag{3.2.12}$$

The following discussion applies to nonlinear as well as to linear transformations, since we shall make no use of equation (3.1.10). The term $\exp(-G)$, stemming from the integration over short wavelengths, is irrelevant to the calculation of correlation functions, because it cancels in the calculation (Section 1.4.1). By contrast, the free energy does depend on this term, which complicates the discussion. Setting

$$Z' = \sum_{[S']} \mathrm{e}^{-H'[S']}, \quad F' = -\ln Z',$$

one obtains the relation ($F' = F(\mu')$),

$$F(\mu) = F(\mu') + G(\mu).$$

Define the free energy per unit volume as

$$f(\mu) = F(\mu)/L^D,$$

where L is the size of the sample, and take for definiteness RGTs with a dilatation factor of 2; we then have

$$f(\mu') = F(\mu')/(L/2)^D$$

and thus

$$f(\mu) = g(\mu) + 2^{-D} f(\mu'). \tag{3.2.13}$$

Let us iterate equation (3.2.13), starting from an initial set of parameters μ_0, representing a system close to the critical surface. We get

$$\begin{aligned} f(\mu_0) &= g(\mu_0) + 2^{-D} f(\mu_1) \\ f(\mu_1) &= g(\mu_1) + 2^{-D} f(\mu_2) \\ &\vdots \\ f(\mu_n) &= g(\mu_n) + 2^{-D} f(\mu_{n+1}) \\ &\vdots \end{aligned}$$

Multiply the second equation by 2^{-D}, the third by $2^{-2D}, \ldots$, the nth by $2^{-nD}, \ldots$, and add. One finds

$$f(\mu_0) = \sum_{n=0}^{\infty} 2^{-nD} g(\mu_n),$$

or reverting to a continuous variable $s = 2^n$,

$$f(\mu_0) = \int_0^{\infty} \frac{\mathrm{d}s}{s} s^{-D} g(\mu(s)).$$

The integral is controlled by the transition region between the critical regime ($s^{y_1}|t| \ll 1$) and the high- (or low-) temperature regime ($s^{y_1}|t| \gg 1$). Thus, as in previous arguments, we have $s \sim |t|^{-1/y_1}$. In this region, $g(\mu(s))$ is slowly varying, since we are in the vicinity of the fixed point, and we obtain

$$f(\mu_0) \sim s^{-D} \sim |t|^{\nu D}. \tag{3.2.14}$$

The specific heat is given by $\sim \mathrm{d}^2 f/\mathrm{d}t^2$, and we see that the critical exponent α depends on ν and D through

$$\alpha = 2 - \nu D \tag{3.2.15}$$

It should be clear that α does not depend on η, as the spin renormalization plays no role in the argument.

3.2.6 Scaling laws and comments

Analysis near the fixed point has enabled us to derive the following scaling laws (i.e. relations between critical exponents):

$$\alpha = 2 - \nu D \tag{3.2.16a}$$

$$\beta = \frac{1}{2}\nu(D - 2 + \eta) \tag{3.2.16b}$$

$$\delta = \frac{D + 2 - \eta}{D - 2 + \eta} \tag{3.2.16c}$$

$$\gamma = \nu(2 - \eta). \tag{3.2.16d}$$

The six basic exponents are not independent: knowledge of η and ν suffices to determine the rest, once the dimension D has been fixed. Let us try to understand why this is so: the exponent η, or equivalently the anomalous dimension d_φ, is tuned in such a way that one obtains a limiting probability distribution, or in other words a fixed point on the critical surface (one can show that the $T \to \infty$ and $T \to 0$ fixed points require different values of d_φ: $d_\varphi = D/2$ for $T \to \infty$ (see the Gaussian model, Section 3.4.2) and $d_\varphi = D$ for $T \to 0$). The exponent ν, or equivalently y_1, indicates how fast one leaves the fixed point in the RGT, and is often called the thermal exponent. On the other hand, ν is related to the scaling behaviour of energy–energy correlations: see Problem 3.11. Since in the GL Hamiltonian the energy operator is to be identified with φ^2, ν is related to the anomalous dimension d_{φ^2} of this operator. Thus the fundamental exponents can be chosen as the anomalous dimension of φ (conjugate to B), and the anomalous dimension of φ^2 (conjugate to the temperature); notice that a second-order transition is obtained by imposing precisely that these two fields vanish: $B = t = 0$.

The scaling laws are satisfied exactly by the $D = 2$ Ising model (if we write $\alpha = 0$ for a logarithmic divergence), and they also appear to be satisfied in all numerical studies that have been made of model systems. Thus the scaling laws represent a remarkable success of the renormalization group. However, one should not accept all the results of the RG on trust. It never yields an explicit solution for a model; in equation (3.2.6) for instance it cannot guarantee that the function $f_\pm$ is nonzero. If this function were zero, then the behaviour of the correlation function would obviously be totally different. Ma (1976, Chapter VI) gives a good example where blind application of renormalization-group results yields incorrect conclusions.

3.3 The Ising model on a triangular lattice, and the approximation by cumulants

In this first example of an application of the renormalization group, we shall use a nonlinear RGT. Such transformations are used chiefly in computer studies. The main problem is the need to truncate the Hamiltonian (3.1.5), keeping only a finite number of terms. This truncation introduces approximations that are difficult to assess, and on the whole results have been disappointing. An alternative method is that of the 'Monte Carlo renormalization group' (Swendsen 1980; Pawley *et al.* 1984).

We shall describe a method of approximation (the cumulant method) which has the merit that the calculations are relatively simple, and which furnishes concrete illustrations of the basic notions of the renormalization group: transformation of the Hamiltonian, fixed point, and calculation of critical exponents. The triangular lattice constitutes the simplest case; the blocks of spins are formed by assigning 3 spins $S_\alpha^{(i)}$ to the apex of a triangle (Fig. 3.6). The dilatation factor is $s = \sqrt{3}$, and the spin of the block is

$$S'_\alpha = \text{sign}(S_\alpha^{(1)} + S_\alpha^{(2)} + S_\alpha^{(3)}) = f(S_\alpha^{(i)}).$$

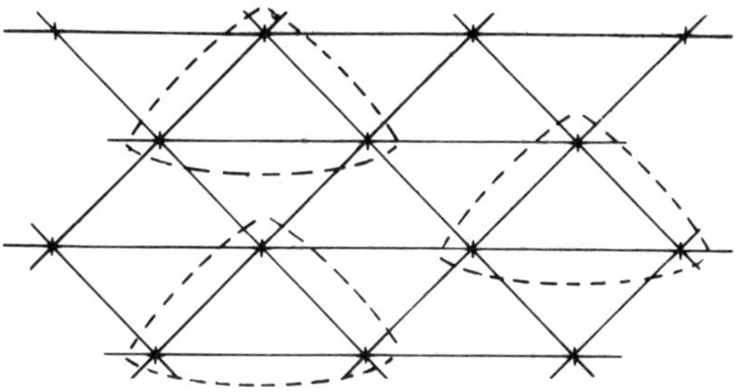

Fig. 3.6 Triangular lattice and the formation of blocks

Accordingly, the transformed Hamiltonian is given by

$$e^{-G-H'[S']} = \sum_{[S]} \prod_\alpha \delta(S'_\alpha - f(S_\alpha^{(i)}))e^{-H[S]}. \tag{3.3.1}$$

One may always write

$$H = H_0 + V,$$

where H_0 embodies the interactions between spins belonging to one and the same block, while V embodies the interactions between spins belonging to different blocks (Fig. 3.6). Rewrite (3.3.1) in the form

$$e^{-G-H'[S']} = \frac{\sum_{[S]} e^{-H_0}\prod \delta(S' - f(S)) \sum_{[S]} e^{-H_0-V} \prod \delta(S' - f(S))}{\sum_{[S]} e^{-H_0}\prod \delta(S' - f(S))}, \tag{3.3.2}$$

and define the average $\langle A \rangle_0$ of any quantity A by

$$\langle A \rangle_0 = \frac{\sum_{[S]} e^{-H_0} A[S] \prod \delta(S' - f(S))}{\sum_{[S]} e^{-H_0}\prod \delta(S' - f(S))}.$$

We stress that $\langle A\rangle_0$ is defined for a given configuration of blocks $[S']$; a more explicit notation would write $\langle A[S']\rangle_0$. Then equation (3.3.2) becomes

$$e^{-G-H'[S']} = \langle e^{-V}\rangle_0 \sum_{[S]} e^{-H_0} \prod \delta(S' - f(S)). \tag{3.3.3}$$

Equation (3.3.3) is exact but does not profit us much for the moment. Next we appeal to an approximation based on the following identity (namely on the expansion in cumulants; see Section 5.2.1):

$$\begin{aligned}\ln\langle e^x\rangle &= \ln[e^{\langle x\rangle}\langle e^{x-\langle x\rangle}\rangle] \\ &= \langle x\rangle + \ln\langle e^{x-\langle x\rangle}\rangle \\ &= \langle x\rangle + \ln\left\langle 1 + (x-\langle x\rangle) + \frac{1}{2!}(x-\langle x\rangle)^2 + \frac{1}{3!}(x-\langle x\rangle)^3 + \dots\right\rangle \\ &= \langle x\rangle + \frac{1}{2}\langle (x-\langle x\rangle)^2\rangle + \frac{1}{6}\langle (x-\langle x\rangle)^3\rangle + \dots .\end{aligned}$$

We shall settle for just the first term of the expansion; calculations have been pushed as far as the third term, but they soon become very complicated. Thus we adopt the approximation

$$\langle e^{-V}\rangle_0 \to e^{-\langle V\rangle_0}.$$

Let us calculate $\sum_{[S]} e^{-H_0} \prod \delta(S' - f(S))$. Since H_0 contains only interactions internal to each block, this term equals $[Z_0(K)]^{N'}$, where $N' = N/3$ is the number of blocks and where

$$Z_0(K) = e^{3K} + 3e^{-K}. \tag{3.3.4}$$

Indeed, for fixed S' there exists one configuration with energy $-3K$, and three others with energy K.

Next we evaluate the term $\langle V\rangle_0$, by considering the interaction between two blocks α and β:

$$-V_{\alpha\beta} = KS_\beta^{(1)}[S_\alpha^{(2)} + S_\alpha^{(3)}].$$

Since H_0 contains no connections between different blocks (Fig. 3.7), we have

$$\langle S_\beta^{(1)} S_\alpha^{(2)}\rangle_0 = \langle S_\beta^{(1)}\rangle_0 \langle S_\alpha^{(1)}\rangle_0 .$$

For instance, if $S'_\alpha = 1$, then

$$\langle S_\alpha^{(2)}\rangle_0 = \frac{1}{Z_0(K)} \sum_{[S_\alpha^{(i)}]} S_\alpha^{(2)} \exp(K(S_\alpha^{(1)}S_\alpha^{(2)} + S_\alpha^{(1)}S_\alpha^{(3)} + S_\alpha^{(2)}S_\alpha^{(3)})).$$

The configurations possible with $S'_\alpha = 1$ are

1 2 3	1 2 3	1 2 3	1 2 3	
↑ ↑ ↑	↑ ↑ ↓	↑ ↓ ↑	↓ ↑ ↑	
e^{3K}	$+e^{-K}$	$-e^{-K}$	$+e^{-K}$	$= e^{3K} + e^{-K};$

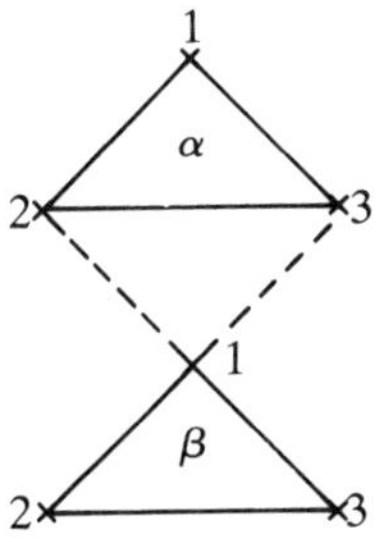

Fig. 3.7

those possible with $S'_\alpha = -1$ are

$$\begin{array}{cccc} 1\ 2\ 3 & 1\ 2\ 3 & 1\ 2\ 3 & 1\ 2\ 3 \\ \downarrow\ \downarrow\ \downarrow & \downarrow\ \downarrow\ \uparrow & \downarrow\ \uparrow\ \downarrow & \uparrow\ \downarrow\ \downarrow \\ -e^{3K} & -e^{-K} & +e^{-K} & -e^{-K} \end{array} = -(e^{3K} + e^{-K}).$$

Thus one finds

$$\langle S_\alpha^{(2)} \rangle_0 = \frac{1}{Z_0(K)} (e^{3K} + e^{-K}) S'_\alpha,$$

whence

$$\langle V_{\alpha\beta} \rangle_0 = 2K \left(\frac{e^{3K} + e^{-K}}{e^{3K} + 3e^{-K}} \right)^2 S'_\alpha S'_\beta = K' S'_\alpha S'_\beta.$$

The relation between K' and K is very simple:

$$K' = 2K \left(\frac{e^{3K} + e^{-K}}{e^{3K} + 3e^{-K}} \right)^2. \tag{3.3.5}$$

Combining equations (3.3.4) and (3.3.5) one finds the transformation law

$$\exp(-G - H'[S']) = \exp\left[N' \ln(e^{3K} + 3e^{-K}) + K' \sum_{\langle\alpha\beta\rangle} S'_\alpha S'_\beta \right], \tag{3.3.6}$$

whence G and H' can be identified immediately.

Accordingly, *at this level of approximation* there is just one parameter. The fixed point K^* satisfies

$$K^* = 2K^* \left(\frac{e^{4K^*} + 1}{e^{4K^*} + 3} \right)^2, \tag{3.3.7}$$

or

$$e^{4K^*} = x = 1 + 2\sqrt{2}; \quad K^* \simeq 0.336.$$

It remains only to determine s^y by calculating $(\mathrm{d}K'/\mathrm{d}K)_{K^*}$:

$$\left.\frac{\mathrm{d}K'}{\mathrm{d}K}\right|_{K^*} = 2\left(\frac{x+1}{x+3}\right)^2 + \frac{32K^*x}{(x+3)^2}\left(\frac{x+1}{x+3}\right), \quad \text{where} \quad \frac{x+1}{x+3} = \frac{1}{\sqrt{2}};$$

this yields

$$s^y \approx 1.634,$$

whence

$$y = \frac{\ln(1.634)}{\ln\sqrt{3}} \simeq 0.894; \quad \nu = \frac{1}{y} \simeq 1.118.$$

These results ($K^* = 0.336$, $\nu = 1.118$) should be compared with the exact values

$$K^* = 0.275, \quad \nu = 1.000.$$

The improvement over mean-field theory is very marked, since the latter approximation (with $q = 6$) yields

$$K_\mathrm{c} = \frac{1}{6} = 0.167, \quad \nu = \frac{1}{2}.$$

On calculating the higher-order terms of the cumulant approximation, they are found to approach the exact values, but the convergence is slow. Already in second order one must introduce three parameters rather than just one.

3.4 The Gaussian model

Instead of defining an RGT by forming blocks of spins, which amounts to integrating over fluctuations with wavelengths between a and sa, one can integrate over these wavelengths directly in Fourier space; in other words one can integrate there over wavenumbers between $\Lambda = 1/a$ and $\Lambda' = \Lambda/s = 1/sa$. From a mathematical viewpoint these two operations are not strictly the same, but they should yield identical end-results if the ideas underlying the RG are right, because their physical inputs are equivalent. This leads us to implement the operations of the renormalization group in Fourier space.

3.4.1 Transformations in Fourier space

We shall consider only *linear* RGTs, and start from a Hamiltonian of the generalized Ginzburg–Landau type. One can consider this Hamiltonian as having been derived from a spin model, by iterating a certain number of linear

RGTs of type (3.1.8). In the course of these iterations the spin has become a continuous variable $\varphi(\boldsymbol{x})$, through integration over all fluctuations with wavenumber $\geqslant \Lambda$; with these integrations done, $\|\boldsymbol{k}\|$ is therefore restricted to

$$\|\boldsymbol{k}\| \leqslant \Lambda. \tag{3.4.1}$$

In other words we have averaged over a domain $a \lesssim \lambda \lesssim \Lambda^{-1} \ll \xi$; equally, (3.4.2) can be viewed as a restricted expansion meant for studying fluctuations around the mean field (Chapter 2). The locality of the theory is reflected by the presence of only a limited number of derivatives of φ. We shall write the Hamiltonian as

$$H = \int \mathrm{d}^D x \left[\frac{1}{2} c(\nabla\varphi)^2 + \frac{1}{2} r_0 \varphi^2 + \frac{1}{4!} u_0 \varphi^4 \right.$$
$$\left. + \frac{1}{6!} u_6 \varphi^6 + \frac{1}{8!} u_8 \varphi^8 + \frac{1}{4!} v_0 \varphi^2 (\nabla\varphi)^2 + \dots \right]. \tag{3.4.2}$$

One recognizes the Ginzburg–Landau Hamiltonian (2.1.19) with a factor $c/2$ instead of $1/2$, and with certain extra terms. Parameter space is the space of these various coefficients or coupling constants from (3.4.2):

$$\mu = \{c, r_0, u_0, u_6, u_8, v_0, \dots\}. \tag{3.4.3}$$

It proves convenient to introduce the Fourier transform $\tilde{\varphi}(\boldsymbol{k})$ of $\varphi(\boldsymbol{x})$ by

$$\tilde{\varphi}(\boldsymbol{k}) = \frac{a^D}{L^{D/2}} \sum_{x} \mathrm{e}^{\mathrm{i}\boldsymbol{k}\cdot\boldsymbol{x}} \varphi(\boldsymbol{x}) \simeq \int \frac{\mathrm{d}^D x}{L^{D/2}} \mathrm{e}^{\mathrm{i}\boldsymbol{k}\cdot\boldsymbol{x}} \varphi(\boldsymbol{x}), \tag{3.4.4}$$

where L is the size of the system.

Note that the normalization of the Fourier transform in (3.4.4) is not the same as in Appendix A (equation (A.8)); the latter normalization applies to correlation functions. The normalization (3.4.4) has been chosen so that

$$\tilde{G}(\boldsymbol{k}) = \langle \tilde{\varphi}(\boldsymbol{k}) \tilde{\varphi}(-\boldsymbol{k}) \rangle \tag{3.4.5}$$

(see Problem 3.2).

The first two terms of (3.4.2) correspond to the Gaussian approximation to this Hamiltonian, and in $\boldsymbol{k}$-space they assume a very simple form. By appeal to Parseval's theorem one finds at once that

$$H = \frac{1}{2} \sum_{k \leqslant \Lambda} (r_0 + ck^2) \tilde{\varphi}(\boldsymbol{k}) \tilde{\varphi}(-\boldsymbol{k}) + \cdots . \tag{3.4.6}$$

The terms written out explicitly in (3.4.6) constitute the Hamiltonian of the *Gaussian model* in Fourier space.

Instead of writing the integration measure for the partition function in the space spanned by the $\varphi(\boldsymbol{x})$, one can write it just as readily in the space spanned by the $\tilde{\varphi}(\boldsymbol{k})$, because the transformation $\varphi(\boldsymbol{x}) \to \tilde{\varphi}(\boldsymbol{k})$ is unitary up to a multiplicative factor (which is irrelevant to correlation functions (see Problem 3.2)):

$$\prod_{\boldsymbol{x}} \mathrm{d}\varphi(\boldsymbol{x}) \to \prod_{k \leqslant \Lambda} \mathrm{d}\tilde{\varphi}(\boldsymbol{k}),$$

$$Z = \int \prod_{k \leqslant \Lambda} \mathrm{d}\tilde{\varphi}(\boldsymbol{k}) \exp\left(-\frac{1}{2} \sum_{\boldsymbol{k}} (r_0 + ck^2) \tilde{\varphi}(\boldsymbol{k}) \tilde{\varphi}(-\boldsymbol{k}) + \ldots \right). \tag{3.4.7}$$

The first two operations in the RGT are, as before, an integration over wavevectors $\Lambda/s \leqslant k \leqslant \Lambda$, followed by a dilatation of the unit of length by a factor s,

$$\boldsymbol{x} \to \boldsymbol{x}' = \boldsymbol{x}/s, \quad \boldsymbol{k} \to \boldsymbol{k}' = s\boldsymbol{k}. \tag{3.4.8}$$

The third operation consists of 'renormalizing' the field variable $\varphi(\boldsymbol{x})$,

$$\varphi(\boldsymbol{x}) \to \varphi'(\boldsymbol{x}') = \lambda(s)\varphi(\boldsymbol{x}) = s^{d_\varphi}\varphi(\boldsymbol{x}); \tag{3.4.9a}$$

or, in $\boldsymbol{k}$-space,

$$\tilde{\varphi}(\boldsymbol{k}) \to \tilde{\varphi}'(\boldsymbol{k}) = \lambda(s) s^{-D/2} \tilde{\varphi}(\boldsymbol{k}) = s^{d_\varphi - D/2} \tilde{\varphi}(\boldsymbol{k}). \tag{3.4.9b}$$

Equation (3.4.9b) follows readily from (3.4.9a) ($L' = L/s$):

$$\begin{aligned} \tilde{\varphi}'(\boldsymbol{k}') &= \int \frac{\mathrm{d}^D x'}{L'^{D/2}} \mathrm{e}^{\mathrm{i}\boldsymbol{k}' \cdot \boldsymbol{x}'} \varphi'(\boldsymbol{x}') \\ &= s^{-D/2} \int \frac{\mathrm{d}^D x}{L^{D/2}} \mathrm{e}^{\mathrm{i}\boldsymbol{k} \cdot \boldsymbol{x}} \lambda(s) \varphi(\boldsymbol{x}) \\ &= \lambda(s) s^{-D/2} \tilde{\varphi}(\boldsymbol{k}). \end{aligned}$$

To summarize, the RGT R_s subdivides into three steps:

(1) Integration over $\boldsymbol{k}$:

$$\Lambda/s \leqslant k \leqslant \Lambda.$$

(2) Dilatation of the unit of length:

$$\boldsymbol{x} \to \boldsymbol{x}' = \boldsymbol{x}/s$$
$$\boldsymbol{k} \to \boldsymbol{k}' = s\boldsymbol{k}.$$

(3) Renormalization of the field:

$$\varphi(\boldsymbol{x}) \to \varphi'(\boldsymbol{x}') = s^{d_\varphi}\varphi(\boldsymbol{x})$$
$$\tilde{\varphi}(\boldsymbol{k}) \to \tilde{\varphi}'(\boldsymbol{k}') = s^{d_\varphi - D/2}\tilde{\varphi}(\boldsymbol{k}).$$

We can, at once, write down a formal relation for the transformed Hamiltonian $H' = R_s H$, namely

$$e^{-H'(\varphi')} = \left[\int \prod_{\Lambda/s \leqslant k \leqslant \Lambda} d\tilde{\varphi}(\boldsymbol{k}) e^{-H(\varphi)} \right]_{\tilde{\varphi}(\boldsymbol{k}) \to s^{(D/2)-d_\varphi} \tilde{\varphi}'(s\boldsymbol{k})}, \tag{3.4.10}$$

where we have dropped the constant G (see (3.2.12).

Let us now establish, for the correlation function, the $\boldsymbol{k}$-space analogue of (3.1.10). Actually this relation is trivial. In fact, if $k < \Lambda/s$, then the probability density $e^{-H'}$ entails the same correlation functions as does the density e^{-H}, because these wavevectors remain unaffected by the integration. The only steps we need take are the change of scale $\boldsymbol{k} \to s\boldsymbol{k}$, and the renormalization of the field; in view of equation (3.4.5) we find for $k < \Lambda/s$ that

$$\tilde{G}(s\boldsymbol{k}, \mu') = s^{2d_\varphi - D} \tilde{G}(\boldsymbol{k}, \mu). \tag{3.4.11a}$$

(see also Problem 3.4(c)).

The relation (3.4.11a) is *exact*; Fourier transformation then yields an approximate relation in $\boldsymbol{x}$-space, valid for $\|\boldsymbol{x}\| \gg 1/\Lambda$:

$$G\left(\frac{\boldsymbol{x}}{s}, \mu'\right) \simeq s^{2d_\varphi} G(\boldsymbol{x}, \mu); \tag{3.4.11b}$$

this is evidently identical to (3.1.10).

3.4.2 The Gaussian model

In order to familiarize oneself with the RGT (3.4.10), it proves useful to consider the case of the Gaussian model (3.4.6); this model is extremely simple, but somewhat too trivial to represent any real physical situation. Parameter space is two-dimensional: $\mu = \{c, r_0\}$. Integration over $d\tilde{\varphi}(\boldsymbol{k})$ yields a constant, since we are dealing with a product of decoupled Gaussian integrals. Accordingly, the new Hamiltonian is

$$\begin{aligned} H' &= \sum_{k \leqslant \Lambda/s} \frac{1}{2} (r_0 + c\boldsymbol{k})^2 |\varphi(\boldsymbol{k})|^2 \\ &= \sum_{k' \leqslant \Lambda} \frac{1}{2} s^{D-2d_\varphi} (r_0 + cs^{-2} \boldsymbol{k}'^2) |\varphi'(\boldsymbol{k}')|^2. \end{aligned}$$

Note that $\varphi(-\boldsymbol{k}) = \varphi^*(\boldsymbol{k})$; henceforth we shall write $\varphi(\boldsymbol{k})$ instead of $\tilde{\varphi}(\boldsymbol{k})$ when no confusion can result.

The Hamiltonian H' has the same form as H, with the transformation law for the parameters given by

$$c' = s^{D-2-2d_\varphi} c; \quad r_0' = s^{D-2d_\varphi} r_0.$$

These equations allow two possible fixed points:

(i) $D - 2d_\varphi = 0$; r_0 arbitrary; $c = 0$;

(ii) $D - 2 - 2d_\varphi = 0$; c arbitrary; $r_0 = 0$.

In (i) c is an irrelevant field, and the fixed point corresponds to an ensemble of decoupled sites, which is not very exciting. In fact this case corresponds to the limit $T \to \infty$, and what we have just found is the $T \to \infty$ fixed point mentioned earlier. Since the Gaussian model does not have a low-temperature phase, we shall not find the $T \to 0$ fixed point; note also that $d_\varphi = D/2$ for the $T \to \infty$ fixed point.

Case (ii) is more interesting. The fixed point is defined by

$$\mu^* = \{c, r_0 = 0\}$$

where c is arbitrary. Equation (3.1.13) implies that $\eta = 0$. On the other hand

$$r_0' = s^2 r_0;$$

this shows that r_0 is a relevant field with $y = 2$, whence $\nu = 1/2$. The parameter r_0 vanishes at the transition, which occurs at $T = T_0$: $r_0 = \bar{r}_0(T - T_0)$. However, for the Gaussian model the low-temperature phase is not defined: when $r_0 < 0$, certain Gaussian integrals over $\varphi(\boldsymbol{k})$ fail to converge. The results $\nu = 1/2$, $\eta = 0$ have already been derived in Section 2.4, by calculating the correlation function directly; recall that in this case the latter equals $(r_0 + ck^2)^{-1}$.

It is instructive to augment (3.4.6) by a term in k^4:

$$H = \sum_{k \leqslant \Lambda} \left(\frac{1}{2} r_0 + \frac{c}{2} k^2 + \frac{w_0}{4} k^4 \right) \varphi(\boldsymbol{k}) \varphi(-\boldsymbol{k}).$$

Under the RGT, the field w_0 transforms according to

$$w_0' = s^{-2} w_0,$$

whence w_0 is irrelevant. The renormalization flow is sketched in Fig. 3.8, on the assumption that w_0 is not too negative; otherwise there is a danger of divergences. The result is interesting, because it explains the 'restoration of rotation invariance' at the critical point. Indeed, on expanding the Hamiltonian of the corresponding lattice model (see Problem 2.5), one meets a term

$$-\frac{1}{12} \sum_{i=1}^{D} k_i^4$$

which violates rotation invariance. However, since this term is irrelevant (in the technical sense), it has no bearing on the long-range behaviour of the correlation functions, which is, therefore, invariant.

Note moreover that the parameter c plays no role at all: one can fix its value at $c = 1$, and this is what we shall do in the sequel. Indeed we can go further, and require that even in the case of the general Hamiltonian (3.4.2) *the coefficient of*

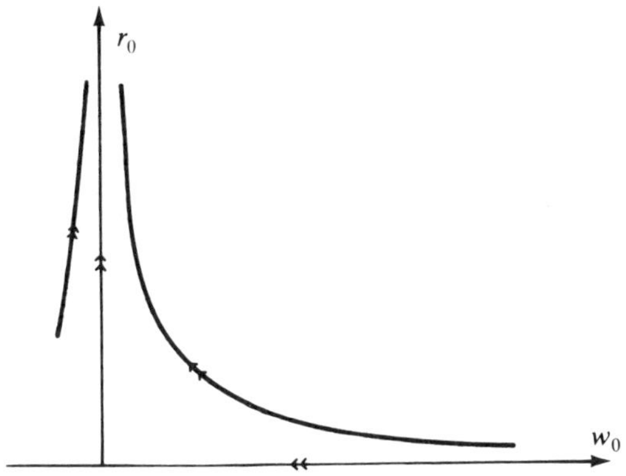

Fig. 3.8 Renormalization flow with a term in k^4

$(\nabla\varphi)^2$ *be always fixed at 1/2 in all iterations of the RGTs.* Certainly this condition is compatible with the existence of a fixed point: one of the parameters of H remains fixed and equal to 1/2. In the context of the perturbative theories of Section 3.5, the condition indeed entails a fixed point.

In fact the results of the Gaussian model are wholly determined by dimensionality. Indeed, requiring $\eta = 0$ (or $d_\varphi = (D/2) - 1$) amounts to requiring that the coefficient either of the term in $k^2|\varphi(\boldsymbol{k})|^2$ or of the term in $(\nabla\varphi)^2$ remain unchanged under any RGT, and equal to 1/2. To ensure this, the change in length scale must be compensated by the change in the normalization of the field, so that

$$\int \mathrm{d}^D x' (\nabla'\varphi')^2 = \int \mathrm{d}^D x \, s^{-D+2+2d_\varphi} (\nabla\varphi)^2$$
$$= \int \mathrm{d}^D x (\nabla\varphi)^2,$$

since $\varphi'(\boldsymbol{x}') = s^{d_\varphi}\varphi(\boldsymbol{x})$ and $\nabla' = s\nabla$; thus $d_\varphi = (D/2) - 1$.

Another way of putting this is to say that if one attributes dimension -1 to length ($x' = s^{-1}x$), then one must attribute dimension $d_\varphi = (D/2) - 1$ to the field φ. Now the invariance of H can be recovered by simple dimensional analysis: since H has dimension zero, it is independent of the length scale. And if H is indeed to have dimension zero, then the field must be assigned dimension $(D/2) - 1$.

The dimension $d_\varphi^0 = (D/2) - 1$ is called the *normal* (*or canonical*) *dimension of the field*: it is the value found by dimensional analysis. In general, for a non-Gaussian Hamiltonian, d_φ will not equal d_φ^0 (equivalently, η will be nonzero); d_φ

is then called the *anomalous dimension of the field.* This anomalous dimensionality stems from the dynamics, and depends on the fixed point under consideration.

Naïve scale invariance corresponds to behaviour governed purely by dimensional analysis. For instance, since $G(\mathbf{k})$ has dimension -2 it must be proportional to k^{-2}; and this indeed is what the Gaussian model gives at $T = T_c(= T_0)$.

It will prove useful to determine the normal dimensions of the coupling constants r_0, u_0, u_6, v_0, etc. in (3.4.2); we write them as $[r_0]$, $[u_0]$, etc. They are easily found, simply by noting that $[H] = 0$ and $[\varphi] = d_\varphi^0 = (D/2) - 1$. One obtains

$$\left.\begin{array}{ll} [r_0] = 2 & [u_0] = 4 - D \\ [u_6] = 6 - 2D & [v_0] = 2 - D. \end{array}\right\} \tag{3.4.12}$$

3.4.3 The Gaussian fixed point

Let us now examine the general Hamiltonian (3.4.2). It contains terms with φ^2, φ^4, etc., referring to just one site. These terms are simple in $\mathbf{x}$-space. By contrast, the term with $(\nabla\varphi)^2$ couples different sites, and this term is simple in Fourier space, being diagonalized by a Fourier transformation. In the Gaussian model the normal modes are decoupled.

On the other hand the term with φ^4 is complicated in Fourier space, because it couples the normal modes with each other:

$$\int \mathrm{d}^D x\, \varphi^4(x) = L^{-D} \sum_{\mathbf{k}_1, \mathbf{k}_2, \mathbf{k}_3} \varphi(\mathbf{k}_1)\varphi(\mathbf{k}_2)\varphi(\mathbf{k}_3)\varphi(-\mathbf{k}_1 - \mathbf{k}_2 - \mathbf{k}_3).$$

It is impossible to find a space where all these terms are simultaneously simple, which is why one must use approximation methods for dealing with the Ginzburg–Landau Hamiltonian. For the time being we confine ourselves to the Hamiltonian (2.1.19); the standard method is the *perturbative expansion in powers of* u_0. The Hamiltonian H is subdivided into a Gaussian term H_0 and an 'interaction' term V:

$$H = H_0 + V, \quad V = \frac{u_0}{4!} \int \mathrm{d}^D x \varphi^4(\mathbf{x}). \tag{3.4.13}$$

Similarly, when applying equation (3.4.10) we must appeal to this subdivision in order to evaluate the integral over the $\mathrm{d}\varphi(\mathbf{k})$.

Write

$$\varphi(\mathbf{x}) = \varphi_1(\mathbf{x}) + \bar{\varphi}(\mathbf{x}), \tag{3.4.14}$$

where $\varphi_1(\mathbf{x})$ has Fourier components in the range $0 \leqslant k \leqslant \Lambda/s$, and $\bar{\varphi}(\mathbf{x})$ in the range $\Lambda/s \leqslant k \leqslant \Lambda$. Accordingly, the integration measure in (3.4.10) is $\mathscr{D}\bar{\varphi}$; and we stress that, in H_0, φ_1 and $\bar{\varphi}$ are decoupled. Disregarding the multiplicative constant ($\exp(-G)$), and disregarding dilatations as well for the moment, we find

$$e^{-H_1'} = e^{-H_0(\varphi_1)} \frac{\int \mathscr{D}\bar{\varphi} \exp(-H_0(\bar{\varphi}) - V(\varphi_1, \bar{\varphi}))}{\int \mathscr{D}\bar{\varphi} \exp(-H_0(\bar{\varphi}))}.$$

If one restricts oneself to first order in u_0, then the new Hamiltonian H_1' is

$$H_1' = H_0(\varphi_1) + \frac{\int \mathscr{D}\bar{\varphi} \exp(-H_0(\bar{\varphi})) V(\varphi_1, \bar{\varphi})}{\int \mathscr{D}\bar{\varphi} \exp(-H_0(\bar{\varphi}))} + O(u_0^2). \tag{3.4.15}$$

The Hamiltonian $H_0(\bar{\varphi})$ is Gaussian, and $V(\varphi_1, \bar{\varphi})$ is a polynomial in $\bar{\varphi}$. Thus our task in (3.4.15) is to evaluate the average of a polynomial over a Gaussian probability distribution. In fact we shall need only $\langle \bar{\varphi}(\boldsymbol{x})\bar{\varphi}(\boldsymbol{y})\rangle_0$, where the subscript 0 identifies averages calculated with the Gaussian Hamiltonian H_0. In order to evaluate this average, note that we already know the result when the integration over k runs from 0 to Λ; then $\langle \bar{\varphi}(\boldsymbol{x})\bar{\varphi}(\boldsymbol{y})\rangle_0$ is the correlation function of the Gaussian model (equation (2.4.7)):

$$\langle \bar{\varphi}(\boldsymbol{x})\bar{\varphi}(\boldsymbol{y})\rangle_0 = \int_{k \leqslant \Lambda} \frac{d^D k}{(2\pi)^D} \frac{e^{-i\boldsymbol{k}\cdot(\boldsymbol{x}-\boldsymbol{y})}}{\boldsymbol{k}^2 + r_0}.$$

In the present case the integral over $\boldsymbol{k}$ runs between the limits $\Lambda/s \leqslant k \leqslant \Lambda$, and the result is simply

$$\langle \bar{\varphi}(\boldsymbol{x})\bar{\varphi}(\boldsymbol{y})\rangle_0 = \bar{G}_0(\boldsymbol{x}-\boldsymbol{y}) = \int_{\Lambda/s \leqslant k \leqslant \Lambda} \frac{d^D k}{(2\pi)^D} \frac{e^{-i\boldsymbol{k}\cdot(\boldsymbol{x}-\boldsymbol{y})}}{\boldsymbol{k}^2 + r_0}. \tag{3.4.16}$$

Reverting now to the calculation of $\langle V(\varphi_1, \bar{\varphi})\rangle_0$, we have

$$\langle(\varphi_1(\boldsymbol{x}) + \bar{\varphi}(\boldsymbol{x}))^4\rangle_0 = \varphi_1^4(\boldsymbol{x}) + 6\varphi_1^2(\boldsymbol{x})\langle \bar{\varphi}(\boldsymbol{x})\bar{\varphi}(\boldsymbol{x})\rangle_0 + \langle \bar{\varphi}^4(\boldsymbol{x})\rangle_0.$$

The final term is a constant and can be dropped. The second term equals $6\varphi_1^2(\boldsymbol{x})\bar{G}_0(0)$. For H_1' we therefore obtain

$$H_1' = \int d^D x \left[\frac{1}{2}(\nabla\varphi_1)^2 + \frac{1}{2} r_0 \varphi_1^2 + \frac{u_0}{4!}\varphi_1^4 + \frac{u_0}{4}\varphi_1^2(x)\bar{G}_0(0)\right].$$

In order to find H' it remains only to implement the dilatations:

$$H' = \int d^D x' s^{D-2d_\varphi-2}\left[\frac{1}{2}(\nabla'\varphi')^2 + \frac{1}{2}s^2\left(r_0 + \frac{u_0}{2}\bar{G}_0(0)\right)\varphi'^2(\boldsymbol{x}) + s^{2-2d_\varphi}\frac{u_0}{4!}\varphi'(\boldsymbol{x})^4\right]. \tag{3.4.17}$$

To this order the integration over $\bar{\varphi}$ has not affected the gradient term. If we wish to keep its coefficient equal to 1/2, we must therefore as in the Gaussian model take $d_\varphi = (D/2) - 1$ and $\eta = 0$. The transformation laws for r_0 and u_0 are

$$\begin{cases} r_0' = s^2\left(r_0 + \dfrac{u_0}{2}\,\bar{G}_0(0)\right), \\ u_0' = s^{4-D}u_0 = s^\varepsilon u_0 \quad (\varepsilon = 4 - D). \end{cases} \tag{3.4.18}$$

Let us now evaluate $\bar{G}_0(0)$. Since $r_0 \to 0$, one can take $r_0 \ll \Lambda/s$ (it is slightly simpler to adopt $H_0 = \frac{1}{2}\int \mathrm{d}^D x(\nabla\varphi)^2$ as one's Hamiltonian: see Ma (1976, Chapter VIII)):

$$\begin{aligned} \bar{G}_0(0) &= K_D \int_{\Lambda/s}^{\Lambda} \frac{k^{D-1}\mathrm{d}k}{k^2} + O(r_0) \\ &= \frac{K_D \Lambda^{D-2}}{D-2}(1 - s^{2-D}) + O(r_0) = 2B(1 - s^{2-D}) + O(r_0). \end{aligned} \tag{3.4.19}$$

Equations (3.4.18) have a fixed point at $r_0 = u_0 = 0$. Linearization in its vicinity is unaffected by the terms that have been dropped from (3.4.19), whose contributions are of order $(u_0 r_0)$ etc. Here, the matrix $T(s)$ (see (3.2.3)) is

$$T(s) = \begin{pmatrix} s^2 & B(s^2 - s^\varepsilon) \\ 0 & s^\varepsilon \end{pmatrix}; \quad B = \frac{K_D \Lambda^{D-2}}{2(D-2)}.$$

Its eigenvalues and eigenvectors are

$$\rho_1 = s^2 \quad (y_1 = 2), \quad e^{(1)} = \begin{pmatrix} 1 \\ 0 \end{pmatrix};$$

$$\rho_2 = s^\varepsilon \quad (y_2 = \varepsilon), \quad e^{(2)} = \begin{pmatrix} -B \\ 1 \end{pmatrix}.$$

For $\varepsilon < 0$, i.e. for $D > 4$, we have one eigenvalue $y_1 = 2$ and another $y_2 = \varepsilon < 0$. This shows that the fixed point is of the type studied in Section 3.2, with one relevant and one irrelevant field. The critical exponents are identically the same as those of the Gaussian model, namely $\nu = 1/2$, $\eta = 0$.

Next we consider parameter space. A point μ in this space is written

$$\mu = r_0 \hat{r}_0 + u_0 \hat{u}_0,$$

where $\hat{r}_0$ and $\hat{u}_0$ are the unit vectors along the r_0 and u_0 axes respectively; since $e^{(1)} = \hat{r}_0$ and $e^{(2)} = -B\hat{r}_0 + \hat{u}_0$, one has

$$\mu = (r_0 + u_0 B)e^{(1)} + u_0 e^{(2)}.$$

The scaling fields are $t_1 = r_0 + u_0 B$ and $t_2 = u_0$, with the critical surface given by $t_1 = r_0 + u_0 B = 0$.

In the linear approximation one finds the critical temperature

$$r_{0c} = \bar{r}_0(T_c - T_0) = -u_0 B = -\frac{u_0 K_D \Lambda^{D-2}}{2(D-2)},$$

which agrees with (2.4.11). The renormalization flow near the critical point is sketched in Fig. 3.9.

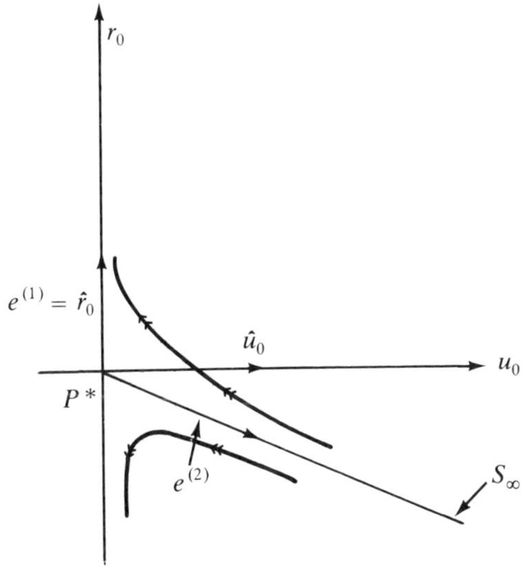

Fig. 3.9 Flow diagram for $D > 4$

What we have shown is that, if one restricts oneself to the parameter space (r_0, u_0), i.e. to the strict Ginzburg–Landau Hamiltonian, then for $D > 4$ the critical exponents are those of the Gaussian model, i.e. of Landau theory. It is easy to generalize this result to arbitrary Hamiltonians of the form (3.4.2). The fact is that under RGTs a term like u_6 transforms according to (3.4.12) into

$$u_6' = s^{6-2D} u_6 + \cdots,$$

(see Problem 3.3); thus the field u_6 is irrelevant just like u_0 (provided always that $D > 4$), because the exponent of s is negative. Starting from any point on the critical surface, iteration of an RGT takes any Hamiltonian of the type (3.4.2) into the Gaussian fixed point. This property enables one to prove the result stated in Chapter 2: *the critical exponents of Landau theory are the correct ones for $D > 4$.* However, care is needed with the exponents α, β, δ (see Ma 1976, p. 185).

3.5 Calculation of the critical exponents to order ε

3.5.1 Non-Gaussian fixed points

For $D < 4$, the fixed point we have just found no longer describes a second-order phase transition, because $y_2 = \varepsilon > 0$. Another fixed point now appears, which does have the appropriate features, namely $y_1 > 0$, $y_2 < 0$; and it is this second fixed point that will determine the critical exponents for $D < 4$. It enters as the continuation of a fixed point which for $D > 4$ is of the wrong type, with two diverging axes; and which moreover is unphysical, because it corresponds to $u_0 < 0$, a case where the integrals over $\varphi(\boldsymbol{x})$ are not defined. The two fixed points interchange their stabilities at $D = 4$ (Fig. 3.10).

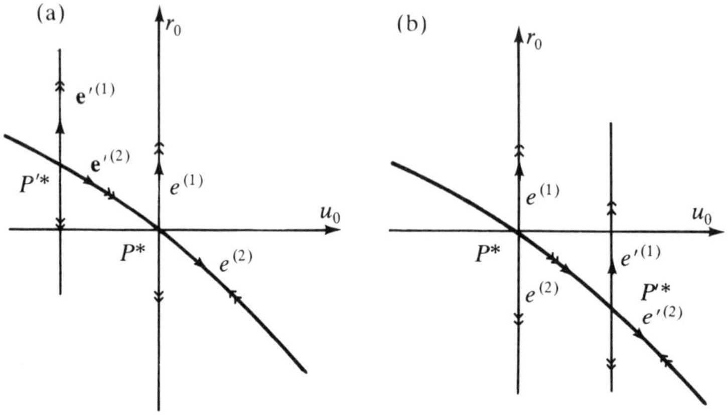

Fig. 3.10 Flow diagram for (a) $D > 4$ and (b) $D < 4$

The reason for the appearance of the 'non-Gaussian' fixed point when $D < 4$ lies in the existence of a nonlinear term in the evolution of u_0. One can show that under an RGT this evolution is of the form

$$u'_0 = s^{\varepsilon}(u_0 - Cu_0^2 \ln s),$$

where C is a constant. The condition that a fixed point exist can be written $\mathrm{d}u'_0/\mathrm{d}\ln s|_{s=1} = 0$, i.e.

$$\left.\frac{\mathrm{d}u'_0}{\mathrm{d}\ln s}\right|_{s=1} = \varepsilon u_0 - Cu_0^2 = 0.$$

Then the fixed point is at $u_0^* = \varepsilon/C$: *u_0^* is of order ε*. In fact calculations based on the principles that we are about to explain, built as they are on a perturbation expansion, apply order by order in an expansion by powers of ε. *The results hold only for 'small' ε*, where the notion 'small ε' is yet to be elucidated.

Obviously it remains to spell out explicitly the transformation laws for r_0 and u_0. The calculation is more complicated than before, because in the perturbation expansion we must now go to order u_0^2. Here we shall settle for quoting the results without proof, because a quicker method will be explained later. One finds (see Ma 1976, Chapter VII)

$$r'_0 = s^2\left[r_0 + \frac{u_0}{16\pi^2}\left(\frac{1}{2}\Lambda^2(1 - s^{-2}) - r_0 \ln s\right)\right], \tag{3.5.1a}$$

$$u'_0 = s^\varepsilon\left[u_0 - \frac{3u_0^2}{16\pi^2}\ln s\right]. \tag{3.5.1b}$$

Equations (3.5.1) are derived through some approximations whose consistency can be checked afterwards, once one has shown that both u_0^* and r_0^* are of order ε.

3.5.2 The differential renormalization equations

It is worth spending some time on the interpretation of equations (3.5.1). Suppose that one has started from an initial Hamiltonian with the values $r_0 = r_0(1)$ and $u_0 = u_0(1)$. After a certain number of iterations one arrives at a Hamiltonian dependent on parameters $r_0(s)$ and $u_0(s)$, and also on other interactions $u_6(s)$ . . . . These interactions greatly complicate the study of the vicinity of the fixed point, unless one confines oneself to order ε, in which case they may be dropped. One pointer to this property is given in Problem 3.7(h). Here we shall simply assume that it is safe to work with the truncated form (3.5.1) of the RG equations. Since the dilatation factor is continuous, we can implement an RGT with the dilatation factor $1 + \delta$, $\delta \to 0$. The relations between $[r_0(s(1 + \delta)), u_0(s(1 + \delta))]$ and $[r_0(s), u_0(s)]$ are given by (3.5.1), if one sets $\ln s = \delta$ ($\ln(1 + \delta) \approx \delta$!). Consequently we can transform (3.5.1) into the differential equations

$$\frac{\mathrm{d}r_0(s)}{\mathrm{d}\ln s} = 2r_0(s) - \frac{u_0(s)r_0(s)}{16\pi^2} + \frac{u_0(s)\Lambda^2}{16\pi^2}, \tag{3.5.2a}$$

$$\frac{\mathrm{d}u_0(s)}{\mathrm{d}\ln s} = \varepsilon u_0(s) - \frac{3u_0^2(s)}{16\pi^2}. \tag{3.5.2b}$$

In fact equations (3.5.1) are correct (subject to the preceding remarks) only if s is small enough, a point to be discussed in detail in Section 3.6: recall that one must *never* implement in just one step an RGT resulting in a dilatation factor $s \gg 1$, but that one must instead factor it into a product of RGTs. The advantage of the differential equations (3.5.2) is that $\ln s$ is infinitesimal. For the parameters

$K_\alpha(s)$ we can in general write differential renormalization equations

$$\frac{\mathrm{d}K_\alpha(s)}{\mathrm{d}\ln s} = \beta_\alpha(K_\beta(s)). \tag{3.5.3}$$

These equations are known as the flow equations, and they have been widely applied to dynamical systems. An excellent introduction can be found in Arnold (1973, 1983).

Iterates of RGTs are given by solutions of these differential equations; i.e. in this formalism a dilatation factor s is achieved as a sequence of iterations of infinitesimal RGTs.

Equations (3.5.2) make it easy to calculate the position of the fixed point,

$$u_0^* = \frac{16\pi^2}{3}\varepsilon; \quad r_0^* = -\frac{\varepsilon}{6}\Lambda^2; \tag{3.5.4}$$

the exponents y_1 and y_2 assume the values

$$y_1 = 2 - \frac{\varepsilon}{3}; \quad y_2 = -\varepsilon < 0 \tag{3.5.5}$$

(Problem 3.4). The fact that $y_2 < 0$ shows that the fixed point really does have the requisite properties. The critical exponent ν is given by

$$\nu = \frac{1}{y_1} = \frac{1}{2} + \frac{\varepsilon}{12} + O(\varepsilon^2); \tag{3.5.6}$$

equation (3.5.6) supplies the order-ε correction to Landau theory. The derivation of equations (3.5.4) and (3.5.5) from (3.5.3) is relegated to Problem 3.4. Equation (3.5.6) gives the first two terms of the 'ε-expansion' of the critical exponent ν. More generally, we write the expansion of a critical exponent ζ as

$$\zeta = \zeta_0 + \zeta_1\varepsilon + \zeta_2\varepsilon^2 + \cdots + \zeta_n\varepsilon^n + \cdots \tag{3.5.7}$$

where ζ_0 is given by Landau theory.

3.5.3 Perturbative calculation of critical exponents

The method just given has the advantage that it proves, explicitly, the existence of a nontrivial fixed point when $D < 4$. However, the calculations are rather laborious; worse, they become flatly unmanageable if one tries to extend them to order ε^2. Indeed the parameter space $\{r_0, u_0\}$ no longer suffices to locate the fixed point, and equations (3.5.1) can no longer account for the RGTs.

For this reason, we turn to a more powerful method for the calculation of critical exponents. We have seen in Section 2.4.3 that one cannot hope to extract the critical exponents from a naïve perturbative expansion, because the effective, dimensionless, expansion parameter is $u_0 r^{-\varepsilon/2}$, where r is the inverse of the susceptibility: $r = \chi^{-1}$, and this parameter tends to infinity at the critical point for $D < 4$.

The GL Hamiltonian depends in fact upon two parameters: in field-theory language (see Chapters 6 and 7), they are the 'bare mass' squared $(r_0 - r_{0c})$ (more precisely the difference between the bare mass squared and the critical mass squared) and the 'bare coupling constant' u_0; in a perturbative approach, r_{0c} is a function of u_0 and of the cut-off $\Lambda \sim a^{-1}$. Two important points need be emphasized:

(i) these parameters are really microscopic parameters: they are defined on a scale of order a, and they are measured in units corresponding to this scale;

(ii) we work at fixed u_0, and we reach the critical point by using $(r_0 - r_{0c})$ as a control parameter, which vanishes at this point.

What we want to do is to reshuffle the perturbative expansion of Chapter 2, which was written as a power series in u_0, by using instead of the two bare parameters r_0 and u_0 two 'renormalized' parameters: the 'renormalized mass' $m = \xi^{-1}$, defined as the inverse of the correlation length ξ, and the 'renormalized coupling constant' g, to be defined later on. It turns out that near $D = 4$, g is small, of order ε, and a perturbative calculation of the critical exponents in powers of g becomes possible. When g is not so small, for example when $D = 3$, one needs more sophisticated techniques in order to obtain reliable results. The point in using renormalized parameters is that they are defined by taking ξ as a basic scale.

We shall use the correlation function

$$\Gamma^{(4)}(\boldsymbol{x}_1, \boldsymbol{x}_2, \boldsymbol{x}_3, \boldsymbol{x}_4) = \frac{\delta^{(4)}\Gamma(M)}{\delta M(\boldsymbol{x}_1)\delta M(\boldsymbol{x}_2)\delta M(\boldsymbol{x}_3)\delta M(\boldsymbol{x}_4)},$$

where Γ is the Gibbs potential. It is related to the 'four-point correlation function'

$$G^{(4)}(\boldsymbol{x}_1, \boldsymbol{x}_2, \boldsymbol{x}_3, \boldsymbol{x}_4) = \langle \varphi(\boldsymbol{x}_1)\varphi(\boldsymbol{x}_2)\varphi(\boldsymbol{x}_3)\varphi(\boldsymbol{x}_4) \rangle$$

and to the correlation functions $G(\boldsymbol{x}_1, \boldsymbol{x}_2)$ (see Section 5.3.3), though we shall not exploit these relations. In fact all we need is the Fourier transform $\tilde{\Gamma}^{(4)}(\boldsymbol{k}_i = 0)$, taken at zero wavevector; strictly speaking one should add that $\tilde{\Gamma}^{(4)}$ is got by extracting a factor $\delta^{(D)}(\sum_{i=1}^{4} \boldsymbol{k}_i)$ (see Section 5.2.4). One obtains this quantity by differentiating the uniform-magnetization Gibbs potential $\Gamma(M)$ of equation (2.4.6) four times; in fact a magnetization M that is uniform possesses just the single Fourier component $\boldsymbol{k} = 0$, and differentiation with respect to M clearly yields $\tilde{\Gamma}^{(4)}(\boldsymbol{k}_i = 0)$.

If at the end of the calculation one sets $M = 0$ $(T > T_c)$, then these differentiations* with respect to M clearly yield the first two terms in an expansion of $\tilde{\Gamma}^{(4)}(0)$ by powers of u_0:

$$\tilde{\Gamma}^{(4)}(0) = u_0 - \frac{3}{2} u_0^2 \int \frac{\mathrm{d}^D k}{(2\pi)^D} \frac{1}{(k^2 + r_0)^2} + O(u_0^3) \tag{3.5.8}$$

(see the argument in Section 2.4.3).

We shall need to determine the transformation law for $\tilde{\Gamma}^{(4)}$ under an RGT. One should note that $\tilde{\Gamma}^{(2)}(0) = 1/\tilde{G}(\boldsymbol{k} = 0)$ (see (1.4.14)) is obtained by twice differentiating $\Gamma(M)$ with respect to M, and that its transformation law is given by (3.4.11a):

$$\tilde{\Gamma}^{(2)}(0; \mu') = s^{D - 2d_\varphi} \tilde{\Gamma}^{(2)}(0; \mu).$$

* One uses $\ln\left(k^2 + r_0 + \dfrac{u_0 M^2}{2}\right) = \ln(k^2 + r_0) + \dfrac{u_0 M^2}{2}(k^2 + r_0)^{-1} + \cdots \; (M \to 0)$

To go from $\tilde{\Gamma}^{(2)}$ to $\tilde{\Gamma}^{(4)}$ we must differentiate twice more with respect to M, and the transformation law of $\tilde{\Gamma}^{(4)}$ reads

$$\tilde{\Gamma}^{(4)}(0; \mu') = s^{D-4d_\varphi}\tilde{\Gamma}^{(4)}(0; \mu).$$

To check, note that according to (3.5.8) the normal dimension of $\tilde{\Gamma}^{(4)}$ is the same as that of u_0, i.e. $4 - D$; and that, if $d_\varphi = d_\varphi^0 = D/2 - 1$, then $D - 4d_\varphi = 4 - D$. The transformation law of $\tilde{\Gamma}^{(4)}$ is equivalent to

$$\tilde{\Gamma}^{(4)}(0) \sim \xi^{4d_\varphi - D} = \xi^{D-4+2\eta}. \tag{3.5.9}$$

At this point we need define accurately the field renormalization, which was already introduced in previous considerations. Since the two-point correlation function $G(\boldsymbol{x})$ behaves as $\exp(-\|\boldsymbol{x}\|/\xi)$ at large distances, its Fourier transform has poles at $k = \pm i\xi^{-1}$ (or more generally a branch cut beginning at $k = \pm i\xi^{-1}$) and can thus be written for $k^2 \lesssim O(\xi^{-2})$

$$\tilde{G}(\boldsymbol{k}) = \frac{Z_3}{k^2 + \xi^{-2} + O(k^4)}, \quad k^2 \lesssim O(\xi^{-2}).$$

The constant Z_3 is by definition the field renormalization constant; the notation Z_3 is chosen in order to be consistent with that of Chapters 6 and 7. An equivalent definition is

$$Z_3^{-1} = \frac{\mathrm{d}}{\mathrm{d}k^2}[\tilde{G}(\boldsymbol{k})]^{-1}|_{k^2=0}.$$

One may notice that Z_3 is the multiplicative factor which relates the susceptibility χ to the correlation length squared, since the fluctuation-response theorem may be written

$$\chi = \tilde{G}(\boldsymbol{k} = 0) = Z_3\xi^2 \sim \xi^{2-\eta}.$$

We thus discover that the renormalization constant Z_3 behaves as $\xi^{-\eta}$; since it is dimensionless, one should rather write $(\xi/a)^{-\eta}$. When $u_0 = 0$, namely in the case of the Gaussian model, we have of course $Z_3 = 1$. Furthermore, we have seen that there is no field renormalization at order u_0, because at this order the coefficient of $(\nabla\varphi)^2$ is not modified in the RG transformations: $Z_3 = 1 + O(u_0^2)$.

Let us now build the dimensionless quantity g from

$$g = \xi^{4-D}Z_3^2\tilde{\Gamma}^{(4)}(0). \tag{3.5.10}$$

From the behaviour (3.5.9) of $\tilde{\Gamma}^{(4)}(0)$ when $\xi \to \infty$, and from that of Z_3, we see that g is finite even at the critical point; g is nothing but the renormalized coupling constant.

Our GL Hamiltonian depended originally on the two bare parameters $(r_0 - r_{0c})$ and u_0. Now, in the vicinity of the critical point, the theory is still parametrized by two quantities: the correlation length ξ and the renormalized coupling constant g. If we are able to reshuffle our old perturbative expansion in powers of $u_0 r^{-\varepsilon/2}$ into an expansion in powers of g, we shall have made a big step forward in the solution of our difficulties, since g remains finite. Let us now examine more closely the relation between g and u_0. We start from equation (3.5.8), and notice that for $u_0 \to 0$, at fixed ξ, we have

$$g \to u_0\xi^\varepsilon \quad (u_0\xi^\varepsilon \to 0),$$

while close to the critical point, at fixed u_0 and $\xi \to \infty$ we have

$$g \to g^* \quad (u_0\xi^\varepsilon \to \infty),$$

where we have called g^* the limit of g at the critical temperature. A fundamental object in what follows will be the β-function $\beta(g, \varepsilon)$, defined as

$$\beta(g,\varepsilon) = -\xi \frac{dg}{d\xi}\bigg|_{u_0}. \tag{3.5.11a}$$

Furthermore it is convenient to define the dimensionless (bare) coupling constant g_0 by

$$g_0 = u_0 \xi^{\varepsilon}.$$

Being dimensionless, g can only be a function of the combination $g_0 = u_0\xi^{\varepsilon}$, so that an alternative expression for $\beta(g, \varepsilon)$ is

$$\beta(g,\varepsilon) = -\varepsilon g_0 \frac{dg}{dg_0}. \tag{3.5.11b}$$

In fact g and $\beta(g, \varepsilon)$ depend also on the ratio a/ξ, and the previous statements are correct only in the limit $a/\xi \to 0$. A calculation analogous to that of Section 7.1.2, with the substitution $\mu \to \xi^{-1}$, shows that g contains terms of order $(a/\xi)^{\varepsilon}$, and $\beta(g, \varepsilon)$ terms of order $(a/\xi)^{2+\varepsilon}$ (this allows us to identify the next-to-leading irrelevant eigenvalue $y_3 = -2 - \varepsilon + O(\varepsilon^2)$). For small enough g_0, $g(g_0) \simeq g_0$, and $\beta(g, \varepsilon) \simeq -\varepsilon g$. For $g_0 \to \infty$, $g(g_0) \to g^*$, so that $g(g_0)$ and $\beta(g, \varepsilon)$ have the qualitative behaviour displayed in Fig. 3.11. One notices that when the (dimensionless) bare coupling constant g_0 varies in the interval $[0, \infty[$, the renormalized one varies in $[0, g^*[$. The solution of the differential equation

$$\frac{dg_0}{dg} = -\frac{\varepsilon g_0}{\beta(g,\varepsilon)}$$

with the boundary condition $g_0(g) \to g$ when $g \to 0$ can be written

$$g_0 = g \exp\left[-\int_0^g dg'\left(\frac{\varepsilon}{\beta(g',\varepsilon)} + \frac{1}{g'}\right)\right].$$

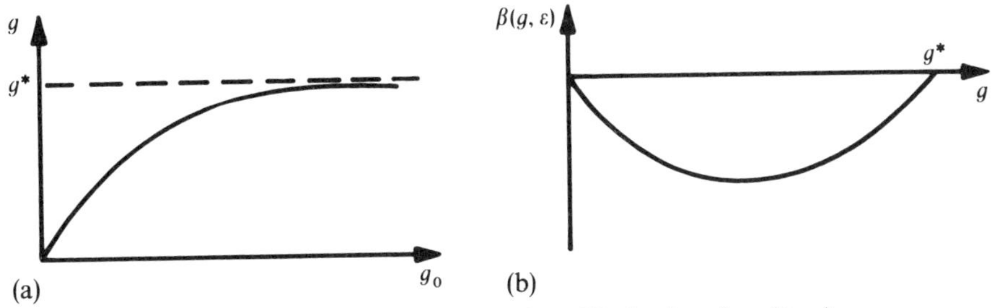

Fig. 3.11 (a) g as a function of g_0; (b) the function $\beta(g, \varepsilon)$

As $\beta(g', \varepsilon) \to -\varepsilon g'$ for $g' \to 0$, the integral converges at $g' = 0$. The critical point corresponds to $g_0 \to \infty$, or to $g \to g^*$; assuming that the zero of $\beta(g, \varepsilon)$ at $g = g^*$ is a simple one, we can parametrize $\beta(g, \varepsilon)$ as

$$\beta(g,\varepsilon) \simeq \omega(g - g^*), \quad g \to g^*,$$

with $\omega = \beta'(g^*, \varepsilon)$. This leads to the behaviour

$$g_0 \sim |g^* - g|^{-\varepsilon/\omega}, \tag{3.5.12a}$$

or

$$\xi \sim |g^* - g|^{-1/\omega}. \tag{3.5.12b}$$

More accurate expressions can be found in Problem 3.10 where it is also shown that $-\omega$ should be identified with the leading irrelevant eigenvalue: $y_2 = -\omega$. It should be clear that g^* corresponds really to a fixed point of RG: the derivative $\xi(\mathrm{d}g/\mathrm{d}\xi)$ vanishes at $g = g^*$. We must now try to compute the critical exponents. We have already defined a renormalization constant Z_3; we need a second one, which we denote by $\bar{Z}$,

$$\bar{Z} = -\frac{1}{2}\xi^3 \frac{\mathrm{d}(r_0 - r_{0c})}{\mathrm{d}\xi}\bigg|_{u_0}. \tag{3.5.13}$$

This definition is not very obvious; notice however that $\bar{Z}$ is dimensionless, and that it is equal to one in the Gaussian model, as $r_0 = \xi^{-2}$ in that case. The definition is also chosen in order to match that of Chapters 6 and 7. The renormalization constants Z_3 and $\bar{Z}$ are functions of g (or g_0) only, since they are dimensionless. From Z_3 and $\bar{Z}$ one defines the functions $\gamma(g)$ and $\bar{\gamma}(g)$,

$$\gamma(g) = -\xi \frac{\mathrm{d}}{\mathrm{d}\xi} \ln Z_3(g)\bigg|_{u_0} = \beta(g, \varepsilon)\frac{\mathrm{d}\ln Z_3}{\mathrm{d}g},$$

$$\bar{\gamma}(g) = -\xi \frac{\mathrm{d}}{\mathrm{d}\xi} \ln \bar{Z}(g)\bigg|_{u_0} = \beta(g, \varepsilon)\frac{\mathrm{d}\ln \bar{Z}}{\mathrm{d}g}.$$

The boundary conditions are $Z_3 = \bar{Z} = 1$ for $g \to 0$, and the solution of the above equations is

$$Z_3(g) = \exp \int_0^g \frac{\gamma(g')}{\beta(g', \varepsilon)}\,\mathrm{d}g',$$

$$\bar{Z}(g) = \exp \int_0^g \frac{\bar{\gamma}(g')}{\beta(g', \varepsilon)}\,\mathrm{d}g'.$$

If we assume that $\gamma(g)$ and $\bar{\gamma}(g)$ do not vanish at $g = g^*$, we get, close to the critical point,

$$Z_3(g) \simeq |g - g^*|^{\gamma(g^*)/\omega} \sim \xi^{-\gamma(g^*)} \tag{3.5.14a}$$

$$\bar{Z}(g) \simeq |g - g^*|^{\bar{\gamma}(g^*)/\omega} \sim \xi^{-\bar{\gamma}(g^*)}. \tag{3.5.14b}$$

The first equation allows us to identify the exponent η,

$$\eta = \gamma(g^*). \tag{3.5.15a}$$

In order to interpret the second one, let us recall that $(r_0 - r_{0c}) \sim (T - T_c)$, and that, from the definition of the exponent ν,

$$r_0 - r_{0c} \sim \xi^{-1/\nu}, \quad \frac{\mathrm{d}(r_0 - r_{0c})}{\mathrm{d}\xi} \sim -\xi^{-\frac{1}{\nu}-1},$$

so that from (3.5.13) and (3.5.14b),

$$\nu = \frac{1}{2 + \bar{\gamma}(g^*)}. \tag{3.5.15b}$$

It remains to evaluate the quantities g^*, $\gamma(g^*)$ and $\bar{\gamma}(g^*)$. We start from (3.5.8) and notice that at order u_0^2 we can set $Z_3 = 1$ and replace r_0 by ξ^{-2} in the integral. Using (B.4) and the definition (3.5.10) of g, we get

$$g\xi^{-\varepsilon} = u_0 - \frac{3}{2} u_0^2 \frac{\Gamma(\varepsilon/2)}{(4\pi)^{D/2}} \xi^{\varepsilon}.$$

Since the integral is convergent, we have taken the infinite cut-off limit: $\Lambda \to \infty$ (or $a \to 0$). Taking the derivative with respect to ξ of both sides of the above equation, and using the definition (3.5.11) of $\beta(g, \varepsilon)$, leads to

$$\beta(g, \varepsilon) = -\varepsilon g + \frac{3}{2} \varepsilon \frac{\Gamma(\varepsilon/2)}{(4\pi)^{D/2}} g^2 + O(g^3).$$

We have thus obtained the first two terms of the perturbative expansion of $\beta(g, \varepsilon)$ in powers of the renormalized coupling constant g. Note that, in the course of the derivation, we have replaced $u_0^2 \xi^{\varepsilon}$ by $g^2 \xi^{-\varepsilon}$, which is correct to this order of perturbation theory: the next term in the expansion of $\tilde{\Gamma}^{(4)}(0)$ is of order $u_0^3 \xi^{2\varepsilon}$, and would lead to a g^3 term in $\beta(g, \varepsilon)$. For small values of ε we can use (B.10) and get

$$\beta(g, \varepsilon) = -\varepsilon g + \frac{3g^2}{16\pi^2}. \tag{3.5.16}$$

The fixed point g^* is thus located at

$$g^* = \frac{16\pi^2 \varepsilon}{3},$$

in agreement with (3.5.4). For small values of ε, the renormalized coupling constant at the critical point is small, of order ε, and this observation lies at the basis of the ε-expansion of the critical exponents.

We have seen that $Z_3 = 1 + O(g^2)$, and we cannot compute the $O(g^2)$ term with the techniques which we have developed so far; we have then to stay with the Landau value of η, $\eta = 0$: the first correction to the Landau value will be computed in Chapter 5. However we can evaluate the first correction to ν, starting from equation (2.4.12); to this order of perturbation theory we can rewrite it as

$$r_0 - r_{0c} = \xi^{-2}\left(1 + \frac{u_0}{2} \int \frac{d^D k/(2\pi)^D}{k^2(k^2 + \xi^{-2})}\right).$$

We have used the relation between $r = \chi^{-1}$ and ξ, and the fact that $Z_3 = 1 + O(u_0^2)$. The integral is easily evaluated thanks to the identity

$$\int_0^{\infty} \frac{u^{\alpha} d\alpha}{(u+1)^{\beta}} = \frac{\Gamma(\alpha+1)\Gamma(\beta-\alpha-1)}{\Gamma(\beta)}.$$

We take immediately the $\varepsilon \to 0$ limit:

$$r_0 - r_{0c} \simeq \xi^{-2}\left(1 + \frac{u_0 \xi^{\varepsilon}}{16\pi^2 \varepsilon}\right),$$

so that from the definitions of $\bar{Z}(g)$ and $\bar{\gamma}(g)$,

$$\bar{Z}(g) = 1 + \frac{g}{16\pi^2 \varepsilon} + O(g^2),$$

$$\bar{\gamma}(g) = -\frac{g}{16\pi^2} + O(g^2).$$

Combining this last equation and (3.5.4) leads to

$$\bar{\gamma}(g^*) = -\frac{\varepsilon}{3} + O(\varepsilon^2)$$

and to

$$\nu = \frac{1}{2} + \frac{\varepsilon}{12} + O(\varepsilon^2)$$

in agreement with (3.5.6).

Instead of looking for an ε-expansion of the critical exponents, we could have worked *at a fixed value of D*, for example $D = 3$. We would then have found $\nu = 0.6$, instead of 0.57 in the ε-expansion (Problem 3.10).

The method we have been following needs some explanation: we began with an infinite-dimensional parameter space, and we ended up by using one parameter only! What happens is that we have followed in parameter space a very peculiar trajectory, which links the origin (trivial Gaussian fixed point) to the non-trivial IR fixed point, when g varies from zero to g^*.

What we have just achieved is the full calculation of a critical exponent to order ε. We add some comments.

(i) The calculations have been done with an order parameter having dimension $n = 1$. It would not be difficult to generalize them to an order parameter having dimension n; the Ginzburg–Landau Hamiltonian for that case reads

$$H = \int \mathrm{d}^D x \left[\frac{1}{2} \sum_{i=1}^{n} (\nabla \varphi_i)^2 + \frac{1}{2} r_0 \left(\sum_{i=1}^{n} \varphi_i^2 \right) + \frac{u_0}{4!} \left(\sum_{i=1}^{n} \varphi_i^2 \right)^2 \right]. \tag{3.5.17}$$

Equation (3.5.6) for ν is then generalized to

$$\nu = \frac{1}{2} + \frac{(n+2)\varepsilon}{4(n+8)} + O(\varepsilon^2) \tag{3.5.18}$$

(Problem 3.6), which displays the explicit dependence of critical exponents on the dimensionality of the order parameter.

(ii) To order ε the critical exponent η is zero. To order ε^2 one finds (see Section 5.5.4)

$$\eta = \frac{(n+2)\varepsilon^2}{2(n+8)^2}.$$

(iii) The calculations given so far depend on perturbation expansions rearranged by means of the renormalization group. The expansion parameter is ε, and the results are trustworthy for $\varepsilon \ll 1$ (in a space of 3.99 dimensions!). Extrapolation to the realistic case $D = 3$ yields reasonably satisfactory results, albeit with some nasty surprises: the results to order ε^3 are rather less good (i.e. further from observations) than those to order ε^2. Expansions like (3.5.7) do not converge; rather they are asymptotic series (like Stirling's approximation to n!).

(iv) However, powerful resummation methods have been applied either to the ε-expansion or to the fixed-D expansion and have produced very accurate values of the critical exponents: to quote Parisi '... the problem of computing critical exponents is essentially solved: if simple computations are performed, approximate results are obtained; if we perform more and more refined computations, we get more and more precise results' (Parisi 1988, chapter 7).

3.6 Marginal fields, the function $\beta(g)$, and logarithmic corrections in 4D

3.6.1 The differential equation for a marginal field

So far we have sidestepped marginal fields. They are particularly important because they constitute the link with the 'old' version of the renormalization group (Stueckelberg and Petermann, and Gell-Mann and Low) which one encounters, nowadays, in the form of the Callan–Symanzik equations (Chapter 7).

For a marginal field (or coupling) the techniques used in Section 3.2.3 cannot directly yield the behaviour of the correlation function. In order to interpret intuitively the somewhat technical discussion that follows, let us try to understand qualitatively the differences in behaviour between RGTs in the presence and in the absence of such coupling.

Consider the renormalization flow in the immediate vicinity of the critical surface (see Fig. 3.12). In case (a) the trajectory converges rapidly to the fixed point, with irrelevant fields decreasing according to a power-law. In case (b) the marginal field remains constant in the linear approximation; we shall show a little later that in fact it varies logarithmically, i.e. very slowly. The trajectory approaches the $e^{(2)}$ axis (which corresponds to the marginal field), and follows it over an appreciable interval of s-values. This explains why the RGTs are

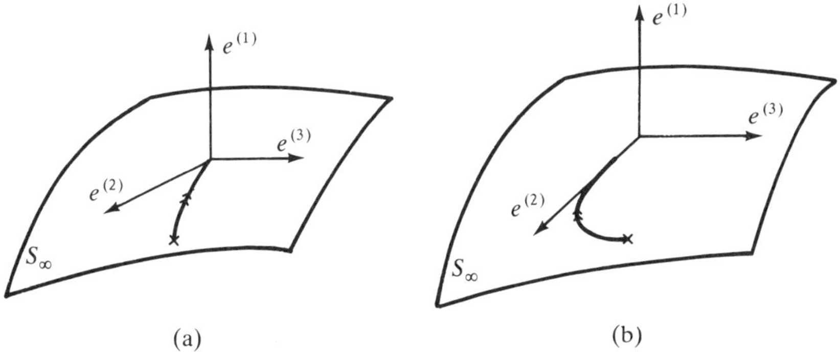

Fig. 3.12 The renormalization flow near the critical surface. (a) No marginal variables: $y_2 < 0$; (b) one marginal variable: $y_2 = 0$

governed at least in part by the evolution of the marginal field; eventually the fixed point regains its influence, but meanwhile the evolution of the marginal field has had time to modify the critical behaviour by logarithmic factors.

The evolution of the marginal field is described by the differential equation (3.6.10), featuring only the marginal field itself. (With several marginal fields we have a system of differential equations.) In what follows, we shall denote the marginal field by g, since it will be identified with the coefficient of φ^4 in the Ginzburg–Landau Hamiltonian: thus the differential equation describes the variation of a coupling constant $g(s)$.

The crucial point is that this differential equation can be ascertained perturbatively, a property that allows the RGTs to be calculated analytically, at least in some approximation. This is why the argument that follows requires us to stay in the perturbative region, i.e. in the region where the various parameters of the problem are small.

We shall use, in what follows, the RG in its differential form (3.5.3), setting $l = \ln s$. Our discussion will rely on Poincaré's theory of normal forms (Arnold 1983). The essential points can be understood by looking at the following simple example, where two functions $x_1(l)$ and $x_2(l)$ obey coupled nonlinear differential equations:

$$\dot{x}_1(l) = \frac{\mathrm{d}x_1}{\mathrm{d}l} = \lambda_1 x_1 + a_{11}x_1^2 + a_{12}x_1x_2 + a_{22}x_2^2, \tag{3.6.1a}$$

$$\dot{x}_2(l) = \frac{\mathrm{d}x_2}{\mathrm{d}l} = \lambda_2 x_2 + b_{11}x_1^2 + b_{12}x_1x_2 + b_{22}x_2^2. \tag{3.6.1b}$$

Let us try to eliminate the nonlinear terms on the RHS of (3.6.1) by the change of variables

$$x_1 = y_1 + h_1(y_1, y_2), \tag{3.6.2a}$$

$$x_2 = y_2 + h_2(y_1, y_2), \tag{3.6.2b}$$

where the functions h_1 and h_2 are given by

$$h_1(y_1, y_2) = \alpha_{11}y_1^2 + \alpha_{12}y_1y_2 + \alpha_{22}y_2^2, \tag{3.6.3a}$$

$$h_2(y_1, y_2) = \beta_{11}y_1^2 + \beta_{12}y_1y_2 + \beta_{22}y_2^2. \tag{3.6.3b}$$

It is readily verified that the new functions $y_1(l)$ and $y_2(l)$ obey approximately uncoupled, linear differential equations (Poincaré's theorem)

$$\dot{y}_1(l) = \lambda_1 y_1(l) + \text{cubic terms},$$

$$\dot{y}_2(l) = \lambda_2 y_2(l) + \text{cubic terms},$$

provided one chooses $\alpha_{11}, \ldots, \beta_{22}$ as follows:

$$\alpha_{11} = \frac{a_{11}}{\lambda_1}, \quad \alpha_{12} = \frac{a_{12}}{\lambda_2}, \quad \alpha_{22} = \frac{a_{22}}{2\lambda_2 - \lambda_1}; \tag{3.6.4a}$$

$$\beta_{11} = \frac{b_{11}}{2\lambda_1 - \lambda_2}, \quad \beta_{12} = \frac{b_{12}}{\lambda_1}, \quad \beta_{22} = \frac{b_{22}}{\lambda_2}. \tag{3.6.4b}$$

The elimination of cubic terms can be achieved through a second change of variables and then that of quartic terms etc. However the elimination of all quadratic terms will not be possible if one of the following 'resonance conditions' is satisfied:

$$\lambda_1 = 0, \quad \lambda_2 = 0, \quad 2\lambda_2 - \lambda_1 = 0, \quad 2\lambda_1 - \lambda_2 = 0. \tag{3.6.5}$$

When one (or several) of the resonance conditions are satisfied, it is still possible to simplify the original equations (3.6.1) (Poincaré–Dulac theorem). For example if $2\lambda_2 - \lambda_1 = 0$, but $\lambda_1 \neq 0$, $\lambda_2 \neq 0$, one finds the normal form

$$\dot{y}_1(l) = \lambda_1 y_1(l) + a_{22} y_2^2 + \text{cubic terms},$$

$$\dot{y}_2(l) = \lambda_2 y_2(l) + \text{cubic terms}.$$

Let us apply the theory of normal forms to a system of two coupled differential equations, involving a relevant and a marginal coupling which we denote by t' and g' respectively. In the GL model, the eigenvalue λ_1 corresponding to the relevant coupling is $\lambda_1 = 2$, and we have of course $\lambda_2 = 0$ for the marginal coupling. The RG differential equations are thus of the form (recall $l = \ln s$)

$$\frac{dt'}{dl} = 2t' + \varphi(t', g'), \tag{3.6.6a}$$

$$\frac{dg'}{dl} = \psi(t', g'), \tag{3.6.6b}$$

where the non-linear functions $\varphi(t', g')$ and $\psi(t', g')$ have power expansions, which can be determined in perturbation theory:

$$\varphi(t', g') = at'^2 + bt'g' + cg'^2 + \cdots, \tag{3.6.7a}$$

$$\psi(t', g') = dt'^2 + et'g' + fg'^2 + \cdots. \tag{3.6.7b}$$

Since $\lambda_2 = 0$, one of the resonance conditions is satisfied, and the nonlinear change of variables

$$t' = t + \frac{1}{2}at^2 - \frac{1}{2}cg^2 + \cdots, \tag{3.6.8a}$$

$$g' = g + \frac{1}{4}dt^2 + \frac{1}{2}etg + \cdots \tag{3.6.8b}$$

allows us to obtain the normal forms

$$\frac{dt}{dl} = 2t + btg + \cdots = t(2 + \bar{\gamma}(g)), \tag{3.6.9a}$$

$$\frac{dg}{dl} = fg^2 + \cdots = -\beta(g). \tag{3.6.9b}$$

We leave to the reader the general proof of (3.6.9).

We encounter for the second time in (3.6.9b) the famous β-function of Callan and Symanzik, which will play a crucial role again in Chapter 7; the notation $\bar{\gamma}(g)$ in (3.6.9a) is introduced for consistency with Subsection 3.5.3 and Chapter 7. The important point is of course that the equations for t and g are now almost entirely decoupled.

Let us first study equation (3.6.9b), which we rewrite

$$\frac{dg}{dl} = -\beta(g) = -\beta_0 g^2 - \beta_1 g^3 - \cdots \tag{3.6.10}$$

We are interested in the $l \to \infty$ behaviour of the coupling $g(l)$, starting from some initial value $g_0 = g(l = 0)$. We can integrate (3.6.10) in the form

$$\int_{g_0}^{g(l)} \frac{\mathrm{d}g'}{\beta(g')} = -l = -\ln s, \tag{3.6.11}$$

which gives $g(l)$ as an implicit function of g_0 and l. Assume now that $\beta(g)$ keeps a constant sign in the interval $[0, g_0]$, which means that it has the sign of β_0 in this interval. If $\beta_0 > 0$, $\mathrm{d}g/\mathrm{d}l$ stays negative and $g(l) \to 0$ as $l \to \infty$. For small enough values of g_0, we need keep only the term $-\beta_0 g^2$ in (3.6.10) and get the explicit solution

$$g(l) = \frac{g_0}{1 + \beta_0 g_0 l} = \frac{g_0}{1 + \beta_0 g_0 \ln s}. \tag{3.6.12}$$

Thus, in the case $\beta_0 > 0$, $g(s)$ behaves as an irrelevant coupling, although it vanishes only as an inverse power of $\ln s$, and not as a power law. Equation (3.6.12) illustrates the power of the renormalization group: the RHS can be expanded in a power series,

$$g(s) = g_0(1 - \beta_0 g_0 \ln s + (\beta_0 g_0 \ln s)^2 - \cdots).$$

This expansion is valid only if $\beta_0 g_0 \ln s < 1$, but the RG allows one to sum the series in a meaningful way, even if $\beta_0 g_0 \ln s$ is large; one never leaves the perturbative regime if the initial value g_0 is small enough. If $\beta_0 < 0$, the present approach admits no conclusion; one interesting situation is the case where $\beta(g)$ has a zero at $g = g^*$. In that case $g(s)$ will be driven to g^* in the limit s (or l) $\to \infty$. More details will be found in Chapter 7.

From now on we assume $\beta_0 > 0$, which will be the case for the GL model. We can then find the large-l behaviour of $t(l)$ by solving (3.6.9a):

$$t(l) = t_0 \exp\left(2l - \int_{g_0}^{g(l)} \frac{\bar{\gamma}(g')\mathrm{d}g'}{\beta(g')}\right). \tag{3.6.13}$$

If we approximate $\bar{\gamma}(g)$ and $\beta(g)$ by the first term of their power expansion,

$$\bar{\gamma}(g) = \bar{\gamma}_0 g + \cdots, \quad \beta(g) = \beta_0 g^2 + \cdots, \tag{3.6.14}$$

then equation (3.6.13) becomes

$$t(s) \approx t_0 s^2 (1 + \beta_0 g_0 \ln s)^{\bar{\gamma}_0/\beta_0}. \tag{3.6.15}$$

If we wish to stay in the perturbative region ($t(s) \ll 1$), it is clear that $t(s)$ should be chosen exponentially small in $\ln s$. The critical surface is obviously defined by $t(s) = 0$. Reverting to the original couplings t' and g', we see that this means (cf. (3.6.8a))

$$t' = -\frac{1}{2} c g'^2 + \cdots \tag{3.6.16}$$

In the previous discussion, we have disregarded the irrelevant couplings. In order to keep the following argument simple, we shall restrict ourselves to a single irrelevant coupling w'. The GL model suggests that the corresponding eigenvalue should be $\lambda_3 = -2$. We must take into account the resonance condition $\lambda_1 + \lambda_3 = 0$, and after a suitable change of variables, we are led to the following RG differential equations in normal form:

$$\frac{dt}{dl} = t(2 + bg + \cdots) \tag{3.6.17a}$$

$$\frac{dg}{dl} = -\beta_0 g^2 + htw + \cdots \tag{3.6.17b}$$

$$\frac{dw}{dl} = w(-2 + kg + \cdots) \tag{3.6.17c}$$

In order to make use of these equations, it is necessary to assume that one is very close to the critical surface, so that the initial value t_0 is exponentially small. Then terms of the form tw in (3.6.17b) can be neglected and one recovers for g a simple differential equation. Starting from an initial value w_0, one sees from (3.6.17c) that w becomes exponentially small. Thus, in the limit $s \to \infty$, the irrelevant coupling $w'(s)$ becomes completely independent of its initial value w'_0 and can be computed as a function of $g'(s)$, by a formula analogous to (3.6.16), which also shows that $w'(s)$ is of order $g'(s)^2$. The only role of the irrelevant coupling is to change the initial condition in the basic differential equation (3.6.9b) for $g(s)$. These remarks will turn out to be extremely important in the discussion of Chapter 7.

3.6.2 Logarithmic corrections to the correlation function

Let us apply the previous considerations to the Ginzburg–Landau Hamiltonian in dimension $D = 4$. The field u_0 is then marginal, and will be denoted by g_0. From the results of Section 3.4.3, the relevant field t_0 is related to r_0 and g_0 by

$$t_0 = r_0 + \frac{\Lambda^2 g_0}{32\pi^2}.$$

The renormalization group equations (3.5.2) become

$$\frac{dt_0(s)}{d\ln s} = 2t_0(s) - \frac{g_0(s)t_0(s)}{16\pi^2} + O(g_0^2(s)) \tag{3.6.18a}$$

$$\frac{dg_0(s)}{d\ln s} = -\frac{3}{16\pi^2}g_0^2(s). \tag{3.6.18b}$$

As explained above, the $O(g_0^2(s))$ term in (3.6.18a) can be absorbed into a redefinition of the relevant coupling t_0. The solution of (3.6.18) has already been written in (3.6.12) and (3.6.15); let us note that, as $s \to \infty$, we can make in (3.6.15) the approximation

$$t_0(s) \simeq t_0 s^2 (\ln s)^{\bar{\gamma}_0/\beta_0}, \tag{3.6.19}$$

where, from (3.6.18),

$$\bar{\gamma}_0 = -\frac{1}{16\pi^2} \qquad \beta_0 = \frac{3}{16\pi^2} \qquad \frac{\bar{\gamma}_0}{\beta_0} = -\frac{1}{3}. \tag{3.6.20}$$

Let us now examine the correlation function in the critical region. We start from (3.4.11a) and define

$$\zeta(s) = s^{D-2}\lambda^{-2}(s). \tag{3.6.21}$$

Equation (3.4.11a) can then be recast into

$$G(\boldsymbol{k}, t_0, g_0, \Lambda) = s^2\zeta(s)G(s\boldsymbol{k}, t_0(s), g(s), \Lambda). \tag{3.6.22}$$

Let us first consider the behaviour of the susceptibility χ, which we obtain by putting $\boldsymbol{k}' = 0$ in (3.6.22):

$$\chi(t_0, g_0, \Lambda) = s^2\zeta(s)\chi(t_0(s), g_0(s), \Lambda). \tag{3.6.23}$$

We shall show later on that $\zeta(s)$ is a constant factor, up to terms of order $(\ln s)^{-1}$. We choose t_0 in the critical region ($|t_0| \to 0$), while taking $|t_0(s)| \approx 1$, so that χ in the RHS of (3.6.23) is regular, since it is evaluated far from the critical region (more precisely we should write $t_0/\Lambda^2 \to 0$ and $t_0(s)/\Lambda^2 \to 1$, as t_0 has dimension 2). From (3.6.19) we get

$$s^2 \simeq \frac{1}{|t_0|}\left(\ln\frac{1}{|t_0|}\right)^{-\bar{\gamma}_0/\beta_0}. \tag{3.6.24}$$

Since $|t_0(s)| \approx 1$, $g_0(s) \to 0$ and $\zeta(s) \approx$ constant, the behaviour of χ for $t_0 \to 0$ is

$$\chi(t_0, g_0, \Lambda) \sim |t_0|^{-1}|\ln|t_0||^{-\bar{\gamma}_0/\beta_0}. \tag{3.6.25}$$

The behaviour of the Landau theory ($\chi \approx |t_0|^{-1}$) is thus modified by the logarithmic factor $|\ln|t_0||^{1/3}$. This is a general rule: in 4D, the critical exponents of the Landau theory are correct, but the behaviour is not given by a pure power law: one finds in addition powers of logarithms.

Let us now justify our choice $\zeta(s) \approx$ constant. We have seen in Section 3.4.3 that the coefficient of the term $\frac{1}{2}(\nabla\varphi)^2$ in the GL Hamiltonian is not modified to first order in g_0. The modification occurs only at order g_0^2, so that (see Problem 3.4(c))

$$\frac{\mathrm{d}\ln\zeta(s)}{\mathrm{d}\ln s} = -\gamma_0 g_0^2(s) + O(g_0^2(s)). \tag{3.6.26}$$

Integrating this equation gives, with the initial condition $\zeta_0 = 1$,

$$\ln\zeta(s) = \frac{\gamma_0}{\beta_0}\int_{g_0}^{g_0(s)} \mathrm{d}g = -\frac{\gamma_0 g_0}{\beta_0} + O\left(\frac{1}{\ln s}\right). \tag{3.6.27}$$

As promised, this expression shows that, up to terms of order $(\ln s)^{-1}$, $\zeta(s)$ is an

s-independent constant. An equivalent statement is that the correlation function obeys the Landau theory at the critical point:

$$G(k) \approx k^{-2}, \quad k \to 0. \tag{3.6.28}$$

This behaviour corresponds to $\eta = 0$, without logarithmic corrections. This property is peculiar to the φ^4-interaction. In general we would expect

$$\frac{\mathrm{d}\ln\zeta(s)}{\mathrm{d}\ln s} = -\gamma(g(s)) = -\gamma_0 g + \cdots$$

and the behaviour (3.6.28) would be modified by powers of $\ln k$.

Let us conclude this chapter with some general remarks on the correlation function at the critical point ($T = T_c$). We have shown that

$$G(\boldsymbol{k}, g_0, \Lambda) \simeq s^2 G(s\boldsymbol{k}, g(s), \Lambda). \tag{3.6.29}$$

In the critical region $k/\Lambda \ll 1$, the perturbative expansion of G on the LHS of (3.6.29) contains terms like $g_0^2 \ln(k/\Lambda)$, $g_0^3 \ln^2(k/\Lambda)$ etc., as will be shown in Chapter 5. Since $\ln(k/\Lambda)$ is arbitrarily large, the perturbative expansion is meaningless, even for small values of g_0. When $sk/\Lambda \approx 1$, a perturbative expansion of the RHS of (3.6.29) becomes possible. This means that for wave numbers $k \approx \Lambda/s$, the coupling constant which should be used in a perturbative expansion is $g_0(s)$, and not g_0: to some scale of wavenumbers corresponds a certain choice of the coupling constant (which is no longer constant!). As we shall see in more detail in Chapter 7, the idea of a coupling constant adapted to each scale of wavenumber is a fundamental one in renormalization theory.

Let us finally emphasize that the RHS in equation (3.6.29) should not be misunderstood: the Hamiltonian to be used on the RHS is not the GL Hamiltonian with a φ^4 interaction only and a coupling constant $g_0(s)$. In fact the Hamiltonian contains many irrelevant couplings, which are generated by the renormalization group transformations. However, as explained in the previous subsection, these irrelevant couplings are calculable as functions of $g_0(s)$.

Problems

3.1 The decimation method

Instead of forming blocks of spins, one can define an RGT by summing over certain spins of the lattice. For instance, one might sum over the spins marked • on the two-dimensional lattice in Fig. 3.13, while those marked x are kept as the spins S'. (Notice that the transformed lattice has undergone a rotation through $\pi/4$ with respect to the original lattice.) Show that if one starts from an Ising Hamiltonian, then (in the notation of equation (3.1.5)) the interaction between the spins S' takes the form

$$H' = A + B \sum_{\langle ij \rangle} S'_i S'_j + C \sum_{\langle\langle ij \rangle\rangle} S'_i S'_j + D \sum_{\langle ijkl \rangle} S'_i S'_j S'_k S'_l,$$

and calculate A, B, C, and D.

(The decimation method gives rather poor results; but the object of the exercise is to show that in an RGT H' takes a form different from H.)

Hint: consider first the case $D = 1$, and show that $\tanh K' = (\tanh K)^2$.

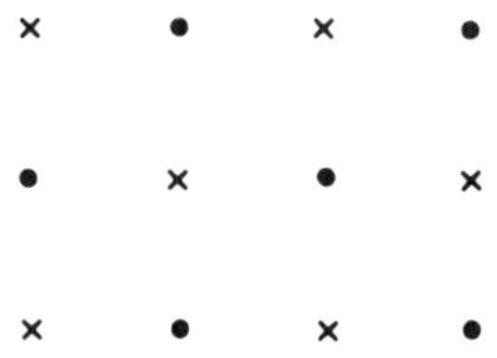

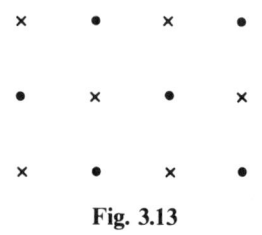

Fig. 3.13

3.2 Fourier transforms

Define the Fourier transform of the field $\varphi(\boldsymbol{x})$ by

$$\varphi(\boldsymbol{k}) = \frac{a^D}{L^{D/2}} \sum_{\boldsymbol{x}} \mathrm{e}^{\mathrm{i}\boldsymbol{k}\cdot\boldsymbol{x}} \varphi(\boldsymbol{x})$$

$$\varphi(\boldsymbol{x}) = \frac{1}{L^{D/2}} \sum_{\boldsymbol{k}} \mathrm{e}^{-\mathrm{i}\boldsymbol{k}\cdot\boldsymbol{x}} \varphi(\boldsymbol{k}).$$

(a) Determine the Gaussian Hamiltonian H_0 as a function of $\varphi(\boldsymbol{k})$.

(b) Show generally that

$G(\boldsymbol{k}) = \langle \varphi(\boldsymbol{k}) \varphi(-\boldsymbol{k}) \rangle.$

(c) The change of variables $\varphi(\boldsymbol{x}) \to \varphi(\boldsymbol{k})$ has the disadvantage that $\varphi(\boldsymbol{k})$ is not real. One needs to introduce two real variables $\alpha(\boldsymbol{k})$ and $\beta(\boldsymbol{k})$:

$$\varphi(\pm \boldsymbol{k}) = \frac{1}{\sqrt{2}} (\alpha(\boldsymbol{k}) \pm \mathrm{i}\beta(\boldsymbol{k})).$$

Evaluate the Jacobian of the transformation

$$\mathscr{D}\varphi(\boldsymbol{x}) \to \prod_{\boldsymbol{k}(k_x > 0)} \mathrm{d}\alpha(\boldsymbol{k})\mathrm{d}\beta(\boldsymbol{k}).$$

(d) From (a) and (c) derive the correlation function $G_0(\boldsymbol{k})$ of the Gaussian model, by direct integration over $\mathrm{d}\alpha(\boldsymbol{k})\mathrm{d}\beta(\boldsymbol{k})$.

3.3 Let the generalized Ginzburg–Landau Hamiltonian be given by (3.4.2). Calculate r_0', u_0', u_6', u_8', and v_0' in an RGT to first order in V, and verify that for $D > 4$ all the fields except r_0 are irrelevant. Calculate also the modification to the term with $(\boldsymbol{\nabla}\varphi)^2$.

3.4 (a) Use the differential equations (3.5.2) to determine the coordinates (r_0^*, u_0^*) of the fixed point.

(b) To linearize near the fixed point one writes

$$r_0 = r_0^* + \delta_{r_0}, \quad u_0 = u_0^* + \delta u_0.$$

Show that

$$\begin{pmatrix} \dfrac{\mathrm{d}\delta r_0}{\mathrm{d}\ln s} \\[2ex] \dfrac{\mathrm{d}\delta u_0}{\mathrm{d}\ln s} \end{pmatrix} = R\begin{pmatrix} \delta r_0 \\ \delta u_0 \end{pmatrix}$$

where R is a 2×2 matrix; and show that y_1 and y_2 are the eigenvalues of the matrix R. Calculate y_1 and y_2.

(c) Starting from the differential renormalization equations (3.5.3), show that the factor $\lambda(s)$ in (3.1.11) obeys a differential equation of the type

$$\frac{\mathrm{d}\ln\lambda(s)}{\mathrm{d}\ln s} = \gamma(K_\alpha(s)),$$

and derive the following expression for $\lambda(s)$:

$$\lambda(s) = \exp\left[\int_1^s \gamma(K_\alpha(s'))\frac{\mathrm{d}s'}{s'}\right].$$

Recover (3.1.11) in the vicinity of the fixed point. How must one amend equation (3.4.11a)?

3.5 Calculation of y_B by the cumulant approximation

(a) One defines

$$a_1 = \frac{\mathrm{e}^{3K} + \mathrm{e}^{-K}}{\mathrm{e}^{3K} + 3\mathrm{e}^{-K}}; \quad a_2 = \frac{\mathrm{e}^{3K} - \mathrm{e}^{-K}}{\mathrm{e}^{3K} + 3\mathrm{e}^{-K}}; \quad a_3 = \frac{\mathrm{e}^{3K} - 3\mathrm{e}^{-K}}{\mathrm{e}^{3K} + 3\mathrm{e}^{-K}}.$$

Show that

$$\langle S_\alpha^{(i)}\rangle_0 = a_1 S'_\alpha; \quad \langle S_\alpha^{(i)} S_\alpha^{(j)}\rangle_0 = a_2; \quad \langle S_\alpha^{(i)} S_\alpha^{(j)} S_\alpha^{(k)}\rangle_0 = a_3 S'_\alpha$$

where the indices i, j, k are all different (the notation is that of Section 3.3).

(b) Introduce an infinitesimally weak B-field:

$$H \to H - B\sum_i S_i.$$

Adopting the approximation $V = 0$, and averaging at fixed $[S']$, show that

$$\sum_i \langle S_\alpha^{(i)}\rangle = 3a_1 \sum_\alpha S'_\alpha$$

and derive y_B in this approximation.

(c) The next approximation consists in writing

$$\mathrm{e}^{-(H_0+V)} \simeq \mathrm{e}^{-H_0}(1 - V).$$

Show that under these conditions $\delta\langle S_\alpha^{(i)}\rangle = \langle S_\alpha^{(i)}\rangle - \langle S_\alpha^{(i)}\rangle_0$ is given by

$$\delta\langle S_\alpha^{(i)}\rangle = \langle V\rangle_0 \langle S_\alpha^{(i)}\rangle_0 - \langle V S_\alpha^{(i)}\rangle_0.$$

Calculate $\langle S_\alpha^{(1)}\rangle$ and from it derive

$$\sum_i \langle S_\alpha^{(i)}\rangle = \sum_\alpha 3a_1[1 + 4K(1 - a_1^2 + 2(a_2 - a_1^2))]S'_\alpha.$$

(*Hint*: consider the conditions needed for $\langle S\rangle_0\langle V\rangle_0 \neq \langle VS\rangle_0$. Remember that V does not link different blocks: see Fig. 3.14.)

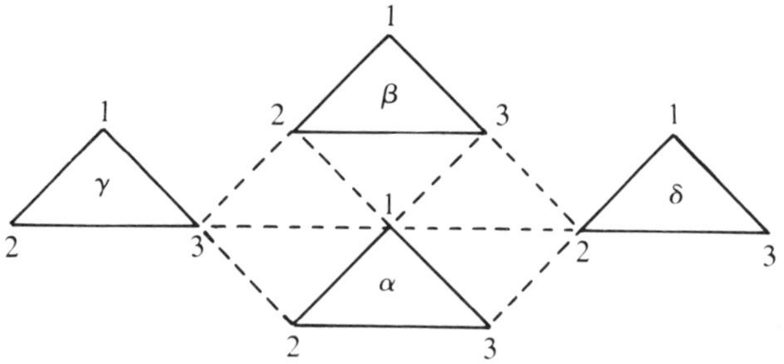

Fig. 3.14

(d) Calculate y_B numerically in this approximation, and likewise α, β, γ, δ, η. Compare them with the experimental values. To this end use the value of ν determined in the text.

3.6 Calculation of the critical exponents with an n-dimensional order parameter

The aim is to calculate the exponent ν starting from a Hamiltonian of the form (3.5.17).

(a) Calculate the thermodynamic potential $\Gamma(M)$ using the method of Section 2.3. Note that the matrix $D(\boldsymbol{x}, \boldsymbol{y})$ now carries internal indices i and j:

$$D_{ij}(\boldsymbol{x}, \boldsymbol{y}) = [(-\nabla_x^2 + r_0)\delta_{ij} + \frac{u_0}{6}(\delta_{ij}\vec{\varphi}^2 + 2\varphi_i\varphi_j)]\delta(\boldsymbol{x} - \boldsymbol{y})$$

where

$$\vec{\varphi}^2 = \sum_{i=1}^{n} \varphi_i^2.$$

Calculate $\ln\det(D_{ij}(\boldsymbol{q}))$ at fixed $\boldsymbol{q}$; it is worth noting that the calculation simplifies if one chooses $M_i = (M, 0, 0, \ldots, 0)$, where $M = (\vec{M}^2)^{1/2}$.

Show that $\Gamma(M)$ may be written

$$L^{-D}\Gamma(M_i) = \frac{1}{2}r_0 M^2 + \frac{u_0}{4!}(\vec{M}^2)^2 + \frac{1}{2}\int \frac{\mathrm{d}^D q}{(2\pi)^D} \ln\left(q^2 + r_0 + \frac{u_0}{2}\vec{M}^2\right)$$
$$+ \frac{n-1}{2}\int \frac{\mathrm{d}^D q}{(2\pi)^D} \ln\left(q^2 + r_0 + \frac{u_0}{6}\vec{M}^2\right).$$

(b) Show that the equations giving $\bar{r}_0(T - T_c)$ and $\tilde{\Gamma}^{(4)}(0)$ are now

$$\bar{r}_0(T - T_c) = r\left(1 + \frac{u_0(n+2)}{6}\int \frac{\mathrm{d}^D k}{(2\pi)^D} \frac{1}{k^2(k^2 + r)}\right),$$

$$\tilde{\Gamma}^{(4)}(0) = u_0 - u_0^2 \frac{(n+8)}{6}\int \frac{\mathrm{d}^D k}{(2\pi)^D} \frac{1}{(k^2 + r)^2}.$$

(c) From this derive the expression (3.5.18) for the critical exponent ν.

3.7 The renormalization group in the limit $n \to \infty$ (Ma 1973)

We aim to study the renormalization group in the limit where the dimensionality n of the order parameter tends to infinity. The Hamiltonian is given by

$$H = \int \mathrm{d}^D x\left[\frac{1}{2}\sum_{i=1}^{n} (\nabla\varphi_i)^2 + U(\vec{\varphi}^2)\right],$$

$$\vec{\varphi}^2 = \sum_{i=1}^{n} \varphi_i^2, \quad U(\vec{\varphi}^2) = \sum_{m=1}^{\infty} \frac{u_{2m}}{m}\left(\frac{\vec{\varphi}^2}{2}\right)^m,$$

$$t(\vec{\varphi}^2) = 2\frac{\mathrm{d}U}{\mathrm{d}\vec{\varphi}^2} = \sum_{m=1}^{\infty} u_{2m}\left(\frac{\vec{\varphi}^2}{2}\right)^{m-1}.$$

(a) Subdivide $\varphi(\boldsymbol{x})$:

$$\varphi(\boldsymbol{x}) = \varphi_1(\boldsymbol{x}) + \bar{\varphi}(\boldsymbol{x}),$$

where $\varphi_1(\boldsymbol{x})$ ($\bar{\varphi}(\boldsymbol{x})$) has Fourier components in the interval $0 \leqslant k \leqslant \Lambda/s$ ($\Lambda/s \leqslant k \leqslant \Lambda$). As $n \to \infty$, terms like

$$\sum_{i=1}^{n} |\bar{\varphi}_i(\boldsymbol{k})|^2$$

are of order n, while

$$\sum_i \sum_{k,k'} \mathrm{e}^{\mathrm{i}(\boldsymbol{k}+\boldsymbol{k}')\cdot \boldsymbol{x}} \bar{\varphi}_i(\boldsymbol{k}'),$$

being a sum of terms with random signs, is of order 1. Exploit this result to show that in the limit $n \to \infty$ the integral over $\bar{\varphi}(\boldsymbol{x})$ may be written

$$\int \prod_{\Lambda/s \leqslant k \leqslant \Lambda} \mathrm{d}N_k \exp\left(-\left\{ \left(\int \mathrm{d}^D x \frac{1}{2} (\nabla \vec{\varphi}_1)^2 \right) + W \right\} \right)$$

where

$$W = \sum_{\Lambda/s \leqslant k \leqslant \Lambda} \left(-\frac{1}{2} n \ln N_k + \frac{1}{2} k^2 N_k \right) + \int \mathrm{d}^D x U(\rho + \vec{\varphi}_1^{\,2}),$$

$$\rho = L^{-D} \sum_{\Lambda/s \leqslant k \leqslant \Lambda} N_k, \quad N_k = \sum_{i=1}^{n} |\bar{\varphi}_i(\boldsymbol{k})|^2.$$

(b) To evaluate the integral, one looks for the maximum of the integrand, and then replaces the integral by this maximum value. Justify this approximation, and show that one must choose ($\varphi_1 \to \varphi$)

$$\bar{N}_k = \frac{n}{k^2 + t(\bar{\rho} + \vec{\varphi}^{\,2})},$$

$$\bar{\rho} = n K_D \int_{\Lambda/s}^{\Lambda} \frac{k^{D-1} \mathrm{d}k}{k^2 + t(\bar{\rho} + \vec{\varphi}^{\,2})}.$$

(c) We now dilate the unit of length, and implement the transformation $\varphi \to s^{-d_\varphi} \varphi(\boldsymbol{x}/s)$. Show that one must choose $\eta = 0$, and that the transformation law $t_s = R_s t$ reads

$$t_s(\vec{\varphi}^{\,2}) = s^2 t(\bar{\rho} + s^{2-D} \vec{\varphi}^{\,2})$$

$$\bar{\rho} = n K_D \int_{\Lambda/s}^{\Lambda} \frac{k^{D-1} \mathrm{d}k}{k^2 + t_s(\vec{\varphi}^{\,2})/s^2}.$$

(d) Show that on the critical surface one is bound to have $t(N_c) = 0$, with $N_c = n K_D \int_0^{\Lambda} k^{D-3} \mathrm{d}k$ (assuming $D > 2$).

(e) Near $N = N_c$ we define ζ and u_c by

$$\bar{\rho} + s^{2-D}\vec{\varphi}^2 = N_c\left(1 + \frac{\zeta}{s^2}\right); \quad u_c = N_c \left.\frac{dt(N)}{dN}\right|_{N=N_c}$$

(one can show that $u_c > 0$). Establish the equation

$$\frac{\vec{\varphi}^2}{N_c} = 1 + \zeta s^{D-4} - (D-2)\Lambda^{2-D}\int_\Lambda^{\Lambda s} dp\, p^{D-1}\left(\frac{1}{p^2 + t_s} - \frac{1}{p^2}\right)$$

and interpret it when $t \in S_\infty$.

(f) For $2 < D < 4$ show that the fixed point is given by

$$\frac{\vec{\varphi}^2}{N_c} = 1 - (D-2)\Lambda^{2-D}\int_\Lambda^{\infty} dp\, p^{D-1}\left(\frac{1}{p^2 + t^*} - \frac{1}{p^2}\right),$$

and explain the precise nature of the fixed point. What is the fixed point for $D > 4$?

Do a numerical calculation for $D = 3$ in order to plot the function $t^*(x)$, $x = \vec{\varphi}^2 N_c$.

(g) Calculate $t_s - t^*$ as a function of $t_1 = t(N_c)$ near the fixed point, and show that, if $2 < D < 4$, the exponent $\nu = 1/y_1$ equals $1/(D-2)$. Likewise, show that $y_2 = D - 4$. What happens if $D > 4$?

(h) If D is close to 4, show that one can find an ε-expansion for t^*; calculate u_2^*, u_4^*, and u_6^*, and show in particular that u_6^* is of order ε^3.

3.8 The example of Bell and Wilson (Bell and Wilson 1975a)

Start from the Hamiltonian

$$H = \frac{1}{2}\int_{q \leqslant \Lambda} \frac{d^D q}{(2\pi)^D}\rho(q)|\varphi(\boldsymbol{q})|^2; \quad \rho(q) = r_0 + cq^2 + wq^4 + \cdots$$

and define the RGT $R_{a,b}$ by

$$R_{a,b}e^{-H} = \int \prod_{\|\boldsymbol{q}\|/2 \leqslant \Lambda} d\varphi(\boldsymbol{q}) \times \exp\left(-\frac{1}{2}a\int\left|\varphi'(\boldsymbol{q}) - b\varphi\left(\frac{\boldsymbol{q}}{2}\right)\right|^2 \frac{d^D q}{(2\pi)^D} - H(\varphi)\right),$$

where $a > 0$, and where the parameter b plays the part of λ in (3.1.8). Thus the transformation $R_{a,b}$ depends on two parameters a and b; note that $\varphi'(\boldsymbol{q}) = 0$ if $\|\boldsymbol{q}\| > \Lambda$.

(a) Show that the expression just given for $R_{a,b}$ does define an RGT having all the requisite properties. What results in the limit $a \to \infty$?

What are the canonical dimensions of $\varphi(\boldsymbol{q})$ (the dimension of c is zero) and of w?

(b) Determine $\rho'(\boldsymbol{q})$ as a function of $\rho(\boldsymbol{q})$; one finds

$$\rho'(\boldsymbol{q}) = \frac{4\rho(\boldsymbol{q}/2)}{2^{D+2}b^2 + 4\rho(\boldsymbol{q}/2)/a}.$$

(c) We write the kth iteration of the RGT in the form

$$\rho^{(k)}(\boldsymbol{q}) = \frac{L_k^2\rho(\boldsymbol{q}/L_k)}{L_k^{D+2}b_k^2 + L_k^2\rho(\boldsymbol{q}/L_k)/a_k}.$$

Establish the recurrence relations for a_k, b_k, and L_k, and then solve them. Show that they yield

$$a_k = \frac{a(1 - 2^D b^2)}{1 - (2^D b^2)^k}, \quad b_k = (b)^k, \quad L_k = (2)^k.$$

(d) Show that a necessary condition for the transformation to have a nontrivial fixed point reads $b^2 = 2^{-(D+2)}$. What (trivial) fixed points are found if $b^2 < 2^{-(D+2)}$ or $b^2 > 2^{-(D+2)}$?

Given the choice $b^2 = 2^{-(D+2)}$, determine the critical exponent η and the fixed point $\rho^*(\boldsymbol{q})$. What does one find in the limit $a \to \infty$, and why?

(e) The fixed point depends on c and on a; c may be chosen arbitrarily, entailing a 'line of fixed points'. The fixed point depends on the choice made for a in the RGT: two physically equivalent RGTs can lead to two different fixed points. Examine the behaviour of r_0 and of w under a sequence of RGTs. Does the critical exponent ν depend on a?

3.9 Cubic anisotropy (Pfeuty and Toulouse 1977, Chapter 8; Amit 1984, Chapter II.5)

To the Ginzburg–Landau Hamiltonian (3.5.17) we add a term with 'cubic anisotropy':

$$H = \int \mathrm{d}^D x \left[\frac{1}{2} \sum_{i=1}^{n} (\nabla\varphi_i)^2 + \frac{1}{2} r_0 \left(\sum_{i=1}^{n} \varphi_i^2 \right) + \frac{1}{4!} u_0 \left(\sum_{i=1}^{n} \varphi_i^2 \right)^2 + \frac{1}{4!} v_0 \sum_{i=1}^{n} \varphi_i^4 \right].$$

(a) Show that the Hamiltonian is positive definite only if both the following conditions are satisfied:

$$u_0 + v_0 > 0; \quad u_0 + \frac{v_0}{n} > 0.$$

Show that for $n = 2$ the set of parameters (u_0, v_0) is equivalent to the set $(u_0 + \frac{3}{2} v_0, - v_0)$.

(b) For $u = u_0/8\pi^2$ and $v = v_0/8\pi^2$ one can derive the following differential renormalization equations (which generalize equation (3.5.2b)):

$$\frac{\mathrm{d}u}{\mathrm{d}\ln s} = \varepsilon u - \frac{n+8}{6} u^2 - uv,$$

$$\frac{\mathrm{d}v}{\mathrm{d}\ln s} = \varepsilon v - 2uv - \frac{3}{2} v^2.$$

Show that these equations imply the existence of four fixed points, called: (1) Ising, (2) Heisenberg, (3) Gaussian, (4) cubic. Justify this nomenclature. For $\varepsilon > 0$, study the stability of these fixed points in the (u, v) plane, and sketch the renormalization flow in this plane. Show that one must distinguish between the cases $n < 4$ and $n > 4$.

3.10 Deviations from scaling

(a) Show that the bare coupling constant u_0 is given as a function of g by

$$u_0 \simeq \xi^{-\varepsilon} g^* A(g^*)(g^*/|g^* - g|)^{\varepsilon/\omega},$$

with

$$A(g) = \exp\left(- \int_0^g \mathrm{d}g' \left(\frac{\varepsilon}{\beta(g', \varepsilon)} + \frac{1}{g'} - \frac{\varepsilon}{\omega(g' - g^*)} \right) \right).$$

(*Hint*: see Section 7.5.1). Deduce from this equation that

$$g \simeq g^*(1 - O(\xi^{-\omega})),$$

and identify the exponent ω with $-y_2$. Compute ω by using the approximation (3.5.16) for $\beta(g, \varepsilon)$.

(b) Repeat the calculations at the end of Section 3.5.3 without taking the small-ε limit. Show that $\bar{\gamma}(g^*) = (D - 4)/3$.

3.11 Energy–energy correlations

In the GL model, one often calls $\varphi^2(\boldsymbol{x})$ the energy 'operator', since it is conjugate to the temperature.

(a) Show that the specific heat C is related to the volume integral of the (connected) energy–energy correlation $\langle \varphi^2(\boldsymbol{x})\varphi^2(\boldsymbol{y})\rangle_c$ (this is another fluctuation-response theorem).

(b) Assume that

$$\langle \varphi^2(\boldsymbol{x})\varphi^2(\boldsymbol{y})\rangle_c = \|\boldsymbol{x}-\boldsymbol{y}\|^{-2\sigma} f\left(\frac{\|\boldsymbol{x}-\boldsymbol{y}\|}{\xi}\right).$$

Show, by comparison with (a), and using the relation between the critical exponents α and ν, that

$$\sigma = D - 1/\nu.$$

$\sigma = d_{\varphi^2}$ is the anomalous dimension of the 'operator' φ^2. Its canonical value is $D-2 = 2d_\varphi^0$, but in general $d_{\varphi^2} \neq 2d_\varphi$.

(c) Consider now the correlation function

$$\langle \varphi^2(\boldsymbol{x})\varphi(\boldsymbol{y})\varphi(\boldsymbol{z})\rangle_c,$$

taking $T \geqslant T_c$ for simplicity. Is it possible to relate the volume integral of this correlation function to the derivative of a thermodynamic function? Show that it is consistent to write, for $T = T_c$,

$$\langle \varphi^2(\boldsymbol{x}/s)\varphi(\boldsymbol{y}/s)\varphi(\boldsymbol{z}/s)\rangle_c = s^{d_{\varphi^2}+2d_\varphi}\langle \varphi^2(\boldsymbol{x})\varphi(\boldsymbol{y})\varphi(\boldsymbol{z})\rangle_c.$$

Try to generalize to any product of φs and φ^2s.

The results obtained in this problem also apply to the Ising model, if the energy 'operator' is taken as the quantity E_i defined in Problem 1.2(f).

Further reading

For an introduction to the basic concepts of the renormalization group see Pfeuty and Toulouse (1977, Chapter 1); Ma (1976, Chapter 1); Shenker (1982, Section 3); Parisi (1988, Chapter 7); and Itzykson and Drouffe (1989, Chapter 4). To supplement these, one might read the instructive though somewhat technical discussions by Kogut and Wilson (1973), and by Bell and Wilson (1975a, b). The calculation of critical exponents is described for instance by Ma (1976, Chapter VI) and by Pfeuty and Toulouse (1977, Chapter 4). The example of a lattice calculation given in Section 3.3 is taken from Niemeijer and Van Leuveen (1975). For the renormalization group in Fourier space and for fixed points for $D > 4$ and $D < 4$, consult Ma (1976, Chapter VII) and Pfeuty and Toulouse (1977, Chapters 4 and 5). The discussion by Wilson (1975), Sections I, II, and III) is outstanding, though somewhat more difficult at first to come to grips with.

4 Two-dimensional models*

The case where the dimensionality D of space equals 2 has interesting peculiarities, because it is a limiting case with the behaviour of a spin system depending qualitatively on the dimensionality n of the order parameter. For $n = 1$ (Ising model) there is a phase transition with spontaneous magnetization. By contrast, for $n \geqslant 2$ one can prove that there is no spontaneous magnetization: but again the case $n = 2$ (XY model) is special: there is an altogether remarkable phase transition without spontaneous magnetization. For $n \geqslant 3$ there is no phase transition at any finite temperature. Nevertheless the $2D$ model is worth studying, because it furnishes a nontrivial illustration of the workings of the renormalization group in the case of a marginal variable. The RG will also be used to analyse the original behaviour of the XY model.

Accordingly, we proceed to study a system of spins $S_{\alpha,i}$, $\alpha = 1, 2, \ldots, n$, located at the sites i of a two-dimensional square lattice, and subject to

$$\sum_{\alpha=1}^{n} S_{\alpha,i}^2 = 1.$$

We start with a heuristic argument for the absence of spontaneous magnetization for $D = 2$ and $n \geqslant 2$ (there is a rigorous proof, due to Mermin and Wagner (1966)). Suppose that near $T = 0$ a spontaneous magnetization does exist, with all the spins parallel to the direction n say:

$$S_{1,i} = S_{2,i} = \cdots = S_{n-1,i} = 0; \quad S_{n,i} = 1;$$

and then study the small fluctuations around this state. The state can be stable only if there are no fluctuations strong enough to destroy it. In particular, as $T \to 0$, the fluctuations must vanish. Suppose therefore that they are small, which allows us to consider only those in a hyperplane normal to n, and to ignore the constraint on $\sum S_{\alpha,i}^2$. Let us write the Hamiltonian describing these fluctuations as

$$H = \tfrac{1}{2} J \sum_{\langle i,j \rangle} \sum_{\alpha=1}^{n-1} (S_{\alpha,i} - S_{\alpha,j})^2,$$

where $\sum_{\langle i,j \rangle}$ denotes a sum over nearest neighbours.

* The rest of the book is independent of this chapter, which may be skipped at a first reading.

At low temperatures one expects that long-wavelength fluctuations dominate, and it is reasonable to take the continuum limit

$$H \to \tfrac{1}{2} J \int d^2 x \sum_{\alpha=1}^{n-1} (\nabla \varphi_\alpha)^2.$$

The Fourier-space correlation function corresponding to this Hamiltonian is $T/J\boldsymbol{k}^2$, and the fluctuation Δ that we require is given by

$$\Delta = \sum_{\alpha=1}^{n-1} \langle \varphi_\alpha^2(\boldsymbol{x}) \rangle = (n-1)\langle \varphi_1^2(\boldsymbol{x}) \rangle = \frac{(n-1)T}{J} \int_{\pi/L}^{\pi/a} \frac{d^2 k}{(2\pi)^2 \boldsymbol{k}^2}, \tag{4.0.1}$$

where a is the lattice spacing and L the size of the lattice. The integration limits in (4.0.1) are determined by the ultraviolet cutoff π/a, and by an infrared cutoff π/L. The maximal wavelength in a lattice of size L is indeed $\sim L$; the $\boldsymbol{k} = 0$ mode describes not a fluctuation, but an overall translation of the variable φ: $\varphi(\boldsymbol{x}) \to \varphi(\boldsymbol{x}) + \text{constant}$. For $D > 2$, the integral (4.0.1) is infrared-convergent, and our assumption, of small fluctuations around the magnetized state as $T \to 0$, is selfconsistent. But for $D = 2$ one has

$$\Delta \sim \frac{(n-1)T}{2\pi J} \ln \frac{L}{a},$$

whence $\Delta \to \infty$ as $L \to \infty$; this reflects the infrared divergence of the integral in (4.0.1), and is a phenomenon characteristic of $D = 2$. The assumption that fluctuations are small is not selfconsistent: in $D = 2$ the long-wavelength fluctuations destabilize the long-range order. One can arrive at the same conclusion by using (4.1.8b) to study $\langle (\varphi(\boldsymbol{x}) - \varphi(0))^2 \rangle$ as $\|\boldsymbol{x}\| \to \infty$.

Section 4.1 is devoted to a qualitative study of the XY model, and to introducing the notion of vortices. Section 4.2 gives the RG analysis, while Section 4.3 studies the case $n = 3$ (or more generally $n \geqslant 3$), which is also called the 'nonlinear σ-model'.

4.1 The *XY* model: qualitative study

Numerical and analytic studies suggest that $n = 2$ is a special case, where apparently one observes a phase transition even though there is no spontaneous magnetization. We shall start with a heuristic argument for such a transition, before the next section proceeds to a more quantitative description by means of the renormalization group. In the XY model the spin at site i is a two-component vector $\boldsymbol{S}$, which may be taken in the plane of the lattice. The Hamiltonian is invariant under rotations in this plane; in other words it has $O(2)$ symmetry, being given by

$$H = -J \sum_{\langle i,j \rangle} \boldsymbol{S}_i \cdot \boldsymbol{S}_j = -J \sum_{\langle i,j \rangle} \cos(\theta_i - \theta_j), \tag{4.1.1}$$

where θ_i is the angle between the spin at site i and some reference direction (e.g. the x-axis). Hence the partition function is

$$Z = \sum_{[\theta_l]} \exp\left(\frac{J}{T}\sum_{\langle i,j\rangle} \cos(\theta_i - \theta_j)\right)$$
$$= \int_0^{2\pi} \prod_l d\theta_l \exp\left(\frac{J}{T}\sum_{\langle i,j\rangle} \cos(\theta_i - \theta_j)\right). \qquad (4.1.2)$$

4.1.1 The high-temperature expansion

The principles of high-temperature expansions have already been explained in Section 1.2.2: as $T \to \infty$ one expands the exponential in (4.1.2) by powers of (J/T), and one attempts to identify the term(s) giving the lowest power.

Let us try to estimate by this method the correlation function between two spins, one at site 0 and the other at site p:

$$\langle \boldsymbol{S}_0 \cdot \boldsymbol{S}_p \rangle = \langle \cos(\theta_0 - \theta_p) \rangle = \langle e^{i(\theta_0 - \theta_p)} \rangle;$$

the last equality here stems from the invariance of H under $\theta_i \to -\theta_i$. Notice that

$$\int_0^{2\pi} d\theta = 2\pi; \quad \int_0^{2\pi} d\theta e^{i\theta} = 0. \qquad (4.1.3)$$

If the integral over θ is not to vanish, then in view of (4.1.3) one must associate with every factor $e^{i\theta}$ another factor $e^{-i\theta}$. This shows that to every nonzero term one can assign a path on the lattice going from site 0 to site p (Fig. 4.1):

$$e^{i\theta_0}(e^{-i\theta_0}e^{i\theta_a})e^{-i\theta_a} \ldots e^{i\theta_h}(e^{-i\theta_h}e^{i\theta_p})e^{-i\theta_p}.$$

Here, the contents of every pair of brackets stem from a factor $\cos(\theta_i - \theta_j)$ associated with a bond. Such a term gives a contribution of order $(J/T)^N$ to the

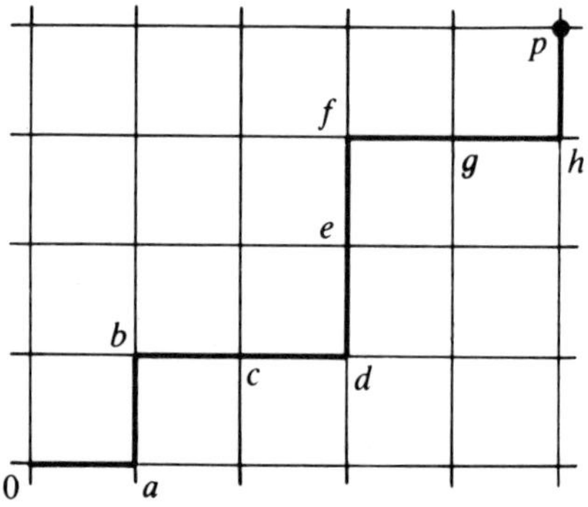

Fig. 4.1 A path from 0 to p

high-temperature expansion, where N is the number of bonds along the path joining the sites 0 and p. The dominant term is found by choosing the shortest path; hence N is approximately r/a, where a is the distance between the sites:

$$\langle e^{i(\theta_0 - \theta_p)} \rangle \simeq \left(\frac{J}{T}\right)^{r/a} = e^{-\frac{r}{a}\ln(T/J)}. \tag{4.1.4}$$

This corresponds to classical exponential behaviour with a correlation length

$$\xi = \frac{a}{\ln(T/J)}.$$

Our argument suggests that at high-enough temperature the system is in a classical disordered phase (high-enough T means the region where the high-T expansion converges, since one can prove that the radius of convergence is nonzero).

4.1.2 The low-temperature expansion

At low temperatures it is reasonable to suppose that the dominant fluctuations are those having long wavelengths, i.e. that one may replace

$$H \to H_0 + \frac{1}{2} J \sum_{\langle i,j \rangle} (\theta_i - \theta_j)^2.$$

In this long-wavelength approximation to fluctuations, θ varies little from one lattice site to the next, and we may replace $\cos(\theta_i - \theta_j)$ by $1 - \frac{1}{2}(\theta_i - \theta_j)^2$. It proves convenient to introduce the notation (see 2.1.12)

$$\partial_\mu \theta_i = \theta_{i+\mu} - \theta_i$$

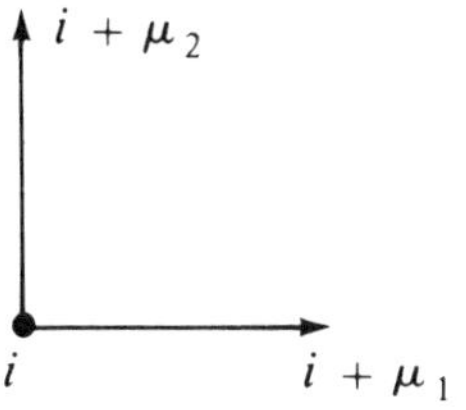

where μ can take the two values μ_1 and μ_2, and $\theta_{i+\mu}$ is a nearest neighbour of θ_i. In this notation the approximate Hamiltonian can be written

$$H = \frac{1}{2} J \sum_i \sum_\mu (\partial_\mu \theta_i)^2. \tag{4.1.5}$$

It is often convenient to go to the continuum limit, adopting the notation

$$i \to \boldsymbol{x}; \quad \theta_i \to \theta(\boldsymbol{x}); \quad \sum_\mu (\partial_\mu \theta_i)^2 \to [\nabla\theta(\boldsymbol{x})]^2;$$

this leads to the following expression for H:

$$H = \frac{1}{2} J \int d^2x (\nabla\theta)^2. \tag{4.1.6}$$

The integral giving $\langle e^{i(\theta_0 - \theta_p)} \rangle$ is Gaussian,

$$\langle e^{i(\theta_0-\theta_p)} \rangle = \frac{1}{Z} \int_{-\infty}^{\infty} \prod_l d\theta_l \exp\left(i(\theta_0 - \theta_p) - \frac{J}{2T} \sum_{i,\mu} (\partial_\mu \theta_i)^2 \right). \tag{4.1.7}$$

Notice that the integration over θ has been taken to run from $-\infty$ to $+\infty$; this point will be discussed later. The correlation function G_{ij} corresponding to the Gaussian Hamiltonian $\frac{1}{2}\sum_{i,\mu}(\partial_\mu \theta_i)^2$ (see Problem 2.5) is given by

$$G_{ij} = a^2 \int \frac{d^2k}{(2\pi)^2} \frac{e^{-ik\cdot(x_i - x_j)}}{4 - 2\cos(k_1 a) - 2\cos(k_2 a)}. \tag{4.1.8a}$$

It should cause no surprise that once again we meet an infrared divergence. To regularize it we subtract 1 from the exponential, defining a function $\bar{G}(\boldsymbol{x})$ by

$$\bar{G}(\boldsymbol{x}) = a^2 \int \frac{d^2k}{(2\pi)^2} \frac{e^{-ik\cdot x} - 1}{4 - 2\cos(k_1 a) - 2\cos(k_2 a)}. \tag{4.1.8b}$$

The Gaussian integral in (4.1.7) yields

$$\langle e^{i(\theta_0-\theta_p)} \rangle = e^{-\frac{T}{2J}[G_{00} + G_{pp} - 2G_{0p}]} = e^{\frac{T}{J}\bar{G}(x)},$$

where $\boldsymbol{x}$ is the vector joining site 0 to site p. In the continuum limit, $\bar{G}(\boldsymbol{x})$ (or $G(\boldsymbol{x})$) satisfies

$$-\nabla^2 \bar{G}(\boldsymbol{x}) = \delta(\boldsymbol{x}), \tag{4.1.9}$$

which is Poisson's equation in two dimensions. To solve it, consider the potential due to a straight and infinitely extended line charge, in a plane normal to the line; this potential satisfies the equation

$$-\nabla^2 G(\boldsymbol{x}) = \lambda \delta(\boldsymbol{x}),$$

where λ is the charge per unit length. It is an elementary exercise in electrostatics to show by means of Gauss's theorem that the corresponding electric field is

$$E(\boldsymbol{x}) = \frac{\lambda}{2\pi \|\boldsymbol{x}\|} = \frac{\lambda}{2\pi r},$$

which leads to the potential $-(\lambda/2\pi)\ln r + \text{constant}$. Accordingly*, the solution of (4.1.9) is

* Direct evaluation of (4.1.8b) shows that at large distances ($r \gg a$) $\bar{G}(\boldsymbol{x})$ indeed behaves according to (4.1.10). Moreover one can compute the constant explicitly as $-(2\pi)^{-1}(\gamma + \frac{3}{2}\ln 2) \approx -\frac{1}{4}$, where $\gamma(=0.577)$ is Euler's constant.

$$\bar{G}(x) = -\frac{1}{2\pi}\ln\frac{r}{a} + \text{constant}, \tag{4.1.10}$$

which yields the long-distance behaviour of the correlation function,

$$\langle e^{i(\theta_0 - \theta_p)}\rangle = e^{-\frac{T}{J}\frac{1}{2\pi}\ln\frac{r}{a}} = \left(\frac{a}{r}\right)^{T/2\pi J}. \tag{4.1.11}$$

The correlation function decreases like a power, with an exponent depending on the temperature: thus we have a *line of critical points*, with a temperature-dependent exponent

$$\eta(T) = \frac{T}{2\pi J}. \tag{4.1.12}$$

Equation (4.1.11) also shows that the XY model yields no spontaneous magnetization: otherwise one would have, in effect,

$$\langle e^{i(\theta_0 - \theta_p)}\rangle = \langle \boldsymbol{S}_0 \cdot \boldsymbol{S}_p\rangle \underset{r\to\infty}{\sim} \langle \boldsymbol{S}_0\rangle\langle \boldsymbol{S}_p\rangle = \boldsymbol{M}^2 \neq 0.$$

If these heuristic arguments are right, then there must exist a transition point where the behaviour changes from power-law to exponential. But one can well ask oneself why, above some eventual transition-point, the low-temperature argument should fail. Since the integral in equation (4.1.7) runs from $-\infty$ to $+\infty$, one could speculate that the periodicity of θ becomes important once the fluctuations become large. This indeed is what happens: the quasi-order obtaining at low temperatures is eventually destroyed by topological excitations, which depend on the periodicity of θ. These excitations are vortices, whose role was first elucidated by Kosterlitz and Thouless.

4.1.3 The role of vortices

Consider a spin configuration such that θ_i equals $\pi/2$ plus the polar angle φ with respect to some origin O (Fig. 4.2). (The choice $\pi/2$ is arbitrary; any other constant value would serve equally well.) In polar coordinates, the gradient of θ is

$$\nabla\theta = \left(0, \frac{1}{r}\right) = \frac{\hat{\varphi}}{r}.$$

If C is a closed contour surrounding the point O, then

$$\oint_C \nabla\theta \cdot d\boldsymbol{l} = 2\pi.$$

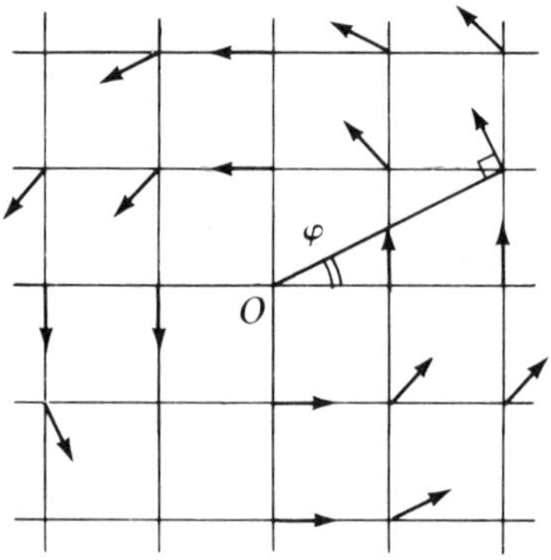

Fig. 4.2 A vortex

Such a spin configuration is one example of a vortex. More generally, because θ is a periodic variable, one has

$$\oint_C \nabla\theta \cdot \mathrm{d}\boldsymbol{l} = 2\pi q, \quad q = 0, \pm 1, \pm 2, \cdots,$$

where q is the vorticity (i.e. the strength of the vortex).

Let us calculate the energy associated with the vortex shown in Fig. 4.2; it is

$$E \simeq \frac{1}{2} J \int (\nabla\theta)^2 \, \mathrm{d}^2 x = \frac{1}{2} J \int \frac{(\hat{\varphi})^2}{r^2} \, \mathrm{d}^2 x = \pi J \ln \frac{L}{a}. \tag{4.1.13}$$

Of course the approximations adopted to obtain (4.1.13) are justified only far enough from the the centre O of the vortex, and (4.1.13) should be augmented by a constant which must be calculated numerically.

Since the centre of the vortex can be chosen anywhere on the lattice, the entropy associated with the creation of a vortex is

$$S = \ln(L/a)^2,$$

seeing that there are $(L/a)^2$ sites. The corresponding free energy is

$$F = E - TS = (\pi J - 2T) \ln \frac{L}{a},$$

and we see that vortices will destabilize the quasi-order when $T > T_c = \pi J/2$. In actual fact vortices are created in pairs, and it would be more correct to reason in terms of pair creation (see Problem 4.1).

We can now summarize the description proposed by Kosterlitz and Thouless for the phases of the system. At low temperatures, the only important configurations are long-wavelength fluctuations, also called spin-waves; correlations decrease according to power-laws, and the system is quasi-ordered, with islands of magnetization of all sizes. There are no 'free' vortices, but one can find pairs having opposite vorticities, whose influence on the system is confined to small distances. As the temperature rises the size of these vortex–antivortex pairs increases; it diverges at $T = T_c$, where free vortices appear. These vortices destabilize the quasi-ordering of the spin waves, and the correlation function then decreases exponentially.

4.2 Renormalization-group analysis

In the XY model the spin waves remain coupled to the vortices, which makes a complete calculation impossible. This is why we now introduce another closely related model, the Villain model, where the spin waves are decoupled from the vortices. The Villain model has the same topological characteristics as the XY model, and it is reasonable to assume (though not rigorously proven) that the phases of the two models are the same.

4.2.1 The Villain model (Villain 1975)

Let us write the Fourier decomposition of a bond as

$$e^{-\beta(1-\cos\theta)} = \sum_{n=-\infty}^{\infty} e^{in\theta} I_n(\beta) e^{-\beta}, \tag{4.2.1}$$

where $\beta = J/T$, and $I_n(\beta)$ is a Bessel function of imaginary argument,

$$I_n(\beta) = \int_0^{2\pi} e^{\beta\cos\theta} e^{in\theta} \frac{d\theta}{2\pi}. \tag{4.2.2}$$

As $\beta \to \infty$ $(T \to 0)$, $e^{-\beta} I_n(\beta)$ may be approximated by $(2\pi\beta)^{-1/2} \exp(-n^2/2\beta)$, whence

$$e^{-\beta(1-\cos\theta)} \simeq \frac{1}{\sqrt{2\pi\beta}} \sum_{n=-\infty}^{\infty} e^{in\theta} e^{-n^2/2\beta}. \tag{4.2.3}$$

This result could have been obtained directly, by noting that for $\beta \to \infty$ the main contribution to the integral (4.2.2) stems from the region $\theta \to 0$, whence $e^{-\beta(1-\cos\theta)} \simeq e^{-\frac{1}{2}\beta\theta^2}$.

The important point in (4.2.3) is that the approximation, even though it applies only for $T \to 0$, nevertheless preserves the periodic nature of the variable θ. On dropping the multiplicative constant $(2\pi\beta)^{-1/2}$ one obtains the Villain model by writing

$$\exp(-\beta(1-\cos(\partial_\mu\theta(\boldsymbol{x})))) \to \sum_{n_\mu(\boldsymbol{x})=-\infty}^{n_\mu(\boldsymbol{x})=+\infty} \exp(in_\mu\partial_\mu\theta(\boldsymbol{x}))\exp(-n_\mu^2/2\beta),$$

where a vector $n_\mu(x)$ having integer components $n_1(x)$ and $n_2(x)$ is associated with every site x. Thus the partition function of the Villain model is

$$Z = \int_0^{2\pi} \prod_x d\theta(x) \prod_{x,\mu} \sum_{n_\mu(x)=-\infty}^{\infty} \exp(in_\mu \partial_\mu \theta(x)) \exp(-n_\mu^2(x)/2\beta), \tag{4.2.4}$$

with no summation over the repeated indices. Consider the integral over $\theta(x)$: in the exponent we find the following terms:

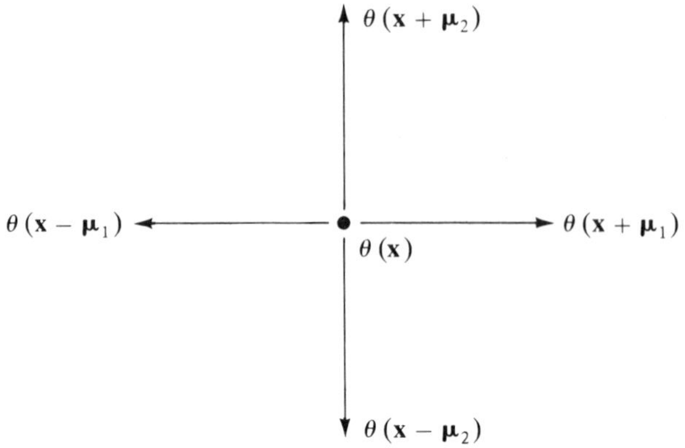

$$n_1(x)[\theta(x+\mu_1) - \theta(x)] + n_2(x)[\theta(x+\mu_2) - \theta(x)]$$
$$+ n_1(x-\mu_1)[\theta(x) - \theta(x-\mu_1)] + n_2(x-\mu_2)[\theta(x) - \theta(x-\mu_2)].$$

The integral over $\theta(x)$ vanishes unless

$$[n_1(x) - n_1(x-\mu_1)] + [n_2(x) - n_2(x-\mu_2)] = 0,$$

i.e. unless

$$\sum_\mu \partial_\mu n_\mu(x) = 0.$$

In other words $n(x)$ is a vector whose 'discrete divergence' is zero. Hence it can be written as a curl (in terms of $\varepsilon_{12} = -\varepsilon_{21} = 1$, $\varepsilon_{11} = \varepsilon_{22} = 0$); thus

$$n_\mu(x) = \sum_\nu \varepsilon_{\mu\nu} \partial_\nu p(x),$$

where $p(x)$ is a scalar field assuming only integer values. Up to a constant factor, the partition function (4.2.4) becomes

$$Z = \sum_{p(x)=-\infty}^{\infty} \exp\left(-\frac{1}{2\beta} \sum_{x,\mu} (\partial_\mu p(x))^2\right). \tag{4.2.5}$$

We appeal to Poisson's summation formula*

$$\sum_{n=-\infty}^{\infty} g(n) = \sum_{m=-\infty}^{\infty} \int_{-\infty}^{\infty} \mathrm{d}\varphi\, g(\varphi)\, \mathrm{e}^{2\mathrm{i}\pi m\varphi}$$

in order to rewrite Z as

$$Z = \int_{-\infty}^{\infty} \prod_{x} \mathrm{d}\varphi(x) \sum_{m(x)=-\infty}^{\infty} \exp\left(-\frac{1}{2\beta}\sum_{x,\mu}(\partial_\mu\varphi)^2 + 2\mathrm{i}\pi\sum_{x} m(x)\,\varphi(x)\right).$$

Here $m(x)$ is a scalar field assuming integer values. One can now integrate over φ:

$$Z = Z_{\mathrm{SW}} \sum_{m(x)=-\infty}^{\infty} \exp\left(-2\pi^2\beta \sum_{x,x'} m(x)G(x-x')m(x')\right), \tag{4.2.6}$$

where Z_{SW}, which is the determinant stemming from the integration, is the partition function of the spin waves. Indeed, if $m = 0$, one recovers precisely this partition function (notice that the integral over the φ now runs from $-\infty$ to $+\infty$). In the Villain model, spin waves and vortices are decoupled, i.e.

$$H = H_{\mathrm{SW}} + H_{\mathrm{V}}$$

and

$$Z = Z_{\mathrm{SW}} Z_{\mathrm{V}}.$$

By contrast, there is no such decoupling in the XY model (and it remains to identify $m(x)$ with the vorticity, as we shall do presently). In the expression (4.2.6)

* Poisson's summation formula reads as follows. Let the function $f(z)$ be holomorphic in the strip $-\rho < \mathrm{Im}\, z < \rho$, and let

$$F(z) = \sum_{\nu=-\infty}^{\infty} f(\nu + z)$$

be likewise holomorphic (the ν are integers). Since $F(z)$ is holomorphic and periodic with period 1, it has a Fourier expansion that converges uniformly:

$$F(z) = \sum_{n=-\infty}^{\infty} A_n \mathrm{e}^{2\mathrm{i}\pi n z},$$

$$A_n = \int_0^1 \mathrm{d}x\, F(x)\mathrm{e}^{-2\mathrm{i}\pi n x} = \sum_{\nu=-\infty}^{\infty} \int_0^1 \mathrm{d}x\, f(\nu + x)\mathrm{e}^{-2\mathrm{i}\pi n x}$$

$$= \sum_{\nu=-\infty}^{\infty} \int_{\nu}^{\nu+1} \mathrm{d}x\, f(x)\mathrm{e}^{-2\mathrm{i}\pi n x} = \int_{-\infty}^{\infty} \mathrm{d}x\, f(x)\mathrm{e}^{-2\mathrm{i}\pi n x},$$

$$F(0) = \sum_{n=-\infty}^{\infty} A_n = \sum_{n=-\infty}^{\infty} \int_{-\infty}^{\infty} f(x)\mathrm{e}^{-2\mathrm{i}\pi n x}\mathrm{d}x.$$

But $F(0)$ can equally well be written as

$$F(0) = \sum_{\nu=-\infty}^{\infty} f(\nu).$$

it is convenient to revert to the regularized correlation function $\bar{G}(\boldsymbol{x})$ of (4.1.8b) by writing

$$G(\boldsymbol{x}-\boldsymbol{x}') = [G(\boldsymbol{x}-\boldsymbol{x}') - G(0)] + G(0)$$
$$= \bar{G}(\boldsymbol{x}-\boldsymbol{x}') + G(0),$$

$$Z = Z_{\mathrm{SW}} \sum_{m(\boldsymbol{x})} \exp\left(-2\pi^2\beta G(0)\left(\sum m(\boldsymbol{x})\right)^2 - 2\pi^2\beta \sum_{\boldsymbol{x},\boldsymbol{x}'} m(\boldsymbol{x})\bar{G}(\boldsymbol{x}-\boldsymbol{x}')m(\boldsymbol{x}')\right).$$

Since $G(0) \sim (1/2\pi)\ln(L/a)$, we see that the only contribution in the limit $L \to \infty$ arises from the term with zero total vorticity:

$$\sum_{\boldsymbol{x}} m(\boldsymbol{x}) = 0.$$

This allows the partition function to be written in its final form

$$Z = Z_{\mathrm{SW}} \sum_{m(\boldsymbol{x})}{}' \exp\left(-2\pi^2\beta \sum_{\boldsymbol{x},\boldsymbol{x}'} m(\boldsymbol{x})\bar{G}(\boldsymbol{x}-\boldsymbol{x}')m(\boldsymbol{x}')\right)$$
$$= Z_{\mathrm{SW}} Z_{\mathrm{V}}, \tag{4.2.7}$$

where $\sum'$ indicates that one must sum only over configurations whose total vorticity is zero. At large distances one can show that, to a good approximation (see the footnote on page 130),

$$\bar{G}(\boldsymbol{x}) \simeq -\frac{1}{2\pi}\ln\frac{r}{a} - \frac{1}{4}. \tag{4.2.8}$$

By virtue of (4.2.8) and of the relation

$$\sum_{\boldsymbol{x}\neq\boldsymbol{x}'} m(\boldsymbol{x})m(\boldsymbol{x}') + \sum_{\boldsymbol{x}} m^2(\boldsymbol{x}) = 0,$$

the partition function may be rewritten as

$$Z = Z_{\mathrm{SW}} \sum_{m(\boldsymbol{x})}{}' \exp\left(-\frac{\pi^2\beta}{2}\sum_{\boldsymbol{x}} m^2(\boldsymbol{x}) + \pi\beta \sum_{\boldsymbol{x}\neq\boldsymbol{x}'} m(\boldsymbol{x})\ln\left\|\frac{\boldsymbol{x}-\boldsymbol{x}'}{a}\right\| m(\boldsymbol{x}')\right). \tag{4.2.9}$$

The term $\exp\left(-\frac{\pi^2\beta}{2}\sum_{\boldsymbol{x}} m^2(\boldsymbol{x})\right)$ can be interpreted as deriving from a chemical potential $\pi^2\beta/2$; it is written conventionally as

$$\exp\left(\ln y \sum_{\boldsymbol{x}} m^2(\boldsymbol{x})\right),$$

so that in the Villain model $y = y_0 = \exp(-\pi^2\beta/2)$. This term enables one to monitor the density of vortices. Thus, as $y \to 0$, the chemical potential discourages vortex formation. As y grows, the number of vortices grows too.

To identify $m(\boldsymbol{x})$ with the vorticity, we must study the law of force between vortices. It is easy to show (Problem 4.1) that two vortices like the one drawn in Fig. 4.2 attract or repel each other (depending on the relative sign of their vorticities) according to a power law determined by the potential (4.2.8). This observation warrants the identification we seek.

It is instructive to notice the analogy with the two-dimensional Coulomb gas: the partition function (4.2.9) is just that of a two-dimensional gas of charges $2\pi\sqrt{J}\,m(\boldsymbol{x})$ situated on the sites of a lattice, with a chemical potential governing the charge density. As $y \to 0$, there are just a few charges, closely bound in pairs, and the medium is an insulator. As y increases, more and more charges appear, the average size of the pairs increases, and eventually one gets free charges, or in other words a plasma.

4.2.2 The renormalization group for the *XY* model

Write the partition function for vortices with given chemical potential y as

$$Z_V = \sum_{m(\boldsymbol{x})}{}' \exp\left(\pi\beta \sum_{\boldsymbol{x} \neq \boldsymbol{z}} m(\boldsymbol{x}) \ln \left\| \frac{\boldsymbol{x} - \boldsymbol{z}}{a} \right\| m(\boldsymbol{z}) + \ln y \sum_{\boldsymbol{x}} m^2(\boldsymbol{x}) \right)$$

(with a change of notation $\boldsymbol{x}' \to \boldsymbol{z}$).

As $y \to 0$, the dominant configurations are those with no vortices, and with two vortices having opposite vorticities ± 1. For instance, we might have one vortex $(+1)$ at the point $\boldsymbol{x}$, and another (-1) at the point $\boldsymbol{z}$. Then the sum over configurations reduces to sums over $\boldsymbol{x}$ and over $\boldsymbol{z}$, and we find

$$Z_V \simeq 1 + e^{2\ln y} \sum_{\boldsymbol{x} \neq \boldsymbol{z}} e^{-2\pi\beta \ln \left\| \frac{\boldsymbol{x} - \boldsymbol{z}}{a} \right\|}.$$

On passage to the continuum limit this becomes

$$Z_V \simeq 1 + \frac{y^2}{a^4} \iint_{(a)} d^2x\, d^2z \left\| \frac{a}{\boldsymbol{x} - \boldsymbol{z}} \right\|^{2\pi\beta}. \tag{4.2.10}$$

The notation $\iint_{(a)}$ prescribes that in (4.2.10) we must have $\|\boldsymbol{x} - \boldsymbol{z}\| > a$; physically, the cutoff on $\|\boldsymbol{x} - \boldsymbol{z}\|$ means that the distance between the centres of the two vortices cannot be less than a. Let us now implement a dilatation of the cutoff, $a \to sa$:

$$Z'_V = 1 + \frac{y^2}{a^4} \iint_{(sa)} d^2x\, d^2z \left\| \frac{a}{\boldsymbol{x} - \boldsymbol{z}} \right\|^{2\pi\beta}.$$

To revert to the original cutoff one sets $x = sx'$, $z = sz'$, whence

$$Z'_V = 1 + \frac{y^2}{a^4} s^{4-2\pi\beta} \iint_{(a)} d^2x' d^2z' \left\| \frac{a}{x' - z'} \right\|^{2\pi\beta}. \tag{4.2.11}$$

Comparing equations (4.2.10) and (4.2.11) we see that under this RG transformation the parameter y transforms into y',

$$y^2 \to y'^2 = s^{4-2\pi\beta} y^2. \tag{4.2.12}$$

For $4 - 2\pi\beta < 0$, i.e. if

$$T < T_c = \frac{\pi J}{2}, \tag{4.2.13}$$

y decreases in such RGTs; hence one reaches the region where the number of vortices tends to zero. By contrast, if $T > T_c$, then the number of vortices increases under an RGT. When $T < T_c$, the long-distance behaviour is described by the situation where $y = 0$, i.e. by pure spin waves: the correlation function decreases like a power. On the other hand, for $T > T_c$ the long-distance behaviour is described by a situation with many vortices: the correlation function falls exponentially (Fig. 4.3).

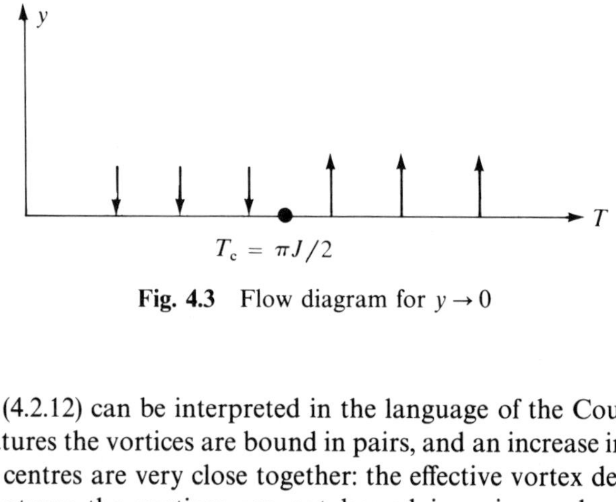

Fig. 4.3 Flow diagram for $y \to 0$

Equation (4.2.12) can be interpreted in the language of the Coulomb gas. At low temperatures the vortices are bound in pairs, and an increase in a eliminates pairs whose centres are very close together: the effective vortex density falls. At high temperatures the vortices are not bound in pairs, and an increase in a increases the effective density of vortices.

The analogy with the Coulomb gas suggests that equation (4.2.12) describes only part of the RG; through their screening effect, tightly-bound pairs weaken the attraction between distant vortices. This effect can be calculated, and leads to the equation

$$\frac{dT}{d \ln s} = y^2.$$

Finally, the complete set of RG equations reads

$$\begin{cases} \dfrac{\mathrm{d}y^2}{\mathrm{d}\ln s} = 2\left(2 - \dfrac{\pi J}{T}\right)y^2, & \text{(4.2.14a)} \\ \dfrac{\mathrm{d}T}{\mathrm{d}\ln s} = y^2. & \text{(4.2.14b)} \end{cases}$$

The point evidently worth attention is located at

$$y = 0, \quad T = T_c = \frac{\pi J}{2}. \tag{4.2.15}$$

Near this point it is easy to verify that

$$\frac{\mathrm{d}}{\mathrm{d}\ln s}\left[\left(2 - \frac{\pi J}{T}\right)^2 - \frac{4}{\pi J}y^2\right] = 0,$$

whence the trajectories under a sequence of RGTs take the form

$$\left(2 - \frac{\pi J}{T}\right)^2 - \frac{4}{\pi J}y^2 = \text{constant}.$$

Their form near the point (4.2.15) can now be traced at once (Fig. 4.4), say by setting $T = T_c + x$.

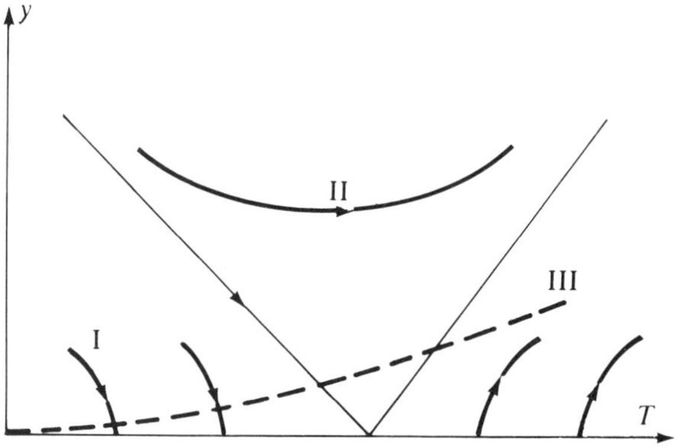

Fig. 4.4 Flow diagrams for the XY model; dashed line: the physical line in Villain's model

As an example, consider a trajectory in region I. Since the long-distance physics is the same for all points on the same trajectory, all points of region I define theories with the same long-distance behaviour as theories having $y = 0$, corresponding to pure spin waves. Thus we meet the line of fixed points once again.

More searching analysis shows that for $T > T_c$ the correlation length $\xi(T)$ behaves like

$$\xi(T) \underset{T \to T_c^+}{\sim} \exp\left[b\left(\frac{T_c}{T - T_c}\right)^{1/2}\right],$$

and that the free energy behaves like

$$\exp\left(-2b\left(\frac{T_c}{T - T_c}\right)^{1/2}\right),$$

where b is a numerical constant.

Accordingly, all derivatives of the free energy with respect to T are continuous at $T = T_c$. Such a phase transition could be described as 'a transition of infinitely high order'.

Finally, at $T = T_c$ the correlation function decreases like $r^{-1/4}$, conformably with the heuristic prediction in Section 4.1.

4.3 Nonlinear σ-models*

When $n \geqslant 3$ one is dealing with the so-called 'nonlinear σ-models' in two dimensions. It is instructive to analyse them by means of the renormalization group, *because temperature in this case is a marginal variable*, and the arguments of Section 3.6 can be illustrated nontrivially. For definiteness we consider the case $n = 3$ (the Heisenberg model); the generalization to arbitrary n is straightforward.

Since the spin $\boldsymbol{S}$ now has three components, one can write the partition function (noting the $O(3)$ symmetry of the Hamiltonian) as

$$Z = \int \prod_x \mathrm{d}\boldsymbol{S}(x) \exp\left(-\frac{1}{2g} \sum_{x,\mu} (\partial_\mu \boldsymbol{S}(x)) \cdot (\partial_\mu \boldsymbol{S}(x))\right), \tag{4.3.1}$$

where the 'coupling constant' g is directly proportional to the temperature:

$$g = T/J.$$

As in the 1D Ising model, there is no phase transition at any finite temperature. But the correlation length diverges as T (or g) tends to zero, and $T = 0$ may be considered as the critical temperature.

As usual it proves convenient to take the continuum limit; however, we wish to retain the condition $\boldsymbol{S}^2 = 1$, and the $O(3)$ symmetry of the original Hamiltonian. Accordingly we obtain a continuum formulation very different from that

* The model is nonlinear because of the constraint $\boldsymbol{S}^2 = 1$.

of Ginzburg and Landau, namely

$$H = \int d^2x \left[\frac{1}{2g} (\nabla \cdot S)^2 + \frac{1}{2f} ((\nabla \cdot S)^2)^2 + \cdots \right] = \int d^2x \, \mathscr{H}(x), \tag{4.3.2}$$

with the notation

$$(\nabla \cdot S)^2 = \sum_{\alpha=1}^{3} (\nabla \cdot S_\alpha)^2 = \sum_{\alpha=1}^{3} \left[\left(\frac{\partial S_\alpha}{\partial x_1} \right)^2 + \left(\frac{\partial S_\alpha}{\partial x_2} \right)^2 \right].$$

In view of the constraint $S^2 = 1$, all the terms in the Hamiltonian density $\mathscr{H}(x)$ (4.3.2) must involve derivatives of S; one cannot simply have monomials in S^2, as in the usual Ginzburg–Landau Hamiltonian. Moreover, $\mathscr{H}(x)$ must be invariant under rotations.

As $g \to 0$, one observes large domains in each of which the orientation of the spins is almost uniform, because near the critical point the system tends to magnetize spontaneously. Of course there are fluctuations around this orientation. We are led to distinguish between long-wavelength fluctuations, corresponding to a slowly rotating magnetization, and short-wavelength fluctuations around such slow evolution. Thus we subdivide S into a slowly-varying component S_l such that S_l is of order 1, and another component S_r fluctuating over short distances. From the reasoning given in the introduction one expects the fluctuations of S_r to be of order g (see (4.0.1)). It proves convenient to parametrize S as follows:

$$S_1 = \sqrt{1 - S_3^2} \cos\theta, \tag{4.3.3a}$$

$$S_2 = \sqrt{1 - S_3^2} \sin\theta, \tag{4.3.3b}$$

with S_l taken in the (1, 2) plane. The parametrization (4.3.3) respects the condition $S^2 = 1$; S_3 fluctuates over short distances, and θ over both short and long distances (θ_r and θ_l). On neglecting higher-order terms in (4.3.2), the Hamiltonian density becomes

$$\mathscr{H} = \frac{1}{2g} \left[(1 - S_3^2)(\nabla\theta)^2 + \frac{(\nabla S_3)^2}{1 - S_3^2} \right]. \tag{4.3.4}$$

Since we expect that $S_3 \sim \sqrt{g}$, we can for $g \to 0$ write

$$\mathscr{H} = \frac{1}{2g} [(\nabla \cdot S_3)^2 + (1 - S_3^2)(\nabla\theta)^2 + S_3^2(\nabla S_3)^2 + \cdots].$$

Finally we change the scale of S_3 so as to eliminate the multiplicative factor $1/g$,

$$S_3(x) \to \sqrt{g}\, h(x),$$

$$\mathscr{H} = \frac{1}{2} \left[(\nabla h)^2 + \left(\frac{1}{g} - h^2 \right) (\nabla\theta)^2 + g h^2 (\nabla h)^2 + \cdots \right]. \tag{4.3.5}$$

The form (4.3.5) of $\mathscr{H}$ is interesting in that the fluctuations of h affect those of θ (via the coefficient of $(\nabla\theta)^2$ in (4.3.5)), while the fluctuations of θ do not affect those of h. As $g \to 0$, the last term of (4.3.5) becomes negligible.

Next, we carry out the integration over fluctuations having wavelengths between Λ/s and Λ; the calculation of Z is organized as follows:

$$Z = \int_{0 \leqslant k \leqslant \Lambda} \mathscr{D}\theta(k) \exp\left(-\frac{1}{2g}\int d^2x (\nabla\theta)^2\right) \int_{0 \leqslant k \leqslant \Lambda/s} \mathscr{D}h(k)$$
$$\times \exp\left(-\frac{1}{2}\int d^2x (\nabla h)^2\right) \int_{\Lambda/s \leqslant k \leqslant \Lambda} \mathscr{D}h(k)$$
$$\times \exp\left(-\frac{1}{2}\int d^2x [(\nabla h)^2 - h^2(\nabla\theta)^2]\right). \tag{4.3.6}$$

Let us evaluate the last integral in (4.3.6), noting that $(\nabla\theta)^2$ is small ($\sim g$):

$$I = \int_{\Lambda/s \leqslant k \leqslant \Lambda} \mathscr{D}h(k) \exp\left(-\frac{1}{2}\int d^2x (\nabla h)^2\right) \exp\left(\frac{1}{2}\int d^2x (\nabla\theta)^2 h^2\right)$$
$$\simeq \int_{\Lambda/s \leqslant k \leqslant \Lambda} \mathscr{D}h(k) \exp\left(-\frac{1}{2}\int d^2x (\nabla h)^2\right)\left(1 + \frac{1}{2}\int d^2x\, h^2 (\nabla\theta)^2 + \cdots\right).$$

Up to a constant, I is given by

$$I = 1 + \frac{1}{2}\int d^2x\, \bar{G}_s(0) (\nabla\theta)^2 \simeq \exp\left(\frac{1}{2}\int d^2x\, \bar{G}_s(0) (\nabla\theta)^2\right) \tag{4.3.7}$$

where

$$\bar{G}_s(0) = \int_{\Lambda/s}^{\Lambda} \frac{d^2k}{(2\pi)^2 k^2} = \frac{1}{2\pi} \ln s, \tag{4.3.8}$$

while the integral over θ_r yields a multiplicative constant. After this integration over short wavelengths, the term in the Hamiltonian featuring $(\nabla\theta)^2$ acquires a coefficient

$$\frac{1}{2g} - \frac{1}{4\pi} \ln s = \frac{1}{2g'},$$

and the new coupling constant g' is determined by

$$\frac{1}{g'} = \frac{1}{g} - \frac{1}{2\pi} \ln s. \tag{4.3.9a}$$

This transforms immediately into a differential equation for the coupling constant,

$$\frac{dg(s)}{d \ln s} = \frac{g^2(s)}{2\pi}. \tag{4.3.9b}$$

On recalling the definition (3.6.10) of the function $\beta(g)$, the interpretation of (4.3.10) is immediate; what we have just calculated is the first term in the perturbative expansion of $\beta(g)$, i.e.

$$\beta(g) = -\frac{g^2}{2\pi} + O(g^3). \tag{4.3.10}$$

Equation (4.3.9) shows that integration over the short-wavelength fluctuations leads to a rise in temperature; in other words this integration establishes a correspondence between the initial system and another system at higher temperature but with the same long-distance behaviour. If successive integrations make the temperature tend to infinity, then the long-range properties will be the same as those of a high-temperature theory, and the correlation function will behave exponentially at all temperatures. For this conclusion to apply, the function $\beta(g)$ must have no zero on the axis $g > 0$ (see Chapter 7); whether it has one or not is a question that obviously cannot be answered by a perturbative calculation.

Quantitatively, the correlation length $\xi(g')$ at temperature $T'(=g'J)$ is given as a function of the correlation length $\xi(g)$ at temperature $T(=gJ)$ by

$$\xi(g') = \frac{1}{s}\xi(g).$$

By virtue of the definition of the function $\beta(g)$ we have

$$s = \exp\left(-\int_g^{g'} \frac{dg''}{\beta(g'')}\right),$$

whence

$$\xi(g') = \xi(g)\exp\int_g^{g'} \frac{dg''}{\beta(g'')}. \tag{4.3.11}$$

Using the expression (4.3.10) for $\beta(g)$, taking $g' \sim 1$, $\xi(g') \sim 1$, and $g \to 0$, one finds

$$\xi(g) \simeq e^{2\pi/g}. \tag{4.3.12}$$

The renormalization group shows that the behaviour of $\xi(g)$ as $g \to 0$ is *nonperturbative*: because of the non-analyticity at $g = 0$, perturbation theory can never give the correct result. Notice too that the behaviour of ξ as $T \to 0$ is analogous to that of the 1D Ising model (see 1.2.11).

The term of order g^3 in the function $\beta(g)$ has also been calculated:

$$\beta(g) = -\beta_0 g^2 - \beta_1 g^3 + O(g^4);$$

it entails for $\xi(g)$ an expression more accurate than (4.3.12) (see Problem 4.2).

We end with a geometric interpretation of (4.3.10): the integration over short-wavelength fluctuations leads to a spin variable $\boldsymbol{S}'$ such that $\|\boldsymbol{S}'\| < 1$. For the Hamiltonian to revert to the form it had initially, we need a change of scale $\boldsymbol{S}' \to \boldsymbol{S}'/\|\boldsymbol{S}'\|$ for the spins, which may be absorbed by modifying the coupling constant, $g \to g/\|\boldsymbol{S}'\|^2$. Thus we rediscover the fact that the operation entails a rise in temperature.

When $n \geqslant 3$, one sees that the only components of the spin contributing to renormalization are those perpendicular to the plane in which the slow evolution occurs. Thus there are $(n-2)$ components contributing effectively to the renormalization of the coupling constant, and equation (4.3.10) becomes

$$\frac{\mathrm{d}g(s)}{\mathrm{d}\ln s} = \frac{(n-2)g^2}{2\pi} + O(g^3). \tag{4.3.13}$$

Nonlinear σ-models have the property of asymptotic freedom (see Section 7.3.2), in that g is a decreasing function of the cutoff. This admits interesting analogies to non-Abelian gauge theories (Chapter 13).

Problems

4.1 We aim to examine a vortex–antivortex pair created in a configuration such that $\theta \to \theta_\infty$ along any direction in the plane.

(a) By examining the configuration drawn in Fig. 4.5 (where $\theta_\infty = \pi$), show that the energy of the vortex-pair is roughly

$$E \approx 2\pi J \ln r,$$

where r is the distance between the two centres.

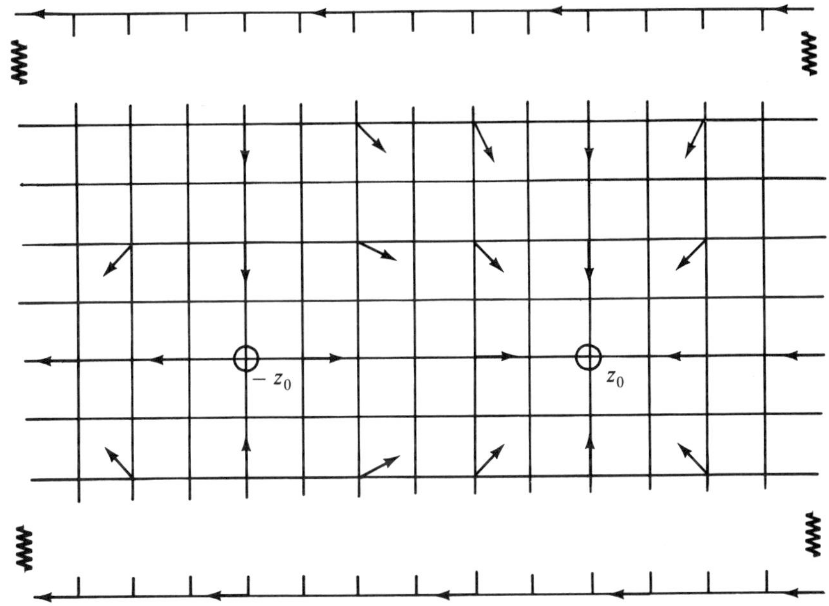

Fig. 4.5 A vortex–antivortex configuration

(b) More quantitatively one proceeds by taking θ as given by the relation (Itzykson and Drouffe 1989, Chapter IV)

$$e^{i\theta} = e^{i\theta_\infty} \left\{ \frac{(z - z_0)\,|z + z_0|}{|z - z_0|\,(z + z_0)} \right\},$$

where $z = x + iy$.

Sketch the general pattern of θ, and show that the expression above does indeed correspond to a vortex $q = +1$ at z_0 plus a vortex $q = -1$ at $-z_0$. (One can choose z_0 on the real axis.)

(c) For evaluating the Hamiltonian it is convenient to define the analytic function

$$F(z) = e^{i\theta_\infty}(z - z_0)/(z + z_0) = e^{i\theta_\infty}\rho e^{i\theta}.$$

To calculate $\int d^2x(\nabla\theta)^2$, we must evidently exclude the singular points, which we enclose in two circles C_i of radius a; this defines a domain D. Show that in D the Cauchy–Riemann conditions on $\ln F$ entail

$$\nabla^2 \ln\rho = \nabla^2\theta = 0; \quad (\nabla\theta)^2 = (\nabla\ln\rho)^2 = \nabla(\ln\rho\,\nabla\ln\rho);$$

thence derive

$$H = \frac{1}{2}J\int d^2x(\nabla\theta)^2 = -\frac{1}{2}J\sum_i\int_{C_i} dl \ln\rho(\boldsymbol{n}\cdot\nabla\ln\rho)$$
$$= -2\pi J\ln|2z_0| - 2\pi\beta\ln a.$$

(d) Use this expression to calculate the free energy of the pair, validating the qualitative argument of Section 4.1.3.

4.2 Given the perturbative expansion

$$\beta(g) = -\beta_0 g^2 - \beta_1 g^3 + O(g^4)$$

for the function $\beta(g)$ of the nonlinear σ-model, show that for $\xi(g)$ one finds

$$\xi(g) = (\text{constant})g^{-\beta_1/\beta_0^2}e^{-1/\beta_0 g}(1 + O(g)).$$

The only source of non-analyticity here lies in the first two terms of $\beta(g)$.

Further reading

The papers by Kogut (1979, Sections VII and VIII.C) and by Shenker (1982, Sections 5 and 6) are particularly clear, and have largely inspired the present discussion. See also the original paper by Kosterlitz and Thouless (1973); Itzykson and Drouffe (1989, Chapter IV); and Nelson (1983). The analogy with the Coulomb gas can likewise be analysed by means of the renormalization group: Kosterlitz (1974); Young (1979); and the book by Ma (1985, Chapter 19). Two-dimensional systems at criticality can be studied thanks to their remarkable property of conformal invariance. The systems can be classified and critical exponents can be computed from group theoretical arguments. This fascinating subject is reviewed in depth by Itzykson and Drouffe (1989, Chapter IX).

II

Perturbation Theory and Renormalization: The Euclidean Scalar Field

5 The perturbation expansion and Feynman diagrams

The aim of this chapter is the systematic development of concepts and techniques that have been introduced in the first three chapters; namely correlation functions and generating functions (Section 1.4.1); Legendre transformations (Section 1.4.5); and perturbation expansions and expansions in the number of loops (Sections 2.4.2, 3.4.3, and 3.5.3). The tools we use are the generating functional and Gaussian integration. In several cases it proves possible to explain the method through an example with just one variable; a preliminary study of such examples then makes it much easier to follow the general discussion.

Section 5.1 establishes Wick's theorem (which underlies the perturbation expansion) on the basis of Gaussian integration. Section 5.2 develops the perturbative expansion of the correlation functions $G^{(2)}$ and $G^{(4)}$, and introduces Feynman diagrams; to every term of the expansion (or rather to every group of terms) one assigns a diagram, and corresponding to every diagram there are specific rules for calculation. Section 5.3 studies the classification of correlation functions, connected, one-particle-irreducible, and proper vertices. The Legendre transformation is used to obtain the generating functional for proper vertices. This generating functional, which generalizes the Gibbs potential, will in Section 5.4 allow us to explicate the ideas of broken symmetry and of loop expansions. Section 5.5 explains how to evaluate Feynman diagrams in practice, while Section 5.6 gives simple but very pregnant arguments for determining how diagrams behave when certain momenta tend to infinity (ultraviolet behaviour) or to zero (infrared behaviour).

The present chapter will employ the vocabulary and the notations of quantum field theory. Accordingly the Ginzburg–Landau Hamiltonian (2.1.19) will be written in a different notation, with $r_0 \to m^2$ and $u_0 \to g$, where m is a mass and g a coupling constant. Wave-vectors will be called momenta ($\boldsymbol{p} = \hbar\boldsymbol{k}$ with $\hbar = 1$). Finally, the vector notation for $\boldsymbol{k}$ and $\boldsymbol{x}$ will be abandoned ($\boldsymbol{k} \to k$ and $\boldsymbol{x} \to x$) unless it is needed to avoid ambiguity.

5.1 Wick's theorem and the generating functional

In this section we revert to and generalize the results of Chapter 1 (Section 4) and of Appendix A. The essential aim is to obtain compact expressions for the

N-point correlation functions $\langle\varphi(x_1)\varphi(x_2)\ldots\varphi(x_N)\rangle$, and the rules for calculating with a Gaussian Hamiltonian. The method can be explained with a single variable.

5.1.1 The generating function with a single variable

Let $P(\varphi)$ be a probability distribution over a random variable; thus $P(\varphi) \geqslant 0$, but we do not assume that it is normalized, whence $\int P(\varphi)\,\mathrm{d}\varphi$ need not equal 1.

The *generating function* $Z(j)$ is defined by

$$Z(j) = \int \mathrm{d}\varphi\, P(\varphi)\mathrm{e}^{j\varphi}; \tag{5.1.1}$$

we assume tacitly that $P(\varphi)$ falls fast enough at infinity for the integral to converge. One is interested in $Z(j)$ because by differentiating it one can obtain the moments $\langle\varphi^n\rangle$ of the probability distribution $P(\varphi)$:

$$\langle\varphi^n\rangle = \frac{\int \mathrm{d}\varphi\, \varphi^n P(\varphi)}{\int \mathrm{d}\varphi\, P(\varphi)} = \frac{1}{Z(0)} \frac{\partial^n Z}{\partial j^n}\bigg|_{j=0}. \tag{5.1.2}$$

Conversely, $Z(j)/Z(0)$ is given as a function of the $\langle\varphi^n\rangle$ by

$$Z(j)/Z(0) = \sum_{n=0}^{\infty} \frac{j^n}{n!}\langle\varphi^n\rangle. \tag{5.1.3}$$

For a Gaussian probability distribution

$$P(\varphi) = \exp\left(-\frac{1}{2}\varphi\frac{1}{A}\varphi\right) \tag{5.1.4}$$

equation (A.2.1) leads to

$$Z(j) = Z(0)\exp\left(\frac{1}{2}jAj\right); \tag{5.1.5}$$

notice that A and A^{-1} have switched roles. Further (with the index 0 indicating averages over a Gaussian distribution) one has

$$\begin{aligned}\langle\varphi^{2n}\rangle_0 &= \frac{\partial^{2n}}{\partial j^{2n}}\exp\left(\frac{1}{2}jAj\right)\bigg|_{j=0} = \frac{\partial^{2n}}{\partial j^{2n}}\frac{1}{n!}\frac{1}{2^n}(jAj)^n \\ &= \frac{(2n)!}{2^n n!}A^n = (2n-1)!!\,A^n.\end{aligned} \tag{5.1.6}$$

The second equality follows on expanding the exponential and noting that only one term of the expansion can contribute, in view of the $2n$ derivatives and of the condition $j = 0$. (We have defined $(2n-1)!! = (2n-1)\cdot(2n-3)\cdot\cdots 3\cdot 1$.) All odd moments of course vanish: $\langle\varphi^{2n+1}\rangle_0 = 0$. Equation (5.1.6) can be rewritten in terms of the second moment $\langle \varphi^2 \rangle_0 = A$:

$$\langle \varphi^{2n} \rangle_0 = (2n-1)!! \langle \varphi^2 \rangle_0^n. \tag{5.1.7}$$

If $P(\varphi)$ is not a Gaussian, but assumes say the form

$$P(\varphi) = \exp\left(-\frac{1}{2A} \varphi^2 + f(\varphi) \right),$$

then

$$Z(j) = \int \mathrm{d}\varphi \exp\left(-\frac{1}{2A} \varphi^2 + f(\varphi) + j\varphi \right).$$

The generating function can be written as

$$Z(j) = \exp\left(f\left(\frac{\partial}{\partial j} \right) \right) \int \mathrm{d}\varphi \exp\left(-\frac{1}{2A} \varphi^2 + j\varphi \right). \tag{5.1.8}$$

This equality follows from

$$f\left(\frac{\partial}{\partial j} \right) \mathrm{e}^{j\varphi} = f(\varphi) \mathrm{e}^{j\varphi},$$

which in turn can be proved through a Taylor expansion of $f\left(\frac{\partial}{\partial j} \right)$ near the origin.

5.1.2 Wick's theorem

By generalizing the argument of the preceding section to N variables we can prove a fundamental theorem for Gaussian integrals which is well known to probabilists. Start from a Gaussian probability distribution for N variables,

$$P(\varphi_1, \ldots, \varphi_N) = \exp\left(-\frac{1}{2} \varphi^{\mathrm{T}} A^{-1} \varphi \right)$$

where

$$\varphi^{\mathrm{T}} A^{-1} \varphi = \sum_{i,j} \varphi_i A_{ij}^{-1} \varphi_j,$$

and define a generating function $Z(j)$,

$$Z(j_1, \ldots, j_N) = \int \prod_{i=1}^{N} \mathrm{d}\varphi_i \exp\left(-\frac{1}{2} \varphi^{\mathrm{T}} A^{-1} \varphi + j^{\mathrm{T}} \varphi \right). \tag{5.1.9}$$

By virtue of equation (A.2.2) we have

$$Z(j) = Z(0) \exp\left(\frac{1}{2} j^{\mathrm{T}} A j \right), \tag{5.1.10}$$

which in turn yields the moments of $P(\varphi)$ on differentiation; thus

$$\langle \varphi_{i_1} \ldots \varphi_{i_{2n}} \rangle_0 = \frac{\partial^{2n}}{\partial j_{i_1} \ldots \partial j_{i_{2n}}} \frac{1}{n!} \frac{1}{2^n} (j^{\mathrm{T}} A j)^n. \tag{5.1.11}$$

From equation (5.1.11) we can prove the generalization of (5.1.6): *all the moments of a Gaussian distribution can be expressed as functions of the second moments alone.* In quantum field theory this result is known as *Wick's theorem.* The proof is straightforward. Start with the second moment

$$\langle \varphi_{i_1} \varphi_{i_2} \rangle_0 = \frac{\partial^2}{\partial j_{i_1} \partial j_{i_2}} \left(\frac{1}{2} \sum_{k,l} j_k A_{kl} j_l \right) = A_{i_1 i_2} = \overbracket{\varphi_{i_1} \varphi_{i_2}}. \tag{5.1.12}$$

The quantity $\overbracket{\varphi_{i_1} \varphi_{i_2}}$ is called the *contraction* of φ_{i_1} and φ_{i_2}. The differentiations in (5.1.11) yield $(2n)!$ terms; we must however divide by $2^n n!$, whence the total number of terms on the right of (5.1.11) is $(2n-1)!!$. But this is simply the number of ways of choosing the pairs $\overbracket{\varphi_{i_1} \varphi_{i_2}}\,\overbracket{\varphi_{i_3} \varphi_{i_4}} \ldots \overbracket{\varphi_{i_{2n-1}} \varphi_{i_{2n}}}$. To see this, note that there are $(2n-1)$ ways of forming the first pair $\overbracket{\varphi_{i_1} \varphi_{i_2}}$, $(2n-3)$ ways of forming the second pair, and so on. Thus one finds

$$\langle \varphi_{i_1} \ldots \varphi_{i_{2n}} \rangle_0 = \overbracket{\varphi_{i_1} \varphi_{i_2}}\,\overbracket{\varphi_{i_3} \varphi_{i_4}} \cdots \overbracket{\varphi_{i_{2n-1}} \varphi_{i_{2n}}} + \text{permutations}, \tag{5.1.13}$$

$$\text{with} \quad \overbracket{\varphi_{i_1} \varphi_{i_2}} = \langle \varphi_{i_1} \varphi_{i_2} \rangle_0 = A_{i_1 i_2}.$$

An example:

$$\langle \varphi_1 \varphi_2 \varphi_3 \varphi_4 \rangle_0 = \overbracket{\varphi_1 \varphi_2}\,\overbracket{\varphi_3 \varphi_4} + \overbracket{\varphi_1 \varphi_3}\,\overbracket{\varphi_2 \varphi_4} + \overbracket{\varphi_1 \varphi_4}\,\overbracket{\varphi_2 \varphi_3}.$$

It is important to realize that *the total number of terms does not change even if some of the indices are the same*:

$$\begin{aligned} \langle \varphi_1 \varphi_2 \varphi_2 \varphi_4 \rangle_0 &= \overbracket{\varphi_1 \varphi_2}\,\overbracket{\varphi_2 \varphi_4} + \overbracket{\varphi_1 \varphi_2}\,\overbracket{\varphi_2 \varphi_4} + \overbracket{\varphi_1 \varphi_4}\,\overbracket{\varphi_2 \varphi_2} \\ &= 2\overbracket{\varphi_1 \varphi_2}\,\overbracket{\varphi_2 \varphi_4} + \overbracket{\varphi_1 \varphi_4}\,\overbracket{\varphi_2 \varphi_2}. \end{aligned}$$

In fact, if all the indices are the same one recovers (5.1.7), i.e.

$$\langle \varphi^{2n} \rangle_0 = (2n-1)!! \langle \varphi^2 \rangle_0^n.$$

5.1.3 The generating functional

The results above can be applied to a continuum theory considered as the limit of a theory on a lattice: see Section 2.1.3. Take the probability density as given by a Hamiltonian (2.1.19) of the Ginzburg–Landau type,

$$P[\varphi] = \exp(-H_{GL}) = \exp\left(- \int \mathrm{d}^D x \left(\frac{1}{2} (\nabla \varphi)^2 + \frac{1}{2} m^2 \varphi^2 + \frac{g}{4!} \varphi^4 \right) \right); \tag{5.1.14}$$

then the 'generating functional' of the correlation functions, $Z(j)$, is defined by

$$Z(j) = \int \mathscr{D}\varphi(x) \exp\left(-H + \int \mathrm{d}^D x \, j(x)\varphi(x) \right). \tag{5.1.15}$$

The function $j(x)$ is called the *source of the field* φ; it plays the same role as did the field $B(x)$. It proves useful to define the *Hamiltonian density* $\mathscr{H}(x)$ by

$$H = \int \mathrm{d}^D x \, \mathscr{H}(x); \quad \mathscr{H}(x) = \frac{1}{2}(\nabla\varphi)^2 + \frac{1}{2} m^2 \varphi^2 + \frac{g}{4!}\varphi^4.$$

The moments of order $(2n)$, which are nothing but the $(2n)$-point correlation functions, are found by differentiating the functional $Z(j)$; thus

$$G^{(2n)}(x_1, \ldots, x_{2n}) = \langle \varphi(x_1) \ldots \varphi(x_{2n}) \rangle = \frac{1}{Z(0)} \left. \frac{\delta^{(2n)} Z(j)}{\delta j(x_1) \ldots \delta j(x_{2n})} \right|_{j=0}. \tag{5.1.16}$$

In Part I, the two-point correlation function

$$G^{(2)}(x_1, x_2) = \langle \varphi(x_1) \, \varphi(x_2) \rangle$$

was simply called 'the correlation function', because it was the only one that entered and because there was no need to specify the number of variables.

It is convenient to subdivide the Hamiltonian H into a Gaussian part H_0 (quadratic in φ),

$$H_0 = \int \mathrm{d}^D x \left(\frac{1}{2}(\nabla\varphi)^2 + \frac{1}{2} m^2 \varphi^2 \right), \tag{5.1.17}$$

plus a so-called *interaction* term V,

$$V = \frac{g}{4!} \int \mathrm{d}^D x \, \varphi^4(x). \tag{5.1.18}$$

More generally one could replace $(g/4!)\varphi^4$ by a polynomial $\mathscr{V}(\varphi)$ in φ, which must be even if one wishes to respect the symmetry $\varphi \to -\varphi$; further, one could introduce so-called *derivative interactions* like $\varphi^2(\nabla\varphi)^2$. However, we shall ignore these interactions for the time being in order to keep the discussion simple. Then, for an interaction

$$V = \int \mathrm{d}^D x \, \mathscr{V}(\varphi) = \int \mathrm{d}^D x \left(\frac{g}{4!}\varphi^4 + \frac{g_6}{6!}\varphi^6 + \cdots \right),$$

the expression in (5.1.8) generalizes to

$$Z(j) = \exp\left(-\int \mathrm{d}^D x \, \mathscr{V}\left(\frac{\delta}{\delta j(x)} \right) \right) \int \mathscr{D}\varphi(x)$$
$$\times \exp\left(-H_0 + \int \mathrm{d}^D x \, j(x)\varphi(x) \right). \tag{5.1.19}$$

The integral over $\varphi(x)$ here is Gaussian; using (5.1.10) we find

$$\int \mathscr{D}\varphi(x) \exp\left(-\int \mathrm{d}^D x \left[\frac{1}{2}(\nabla\varphi)^2 + \frac{1}{2}m^2\varphi^2 - j(x)\varphi(x)\right]\right)$$

$$= Z_0(j=0)\exp\left(\frac{1}{2}\int \mathrm{d}^D x\, \mathrm{d}^D y\, j(x) G_0(x-y) j(y)\right), \tag{5.1.20}$$

where $G_0(x-y)$ is the two-point correlation function of the Gaussian model ($G_0(x, y)$ now plays the role of A_{ij}):

$$G_0(x) = \int \frac{\mathrm{d}^D k}{(2\pi)^D} \frac{\mathrm{e}^{-\mathrm{i}k \cdot x}}{k^2 + m^2} = \langle \varphi(x)\, \varphi(0) \rangle_0. \tag{5.1.21}$$

Accordingly, up to a multiplicative constant the generating functional $Z(j)$ may be written as

$$Z(j) = \mathscr{N} \exp\left(-\int \mathrm{d}^D x\, \mathscr{V}\left(\frac{\delta}{\delta j(x)}\right)\right)$$

$$\times \exp\left(\frac{1}{2}\int \mathrm{d}^D x\, \mathrm{d}^D y\, j(x) G_0(x-y) j(y)\right). \tag{5.1.22}$$

Throughout the following, $\mathscr{N}$ represents a normalization constant for generating functionals. In general such constants do not matter and need not be specified. The form (2.1.22) of $Z(j)$, however compact, cannot be evaluated directly. We shall have to expand it in a perturbation series.

Before doing so, we recall that to describe a physical system whose order parameter is n-dimensional, we had to generalize the Ginzburg–Landau Hamiltonian by introducing an n-component field $\varphi_i(x)$ ($i = 1, 2, \ldots, n$):

$$H = \int \mathrm{d}^D x \left[\frac{1}{2}\sum_{i=1}^{n} (\nabla\varphi_i)^2 + \frac{1}{2}m^2 \sum_{i=1}^{n} \varphi_i^2 + \frac{g}{4!}\left(\sum_{i=1}^{n} \varphi_i^2\right)^2\right]. \tag{5.1.23}$$

H is invariant under rotations in the n-dimensional space of the 'internal indices' i, depending as it does only on the 'length' of the field in this space, namely on

$$\vec{\varphi}^2 = \sum_{i=1}^{n} \varphi_i^2,$$

and perhaps on terms containing derivatives, but likewise invariant under such rotations. It is called the '$O(n)$-symmetric Ginzburg–Landau Hamiltonian', because the group of rotations in an n-dimensional space is commonly denoted by $O(n)$ ($O(n)$ contains in addition the symmetry operations $\varphi_i \to -\varphi_i$, which also leave H invariant).

5.2 The perturbation expansion of $G^{(2)}$ and of $G^{(4)}$: Feynman diagrams

This section aims to establish general rules for the perturbative calculation of correlation functions, rules designed to yield the result in the form of an expansion in powers of g,

$$G = G_0 + gG_1 + g^2G_2 + \cdots + g^nG_n + \cdots,$$

where G_0 is the correlation function of the Gaussian model. These rules are simple to express in diagrammatic form, namely in terms of the famous 'Feynman diagrams'. We start as before with just one variable, in order to explicate the principle underlying the expansion. As our example we adopt the interaction φ^4, but the results are easily generalized to arbitrary polynomial interactions $\mathscr{V}(\varphi)$.

5.2.1 The perturbation expansion in a single variable

Consider the probability distribution

$$P(\varphi) = \mathrm{e}^{-\frac{1}{2A}\varphi^2 - \frac{g}{4!}\varphi^4},$$

and try to calculate its moments. It is impossible to find an exact closed formula for $Z(0)$, but if g is small one can expand $\exp(-g\varphi^4/4!)$, obtaining

$$\begin{aligned} Z(0) &= \int \mathrm{d}\varphi\, \mathrm{e}^{-\frac{1}{2A}\varphi^2}\left(1 - \frac{g}{4!}\varphi^4 + \frac{g^2}{2!(4!)^2}\varphi^8 + \cdots\right) \\ &= \sqrt{2\pi A}\left(1 - \frac{1}{8}gA^2 + \frac{35}{384}g^2A^4 + \cdots\right). \end{aligned} \tag{5.2.1}$$

Now calculate $\langle\varphi^2\rangle$, the single-variable analogue of the two-point correlation function:

$$\begin{aligned} \int \mathrm{d}\varphi\, \varphi^2 P(\varphi) &= \int \mathrm{d}\varphi\, \varphi^2 \mathrm{e}^{-\frac{1}{2A}\varphi^2}\left(1 - \frac{g}{4!}\varphi^4 + \frac{g^2}{2!(4!)^2}\varphi^8 + \cdots\right) \\ &= \sqrt{2\pi A}\left(A - \frac{5}{8}gA^3 + \frac{105}{128}g^2A^5 + \cdots\right). \end{aligned} \tag{5.2.2}$$

$\langle\varphi^2\rangle$ is found by dividing (5.2.2) by (5.2.1), which yields

$$\langle\varphi^2\rangle = A\left(1 - \frac{1}{2}gA^2 + \frac{2}{3}g^2A^4 + \cdots\right). \tag{5.2.3}$$

This is an elementary example of a perturbation series. Notice that, if $g = 0$, then one recovers simply the Gaussian value $\langle\varphi^2\rangle = A$. Note also that the factors $\sqrt{2\pi A}$ cancel from the ratio: thus our special example bears out that the constant stemming from the Gaussian integral is irrelevant to the correlation function.

5.2.2 The calculation of $G^{(2)}$ to order g

Reverting now to a Ginzburg–Landau theory we calculate the term of order g in the correlation function $G^{(2)}(x-y)$. One must evaluate the integral

$$I(x, y) = \int \mathscr{D}\varphi\, \varphi(x)\varphi(y)\mathrm{e}^{-H}$$
$$= \int \mathscr{D}\varphi\, \varphi(x)\varphi(y)\mathrm{e}^{-H_0}\left[1 - \frac{g}{4!}\int \mathrm{d}^D z\, \varphi^4(z) + \cdots\right].$$

The first term in the square brackets merely yields

$$\mathscr{N}\langle\varphi(x)\,\varphi(y)\rangle_0 = \mathscr{N}G_0(x-y),$$

where the constant $\mathscr{N}$ is given by

$$\mathscr{N} = \int \mathscr{D}\varphi\, \mathrm{e}^{-H_0} = Z_0(j=0).$$

To evaluate the integral in the second term,

$$\int \mathscr{D}\varphi\, \varphi(x)\varphi(y)\mathrm{e}^{-H_0}\int \mathrm{d}^D z\, \varphi^4(z), \tag{5.2.4}$$

we use Wick's theorem (5.1.13). The number of contractions in (5.2.4) equals the number of contractions of $\varphi(x)$, $\varphi(y)$, and $\varphi^4(z)$. It proves convenient to represent these contractions on a diagram, by drawing two 'external' points x and y ('external' means that they refer to the arguments of the correlation function), marked by crosses; and an 'internal' point or 'vertex' z, which stems from the expansion of $\exp(-V)$, and over which we shall integrate. Because $\varphi(z)$ enters through its fourth power, at first z is drawn as four separate points. Every contraction is represented by a line joining arguments of φ, as for instance in

$$\overline{\varphi(x)\varphi(y)} \rightarrow \underset{x}{\times}\!\!-\!\!\!-\!\!\!-\!\!\underset{y}{\times}\,.$$

Two types of terms are possible (Fig. 5.1):

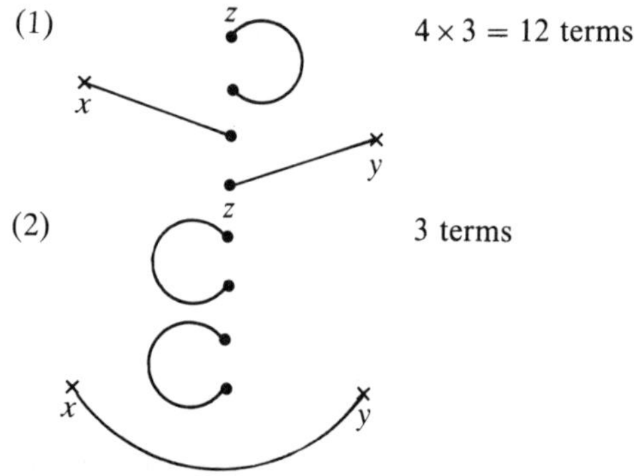

Fig. 5.1

One checks that $12 + 3 = 15 = (6 - 1)!!$. In order to simplify the diagrams, the four points z are merged into a single point, with the result shown in Fig. 5.2:

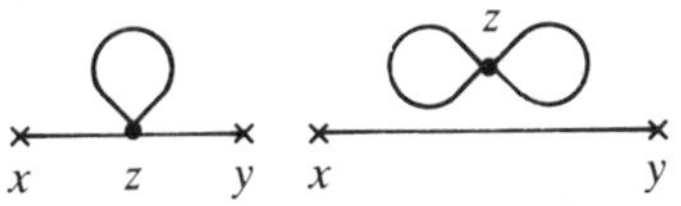

Fig. 5.2 The two diagrams of order g

These diagrams are called *Feynman diagrams* (*or graphs*); one such diagram corresponds to every term (or rather to every group of terms) of the perturbation expansion. The integral I reads

$$I(x, y) = \mathcal{N}\left[G_0(x-y) - \frac{1}{2}g\int d^D z\, G_0(x-z)G_0(0)G_0(z-y) - \frac{1}{8}gG_0(x-y)(G_0(0))^2\int d^D z\right]. \tag{5.2.5}$$

In order to obtain the correlation function, we must divide by $Z(0)$, just as in the example with a single variable:

$$\begin{aligned} Z(0) &= \int \mathcal{D}\varphi\, e^{-H_0}\left(1 - \frac{g}{4!}\int d^D z\, \varphi^4(z) + \cdots\right) \\ &= \mathcal{N}\left[1 - \frac{g}{8}(G_0(0))^2\int d^D z + \cdots\right]. \end{aligned} \tag{5.2.6}$$

The second term in the square brackets is represented by the diagram of Fig. 5.3.

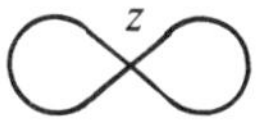

Fig. 5.3 A vacuum-fluctuation diagram

To obtain the correlation function to order g, we must divide (5.2.5) by (5.2.6), and find

$$G^{(2)}(x-y) = \frac{I(x, y)}{Z(0)} = G_0(x-y) - \frac{1}{2}g\int d^D z\, G_0(x-z)G_0(0)G_0(z-y) + O(g^2).$$

The graph (2) from Fig. 5.2 does not feature in the perturbation expansion of G. Diagrams of this type contain parts called 'vacuum fluctuations' or 'vacuum to vacuum' (sub)diagrams (a terminology borrowed from quantum field theory),

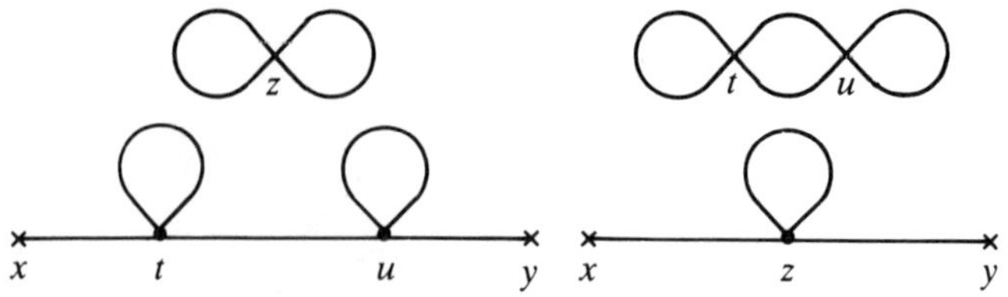

Fig. 5.4 Diagrams not contributing to $G^{(2)}(x - y)$

meaning a *subgraph that is completely disconnected from the 'external' points x and y* (see Fig. 5.2). Further examples of diagrams of this type are shown in Fig. 5.4. Such graphs never enter the perturbative expansion of correlation functions. The sum of all vacuum-fluctuation diagrams equals $Z(0) = \int \mathscr{D}\varphi\, e^{-H}$, its single-variable analogue being (5.2.1). Division by $Z(0)$ cancels all the graphs containing 'vacuum-fluctuation' parts disconnected from the rest of the diagram. This can be proved for the general case without too much trouble.

Consider a diagram featuring the interaction Hamiltonian to order $(p + q)$, and containing a vacuum-fluctuation part of order q; Fig. 5.5 shows an example for the case where $\mathscr{V} = g\varphi^4/4!$. If $n = p + q$, there are C_n^q ways of choosing the q factors V needed to construct the vacuum-fluctuation diagrams. On the other hand, the expansion of the exponential $\exp(-V)$ supplies a factor $1/n!$. Hence a graph with given topology enters with a factor $C_n^q/n! = 1/q!p!$. In view of

$$\sum_n \sum_{(p+q)=n} = \sum_{p,q},$$

the numerator in the calculation of the correlation function reads

$$\sum_{p,q} \frac{1}{p!q!} [\text{vac. fluct. } (q)]\, [\text{connected}(p)]$$

$$\left(\sum_p \frac{1}{p!} [\text{connected}(p)]\right)\left(\sum_q \frac{1}{q!} [\text{vac. fluct. } (q)]\right).$$

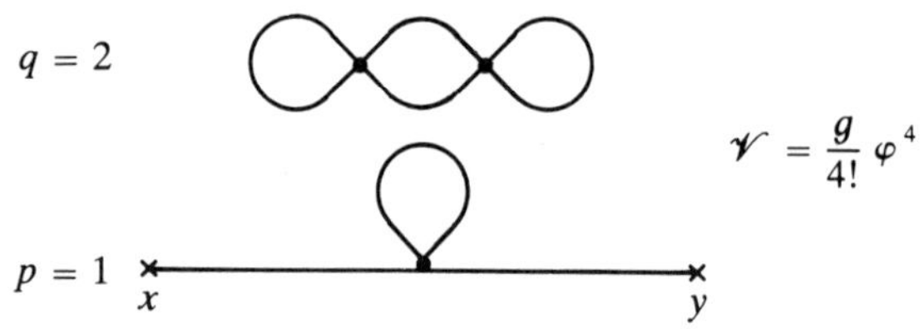

Fig. 5.5

But the second factor is just $Z(0)$, and the correlation function is given by

$$\sum_p \frac{1}{p!} [\text{connected}(p)].$$

The calculation just described enables one to begin guessing the 'Feynman rules', i.e. the rules for associating diagrams with the perturbation expansion; and also the prescriptions for evaluating these diagrams (a complete proof will be given in Section 5.2.5). To order g, we have determined the expression for $G^{(2)}(x-y)$, namely

$$G^{(2)}(x-y) = G_0(x-y) - \frac{1}{2} g \int \mathrm{d}^D z\, G_0(x-z) G_0(0) G_0(z-y), \tag{5.2.8}$$

which is represented graphically by Fig. 5.6:

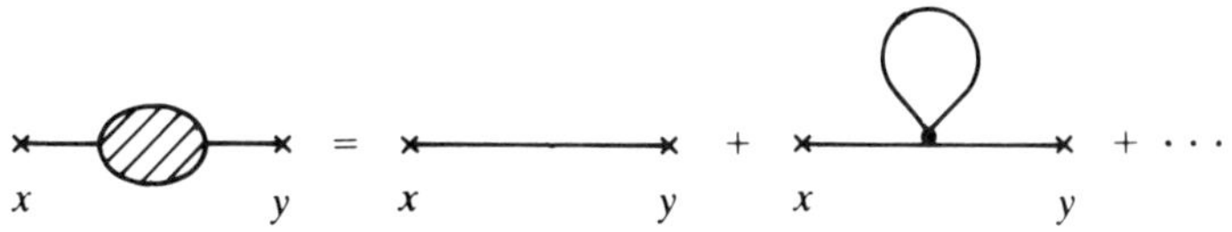

Fig. 5.6 Diagrammatic expansion of $G^{(2)}(x-y)$

The analytic expression corresponding to a graph is evaluated according to the following rules (these are the 'Feynman rules in x-space'):

1. To every internal point (i.e. vertex) we assign a factor $-g$.
2. To every line joining two points x_i and x_j we assign a factor $G_0(x_i - x_j)$, often called a '*propagator*' (again a term borrowed from quantum field theory).
3. We integrate over all internal points z_i: $\int \mathrm{d}^D z_i$.
4. To every graph we assign a multiplicative numerical factor, called a '*symmetry factor*'. In the case considered above, this factor equals 1/2.

On taking its Fourier transform, equation (5.2.8) in k-space becomes

$$G^{(2)}(k) = G_0(k) - \frac{1}{2} g G_0(k) \left[\int \frac{\mathrm{d}^D q}{(2\pi)^D} G_0(q) \right] G_0(k). \tag{5.2.9}$$

This expression suggests the following 'Feynman rules in k-space' (fully proved in Section 5.2.5):

1. To every vertex we assign a factor $-g$.
2. To every line we assign a factor $G_0(k)$.
3. To every independent loop (this notion will be explained below) there corresponds an integration $\int \mathrm{d}^D q/(2\pi)^D$.
4. Finally, every graph is multiplied by a symmetry factor.

The diagrammatic expansion and the Feynman rules follow immediately from Wick's theorem (5.1.13): every term of the perturbation expansion is a product of factors $G_0(x_i - x_j)$. The only factor which is not completely selfevident is the symmetry factor. Moreover, we must still specify the notion of 'independent loops'. In order to become familiar with the perturbation expansion and with the Feynman rules, it is worth pushing the calculation of $G^{(2)}$ to order g^2.

5.2.3 The calculation of $G^{(2)}$ to order g^2

We use Wick's theorem to compute the expression

$$\left\langle \varphi(x)\varphi(y) \int \mathrm{d}^D z\, \mathrm{d}^D u\, \varphi^4(z)\varphi^4(u) \right\rangle_0 .$$

Eliminating the terms that contain vacuum-fluctuation parts, one finds the three types of graphs shown in Fig. 5.7, with their symmetry factors given in brackets (Problem 5.1):

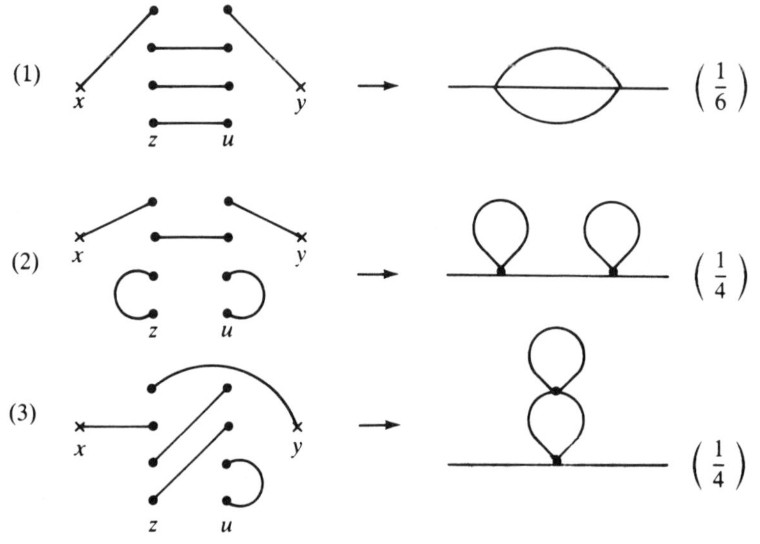

Fig. 5.7 Diagrams of order g^2

We must stress that the vertices z and u may be permuted, which yields a multiplicative factor 2!; however, this is exactly cancelled by the factor 1/2! from the expansion of the exponential. It is the general rule, when considering a term in g^n, that the factor $n!$ that stems from the possibility of permuting the vertices is cancelled by the factor $1/n!$ that stems from the expansion of the exponential.

We shall settle for examining the contribution $\bar{G}(x-y)$ to the correlation function from graph (1) in Fig. 5.7, and leave the reader to write down the analytic expressions for the two other graphs. Thus

$$\bar{G}(x-y)=\frac{1}{6}g^2\int \mathrm{d}^D z\,\mathrm{d}^D u\,G_0(x-z)[G_0(z-u)]^3 G_0(u-y).$$

Let us write $\bar{G}(x-y)$ as a Fourier transform, by replacing every factor G_0 by its Fourier representation

$$\bar{G}(x-y)=\frac{1}{6}g^2\int \mathrm{d}^D z\,\mathrm{d}^D u\,\frac{\mathrm{d}^D k}{(2\pi)^D}\frac{\mathrm{d}^D k'}{(2\pi)^D}\prod_{l=1}^{3}\left\{\int\frac{\mathrm{d}^D q_l}{(2\pi)^D}\,\mathrm{e}^{-\mathrm{i}\sum_{l=1}^{3} q_l\cdot(z-u)}\right\}$$
$$\times\,\mathrm{e}^{-\mathrm{i}k\cdot(x-z)}\mathrm{e}^{-\mathrm{i}k'\cdot(u-y)}G_0(k)G_0(k')\prod_{l=1}^{3}G_0(q_l).$$

The integrations over z and u yield a product of two delta-functions, namely

$$(2\pi)^D\delta^{(D)}(k-q_1-q_2-q_3)\times(2\pi)^D\delta^{(D)}(k'-q_1-q_2-q_3),$$

whence

$$\bar{G}(x-y)=\frac{1}{6}g^2\int\frac{\mathrm{d}^D k}{(2\pi)^D}\,\mathrm{e}^{-\mathrm{i}k\cdot(x-y)}[G_0(k)]^2$$
$$\times\int\frac{\mathrm{d}^D q_1}{(2\pi)^D}\frac{\mathrm{d}^D q_2}{(2\pi)^D}G_0(q_1)G_0(q_2)G_0(k-q_1-q_2).$$

The last expression shows that $\bar{G}(x-y)$ is the Fourier transform of a function $\bar{G}(k)$,

$$\bar{G}(k)=\frac{1}{6}g^2G_0(k)\left[\int\frac{\mathrm{d}^D q_1}{(2\pi)^D}\frac{\mathrm{d}^D q_2}{(2\pi)^D}G_0(q_1)G_0(q_2)G_0(k-q_1-q_2)\right]G_0(k),\tag{5.2.10}$$

represented diagrammatically in Fig. 5.8. The graph shown there features two external propagators $G_0(k)$, and three internal propagators; because of the two delta-functions $\delta^{(D)}(\ldots)$, only two of the three internal lines are independent. The diagram in Fig. 5.8 allows us to introduce the notion of independent loops.

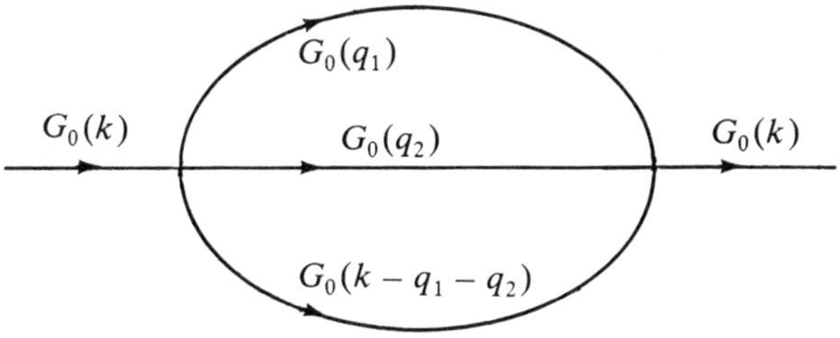

Fig. 5.8 Diagrammatic representation of (5.2.10)

By following the internal propagators one can describe three different closed loops, but on account of the delta-functions $\delta^{(D)}$ only two of these loops are independent; in other words there are only two integration variables in (5.2.10).

To complete this discussion of the perturbative expansion of $G^{(2)}$, we investigate the symmetry factor, starting with the case of the simple Ginzburg–Landau Hamiltonian. To order (p) in perturbation theory, one draws all topologically inequivalent Feynman diagrams, and evaluates each diagram using the rules given above. In principle, the 4! permutations of the four points of each vertex cancel the factor 1/4! multiplying g. However, some of these permutations may correspond to one and the same term in Wick's theorem, and we must avoid double-counting. One can try to formulate general rules for calculating this factor (see Itzykson and Zuber 1980, pp. 265–568). Instead, we shall settle for calculating it explicitly case by case*.

Next, we generalize to a Ginzburg–Landau Hamiltonian with $O(n)$ symmetry (see equation (5.1.23)). For a Hamiltonian of this kind, the Gaussian correlation function is

$$\langle \varphi_i(x)\varphi_j(y)\rangle_0 = G_{0;\,ij}(x-y) = \delta_{ij}G_0(x-y). \tag{5.2.11}$$

To evaluate the correction of order g, we use Wick's theorem to compute

$$\left\langle \int \mathrm{d}^D z \left(\sum_{k=1}^{n} \varphi_k^2(z) \right)^2 \varphi_i(x)\varphi_j(y) \right\rangle_0 .$$

Note the relation

$$\left(\sum_{k=1}^{n} \varphi_k^2(z) \right)^2 = \sum_{k,\,l} \varphi_k^2(z)\varphi_l^2(z),$$

and the fact that the 'fourfold' point z can be split into a pair of points with index k, and another pair with index l. Hence there are two kinds of graphs, as in Figs. 5.9 and 5.10 (where the broken lines link the group of indices (k) with the group of indices (l)):

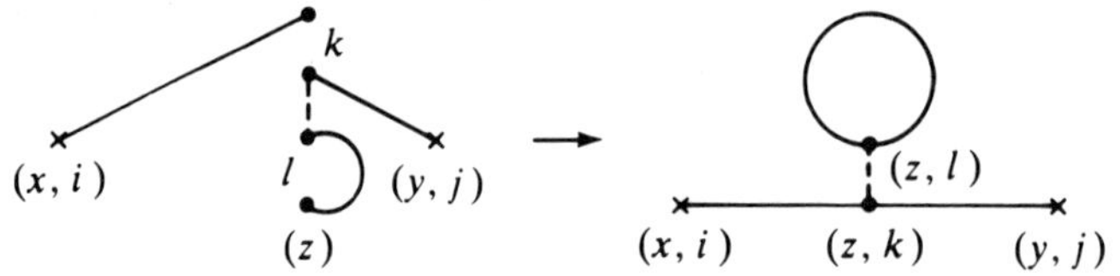

Fig. 5.9

* When $n = 1$, the symmetry factor is given by the reciprocal of $y\Pi 2^{\beta}(n!)^{\alpha_n}$ where α_n is the number of *pairs* of vertices connected by identical self-conjugate lines; β is the number of lines connecting a vertex to itself; and y is the number of permutations of vertices that leave the diagram invariant, with external lines fixed.

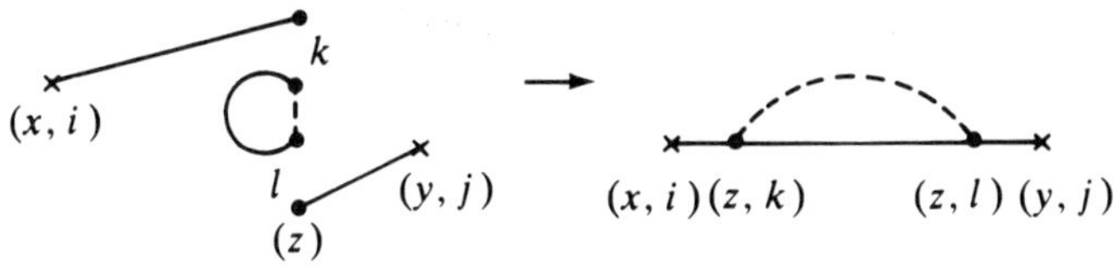

Fig. 5.10

Since $\sum_{k,l} \delta_{ik}\delta_{jk}\delta_{ll} = n\delta_{ij}$, the symmetry factor is $(4n)/4! = n/6$. The symmetry factor of the graph in Fig. 5.10 is $8/4! = 1/3$. The sum of the two graphs (Fig. 5.11) has a symmetry factor $(n + 2)/6$ ($=1/2$ if $n = 1$).

Fig. 5.11

The three graphs of order g^2, shown in Fig. 5.12, have symmetry factors $(n + 2)/18$, $(n + 2)^2/36$, and $(n + 2)^2/36$ respectively (Problem 5.4).

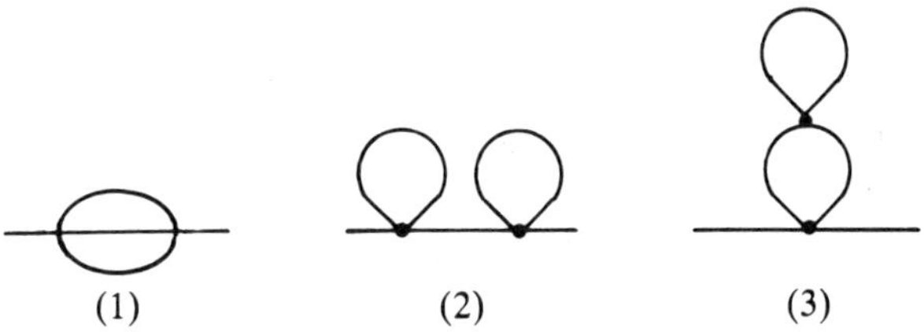

Fig. 5.12 $G^{(2)}$ to order g^2

5.2.4 The four-point correlation function $G^{(4)}$

We carry on with the Hamiltonian (5.1.23). The four-point correlation function $G^{(4)}_{ijkl}(x_1, x_2, x_3, x_4)$ is given by

$$G^{(4)}_{ijkl}(x_1, x_2, x_3, x_4) = \langle \varphi_i(x_1)\varphi_j(x_2)\varphi_k(x_3)\varphi_l(x_4)\rangle.$$

By translation-invariance, this function depends only on three coordinate differences, e.g. on $(x_1 - x_2)$, $(x_2 - x_3)$, $(x_3 - x_4)$.

To order g^0, $G^{(4)}_{0;\,ijkl}$ is the sum of three disconnected graphs (Fig. 5.13):

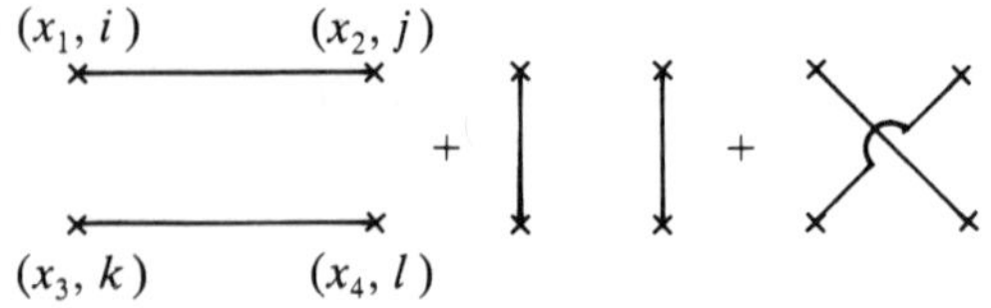

Fig. 5.13 $G^{(4)}$ to order g^0

$$G^{(4)}_{0;\,ijkl}(x_1, x_2, x_3, x_4) = G_{0;\,ij}(x_1 - x_2)G_{0;\,kl}(x_3 - x_4)$$
$$+ G_{0;\,ik}(x_1 - x_3)G_{0;\,jl}(x_2 - x_4) + G_{0;\,il}(x_1 - x_4)G_{0;\,jk}(x_2 - x_3).$$

To order g there are three types of graphs (Fig. 5.14):

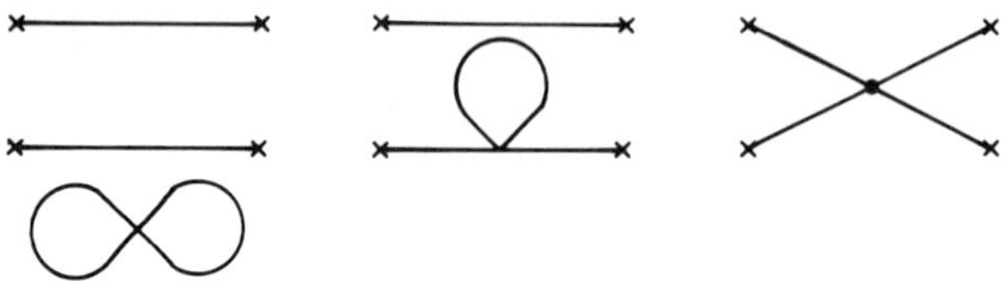

Vacuum fluctuation (1)   Disconnected (2)   Connected (3)

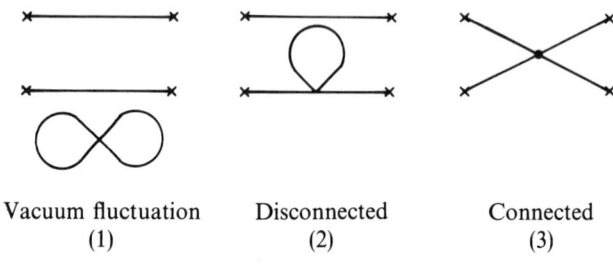

Fig. 5.14 $G^{(4)}$ to order g

The graphs of type (1) contain vacuum-fluctuation parts, and are cancelled on division by $Z(0)$. The graphs of type (2) can be written as products of two independent factors; these are examples of *disconnected graphs*. Their expressions are known, since they involve only two-point correlation functions, which have been calculated already. The graphs of type (3) are the most interesting. The analytic expression for graph (3) in Fig. 5.14 reads

$$\bar{G}_{ijkl} = -\bar{\Gamma}_{ijkl}\int \mathrm{d}^D z\, G_0(x_1 - z)G_0(x_2 - z)G_0(x_3 - z)G_0(x_4 - z). \tag{5.2.12}$$

(Note the minus sign in (5.2.12).) $\bar{\Gamma}_{ijkl}$ can be evaluated immediately:

$$\begin{cases} \bar{\Gamma}_{ijkl} = \dfrac{1}{3} g S_{ijkl}, \\ S_{ijkl} = \delta_{ij}\delta_{kl} + \delta_{ik}\delta_{jl} + \delta_{il}\delta_{jk}. \end{cases} \tag{5.2.13}$$

If $n = 1$, we have simply $\bar{\Gamma} = g$, since $S_{ijkl} = 3$. The minus sign in (5.2.12) ensures that $\bar{\Gamma}$ equals $+g$ rather than $-g$ (see Section 5.3.3).

Let us now replace the G_0s in (5.2.12) by their Fourier representations:

$$\bar{G}_{ijkl} = -\bar{\Gamma}_{ijkl}\int \mathrm{d}^D z\left\{\prod_{i=1}^{4}\frac{\mathrm{d}^D k_i}{(2\pi)^D}\,\mathrm{e}^{-\mathrm{i}k_i(x_i-z)}G_0(k_i)\right\}$$

$$= -\bar{\Gamma}_{ijkl}\int\prod_{i=1}^{4}\left\{\frac{\mathrm{d}^D k_i}{(2\pi)^D}\,\mathrm{e}^{-\mathrm{i}k_i x_i}G_0(k_i)\right\}(2\pi)^D\delta^{(D)}\left(\sum_{i=1}^{4}k_i\right). \qquad (5.2.14)$$

Here one notes the factor $(2\pi)^D\delta^{(D)}(\sum k_i)$; it stems from the translation-invariance of $G^{(4)}(x_i)$. Let us establish its presence quite generally, i.e. without appeal to the perturbation expansion. To this end we change variables in the integral

$$\int\left[\prod_{i=1}^{4}\mathrm{d}^D x_i\,\mathrm{e}^{\mathrm{i}k_i x_i}\right]G^{(4)}(x_i)$$

as follows:

$$y_1 = x_1 - x_4,\quad y_2 = x_2 - x_4,\quad y_3 = x_3 - x_4,\quad y_4 = x_4,$$

$$\sum_{i=1}^{4}k_i x_i = \sum_{i=1}^{3}k_i y_i + \left(\sum_{i=1}^{4}k_i\right)y_4.$$

By translation invariance, $G^{(4)}(x_i)$ can be written as a function only of y_1, y_2, y_3; the integral over $y_4 = x_4$ indeed yields $(2\pi)^D\delta^{(D)}(\sum_{i=1}^{4}k_i)$. *Thus, in the definition of the Fourier transform $G^{(4)}(k)$ one can always identify a factor* $(2\pi)^D\delta^{(D)}(\sum_{i=1}^{4}k_i)$, so that

$$(2\pi)^D\delta^{(D)}\left(\sum_{i=1}^{4}k_i\right)G^{(4)}(k_i) = \int\left[\prod_{i=1}^{4}\mathrm{d}^D x_i\,\mathrm{e}^{\mathrm{i}k_i x_i}\right]G^{(4)}(x_i). \qquad (5.2.15)$$

In an N-point correlation function, one can identify a factor $(2\pi)^D\delta^{(D)}(\sum_{i=1}^{N}k_i)$. The reader can readily verify that this convention has already been adopted tacitly for $G^{(2)}$. If discrete variables k_i are used, one identifies a factor $L^D\delta_{\sum_i^N k_i}$, using in these cases a Kronecker delta (L is the size of the system under study).

The examples given above have enabled us to guess the form of the Feynman rules; it remains only to validate these rules in general. We shall do this only in k-space, which is the more useful in practice.

5.2.5 The Feynman rules in *k*-space

Define the Fourier transform of $j(x)$ by

$$j(k) = \int\frac{\mathrm{d}^D x}{(2\pi)^{D/2}}\,\mathrm{e}^{\mathrm{i}k\cdot x}j(x).$$

By virtue of Parseval's theorem one has

$$\int\mathrm{d}^D x\, j(x)\varphi(x) = \int\mathrm{d}^D k\,\varphi(k)j(-k).$$

The Gaussian generating functional in k-space reads

$$Z_0(j) = \int \mathscr{D}\varphi(k) \exp\left(-\frac{1}{2} \int d^D k (m^2 + k^2) \varphi(k) \varphi(-k) + \int d^D k \, \varphi(k) j(-k) \right)$$

$$= Z_0(0) \exp\left(\frac{1}{2} \int d^D k \, j(k) G_0(k) j(-k) \right), \tag{5.2.16}$$

and the interaction becomes

$$V = \frac{g}{4!} \int \left[\prod_{i=1}^{4} \frac{d^D q_i}{(2\pi)^{D/2}} \varphi(q_i) \right] (2\pi)^D \delta^{(D)}(\textstyle\sum q_i). \tag{5.2.17}$$

Let us now relate the Fourier transform of the N-point correlation function to $\langle \varphi(k_1) \ldots \varphi(k_N) \rangle$:

$$G^{(N)}(x_1, \ldots, x_N) = \langle \varphi(x_1) \ldots \varphi(x_N) \rangle$$
$$= \int \left\{ \prod_{i=1}^{N} \frac{d^D k_i}{(2\pi)^{D/2}} e^{-ik_i \cdot x_i} \right\} \langle \varphi(k_1) \ldots \varphi(k_N) \rangle.$$

But by virtue of equation (5.2.15) adapted for the general case, we can equally well write

$$\int \left\{ \prod_{i=1}^{N} \frac{d^D k}{(2\pi)^D} e^{-ik_i \cdot x_i} \right\} (2\pi)^D \delta^{(D)}\left(\sum_1^N k_i \right) G^{(N)}(k_1 \ldots k_N),$$

whence

$$(2\pi)^D \delta^{(D)}\left(\sum_1^N k_i \right) G^{(N)}(k_1, \ldots, k_N) = (2\pi)^{ND/2} \langle \varphi(k_1) \ldots \varphi(k_N) \rangle. \tag{5.2.18}$$

Note the special case

$$(2\pi)^D \delta^{(D)}(k_1 + k_2) G^{(2)}(k_1) = (2\pi)^D \langle \varphi(k_1) \varphi(k_2) \rangle,$$

or in other words

$$G^{(2)}(k) = \langle \varphi(k) \varphi(-k) \rangle.$$

Combining (5.2.16), (5.2.17),and (5.2.18), we find for $G^{(N)}(k_1, \ldots, k_N)$ the expression

$$(2\pi)^D \delta^{(D)}\left(\sum_1^N k_i \right) G^{(N)}(k_1, \ldots, k_N)$$
$$= \mathscr{N} (2\pi)^{ND/2} \exp\left(-V\left(\frac{\delta}{\delta j(-q)} \right) \right) \frac{\delta^N}{\delta j(-k_1) \ldots \delta j(-k_N)}$$
$$\times \exp\left(\frac{1}{2} \int d^D k \, j(k) G_0(k) j(-k) \right). \tag{5.2.19}$$

We see that

$$\frac{\delta^2}{\delta j(-k_1)\delta j(-k_2)}\int \mathrm{d}^D k\, j(k) G_0(k) j(-k) = 2\delta^{(D)}(k_1 + k_2) G_0(k_1).$$

To every line, internal or external, one assigns a factor $\delta^{(D)}(k_1 + k_2)G_0(k_1)$; on the other hand, by virtue of the factor $(2\pi)^D\delta^{(D)}(\sum k_i)$ in (5.2.17), the sum of all the momenta entering a vertex must be zero. On every line of the diagram (internal or external) we can choose a definite direction for the momentum carried by the line. Four lines join at every vertex; they carry momenta k_i, counted as positive if they enter and as negative if they leave the vertex. For every vertex the delta-function in (5.2.17) entails the conservation law

$$\sum_{i=1}^{4} (\pm k_i) = 0.$$

Often it proves convenient to adopt the *convention* that all the external momenta $k_1, \ldots, k_N$ are directed towards the diagram. Then, by virtue of the factor $\delta^{(D)}(k_1 + k_2)$, a line arriving at a vertex with momentum k must either arrive at the diagram with momentum k, or it must enter another vertex with momentum $-k$. Figure 5.15 shows an example where

$$k_1 + k_2 - q_1 - q_2 = 0, \quad q_1 + q_2 + k_3 + k_4 = 0.$$

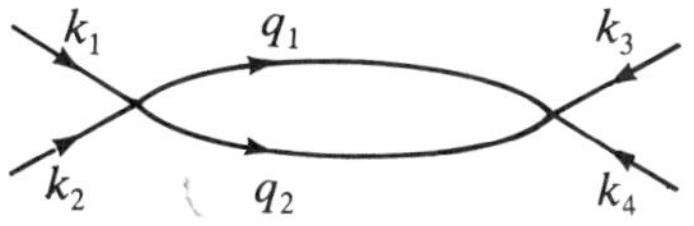

Fig. 5.15 A contribution to $G^{(4)}$

The conservation of momentum at every vertex ensures overall momentum conservation: $k_1 + k_2 + k_3 + k_4 = 0$.

Finally, let us determine the factors (2π). The correlation function has N external lines, and to order V of perturbation theory it has V vertices. Therefore (see (5.2.17) and (5.2.19)) in this order we have a factor

$$(2\pi)^{ND/2}(2\pi)^{-4VD/2}(2\pi)^{VD}.$$

But in a connected diagram we also have the relation

$$4V = 2I + N,$$

where I is the number of internal lines; to see this, notice that if every internal line is cut, then four lines arrive at every vertex, while the total number of lines is $2I + N$. Figure 5.16 shows an example.

Accordingly, the multiplicative factor is

$$(2\pi)^{VD}(2\pi)^{(N-4V)D/2} = (2\pi)^{VD}(2\pi)^{-ID}.$$

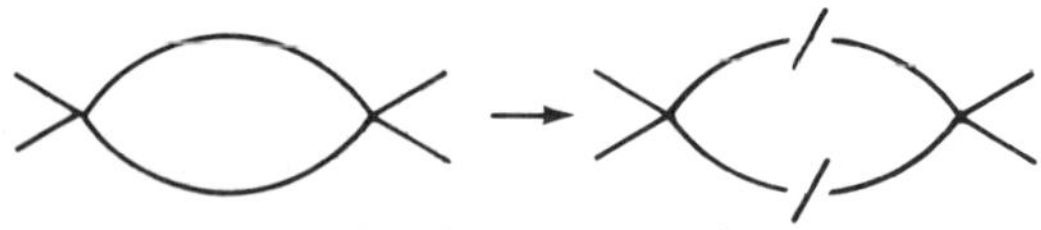

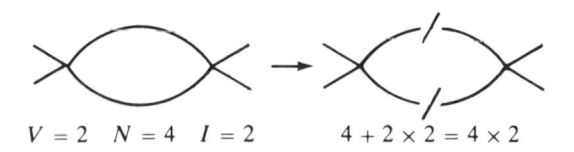

$V = 2 \quad N = 4 \quad I = 2 \qquad\qquad 4 + 2 \times 2 = 4 \times 2$

Fig. 5.16

The Feynman rules for $G^{(N)}(k_1 \ldots k_N)$ in k-space can be summarized in two equivalent forms. (For disconnected diagrams there may be some trivial modifications; as an exercise one might write down the contribution to $G^{(4)}$ from the diagrams of Fig. 5.13.) The first alternative is as follows:

(1) Draw all topologically inequivalent diagrams (to a given order of perturbation theory).

(2) To every line of the diagram assign a factor $G_0(q)$.

(3) To every vertex assign a factor $-g(2\pi)^D\delta^{(D)}(\sum_1^4 q_i)$, and extract the momentum-conserving factor $(2\pi)^D\delta^{(D)}(\sum_1^N k_i)$.

(4) Integrate over every internal line with the integration measure $\int \mathrm{d}^Dq/(2\pi)^D$.

(5) Multiply by the symmetry factor.

The second alternative is found by using the notion of independent loops. There are

$$L = I - V + 1$$

independent integration variables in a diagram, or in other words L *independent loops*; every vertex supplies a delta-function, but one of these is already accounted for by overall momentum conservation. The factor $(2\pi)^{(-I+V)D}$ may be written

$$(2\pi)^{-ID}(2\pi)^{VD} = (2\pi)^{-LD}(2\pi)^D.$$

The factor $(2\pi)^D$ is to be associated with the function $\delta(\ldots)$ responsible for overall momentum conservation in the diagram. Accordingly we can replace (3) and (4) by

(3a) To every vertex assign a factor $-g$.

(4a) Choose internal momenta so that momentum is conserved at every vertex, and integrate over all independent loops with the integration measure $\mathrm{d}^Dq/(2\pi)^D$.

An example applying these rules is given in Fig. 5.17. The contribution of this graph to $G^{(2)}$ reads

$$\frac{1}{6}g^2G_0(k_1)\left[\int \frac{\mathrm{d}^Dq_1}{(2\pi)^D}\frac{\mathrm{d}^Dq_2}{(2\pi)^D}\,G_0(q_1)G_0(q_2)G_0(k_1 - q_1 - q_2)\right]G_0(k_1),$$

conformably with (5.2.10).

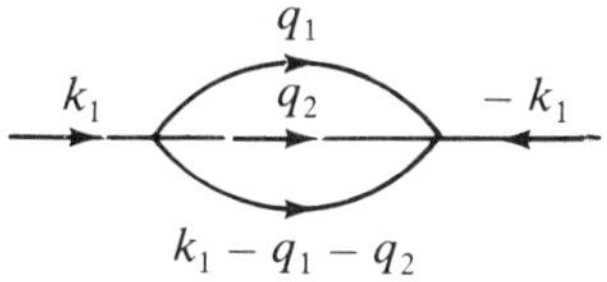

Fig. 5.17 A contribution to $G^{(2)}$

5.3 Connected correlation functions and proper vertices

We have seen that some of the contributions to the correlation functions can be written as products of two or more independent factors; for instance, the contribution to $G^{(4)}$ from the diagram in Fig. 5.18 becomes a product of two such factors. In order to limit the number of diagrams it proves useful to define *connected diagrams*, which are those that cannot be separated into two or more disjoint parts without cutting at least one line. Decomposition into connected diagrams generalizes the expression of the moments of a probability distribution in terms of *cumulants*.

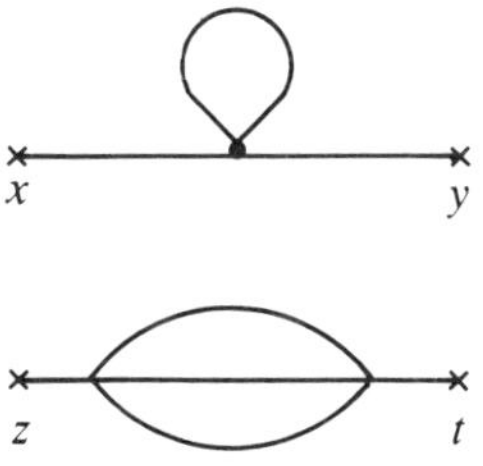

Fig. 5.18 A disconnected contribution to $G^{(4)}$

5.3.1 The cumulants of a probability distribution

Let $Z(j)$ be the generating function of a probability distribution $P(\varphi)$:

$$Z(j) = Z(0) \sum_{n=0}^{\infty} \frac{1}{n!} j^n \langle \varphi^n \rangle.$$

We write $Z(j) = Z(0)\exp[W(j)]$ or $W(j) = \ln(Z(j)/Z(0))$ (W is analogous to a free energy), and define the cumulants $\langle \varphi^n \rangle_c$ of order n by

$$W(j) = \sum_{n=1}^{\infty} \frac{j^n}{n!} \langle \varphi^n \rangle_c,$$

$$\langle \varphi^n \rangle_c = \left. \frac{\partial^n W(j)}{\partial j^n} \right|_{j=0}.$$

By comparison one can identify

$$\langle\varphi\rangle_c = \langle\varphi\rangle,$$

$$\langle\varphi^2\rangle_c = \langle\varphi^2\rangle - \langle\varphi\rangle^2 = \langle(\varphi - \langle\varphi\rangle)^2\rangle,$$

$$\langle\varphi^3\rangle_c = \langle\varphi^3\rangle - 3\langle\varphi\rangle\langle\varphi^2\rangle + 2\langle\varphi\rangle^3 = \langle(\varphi - \langle\varphi\rangle)^3\rangle,$$

$$\langle\varphi^4\rangle_c = \langle\varphi^4\rangle - 4\langle\varphi\rangle\langle\varphi^3\rangle + 3\langle\varphi^2\rangle^2 + 12\langle\varphi\rangle^2\langle\varphi^2\rangle - 6\langle\varphi\rangle^4$$
$$= \langle(\varphi - \langle\varphi\rangle)^4\rangle - 3\langle(\varphi - \langle\varphi\rangle)^2\rangle^2.$$

The cumulants of a Gaussian distribution vanish for $n \geqslant 3$. Indeed from (5.1.5), $Z(j) \sim \exp(\frac{1}{2}Aj^2)$, and $W(j) = \frac{1}{2}Aj^2$. Even if the Gaussian distribution is not centred on the origin, this results only in the addition to $F(j)$ of a term linear in j: $\langle\varphi\rangle_c$ and $\langle\varphi^2\rangle_c$ are nonzero, but $\langle\varphi^n\rangle_c = 0$ if $n \geqslant 3$. We shall generalize what has just been stated to correlation functions, through the correspondence (correlation function) → (moment), and (connected correlation function) → (cumulant).

5.3.2 The generating functional in terms of connected diagrams

We start with an example, by investigating the correlation function $G^{(4)}$. It subdivides into one connected and three disconnected diagrams,

$$G^{(4)}(1, 2, 3, 4) = G_c^{(4)}(1, 2, 3, 4) + \{G_c^{(2)}(1, 2)G_c^{(2)}(3, 4) + \text{permutations}\},$$

where G_c denotes a connected correlation function; notice that $G_c^{(2)} = G^{(2)}$. In terms of graphs one obtains the equation of Fig. 5.19:

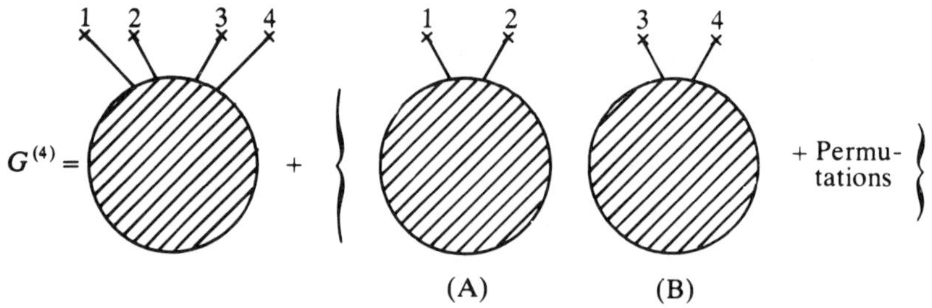

Fig. 5.19

The number of disconnected terms is $3 = 4!/[(2!)^2 \times (2!)]$. $4!$ is the number of permutations of the external points (1, 2, 3, 4); but the result is unaffected by permuting (1, 2), or (3, 4), or the two bubbles (A) and (B), whence the factor $(2!)^2 \times 2!$.

The case of φ^4 theory is somewhat peculiar, in that all correlation functions with odd N vanish: $G^{(2k+1)} = 0$*. For complete generality, we shall assume that the interaction contains terms in φ^{2p+1}, e.g. φ^3, so that $G^{(2k+1)} \neq 0$. Consider a disconnected diagram of $G^{(N)}$ corresponding to the subdivision of Fig. 5.20 into connected diagrams:

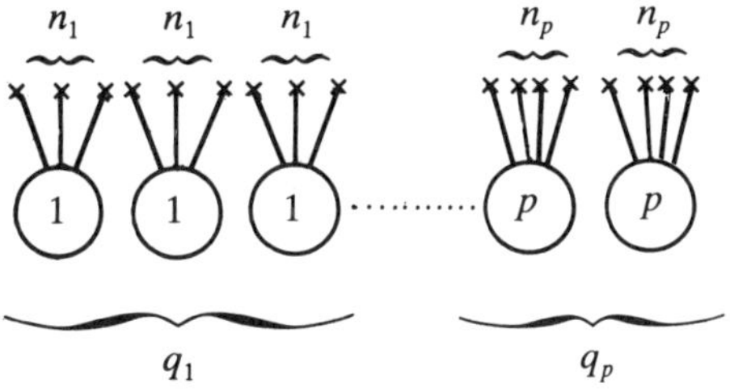

Fig. 5.20

There are q_l bubbles connected to n_l external points, . . . , q_p bubbles connected to n_p external points, with

$$q_1 n_1 + \cdots + q_p n_p = N.$$

The number of independent terms is

$$\frac{N!}{[(n_1!)^{q_1} q_1!] \ldots [(n_p!)^{q_p} q_p!]}. \tag{5.3.1}$$

The generating functional $Z(j)$ reads

$$\begin{aligned}
\frac{Z(j)}{Z(0)} &= \sum_{N=0}^{\infty} \frac{1}{N!} \int \mathrm{d}x_1 \ldots \mathrm{d}x_N j(x_1) \ldots j(x_N) \\
&\quad \times \sum_{q_1 n_1 + \cdots + q_p n_p = N} G_c^{(n_1)}(x_1 \ldots x_{n_1}) \ldots G_c^{(n_p)}(\ldots x_N) \\
&= \sum_{N=0}^{\infty} \sum_{q_1 n_1 + \cdots + q_p n_p = N} \prod_{i=1}^{p} \frac{1}{q_i!} \\
&\quad \times \left[\frac{\int \mathrm{d}x_1 \ldots \mathrm{d}x_{n_i} \, j(x_1) \ldots j(x_{n_i}) G_c^{(n_i)}(x_1 \ldots x_{n_i})}{n_i!} \right]^{q_i} \\
&= \sum_{q_i} \prod_i \frac{1}{q_i!} \left[\frac{\int \mathrm{d}x_1 \ldots \mathrm{d}x_{n_i} \, j(x_1) \ldots j(x_{n_i}) G_c^{(n_i)}(x_1 \ldots x_{n_i})}{n_i!} \right]^{q_i} \\
&= \exp \sum_{N=1}^{\infty} \frac{1}{N!} \int \mathrm{d}x_1 \ldots \mathrm{d}x_N j(x_1) \ldots j(x_N) G_c^{(N)}(x_1, \ldots, x_N).
\end{aligned}$$

* $G^{(1)} = \langle \varphi \rangle \neq 0$ is possible in cases of broken symmetry, or if $\mathscr{V}(\varphi)$ contains terms odd in φ, e.g. φ^3.

In going from the first to the second equality we have used (5.3.1), and the symmetry of G_c with respect to its arguments. Thus we find that *the generating functional $W(j)$ of the connected correlation functions is* $\ln[Z(j)/Z(0)]$,

$$W(j) = \ln\frac{Z(j)}{Z(0)} = \sum_{N=1}^{\infty} \frac{1}{N!}\int dx_1 \ldots dx_N\, j(x_1)\ldots j(x_N) \times G_c^{(N)}(x_1, \ldots, x_N). \tag{5.3.2}$$

Note that $G^{(N)}$ may be written

$$G^{(N)} = G_c^{(N)}(x_1, \ldots, x_N) + \sum_{UI_\alpha = I} \prod_\alpha G_c(I_\alpha),$$

where $I = \{x_1, \ldots, x_N\}$, and I_α is a partition of I: $\bigcup(I_\alpha) = I$. On the right, every term appears once and only once. Notice also that for a Gaussian Hamiltonian the connected correlation functions vanish for $N \geqslant 3$.

5.3.3 Proper vertices and the generating functional

We can find one further simplification. In k-space, a diagram like that of Fig. 5.21 effectively reads $\hat{G}_1 G_0(k)\hat{G}_2$.

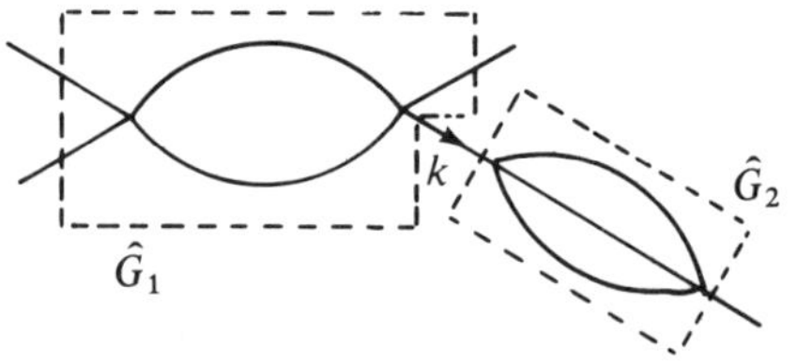

Fig. 5.21 A 1-particle reducible diagram

In order to obtain the expression for this diagram it is therefore enough to know how to calculate $\hat{G}_1$ and $\hat{G}_2$ independently. Such a diagram is called 1-particle reducible (1-PR). *We shall call any connected correlation function 1-particle irreducible (1-PI) if it cannot be separated into two independent parts by cutting just one internal line* (Fig. 5.22). Finally, *a 1-PI correlation function shorn of its external lines is called a proper vertex.*

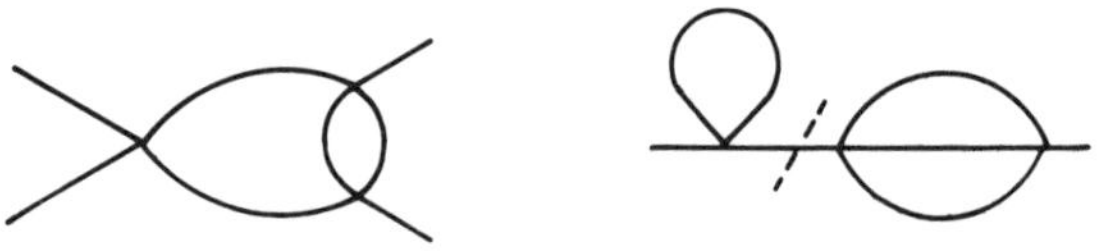

1-*PI* 1-*PR*.

Fig. 5.22

For example, the expression for the proper vertex in Fig. 5.23 is

$$-\frac{g^2}{2}\int\frac{\mathrm{d}^D q}{(2\pi)^D}\frac{1}{(q^2+m^2)}\frac{1}{((k-q)^2+m^2)}.$$

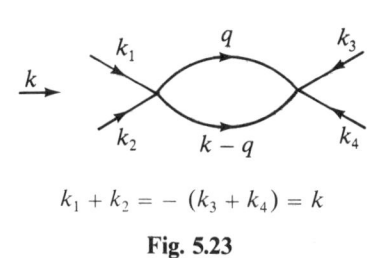

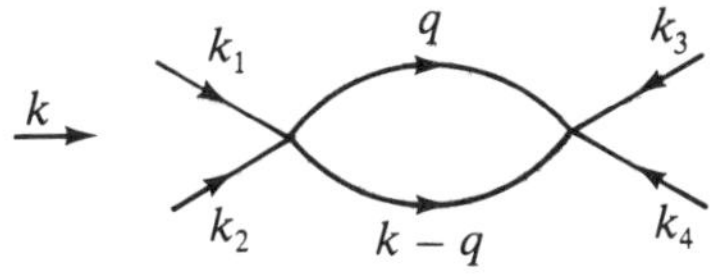

$k_1 + k_2 = -\ (k_3 + k_4) = k$

Fig. 5.23

No factor $\prod_{i=1}^{4} G_0(k_i)$ is associated with the external lines. Proper vertices in general will be written Γ. We prove the following remarkable result:

Theorem: The generating functional of proper vertices is the Legendre transform of $W(j)$.

To ease the notation in the proof that follows, it is convenient to write

$$\frac{\delta}{\delta j(x)}\to\frac{\delta}{\delta j_i},\quad \int\mathrm{d}^D x\to\sum_i.$$

Let $\bar{\varphi}_i = \delta W/\delta j_i$ be the mean value of the field, and Γ the Legendre transform of W; in field theory, the Gibbs potential Γ is called the *effective action*. (Notice that $j(x)$ now plays the role of the B-field, and $\bar{\varphi}(x)$ that of the magnetization; the equations below generalize those of Section 1.4.5.) Thus

$$\Gamma = \mathrm{Max}|_j\left(\sum_i \bar{\varphi}_i j_i - W(j)\right),\quad \frac{\delta\Gamma}{\delta\bar{\varphi}_k} = j_k. \tag{5.3.3}$$

In Section 1.4.5 we have already met the identity

$$\sum_l \frac{\delta^2 W}{\delta j_i \delta j_l}\frac{\delta^2\Gamma}{\delta\bar{\varphi}_l\delta\bar{\varphi}_k} = \sum_l G_{il}^{(2)}\Gamma_{lk}^{(2)} = \delta_{ik}, \tag{5.3.4}$$

which shows that $\Gamma_{kl}^{(2)} = \delta^2\Gamma/\delta\bar{\varphi}_k\delta\bar{\varphi}_l$ is the inverse of the (connected) correlation function $G_{kl}^{(2)}$. With continuous variables, equation (5.3.4) reads

$$\int\mathrm{d}^D z\, G^{(2)}(x-z)\Gamma^{(2)}(z-y) = \delta^{(D)}(x-y),$$

and, in Fourier space,

$$G^{(2)}(k) = 1/\Gamma^{(2)}(k).$$

It proves useful to define the *self-energy** $\Sigma(k)$; $-\Sigma(k)$ is the sum of all two-point 1-PI diagrams shorn of their external lines (i.e. it is the sum of all proper two-point vertices):

$-\Sigma(k) =$

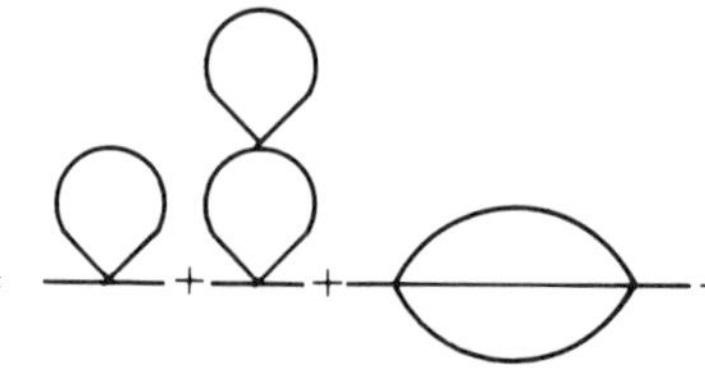

$+\ O(g^3).$

The correlation function $G^{(2)}(k)$ can be written in terms of $\Sigma(k)$ as

$$G^{(2)}(k) = G_0(k) - G_0(k)\Sigma(k)G_0(k) + \cdots$$

$$= G_0(k)\left(1 + \sum_{n=1}^{\infty} [-\Sigma(k)G_0(k)]^n\right) = [G_0^{-1}(k) + \Sigma(k)]^{-1}.$$

For $G^{(2)}(k)$ and $\Gamma^{(2)}(k)$ this entails the expressions

$$G^{(2)}(k) = \frac{1}{k^2 + m^2 + \Sigma(k)}, \quad \Gamma^{(2)}(k) = k^2 + m^2 + \Sigma(k). \tag{5.3.5}$$

We now proceed to correlations of higher order, by differentiating the identity (5.3.4) with respect to j_m:

$$\sum_l \frac{\delta^3 W}{\delta j_i \delta j_l \delta j_m} \frac{\delta^2 \Gamma}{\delta \bar{\varphi}_l \delta \bar{\varphi}_k} + \sum_l \frac{\delta^2 W}{\delta j_i \delta j_l} \frac{\delta^3 \Gamma}{\delta j_m \delta \bar{\varphi}_l \delta \bar{\varphi}_k} = 0. \tag{5.3.6a}$$

Since Γ is a function of the $\bar{\varphi}_i$, we must transform the second derivative in (5.3.6a); we do this for the general case ($\Gamma^{(N)}_{i_1 \cdots i_N} = \delta^{(N)}\Gamma/\delta\varphi_{i_1} \ldots \delta\varphi_{i_N}$):

$$\frac{\delta}{\delta j_m} \Gamma^{(N)}_{i_1 \cdots i_N} = \sum_n \frac{\delta \bar{\varphi}_n}{\delta j_m} \frac{\delta \Gamma^{(N)}_{i_1 \cdots i_N}}{\delta \bar{\varphi}_n} = \sum_n G^{(2)}_{mn} \Gamma^{(N+1)}_{n i_1 \cdots i_N}. \tag{5.3.6b}$$

Equations (5.3.6) can be put into graphical form if we represent the $\Gamma^{(N)}$ by shaded bubbles (Fig. 5.24):

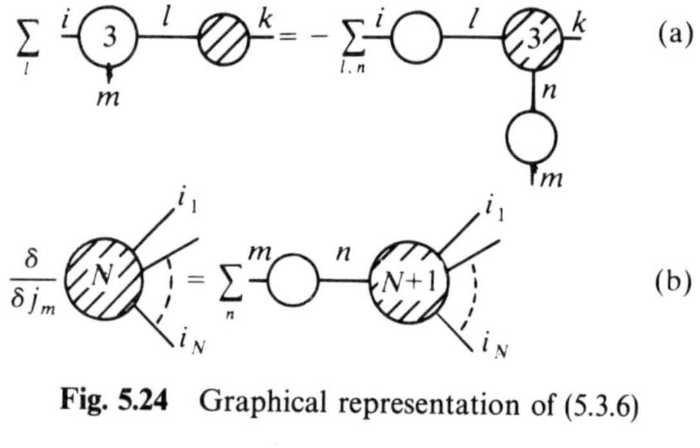

Fig. 5.24 Graphical representation of (5.3.6)

* The name is borrowed from quantum field theory. One must be careful of the fact that $\Sigma(k)$ is given different signs by different writers.

(Here we have used (5.3.6b) in order to transform (5.3.6a).) Multiplying the two terms in (5.3.6a) from the right by G_{kp} and summing over k we find the relation between $G_c^{(3)}$ and $\Gamma^{(3)}$,

$$-G_c^{(3)}(i, m, p) = \sum_{l, k, n}$$

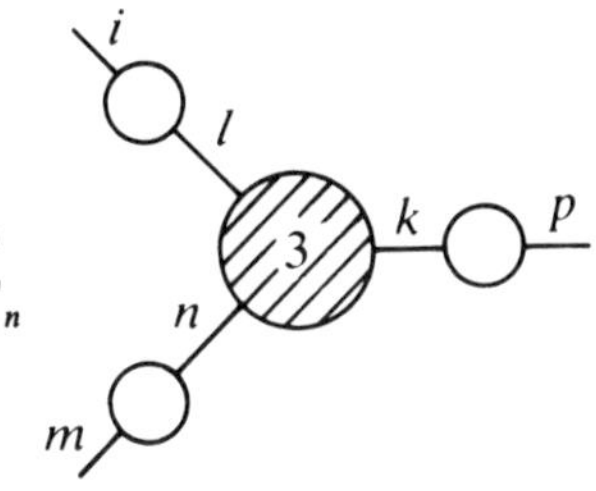

(5.3.7)

Equation (5.3.7) allows one to identify $\Gamma^{(3)}_{kln} = \delta^{(3)}\Gamma/\delta\bar{\varphi}_k\delta\bar{\varphi}_l\delta\bar{\varphi}_n$ with the proper vertex $\bar{\Gamma}^{(3)}_{kln}$; indeed $\Gamma^{(3)}_{kln}$ is just the connected correlation function deprived of its full external propagators, which is precisely a 1-PI correlation function.

We continue the process by differentiating equation (5.3.7) once more with respect to j_l. Using (5.3.6a) and (5.3.6b) we obtain the relation between $G^{(4)}$ and $\Gamma^{(4)}$ (Fig. 5.25):

$$\frac{-\delta}{\delta_{je}} G_c^{(3)}(i, m, p) = -G_c^{(4)}(i, m, p, l) =$$

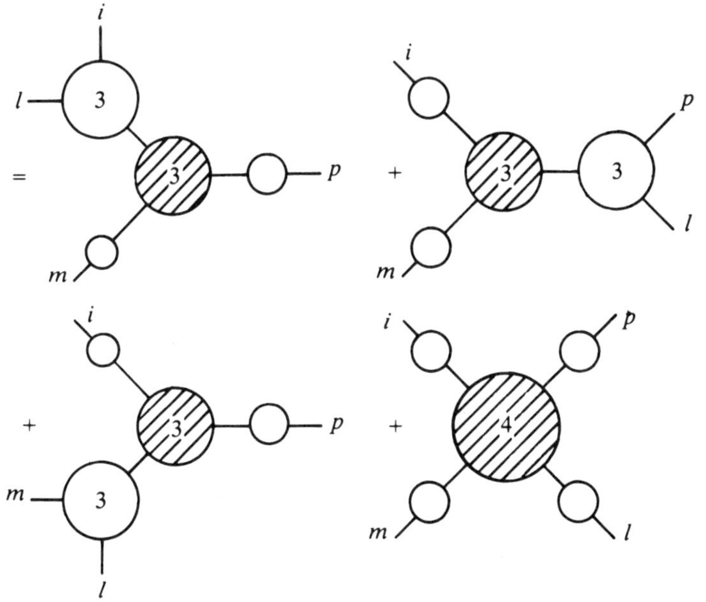

Fig. 5.25

The first three terms can be rewritten by appeal to the equation of Fig. 5.26:

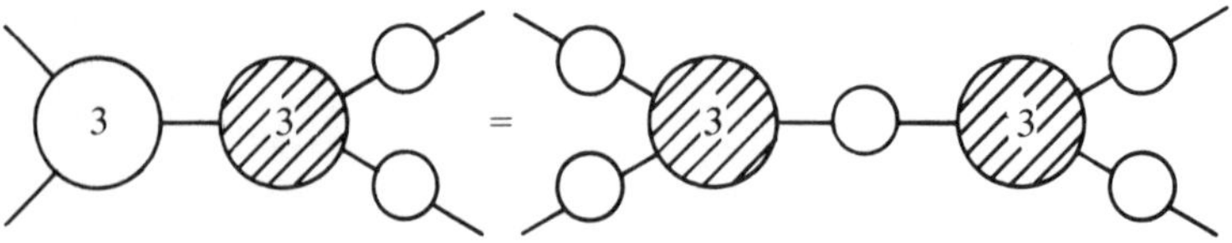

Fig. 5.26

In a φ^4 theory, and without broken symmetry, the end-result is very simple, because there is only one nonzero term (Fig. 5.27):

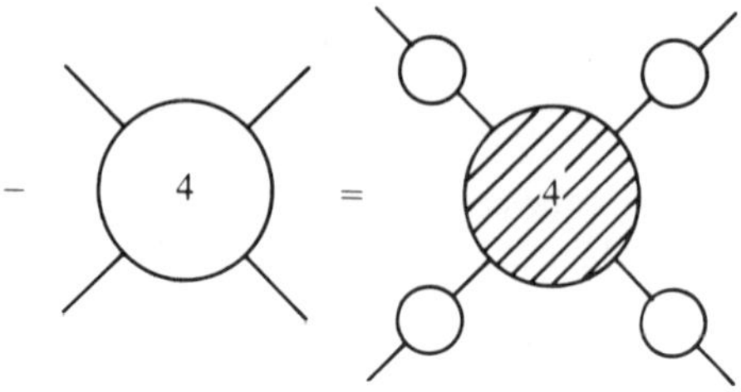

Fig. 5.27

As before, Fig. 5.25 (or Fig. 5.27 in the simplest case) allows one to identify $\delta^{(4)}\Gamma/(\delta\bar{\varphi})^4$ with the proper vertex $\bar{\Gamma}^{(4)}$.

The special cases $N = 3$ and $N = 4$ which we have just studied enable us to see how one can set out to prove the theorem formulated above. One need merely proceed by induction, assuming that an equation like that of Fig. 5.25 can indeed be written to order N, and that the proper vertex $\bar{\Gamma}^{(N)}$ can indeed be identified with the Nth derivative of the Gibbs potential Γ. Differentiating this equation with respect to j_l, one finds

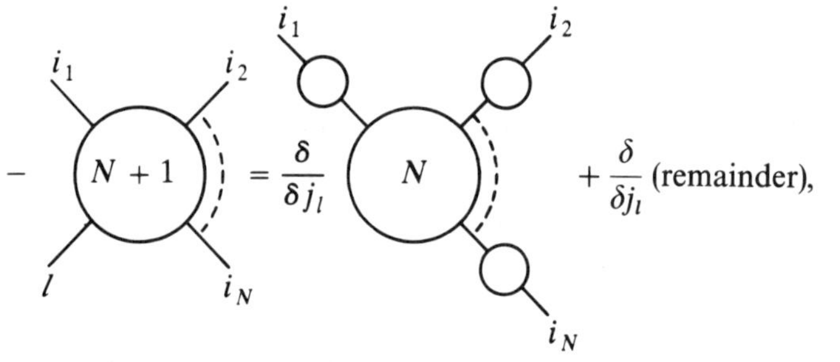

where the 'remainder' does not contain $\Gamma^{(N)}$, but only $\Gamma^{(N-1)}$, $\Gamma^{(N-2)}$, etc. (see Fig. 5.25). Hence we obtain

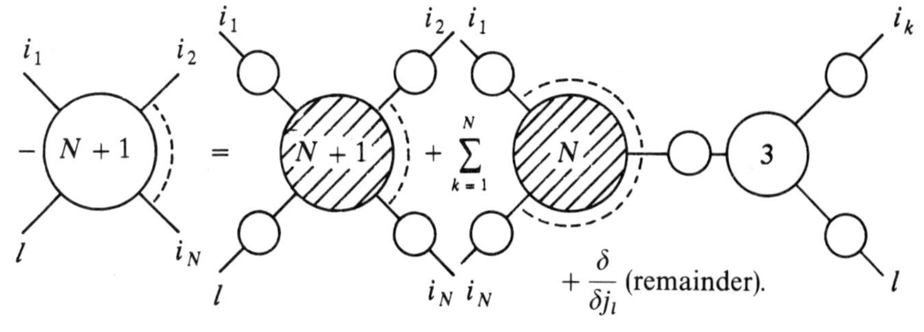

On removing the full external propagators from both sides, we can identify $\bar{\Gamma}^{(N+1)}$ on the left with $\Gamma^{(N+1)}$ on the right; in fact, such amputation leaves us with just a single 1-PI graph on each side.

5.4 The effective potential; the loop expansion

Bearing in mind the results established in the two preceding sections, it proves useful to return to those of Section 2.4, and to reinterpret them in the light of the perturbation expansion. Recall that the external field (or source) is $j(x)$, that it plays the same role as did the field $B(x)$ in Chapter 2, and that the magnetization is written as $\bar{\varphi}(x)$.

5.4.1 Broken symmetry and the effective potential

When $T < T_c$, the mean value of $\varphi(x)$ is nonzero even without an external field:

$$\lim_{j\to 0}\lim_{L^D\to\infty} \bar{\varphi}(x) = v \neq 0.$$

We shall consider only the simple Ginzburg–Landau Hamiltonian, i.e. $n = 1$, even though $n > 1$ introduces the interesting new feature of Goldstone bosons (see Problem 2.4 and Section 13.3.1).

In zero external field ($j = 0$) we have

$$\Gamma^{(1)} = \left.\frac{\delta\Gamma}{\delta\bar{\varphi}}\right|_{j=0} = 0;$$

since $j = 0$ implies $\bar{\varphi} = v$, $\Gamma(\bar{\varphi}(x))$ can be expanded as

$$\Gamma[\bar{\varphi}] = \sum_{N=2}^{\infty} \frac{1}{N!}\int \mathrm{d}^D x_1 \ldots \mathrm{d}^D x_N \Gamma^{(N)}(x_1, \ldots, x_N; v)$$
$$\times(\bar{\varphi}(x_1) - v)\ldots(\bar{\varphi}(x_N) - v), \tag{5.4.1}$$

where the proper vertex $\Gamma^{(N)}(x_1, \ldots, x_N; v)$ is evaluated in the presence of spontaneous symmetry breaking. We recall that a function $f(\bar{\varphi})$ can be Taylor-expanded either near $\bar{\varphi} = v$,

$$f(\bar{\varphi}) = \sum_{N=0}^{\infty} \frac{1}{N!} f^{(n)}(v)(\bar{\varphi} - v)^N,$$

or near $\bar{\varphi} = 0$,

$$f(\bar{\varphi}) = \sum_{N=0}^{\infty} \frac{1}{N!} f^{(N)}(0)\bar{\varphi}^N.$$

Equation (5.4.1) can therefore be transformed into

$$\Gamma[\bar{\varphi}] = \sum_{N=2}^{\infty} \frac{1}{N!} \int \mathrm{d}^D x_1 \ldots \mathrm{d}^D x_N \, \Gamma^{(N)}(x_1, \ldots, x_N)\bar{\varphi}(x_1) \ldots \bar{\varphi}(x_N), \qquad (5.4.2)$$

where

$$\Gamma^{(N)}(x_1, \ldots, x_N) = \Gamma^{(N)}(x_1, \ldots, x_N; v = 0).$$

In the expression (5.4.2) for $\Gamma(\bar{\varphi})$, *the proper vertices are, accordingly, calculated without symmetry breaking* (i.e. for a vanishing mean value of the field).

If $\bar{\varphi}(x)$ is independent of x (i.e. if $\bar{\varphi}(x) = \bar{\varphi}$), then (5.4.2) yields

$$\Gamma(\bar{\varphi}) = \sum_{N=2}^{\infty} \frac{\bar{\varphi}^N}{N!} \int \mathrm{d}^D x_1 \ldots \mathrm{d}^D x_N \, \Gamma^{(N)}(x_1, \ldots, x_N).$$

Next we recall that, according to the definition of the Fourier transform of a correlation function, one extracts a factor $(2\pi)^D \delta^{(D)}(\sum k_i)$, or $L^D \delta_{\sum k_i}$; hence the Gibbs potential per unit volume becomes

$$\boldsymbol{L^{-D}\Gamma(\bar{\varphi}) = \sum_{N=2}^{\infty} \frac{\bar{\varphi}^N}{N!} \Gamma^{(N)}(k_i = 0).} \qquad (5.4.3)$$

The quantity $L^{-D}\Gamma(\bar{\varphi})$ is often called the *effective potential* $U(\bar{\varphi})$. The equation of state (in a uniform external field) reads

$$\mathrm{d}U/\mathrm{d}\bar{\varphi} = j.$$

In zero external field, the mean value v is found by locating the minimum of the effective potential $U(\bar{\varphi})$. We saw this in detail in the framework of the Landau approximation to $\Gamma(\bar{\varphi})$ (see Section 2.2).

At this point it is instructive to relate the exact expression (5.4.3) to the approximate expressions we found previously (Landau approximation: Section 2.2; the first Landau correction: Section 2.4). We shall establish this connection with the help of the loop expansion.

5.4.2 The loop expansion

We replace H by $H/\hbar$, and look for an expansion of the generating functional $Z(j)$ in powers of $\hbar$:

$$Z(j) = \mathcal{N} \exp\left(-\int \frac{1}{\hbar} V\left(\frac{\delta}{\delta j}\right)\right) \int \mathcal{D}\varphi \exp\left(-\frac{1}{2}\int \varphi \frac{1}{\hbar} G^{(0)-1}\varphi + \int j\varphi\right)$$

$$= \mathcal{N}' \exp\left(-\int \frac{1}{\hbar} V\left(\frac{\delta}{\delta j}\right)\right) \exp\left(\frac{1}{2}\hbar \int jG^{(0)}j\right).$$

One finds the same perturbation expansion as before, except that

- every interaction is multiplied by $1/\hbar$;
- every line is multiplied by $\hbar$.

To a given order of perturbation theory, every proper vertex is therefore multiplied by

$$\hbar^{I-V} = \frac{1}{\hbar}\hbar^{I-V+1} = \frac{1}{\hbar}\hbar^{L},$$

where I is the number of internal lines (a proper vertex has no external lines), V is the number of interactions, and L is the number of (closed) loops.

In Chapter 2 we calculated $\Gamma(\bar{\varphi})$ directly in the form of an expansion by powers of $\hbar$; the remarks just made show that, automatically, *this is also an expansion in the number of loops*. In particular,

$L = 0$ corresponds to the Landau approximation;

$L = 1$ corresponds to the first correction calculated in Section 2.4.

For a *monomial* interaction ($\mathcal{V}(\varphi) = c\varphi^n$) there is, *at fixed* N, a one-to-one correspondence between the order of perturbation theory and the number of loops; thus, if $\mathcal{V} = c\varphi^4$, then $L = p + 1 - N/2$, where p is the order of perturbation theory. This remark confirms the reasoning in Section 2.4.2.

Hence we have two methods available for determining the effective potential:

- either, as in Section 2.4, through a direct calculation to each order in $\hbar$;
- or by using equation (5.4.3) to calculate all the $\Gamma^{(N)}$ perturbatively, obtaining $\Gamma^{(N)}$ to order $L = 0$, $L = 1$, etc.

In Problem 5.6 one shows explicitly that the two methods give the same result when $L = 0$ and $L = 1$. Note that the Landau approximation $L = 0$ is often called the *tree approximation* (Fig. 5.28), or the *classical approximation* ($\hbar \to 0$: see the footnote on page 53).

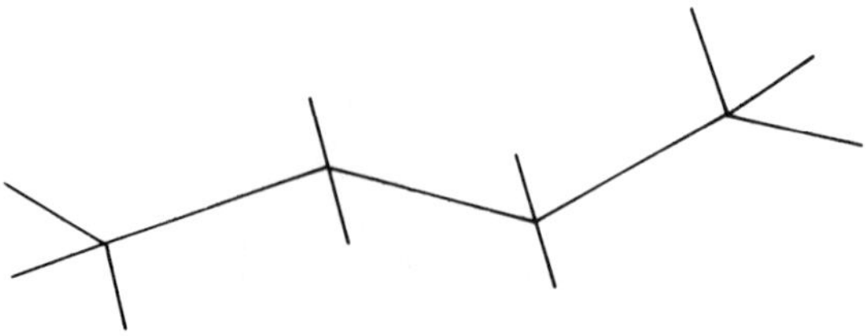

Fig. 5.28 A tree diagram

5.5 The evaluation of Feynman diagrams

Except for the case of just one loop ($L = 1$), the evaluation of Feynman diagrams is complicated, and an exact analytic calculation impossible as a rule. It often happens however that one does not need the full expression for the graph, but only its behaviour in certain limits. There are no general prescriptions for such calculations, and techniques must be evolved case by case. In this book there can evidently be no question of reviewing all the various evaluation methods developed by physicists over the last forty years. We shall confine ourselves to describing the calculations for a single loop in some detail; and to a parametric representation which may or may not be convenient to evaluate in practice, but which does at least allow one to demonstrate many general properties.

5.5.1 An elementary example

Consider first a very simple case, namely the correction of order g to $G^{(2)}(k)$; Fig. 5.29 is called the 'tadpole graph'. (The name stems from φ^3 theory. A typical

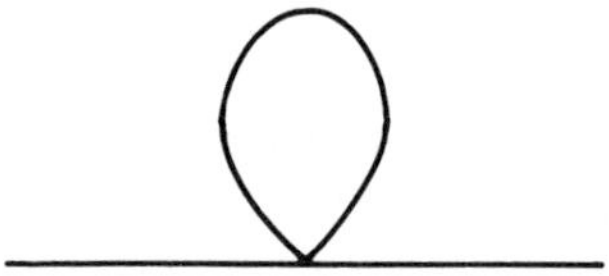

(a)

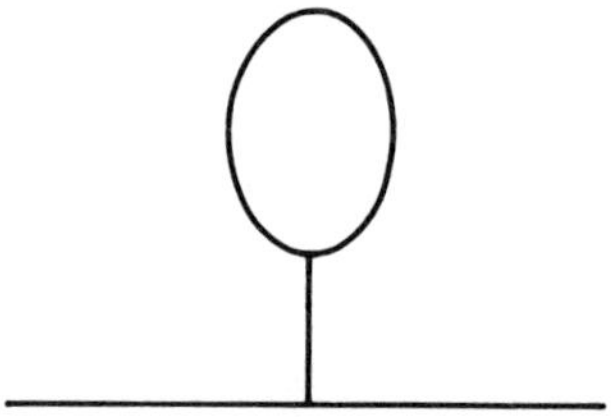

(b)

Fig. 5.29

'tadpole graph' in that theory is drawn in Fig. 5.29(b), which with some exercise of the imagination can be recognized as a tadpole.) Technically speaking, a tadpole is a graph whose expression is independent of the external momenta. The analytic expression for the proper vertex in Fig. 5.29 is

$$\bar{\Sigma} = g\frac{1}{2}\int \frac{d^Dq}{(2\pi)^D}\frac{1}{q^2+m^2} = \frac{1}{2}gK_D\int_0^\Lambda \frac{q^{D-1}\,dq}{q^2+m^2}.$$

The integral converges if the cutoff Λ is finite. When $\Lambda/m \gg 1$, and if $D > 2$, the integral behaves like Λ^{D-2}. In quantum field theory one would wish to take the limit $\Lambda \to \infty$, and the divergences of the Feynman integrals would need interpretation. Those that stem from the region $q \to \infty$ are called *ultraviolet divergences*. When $m^2 = 0$, and if $D \leqslant 2$, the integral diverges on account of the region $q \to 0$; here one is dealing with an *infrared divergence.*

Suppose that $D < 2$, and let Λ tend to infinity; for $\bar{\Sigma}$ one can then obtain the analytic expression

$$\bar{\Sigma} = \frac{1}{4}gK_Dm^{D-2}\int_0^\infty \frac{x^{D/2-1}\,dx}{x+1} = \frac{1}{4}gK_Dm^{D-2}\Gamma\left(\frac{D}{2}\right)\Gamma\left(1-\frac{D}{2}\right). \tag{5.5.1}$$

The pole of the function $\Gamma(1 - D/2)$ at $D = 2$ reflects the ultraviolet divergence for $D \geqslant 2$. The expression can however be used to define $\bar{\Sigma}$ for values of D other than 2, 4, 6, ... by analytic continuation.

In the limit $m^2 = 0$, and for values of $D > 2$, $D \neq 4, 6, \ldots$, $\bar{\Sigma}$ vanishes. This suggests (but does not prove!) that if one defines Feynman integrals for $\Lambda \to \infty$ by analytic continuation (called 'dimensional regularization' in the next chapter), then one can simply ignore tadpoles, even though strictly speaking the integral is not defined for any value of D when $m = 0$! This conclusion is justified for instance in the book by Collins (1983, Chapter 4). Here we shall settle for the heuristic argument just given.

5.5.2 The method using the Feynman identity

For one-loop diagrams one generally exploits the Feynman identity

$$\frac{1}{ab} = \int_0^1 \frac{dx}{[ax + b(1-x)]^2} \tag{5.5.2}$$

(Problem 5.8). Let us apply it to calculate the contribution of Fig. 5.30 to $\Gamma^{(4)}_{ijkl}$, given by

$$\bar{\Gamma}^{(4)}_{ijkl} = \frac{g^2}{18}\left[\delta_{ij}\delta_{kl}(n+4) + 2\delta_{ik}\delta_{jl} + 2\delta_{il}\delta_{jk}\right] I(k);$$

for the factor in square brackets see Problem 5.3. For the integral $I(k)$ one finds

$$I(k) = \int \frac{d^Dq}{(2\pi)^D}\frac{1}{(q^2+m^2)((k-q)^2+m^2)};$$

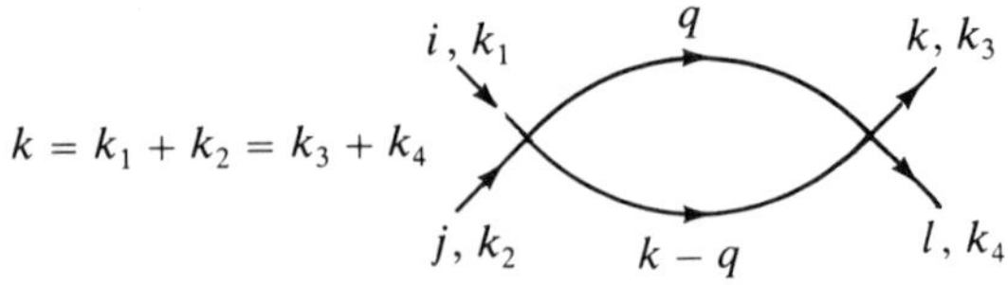

Fig. 5.30

by virtue of the identity (5.5.2) this yields

$$I(k) = \int_0^1 \mathrm{d}x \int \frac{\mathrm{d}^D q}{(2\pi)^D} \frac{1}{[(1-x)(q^2+m^2) + x[(k-q)^2+m^2]]^2}.$$

A change of variables $q \to q' + xk$ turns $I(k)$ into

$$I(k^2) = \int_0^1 \mathrm{d}x \int \frac{\mathrm{d}^D q}{(2\pi)^D} \frac{1}{(q^2 + x(1-x)k^2 + m^2)^2}.$$

This form shows that in fact I is a function only of k^2. The integration over q can be carried out in polar coordinates:

$$I(k^2) = K_D \int_0^1 \mathrm{d}x \int_0^\infty \frac{q^{D-1}\,\mathrm{d}q}{(q^2 + x(1-x)k^2 + m^2)^2}.$$

One sets $u = q^2$ and uses

$$\int_0^\infty \frac{u^\alpha\,\mathrm{d}u}{(u+r)^\beta} = r^{\alpha+1-\beta} \frac{\Gamma(\alpha+1)\Gamma(\beta-\alpha-1)}{\Gamma(\beta)},$$

whence

$$I(k^2) = \frac{\Gamma(2-D/2)}{(4\pi)^{D/2}} \int_0^1 \mathrm{d}x [m^2 + x(1-x)k^2]^{(D/2)-2}. \tag{5.5.3}$$

In the integration over q, we have assumed that $\Lambda \to \infty$. This is possible only if $D < 4$, because otherwise the integral is ultraviolet divergent, i.e. it diverges in the region $q \to \infty$. The divergence at $D = 4$ is reflected by the pole of the function $\Gamma(2 - D/2)$ at $D = 4$. The expression (5.5.3) for $I(k^2)$ can be used to define it for arbitrary values of D (other than $D = 4, 6, 8, \ldots$).

When $m^2 \neq 0$, the result of the integration over x can be expressed in terms of hypergeometric functions, though this is not particularly illuminating. It is a general rule that Feynman diagrams are simpler to evaluate in the massless case; indeed, for $m^2 = 0$ we use

$$\int_0^1 \mathrm{d}u\, u^{\alpha-1}(1-u)^{\beta-1} = B(\alpha,\beta) = \frac{\Gamma(\alpha)\Gamma(\beta)}{\Gamma(\alpha+\beta)}$$

and find

$$I(k^2) = \frac{\Gamma(2 - D/2)}{(4\pi)^{D/2}} \frac{\left[\Gamma\left(\frac{D}{2} - 1\right)\right]^2}{\Gamma(D - 2)} (k^2)^{(D/2)-2}. \tag{5.5.4}$$

The factor $(k^2)^{(D/2)-2}$ enters for purely dimensional reasons; the dimension of I is $D - 4$, and k is the only dimensional quantity at our disposal.

On the other hand, a new divergence appears for $D = 2$, on account of the factor $[\Gamma(D/2 - 1)]^2/\Gamma(D - 2)$. It stems from the region $q \to 0$, and is another example of an infrared divergence.

5.5.3 The general parametric representation

Let us set up a parametric representation of Feynman integrals in general, following Itzykson and Zuber (1980). Consider a proper vertex containing*

- E external lines k_s: $1 \leqslant s \leqslant E$;
- I internal lines p_i: $1 \leqslant i \leqslant I$, with masses m_i;
- L independent integration variables q_r: $1 \leqslant r \leqslant L$.

All the external momenta are taken to be directed inwards (into the diagram); the directions of the internal lines are chosen arbitrarily. Consider a vertex v and an internal line i, and choose a factor ε_{vi} as follows:

$\varepsilon_{vi} = +1$ if the line i is directed away from the vertex v:

$\varepsilon_{vi} = -1$ if the line i is directed towards the vertex v;

$\varepsilon_{vi} = 0$ if the line i does not (directly) connect with the vertex v.

Examples:

(E1) 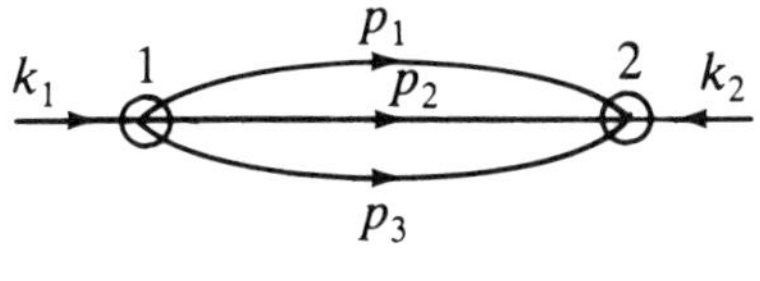

$$\varepsilon_{vi} = \begin{pmatrix} 1 & 1 & 1 \\ -1 & -1 & -1 \end{pmatrix} \quad (i \to,\ v \downarrow) \qquad \text{(E1)}$$

(E2) 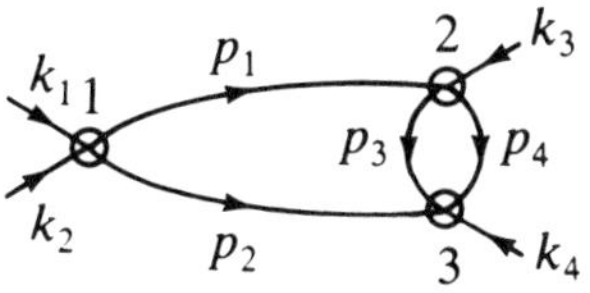

$$\varepsilon_{vi} = \begin{pmatrix} 1 & 1 & 0 & 0 \\ -1 & 0 & 1 & 1 \\ 0 & -1 & -1 & -1 \end{pmatrix} \quad (i \to,\ v \downarrow) \qquad \text{(E2)}$$

The integral reads

$$J' = \int \prod_{i=1}^{I} \left\{ \frac{\mathrm{d}^D p_i}{(2\pi)^D} \frac{1}{p_i^2 + m_i^2} \right\} \prod_{v=1}^{V} (2\pi)^D \delta^{(D)}\left(k_v - \sum_i \varepsilon_{vi} p_i\right), \tag{5.5.5}$$

* To the end of this chapter, the number of external lines is called E rather than N.

where k_v is the sum of all the external momenta running into the vertex v ($k_v = 0$ if no external lines connect directly with the vertex).

We shall use the 'Schwinger representation' of the propagator,

$$\frac{1}{p^2+m^2} = \int_0^\infty \mathrm{d}\alpha\, \mathrm{e}^{-\alpha(p^2+m^2)}. \tag{5.5.6}$$

It has two advantages:

- it reduces the evaluation of (5.5.5) to Gaussian integrations;
- it lends itself readily to the introduction of a cutoff, through

$$\frac{1}{p^2+m^2} \to \int_{\Lambda^{-2}}^\infty \mathrm{d}\alpha\, \mathrm{e}^{-\alpha(p^2+m^2)} = \frac{\mathrm{e}^{-(p^2+m^2)/\Lambda^2}}{p^2+m^2}. \tag{5.5.7}$$

This way of introducing a cutoff ('Schwinger's regularization') is far more elegant than the brute-force restriction of the integration over q to $q \leqslant \Lambda$; once $L > 1$, brute-force cutoffs become impracticable. By virtue of (5.5.6) and of the exponential representation of the delta-function, J' may be written as

$$J' = \int \prod_{i=1}^{I} \left\{ \frac{\mathrm{d}^D p_i}{(2\pi)^D} \mathrm{d}\alpha_i \exp(-\alpha_i(p_i^2 + m_i^2)) \right\} \prod_{v=1}^{V} \left\{ \mathrm{d}^D y_v \exp\left(-\mathrm{i} y_v \cdot \left(k_v - \sum_i \varepsilon_{vi} p_i \right) \right) \right\}.$$

The integral over p is the Fourier transform of a Gaussian,

$$\int \frac{\mathrm{d}^D p_i}{(2\pi)^D} \exp\left(-\alpha_i p_i^2 + \mathrm{i} p_i \cdot \sum_v \varepsilon_{vi} y_v \right) = \frac{1}{(4\pi\alpha_i)^{D/2}} \exp\left(-\frac{1}{4\alpha_i} \left(\sum_v \varepsilon_{vi} y_v \right)^2 \right),$$

$$J' = \int \prod_{i=1}^{I} \left\{ \frac{\mathrm{d}\alpha_i\, \mathrm{e}^{-\alpha_i m_i^2} \exp\left(-\frac{1}{4\alpha_i} \left(\sum_v \varepsilon_{vi} y_v \right)^2 \right)}{(4\pi\alpha_i)^{D/2}} \right\} \prod_{v=1}^{V} \{ \mathrm{d}^D y_v \mathrm{e}^{-\mathrm{i} k_v \cdot y_v} \}.$$

Next, we perform the integrations over the variables y_v, numbering the vertices from 1 to V, and changing variables (with Jacobian equal to 1):

$$y_1 = z_1 + z_V; \quad y_2 = z_2 + z_V; \ldots; \quad y_V = z_V.$$

Notice that $\sum_v \varepsilon_{vi} = 0$, because, with fixed i, only two ε_{vi} are nonzero, one being equal to $+1$ and the other to -1.

$$\sum_{v=1}^{V} \varepsilon_{vi} y_v = \sum_{v=1}^{V-1} \varepsilon_{vi} z_v + \sum_{v=1}^{V} \varepsilon_{vi} z_V = \sum_{v=1}^{V-1} \varepsilon_{vi} z_v.$$

The integral over z_V yields a factor $(2\pi)^D \delta^{(D)}(\sum_1^V k_v)$:

$$J' = (2\pi)^D \delta^{(D)}\left(\sum_{s=1}^{E} k_s \right) J.$$

By convention this factor $(2\pi)^D \delta(\ldots)$ is to be removed from $\Gamma^{(E)}$; then J becomes

$$J = \int \prod_{i=1}^{I} \left\{ \frac{\mathrm{d}\alpha_i \mathrm{e}^{-\alpha_i m_i^2} \exp\left(-\frac{1}{4\alpha_i} \left(\sum_{v=1}^{V-1} \varepsilon_{vi} z_v \right)^2 \right)}{(4\pi\alpha_i)^{D/2}} \right\} \prod_{v=1}^{V-1} \{ \mathrm{d}^D z_v \mathrm{e}^{-\mathrm{i} k_v \cdot z_v} \}.$$

Here we introduce the symmetric $(V-1)\times(V-1)$ matrix $A_{st}(\alpha)$,

$$A_{st}(\alpha) = \sum_i \varepsilon_{si}\frac{1}{\alpha_i}\varepsilon_{ti}.$$

It can be shown that this matrix is nonsingular and positive. Again the integral over the variables z_V is Gaussian:

$$\int\left(\prod_{v=1}^{V-1} \mathrm{d}^D z_v \mathrm{e}^{-\mathrm{i}k_v\cdot z_v}\right)\exp\left(-\frac{1}{4}\sum_{s,t} z_s A_{st}(\alpha) z_t\right)$$
$$= \frac{(4\pi)^{\frac{D}{2}(V-1)}}{[\det A(\alpha)]^{D/2}}\exp\left(-\sum_{s,t=1}^{V-1} k_s A_{st}^{-1}(\alpha) k_t\right).$$

Note that $I - V + 1 = L$ is the number of independent loops. Finally one obtains

$$J = (4\pi)^{-LD/2}\int\prod_{i=1}^{I}\{\mathrm{d}\alpha_i \mathrm{e}^{-\alpha_i m_i^2}\}\left[\frac{1}{P(\alpha)}\right]^{D/2}\mathrm{e}^{-k^{\mathrm{T}}A^{-1}k}$$
$$A_{st} = \sum_i \varepsilon_{si}\frac{1}{\alpha_i}\varepsilon_{ti};\quad P(\alpha) = \alpha_1\ldots\alpha_I \det A(\alpha). \tag{5.5.8}$$

Let us write down explicitly the integrals corresponding to the examples (E1) and (E2):

(E1) $A_{11} = \dfrac{1}{\alpha_1} + \dfrac{1}{\alpha_2} + \dfrac{1}{\alpha_3} = \dfrac{\alpha_1\alpha_2 + \alpha_1\alpha_3 + \alpha_2\alpha_3}{\alpha_1\alpha_2\alpha_3}$,

$$P(\alpha) = \alpha_1\alpha_2 + \alpha_1\alpha_3 + \alpha_2\alpha_3,$$

$$J = (4\pi)^{-D}\int\prod_{i=1}^{3}\{\mathrm{d}\alpha_i\,\mathrm{e}^{-\alpha_i m_i^2}\}\,[P(\alpha)]^{-D/2}\,\mathrm{e}^{-k^2\frac{\alpha_1\alpha_2\alpha_3}{P(\alpha)}};$$

(E2) $A_{11} = \dfrac{1}{\alpha_1} + \dfrac{1}{\alpha_2}$, $\quad A_{12} = A_{21} = \dfrac{-1}{\alpha_1}$, $\quad A_{22} = \dfrac{1}{\alpha_1} + \dfrac{1}{\alpha_3} + \dfrac{1}{\alpha_4}$,

$$P(\alpha) = \alpha_1\alpha_3 + \alpha_1\alpha_4 + \alpha_2\alpha_3 + \alpha_2\alpha_4 + \alpha_3\alpha_4.$$

It is possible to formulate general rules for writing down $P(\alpha)$ and $A^{-1}(\alpha)$ for an arbitrary graph: see Itzykson and Zuber (1980, p. 297).

The final step is to introduce homogeneous variables

$$\lambda = \alpha_1 + \cdots + \alpha_I,\quad \alpha_i = \lambda x_i,\quad 0 \leqslant x_i \leqslant 1$$
$$\int\prod \mathrm{d}\alpha_i = \int\prod \mathrm{d}x_i\frac{\mathrm{d}\lambda}{\lambda^{-I+1}}\delta\left(1 - \sum x_i\right).$$

On the other hand

$$A(\alpha) = \frac{1}{\lambda}A(x),$$

$$P(\alpha) = \alpha_1\ldots\alpha_I \det A(\alpha) = \lambda^{I-(V-1)}(x_1\ldots x_I \det A(x))$$
$$= \lambda^{LD/2}P(x),$$

whence the end-result reads

$$J = (4\pi)^{-LD/2} \int \frac{\prod \mathrm{d}x_i \delta(1 - \sum x_i)}{[P(x_i)]^{D/2}} \int \frac{\mathrm{d}\lambda}{\lambda^{1-I+LD/2}} \times \exp\left(-\lambda\left(\sum_i x_i m_i^2 + k^{\mathrm{T}} A^{-1}(x) k\right)\right). \tag{5.5.9}$$

If $m_i = m$, then $\sum m_i^2 x_i = m^2$; on the other hand, if the integral over λ converges at $\lambda = 0$, then

$$J = \frac{\Gamma(I - LD/2)}{(4\pi)^{LD/2}} \int \frac{\prod \mathrm{d}x_i \delta(1 - \sum x_i)}{[P(x_i)]^{D/2}} (m^2 + k^{\mathrm{T}} A^{-1}(x) k)^{(LD/2)-I}. \tag{5.5.10}$$

For the case $I = 2$, $L = 1$ we evidently recover (5.5.3); it is easy to check that $P(x) = 1$ and $A^{-1}(x) = x(1 - x)$. The exponent $(LD/2) - I$ can be obtained by a dimensional argument: the dimension of J is indeed $LD - 2I$.

The apparent simplicity of equation (5.5.10) should be mistrusted, because generally there are hidden divergences in the integrals over the x_i. In fact for a single loop there are none, but complications do set in as soon as $L > 1$. We proceed to two explicitly evaluated examples.

5.5.4 The calculation of η to order ε^2

The divergences of Feynman integrals near $D = 4$ will play a crucial role in the next two chapters. We start by studying a simple case, namely $\Gamma^{(4)}$, reverting to the form (5.5.3) of $I(k^2)$, and evaluating it near $\varepsilon = 0$, where $\varepsilon \equiv (4 - D)$:

$$I(k^2) = \frac{\Gamma(\varepsilon/2)}{(4\pi)^{D/2}} \int_0^1 \mathrm{d}x [m^2 + x(1 - x)k^2]^{-\varepsilon/2}.$$

Using

$$X^{-\varepsilon} \simeq 1 - \varepsilon \ln X, \quad \Gamma(\varepsilon) \simeq \frac{1}{\varepsilon}$$

one finds that to order $(\varepsilon)^0$

$$(4\pi)^2 I(k^2) = \frac{2}{\varepsilon} - \int_0^1 \mathrm{d}x \ln[m^2 + x(1 - x)k^2] + \text{constant}. \tag{5.5.11}$$

The pole at $\varepsilon = 0$ reflects the ultraviolet divergence at $D = 4$.

Our work on example (E1) will enable us to calculate the critical exponent η to order ε^2. The graph in this example contributes to the self-energy $\Sigma(k^2)$, and this time we shall calculate directly in $D = 4$ dimensions, with a cutoff Λ; then

$$I(k^2) = (4\pi)^{-4} \int \frac{\prod_1^3 \mathrm{d}x_i \delta(1 - \sum x_i)}{[P(x_i)]^2} \int_{c\Lambda^{-2}}^{\infty} \frac{\mathrm{d}\lambda}{\lambda^2} \exp(-\lambda(m^2 + k^2 f(x_i)))$$

$$P(x_i) = x_1 x_2 + x_1 x_3 + x_2 x_3; \quad f(x_i) = (x_1 x_2 x_3)/P(x_i),$$

where c is a constant.

The integral involves two loops, and contains overlapping divergences, reflected by the divergence of the integral over the x_i.

These divergences occur as some of the x_i tend to zero; in actual fact the cutoff prevents any of them from vanishing, but keeping track of the cutoff is complicated. Fortunately, all we need is the derivative

$$\frac{\mathrm{d}I}{\mathrm{d}k^2} = -\frac{1}{(4\pi)^4} \int \frac{\prod \mathrm{d}x_i \delta(1 - \sum x_i) f(x_i)}{[P(x_i)]^2} \int_{c\Lambda^{-2}}^{\infty} \frac{\mathrm{d}\lambda}{\lambda} \exp(-\lambda(m^2 + k^2 f(x_i))),$$

where the integral over the x_i converges. In fact, from

$$P(x_i) = x_1 x_2 + x_1 x_3 + x_2 x_3$$

we see that the dangerous region is the region where (x_1, x_2), or (x_1, x_3), or (x_2, x_3) tend to zero *simultaneously*; for instance, if $x_3 \to 1$, then $P(x_i) \approx (x_1 + x_2)$, and

$$\int \frac{\mathrm{d}x_1\, \mathrm{d}x_2}{(x_1 + x_2)^2}$$

diverges. By contrast,

$$\int \frac{x_1 x_2\, \mathrm{d}x_1\, \mathrm{d}x_2}{(x_1 + x_2)^3}$$

is perfectly convergent. The integral over λ yields a factor ($\ln \Lambda^2$ + nondivergent terms), and one finds

$$\frac{\mathrm{d}I}{\mathrm{d}k^2} \simeq -\frac{\ln \Lambda^2}{(4\pi)^4} \int \frac{\prod \mathrm{d}x_i \delta(1 - \sum x_i) f(x_i)}{[P(x_i)]^2}.$$

To evaluate the integral, we change variables:

$$x_1 = \rho x; \quad x_2 = x(1 - \rho); \quad x_3 = 1 - x.$$

A tedious but straightforward calculation shows that the integral equals 1/2,

$$\frac{\mathrm{d}I}{\mathrm{d}k^2} = -\frac{1}{2(4\pi)^4} \ln \Lambda^4 + \text{nondivergent terms.} \tag{5.5.12}$$

As an application, we calculate the critical exponent η to order ε^2. To order g^2, the graph we have just studied is the only one to give a contribution that depends on k. The other 1-PI graph (Fig. 5.31) also contributes, but independently of k. When $m^2 = 0$, the logarithmic term of $\Sigma\,(k^2)$ is proportional to $\ln(\Lambda^2/k^2)$, because the argument of the logarithm must be dimensionless. On restoring all the factors, the contribution to $\mathrm{d}\Sigma(k^2)/\mathrm{d}k^2$ that we have just calculated reads

$$-\frac{g^2}{12(4\pi)^4} \ln \frac{k^2}{\Lambda^2};$$

for the derivative of the inverse of the correlation function, this yields

$$\frac{\mathrm{d}}{\mathrm{d}k^2}\Gamma(k^2) = 1 - \frac{g^2}{12(4\pi)^4}\ln\frac{k^2}{\Lambda^2} + O(g^2 \times 1, g^3).$$

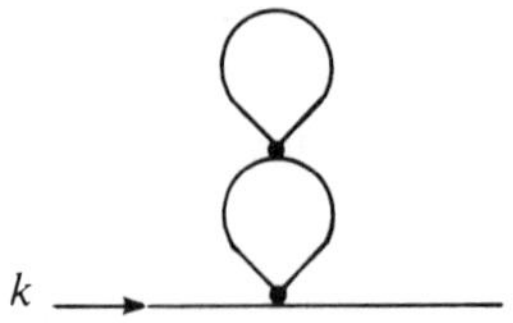

Fig. 5.31

On the other hand, we know that, at $T = T_c$,

$$\Gamma(k^2) = k^{2-\eta} = k^2\left(1 - \frac{\eta}{2}\ln k^2 + \cdots\right), \quad \Gamma'(k^2) \simeq 1 - \frac{\eta}{2}\ln k^2.$$

Further, we know that for g at the fixed point (see (3.5.4)) one must choose

$$g = g^* = \frac{(4\pi)^2\varepsilon}{3}.$$

By comparing the terms in $\ln k^2$, one obtains for η the end-result

$$\eta = \frac{(g^*)^2}{6(4\pi)^4} = \frac{\varepsilon^2}{54}. \tag{5.5.13}$$

The above reasoning, which relies on the so-called 'matching method', is somewhat cavalier. The scrupulous reader will rederive (5.5.13) by going through the details of the method described in Section 3.5.3.

5.6 Power-counting: ultraviolet and infrared divergences

In factors like $\lambda^{-1+I-LD/2}$ in (5.5.9) or $(m^2 + k^{\mathrm{T}}A^{-1}k)^{LD/2-I}$ in (5.5.10), the exponents have a purely dimensional origin: schematically speaking, the integrand is $\int (\mathrm{d}^D q)^L (p^2)^{-I}$, and its dimension is $LD - 2I$. One defines a quantity $\omega(G)$ called the *superficial degree (or index) of divergence of the diagram*, which determines the *degree of the ultraviolet divergence* ($q \to \infty$), namely

$$\omega(G) = LD - 2I.$$

Schematically, the integral over q behaves like $\int^{\Lambda} q^{\omega(G)-1}\mathrm{d}q$: if $\omega(G) > 0$, then the integral diverges like $\Lambda^{\omega(G)}$ (more precisely, like $\Lambda^{\omega(G)}$ $(\ln\Lambda)^p$, where p is an integer); if $\omega(G) = 0$, then the integral diverges like $(\ln\Lambda)^p$. If $\omega(G) < 0$, then the graph is superficially convergent ('superficially' because sub-integrations could still diverge). We shall derive an interesting expression for $\omega(G)$, featuring the number of external lines of the diagram, by appeal first to a topological and then to a dimensional argument.

5.6.1 The topological argument

Consider a completely general type of interaction, admitting terms in φ^3, φ^4, etc., and even derivative couplings $\varphi^2(\nabla\varphi)^2$ etc. In derivative interactions, every ∇ multiplies the vertex by a factor ik; this is evident from Wick's theorem,

$$\overbrace{\varphi(x)\nabla\varphi}(z) \to \nabla_z G_0(x-z) = \text{F.T.}(ikG_0(k)).$$

Given a vertex of type (i) with an interaction containing δ_i derivatives, the integrand of the diagram will have to be multiplied by a factor $(k)^{\delta_i}$. For derivative couplings we therefore have

$$\omega(G) = LD - 2I + \sum_i \delta_i.$$

Start by using $L = I - V + 1$:

$$\omega(G) - D = I(D-2) + \sum_i (\delta_i - D).$$

There are n_i lines running into the vertex (i) if it corresponds to an interaction $(\varphi)^{n_i}$, possibly with derivatives: $n_i = n_i^{(\text{int})} + n_i^{(\text{ext})}$, and

$$I = \frac{1}{2}\sum_i n_i^{(\text{int})}, \quad E = \sum_i n_i^{(\text{ext})}.$$

Define an index

$$\omega_i = n_i\left(\frac{D}{2} - 1\right) + \delta_i, \tag{5.6.1}$$

to be interpreted presently. The expression for $\omega(G)$ becomes

$$\begin{aligned}\omega(G) - D &= \left(\frac{D}{2} - 1\right)\sum_i n_i^{(\text{int})} + \sum_i (\delta_i - D)\\ &= \left(\frac{D}{2} - 1\right)\sum_i n_i - \left(\frac{D}{2} - 1\right)\sum_i n_i^{(\text{ext})} + \sum_i (\delta_i - D)\\ &= \sum_i (\omega_i - D) - E\left(\frac{D}{2} - 1\right),\end{aligned}$$

whence

$$\omega(G) - D = \sum_i (\omega_i - D) - E\left(\frac{D}{2} - 1\right). \tag{5.6.2}$$

Example: A theory with $g\varphi^4$: $\omega = 4(D/2 - 1) = 2D - 4$,

$$E = 2: \quad \omega(G) = V(D-4) + 2,$$

$$E = 4: \quad \omega(G) = V(D-4) + 4 - D,$$

$$E = 6: \quad \omega(G) = V(D-4) + 6 - 2D.$$

Equation (5.6.1) shows quite generally that for $D > 2$ (otherwise the theory is ill defined), $\omega(G)$ *falls as the number of external lines rises.* For a theory with $g\varphi^4$, and for $D = 4$, only the diagrams with $E = 2$ ($\omega(G) = 2$) and $E = 4$ ($\omega(G) = 0$) are superficially divergent.

5.6.2 The dimensional argument

In Section 3.4.2 we saw that the dimension of the field φ is $[\varphi] = D/2 - 1$. A general interaction term may be written

$$g_i \int \mathrm{d}^D x (\nabla)^{\delta_i} (\varphi)^{n_i}.$$

It must be dimensionless, whence

$$[g_i] - D + \delta_i + n_i \left[\frac{D}{2} - 1\right] = 0,$$

or

$$[g_i] = D - \omega_i. \tag{5.6.3}$$

Accordingly, the dimension $[g_i]$ of the coupling constant is related very directly to the index ω_i defined in (5.6.1).

To recover equation (5.6.2), let us first determine the dimension of $\Gamma^{(E)}$ in k-space; $\Gamma^{(E)}(k_i = 0)$ is found by differentiating the (dimensionless) Gibbs potential E times with respect to a uniform magnetization $\bar{\varphi}$ with dimension $(D/2 - 1)$, and multiplying by an inverse volume (dimension $-D$):

$$[\Gamma^{(E)}] = D - E\left(\frac{D}{2} - 1\right). \tag{5.6.4}$$

The quantity

$$\left(\prod_{i=1}^{V} g_i\right) \int k^{\omega(G)-1}\, \mathrm{d}k$$

must have dimension $D - E(D/2 - 1)$; in view of (5.6.3) this yields the desired result (5.6.2).

5.6.3 Infrared divergences (φ^4 theory)

When $m^2 = 0$, infrared divergences can arise in certain cases. By contrast to the ultraviolet, *infrared divergences occur only for special configurations of the external momenta, at least when $D = 4$.* For instance, with $D = 4$, $m^2 = 0$, the

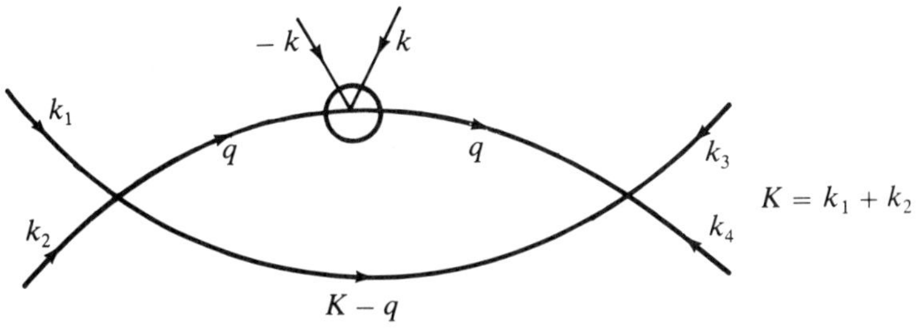

Fig. 5.32

contribution to $\Gamma^{(6)}$ from Fig. 5.32 is given up to a factor by the integral

$$\int \frac{d^4 q}{q^4 (K-q)^4},$$

which is infrared divergent. This divergence stems from the entry, at the circled vertex, of an external momentum with value zero. In general, a configuration of external momenta is called a *non-exceptional configuration* if no partial sum of the k_i vanishes:

$$\sum_{i \in I} k_i \neq 0 \quad \text{for any } I,$$

where I is any arbitrary subset of the indices $(1, \ldots, E)$ of the E external momenta $k_i \ldots$. In a non-exceptional configuration, all the external momenta can be linked by internal lines of nonzero momentum; such lines are called 'hard'. Indeed, if this were not possible, then the diagram could be split into two parts by cutting only internal lines of zero momentum (called 'soft'). However, one of these parts receives momentum $\sum_{i=1}^{E_i} k_i \neq 0$, while the momentum removed by the internal lines vanishes.

Consider for instance the graph of Fig. 5.33, taken from Itzykson and Zuber (1980), showing a possible scheme for the 'hard momentum flow'. Imagine that

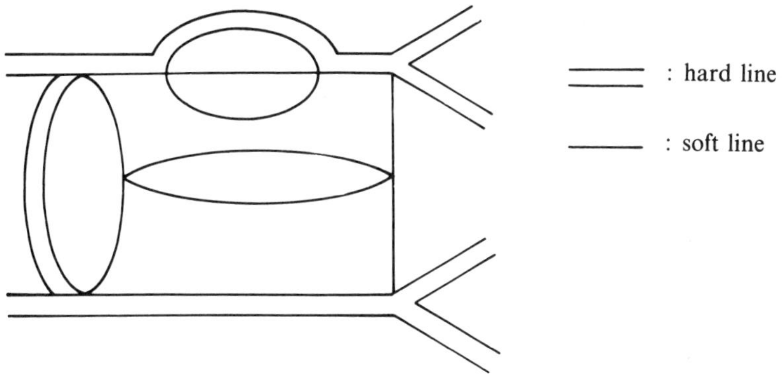

Fig. 5.33

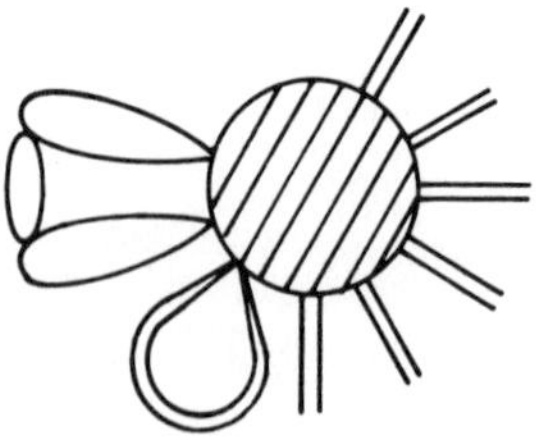

Fig. 5.34

we contract all the hard internal lines into just one vertex (Fig. 5.34); let I, L, and V be the numbers of loops, internal lines and vertices of the *contracted* diagram; and let n be the number of soft internal lines attached to the contracted vertex. We have the relations

$$L = I - (V + 1) + 1 = I - V,$$

$$2I = 4V + n.$$

The contracted diagram has the superficial degree of divergence

$$\omega = DL - 2I = L(D - 4) + 4L - 2I,$$

$$\omega = n + L(D - 4). \tag{5.6.5}$$

If all internal loops have near-zero momenta, then the contracted diagram is homogeneous of order ω. When $D = 4$, this semi-heuristic argument shows that *in a non-exceptional configuration the diagram is infrared-convergent*; indeed we have $n \geqslant 2$ because the diagram is 1-PI.

By contrast, for $D < 4$ one can easily find infrared-divergent diagrams even in non-exceptional configurations. For example, the superficial degree of divergence of Fig. 5.35 is

$$\omega = 4 + 3(D - 4) = 3D - 8.$$

Hence it is infrared divergent even in non-exceptional configurations, provided $D \leqslant 8/3$.

The reasoning just given shows that in the infrared region the φ^4 interaction is the most singular (the φ^3 interaction being excluded by the symmetry $\varphi \to -\varphi$).

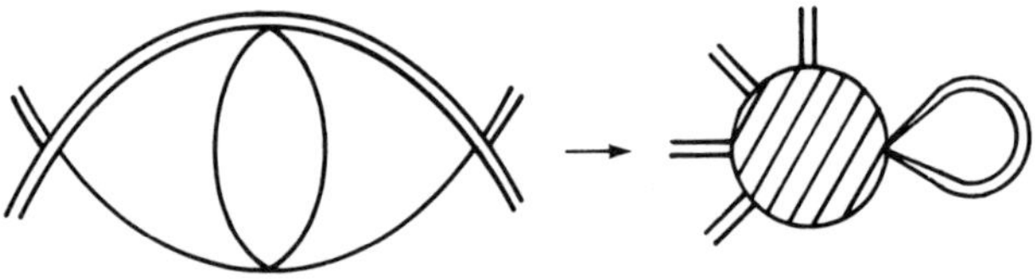

Fig. 5.35

Heuristically at least this justifies the use of the Ginzburg–Landau Hamiltonian (2.1.19) for studying critical behaviour.

Weinberg's theorem (Weinberg 1960; Hahn and Zimmermann 1968) allows one to determine the asymptotic behaviour of a graph when all its external momenta tend to infinity at the same rate: $k_i \to \lambda k_i$, $\lambda \to \infty$. Consider φ^4 theory in 4 dimensions, and assume that the integral in (5.5.10) is ultraviolet convergent; if its $m = 0$ limit exists, as it will if the configuration $[k_i]$ is non-exceptional, then

$$J(\lambda k_i) \underset{\lambda \to \infty}{\sim} \lambda^{\omega(G)}. \tag{5.6.6}$$

If $J(k)$ has to be renormalized on account of ultraviolet divergences (Chapter 6), then

$$J(\lambda k_i) \underset{\lambda \to \infty}{\sim} \lambda^{\omega(G)}(\ln \lambda)^p,$$

where the integer p depends on the graph in question.

Problems

5.1 When $n = 1$, what are the symmetry factors for the graphs in Fig. 5.36?

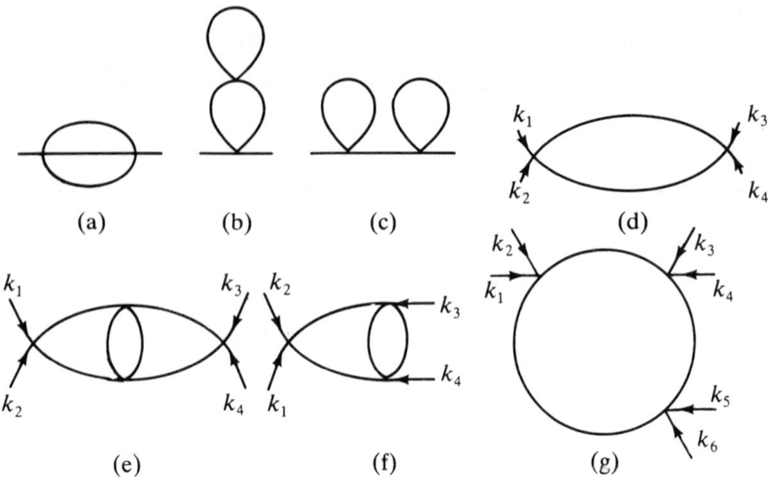

Fig. 5.36

5.2 What are the symmetry factors for the vacuum-fluctuation diagrams of Fig. 5.37 (when $n = 1$)? Check your results by using (5.2.1).

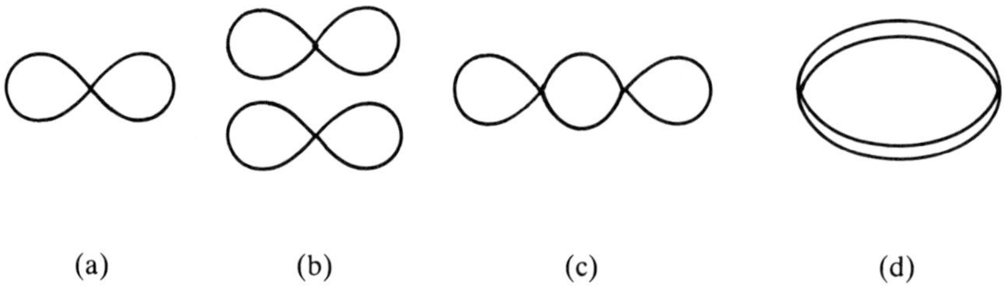

Fig. 5.37

5.3 For arbitrary n, determine the symmetry factor of the graph in Fig. 5.38.

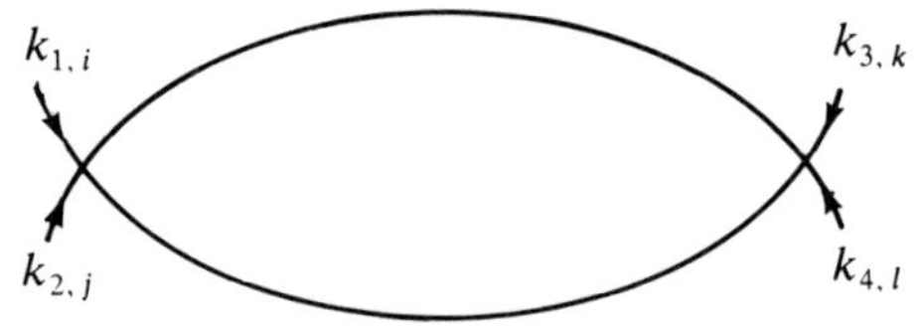

Fig. 5.38

5.4 For arbitrary n, what are the symmetry factors of the graphs in Fig. 5.39?

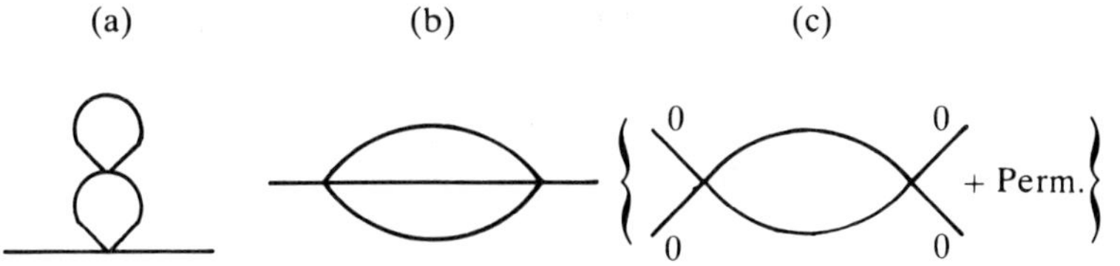

Fig. 5.39

5.5 (a) For the interaction $(g/3!)\varphi^3$, draw

- the vacuum-fluctuation diagrams of orders g^2 and g^4;
- $G^{(1)}$ to orders g and g^3;
- $G^{(2)}$ to orders g^2 and g^4;
- $G^{(3)}$ to orders g and g^3;
- $G^{(4)}$ to orders g^2 and g^4.

(b) To these orders, what are the 1-PI graphs for $G^{(2)}$, $G^{(3)}$, and $G^{(4)}$?

5.6 (a) What is the symmetry factor for the graph of Fig. 5.40, which contributes to $\Gamma^{(2N)}(k_i = 0)$? (Restrict yourself to $n = 1$.)

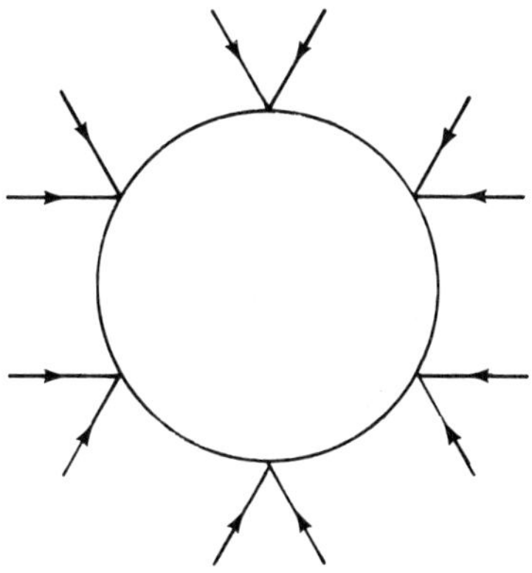

Fig. 5.40

(b) By summing over all graphs with one loop or none, recover the expression (2.4.6) for the effective potential to order $\hbar$.

5.7 When $n = 1$, obtain the expression of type (5.5.8) for the graph of Fig. 5.41, numbering the internal lines as in the figure. Check your result by using the rules given by Itzykson and Zuber (1980, p. 297).

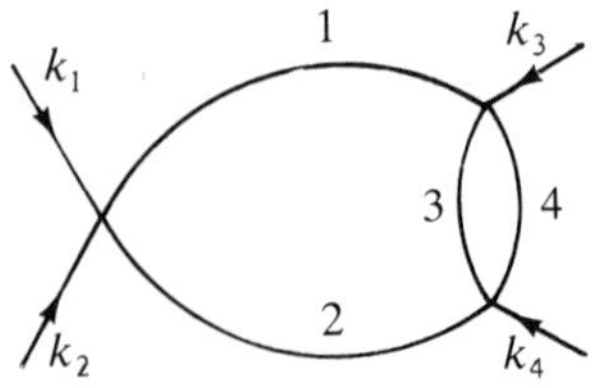

Fig. 5.41

5.8 The Feynman identity

Prove the identity

$$\frac{1}{a^{\alpha_1}a^{\alpha_2}} = \frac{\Gamma(\alpha_1 + \alpha_2)}{\Gamma(\alpha_1)\Gamma(\alpha_2)} \int_0^1 \frac{\mathrm{d}x_1\, \mathrm{d}x_2\, x_1^{\alpha_1 - 1} x_2^{\alpha_2 - 1} \delta(1 - x_1 - x_2)}{(x_1 a_1 + x_2 a_2)^{\alpha_1 + \alpha_2}}.$$

Hint:

$$\int \mathrm{d}x\, x^n \mathrm{e}^{-ax} = \frac{\Gamma(n + 1)}{a^{n+1}}.$$

Generalize to $(a_1^{\alpha_1} \ldots a_n^{\alpha_n})^{-1}$ (see equation (B.1)).

5.9 Evaluate the integral

$$\int \frac{\prod_{i=1}^{3} \mathrm{d}x_i\, \delta(1 - \sum x_i) x_1 x_2 x_3}{(x_1 x_2 + x_1 x_3 + x_2 x_3)^3}.$$

5.10 By direct calculation in $(4 - \varepsilon)$ dimensions, show that

$$\bar{\Sigma}'(k^2) = -\frac{\mathrm{d}}{\mathrm{d}k^2}$$  $$= \frac{g^2}{12(4\pi)^4} \frac{1}{\varepsilon} + O(\varepsilon)^0.$$

5.11 Calculate $\bar{\Sigma}(k^2)$ in x-space (for $D = 4$) (Ma 1976, Chapter 9).

(a) Show that

$$\bar{\Sigma}(k^2) = (\text{constant}) \int \mathrm{d}^4x\, \mathrm{e}^{\mathrm{i}k \cdot x} [G_0(x)]^3.$$

(b) Determine the coefficient of $k^2 \ln(k^2/\Lambda^2)$ by using the expression for $G_0(x)$ with a cutoff Λ (see Problem 2.6):

$$G_0(x) = \frac{1}{(2\pi)^2 \|\boldsymbol{x}\|^2} (1 - J_0(\Lambda \|\boldsymbol{x}\|)).$$

5.12 Successive integration over loops

One method for evaluating Feynman integrals is to integrate over the loops one after the other. This method is well suited to the case $m = 0$ and to dimensional regularization.

(a) Denote by $I(k^2)$ the integral of the graph

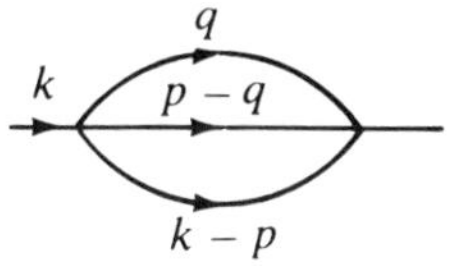

$$= \int \frac{\mathrm{d}^D p}{(2\pi)^D} \frac{1}{(p-k)^2} \int \frac{\mathrm{d}^D q}{(2\pi)^D} \frac{1}{q^2 (p-q)^2}$$

$$= \int \frac{\mathrm{d}^D p}{(2\pi)^D} \frac{1}{(p-k)^2} \Pi(p^2) = I(k^2).$$

The expression for $\Pi(p^2)$ can be inferred from the calculation leading to (5.5.4). Use the identity given in Problem 5.8 to evaluate $I(k^2)$. Do not perform the x-integration.

(b) Show that the calculation simplifies if one requires only the part of $I(k^2)$ that diverges (like $1/\varepsilon$), and recover the result of Problem 5.10.

5.13 Expansion in $1/n$ (Ma 1976, Chapter 9)

We aim to calculate the $1/n$-proportional corrections to the results of Problem 3.7, using a diagrammatic method exploiting the representation based on Figs. 5.9 to 5.11. We shall take the coupling constant u_0 in the Ginzburg–Landau Hamiltonian with $O(n)$ symmetry to be of order $1/n$; in other words we take $u_0 \to u_0/n$.

(a) Show that the graphs dominating $\Sigma(k^2, r_0)$ in the limit $n \to \infty$ are those drawn in Fig. 5.42; thence derive for $\Sigma(0, r_0)$ the relation

$$\Sigma(0, r_0) = \frac{u_0}{6} \int \frac{\mathrm{d}^D p}{(2\pi)^D} \frac{1}{p^2 + r_0 + \Sigma(0, r_0)}.$$

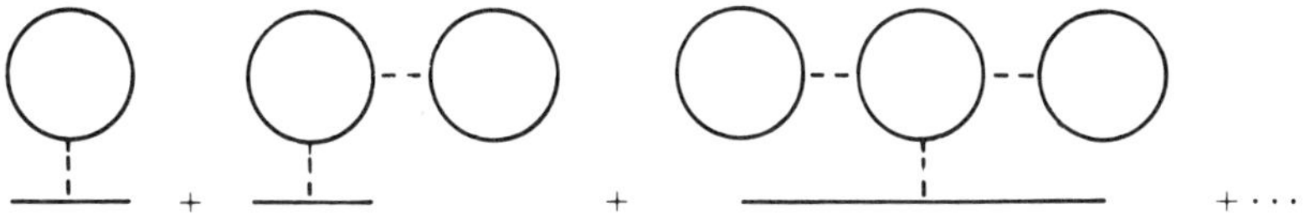

Fig. 5.42

From this expression recover the results of Problem 3.7:

$$\eta = 0, \quad \nu = \frac{1}{D-2} \quad \text{for } 2 < D < 4.$$

(b) The dominant contribution to $\Sigma(k^2)$ that depends on k^2 is of order $1/n$. Show that it is given by the graphs of Fig. 5.43.

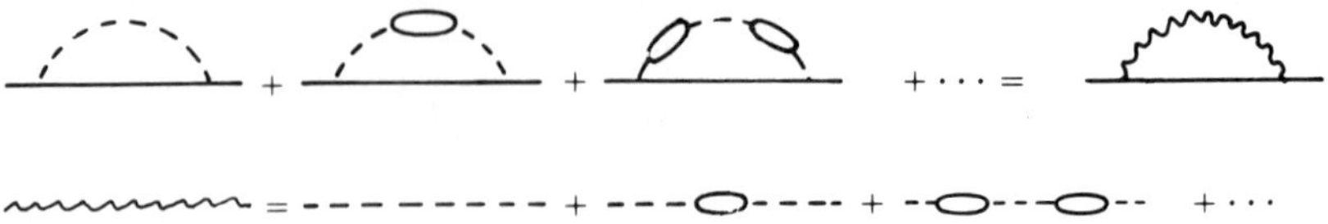

Fig. 5.43

Thence derive for $\Sigma(k^2, 0)$ to order $1/n$ the expression

$$\Sigma(k^2, 0) = \frac{u_0}{3n} \int \frac{\mathrm{d}^D p}{(2\pi)^D} \frac{1}{\left(1 + \frac{u_0}{6} \Pi(p^2)\right)(p + k)^2},$$

where

$$\Pi(p^2) = \int \frac{\mathrm{d}^D q}{(2\pi)^D} \frac{1}{q^2 (q + p)^2}.$$

(c) Determine η by using the 'matching method' of Section 5.5.4,

$$\Gamma^{(2)} \simeq k^2 \left(1 - \frac{\eta}{2} \ln k^2 + \cdots\right),$$

identifying the coefficient of $\ln k^2$ (one must notice that $y_2 \approx D - 4$: see Problem 3.7(g)). Show that the $\ln k^2$ singularity of $\Sigma(k^2)$ stems from the integration region $p \to 0$, and calculate the coefficient of $\ln k^2$ by using the expression (5.5.4) for $\Pi(p^2)$. *Hint*: calculate $\partial^2 \Sigma / (\partial k^2)^2$, and identify the coefficient of $1/k^2$. The result reads

$$\eta = \frac{2(4 - D)\Gamma(D - 2)}{n\Gamma\left(\frac{D}{2} + 1\right)\Gamma\left(2 - \frac{D}{2}\right)\left(\Gamma\left(\frac{D}{2} - 1\right)\right)^2}.$$

5.15 Revert to the reasoning in Section 5.6.3 for the case of a soft external line. Show that in (5.6.5) one must replace n by $(n - 1)$.

Further reading

Wick's theorem and Feynman diagrams are explained by Amit (1984, Chapters 2 and 4) and by Zinn-Justin (1989, Chapter 4). For the generating functional for proper vertices, and for the loop expansion, see Amit (1984, Chapter 5), Itzykson and Zuber (1980, Chapters 6 and 9), Abers and Lee (1973, Section 16) or Zinn-Justin (1989, Chapter 5). The parametric representation of Feynman diagrams is described in detail by Itzykson and Zuber (1980, Chapter 6); they also discuss the divergence problem, as do Amit (1984, Chapter 7) and Zinn-Justin (1989, Chapter 6).

6 Renormalization

We saw in the last chapter that, if the dimensionality of space is high enough, then correlation functions often suffer from 'ultraviolet divergences' as the momenta q circulating round closed loops tend to infinity. Hence one must subject these momenta to a cutoff Λ, at least at intermediate stages. In quantum field theory (recalling that $\Lambda \sim 1/a$, where a is the lattice spacing, while ordinary space is a continuum) we want to allow Λ to tend to infinity, while keeping the physical masses and the external momenta k_i fixed: $m_i/\Lambda \to 0$, $k_i/\Lambda \to 0$. This is possible only if one can eliminate the divergences in one way or another, which is the object of renormalization: we shall see that in renormalizable theories one obtains finite results once *the infinities are absorbed through a redefinition of the mass, the coupling constant, and the normalization of the fields.* In statistical physics it should be unnecessary to trouble with renormalization, because the cutoff is finite. Nevertheless, in the physics of critical phenomena one does study the limit $m/\Lambda \to 0$ $(m = \xi^{-1})$, $k_i/\Lambda \to 0$, and this does naturally lead one back to the problem just described. It must be added that the relevance of the renormalized theory to critical phenomena, for $D < 4$, is somewhat subtle (see Chapter 7).

The advantage of renormalization theory over the approach of Chapter 3 is that it is built on rigorously proved theorems, which fully justify all the manipulations required in order to make the theory finite. However, these theorems are complicated to prove, and we shall settle for stating them without proof; our overriding aim is to demonstrate the mechanism of renormalization, with simple examples to help us.

The results described in this chapter suffice for applications to statistical mechanics. In quantum field theory they need extending in order to deal with the complications due to particle spins. Moreover it is essential to show that renormalization preserves properties like causality and the unitarity of the S-matrix. Finally, field theories with local gauge symmetries present problems of their own; we shall see these in Chapter 12 for quantum electrodynamics, and in Chapter 13 for non-Abelian gauge theories.

Section 6.1 introduces some general ideas, namely the classification of field theories into renormalizable and nonrenormalizable, and regularization. One essential remark must be made before tackling examples at one-loop order in Section 6.2. The Ginzburg–Landau Hamiltonian (5.1.17–18) depends on three parameters: a mass m, a coupling constant g, and a cutoff Λ. But we shall see that under our definitions m and g enter the calculation only as intermediaries, called *the bare mass and the bare coupling constant, which we shall write as* m_0 *and* g_0*.

The first version of renormalization consists of calculating the correlation functions as functions of m_0, g_0, and Λ, and then eliminating the latter in favour of physical (renormalized) parameters m and g, which will be defined through the values of the correlation functions at certain special points. Simple examples illustrating this procedure are given in Sections 6.2 and 6.3. A second and alternative procedure avoids introducing bare parameters, by adding to the Hamiltonian certain counterterms expressed in terms of the physical parameters m and g, and dependent on the cutoff. The two approaches are equally useful, and it is essential to be able to switch between them; the requisite connection is explained in Section 6.4.

Some problems lead one to introduce correlation functions that depend on the products of fields (or of their derivatives) at a common point x, like $\varphi^2(x)$, $\varphi^2(x)(\nabla\varphi(x))^2$, etc. Such products are called *composite operators*, and their renormalization too has to be studied. The simplest case, that of the composite operator $\varphi^2(x)$, is considered in Section 6.5, which shows in particular that this operator requires a renormalization independent of that of the field $\varphi(x)$. Finally, the 'minimal subtraction scheme', which is very useful in calculations, is illustrated in Section 6.6 to two-loop order.

6.1 Introduction

6.1.1 Classification of the theories

We saw in the last Chapter (equation (5.6.2)) that, for a monomial interaction, the superficial degree of divergence of a graph G contributing to a proper vertex $\Gamma^{(E)}$ with E external lines is

$$\omega(G) - D = V(\omega - D) - E\left(\frac{D}{2} - 1\right), \tag{6.1.1}$$

where $D - \omega = [g]$ is the dimension of the coupling constant. At fixed E we notice that

(i) If $\omega > D$, then the degree of divergence grows with the order V of perturbation theory.

(ii) If $\omega = D$, then the degree of divergence is independent of the order of perturbation theory.

(iii) If $\omega < D$, then the degree of divergence decreases with increasing order of perturbation theory, and there is only a *finite* number of divergent graphs.

* In statistical mechanics, m_0 and g_0 are parameters of the Ginzburg–Landau Hamiltonian; in quantum field theory they have no physical significance.

Case (i) corresponds to *nonrenormalizable* theories. One can make no sense of these in perturbation theory, because one would need to fix infinitely many parameters in order to make such a theory finite. Recall that in this case the coupling constant has dimension less than zero; powers of g must be compensated by powers of Λ. In certain very special cases (massive electrodynamics: see Chapter 12; or spontaneously broken gauge theory quantized in a unitary gauge) it can happen that the theory contains nonrenormalizable interactions, but the divergences cancel for all physical quantities, rather than for off-shell Green's functions; then in practice the theory is equivalent to a renormalizable one. But as a general rule one can ascribe no significance in perturbation theory to nonrenormalizable theories, though this does not mean that they are of no interest at all!* General Relativity is one good example of a nonrenormalizable theory. We can handle it on the classical but not on the quantum level. Perhaps the solution will be supplied by a nonperturbative method; perhaps General Relativity is the low-energy limit of a more complex theory;. . . or perhaps there exists some completely unexpected solution.

Case (iii) is that of super-renormalizable theories. Though appealing at first sight (only a finite number of graphs are divergent), it seems that such theories are pathological in 4 dimensions, and so far they have found no application in physics.

Case (ii), that of *renormalizable* theories, is the most interesting. Note that $\omega = D$ corresponds to $[g] = 0$: in a renormalizable theory the coupling constant is dimensionless. In such a theory the divergences can be absorbed by fixing a *finite* number of parameters, and by calculating the correlation functions in terms of these.

6.1.2 The divergent diagrams in a renormalizable theory

To be definite, consider a φ^4 interaction, where $\omega = 2D - 4$:

$$\omega - D = D - 4; \quad [g_0] = 4 - D.$$

This entails that one has for

(i) $D > 4$: a non-renormalizable theory;

(ii) $D = 4$: a renormalizable theory;

(iii) $D < 4$: a super-renormalizable theory.

When $D = 4$, the dimension of g_0 is zero (as we saw already), and $\omega(G)$ is given by

$$\omega(G) = D - E\left(\frac{D}{2} - 1\right) = 4 - E.$$

* In some cases, it seems possible to give a meaning to nonrenormalizable theories considered as limits of theories with cutoffs, thanks to nonperturbative methods: see Kupiainen (1986).

For the graphs G contributing to $\Gamma^{(E)}$ this yields

$$E = 2(\Gamma^{(2)}): \quad \omega(G) = 2$$

$$E = 4(\Gamma^{(4)}): \quad \omega(G) = 0$$

$$E \geqslant 6(\Gamma^{(E)}): \quad \omega(G) < 0.$$

Only the two-point and the four-point correlation functions are divergent. For $E \geqslant 6$, the correlation functions are superficially convergent. Nevertheless one can encounter divergences due to sub-integrations (Fig. 6.1).

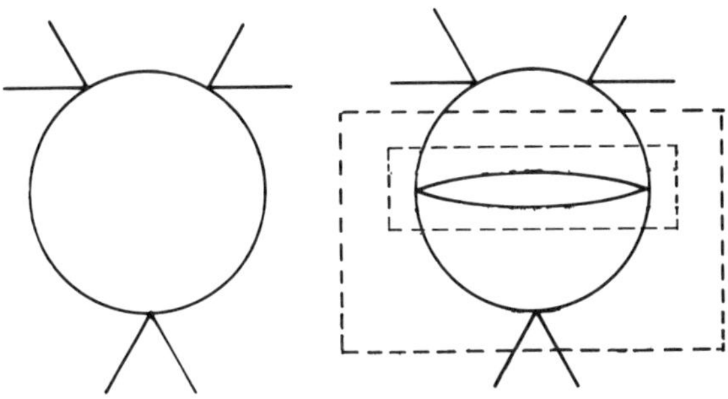

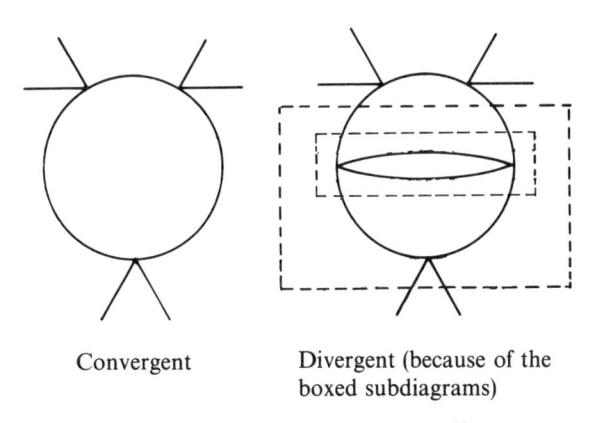

Fig. 6.1 Two contributions to $\Gamma^{(6)}$

Let γ be a sub-diagram of G, namely a set of vertices belonging to G plus lines joining them. When $\omega(G) < 0$, one can prove the following theorem:

First convergence theorem: If all connected 1-PI subdiagrams γ of a diagram G (including G itself) are such that $\omega(\gamma) < 0$, then the Feynman integral of G is absolutely convergent.

The theorem is proved by using the parametric representation (5.5.8), and separating the integration region in α into subregions $0 \leqslant \alpha_{\pi_1} \leqslant \alpha_{\pi_2} \leqslant \ldots \leqslant \alpha_{\pi_I}$, where $\{\pi_1, \ldots, \pi_I\}$ is a permutation of $\{1, \ldots, I\}$ (see Itzykson and Zuber 1980, Chapter 8).

6.1.3 Regularization

To manipulate integrals that a priori are divergent, it proves useful to regularize them, or in other words to make them finite at intermediate stages of the calculation. Regularization should not be confused with renormalization: the renormalized theory must be independent of the regularization procedure, which disappears completely from the end-result. The main regularization methods are the following:

(a) Brute-force cutoff: integrals over q are restricted to $\|\boldsymbol{q}\| < \Lambda$. This kind of cutoff is useful only in heuristic arguments, and is impracticable beyond one-loop order.

(b) Schwinger regularization:

$$\frac{1}{q^2 + m^2} \to \int_{\Lambda^{-2}}^{\infty} \mathrm{d}\alpha\, \mathrm{e}^{-\alpha(q^2 + m^2)}.$$

(c) Pauli-Villars regularization:

$$\frac{1}{q^2 + m^2} \to \frac{1}{q^2 + m^2} - \frac{1}{q^2 + \Lambda^2}.$$

(d) Dimensional regularization: one evaluates the integrals for some small enough value of D. The divergences enter as poles in $(2 - D)$, $(4 - D)$, . . . , and away from these poles the integrals can be defined through analytic continuation in D.

(e) Lattice regularization: one 'puts the theory on a lattice' by discretizing space. The field variable $\varphi(x)$ is replaced by a lattice variable φ_i:

$$\varphi_i \simeq \frac{1}{a^D} \int \varphi(x) \mathrm{d}^D x,$$

where the integral runs over a volume a^D centred on site i. The cutoff is $\Lambda \sim \pi/a$.

The regularization methods popular at the time of writing were (d) and (e). But we shall start off with regularization by cutoff, because it seems more intuitive; moreover, for certain arguments in statistical mechanics it is essential to retain the cutoff (or the lattice).

Some regularization procedures can prove incompatible with symmetries that one wishes to preserve; for example, lattice regularization breaks rotation and translation invariance. Though this can be a serious nuisance, it is not fatal; at the end of the calculation one must show that the renormalized theory does have the requisite symmetries, even if they were absent at some intermediate stage. It can also happen that renormalization is incapable of preserving a symmetry of the classical theory, i.e. of the tree approximation; in such cases one has to do with an *anomaly*.

6.2 The renormalization of mass and coupling constant

Consider φ^4 theory in $D = 4$ dimensions, with an order parameter of dimension $n = 1$. The Hamiltonian depends on the bare mass m_0 and on the bare coupling constant g_0, as well as on a cutoff Λ. Here we shall study the proper vertices $\Gamma^{(2)}$

and $\Gamma^{(4)}$ at one-loop order, relegating two-loop order to Section 6.3. It proves useful to define the following integrals, which enable one to write down the one-loop and two-loop contributions to $\Gamma^{(2)}$, as well as the one-loop contribution to $\Gamma^{(4)}$:

$$I_\Lambda(m_0^2) = \int_\Lambda \frac{d^4 q}{(2\pi)^4} \frac{1}{q^2 + m_0^2}: \tag{6.2.1}$$

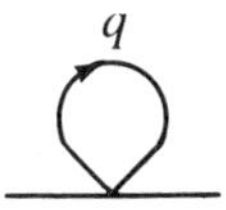

$$J_\Lambda(k_i, m_0^2) = \int_\Lambda \frac{d^4 q}{(2\pi)^4} \frac{1}{(q^2 + m_0^2)((k-q)^2 + m_0^2)}: \tag{6.2.2}$$

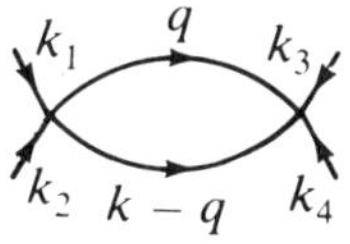

$$K_\Lambda(k^2, m_0^2) = \int_\Lambda \frac{d^4 q\, d^4 q'}{(2\pi)^4 (2\pi)^4} \frac{1}{(q^2 + m_0^2)(q'^2 + m_0^2)[(k - q - q')^2 + m_0^2]}:$$

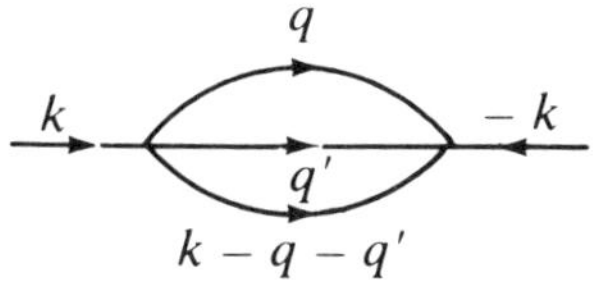

(6.2.3)

These integrals are regularized by a cutoff Λ whose precise form is not needed for the moment. Power-counting shows that $I_\Lambda \sim \Lambda^2$, $J_\Lambda \sim \ln \Lambda$, $K_\Lambda \sim \Lambda^2$; K_Λ contains terms in $(\ln \Lambda)^2$, $\ln \Lambda$, plus (as we saw in Chapter 5) terms in $k^2 \ln \Lambda$. Note that the integrals are defined without the symmetry factors of the corresponding graphs.

6.2.1 $\Gamma^{(2)}$ in one-loop order: mass renormalization

At one-loop order $\Gamma^{(2)}$ is given by

$$\Gamma^{(2)}(k^2) = k^2 + m_0^2 + \frac{1}{2} g_0 I_\Lambda(m_0^2). \tag{6.2.4}$$

We want it to be finite as $\Lambda \to \infty$; in particular, we want $\Gamma^{(2)}(k^2 = 0)$ to be finite,

$$\Gamma^{(2)}(k^2 = 0) = m^2, \tag{6.2.5}$$

where m is a finite parameter with the dimensions of mass, called the renormalized mass:

$$m^2 = m_0^2 + \frac{1}{2} g_0 I_\Lambda(m_0^2). \tag{6.2.6}$$

Next, we reexpress m_0^2 in (6.2.4) as a function of m^2:

$$\Gamma^{(2)}(k^2) = k^2 + m^2; \tag{6.2.7}$$

$\Gamma^{(2)}$ is finite for all k^2. The divergences have been absorbed into a redefinition of the mass ($m_0^2 \to m^2$). The bare mass m_0 is a function of g_0, Λ, and m (see equation (6.2.6)).

6.2.2 $\Gamma^{(4)}$ to one-loop order: coupling-constant renormalization

The results of Chapter 5 yield $\Gamma^{(4)}$ to one-loop order,

$$\Gamma^{(4)}(k_i) = g_0 - \frac{1}{2} g_0^2 [J_\Lambda(k_i, m_0^2) + \text{perm.}]. \tag{6.2.8}$$

Once again we require $\Gamma^{(4)}$ to be finite at some chosen point, e.g. at $k_i = 0$; this defines a second finite parameter g, called the renormalized coupling constant:

$$\Gamma^{(4)}(k_i = 0) = g. \tag{6.2.9}$$

Equation (6.2.9) yields g as a function of g_0,

$$g = g_0 - \frac{3}{2} g_0^2 J_\Lambda(0, m_0^2);$$

conversely

$$g_0 = g + \frac{3}{2} g^2 J_\Lambda(0, m^2). \tag{6.2.10}$$

In writing (6.2.10) we have ignored all terms of order g^3; because we have calculated only the one-loop diagrams, we do not in fact know the terms of order g_0^3. Hence it is perfectly consistent to replace g_0^2 by g^2 and $J_\Lambda(0, m_0^2)$ by $J_\Lambda(0, m^2)$: the error involved is of order g^3. It is very important to realize quite generally that one may legitimately carry out all manipulations to a given order in g_0, even if the higher order terms ignored in the process have coefficients that diverge as $\Lambda \to \infty$. In the light of these remarks, we substitute into (6.2.8) the value (6.2.10) of g_0, and find

$$\Gamma^{(4)}(k_i) = g - \frac{1}{2} g^2 [(J_\Lambda(k_i, m^2) - J_\Lambda(0, m^2)) + \text{perm.}]. \tag{6.2.11}$$

We can see at once that the combination

$$J_\Lambda(k_i, m^2) - J_\Lambda(0, m^2)$$

remains finite as $\Lambda \to \infty$ (Problem 6.1). The expression (6.2.11) for $\Gamma^{(4)}(k_i)$ has a finite limit for all k_i as $\Lambda \to \infty$, and has been written in terms of the *finite* parameters g and m^2.

To sum up at one-loop order, the choice of two finite parameters m^2 and g has enabled us to make finite the correlation functions $\Gamma^{(2)}$ and $\Gamma^{(4)}$; the divergences have been absorbed into a redefinition of the mass ($m_0^2 \to m^2$) and of the coupling constant ($g_0 \to g$).

From the one-loop calculation we can envisage a strategy for the general case. To a given order of perturbation theory, the conditions (6.2.5) and (6.2.9) enable one to express m_0^2 and g_0 as functions of m^2 and g. Substituting these expressions into the proper vertices, one can in the limit $\Lambda \to \infty$ hope for finite formulae featuring m^2 and g. But another supplementary operation is needed first, namely field renormalization, which we proceed to illustrate for the case of the proper vertex $\Gamma^{(2)}$ in two-loop order.

6.3 Field renormalization; counterterms

6.3.1 $\Gamma^{(2)}$ in two-loop order: field renormalization

The diagrammatic expansion of $\Gamma^{(2)}$ to two-loop order is given in Fig. 6.2. In analytic form this reads

$$\Gamma^{(2)}(k^2) = k^2 + m_0^2 + \frac{1}{2} g_0 I_\Lambda(m_0^2)$$
$$- \frac{1}{4} g_0^2 I_\Lambda(m_0^2) J_\Lambda(0, m_0^2) - \frac{1}{6} g_0^2 K_\Lambda(k^2, m_0^2). \tag{6.3.1}$$

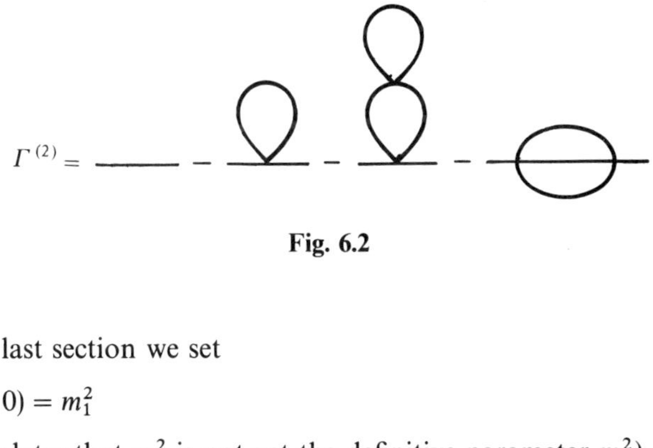

Fig. 6.2

As in the last section we set

$$\Gamma^{(2)}(k^2 = 0) = m_1^2 \tag{6.3.2}$$

(we shall see later that m_1^2 is not yet the definitive parameter m^2). In the terms proportional to g_0^2, one may replace m_0^2 by m_1^2; but to treat the term explicitly proportional to g_0 consistently to order g_0^2, one must take note of (6.2.6):

$$I_\Lambda(m_0^2) = \int \frac{d^4 q}{(2\pi)^4} \frac{1}{q^2 + m_0^2} = \int \frac{d^4 q}{(2\pi)^4} \frac{1}{q^2 + m_1^2 - \frac{1}{2} g_0 I_\Lambda(m_1^2)}$$

$$= I_\Lambda(m_1^2) + \frac{1}{2} g_0 I_\Lambda(m_1^2) J_\Lambda(0, m_1^2) + O(g_0^2).$$

Substituting into (6.3.1) and setting $k^2 = 0$, we find m_0^2 as a function of m_1^2:

$$m_0^2 = m_1^2 - \frac{1}{2} g_0 I_\Lambda(m_1^2) + \frac{1}{6} g_0^2 K_\Lambda(0, m_1^2). \tag{6.3.3}$$

On substitution into (6.3.1) this yields

$$\Gamma^{(2)}(k^2) = k^2 + m_1^2 - \frac{1}{6} g^2 [K_\Lambda(k^2, m_1^2) - K_\Lambda(0, m_1^2)], \tag{6.3.4}$$

once we notice that g_0 may now be replaced by g, since the resultant error is only of order g^3. Unfortunately, (6.3.4) still diverges as $\Lambda \to \infty$. In fact we have already seen that $K_\Lambda(k^2, m^2)$ contains a term proportional to $k^2 \ln \Lambda$, which is not eliminated on subtracting $K_\Lambda(0, m_1^2)$. The one operation that is still needed is a change in the normalization of the field, also called field renormalization.

This renormalization leads us to distinguish between the bare correlation functions $\Gamma^{(E)}$ and the renormalized (and finite) correlation functions $\Gamma_{\mathrm{R}}^{(E)}$; they can be related through the *renormalization constant* Z_3 (the notation Z_3 is conventional). For instance, as regards $\Gamma_{\mathrm{R}}^{(2)}$ one writes

$$\Gamma_{\mathrm{R}}^{(2)}(k^2, m^2, g) = Z_3 \Gamma^{(2)}(k^2, m_0^2, g_0, \Lambda). \tag{6.3.5}$$

We stress that $\Gamma_{\mathrm{R}}^{(2)}$ depends on the renormalized (and finite) parameters m^2 and g, while $\Gamma^{(2)}$ depends on the bare parameters m_0^2 and g_0, and also on Λ; and that on dimensional grounds Z_3 can be a function only of g and of Λ/m:

$$Z_3 = Z_3(g, \Lambda/m).$$

For $\Gamma_{\mathrm{R}}^{(2)}$ to be finite, we need one further condition

$$\left.\frac{\mathrm{d}}{\mathrm{d}k^2} \Gamma_{\mathrm{R}}^{(2)}(k^2)\right|_{k^2=0} (= \Gamma_{\mathrm{R}}^{\prime(2)}(k^2 = 0)) = 1. \tag{6.3.6}$$

The choice of the point $k^2 = 0$ in (6.3.6) is completely arbitrary, but convenient. Equation (6.3.5) entails

$$Z_3^{-1} = \Gamma^{\prime(2)}(k^2 = 0) = 1 - \frac{1}{6} g^2 K_\Lambda'(0, m_1^2); \tag{6.3.7}$$

substitution into (6.3.4) yields the renormalized correlation function

$$\Gamma_{\mathrm{R}}^{(2)}(k^2) = k^2 + Z_3 m_1^2 - \frac{1}{6} g^2 [K_\Lambda(k^2, m_1^2) - K_\Lambda(0, m_1^2) - k^2 K_\Lambda'(0, m_1^2)]. \tag{6.3.8}$$

The terms between square brackets are finite (Problem 6.1); for $\Gamma_{\mathrm{R}}^{(2)}$ to be finite we need merely fix the parameter m_1^2 by

$$m^2 = Z_3 m_1^2,$$

and the end-result reads

$$\Gamma_{\mathrm{R}}^{(2)}(k^2) = k^2 + m^2 - \frac{1}{6}g^2[K_\Lambda(k^2, m^2) - K_\Lambda(0, m^2) - k^2 K'_\Lambda(0, m^2)], \quad (6.3.9)$$

the replacement of m_1^2 by m^2 in the square brackets being warranted to this order of perturbation theory. To summarize, the divergences in $\Gamma^{(2)}$ have been absorbed by imposing the following *normalization conditions* on the renormalized correlation functions:

(1) $\Gamma_{\mathrm{R}}^{(2)}(k^2 = 0) = m^2$, (6.3.10a)

(2) $\left.\dfrac{\mathrm{d}}{\mathrm{d}k^2}\Gamma_{\mathrm{R}}^{(2)}(k^2)\right|_{k^2=0} = 1$, (6.3.10b)

(3) $\Gamma_{\mathrm{R}}^{(4)}(k_i = 0) = g$. (6.3.10c)

Several remarks are in order at this point.

(a) Equation (6.3.5) determines the relation between $\Gamma_{\mathrm{R}}^{(2)}$ and $\Gamma^{(2)}$; $\Gamma^{(2)}$, a bare correlation function, is written in terms of the bare parameters m_0^2, g_0, and of the cutoff Λ. These parameters must be reexpressed as functions of the physical (renormalized) parameters m^2 and g. Provided this operation is supplemented by a change in the normalization of the field (by a factor Z_3), $\Gamma_{\mathrm{R}}^{(2)}$ expressed as a function of m^2 and g is finite in the limit $\Lambda \to \infty$. The divergences have been absorbed by a redefinition of the mass and of the coupling constant, and by a change in the normalization of the field. Note that $\Gamma_{\mathrm{R}}^{(2)}$ depends on only *one* dimensional parameter, namely on m, while $\Gamma^{(2)}$ depends on m_0 and on Λ (see point (c) below).

(b) The parameters m^2 and g need not be *directly* accessible to experiment. On the other hand, all renormalized correlation functions are expressed as functions of the two parameters m^2 and g, and of these two alone. In this sense the theory is predictive: two independent experiments suffice to determine m^2 and g.

(c) As regards critical phenomena, the parameter m is related to the correlation length: $m \approx \xi^{-1}$. Indeed, for $k \to 0$ we find

$$\Gamma_{\mathrm{R}}^{(2)} \approx k^2 + m^2,$$

and $G_{\mathrm{R}}^{(2)}$ has a pole at $k = \pm im$. On calculating the Fourier transform, one finds that at large distances it behaves like

$$\mathrm{e}^{-mr} = \mathrm{e}^{-r/\xi}$$

(see the calculation of the correlation function in the Landau approximation, Section 2.3). The critical region corresponds to $m \ll \Lambda$ ($\xi \gg a$); for generic values of the parameters m_0, g_0, and Λ, the parameter m is expected to be of order Λ. To be in the critical region one therefore needs an appropriate relation between m_0, g_0, and Λ. In other words, *the construction of a renormalized theory presupposes the existence of a critical point.*

(d) The condition (6.3.10c) reads $\Gamma_R^{(4)}(0) = g$, while the preceding section had $\Gamma^{(4)}(0) = g$. There is no contradiction however, because $\Gamma_R^{(4)} = Z_3^2 \Gamma^{(4)}$ (as we shall show later), and because to this order of perturbation theory one can take $Z_3 = 1$, given that $Z_3 = 1 + O(g^2)$. This is peculiar to φ^4 theory; for instance, φ^3 theory in six dimensions does need the factor Z_3 in order to define g (Problem 6.2).

(e) It proves useful to define a renormalization constant Z_1 by

$$Z_1^{-1} g_0 = \Gamma^{(4)}(0, g_0, m_0^2, \Lambda). \tag{6.3.11}$$

Then g_0 and g are related through

$$g = Z_3^2 Z_1^{-1} g_0. \tag{6.3.12}$$

(f) The normalization conditions (6.3.10) are largely arbitrary. The only condition is that they are satisfied at tree level. We shall comment further on this point at the end of Section 6.4.

6.3.2 Counterterms

Since, ultimately, we aim to calculate the renormalized correlation functions in terms of the parameters m^2 and g, it makes sense to try and avoid introducing the bare quantities m_0^2 and g_0 in the first place. Renormalization then follows the strategy of adding 'counterterms', i.e. adding a supplementary Hamiltonian δH; this will be treated as an interaction Hamiltonian, whose coefficients are to be determined by the norming conditions (6.3.10) order by order in perturbation theory, or rather in the number of loops. Since in the limit $\Lambda \to \infty$ these norming conditions determine the correlation functions uniquely, this is equivalent to our previous procedure, but has the advantage that all calculations are carried out in terms of the renormalized parameters m^2 and g. Accordingly we write the Hamiltonian density as

$$\mathscr{H}_0(x) = \mathscr{H}_\Lambda(x) + \delta\mathscr{H}_\Lambda(x), \tag{6.3.13a}$$

$$\mathscr{H}_\Lambda(x) = \frac{1}{2}(\nabla\varphi)^2 + \frac{1}{2}m^2\varphi^2 + \frac{1}{4!}g\varphi^4, \tag{6.3.13b}$$

$$\delta\mathscr{H}_\Lambda(x) = \frac{1}{2}(Z_3 - 1)(\nabla\varphi)^2 + \frac{1}{2}(Z_3 m_0^2 - m^2)\varphi^2 + \frac{1}{4!}g(Z_1 - 1)\varphi^4, \tag{6.3.13c}$$

where the index Λ specifies that one is calculating with a cutoff Λ, and that δm^2, Z_1 and Z_3 depend on Λ.

Notice that the counterterms have the same form as those in the initial Hamiltonian. This is sometimes used as a criterion for renormalizability. It can happen however (as in scalar electrodynamics: see Problem 12.8) that one needs to introduce counterterms different in form from those of the initial Hamiltonian. In practice the theory is renormalizable as long as the number of counterterms is finite.

Since $\delta\mathscr{H}_\Lambda$ is treated perturbatively, the counterterms can be represented graphically:

 $: -(Z_3 m_0^2 - m^2) = -\delta m^2 = -[\delta m^{2(1)} + \cdots + \delta m^{2(l)} + \cdots],$

 $: -(Z_3 - 1)q^2 = -q^2[Z_3^{(2)} + \cdots + Z_3^{(l)} + \cdots],$

 $: -g(Z_1 - 1) = -g[Z_1^{(1)} + \ldots + Z_1^{(l)} + \cdots],$

where $\delta m^{2(l)}$, $Z_1^{(l)}$, and $Z_3^{(l)}$ are calculated to l-loop order.

The counterterms are calculated recursively: once they have been determined to l-loop order, they are reinjected into the $(l+1)$-loop calculation. By applying the norming conditions (6.3.10) one can then determine the counterterms to order $(l+1)$, and so on. The expansions of δm^2, Z_3, Z_1 in the number of loops are also expansions in powers of g: $Z_3^{(l)} \propto g^l$. Let us now summarize the results already established.

One-loop order (Fig. 6.3):

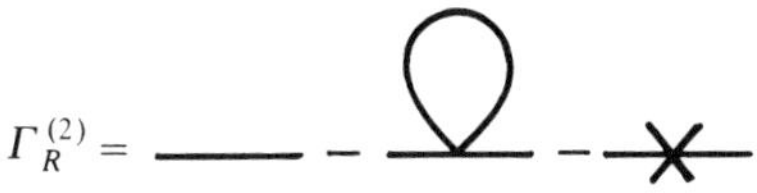

Fig. 6.3

$$\Gamma_R^{(2)} = k^2 + m^2 + \frac{1}{2} g I_\Lambda(m^2) + \delta m^{2(1)}$$

and, by virtue of (6.3.10a),

$$\delta m^{2(1)} = -\frac{1}{2} g I_\Lambda(m^2) \tag{6.3.14}$$

$-\Gamma_R^{(4)} =$  $+ \{$  $+$ perm.$\} +$ 

$$\Gamma_R^{(4)} = g - \frac{1}{2} g^2 [J_\Lambda(k_i, m^2) + \text{perm.}] + g Z_1^{(1)},$$

where $Z_1 = 1 + Z_1^{(1)} + Z_1^{(2)} + \cdots$. By virtue of (6.3.10c),

$$Z_1^{(1)} = \frac{3}{2} g J_\Lambda(0, m^2). \tag{6.3.15}$$

The expression for $\Gamma_R^{(4)}$ is indeed the same as (6.2.11).

Two-loop order: for $\Gamma_R^{(4)}$ we find (Fig. 6.4):

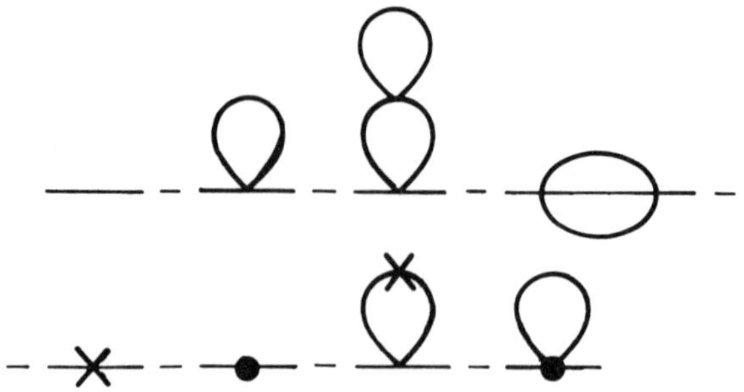

Fig. 6.4

$$\Gamma_R^{(2)} = k^2 + m^2 + \frac{1}{2} g I_\Lambda(m^2) - \frac{1}{4} g^2 I_\Lambda(m^2) J_\Lambda(0, m^2)$$
$$- \frac{1}{6} g^2 K_\Lambda(k^2, m^2) + \delta m^{2(1)} + \delta m^{2(2)} + Z_3^{(2)} k^2$$
$$- \frac{1}{2} g \delta m^{2(1)} J_\Lambda(0, m^2) + \frac{1}{2} g Z_1^{(1)} I_\Lambda(m^2).$$

The condition (6.3.10a) yields

$$\delta m^{2(2)} = -\frac{3}{4} g^2 I_\Lambda(m^2) J_\Lambda(0, m^2) + \frac{1}{6} g^2 K_\Lambda(0, m^2),$$

while (6.3.10b) becomes

$$Z_3^{(2)} = \frac{1}{6} g^2 K'_\Lambda(0, m^2),$$

in agreement with (6.3.7). The final expression for $\Gamma_R^{(2)}$ is, as it should be, the same as (6.3.9).

6.4 The general case

6.4.1 $\Gamma^{(4)}$ to two-loop order

By studying $\Gamma^{(4)}$ to two-loop order one can illustrate how the counterterms work, and understand intuitively how this mechanism may be generalized to all orders of perturbation theory. The two-loop graphs contributing to $\Gamma^{(4)}$ are

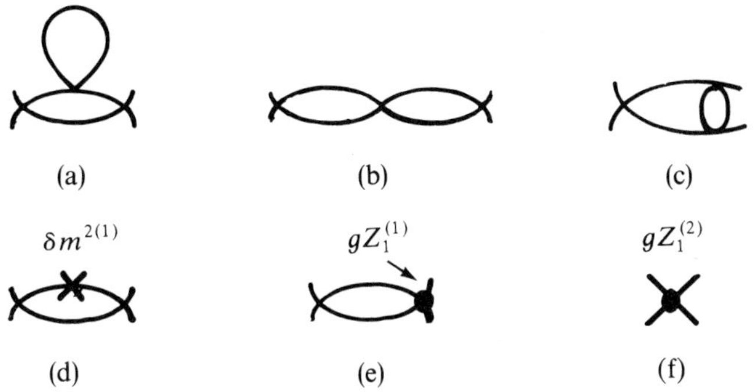

Fig. 6.5

drawn in Fig. 6.5(a–c), together with the graphs constructed from the counterterms (d–f).

The relation (6.3.14) giving $\delta m^{2(1)}$ shows that graphs (a) and (d) sum to zero. Graph (b) is easy to calculate, being the product of two single loops. Graph (c) is more complicated, but also more interesting, and we shall examine it in detail. Using the parametric representation (5.5.8) together with the results of Problem (5.7), one finds the analytic expression for this graph in the form an integral over four parameters α_i (suppressing the multiplicative factor $g^3/2(4\pi)^4$ which plays no part in the argument that follows). Thus

$$I(k_i) = \int \frac{\prod_{i=1}^{4} d\alpha_i}{[P(\alpha_i)]^2} \exp\left(-\left[m^2\left(\sum_{i=1}^{4} \alpha_i\right) + Q(\alpha_i, k_i)\right]\right), \qquad (6.4.1)$$

where

$$P(\alpha_i) = (\alpha_1 + \alpha_2)(\alpha_3 + \alpha_4) + \alpha_3\alpha_4,$$

$$Q(\alpha_i, k_i) = [P(\alpha_i)]^{-1}[\alpha_1\alpha_2(\alpha_3 + \alpha_4)k^2 + \alpha_3\alpha_4(\alpha_1 k_3^2 + \alpha_2 k_4^2)],$$

and $k^2 = (k_1 + k_2)^2$. The labelling of the internal lines is shown in Fig. 6.6.

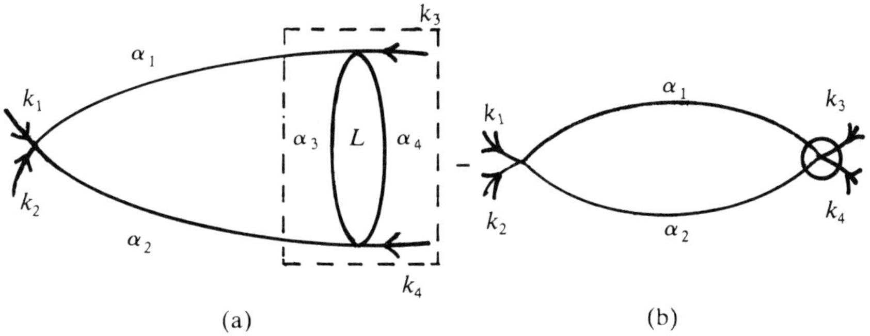

Fig. 6.6 The graph of Fig. 5(c) and its counterterm

The integral (6.4.1) diverges first of all on account of the sub-integration over the loop L shown boxed in Fig. 6.6. The divergence stems from the region where α_3 and α_4 tend to zero simultaneously, while α_1 and α_2 remain finite. The counterterm is defined by the graph 6.6(b), where the circled vertex corresponds to a loop L, evaluated for vanishing external momenta. One should note that the counterterm belongs to a particular graph, and not to the graph (e) (Fig. 6.5) as a whole.

A second divergence stems from the region where all the α_i vanish simultaneously, or in other words from the region $\lambda \to 0$, where λ is the homogeneity parameter in equation (5.5.9); this divergence therefore is linked to the superficial degree of divergence of the graph: $\omega(G) = 0$.

The divergence stemming from the sub-integration is cancelled by the counterterm, so that the overall contribution of Fig. 6.6 reads

$$\bar{I}(k_i) = \int \prod_{i=1}^{4} \mathrm{d}\alpha_i \left\{ \frac{1}{[P(\alpha_i)]^2} \exp\left(-\left[m^2 \left(\sum_{i=1}^{4} \alpha_i \right) + Q(\alpha_i, k_i) \right] \right) \right.$$
$$\left. - \frac{1}{(\alpha_1 + \alpha_2)^2 (\alpha_3 + \alpha_4)^2} \exp\left(-\left[m^2 \left(\sum_{i=1}^{4} \alpha_i \right) + \frac{\alpha_1 \alpha_2}{\alpha_1 + \alpha_2} k^2 \right] \right) \right\}. \tag{6.4.2}$$

In fact, in the region where α_1 and α_2 are finite while $\alpha_3, \alpha_4 \to 0$, one can make the replacements

$$P(\alpha_i) \to (\alpha_1 + \alpha_2)(\alpha_3 + \alpha_4),$$

$$Q(\alpha_i, k_i) \to \frac{\alpha_1 \alpha_2}{\alpha_1 + \alpha_2} k^2$$

(Problem 6.4 supplies a full proof); this shows that the singularity indeed cancels from (6.4.2). The integral $\bar{I}(k_i)$ is superficially divergent, but its divergence is *independent of* k^2, while $I(k_i)$ contains a term in $(\ln \Lambda)(\ln k^2)$; to see this, we need merely differentiate (6.4.2) with respect to k^2: $\partial \bar{I}/\partial k^2$ is given by a convergent integral, with $\omega(G) = -2$. If we now add the graphs from Figs. 6.6(a) and 6.6(b), and subtract the value of the sum-total at $k_i = 0$, then we obtain a convergent result; the subtracted term yields a contribution to $Z_1^{(2)}$ containing terms in $(\ln \Lambda)^2$ and $(\ln \Lambda)$.

From this example one can understand the mechanism of the BPHZ renormalization scheme (due to Bogoliubov, Parasiuk, Hepp, and Zimmermann). Here every graph is dealt with individually, by contrast to the norming conditions (6.3.10) which feature the set of all graphs contributing to $\Gamma_{\mathrm{R}}^{(2)}$ and to $\Gamma_{\mathrm{R}}^{(4)}$ in a given order of perturbation theory. To one-loop order, one subtracts from $\Gamma_{\mathrm{R}}^{(4)}$ ($\Gamma_{\mathrm{R}}^{(2)}$) the first (the first and the second) term of their Taylor expansion about $k_i = 0$; this yields convergent Feynman integrals, and dispenses with regularization altogether, since the subtraction can be made directly in the *integrands*. Similarly, to two-loop order the sum of graphs 6.6(a) and 6.6(b) is made convergent by subtracting the value of the integrand at $k_i = 0$. One can see that the graph 6.6(b) (with counterterm) has been defined recursively, because from the integrand of the loop L we have subtracted its value corresponding to vanishing external momenta.

Consider next the general case of a graph with $(l + 1)$ loops. The divergences stemming from sub-integrations are cancelled by counterterms of order $\leqslant l$; the general prescription for writing down these counterterms is Bogoliubov's recurrence formula, or rather its solution, namely Zimmermann's 'forest formula' (see Itzykson and Zuber 1980, Chapter 8, or Collins 1983, Chapter 5). There are two cases:

(i) The graph in question is superficially convergent ($\omega(G) < 0$); then the sum of graph and counterterms to order $\leqslant l$ is given by an absolutely convergent integral.

(ii) The graph is superficially divergent ($\omega(G) \geqslant 0$); then the sum of graph and counterterm diverges, but a convergent result can be secured merely by one subtraction from the integrand ($\Gamma_{\mathrm{R}}^{(4)}$: $\omega(G) = 0$), or by two subtractions ($\Gamma_{\mathrm{R}}^{(2)}$: $\omega(G) = 2$). Thus the overall divergence is proportional to a polynomial of degree $\omega(G)$ in the external momenta, corresponding to a local interaction in x-space (in other words to an interaction expressible in terms of $\varphi(x)$ and of a finite number of its derivatives).

By means of a complicated combinatorial argument one can show that this procedure is indeed equivalent to the alternative consisting in the construction of counterterms by appeal to the norming conditions (6.3.10). For instance, the full expression for $Z_1^{(2)}$ is given by the equation from Fig. 6.7:

Fig. 6.7

Though it is clear that our discussion has been purely descriptive, full proofs of all its assertions do emerge from arguments along the same lines. Before leaving the BPHZ scheme we stress once more the fact that the systematic subtraction procedure for the integrands frees one from reliance on intermediate regularizations. Because one always has the option of applying the BPHZ scheme instead of such regularizations, the $\Gamma_{\mathrm{R}}^{(N)}$ are indeed independent of the regularization method, seeing that all the integrals converge absolutely; the renormalized correlation functions actually depend on nothing but the norming conditions.

Nevertheless, in spite of its great theoretical interest, the BPHZ scheme is generally not the most convenient for practical calculation, and it leads to complications in the zero-mass case. That is why we now revert to our original point of view, establishing the links between the bare and the renormalized correlation functions.

6.4.2 The relation between bare and renormalized correlation functions

Let us combine $\mathscr{H}_\Lambda$ and $\delta\mathscr{H}_\Lambda$ as in (6.3.13),

$$\mathscr{H}_0(x) = \frac{1}{2} Z_3 (\nabla\varphi)^2 + \frac{1}{2} Z_3 m_0^2 \varphi^2 + \frac{1}{4!} g Z_1 \varphi^4,$$

and change the normalization of the field,

$$\varphi_0(x) = Z_3^{1/2} \varphi(x). \tag{6.4.3}$$

By virtue of (6.3.12), $\mathscr{H}_0$ becomes

$$\mathscr{H}_0(x) = \frac{1}{2} (\nabla\varphi_0)^2 + \frac{1}{2} m_0^2 \varphi_0^2 + \frac{1}{4!} g_0 \varphi_0^4. \tag{6.4.4}$$

Next, consider the generating functional

$$Z(j) = \exp\left(- \int d^4x [\mathscr{H}_\Lambda + \delta\mathscr{H}_\Lambda + j(x)\varphi(x)] \right)$$

$$= \exp\left(- \int d^4x [\mathscr{H}_0 + Z_3^{-1/2} j(x)\varphi_0(x)] \right).$$

Functional differentiation shows at once that

$$G_R^{(E)}(x_1, \ldots, x_E; g, m^2) = Z_3^{-E/2} G^{(E)}(x_1, \ldots, x_E; g_0, m_0^2, \Lambda).$$

In general this relation is written for proper vertices, after going over to k-space; in view of the definition of such vertices (Section 5.3.3) one finds

$$\Gamma_R^{(E)}(k_1, \ldots, k_E; g, m^2) = Z_3^{E/2} \Gamma^{(E)}(k_1, \ldots, k_E; g_0, m_0^2, \Lambda). \tag{6.4.5}$$

Note the factor $Z_3^{E/2}$ (instead of $Z_3^{-E/2}$), which stems from division by the full external propagators.

We are now in a position to formulate the *second convergence theorem*. With $\Gamma_R^{(2)}(0)$, $\Gamma_R'^{(2)}(0)$, and $\Gamma_R^{(4)}(0)$ fixed by the norming conditions (6.3.10), one can define the counterterms δm^2, Z_1, and Z_3. Then the integrals giving $\Gamma_R^{(E)}$ are absolutely convergent:

$$\Gamma_R^{(E)}(k_1, \ldots, k_E; g, m^2) = \lim_{\Lambda \to \infty} Z_3^{E/2} \Gamma^{(E)}(k_1, \ldots, k_E; g_0, m_0^2, \Lambda)$$

is finite and independent of the regularization procedure.

It is essential to understand the meaning of equation (6.4.5). On the right, g_0, m_0/m, and Z_3 are functions of g and of Λ/m: $g_0 = g_0(g, \Lambda/m)$, $m_0/m = f(g, \Lambda/m)$,

$Z_3 = Z_3(g, \Lambda/m)$, governed by the norming conditions (6.3.10). With bare correlation functions $\Gamma^{(E)}$ calculated using a cutoff Λ, the limit $\Lambda \to \infty$ of $Z_3^{E/2}\Gamma^{(E)}$ defines the renormalized correlation functions $\Gamma_R^{(E)}$. When Λ is large but finite, the $\Gamma_R^{(E)}$ do still depend on it though only weakly, through terms $\sim (k/\Lambda)^2$ and $(m/\Lambda)^2$.

6.4.3 Zero mass

The zero-mass case presents the special problem that one cannot use the norming conditions (6.3.10); if the condition $\Gamma_R^{(2)}(k^2 = 0) = 0$ defines the (renormalized) mass as zero, then $\Gamma_R'^{(2)}(k^2)$ and $\Gamma_R^{(4)}(k_i)$ are infrared-divergent at $k^2 = 0$ and $k_i = 0$ (see Section 5.6.3). One must arm oneself with an auxiliary mass μ, whose value is arbitrary, but which is indispensable for writing the normalization conditions. This mass breaks the scale-invariance of the classical approximation, which has no dependence on any dimensional parameter. The norming conditions can then be chosen as

$$\Gamma_R^{(2)}(k^2 = 0) = 0, \tag{6.4.6a}$$

$$\left.\frac{d}{dk^2}\Gamma_R^{(2)}(k^2)\right|_{k^2 = \mu^2} = 1, \tag{6.4.6b}$$

$$\Gamma_R^{(4)}(k_{i0}) = g; \tag{6.4.6c}$$

here k_{i0}, called the *subtraction point*, is defined as symmetrically as possible, namely by

$$k_{i0} \cdot k_{j0} = \frac{1}{4}\mu^2(4\delta_{ij} - 1). \tag{6.4.7}$$

One can understand this choice by noticing that if $k_1 + k_2 = -(k_3 + k_4)$ in $\Gamma^{(4)}$, then $k_i^2 = k^2$ entails

$$s = (k_1 + k_2)^2 = 2k^2 + 2k_1 \cdot k_2,$$

$$t = (k_1 + k_3)^2 = 2k^2 + 2k_1 \cdot k_3,$$

$$u = (k_1 + k_4)^2 = 2k^2 + 2k_1 \cdot k_4,$$

and

$$s + t + u = 4k^2.$$

The most symmetric choice is

$$s = t = u = \mu^2; \quad k^2 = \frac{3}{4}\mu^2; \quad k_i \cdot k_j = -\frac{1}{4}\mu^2,$$

conformably with (6.4.7).

Let us spell out once more that the conditions (6.4.6) are largely arbitrary (except for $\Gamma_R^{(2)}(0) = 0$, which is just the zero-mass condition). For instance, we could adopt $\Gamma_R'^{(2)}(2\mu^2) = 1$, or $k_{i0} \cdot k_{j0} = (\mu^2/\sqrt{2})(4\delta_{ij} - 1)$, etc. One point of importance is that the renormalized theory depends on *two* finite parameters, g and μ^2. However, these two parameters are not independent, as we shall see in the next chapter.

In general, it is only the divergent parts of the counterterms that are uniquely fixed by the norming conditions. A change in these conditions changes the counterterms by *finite* amounts, and is absorbed by a redefinition of the mass, of the coupling constant, and of the normalization of the field; the theory itself is unchanged, but is parametrized differently. This invariance of the theory under reparametrization underlies the (first version of the) renormalization group, which will be discussed in the next chapter; a simple example of reparametrization is given in Problem 6.7.

6.5 Composite operators and their renormalization

One calls composite any operator $O(x)$ that is a function of the field $\varphi(x)$ and of its derivatives. We restrict ourselves to the only case we shall need later, namely to

$$O(x) = \varphi^2(x).$$

This section is devoted to studying the correlation functions constructed from $\varphi(x)$ and $\varphi^2(x)$:

$$G^{(N,L)} = \left(\frac{1}{2}\right)^L \langle \varphi(x_1) \ldots \varphi(x_N)\varphi^2(y_1) \ldots \varphi^2(y_L)\rangle. \tag{6.5.1}$$

At first sight one might be surprised at the need to study such objects, since, after all, $\varphi^2(y) = \varphi(y)\varphi(y)$. The problem lies in the fact that, taken inside a correlation function, the limit

$$\lim_{\|\varepsilon\| \to 0} \varphi(y + \varepsilon)\varphi(y)$$

is singular as $\Lambda \to \infty$. For instance, in the Gaussian approximation

$$\langle \varphi(0)\varphi(0)\rangle = \int_{q \leqslant \Lambda} \frac{d^4 q}{(2\pi)^4} \frac{1}{q^2 + m_0^2}$$

diverges as $\Lambda \to \infty$. Because of the process of renormalization, it is essential to define $\varphi^2(x)$ as an object independent of $\varphi(x)$.

6.5.1 The generating functional

A generating functional for the correlation functions (6.5.1) can be written down at once:

$$\hat{Z}(j, t; m_0^2) = \int \mathscr{D}\varphi(x) \exp\left(- H + \int \left[j(x)\varphi(x) - \frac{1}{2} t(x)\varphi^2(x) \right] \mathrm{d}^4 x \right) \tag{6.5.2}$$

and

$$G^{(N,L)}(x_1, \ldots, x_N; y_1, \ldots, y_L; m_0^2) = \frac{(-1)^L}{Z(0; m_0^2)} \left[\frac{\delta^{N+L} \hat{Z}(j, t; m_0^2)}{\delta j(x_1) \ldots \delta j(x_N) \delta t(y_1) \ldots \delta t(y_L)} \right]_{j=t=0}. \tag{6.5.3}$$

If $t(x)$ in (6.5.2) is independent of x,

$$t(x) = \mu^2,$$

then the source term featuring $\varphi^2(x)$ can be combined with the mass term $\frac{1}{2} m_0^2 \varphi^2$ in H, and one obtains a relation between the generating functionals Z and $\hat{Z}$:

$$\hat{Z}(j, \mu^2; m_0^2) = Z(j; m_0^2 + \mu^2). \tag{6.5.4}$$

From this one can derive

$$G^{(N)}(x_1, \ldots, x_N; m_0^2 + \mu^2) = \sum_{L=0}^{\infty} \frac{(-1)^L \mu^{2L}}{L!} \times \int \mathrm{d}^4 y_1 \ldots \mathrm{d}^4 y_L \, G^{(N,L)}(x_1, \ldots, x_N; y_1, \ldots, y_L; m_0^2) \tag{6.5.5}$$

where the $G^{(N,L)}$ are connected with respect to the φ^2 (see Problem 6.5). Taking the limit $\mu^2 \to 0$, one calculates the derivative of $G^{(N)}$ with respect to m_0^2:

$$\frac{\partial}{\partial m_0^2} G^{(N)}(x_1, \ldots, x_N; m_0^2) = - \int \mathrm{d}^4 y \, G^{(N,1)}(x_1, \ldots, x_N; y; m_0^2). \tag{6.5.6}$$

We define the Fourier transform just as for ordinary correlation functions, by

$$(2\pi)^4 \delta^{(4)}\left(\sum_1^N k_i + q \right) G^{(N,1)}(k_1, \ldots, k_N; q; m_0^2) = \int \left(\prod_{i=1}^N \mathrm{d}^4 x_i \mathrm{e}^{\mathrm{i} k_i \cdot x_i} \right) \mathrm{d}^4 y \, \mathrm{e}^{\mathrm{i} q \cdot y} G^{(N,1)}(x_1, \ldots, x_N; y; m_0^2),$$

where the momentum q is associated with φ^2. Then (6.5.6) reduces to

$$\frac{\partial}{\partial m_0^2} G^{(N)}(k_1, \ldots, k_N; m_0^2) = - G^{(N,1)}(k_1, \ldots, k_N; q = 0; m_0^2). \tag{6.5.7}$$

The operation of taking the derivative with respect to m_0^2 is often called a '*mass insertion*'.

By virtue of Wick's theorem we can formulate Feynman rules for the $G^{(N,L)}$ exactly as for the $G^{(N)}$. Consider as an example the first few terms of $G^{(2,1)}(x, y; z)$ (Fig. 6.8):

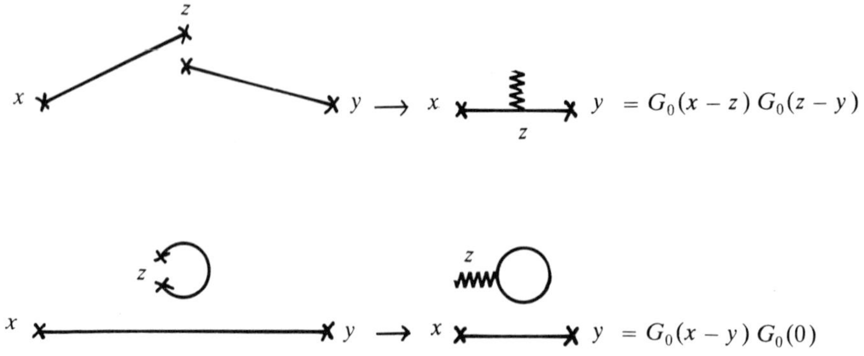

Fig. 6.8 Contributions to $G^{(2,1)}$

The φ^2 insertion at the point z is represented by a wavy line. With respect to φ^2 the second term $(G_0(x - y)G_0(0))$ is not connected, and must not be reckoned in (6.5.5). For the first term, if $k_1 + k_2 = q = 0$, $k_1 = -k_2 = k$, one finds in k-space that

$$G^{(2,1)}(k, -k; q = 0) = \frac{1}{(k^2 + m_0^2)^2} = -\frac{\partial}{\partial m_0^2}\frac{1}{k^2 + m_0^2},$$

conformably with (6.5.7).

As in the case of ordinary correlation functions $G^{(N)}$, one defines connected correlation functions $G_c^{(N,L)}$ and proper vertices $\Gamma^{(N,L)}$ through a Legendre transformation with respect to j.

Consider for instance the proper vertex $\Gamma^{(2,1)}$ found by dividing $G^{(2,1)}$ by two propagators:

$$\Gamma^{(2,1)}(k_1, k_2; q = -(k_1 + k_2)) = \frac{1}{G^{(2)}(k_1)} G^{(2,1)}(k_1, k_2; q) \frac{1}{G^{(2)}(k_2)}.$$

When $q = 0$, $\Gamma^{(2,1)}$ is easily related to the derivative of $G^{(2)}$ with respect to m_0^2,

$$\frac{\partial}{\partial m_0^2}\frac{1}{G^{(2)}(k)} = \frac{1}{G^{(2)}(k)} G^{(2,1)}(k, -k; 0) \frac{1}{G^{(2)}(-k)},$$

whence

$$\frac{\partial}{\partial m_0^2}\Gamma^{(2)}(k) = \Gamma^{(2,1)}(k, -k; q = 0). \tag{6.5.8}$$

6.5.2 Example: $\Gamma^{(2,1)}$ to one-loop order

To show that the renormalizations described in the preceding sections are insufficient to make $\Gamma^{(N,L)}$ finite, we need merely calculate $\Gamma^{(2,1)}$ to one-loop order (Fig. 6.9):

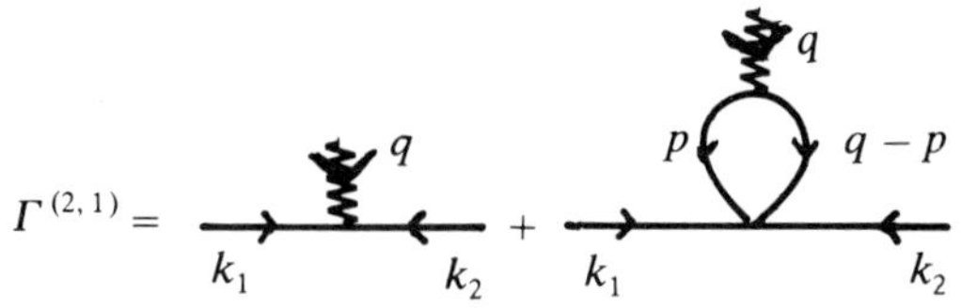

Fig. 6.9 Graphs contributing to $\Gamma^{(2,1)}$

$$\Gamma^{(2,1)}(k_1, k_2; q) = 1 - \frac{1}{2} g_0 \int \frac{\mathrm{d}^4 p}{(2\pi)^4} \frac{1}{(p^2 + m_0^2)((q-p)^2 + m_0^2)}. \tag{6.5.9}$$

It is instructive to check (6.5.8) by using

$$\Gamma^{(2)}(k) = k^2 + m_0^2 + \frac{1}{2} g_0 \int \frac{\mathrm{d}^4 p}{(2\pi)^4} \frac{1}{p^2 + m_0^2}.$$

The integral in (6.5.9) diverges logarithmically, and cannot be made finite by any of the renormalizations discussed so far; nor, equivalently, by any of the counterterms in (6.3.13c). One must introduce an additional renormalization constant $\bar{Z}$ through the relation

$$\Gamma_{\mathrm{R}}^{(2,1)}(k_1, k_2; q; g, m^2) = \bar{Z} Z_3 \Gamma^{(2,1)}(k_1, k_2; q; g_0, m_0^2, \Lambda), \tag{6.5.10}$$

where $\bar{Z}$ is determined by the norming condition

$$\Gamma_{\mathrm{R}}^{(2,1)}(k_i = 0; q = 0) = 1 \tag{6.5.11}$$

provided the mass is nonzero. For zero mass, the integral in (6.5.9) is infrared divergent at $q = 0$, and one must choose a different subtraction point. To one-loop order, and taking into account that $Z_3 = 1 + O(g^2)$, $\bar{Z}$ can be found immediately:

$$\bar{Z} = 1 + \frac{1}{2} g \int \frac{\mathrm{d}^4 p}{(2\pi)^4} \frac{1}{(p^2 + m^2)^2}. \tag{6.5.12}$$

To understand the general case, one proceeds by power-counting, as for ordinary correlation functions.

6.5.3 Power-counting and counterterms

Consider the superficial degree of divergence of a diagram with N external φ lines, L external φ^2 lines, and of order V in perturbation theory. By (5.6.1), the factor ω corresponding to a φ^2 insertion is given by

$$\omega - D = 2\left(\frac{D}{2} - 1\right) - D = -2; \tag{6.5.13}$$

this could have been seen just as well by noticing that the coupling constant for a φ^2 interaction has dimension 2. The superficial degree of divergence $\omega(N, L)$ of the proper vertex $\Gamma^{(N,L)}$ is governed by (5.6.2),

$$\omega(N, L) = D - 2L + V(D - 4) - \frac{N}{2}(D - 1),$$

whence, if $D = 4$,

$$\omega(N, L) = 4 - 2L + N. \tag{6.5.14}$$

One could derive this also by noting that every φ^2 insertion is equivalent to a differentiation with respect to m_0^2, and accordingly diminishes the dimension of the integrand by 2. Thus the only two superficially divergent proper vertices correspond to

$$N = 0,\ L = 2:\quad \omega(0, 2) = 0;$$

$$N = 2,\ L = 1:\quad \omega(2, 1) = 0.$$

The first case, which we shall ignore, becomes relevant only if one wants to study the free energy. Since $\Gamma^{(2,1)}$ is the only superficially divergent graph, all the $\Gamma^{(N,L)}$ can be made finite merely by renormalizing φ^2. For instance, one can use the following normalization conditions.

Nonzero mass:

$$\Gamma_{\mathrm{R}}^{(2,1)}(k_1 = k_2 = q = 0) = 1. \tag{6.3.10d}$$

Zero mass:

$$\Gamma_{\mathrm{R}}^{(2,1)}(k_{i0}; q_0) = 1, \tag{6.4.6d}$$

where

$$k_{10}^2 = k_{20}^2 = \frac{3}{4}\mu^2, \quad k_1 \cdot k_2 = -\frac{1}{4}\mu^2.$$

Just as for $\Gamma_{\mathrm{R}}^{(2)}$ and $\Gamma_{\mathrm{R}}^{(4)}$, this choice of normalization is largely arbitrary.

After renormalization, inserting a φ^2 is no longer equivalent to inserting a product of two φ's; to stress this distinction, we shall (provisionally) write a φ^2 insertion as $[\varphi^2]$. The correlation functions are calculated from the Hamiltonian H_Λ, the counterterm δH_Λ, and the source term

$$\int \left[j(x)\varphi(x) - \frac{1}{2} t(x) [\varphi^2] \right] \mathrm{d}^4 x.$$

The renormalization constant $\bar{Z}$ relates $[\varphi^2(x)]$ to $\varphi_0^2(x)$; by definition we have

$$[\varphi^2(x)] = \bar{Z}Z_3\varphi^2(x) = \bar{Z}\varphi_0^2(x). \tag{6.5.15}$$

After the change in the normalization of the field, the source term becomes

$$\int\left[Z_3^{-1/2}j(x)\varphi_0(x) - \frac{1}{2}\bar{Z}t(x)\varphi_0^2(x)\right]\mathrm{d}^4x,$$

whence one derives the relation between the bare and the renormalized correlation functions,

$$G_{\mathrm{R,c}}^{(N,L)}(k_i, q_j; g, m^2) = Z_3^{-N/2}\bar{Z}^L G_{\mathrm{c}}^{(N,L)}(k_i, q_j; g_0, m_0^2, \Lambda)$$

or

$$\Gamma_{\mathrm{R}}^{(N,L)}(k_i, q_j; g, m^2) = Z_3^{N/2}\bar{Z}^L \Gamma^{(N,L)}(k_i, q_j; g_0, m_0^2, \Lambda). \tag{6.5.16}$$

(One must remember that the Legendre transformation is implemented only with respect to j.) Equation (6.5.6) is of course a special case of (6.5.16).

Notice also that

$$\frac{1}{2}t(x)[\varphi^2(x)] = \frac{1}{2}t(x)\varphi^2(x) + \frac{1}{2}t(x)(Z_3\bar{Z} - 1)\varphi^2(x). \tag{6.5.17}$$

The first term can be assigned to $\mathscr{H}_\Lambda$; the second can be assigned to $\delta\mathscr{H}_\Lambda$, since it acts as a counterterm (Fig. 6.10), and can therefore be dealt with by the same rules that apply to counterterms generally. It is determined recursively through the norming condition (6.3.10d) (or (6.4.6d) for zero mass).

Fig. 6.10 The counterterm associated with a φ^2 insertion

Finally we point out that the renormalization of $\varphi^2(x)$ in $g\varphi^4$ theory constitutes a particularly simple case. In general, renormalization couples insertions of a composite operator to insertions of all other operators having dimension less than or equal to its own, unless there are reasons to the contrary from symmetry (see Problem 6.9).

6.6 The minimal subtraction scheme (MS)

We conclude this chapter by describing the minimal renormalization scheme, which often proves very convenient in actual calculations. This scheme works only with dimensional regularization; instead of using norming conditions like (6.3.10) or (6.4.6), one constructs counterterms by removing the ε-poles from the superficially divergent correlation functions, having first regularized the latter dimensionally. We stress that dimensional regularization is perfectly possible *with* the norming conditions (6.3.10) or (6.4.6); the renormalized correlation functions would then be identically the same as those found by regularization with a cutoff: they depend only on the norming conditions, and not on the regularization procedure. By contrast, the renormalized correlation functions of the minimal scheme *differ* from those found by starting from (6.3.10) or (6.4.6). Nevertheless their physical predictions are the same, provided one sums all the terms of the perturbation series. The differences are effectively absorbed by a redefinition of the mass and of the coupling constant; if g is the (renormalized) coupling constant derived from (6.3.10) or (6.4.6), and g' that of the minimal scheme, then g' can be expressed as a perturbative expansion in powers of g:

$$g' = g + c_1 g^2 + c_2 g^3 + \cdots . \tag{6.6.1}$$

(The calculation is called for in Problem 6.7.) Of course these properties merely constitute a special case of the invariance under reparametrization evoked earlier. However, one knows only a finite number of the terms of the perturbation series (generally two or three), whence in practice the predictions of the two schemes do differ. Suppose for example that a correlation function has been calculated to order g^3 in both schemes: the two expressions will then differ by terms of order g^4.

Choosing the 'best renormalization scheme' is important, as for instance in quantum chromodynamics. Several prescriptions have been proposed, but arguments about 'optimizing the renormalization scheme' can never be more than heuristic.

We illustrate the mechanism of the minimal scheme by renormalizing to two-loop order for zero mass, with the appropriate Hamiltonian

$$H = \int \mathrm{d}^D x \left(\frac{1}{2}(\nabla\varphi)^2 + \frac{\mu^\varepsilon g}{4!}\varphi^4 \right) + \text{counterterms}. \tag{6.6.2}$$

Because dimensional regularization is to be used, the Hamiltonian has been written for D-dimensional space, with $\varepsilon = 4 - D$ as usual. Since the coupling constant is dimensional, it proves convenient to define a dimensionless constant g and a mass μ: the combination $\mu^\varepsilon g$ then has the right dimensions for a coupling constant. We shall need the results calculated for the following graphs:

$$= -\frac{1}{2\varepsilon}\left(1 - \varepsilon \ln\left(\frac{k^2}{\mu^2}\right) + \varepsilon \times \text{constant} + O(\varepsilon^2)\right) \tag{6.6.3a}$$

$$= \frac{2}{\varepsilon}(1 + \varepsilon a_1 + O(\varepsilon^2)) \tag{6.6.3b}$$

$$= \frac{4}{\varepsilon^2}(1 + 2\varepsilon a_1 + O(\varepsilon^2)) \tag{6.6.3c}$$

$$= \frac{2}{\varepsilon^2}\left(1 + 2\varepsilon a_1 + \frac{\varepsilon}{2} + O(\varepsilon^2)\right) \tag{6.6.3d}$$

$$a_1 = 1 - \frac{1}{2}\gamma - \frac{1}{2}\ln\frac{(k_1 + k_2)^2}{\mu^2}. \tag{6.6.3e}$$

Equations (6.6.3) have been written somewhat schematically. From $\Gamma^{(4)}$ we have dropped a multiplicative factor $g\mu^{\varepsilon}$; moreover, every graph must still be multiplied by a symmetry factor and by α^l, where $\alpha = g/(4\pi)^{D/2}$, with l the number of loops. For instance, (6.6.3a) should be multiplied by $\alpha^2/6$. In (6.6.3e), $\gamma(=0.577\ldots)$ is Euler's constant. The proof of (6.6.3a) is asked for in Problems 5.10 or 5.12. The proof of (6.6.3b)–(6.6.3d) is relegated to Problem 6.6. Let us start with Z_3, by requiring the cancellation of the pole of (6.6.3a)* at $\varepsilon = 0$:

$$\frac{\partial}{\partial k^2}$$

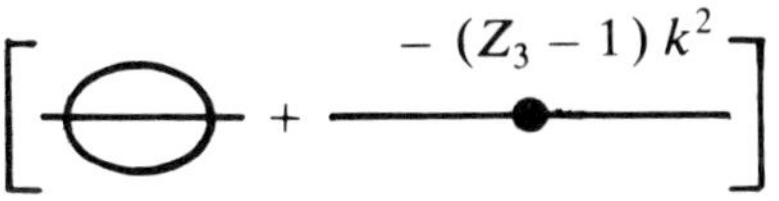

has no pole at $\varepsilon = 0$,

whence

$$Z_3 = 1 - \frac{\alpha^2}{12\varepsilon} + O(\alpha^3). \tag{6.6.4a}$$

The advantage of the minimal scheme is that Z_3 is independent of the constant in (6.6.3b); therefore the constant need not be calculated, as it would have to be if one were to use the norming condition (6.4.6b) to find Z_3.

Next we determine Z_1 in one loop order from (6.6.3b):

$$Z_1^{(1)} = \frac{3\alpha}{\varepsilon}. \tag{6.6.5}$$

To calculate Z_1 in two-loop order, one must remember to add the one-loop graph constructed with the counterterm, namely

$$= \frac{1}{2}\alpha g Z_1^{(1)}\frac{2}{\varepsilon}(1 + \varepsilon a_1). \tag{6.6.3f}$$

Taking symmetry factors and permutations into account one finds

$$Z_1^{(2)} = -\frac{3\alpha^2}{\varepsilon^2}(1 + 2\varepsilon a_1) - \frac{6\alpha^2}{\varepsilon^2}\left(1 + 2\varepsilon a_1 + \frac{\varepsilon}{2}\right) + \frac{18\alpha^2}{\varepsilon^2}(1 + \varepsilon a_1), \tag{6.6.6}$$

where the three terms correspond to the graphs (6.6.3c), (6.6.3d), and (6.6.3f) respectively. For $Z_1^{(2)}$ this equation yields

$$Z_1^{(2)} = \frac{9\alpha^2}{\varepsilon^2} - \frac{3\alpha^2}{\varepsilon}, \tag{6.6.7}$$

* The coefficient of the pole has the form $f(D)/\varepsilon$. The strict minimal scheme would subtract $f(4)/\varepsilon$, i.e. it would expand $(4\pi)^{-D/2}$ in powers of ε. In fact we use a variant which consists of choosing the mass $\mu' = \mu(4\pi)^{1/2}$ instead of μ. Yet another variant is the scheme $\overline{\text{MS}}$, much employed in quantum chromodynamics, which consists of choosing $\bar{\mu} = \mu[e^{-\gamma}(4\pi)]^{1/2}$.

whence to two-loop order

$$Z_1 = 1 + \frac{3\alpha}{\varepsilon} + \frac{9\alpha^2}{\varepsilon^2} - \frac{3\alpha^2}{\varepsilon}. \tag{6.6.4b}$$

Equations (6.6.4) yield the two renormalization constants Z_1 and Z_3 to two-loop order. The relation between g_0 and g defines the renormalization constant Z (see (6.3.12)) through

$$g_0 = Z_1 Z_3^{-1} g = Zg, \tag{6.6.8}$$

where

$$Z = 1 + \frac{3\alpha}{\varepsilon} + \frac{9\alpha^2}{\varepsilon^2} - \frac{17\alpha^2}{6\varepsilon} + O(\alpha^3). \tag{6.6.4c}$$

We conclude with some observations. The coefficient of $1/\varepsilon^2$ in (6.6.4c) is the square of the coefficient of $1/\varepsilon$ in (6.6.4b); this is not a coincidence, but stems from the renormalizability of the theory (see Chapter 7). On the other hand, from equation (6.6.6) the terms in $(1/\varepsilon)\ln[(k_1 + k_2)^2/\mu^2]$ have cancelled; fortunately so, because otherwise the counterterms would not be local in x, and the entire renormalization scheme would crumble. This cancellation is evidently related to the cancellation of the terms with $\ln \Lambda \ln k^2$ between the graph of Fig. 6.6(a) and its counterterm in Fig. 6.6(b). As in other schemes, the counterterms therefore appear as polynomials of order $\leqslant \omega(G)$ in the external momenta, being thereby local in x; to l-loop order the coefficients contain poles in $\varepsilon^{-1}, \ldots, \varepsilon^{-l}$, but no constant ($\varepsilon^0$) terms.

Recall finally that we have been calculating with zero mass. However, *in the minimal scheme* the renormalization constants are independent of the renormalized mass m, which is related to the bare mass m_0 by $m_0^2 = Z_m m^2$. Indeed, by dimensional analysis Z_1, Z_3, and Z_m are functions of g, ε, and μ/m. But the mass μ enters only through μ^ε, whose expansion is $(1 + \varepsilon \ln \mu + \cdots)$. Hence the renormalization constants can be functions only of $\ln(\mu/m)$; but, being regular at $m = 0$, they can have no such dependence; therefore in the minimal scheme Z_1, Z_3, and Z_m depend only on g and on ε. This would not be the case under renormalization conditions of the type (6.4.6).

Problems

6.1 (a) Show that $J_\Lambda^{\mathrm{R}} = [J_\Lambda(k_i, m^2) - J_\Lambda(0, m^2)]$ (see (6.2.11)) is finite in the limit $\Lambda \to \infty$. Show likewise that J_Λ^{R} differs from J_∞^{R} by terms of order k^2/Λ^2.

(b) Show that $[K_\Lambda(k^2, m^2) - K_\Lambda(0, m^2) - k^2 K'_\Lambda(0, m^2)]$ (see (6.3.9)) is finite in the limit $\Lambda \to \infty$ (use the parametric representation from Section 5.5.3).

6.2 Renormalization of φ^3 theory

Let H be the Hamiltonian of 'φ^3 theory',

$$H = \int \mathrm{d}^D x \left[\frac{1}{2}(\nabla\varphi)^2 + \frac{1}{2}m^2\varphi^2 + \frac{1}{3!}g\mu^{\varepsilon/2}\varphi^3\right] + \text{counterterms},$$

with $\varepsilon = 6 - D$. The factor $\mu^{\varepsilon/2}$ ensures that the coupling constant g is dimensionless.

(a) Show that φ^3 theory is renormalizable when $D = 6$. Which are the superficially divergent correlation functions?

(b) One gets rid of tadpoles (see Section 5.5.1) with the help of a counterterm proportional to φ. Explain why this operation (which consists in requiring $\langle\varphi\rangle = 0$) is possible. For $D = 6$, the only superficially divergent correlation functions one is left with are $\Gamma^{(2)}$ and $\Gamma^{(3)}$.

(c) Calculate $\Gamma^{(2)}$ to one-loop order by using

(c.1) Schwinger regularization;

(c.2) dimensional regularization.

Calculate $\Gamma_{\mathrm{R}}^{(2)}$ by using the norming conditions (6.3.10), and check (without evaluating the integral over the Feynman parameter) that the result is independent of the regularization. For case (c.1) use identities like

$$\int_0^\infty \frac{\mathrm{d}\lambda}{\lambda}(\mathrm{e}^{-\lambda x} - \mathrm{e}^{-\lambda x_0}) = -\ln(x/x_0).$$

Solution:

$$\Gamma_{\mathrm{R}}^{(2)} = k^2 + m^2 + \frac{1}{12}\alpha k^2 - \frac{1}{2}\alpha\int_0^1 \mathrm{d}x\, f(x, k^2)\ln[f(x, k^2)/m^2],$$

where

$$\alpha = g^2/(4\pi)^3 \quad \text{and} \quad f(x, k^2) = m^2 + x(1-x)k^2.$$

Determine also the divergent parts of δm^2 and Z_3.

(d) Determine the divergent part of $\Gamma^{(3)}$; thence derive the divergent part of Z_1 to one-loop order. Show that in the minimal scheme $Z = Z_1 Z_3^{-3/2}$ is given by

$$Z = 1 - \frac{3\alpha}{4\varepsilon}.$$

6.3 Renormalization constants from regularization by cutoff

(a) Using Schwinger regularization, calculate $\Gamma^{(4)}(0)$ in one-loop order, keeping track of constant terms (i.e. those remaining finite as $\Lambda \to \infty$). Thence derive

$$Z_1^{(1)} = \frac{3}{2}\frac{g}{(4\pi)^2}\left[\ln\frac{\Lambda^2}{m^2} - (\gamma + \ln 2 + 1)\right],$$

where γ is Euler's constant. One must pay attention to the limits of the integrals over the parameters α_i, and use the representation

$$\gamma = -\left[\int_0^1 \frac{du}{u}(e^{-u} - 1) + \int_0^\infty \frac{du}{u} e^{-u}\right].$$

(b) We aim to calculate the integral corresponding to the graph (e) in Fig. 6.6, for $k_i = 0$. Use the parametric representation (5.5.9), integrating first over the variables x_i (and being careful with the integration limits). Restrict yourself to calculating the divergent terms, of orders $(\ln \Lambda)^2$ and $(\ln \Lambda)$. Show that the result reads

$$I = \frac{1}{2}\ln^2\frac{\Lambda^2}{m^2} - (\gamma + \ln 2)\ln\frac{\Lambda^2}{m^2}.$$

This must be multiplied by $g\alpha$, where $\alpha = g/(4\pi)^2$, and by the symmetry factor.

(c) Show that to two-loop order the renormalization constant Z_1 is given by

$$Z_1 = 1 + \frac{3}{2}\alpha\left[\ln\frac{\Lambda^2}{m^2} - (\gamma + \ln 2 + 1)\right]$$
$$+ \alpha^2\left\{\frac{9}{4}\ln^2\frac{\Lambda^2}{m^2} - \left[\frac{9}{2}(\gamma + \ln 2) + \frac{15}{2}\right]\ln\frac{\Lambda^2}{m^2}\right\}.$$

From equation (5.5.12) calculate $d\Sigma/dk^2$ and

$$Z_3 = 1 - \frac{\alpha^2}{12}\ln\frac{\Lambda^2}{m^2}.$$

Use these formulae for Z_1 and Z_3 to evaluate the renormalization constant $Z = Z_1 Z_3^{-2}$ to two-loop order.

6.4 To determine rigorously the behaviour of the integral (6.4.1) as $\alpha_i \to 0$, one splits the integration region into sectors like

$$\alpha_1 \leqslant \alpha_2 \leqslant \alpha_3 \leqslant \alpha_4,$$

and (in this particular sector) one makes the change of variables ($0 \leqslant \beta_i \leqslant 1$, $i \leqslant 3$; $0 \leqslant \beta_4 < \infty$):

$$\alpha_1 = \beta_1\beta_2\beta_3\beta_4 \qquad \alpha_2 = \beta_2\beta_3\beta_4$$

$$\alpha_3 = \beta_3\beta_4 \qquad \alpha_4 = \beta_4 .$$

Justify the assertions of Section 6.4.1 about the behaviour of $I(k_i)$, $\partial I/\partial k^2$, $\bar{I}(k_i)$, $\partial \bar{I}/\partial k^2$. You are asked only to examine the sectors $\alpha_1 \leqslant \alpha_2 \leqslant \alpha_3 \leqslant \alpha_4$ and $\alpha_3 \leqslant \alpha_4 \leqslant \alpha_1 \leqslant \alpha_2$ as examples.

The subdivision into sectors is essential in order to avoid mishaps due to integrals like

$$\int_0^1 \mathrm{d}\alpha_1 \int_0^1 \mathrm{d}\alpha_2 \frac{2\alpha_1}{[\alpha_1^2 + \alpha_2]^2}$$

(an example borrowed from Itzykson and Zuber 1980). This integral appears convergent on examining the regions ($\alpha_1 \to 0$, α_2 finite), ($\alpha_2 \to 0$, α_1 finite), ($\alpha_1, \alpha_2 \to 0$ at the same rate); nevertheless it diverges.

6.5 By taking the logarithm of equation (6.5.4), prove the relation (6.5.5) for the connected correlation functions $G_c^{(N,L)}$. Thence derive (6.5.5) for 'φ^2-connected' correlation functions, where the φ^2 insertions are linked to external points φ (one diagram that is not 'φ^2-connected' is drawn in Fig. 6.8).

6.6 Two-loop graphs under dimensional regularization

(a) Calculate the one-loop graph contributing to $\Gamma^{(4)}$, for zero mass, and using dimensional regularization. Evaluate both the divergent part and the part that remains finite in the limit $\varepsilon \to 0$. (Solution: see (6.6.3b).)

(b) Calculate the two-loop graph of Fig. 6.5(c), still for zero mass and with dimensional regularization. Keep only terms of orders $1/\varepsilon^2$ and $1/\varepsilon$.

Use either of the two following techniques:

(i) start from the parametric representation (5.5.9);

(ii) integrate over the two loops in turn (see Problem 5.12).

The second method is probably quicker. We indicate some intermediate results. The integral over $d^D q$ yields (with $k = k_1 + k_2$)

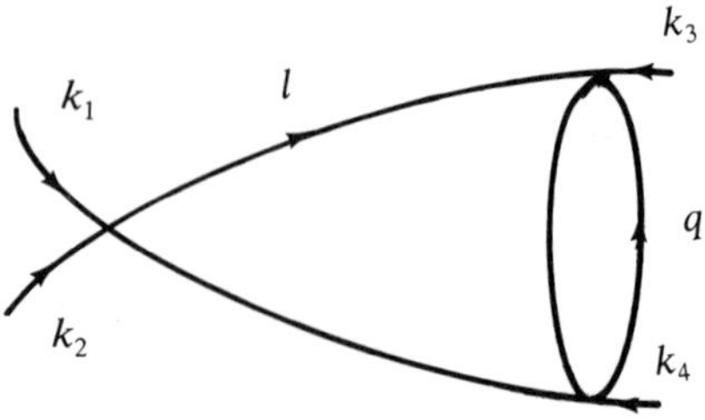

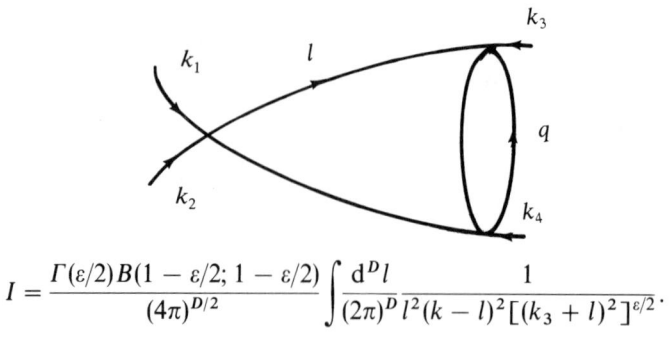

$$I = \frac{\Gamma(\varepsilon/2)B(1-\varepsilon/2; 1-\varepsilon/2)}{(4\pi)^{D/2}} \int \frac{d^D l}{(2\pi)^D} \frac{1}{l^2(k-l)^2[(k_3+l)^2]^{\varepsilon/2}}.$$

The three denominators are combined via Feynman's identity (B.1), and then one integrates over l:

$$I = \frac{\Gamma(\varepsilon)B(1-\varepsilon/2; 1-\varepsilon/2)}{(4\pi)^D}$$

$$\times \int \frac{\left(\prod_1^3 dx_i\right)\delta\left(1-\sum_1^3 x_i\right)x_3^{\varepsilon/2-1}}{[x_2(1-x_2)k^2 + x_3(1-x_3)k_3^2 + 2x_2x_3 k\cdot k_3]^{\varepsilon}}.$$

To find the terms of order $1/\varepsilon^2$ and $1/\varepsilon$ we note that

$$\int_0^{x_2} dx_3\, x_3^{\varepsilon/2-1}(f(x_3))^{\varepsilon} = (f(0))^{\varepsilon}\int_0^{x_2} dx_3\, x_3^{\varepsilon/2-1} + O(\varepsilon).$$

The result is given in (6.6.3d).

6.7 To order g^2, find the relation between the renormalized coupling constant g_{MS} of the minimal scheme (defined in Section 6.6), and the constant $g(\mu)$ found by starting from the normalization conditions (6.4.6). Calculate the renormalization constant Z_1 in both schemes, for the same value of the bare coupling constant.

Solution:

$$g(\mu) = g_{\mathrm{MS}} - \frac{3g_{\mathrm{MS}}^2}{(4\pi)^2}\left(1 - \frac{1}{2}\gamma\right).$$

6.8 Renormalization of the effective action

We start from the expression (2.4.6) for the effective potential to one-loop order in $D = 4$ dimensions:

$$V(\bar{\varphi}) = \frac{1}{2}m^2\bar{\varphi}^2 + \frac{1}{4!}g\bar{\varphi}^4 + \frac{1}{2}\int \frac{d^4 q}{(2\pi)^4}\ln\left(1 + \frac{g\bar{\varphi}^2/2}{q^2+m^2}\right)$$

(explain the difference between this and (2.4.6)). The integral over q diverges. To renormalize the effective potential one uses the conditions (6.3.10), recalling that

$$\Gamma^{(2)}(0) = \left.\frac{\mathrm{d}^2 V}{\mathrm{d}\bar{\varphi}^2}\right|_{\bar{\varphi}=0}, \quad \Gamma^{(4)}(0) = \left.\frac{\mathrm{d}^4 V}{\mathrm{d}\bar{\varphi}^4}\right|_{\bar{\varphi}=0}.$$

(a) Show that the conditions (6.3.10) are satisfied if the renormalized effective potential $V_{\mathrm{R}}(\bar{\varphi})$ is given by

$$V_{\mathrm{R}}(\bar{\varphi}) = \frac{1}{2}m^2\bar{\varphi}^2 + \frac{1}{4!}g\bar{\varphi}^4 + \frac{1}{2}\int\frac{\mathrm{d}^4 q}{(2\pi)^4}\left\{\ln\left(1 + \frac{g\bar{\varphi}^2/2}{q^2+m^2}\right) - \frac{g\bar{\varphi}^2/2}{q^2+m^2} + \frac{1}{2}\frac{(g\bar{\varphi}^2/2)^2}{(q^2+m^2)^2}\right\}.$$

Check that the integral converges.

(b) Evaluate $V_{\mathrm{R}}(\bar{\varphi})$ explicitly; it should yield

$$V_{\mathrm{R}}(\bar{\varphi}) = \frac{1}{2}m^2\bar{\varphi}^2 + \frac{1}{4!}g\bar{\varphi}^4 + \frac{1}{(8\pi)^2}\left[\left(m^2 + \frac{g\bar{\varphi}^2}{2}\right)^2\ln\left(1 + \frac{g\bar{\varphi}^2}{2m^2}\right) - \frac{g\bar{\varphi}^2}{2}\left(\frac{3}{2}\frac{g\bar{\varphi}^2}{2} + m^2\right)\right].$$

(c) If $m = 0$, the method just given yields an infrared-divergent integral. Then one defines a new coupling constant

$$g(\mu) = \left.\frac{\mathrm{d}^4 V(\bar{\varphi})}{\mathrm{d}\bar{\varphi}^4}\right|_{\bar{\varphi}=\mu}.$$

Calculate $V(\bar{\varphi})$ using a cutoff Λ, and show that the definition of $g(\mu)$ and the condition $\mathrm{d}^2 V(\bar{\varphi})/\mathrm{d}\bar{\varphi}|_{\bar{\varphi}=0} = 0$ lead to the following expression for $V_{\mathrm{R}}(\bar{\varphi})$ due to Coleman and Weinberg (1973):

$$V_{\mathrm{R}}(\bar{\varphi}) = \frac{1}{4!}g(\mu)\bar{\varphi}^4 + \frac{g^2(\mu)\bar{\varphi}^4}{(16\pi)^2}\left(\ln\frac{\bar{\varphi}^2}{\mu^2} - \frac{25}{6}\right).$$

Sketch $V_{\mathrm{R}}(\bar{\varphi})$.

6.9 Renormalization of the operator φ^2 in φ^3 theory (Collins 1983, Chapter 6)

Consider φ^3 theory in six dimensions (see Problem 6.2), and study the renormalization of the correlation function

$$G_{\mathrm{c}}^{(2,1)}(x, y; z) = \left\langle \varphi(x)\varphi(y)\frac{1}{2}\varphi^2(z)\right\rangle_{\mathrm{c}}$$

or of the proper vertex $\Gamma^{(2,1)}$ which is associated with it.

(a) Show that to one-loop order the divergent diagrams of $\Gamma^{(2,1)}$ are

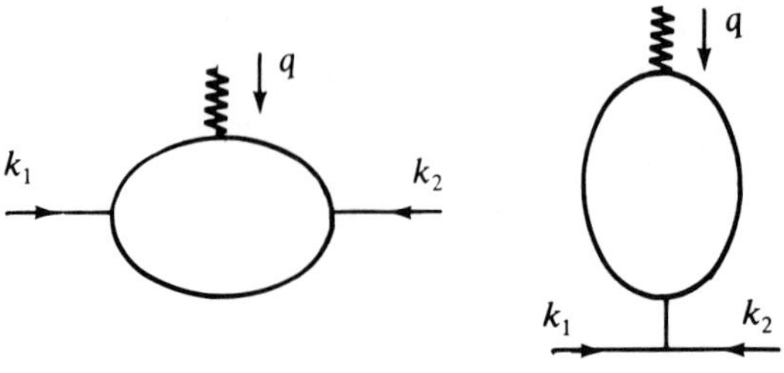

(a) (b)

Thence derive the corresponding counterterms,

(a) $-\dfrac{\alpha}{\varepsilon}$, with $\alpha = g^2/(4\pi)^3$

(b) $-\dfrac{\alpha}{\varepsilon g}\left(m^2 + \dfrac{1}{6}q^2\right)$.

(b) Show that φ^2 is renormalized according to

$$\frac{1}{2}[\varphi^2] = \frac{1}{2}Z_a\varphi^2 + \mu^{-\varepsilon/2}Z_b m^2\varphi + \mu^{-\varepsilon/2}Z_c\nabla^2\varphi,$$

where, to one-loop order,

$$Z_a = 1 - \frac{\alpha}{\varepsilon}; \quad Z_b = -\frac{\alpha}{\varepsilon g}; \quad Z_c = \frac{-1}{6}Z_b.$$

Further reading

An elementary discussion of renormalization is given by Coleman (1970). The treatments by Brézin *et al.* (1976, Section III) and by Amit (1984, Chapters 6 and 7) are at roughly the same level as this chapter. At a more advanced level, see Itzykson and Zuber (1980, Chapter 8), Collins (1983, Chapters 3 and 5) and Zinn-Justin (1989, Chapters 6–9). Dimensional regularization is explained in detail by Collins (1983, Chapter 4). For the renormalization of composite operators, see Zinn-Justin (1989, Chapter 10).

7 The Callan–Symanzik equations

In Chapter 3 we met the word 'renormalization' through the expression 'renormalization group'; recall that our strategy was to follow the evolution of the parameters defining the Hamiltonian through successive contractions of the cutoff Λ, aiming to obtain the long-distance ($r \gg 1/\Lambda$) behaviour of the theory. By contrast, the last chapter seemed to feature renormalization in the completely different context of eliminating a nonphysical (i.e. not directly applicable) bare theory in favour of a renormalized theory depending on just a few parameters. But we can observe already at this point that the renormalized theory, being finite as $\Lambda \to \infty$, is approximately independent of Λ when Λ is large but still finite; at least this is so in the long-distance region $r \gg 1/\Lambda$. Therefore one may hope to find a connection between the theory found by integrating over short-wavelength fluctuations, and the renormalized theory: both can be characterized as theories for long distances.

In quantum field theory too there has been a progressive change in our views about what a renormalized theory signifies. Up to the early 1970s, the bare theory was considered as a mathematical artefact without physical meaning. Today one is more inclined to think of a renormalized theory as a long-distance approximation to a more complex theory not yet known. At long distances however (i.e. for $r \sim 10^{-18}$ m in this case!) the details of this unknown short-distance theory are unimportant, and all we need to know are the few parameters that define the renormalized theory. It is clear that the investigation of critical phenomena has had a strong influence on this change in our point of view.

The results of Chapter 3, Sections 5 and 6, will now be rederived by exploiting a feature of the renormalized theory which we have already noted: namely its invariance under reparametrizations of the mass and of the coupling constant, which stems from the arbitrariness in choosing the finite parts of the counterterms. In fact we shall restrict ourselves to the class of reparametrizations corresponding to variations of the renormalization mass μ (see (6.4.6)). The mathematical expression of such invariance then takes the form of the Callan–Symanzik (CS) equations*. Under certain conditions (i.e. if the function

* In fact the equation originally written by Callan and Symanzik is that appearing in Problem 7.2. However, all equations of a similar form are usually called CS equations, although their meaning may be somewhat different.

$\beta(g)$ has a zero with appropriate properties) one can from these equations derive either the long-distance behaviour of the theory ($r \gg \mu^{-1}$), or its short-distance behaviour ($\Lambda^{-1} \ll r \ll \mu^{-1}$), depending on the properties of this zero. In this way one recovers the property of scale invariance, with exponents governed by anomalous dimensions, or with logarithmic corrections.

Some of the physical content of the CS equations has been discussed already, in Chapter 3, Sections 5 and 6. But we need to examine the equations more closely, for several reasons.

(i) The equations are exact, and follow directly from the existence of a renormalized theory, while the method of Chapter 3 is semi-quantitative at best (see however Polchinski (1984), Kupiainen (1986) and the references cited there).

(ii) They speak the language of the 'first version' of the renormalization group, that of Stueckelberg and Petermann (1953), and of Gell-Mann and Low (1954). This is the language generally used by field theorists.

(iii) They furnish an almost automatic procedure for deriving the consequences of the renormalization group, and permit unambiguous calculations to higher orders in perturbation theory.

(iv) On the debit side, Wilson's method applies to a much wider class of physical systems, since it does not rely on perturbation theory.

This chapter is arranged as follows. In Section 7.1, we make use of the arbitrariness in the renormalization mass μ to derive CS equations for renormalized correlation functions at the critical point (massless case). We introduce the β-function, which describes the coupling-constant flow, and study various possibilities for fixed points. Finally we prove the power law behaviour (scale invariance) of the correlation functions in the low-momentum limit, and compute the anomalous dimensions.

In Section 7.2 we construct the theory at finite T ($T > T_c$) from that previously obtained at the critical temperature. This allows us to derive a CS equation, from which we get the behaviour of correlation functions and the critical indices η and ν.

Section 7.3 is devoted to a comparison between the CS approach and that followed in Chapter 3. We first derive CS equations for the bare theory, by observing that the renormalized one is approximately cutoff-independent in the large-distance ($r \gg \Lambda^{-1}$), or low-momentum ($k \ll \Lambda$) region. These CS equations allow one to study the response of the correlation functions to variations of the cutoff, which is very close in spirit to what was done in Chapter 3. Then a qualitative study of the renormalization-group flow in Wilson's approach will allow us to understand the relation between this approach and the standard one, which was described in Chapter 6. The last two sections contain some technical details on the calculations of the functions $\beta(g)$, $\gamma(g)$, . . . , as well as a computation of critical exponents to order ε^2.

In Sections 7.1 and 7.2, we describe the usual field-theoretical approach to the RG. We have thus attempted to make these two sections independent of the developments of Chapter 3, while they are of course closely related to the material of Chapter 6. By contrast, Section 7.3, which can be omitted in a first reading, relies on the discussions in Chapter 3, particularly those of Sections 3.5 and 3.6. Because of the presentation we have been following, there is some unavoidable duplication between the present Chapter and Chapter 3.

7.1 The Callan–Symanzik equations at the critical temperature

7.1.1 Introduction

We want to derive the Callan–Symanzik (CS) equations in a formalism which can be applied to both critical phenomena and elementary-particle physics. Although the equations look superficially similar in the two cases, one must be aware of some subtleties. In the case of critical phenomena, we are interested in going below the dimension where the theory is renormalizable ($D = 4$), while in the case of elementary-particle physics we wish to stick to the $D = 4$ case. (As will be explained in Chapter 10, $D = 4$ is really the dimension of *space-time*, but an analytical continuation allows one to work in a Euclidean space with dimension $D = 4$).

Let us write the Ginzburg–Landau Hamiltonian in the case $D < 4$ and $n = 1$ in the form

$$H = \int \mathrm{d}^D x \left(\frac{1}{2} (\nabla \varphi)^2 + \frac{1}{2} m_{0c}^2 \varphi^2 + \frac{\Lambda^\varepsilon g_0}{4!} \varphi^4 \right), \tag{7.1.1}$$

where Λ is the cutoff; $\varepsilon = 4 - D$; and as the bare coupling constant u_0 is dimensional, we have introduced a dimensionless g_0 through

$$u_0 = \Lambda^\varepsilon g_0. \tag{7.1.2}$$

The bare mass m_{0c} has been chosen in such a way that the system is at the critical point $T = T_c$, or, equivalently, that the renormalized mass m is zero. m_{0c} is not an independent parameter: it can be computed (at least in principle) as a function of g_0 and Λ. For purely dimensional reasons, u_0 will be large for large values of Λ (see Problem 2.1), and the application of the perturbative methods described in Chapter 5 looks problematic. Another (but not unrelated) difficulty is linked to infrared divergences. At least in perturbation theory, setting $m = 0$ means that we are bound to encounter IR divergences which appear to be catastrophic, since Weinberg's theorem implies that for any value of $D < 4$ they may occur even for non-exceptional values of the external momenta if we go far enough in the perturbative expansion. An example was already provided in Chapter 5. Let us look at another instructive case, and examine the $(n + 1)$-loop graph of Fig. 7.1(a).

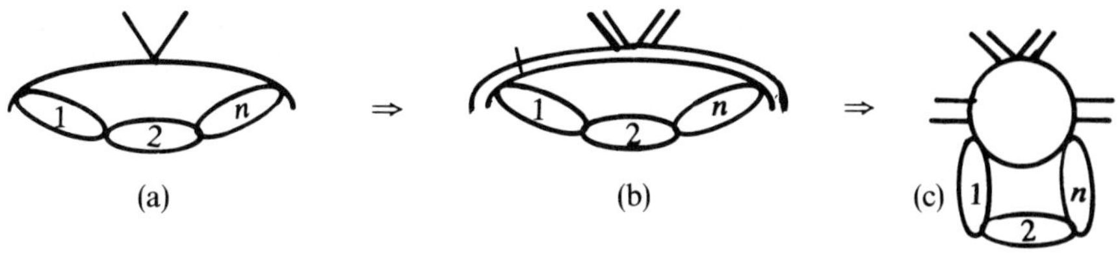

Fig. 7.1 An example of infrared divergences

A possible configuration of the flow of hard momenta is given in Fig. 7.1(b), while the corresponding contracted diagram is drawn in Fig. 7.1(c). As four external lines are connected to the contracted vertex, the superficial degree of divergence is given by (5.6.5):

$$\omega = 4 + (n + 1)(D - 4) = D - n\varepsilon.$$

At fixed ε, the diagram is divergent even in a non-exceptional configuration for $n > D/\varepsilon$. A simple way to rederive this result is as follows: an individual loop behaves, from dimensional analysis, as $k^{-\varepsilon}$, so that the string of loops in Fig. 7.1(a) can be replaced by $k^{-n\varepsilon}$. If we rescale k, $k \to \lambda k$, the Feynman integrand scales as $\lambda^{D-n\varepsilon}$ and IR divergences occur for $D - n\varepsilon < 0$.

The most natural, and probably most satisfactory way out, is to keep a non-zero renormalized mass m in order to regulate the divergences. This method, which is sketched in Problem 7.2, and which is also implicit in the approach of Section 3.5.3, is described in full detail by Itzykson and Drouffe (1989). There are two drawbacks however:

(1) The CS equations derived in this method (which are in fact the equations originally derived by Callan and Symanzik) are inhomogeneous, and the treatment of the RHS is not precisely straightforward.

(2) The equations become singular at $m = 0$, since they rely on the use of the finite-mass normalization conditions (6.3.10), and the continuation to $T = T_c$ and $T < T_c$ is problematic.

We shall follow what seems to be the simplest approach by working directly at $m = 0$. This will lead to homogeneous CS equations, which are easy to solve. Moreover, in the next section, we shall show how to deal with the $m \neq 0$ case, while using results of the massless theory: this is an interesting technical advantage, as Feynman diagrams are easier to evaluate in the massless case. However, there is a price to be paid: as explained above, it is not possible to perform a perturbative expansion for fixed $D < 4$, or $\varepsilon > 0$; it will be necessary to work with a double expansion in powers of ε and g, so that all Feynman integrals will be well defined at any finite order in perturbation theory, and will be guaranteed to be IR finite at nonexceptional points.

In both approaches, the difficulty related to the large value of u_0 will be bypassed by turning to the renormalized theory, where the expansion parameter—the renormalized coupling constant—is small, at least in the vicinity of $D = 4$. This corresponds to a reshuffling of the original perturbative expansion, where the renormalized coupling constant plays the role of a small parameter.

To conclude, it is fair to say that the strategy in which m is kept nonzero is more convincing and more general—as it allows one to work at fixed D—than that in which one sets $m = 0$. However, the latter strategy is simpler to follow and has more direct applications to elementary particle physics, where $D = 4$ and the mass is not a free parameter.

7.1.2 Derivation of the CS equations

We have seen in the previous Chapter that it was possible to obtain, for $D \leqslant 4$ and it the vicinity of $D = 4$, a well-defined renormalized, cutoff-independent theory. Since $m = 0$, the renormalized proper vertices $\Gamma_R^{(N)}$ are specified by the normalization conditions (6.4.6) for $\Gamma_R^{(2)}$ and $\Gamma_R^{(4)}$; as we shall need correlation functions involving the composite operator φ^2, we also recall the normalization conditions for $\Gamma_R^{(2,1)}$. The subtraction point k_{i0} depends on a mass scale $\mu \ll \Lambda$. One notes in (7.1.3c) the factor μ^ε, which ensures that the renormalized coupling constant g is dimensionless. We have

$$\Gamma_R^{(2)}(k^2 = 0) = 0, \tag{7.1.3a}$$

$$\frac{\partial}{\partial k^2} \Gamma_R^{(2)}(k^2 = \mu^2) = 1, \tag{7.1.3b}$$

$$\Gamma_R^{(4)}(k_{i0}) = \mu^\varepsilon g, \tag{7.1.3c}$$

$$\Gamma_R^{(2,1)}(k_{i0}, q_0) = 1. \tag{7.1.3d}$$

The bare proper vertices $\Gamma^{(N)}$ are related to the renormalized ones $\Gamma_R^{(N)}$ through a multiplicative factor $Z_3^{-N/2}(g, \Lambda/\mu)$,

$$\Gamma^{(N)}(k_i, u_0, \Lambda) = Z_3^{-N/2}\left(g, \frac{\Lambda}{\mu}\right) \Gamma_R^{(N)}(k_i, g, \mu).$$

As the bare theory is defined without any reference to the scale μ, it is independent of μ at fixed u_0 and Λ (or g_0 and Λ):

$$\frac{\mathrm{d}}{\mathrm{d}\ln\mu} \Gamma^{(N)}(k_i, u_0, \Lambda)|_{u_0, \Lambda} = 0. \tag{7.1.4}$$

Let us define the functions $\beta(g, \varepsilon)$ and $\gamma(g, \varepsilon)$ through

$$\beta(g, \varepsilon) = \left.\frac{\mathrm{d}g}{\mathrm{d}\ln\mu}\right|_{u_0, \Lambda}, \tag{7.1.5}$$

$$\gamma(g, \varepsilon) = \left.\frac{\mathrm{d}\ln Z_3}{\mathrm{d}\ln\mu}\right|_{u_0, \Lambda}. \tag{7.1.6}$$

The notations in equations (7.1.5) and (7.1.6) anticipate the fundamental property of β and γ: being dimensionless, β and γ can only depend on g, ε, and μ/Λ. The crucial result is that they have finite limits when $\Lambda \to \infty$, as we shall show shortly. We can transform the total derivative in (7.1.4) into partial derivatives thanks to the definitions (7.1.5) and (7.1.6), and we get a partial differential equation, which is nothing but the CS equation obeyed by the proper vertex $\Gamma_R^{(N)}(k_i, g, \mu)$*:

$$\left[\frac{\partial}{\partial\ln\mu} + \beta(g, \varepsilon)\frac{\partial}{\partial g} - \frac{1}{2}N\gamma(g, \varepsilon)\right]\Gamma_R^{(N)}(k_i, g, \mu) = 0. \tag{7.1.7}$$

One should keep in mind the physical interpretation of (7.1.7): although necessary, the normalization scale μ is arbitrary, and the theory should be independent of it. A variation in the renormalization scale must be compensated by a redefinition of the coupling ($\beta(g)$) and of the field normalization ($\gamma(g)$).

In order to show that β and γ are indeed finite in the limit $\Lambda \to \infty$, let us derive a CS equation for an invariant charge; an *invariant charge* is a combination of correlation functions whose renormalization does not require Z_3-factors. For example,

$$\hat{\Gamma}_R(k_i, g, \mu) = \frac{\Gamma_R^{(4)}(k_i, g, \mu)\left(\prod_{i=1}^{4} k_i^2\right)^{1/2}}{\left(\prod_{i=1}^{4} \Gamma_R^{(2)}(k_i, g, \mu)\right)^{1/2}} \tag{7.1.8}$$

is such an invariant charge; notice that it is simply equal to $\mu^\varepsilon g$ at the tree level. The renormalized invariant charge is equal to the bare one,

$$\hat{\Gamma}_R(k_i, g, \mu) = \hat{\Gamma}_R(k_i, g_0, \Lambda), \tag{7.1.9}$$

and it obeys the simple equation

$$\left(\frac{\partial}{\partial\ln\mu} + \beta(g, \varepsilon)\frac{\partial}{\partial g}\right)\hat{\Gamma}_R(k_i, g, \mu) = 0, \tag{7.1.10}$$

* Note the difference between (7.1.7) and the inhomogeneous equation of Problem 7.2. In the latter case, m is a physical parameter.

so that

$$\beta(g, \varepsilon) = -\left[\frac{\partial}{\partial \ln \mu}\hat{\Gamma}_{\mathrm{R}}(k_i, g, \mu)\right]\left[\frac{\partial}{\partial g}\hat{\Gamma}_{\mathrm{R}}(k_i, g, \mu)\right]^{-1}. \tag{7.1.11}$$

The RHS of (7.1.11) is independent of Λ and finite, as it involves the ratio of renormalized quantities, and so must be the LHS. It is then straightforward to show that γ is also Λ-independent and finite. Note that any physical quantity obeys, at the critical point, an equation identical to (7.1.10), since a physical quantity must be μ-independent.

7.1.3 The β-function to one-loop order

Before trying to solve the CS equation, it is instructive to compute the β-function to the lowest nontrivial order, in order to acquire some familiarity with this object and to make some guesses about its properties. If we wish to use the definition (7.1.5) of the β-function, we have to compute the renormalized coupling g as a function of g_0 (or u_0) and Λ. Let us do it to one-loop order. From the normalization condition (7.1.3) and the results of Section 5.5.2 we get

$$g\mu^{\varepsilon} = g_0\Lambda^{\varepsilon} - \frac{1}{2}g_0^2\Lambda^{2\varepsilon}\left[\sum_{i=2}^{4}\int_{q \leqslant \Lambda}\frac{\mathrm{d}^D q}{(2\pi)^D}\frac{1}{q^2(k_1 + k_i - q)^2}\right]_{k_i = k_{i0}}.$$

We have made use of a sharp cutoff in momentum space, but Schwinger's regularization would lead to the same conclusions, and we have taken into account the fact that $Z_3 = 1$ to this order of perturbation theory. Introducing a Feynman parameter x, we get

$$g\mu^{\varepsilon} = g_0\Lambda^{\varepsilon} - \frac{3}{2}g_0^2\Lambda^{2\varepsilon}K_D\int_0^1 \mathrm{d}x\int_0^{\Lambda}\frac{q^{D-1}\mathrm{d}q}{(q^2 + x(1-x)\mu^2)^2}.$$

The q-integral is convergent at infinity if $\varepsilon > 0$; we write

$$\int_0^{\Lambda} = \int_0^{\infty} - \int_{\Lambda}^{\infty},$$

and expand the denominator in the second integral by powers of $(\mu/q)^2$; this allows us to write our final result as an expansion in powers of $(\mu/\Lambda)^2$:

$$g\mu^{\varepsilon} = g_0\Lambda^{\varepsilon} - \frac{3}{2}g_0^2\frac{\Gamma(\varepsilon/2)f(\varepsilon)}{(4\pi)^2}\left(\frac{\Lambda^2}{\mu}\right)^{\varepsilon} + \frac{3}{2\varepsilon}g_0^2K_D\Lambda^{\varepsilon} - \frac{g_0^2}{4(1+\varepsilon/2)}K_D\Lambda^{\varepsilon}\left(\frac{\mu}{\Lambda}\right)^2 + \cdots. \tag{7.1.12}$$

In this equation $f(\varepsilon)$ has been defined in such a way that $f(0) = 1$:

$$f(\varepsilon) = (4\pi)^{\varepsilon/2}\int_0^1 \mathrm{d}x[x(1-x)]^{-\varepsilon/2}.$$

We take the derivative of both sides of (7.1.12) with respect to $\ln \mu$:

$$\varepsilon g \mu^{\varepsilon} + \mu^{\varepsilon} \beta(g, \varepsilon) = 3g_0^2 \frac{\Gamma(1 + \varepsilon/2) f(\varepsilon)}{(4\pi)^2} \left(\frac{\Lambda^2}{\mu}\right)^{\varepsilon} - \frac{g_0^2 K_D}{2(1 + \varepsilon/2)} \Lambda^{\varepsilon} \left(\frac{\mu}{\Lambda}\right)^2 + \cdots .$$

On the RHS of this equation, we can replace g_0 by $g(\mu/\Lambda)^{\varepsilon}$ to this order of perturbation theory, so that we get the following explicit form of $\beta(g, \varepsilon)$:

$$\beta(g, \varepsilon) = -\varepsilon g + \frac{3g^2}{(4\pi)^2} \Gamma\left(1 + \frac{\varepsilon}{2}\right) f(\varepsilon) - \frac{g^2 K_D}{2(1 + \varepsilon/2)} \left(\frac{\mu}{\Lambda}\right)^{2+\varepsilon} + \cdots . \tag{7.1.13}$$

The derivation of (7.1.13), as well as the final result, are extremely instructive and the following points need to be emphasized.

(i) The coefficient of the second term in (7.1.12) is singular when $\varepsilon \to 0$, since $\Gamma(\varepsilon/2) \sim 2/\varepsilon$; however, the differentiation with respect to μ in the calculation of $\beta(g, \varepsilon)$ brings down a factor of ε, which cancels the singularity. Thus the expression giving g as a function of g_0 is singular when $\varepsilon \to 0$, while the β-function is finite.

(ii) For $\varepsilon > 0$, it is possible to take the limit $\Lambda \to \infty$, but we must then modify the relation between u_0 and g_0 and define a different bare coupling constant g_0',

$$u_0 = \mu^{\varepsilon} g_0'. \tag{7.1.14}$$

The relation between g_0' and g is then much simpler:

$$g = g_0' - \frac{3}{2} g_0'^2 \frac{\Gamma(\varepsilon/2) f(\varepsilon)}{(4\pi)^2}.$$

In Section 7.5 we shall show how to compute the β-function from this expression.

(iii) When Λ/μ is large, but finite, and $\varepsilon \to 0$, we can make an ε-expansion of (7.1.13) by using

$$\left(\frac{\Lambda}{\mu}\right)^{\varepsilon} = 1 + \varepsilon \ln \frac{\Lambda}{\mu} + O(\varepsilon^2);$$

this leads to

$$g = g_0 \left(1 + \varepsilon \ln \frac{\Lambda}{\mu}\right) - \frac{3g_0^2}{16\pi^2} \left(\ln \frac{\Lambda}{\mu} + \text{constant}\right),$$

whence we get the ε-expansion of $\beta(g, \varepsilon)$,

$$\beta(g, \varepsilon) = -\varepsilon g + \frac{3g^2}{16\pi^2}. \tag{7.1.15}$$

This can also be obtained straightforwardly from (7.1.13). One also checks that the $D = 4$ limit of $\beta(g, \varepsilon)$ is obtained smoothly from $D < 4$. Setting $\varepsilon = 0$ in the above two equations gives the four-dimensional result.

(iv) One can see that for Λ large, but finite, the dependence of the β-function on Λ is of order $(\mu/\Lambda)^{2+\varepsilon}$.

(v) Finally we have to comment on the *renormalization-scheme dependence* of the β-function. Indeed the β-function is not uniquely defined; it depends on the details of the renormalization procedure. For simplicity, let us limit ourselves to $D = 4$ and write the perturbative expansion of $\beta(g)$,

$$\beta(g) = \beta_0 g^2 + \beta_1 g^3 + \cdots . \tag{7.1.16}$$

The remarkable result is that β_0 and β_1 are *independent of the renormalization scheme*, and thus universal: they depend only on the field theory under consideration. To prove this result, let us remark that, if g' is the coupling in another renormalization scheme, we have

$$g' = g + cg^2 + \cdots,$$

or conversely

$$g = g' - cg'^2 + \cdots .$$

Let us take the derivative of the second equation with respect to $\ln \mu$ and use the expansion (7.1.16):

$$\beta_0 g^2 + \beta_1 g^3 + \cdots = \beta'_0 g'^2 + \beta'_1 g'^3 - 2cg'(\beta_0 g' + \cdots).$$

Expressing g' as function of g on the RHS of the above equation, one gets immediately

$$\beta'_0 = \beta_0, \quad \beta'_1 = \beta_1 .$$

7.1.4 Fixed points

The expression (7.1.15) of the β-function to one-loop order is drawn in Fig. 7.2: the most remarkable feature of $\beta(g, \varepsilon)$ is its zero at

$$g = g^* = \frac{16\pi^2 \varepsilon}{3}. \tag{7.1.17}$$

This zero is a *fixed point* of the coupling-constant flow, since $\mathrm{d}g/\mathrm{d}\ln\mu$ vanishes at that point. Another fixed point is of course the origin $g = 0$. It is not enough to know about the existence of a fixed point: one must also study its stability. Let us consider the point $g = g^*$ and start from an initial value $g(\mu) = g$; we then rescale the renormalization mass μ by a factor s: $\mu \to \mu s$, and let s go to zero ($\ln s \to -\infty$). To the renormalization mass μs will correspond a coupling constant $\bar{g}(s, g)$, such that $\bar{g}(1, g) = g$. For simplicity we write $g(s) = \bar{g}(s, g)$ and $g(1) = g(\mu) = g$. Notice that, in order to avoid too many different notations, we may take the argument of g to be dimensionless (s) or dimensional (μ), when

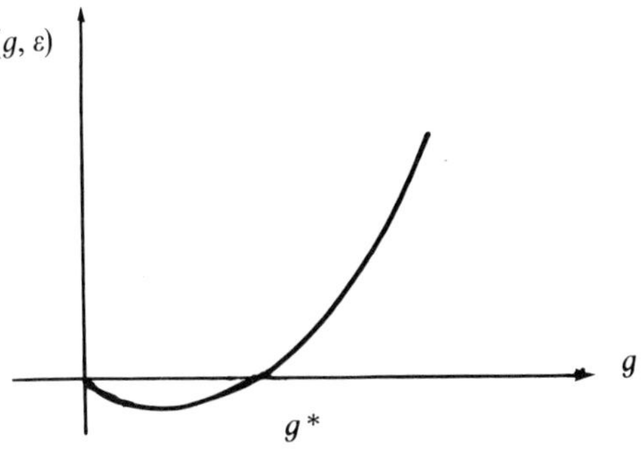

Fig. 7.2 The function $\beta(g, \varepsilon)$ to one-loop order

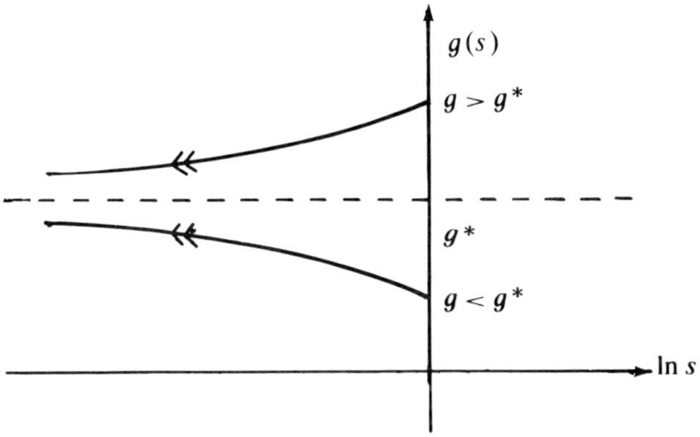

Fig. 7.3 Infrared stability

there is no possible ambiguity. We have*

$$\frac{\mathrm{d}g(s)}{\mathrm{d}\ln s} = \beta(g(s)); \quad g(1) = g, \tag{7.1.18}$$

or

$$\ln s = \int_g^{g(s)} \frac{\mathrm{d}g'}{\beta(g')}. \tag{7.1.19}$$

(The careful reader will notice a minus sign with respect to (3.6.10); the reason is that, in the renormalized theory, μ plays the role of a cutoff. In field theory, it is customary to dilate μ, while in Chapter 3 we contracted the cutoff Λ by s.) Assume first that $g(1) = g < g^*$, so that $\beta(g, \varepsilon) < 0$. When s decreases,

* In Sections (7.1.1) to (7.1.3), we sometimes omit the ε-dependence of β and γ.

$\mathrm{d}g(s)/\mathrm{d}\ln s < 0$ and $g(s)$ is driven to a fixed point g^* (Fig. 7.3). Assume next that the initial value $g > g^*$: $\beta(g, \varepsilon)$ is now positive and $g(s)$ is again driven to g^*.

We thus discover that g^* is a stable fixed point when the renormalization scale tends to zero: one speaks then of an *infrared-stable fixed point*. By contrast, if $0 < g(1) = g < g^*$, it is easy to see that the origin is a stable fixed point for $s \to \infty$: one then speaks of an *ultraviolet-stable fixed point*. Naturally an UV-stable fixed point need not be located at the origin (Fig. 7.4).

To summarize: to a zero g^* of the β-function, $\beta(g^*, \varepsilon) = 0$, there corresponds a fixed point for the flow of the coupling constant. If the derivative $\beta'(g^*, \varepsilon)$ is positive, the fixed point is IR-stable; if it is negative, the fixed point is UV-stable (Fig. 7.5).

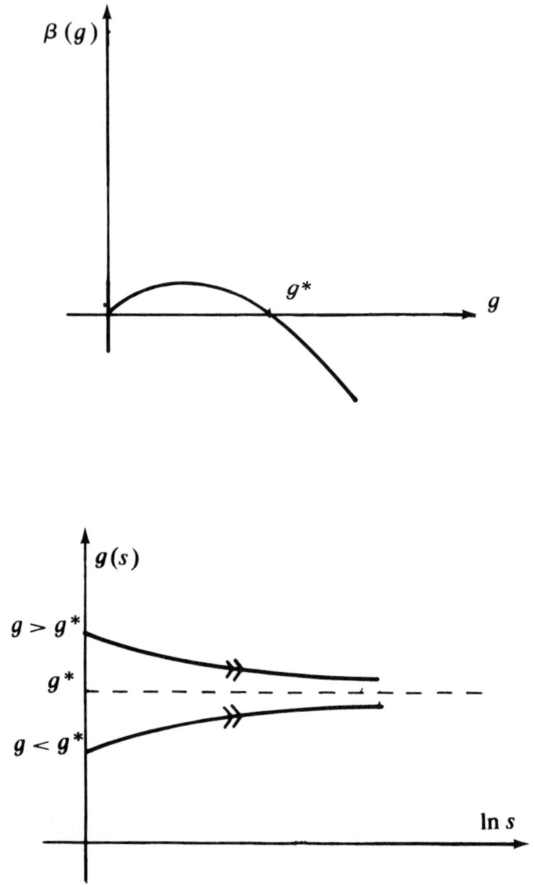

Fig. 7.4 Ultraviolet stability

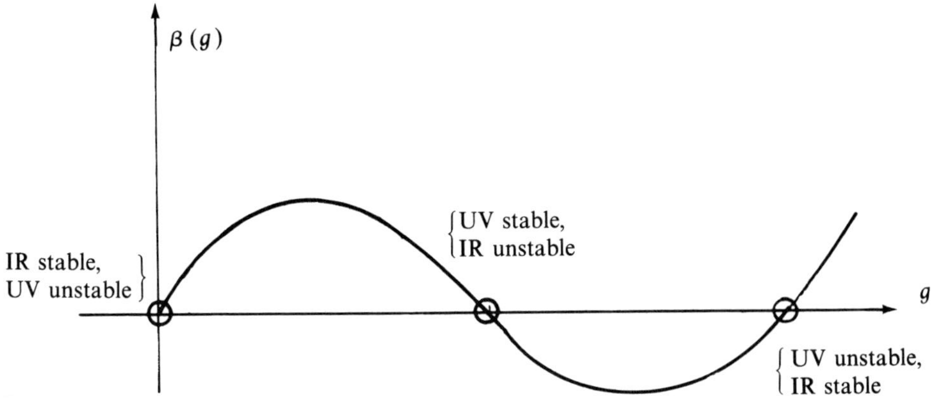

Fig. 7.5 A hypothetical configuration of fixed points

7.1.5 Solution of the Callan–Symanzik equation

Before trying to solve the full CS equation (7.1.7), let us examine the simpler case of an invariant charge (7.1.8). Equation (7.1.10) expresses the μ-independence of $\hat{\Gamma}_R(k_i, g, \mu)$ and is equivalent to

$$\frac{\mathrm{d}}{\mathrm{d}\ln s}\hat{\Gamma}_R(k_i, g(s), \mu(s)) = 0,$$

where $\mu(s) = \mu s$ and $g(s)$ has been defined in (7.1.19). Thus the solution of (7.1.10) is simply

$$\hat{\Gamma}_R(k_i, g, \mu) = \hat{\Gamma}_R(k_i, g(s), \mu s). \tag{7.1.20}$$

A more detailed way of solving (7.1.10) is given in Problem 7.1. It is now straightforward to check that the solution of (7.1.7) reads

$$\Gamma_R^{(N)}(k_i, g(s), \mu(s)) = \exp\left(\frac{N}{2}\int_g^{g(s)} \frac{\gamma(g')\mathrm{d}g'}{\beta(g')}\right)\Gamma_R^{(N)}(k_i, g, \mu), \tag{7.1.21}$$

since both the partial differential equation and the boundary condition at $s = 1$ are satisfied. It is often useful to write (7.1.21) by rescaling the momenta rather than the renormalization mass μ. This is easily done from ordinary dimensional analysis, which tells us that

$$\Gamma_R^{(N)}(k_i, g, \mu) = \mu^{d_N} F^{(N)}\left(\frac{k_i}{\mu}, g\right), \tag{7.1.22}$$

where

$$d_N = D - N\left(\frac{D}{2} - 1\right) \tag{7.1.23}$$

is nothing but the canonical dimension of $\Gamma^{(N)}$ (cf. (5.6.4)). From (7.1.22) we get

$$\Gamma_{\mathrm{R}}^{(N)}(sk_i, g, \mu s) = s^{d_N}\Gamma_{\mathrm{R}}^{(N)}(k_i, g, \mu), \tag{7.1.24}$$

and combining (7.1.21) and (7.1.24) yields

$$\Gamma_{\mathrm{R}}^{(N)}(sk_i, g, \mu) = s^{D-N(D/2-1)}\exp\left(-\frac{1}{2}N\int_g^{g(s)}\frac{\gamma(g')\mathrm{d}g'}{\beta(g')}\right) \times \Gamma_{\mathrm{R}}^{(N)}(k_i, g(s), \mu). \tag{7.1.25}$$

7.1.6 Application to critical phenomena and asymptotic freedom

In the theory of critical phenomena, the interesting region is that of low momenta (or long wavelengths), $k \to 0$; since, in the renormalized theory, μ is the only scale at our disposal, we have to look at the region $k_i \ll \mu$, and from (7.1.25) we see that it can be reached if $s \to 0$. We now make the crucial assumption that the β-function keeps the same shape as suggested by perturbation theory to one loop order, with an IR-stable zero as displayed in Fig. 7.2. This will lead later on to typically non-perturbative results. In the limit $s \to 0$, the coupling $g(s)$ is driven to the IR fixed point g^*. For simplicity, we restrict ourselves to the $N = 2$ case, where $d_N = 2$. The integral in the exponential of (7.1.25) is dominated by the region $g \to g^*$, or $\ln s \to -\infty$; thus

$$\int_g^{g(s)}\frac{\gamma(g')\mathrm{d}g'}{\beta(g')} = \int_0^{\ln s}\gamma(g(s'))\mathrm{d}\ln s' \approx \gamma(g^*, \varepsilon)\ln s, \tag{7.1.26}$$

so that in the $N = 2$ case (7.1.25) becomes

$$\Gamma_{\mathrm{R}}^{(2)}(sk_i, g, \mu) \underset{s\to 0}{\sim} s^{2-\gamma(g^*, \varepsilon)}\Gamma_{\mathrm{R}}^{(2)}(k_i, g^*, \mu). \tag{7.1.27}$$

We have assumed that the zero of the β-function at $g = g^*$ is a simple one and that $\gamma(g^*, \varepsilon)$ is nonzero. Equation (7.1.27) allows one to identify the critical exponent η, although we have worked with renormalized correlation functions, while the physical ones are the bare correlation functions $\Gamma^{(N)}$. However $\Gamma_{\mathrm{R}}^{(N)}$ and $\Gamma^{(N)}$ are related through a multiplicative, k-independent factor, so that (7.1.27) also holds for $\Gamma^{(2)}$, which thus behaves for small k as $k^{2-\gamma(g^*, \varepsilon)}$. The critical exponent η is then

$$\eta = \gamma(g^*, \varepsilon). \tag{7.1.28}$$

As in Chapter 3, this equation can be interpreted by attributing to the field φ an anomalous dimension

$$d_\varphi = \frac{D}{2} - 1 + \frac{\eta}{2}. \tag{7.1.29}$$

The occurrence of anomalous dimensions ($\eta \neq 0$) can be traced back to the necessity of introducing a mass scale μ, which breaks naïve scale invariance. Let us now examine the UV behaviour $s \to \infty$. For $D < 4$, the origin is a UV-stable fixed point. The flow equation shows that as $s \to \infty$ (more precisely for $\mu \ll sk_i \ll \Lambda$), $g(s)$ behaves as a power,

$$g(s) \underset{s \to \infty}{\sim} s^{-\varepsilon},$$

since the integral in (7.1.19) is controlled by the small g region, where $\beta(g, \varepsilon) \approx -\varepsilon g$. We can even relate $g(s)$ to the bare coupling constant u_0 by

$$g(s) \approx u_0 \mu^{-\varepsilon} s^{-\varepsilon}.$$

Since $g(s) \to 0$, the exponential in (7.1.25) can be shown to be equal to $Z_3^{N/2}$. If we set for example $N = 4$ in equation (7.1.27), we get for the *bare* proper vertex

$$\lim_{s \to \infty} \Gamma^{(4)}(sk_i) = u_0.$$

In the limit of large momenta, the bare correlation functions tend to their tree-level expression. This is related to the fact that the φ^4-theory is superrenormalizable in $D < 4$, and can be shown to have a trivial UV behaviour. However, as we shall see later on, the field theory which is relevant for critical phenomena behaves as a renormalizable theory, because in this case g is fixed at g^*.

In dimension four, and more generally in the dimension where the field theory under consideration is renormalizable, we have two possibilities:

(i) as in the case of the φ^4-theory, or of quantum electrodynamics (Chapter 12), the origin is an IR-stable fixed point;

(ii) as in the case of the φ^3-theory in $D = 6$ (Problem 7.4) or of non-Abelian gauge theories (Chapter 13) the origin is a UV fixed point (Fig. 7.6).

Let us look first at the former case: then $\lim_{s \to 0} g(s) = 0$, and $g(s) \sim |\ln s|^{-1}$ for small s. The proper vertex on the RHS of (7.1.25) can be calculated perturbatively, while the scaling behaviour (s^{d_N}) is modified by powers of $\ln s$, as was already shown in Section 3.6. In dimension four, the critical behaviour is not given by pure power laws; there are in addition powers of logarithms. On the other hand, the theory seems to be in bad shape for large momenta. Unless there exists another fixed point, or unless $\beta(g)$ grows no faster than g as $g \to \infty$, the coupling $g(s)$ grows without bounds for a *finite* value of s, and perturbation theory becomes meaningless. The existence of another fixed point for the φ^4-theory has not been rigorously excluded, but seems unlikely from numerical simulations and other arguments. Indeed it is widely believed that a nontrivial φ^4-theory cannot be defined in $D = 4$ in the limit of infinite cutoff (but the φ^4-theory can nevertheless be relevant for physics in a wide range of momenta for large, but finite Λ). This 'φ^4-triviality' will be examined further in Section 7.3.

By contrast, the theory is under control when the origin is a UV-stable fixed point. The coupling constant $g(s)$ tends to zero as $(\ln s)^{-1}$ when $s \to \infty$: such

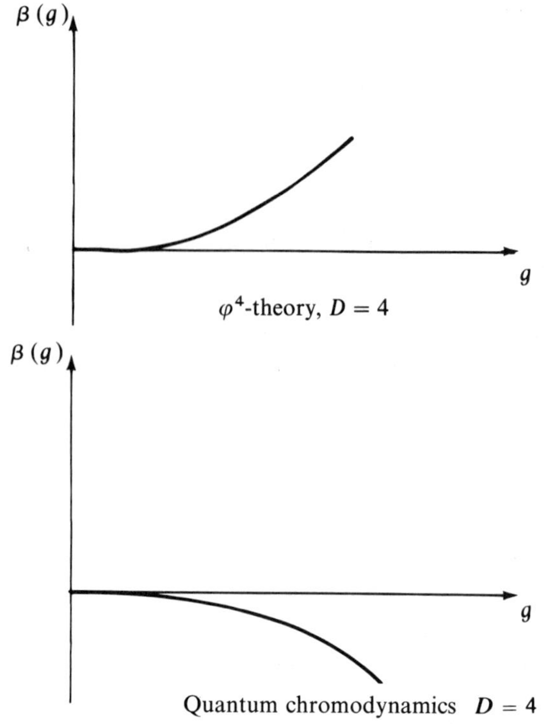

Fig. 7.6 Fixed points at the origin in $D = 4$

a theory is called *asymptotically free*. The large-momentum behaviour of correlation functions is calculable from perturbation theory (or more precisely from RG-improved perturbation theory: see Section 7.4). As we shall see in Chapter 13, asymptotic freedom in the case of quantum chromodynamics has allowed, for the first time, reliable calculations of strong interaction processes.

7.2 The Callan–Symanzik equations for $T > T_c$

When T is larger than the critical temperature, or, equivalently, when the renormalized mass m is nonzero, one can use m as a control parameter and study the variations of correlation functions with m. One starts from the normalization conditions (6.3.10) in order to define properly the renormalized correlation functions, and one derives (inhomogeneous) CS equations (Problem 7.2). As already mentioned, this method is probably the most satisfactory one; however, since the normalization conditions (6.3.10) cannot be used when $m = 0$, it is not possible to reach continuously the critical temperature and the region $T < T_c$. We shall find it preferable to construct the $T \neq T_c$ theory from that at $T = T_c$, using the normalization conditions (7.1.3), which are, by construction, regular at $T = T_c$. Another (technical) advantage will be that the renormalization constants Z_3, Z and $\bar{Z}$ will be computed in the massless theory, where Feynman diagrams are easier to evaluate.

7.2.1 Derivation of the CS equations for $T > T_c$

We start from the following (bare) Hamiltonian density (see (6.5.15))

$$\mathscr{H}_0(x) = \frac{1}{2}(\nabla\varphi_0)^2 + \frac{\Lambda^\varepsilon}{4!} g_0 \varphi_0^4 + \frac{1}{2} m_{0c}^2 \varphi_0^2 + \frac{1}{2} t\bar{Z}\varphi_0^2. \tag{7.2.1}$$

Let us remark once more that m_{0c} is not an independent parameter, but is determined by g_0 and Λ; if one uses a dimensional regularization ($\Lambda \to \infty$), m_{0c} is simply zero. The renormalization constant $\bar{Z}$ is associated with the renormalization of the composite operator φ^2 (see Section 6.5.2). The coefficient of $\frac{1}{2}\varphi^2$ is by definition the bare mass squared, m_0^2, and is given by

$$m_0^2 = m_{0c}^2 + t\bar{Z} \tag{7.2.2}$$

or

$$\bar{Z}t = m_0^2 - m_{0c}^2. \tag{7.2.3}$$

Then the parameter t can be interpreted as a renormalized temperature difference $T - T_c$; let us recall that $m_0^2 - m_{0c}^2$ ($= r_0 - r_{0c}$ with the notations of statistical physics) is proportional to $T - T_c$ (see (2.1.11)).

Thanks to the identity (6.5.5) one can write for the bare correlation functions

$$\Gamma^{(N)}(m_{0c}^2 + \bar{Z}t) = \sum_L \frac{\bar{Z}^L t^L}{L!} \int \Gamma^{(N,L)}(\ldots; m_{0c}^2)\mathrm{d}y_1 \ldots \mathrm{d}y_L. \tag{7.2.4}$$

Actually we are ignoring a difficulty: because of infrared divergences, one should begin by choosing instead of a constant t a function $t(x)$ vanishing at infinity. A full treatment can be found in the references. We now multiply both sides of (7.2.4) by $Z_3^{N/2}$ and use equation (6.5.16):

$$\Gamma_R^{(N)}(k_i, t, g) = \sum_L \frac{t^L}{L!} \Gamma_R^{(N,L)}(k_i, q_j = 0, t = 0, g). \tag{7.2.5}$$

The RHS of (7.2.5) contains finite (renormalized) correlation functions $\Gamma_R^{(N,L)}$; if the sum over L converges, then $\Gamma_R^{(N)}(k_i, t, g)$ is a finite quantity, while the renormalization constants Z_3 and $\bar{Z}$ have been evaluated in the *massless* theory.

The function $\Gamma_R^{(N,L)}$ obeys a CS equation

$$\left[\frac{\partial}{\partial \ln \mu} + \beta(g, \varepsilon)\frac{\partial}{\partial g} - \frac{1}{2}N\gamma(g, \varepsilon) - L\bar{\gamma}(g, \varepsilon)\right]\Gamma_R^{(N,L)} = 0,$$

with

$$\bar{\gamma}(g, \varepsilon) = \left.\frac{\mathrm{d}\ln \bar{Z}}{\mathrm{d}\ln \mu}\right|_{u_0,\, \Lambda}. \tag{7.2.6}$$

Since

$$\sum_L \frac{Lt^L}{L!} f^{(L)}(0) = \frac{\partial f(t)}{\partial \ln t},$$

one obtains the following CS equation for $\Gamma_{\mathrm{R}}^{(N)}$ at $T \neq T_{\mathrm{c}}$:

$$\left[\frac{\partial}{\partial \ln \mu} + \beta(g, \varepsilon)\frac{\partial}{\partial g} - \frac{1}{2}N\gamma(g, \varepsilon) - \bar{\gamma}(g, \varepsilon)\frac{\partial}{\partial \ln t}\right]\Gamma_{\mathrm{R}}^{(N)}(k_i, t, g, \mu) = 0. \tag{7.2.7}$$

7.2.2 Critical exponents

A simple generalization of the method used in the previous section allows us to write the solution of (7.2.7) (see Problem 7.1) as

$$\begin{aligned}\Gamma_{\mathrm{R}}^{(N)}(k_i, t, g, \mu) = {} & s^{D-N(D/2-1)}\exp\left(-\frac{1}{2}N\int_0^{\ln s}\gamma(g(s'))\mathrm{d}\ln s'\right)\\ & \times \Gamma_{\mathrm{R}}^{(N)}\left(\frac{k_i}{s}, \frac{t(s)}{s^2}, g(s), \mu\right),\end{aligned} \tag{7.2.8}$$

where $t(s)$ is defined by

$$t(s) = t\exp\left(-\int_0^{\ln s}\bar{\gamma}(g(s'))\mathrm{d}\ln s'\right). \tag{7.2.9}$$

The reader will note that $t(s)$ as defined in (7.2.9) differs by a factor s^2 from $t(s)$ defined in (3.6.13). Let us now choose s so that

$$\frac{t(s)}{s^2\mu^2} = 1. \tag{7.2.10}$$

The physical meaning of this choice is as follows: we are interested in the critical region $t \ll \mu^2$, which cannot be reached in perturbation theory. However, on the RHS of (7.2.8) the temperature parameter $t(s)/s^2$ is equal to μ^2, so that we are very far from the critical region and the correlation function is regular. From (7.2.9) we can rewrite (7.2.10) as

$$\frac{t}{\mu^2} = \exp\left(\int_0^{\ln s}(2 + \bar{\gamma}(g(s'))\mathrm{d}\ln s')\right).$$

In the critical region $t/\mu^2 \to 0$, $s \to 0$ and $g(s) \to g^*$. The integral is dominated by large and negative values of $\ln s'$, so that

$$\frac{t}{\mu^2} \simeq e^{(2 + \bar{\gamma}(g^*, \varepsilon))\ln s} = s^{2 + \bar{\gamma}(g^*, \varepsilon)}.$$

Let us set

$$\nu = \frac{1}{2 + \bar{\gamma}(g^*, \varepsilon)}, \tag{7.2.11}$$

where the identification of ν with a critical exponent will be made shortly. This gives

$$s \simeq \left(\frac{t}{\mu^2}\right)^{\nu},$$

and (7.2.8) can be transformed into

$$\Gamma_{\mathrm{R}}^{(N)}(k_i, t, g, \mu) = \left[\frac{t}{\mu^2}\right]^{\nu[D - N(D/2 - 1) - (1/2)N\gamma(g^*, \varepsilon)]}$$
$$\times \Gamma_{\mathrm{R}}^{(N)}\left(\frac{k_i}{(t/\mu^2)^{\nu}}, \mu^2, g^*, \mu\right). \tag{7.2.12a}$$

Specializing to $N = 2$, we find that $\Gamma_{\mathrm{R}}^{(2)}$ behaves as

$$\Gamma_{\mathrm{R}}^{(2)}(k, t, g, \mu) \simeq \mu^2 \left(\frac{t}{\mu^2}\right)^{\nu(2 - \eta)} F\left(\frac{k}{\mu}\left(\frac{t}{\mu^2}\right)^{-\nu}\right), \tag{7.2.12b}$$

where $\eta = \gamma(g^*, \varepsilon)$ and F is a dimensionless function. One notices the following points.

(a) The quantity with dimension 1, $\mu(t/\mu^2)^{\nu}$, must be identified with the renormalized mass $m = \xi^{-1}$, since the function F in (7.2.12b) depends only on the ratio $k/m = k\xi$. The mass defined in this way differs at most by a finite renormalization from that used in the normalization conditions (6.3.10).

(b) Equation (7.2.12b) can of course be written as

$$\Gamma_{\mathrm{R}}^{(2)} \simeq \mu^{\eta} k^{2-\eta} G(k\xi).$$

Comparison with (1.4.9) allows us to complete the identification of the critical exponents η and ν:

$$\eta = \gamma(g^*, \varepsilon); \quad \nu = [2 + \bar{\gamma}(g^*, \varepsilon)]^{-1}. \tag{7.2.13}$$

These equations, as in Chapter 3, allow an explicit check on the property of *universality*: η and ν depend only on D (or ε), and on the dimension n of the order parameter (we have always chosen $n = 1$, but it would be straightforward to treat the general case). The critical exponents do not depend on details of the GL Hamiltonian, for example the value of the bare coupling constant. The exponent ν is of course related to the anomalous dimension d_{φ^2} of the composite operator φ^2:

$$d_{\varphi^2} = D - \frac{1}{\nu}.$$

Let us conclude this section with the following comments.

(i) In the renormalized theory, the parameter μ fixes the mass scale, and the critical theory is defined by $k_i \ll \mu, t \ll \mu^2$. Thus μ plays the role of Λ in the bare theory.

(ii) By construction, the proper vertices in (7.2.12) have a finite limit when $t \to 0$. Let us consider the case $N = 2$, although the results are valid for all N. If we fix k and let t go to zero, the argument of the function F in (7.2.12b) tends to infinity. The factor $t^{\nu(2-\eta)}$ must be compensated in order to yield a finite limit, which implies

$$\lim_{x \to \infty} F(x) = x^{2-\eta}$$

and

$$\Gamma_{\mathrm{R}}^{(2)}(k, t, g, \mu) \underset{k \gg m}{\simeq} k^{2-\eta}. \tag{7.2.14}$$

One thus recovers scale invariance in the *ultraviolet* region $k/m \gg 1$, or $r \ll \xi$ in configuration space. One should remark that the UV behaviour of the *massive* theory is controlled by the same exponent as that found in the IR behaviour ($k/\mu \ll 1$) of the *massless* theory. This looks paradoxical, as one would expect the UV behaviour of the massive theory to be controlled by the UV stable fixed point at the origin. However, the limits $s = k/m \to \infty$ and $g \to g^*$ do not commute, and the critical behaviour is controlled by the latter. Indeed, for Λ large, m and g are related to the bare coupling constant u_0 by (see Problem 7.2)

$$u_0 = \Lambda^{\varepsilon} g_0 = \frac{(\text{constant}) m^{\varepsilon}}{(g^* - g)^{\varepsilon/\omega}}, \tag{7.2.15}$$

where $\omega = \beta'(g^*) = \varepsilon + O(\varepsilon^2)$. Letting now $\Lambda \to \infty$ at fixed g_0 and m shows that $g \to g^*$: at the critical point, the coupling constant of the renormalized theory is $g = g^*$. Comparison with Section 3.5.3 shows that, while we had in that case Λ finite, but $m \to 0$, in both cases the effective bare coupling constant goes to ∞. Further comments on this point can be found in Sections 7.3.3 and 7.5.1.

(iii) A slightly more precise definition of m would be

$$m = s\mu = \mu \exp\left[\int_g^{g(s)} \frac{\mathrm{d}g'}{\beta(g')}\right] \simeq \mu(t/\mu^2)^{\nu},$$

while the solution (7.2.8) of the CS equation can also be written

$$\Gamma_{\mathrm{R}}^{(N)}(k_i, t, g, \mu) = \tilde{Z}_3^{-N/2}\left(\frac{m}{\mu}, g(s)\right)\tilde{\Gamma}_{\mathrm{R}}^{(N)}(k_i, g(s), m),$$

with

$$\tilde{Z}_3\left(\frac{m}{\mu}, g(s)\right) = \exp\left(\int_g^{g(s)} \frac{\gamma(g')}{\beta(g')}\mathrm{d}g'\right) \simeq \left(\frac{m}{\mu}\right)^{\eta}.$$

The proper vertex $\tilde{\Gamma}_{\mathrm{R}}^{(N)}$ differs at most by a finite renormalization from that defined by the renormalization conditions (6.3.10). This proper vertex is singular in the limit $m \to 0$, as it should be, since $\Gamma_{\mathrm{R}}^{(N)}$ is regular in this limit and $\tilde{Z}_3$ is singular.

(iv) The critical exponents α, β, δ, and γ are obtained from η and ν by using the scaling laws (3.2.16). These scaling laws could be rederived, in the formalism of the present chapter, by studying the free energy and the equation of state for $T < T_c$ (see the references).

7.2.3 The role of irrelevant operators (Brézin 1980)

The previous discussion relied on the use of a GL Hamiltonian with a φ^4-coupling only, and one may wonder about the role of non-renormalizable interactions like φ^6, $\varphi^2(\nabla\varphi)^2$ etc. Indeed, a term like $u_6^0\varphi^6$ has been excluded on purely dimensional grounds: in dimension four for example, the coupling constant u_6^0 has dimension -2, and must be written in the bare theory as $\Lambda^{-2}g_6^0$, where g_6^0 is dimensionless. One can then hope that a φ^6-interaction is negligible for large values of Λ. However, this expectation is too naïve: when a composite operator like φ^6 is inserted in a correlation function, it induces further divergences and the global effect is not evident. Let us for instance insert φ^6 in $\Gamma^{(2)}$ and $\Gamma^{(4)}$ (Fig. 7.7):

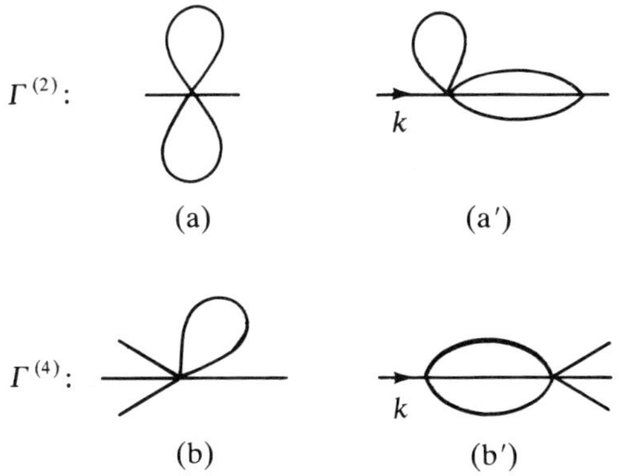

Fig. 7.7 Insertion of φ^6 in $\Gamma^{(2)}$ and $\Gamma^{(4)}$

Graph 7.7(a) diverges as $\Lambda^4 \times$ constant, while graph 7.7(a′) contains a divergent piece $\Lambda^4 \times$ constant and another one $\sim \Lambda^2 \ln(k/\Lambda)$. Multiplying by $u_6^0 \sim \Lambda^{-2}$, we see that these graphs are going to shift the critical temperature and to change the field normalization. Graph 7.7(b) contains a divergent piece $\sim \Lambda^2 \times$ constant, while 7.7(b′) contains $\Lambda^2 \times$ constant and $\ln(k/\Lambda)$ divergences. Multiplying by u_6^0, we see that the contribution of these diagrams can be absorbed into a redefinition of the renormalized coupling constant. These redefinitions lead to corrections to scaling laws $\sim (k/\Lambda)^{\varepsilon}$ (Problem 7.8), which correspond precisely to the existence of an eigenvalue $y_2 = -\varepsilon$ in the formalism of Section 3.2.3. This heuristic discussion is confirmed by a rigorous analysis, which relies on the renormalization of composite operators like φ^6, $\varphi^2(\nabla\varphi)^2$ etc., inserted into correlation functions. However, this analysis is cumbersome, because all these operators are coupled by renormalization: the renormalization constants form a matrix. Once one has taken care of the redefinition of the field and of the coupling constant as well as of the displacement of the critical temperature, all corrections due to irrelevant operators behave as $\Lambda^{-2+O(\varepsilon)}$. Their influence on the critical behaviour is negligible, which gives a further confirmation of the property of universality.

7.3 Renormalization and the renormalization group*

The previous sections relied heavily on the use of the renormalized theory. While this approach is quite natural if one has in mind applications to elementary-particle physics, where the renormalized correlation functions are the physical ones, this is not the case in applications to critical phenomena, where the bare correlation functions are the physical ones. In the present section, we shall explore the possibility of working directly with the bare theory: in Section 7.3.1, it will be shown that the bare correlation functions obey CS equations, which allow one to recover the results already obtained in the renormalized theory. However, we also want to go further, by exploring the relationship between the CS version of the RG group and Wilson's version, which was described in Chapter 3. This will allow us to understand in a deeper fashion the connection between the bare and the renormalized theory.

7.3.1 Callan–Symanzik equations for the bare correlation functions

As already emphasized at the beginning of this Chapter, our basic assumption is the existence of a renormalized theory valid at large distances, or low momenta: for $k_i \ll \Lambda$, the renormalized correlation functions are approximately Λ-independent (with an accuracy $\sim (k/\Lambda)^2$ in $D = 4$: Problem 6.1) and tend to a finite limit when $\Lambda \to \infty$. For simplicity, we consider the massless case ($T = T_c$) and use the normalization conditions

* This section can be omitted in a first reading.

(7.1.3). The Λ-independence of $\Gamma_{\mathrm{R}}^{(N)}$ is expressed by

$$\frac{\mathrm{d}}{\mathrm{d}\ln\Lambda}\Gamma_{\mathrm{R}}^{(N)}|_{g,\mu} = O\left(\frac{(\ln\Lambda)^p}{\Lambda^2}\right), \tag{7.3.1}$$

where p depends on the order of the perturbation expansion. However, the bound in (7.3.1) is uniform for $D \leqslant 4$, including $D = 4$. If we assume that powers of logarithms do not sum up to compensate for the Λ^{-2} factor, we can write

$$\frac{\mathrm{d}}{\mathrm{d}\ln\Lambda}\left[Z_3^{N/2}\left(g, \frac{\Lambda}{\mu}\right)\Gamma^{(N)}(k_i, g_0, \Lambda)\right]_{g,\mu} \simeq 0.$$

As in Section 7.1, this equation is transformed into a partial differential equation by introducing two functions $\beta(g_0, \varepsilon)$ and $\gamma(g_0, \varepsilon)$,

$$\beta(g_0, \varepsilon) = \left.\frac{\mathrm{d}g_0}{\mathrm{d}\ln\Lambda}\right|_{g,\mu}, \tag{7.3.2}$$

$$\gamma(g_0, \varepsilon) = -\left.\frac{\mathrm{d}\ln Z_3}{\mathrm{d}\ln\Lambda}\right|_{g,\mu}. \tag{7.3.3}$$

Although these functions differ from those defined in (7.1.5) and (7.1.6), we keep the same notation. We get the CS equation for the bare correlation functions

$$\left[\frac{\partial}{\partial\ln\Lambda} + \beta(g_0, \varepsilon)\frac{\partial}{\partial g_0} - \frac{1}{2}N\gamma(g_0, \varepsilon)\right]\Gamma^{(N)}(k_i, g_0, \Lambda) = 0. \tag{7.3.4}$$

To be consistent with the notations of Chapter 3, we define a cutoff *contraction* $\Lambda \to \Lambda/s$, so that (7.3.2) becomes

$$\frac{\mathrm{d}g_0(s)}{\mathrm{d}\ln s} = -\beta(g_0(s), \varepsilon), \quad g_0(1) = g_0, \tag{7.3.5a}$$

or, equivalently,

$$\ln s = -\int_{g_0}^{g_0(s)} \frac{\mathrm{d}g'}{\beta(g', \varepsilon)}. \tag{7.3.5b}$$

As in Section 7.1, the crucial property of $\beta(g_0, \varepsilon)$ and $\gamma(g_0, \varepsilon)$ is that they are Λ-independent, within an accuracy $\sim(\mu/\Lambda)^2$. The proof is identical to that given previously. We can write without further comments the solution of (7.3.4) as

$$\Gamma^{(N)}(k_i, g_0, \Lambda) = \exp\left(-\frac{1}{2}N\int_{g_0}^{g_0(s)}\frac{\gamma(g')\mathrm{d}g'}{\beta(g', \varepsilon)}\right)\Gamma\left(k_i, g_0(s), \frac{\Lambda}{s}\right), \tag{7.3.6}$$

which can also be rewritten in the form

$$\Gamma^{(N)}(k_i, g_0, \Lambda) = \frac{Z_3^{N/2}(g, \Lambda/\mu s)}{Z_3^{N/2}(g, \Lambda/\mu)}\Gamma^{(N)}\left(k_i, g_0(s), \frac{\Lambda}{s}\right).$$

This equation is clearly reminiscent of the equation relating bare and renormalized correlation functions. Further comments on this point will be found in the next Section. The critical behaviour is obtained by noticing that $g_0(s)$ is driven to the fixed point g_0^*, which, to one loop order, is given by the same value as was found previously for g^* (see also Chapter 3),

$$g_0^* = \frac{16\pi^2\varepsilon}{3}.$$

Then, following the same reasoning as in Section 7.1, one easily derives that the correlation function $\Gamma^{(2)}$ behaves as a power of k and one identifies the critical exponent η,

$$\eta = \gamma(g_0^*, \varepsilon).$$

Details of the proof are left to the reader. However, before leaving this topic, let us remark that the derivation of (7.3.4) is quite close in spirit to Wilson's method, as we study the variation of the correlation functions when the cutoff is contracted. This leads us naturally to revert to Wilson's version of the RG.

7.3.2 Bare and renormalized coupling constants

The discussion in the present section is limited to the case $D = 4$, and for simplicity to $T = T_c$. As we have shown in Section 3.6, the RHS of the flow equations depends, in this case, only on the marginal coupling $g_0(s)$. The transformation law of the proper vertex $\Gamma^{(N)}$ (chosen in a nonexceptional configuration) is given by the generalization of (3.6.22) with $t_0 = t_0(s) = 0$:

$$\Gamma^{(N)}(sk_i, g_0(s), \Lambda) = s^{D-N(D/2-1)}[\zeta(s)]^{-N/2}\Gamma^{(N)}(k_i, g_0, \Lambda). \tag{7.3.7}$$

The exponent $d_N = D - N(D/2 - 1)$ is, as previously, the canonical dimension of $\Gamma^{(N)}$. Let us recall that, in Chapter 3, we rescaled the unit of length by a factor s, in order to keep Λ, or the lattice spacing a, at a fixed value. This was convenient, but not strictly necessary. In the present section, it will be convenient, in order to compare with previous results, not to make this rescaling, and to compare two systems with the same correlation length on lattices with different spacings (Fig. 7.8).

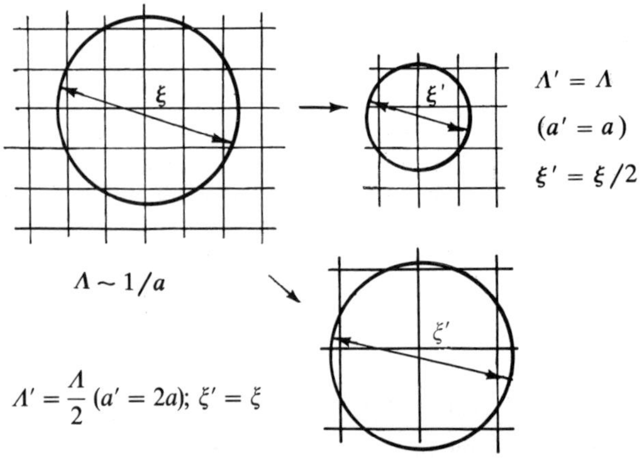

Fig. 7.8 Two ways of looking at a renormalization-group transformation

Using dimensional analysis, it is straightforward to rewrite (7.3.7) as

$$\Gamma^{(N)}\left(k_i, g_0(s), \frac{\Lambda}{s}\right) = [\zeta(s)]^{-N/2}\Gamma^{(N)}(k_i, g_0, \Lambda). \tag{7.3.8}$$

Let us recall the physical interpretation of (7.3.8): a perturbative expansion of the RHS is not possible, because of large logarithms $\ln(k/\Lambda)$; the RGT allows one to decrease the cutoff down to a scale $\mu = \Lambda/s \gtrsim k$, by integrating out the short-wavelength fluctuations in the interval $\Lambda^{-1} \leqslant \lambda \leqslant \mu^{-1}$. Then the LHS of (7.3.8) has a perturbative expansion, as $\ln(k/\mu) \sim 1$. Moreover, from (3.6.12), $g_0(s) \to 0$ as $(\ln s)^{-1}$ for large values of s, unless there exists another zero g_0^* of the β-function: in such cases one has only to choose an initial value $g_0(1) < g_0^*$. The critical behaviour is contained in the prefactor $[\zeta(s)]^{-N/2}$.

Let us first consider the transformation law (7.3.8) in the particular case of an invariant charge (7.1.8):

$$\hat{\Gamma}\left(k_i, g_0\left(\frac{\Lambda}{\mu}\right), \mu\right) = \hat{\Gamma}(k_i, g_0, \Lambda).$$

Physics at large distances $k_i \lesssim \mu \ll \Lambda$ is described by a correlation function $\hat{\Gamma}(k_i, g_0(\Lambda/\mu), \mu)$ which can be expanded perturbatively, without large logarithms, in powers of $g_0(s) = g_0(\Lambda/\mu)$. If we define the renormalized coupling $g = g(\mu)$ by

$$\hat{\Gamma}\left(k_{i0}, g_0\left(\frac{\Lambda}{\mu}\right), \mu\right) = g,$$

which differs from the usual definition (7.1.3c) by a finite renormalization, we see that, at the tree level, $g = g_0(\Lambda, \mu)$, and that g can be expanded perturbatively in powers of $g_0(\Lambda/\mu)$:

$$g = g_0\left(\frac{\Lambda}{\mu}\right) + c g_0^2\left(\frac{\Lambda}{\mu}\right) + \cdots . \tag{7.3.9}$$

In other words, g and $g_0(\Lambda/\mu)$ *differ at most by a finite renormalization*; as the difference can always be absorbed into a change of renormalization scheme, $g_0(\Lambda/\mu)$ can be identified with the renormalized coupling constant. The renormalized coupling constant is equal to the bare one when the renormalization mass is of the order of the cutoff. Let us now identify the field-renormalization constant Z_3; since $\zeta(s)$ obeys the differential equation

$$\frac{\mathrm{d}\ln\zeta(s)}{\mathrm{d}\ln s} = \gamma(g_0(s)), \quad \zeta(1) = 1,$$

we have

$$\zeta(s) = \exp\left(-\int_{g_0}^{g_0(s)} \frac{\gamma(g')\mathrm{d}g'}{\beta(g')}\right),$$

and we can recast the transformation law (7.3.8) into

$$\Gamma^{(N)}\left(k_i, g_0\left(\frac{\Lambda}{\mu}\right), \mu\right) = \exp\left(\frac{1}{2}N\int_{g_0}^{g_0(\Lambda/\mu)} \frac{\gamma(g')\mathrm{d}g'}{\beta(g')}\right)\Gamma^{(N)}(k_i, g_0, \Lambda). \tag{7.3.10}$$

Using the same arguments as before, we can identify $\Gamma^{(N)}(k_i, g_0(\Lambda/\mu), \mu)$ with the renormalized proper vertex, and comparison with (6.4.5) yields

$$Z_3 = \zeta\left(\frac{\Lambda}{\mu}\right)^{-1} = \exp\left(\int_{g_0}^{g_0(\Lambda/\mu)} \frac{\gamma(g')\mathrm{d}g'}{\beta(g')}\right), \tag{7.3.11}$$

where $g_0(\Lambda/\mu) \simeq g(\mu)$.

7.3.3 Renormalization group flow and renormalization

We pursue Wilson's RG, and examine the possibility of taking the limit $\Lambda \to \infty$ in order to define a Λ-independent renormalized theory. The discussion will rely on a qualitative picture of the RG flow; for clarity, we consider only two couplings which we choose dimensionless. The first one, $\bar{t}_0$ ($\bar{t}_0 = t_0/\Lambda^2$) is relevant, the second one, g_0, is marginal; we also assume that the fixed point P^* is IR-stable. We wish to characterize the renormalized theory by a mass m, say $m = 1$ GeV, and we call $\bar{\xi}(\bar{t}_0, g_0)$ the dimensionless correlation length, i.e. the correlation length measured in units of Λ^{-1}. We finally choose a fixed value g_0 of the coupling constant, and determine Λ from the equation

$$\Lambda = m\bar{\xi}(\bar{t}_0, g_0).$$

When $\bar{t}_0$ approaches the critical manifold, Λ goes to infinity (Fig. 7.9); when $\bar{t}_0$ varies, each point on the AQ line corresponds to a bare Hamiltonian $H_0(m, \Lambda)$. We can apply RGTs

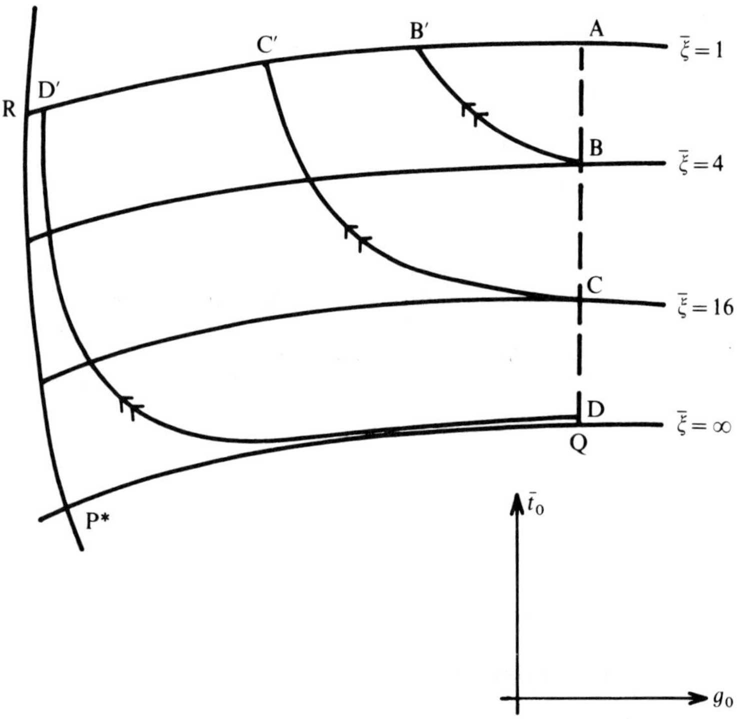

Fig. 7.9 The renormalization-group flow

to these Hamiltonians, and define $H_{\text{eff}}(m, \Lambda)$ as the transformed Hamiltonian having $\bar{\xi} = 1$ (points B′, C′, D′ on Fig. 7.9). Note that C and C′ for example describe the same long-distance physics. All effective Hamiltonians correspond to the same renormalized mass, and we can take the limit $\Lambda \to \infty$, obtaining an effective renormalized Hamiltonian $H_{\text{eff}}(m, \infty)$. This effective Hamiltonian, which corresponds to point R on Fig. 7.9, is unique and depends only on the renormalized mass m; the other couplings, and in particular the renormalized coupling constant g, are completely fixed by m, and moreover $H_{\text{eff}}(m, \infty)$ is completely independent of the initial value g_0. Notice that in obtaining $H_{\text{eff}}(m, \infty)$, we have effectively pushed the cutoff to infinity, or the lattice spacing to zero, while keeping the renormalized mass m at a fixed value.

In general, one can see that the number of independent parameters is equal to the number of relevant couplings, which is the same as the number of independent directions of instability at the fixed point P*. In order to prove the renormalizability of a field theory, one must be able to control the renormalization flow for large values of s, the divergences as $\Lambda \to \infty$ being reflected in the sensitive dependence of the global flow on initial conditions. If the flow is found to be insensitive to the initial conditions, then the qualitative picture of Fig. 7.9 is valid. This point has been recently under active investigation (Polchinski 1984; Kupiainen 1986), and rigorous results have been obtained for simple, albeit unrealistic field-theoretic models. It appears that perturbative renormalizability is neither sufficient (φ^4-case), nor necessary (Gross–Neveu model in $2 + \varepsilon$ dimension) in order for a field theory to be renormalizable in this nonperturbative approach.

Let us now look in some detail at an example, borrowed from Polchinski (1984), which will clarify the role of irrelevant variables. We take a toy model, inspired by the φ^4-theory, with a marginal coupling g_0 and an irrelevant coupling u_6^0, which can be thought of as the coefficient of a nonrenormalizable φ^6 interaction. We shall study the flow on the critical surface $m = 0$. The coupling u_6^0 has dimension -2, and we define a dimensionless $g_6^0 = \Lambda^2 u_6^0$; if we scale down the cutoff, $\Lambda \to \Lambda/s$, we shall have $g_6^0(s) = (\Lambda/s)^2 u_6^0(s)$. Let us consider in the (g, g_6) plane two trajectories $(g_0(s), g_6^0(s))$ and $(\bar{g}_0(s), \bar{g}_6^0(s))$, issuing from points A and B respectively (see Fig. 7.10). They correspond to the same initial value of g_0, but to two different values of g_6^0, ($g_6^0 = 0$ and $g_6^0 \neq 0$). In order to avoid a proliferation of indices, we use the notations of Section 3.6.1:

$$g_0 \to g' \qquad g_6^0 \to w'.$$

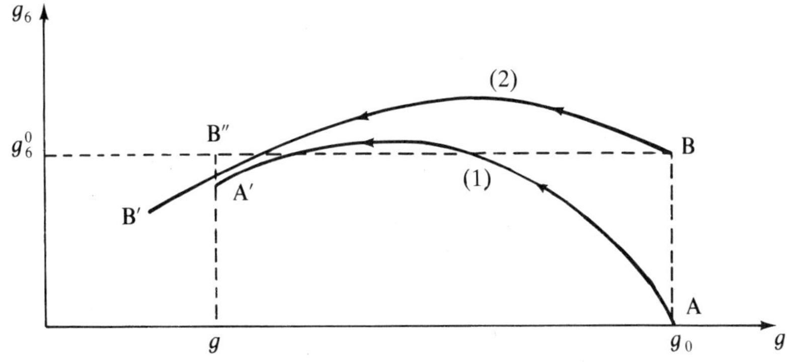

Fig. 7.10 Renormalization flow in the (g, g_6) plane

The flow equations read

$$\frac{dg'}{d\ln s} = \beta_4(g', w') = ag'^2 + bg'w' + cw'^2 + \cdots , \tag{7.3.12}$$

$$\frac{dw'}{d\ln s} = -2w' + \beta_6(g', w') = -2w' + dg'^2 + eg'w' + f'w'^2 + \cdots . \tag{7.3.13}$$

The nonlinear change of variables

$$g' = g - \frac{1}{2}bgw - \frac{1}{4}cw^2, \tag{7.3.14}$$

$$w' = w + \frac{1}{2}dg^2 - \frac{1}{2}fw^2 \tag{7.3.15}$$

allows us to recast equations (7.3.12) and (7.3.13) into their normal form:

$$\frac{dg}{d\ln s} = ag^2 + \text{cubic terms}, \tag{7.3.16}$$

$$\frac{dw}{d\ln s} = -2w + egw + \text{cubic terms}. \tag{7.3.17}$$

More generally we have

$$\frac{dg}{d\ln s} = -\beta(g), \tag{7.3.18}$$

$$\frac{dw}{d\ln s} = (-2 + \lambda(g))w. \tag{7.3.19}$$

Now the solution of (7.3.19) reads

$$w(s) = w(s=1)s^{-2}\exp\left(\int_{g(1)}^{g(s)} \frac{\lambda(g')dg'}{\beta(g')}\right).$$

Unless the canonical dimensions are overwhelmed by the anomalous ones, $w(s)$ will decrease as a power; indeed we expect the s^{-2} behaviour to be corrected by powers of $\ln s$.

The different initial conditions on $w'(\equiv g_6^0)$ are reflected by different initial conditions $g(1)$ and $\bar{g}(1)$ for $g(s)$. The trajectories issuing from A and B will land, for the same value of s, on two different points A′ and B′. However, there exists a value $\bar{s}$ such that $g(s) = \bar{g}(\bar{s})$, which implies, within an accuracy $\sim s^{-2}$, that $g_0(s) = \bar{g}_0(\bar{s})$. This choice corresponds to point B″ on the BB′ trajectory. Moreover we also have $g_6^0(s) = \bar{g}_6^0(\bar{s})$, within accuracy $\sim s^{-2}$.

If $s = \Lambda/\mu$, where μ is the renormalization scale, we see that, with an accuracy $\sim(\mu/\Lambda)^2$, the renormalized theory is entirely determined by the knowledge of $g = g(\mu)$, independently of the initial conditions. It is always possible, as was done in Chapter 6, to restrict oneself to renormalizable interactions in the bare theory, that is, to start from points located on the g_0-axis of Fig. 7.10. If one adds nonrenormalizable interactions, like φ^6, it is however necessary to choose u_6^0 proportional to Λ^{-2}, and not to some inverse power of a lower scale like μ^{-2}.

The toy model illustrates the previous general discussion: all trajectories of the renormalization flow converge, independently on the initial conditions, towards an n-dimensional manifold (a curve, $n = 1$, in this case) where n is the number of independent renormalized couplings in the theory. The irrelevant couplings are not zero, but they are calculable as functions of the renormalized ones, and from (7.3.15) one sees that $g_6^0(s)$ is—at least—of order $g_0^2(s)$. It should be emphasized once more that the notation $\Gamma^{(N)}(k_i, g_0(s), \Lambda/s)$ should not be misunderstood: $\Gamma^{(N)}$ depends also on irrelevant couplings, but they are calculable as functions of $g_0(s)$, within an accuracy $\sim s^{-2}$. One must also be aware that the true cutoff is Λ, and not $\mu = \Lambda/s$: when Λ is finite, the corrections to $\Gamma^{(N)}$ are $\sim(k_i/\Lambda)^2$, and not $\sim(k_i/\mu)^2$. Irrelevant couplings are of course crucial in order to enforce this property.

The careful reader will have noticed that the discussion of this toy model is valid only for large but finite Λ, and that it is not possible to push Λ to infinity. If we fix g_0 and push Λ to infinity, $g \to 0$, as the origin is an IR-stable fixed point, and the renormalized theory is a free theory. The coupling g is marginal, but it behaves as if it were irrelevant, and it is thus calculable: it is in fact zero! This is the triviality-problem of the φ^4-theory.

Things look much better in an asymptotically free theory, whose flow diagram is drawn schematically in Fig. 7.11. The unstable manifold, which is also called the renormalized trajectory, is the line issuing from the fixed point P^*.

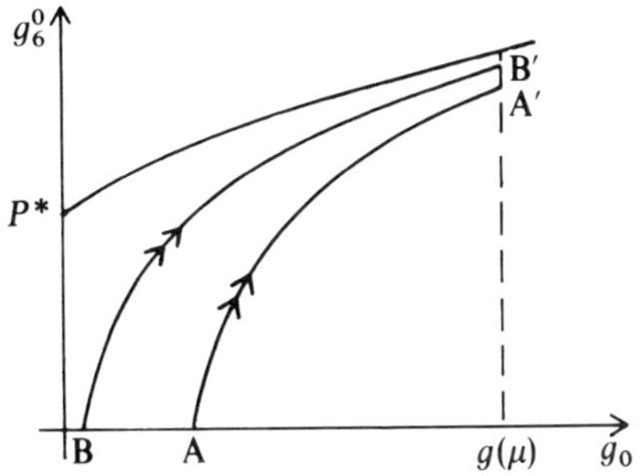

Fig. 7.11 Flow diagram in the (g, g_6^0) plane for an asymptotically free theory

The trajectories (AA′) and (BB′) correspond to cutoffs $\Lambda = s\mu$ and $\bar{\Lambda} = \bar{s}\mu$ respectively, with $\bar{s} > s$. In the limit $\Lambda \to \infty$, $g(\mu)$ fixed, the bare coupling $g_0(\Lambda)$ tends to zero. The use of perturbation theory for the bare Hamiltonian is thus entirely justified. As before, the irrelevant couplings $g_6(\mu)$ etc. can be calculated from $g(\mu)$ within an accuracy $\sim(\mu/\Lambda)^2$, and the renormalized theory becomes Λ-independent in the large Λ limit.

The same picture obtains in the case of a UV-stable fixed point which is not located at the origin (Fig. 7.4). However, the bare coupling constant is not zero for $\Lambda \to \infty$, but is equal to its fixed-point value; the validity of perturbation theory is then problematic. On the other hand, if the situation is that of Fig. 7.2, the flow diagram shows that the renormalized coupling constant is equal to g^*, independently of g_0: as already mentioned, g in this case is not arbitrary, but is fixed at g^*. A particular case is of course the φ^4-theory in $D = 4$, where $g^* = 0$.

Let us now discuss the relation between the β-functions $\beta(g)$ and $\beta(g_0)$: as it is now necessary to distinguish between them, we denote the latter by $\bar{\beta}(g_0)$. We have previously identified g_0 at scale μ with the renormalized $g(\mu)$, within a finite renormalization. Thus we have the case of two different renormalization schemes, and if we write

$$\bar{\beta}(g_0) = \bar{\beta}_0 g_0^2 + \bar{\beta}_1 g_0^3 + \cdots,$$

then we must have, comparing with (7.1.16),

$$\beta_0 = \bar{\beta}_0, \quad \beta_1 = \bar{\beta}_1. \tag{7.3.20}$$

An explicit proof is easily obtained from

$$\ln\frac{\Lambda}{\mu} = \int_{g_0(\mu)}^{g_0(\Lambda)} \frac{dg'}{\bar{\beta}(g')},$$

and

$$g_0(\mu) = g(\mu) - cg^2(\mu) + \cdots .$$

7.4 The renormalization group in $D = 4$ dimensions

$D = 4$ is the case that is important in quantum field theory; this section establishes some useful results.

7.4.1 The calculation of $\beta(g)$

If one calculates in 4 dimensions, one must use a cutoff, though $\beta(g)$ can of course be obtained also by dimensional regularization (see Section 7.5). But for the moment we suppose that the renormalization constants Z_1 and Z_3 have been calculated by using counterterms which are functions of g and of Λ. Recall that the bare and the renormalized coupling constants are related through

$$g_0 = Z_1 Z_3^{-2} g = Zg.$$

Z is a function of g and of Λ/μ, or of g and $X = \ln(\Lambda/\mu)$. Equation (7.1.5) defining $\beta(g)$ entails

$$\beta(g) = \left.\frac{dg}{d\ln\mu}\right|_{g_0,\Lambda} = -\left.\frac{dg}{dX}\right|_{g_0} = -g_0\frac{dZ^{-1}}{dX} = -gZ\frac{dZ^{-1}}{dX}$$
$$= g\left.\frac{d\ln Z}{dX}\right|_{g_0}.$$

Now express the total in terms of the partial derivatives:

$$\left.\frac{d\ln Z}{dX}\right|_{g_0} = \left.\frac{\partial\ln Z}{\partial g}\frac{dg}{dX}\right|_{g_0} + \frac{\partial\ln Z}{\partial X} = -\beta(g)\frac{\partial\ln Z}{\partial g} + \frac{\partial\ln Z}{\partial X},$$

which yields

$$\beta(g) = \frac{g\,\partial\ln Z/\partial X}{1 + g\,\partial\ln Z/\partial g}. \tag{7.4.1}$$

The interesting point about (7.4.1) is that all the calculations are performed in terms of g and X: there is no need to revert to g_0. Suppose that we have calculated Z to one-loop order:

$$Z(g, X) = 1 + g(a_1 X + a_0) + O(g^2).$$

Then (7.4.1) immediately leads to

$$\beta(g) = a_1 g^2 + O(g^3), \tag{7.4.2}$$

whence the coefficient β_0 equals a_1: $\beta_0 = a_1$. Next, we show that the structure of $Z(g)$ is constrained by (7.4.1) and by the perturbative expansion

$$\beta(g) = \beta_0 g^2 + \beta_1 g^3 + \cdots \tag{7.4.3}$$

of $\beta(g)$. Write the order g^2 term of $Z(g)$ in the form $g^2 f(X)$:

$$Z(g, X) = 1 + g(a_1 X + a_0) + g^2 f(X) + O(g^3),$$

$$\ln Z(g, X) = g(a_1 X + a_0) + g^2\left[f(X) - \frac{1}{2}(a_1 X + a_0)^2\right] + \cdots.$$

Comparing the terms of order g^3 we find for $f(X)$ the differential equation

$$f'(X) = 2a_1^2 X + \beta_1 + 2a_1 a_0,$$

whence

$$f(X) = a_1^2 X^2 + (\beta_1 + 2a_1 a_0)X + b_2.$$

Thus the coefficient of g^2 contains a term in $\ln^2(\Lambda/\mu)$ which is fully determined by the calculation to order g, or rather to one-loop order; a two-loop calculation is needed to determine the coefficient β_1. In general, one could show, recursively, that

$$Z(g) = 1 + \sum_{n=1}^{\infty} \sum_{p=0}^{n} g^n c_{np} (\ln \Lambda/\mu)^p. \tag{7.4.4}$$

The highest power of the logarithm in the term of order g^n *is* $(\ln \Lambda/\mu)^n$: this dominant term is called *the leading logarithm.* The term in $(\ln \Lambda/\mu)^{n-1}$ is subdominant, and is called the next-to-leading logarithm. The dominant logarithmic term is wholly determined by the one-loop calculation; the subdominant by the two-loop calculation; and so on. We have already seen a specific example of this phenomenon in the last chapter.

The structure of equation (7.4.4) generalizes to the other renormalization constants, and to the correlation functions. Of course there is nothing magic about this structure: it follows from the renormalization group. For instance, it could be found through perturbative expansions of expressions like (7.3.11) for Z_3 or like (7.1.25) for the solution of the Callan–Symanzik equations.

7.4.2 RG-improved perturbation theory

Suppose we wish to calculate a correlation function $\Gamma_{\mathrm{R}}^{(N)}$ for values of the momenta $p_i \sim p \gg \mu$. Assuming that $m = 0$, or that $\mu \gg m$, a direct calculation in the form of a perturbative expansion would yield (with normalization conditions of the kind (6.4.6))

$$\Gamma_{\mathrm{R}}^{(N)} = g^{N/2-1}A(p_i)\left(1 + g\left(\ln\frac{p}{\mu} + c\right) + \cdots\right). \tag{7.4.5}$$

If $\ln(p/\mu)$ is large, the perturbation series converges slowly, or not at all. The case of an IR-free theory is certainly irredeemable; by contrast, for a theory that is asymptotically free in the ultraviolet, we can get sensible results. One defines a dilatation factor s by $s = p/\mu \gg 1$ and $k_i = p_i/s$. Equation (7.3.6) then yields

$$\Gamma_{\mathrm{R}}^{(N)}(p_i, g, \mu) = s^{D-N(D/2-1)}\exp\left(-\frac{N}{2}\int_g^{g(s)}\frac{\gamma(g')\mathrm{d}g'}{\beta(g')}\right)$$
$$\times\,\Gamma_{\mathrm{R}}^{(N)}(k_i, g(s), \mu).$$

In view of $k_i/\mu \sim 1$, the perturbative expansion of the right-hand side takes the form

$$\Gamma_{\mathrm{R}}^{(N)}(k_i, g(s), \mu) = g(s)^{N/2-1}A(k_i)\left(1 + g(s)f\left(\frac{k_i\cdot k_j}{\mu^2}\right) + \cdots\right),$$

where $f(k_i\cdot k_j/\mu^2) \sim 1$. Since the dimension of $A(k_i)$ is the canonical dimension of $\Gamma_{\mathrm{R}}^{(N)}$, one finds eventually that

$$\Gamma_{\mathrm{R}}^{(N)} = [g(s)]^{N/2-1}A(p_i)\exp\left(-\frac{1}{2}N\int_g^{g(s)}\frac{\gamma(g')\mathrm{d}g'}{\beta(g')}\right)$$
$$\times\left(1 + g(s)f\left(\frac{k_i\cdot k_j}{\mu^2}\right) + \cdots\right). \tag{7.4.6}$$

For a theory that is asymptotically free in the ultraviolet, $g(s)\to 0$, and the perturbative expansion of the right-hand side of (7.4.6) converges rapidly; by applying the RG we have succeeded in 'summing the large logarithms' in equation (7.4.5).

An equivalent procedure consists in defining the renormalized coupling constant by choosing the subtraction point at $s\mu$ instead of μ. Suppose we had chosen the invariant charge of equation (7.1.8) to define g. Then we would have

$$\hat{\Gamma}_{\mathrm{R}}(p_{i0}, g, \mu) = \hat{\Gamma}_{\mathrm{R}}(k_{i0}, g(s), \mu), \tag{7.4.7}$$

where k_{i0} is the subtraction point (6.4.7), and $p_{i0} = sk_{i0}$. By definition the right-hand side of (7.4.7) equals $g(s)$, and

$$\hat{\Gamma}_{\mathrm{R}}(p_{i0}, g, \mu) = g(s).$$

In order to avoid large logarithms in the perturbation series when all the momenta are $\sim p$, it is therefore advisable to choose the subtraction point at $\mu \sim p$. The troublesome case is $p_i/p_j \gg 1$: the renormalization group does not then afford a grip on the logarithms $\ln(p_i/p_j)$ that appear in the perturbation series.

7.5 The renormalization group in dimensions $D < 4$

For application to critical phenomena and to the minimal renormalization scheme, it proves interesting to calculate the functions β and γ in $D < 4$. The expressions one finds depend on $\varepsilon = 4 - D$, but in order to obtain the results in $D = 4$ we need merely let ε tend to zero.

7.5.1 An equation for $\beta(g, \varepsilon)$

When $D < 4$, one can let the cutoff tend to infinity, so that the only dimensional parameter which is left is μ; this allows us to write for the bare coupling u_0 an equation of the form

$$u_0 = \mu^\varepsilon f(g) = \mu^\varepsilon g_0'.$$

Taking the derivative with respect to μ of this equation at fixed u_0, we get from the definition (7.1.5)

$$\left(\varepsilon + \beta(g, \varepsilon)\frac{\mathrm{d}}{\mathrm{d}g}\right) f(g) = 0. \qquad (7.5.1)$$

This differential equation can be solved explicitly if we take into account the boundary condition $u_0 \to \mu^\varepsilon g$ when $g \to 0$:

$$u_0 = \mu^\varepsilon g \exp\left[-\int_0^g \mathrm{d}g' \left(\frac{\varepsilon}{\beta(g', \varepsilon)} + \frac{1}{g'}\right)\right]. \qquad (7.5.2)$$

The integral converges when $g' \to 0$, because $\beta(g', \varepsilon) \simeq -\varepsilon g'$ in this region. Assume now that $\beta(g, \varepsilon)$ has the shape drawn in Fig. 7.2, with an IR-stable fixed point at $g = g^*$, so that in the vicinity of $g = g^*$,

$$\beta(g, \varepsilon) \simeq \omega(g - g^*), \quad \omega = \beta'(g^*, \varepsilon).$$

We can write the integral in (7.5.2) as

$$\int_0^g \mathrm{d}g' \frac{\varepsilon}{\omega(g' - g^*)} + \int_0^g \mathrm{d}g' \left(\frac{\varepsilon}{\beta(g', \varepsilon)} + \frac{1}{g'} - \frac{\varepsilon}{\omega(g' - g^*)}\right).$$

The last integral converges at $g = 0$ and $g = g^*$. We can write it as $-\ln A(g)$, where $A(g) = 1 + cg + \cdots$ (note that $A(g) = 1$ in the one-loop approximation (7.1.15)), and we obtain for u_0 the equation

$$u_0 = \mu^\varepsilon g A(g) \left|\frac{g^*}{g^* - g}\right|^{\varepsilon/\omega}. \qquad (7.5.3)$$

This equation shows that u_0 varies between 0 and ∞ when g varies in the range $[0, g^*[$. Since $g^* \to 0$ when $D \to 4$, the region $g > 0$ cannot be reached when $D = 4$, without an analytical continuation where u_0 becomes complex.

This is another indication that it is impossible to construct a nontrivial renormalized theory in the limit of infinite cutoff. For Λ finite, there exists an acceptable domain $[0, \ln^{-1}\Lambda]$ for g.

Similarly for $\gamma(g, \varepsilon)$ we find

$$\gamma(g, \varepsilon) = \mu\left(\frac{\partial \ln Z_3}{\partial \mu}\right)_{u_0} = \mu\left(\frac{\partial g}{\partial \mu}\right)_{u_0}\left(\frac{\partial \ln Z_3}{\partial g}\right)_{u_0}.$$

In the limit $\Lambda \to \infty$, $Z_3 = Z_3(g, \varepsilon)$, and

$$\gamma(g, \varepsilon) = \beta(g, \varepsilon)\frac{\partial \ln Z_3(g, \varepsilon)}{\partial g}. \tag{7.5.4}$$

This equation can be solved for Z_3, thanks to the boundary condition $Z_3(g, \varepsilon) \to 1$ as $g \to 0$:

$$Z_3(g, \varepsilon) = \exp\int_0^g \frac{\gamma(g', \varepsilon)}{\beta(g', \varepsilon)}\,dg'.$$

For $g \to g^*$ (namely for the critical theory) this leads to

$$Z_3(g, \varepsilon) \simeq \left|\frac{g^*}{g^* - g}\right|^{-\eta/\omega},$$

or, using (7.5.3), to

$$Z_3(g, \varepsilon) \simeq \mu^\eta (g^*/u_0)^{\eta/\varepsilon}.$$

Using equation (7.1.27) and dimensional analysis, which tell us that in the low k region

$$\Gamma_R^{(2)} \sim \mu^\eta k^{2-\eta},$$

we obtain the bare correlation function in the form

$$\Gamma^{(2)} \sim (u_0/g^*)^{\eta/\varepsilon} k^{2-\eta}.$$

Besides the scaling behaviour of $\Gamma^{(2)}$, the most interesting feature of this equation is the nonanalytic behaviour with respect to the bare coupling constant u_0 (or g_0'). It is related to the breakdown of the perturbative expansion due to IR singularities. This nonanalytic behaviour has been obtained from the assumption that the β-function has an IR-stable zero beyond the reach of perturbation theory. The above derivation could of course be criticized, since we have used tools, like renormalization theory, which rely on the perturbative expansion. However, since we have always worked in the case $m = 0$, this expansion does not exist at fixed $D < 4$. Nevertheless the use of the renormalization group leads to the correct result, as can be shown by keeping $m \neq 0$ and taking the limit $m \to 0$ (see Itzykson and Drouffe 1989, or Parisi 1988).

Let us also remark that, since $u_0 = \Lambda^\varepsilon g_0$, a Λ-dependence is hidden in the equations for Z_3 and $\Gamma^{(2)}$. This shows that naïve scaling ($\Gamma^{(2)} \sim k^2$) does not hold, because the cutoff is still relevant at low k. Furthermore the behaviour

$$Z_3(g, \varepsilon) \approx (\mu/\Lambda)^{\eta}$$

shows that the field-renormalization constant is not finite at the critical point in the limit $\Lambda \to \infty$: the critical theory behaves as a renormalizable field theory, although naïve power counting would have led us to conclude that it should be renormalizable and that Z_3 should thus be finite.

Finally, it will be convenient to transform (7.5.1) into an equation for $Z(g, \varepsilon)$, which is defined by $g_0' = Z(g, \varepsilon)g$; we obtain

$$\left[\beta(g, \varepsilon) + \varepsilon g + g\beta(g, \varepsilon)\frac{\partial}{\partial g}\right] Z(g, \varepsilon) = 0, \tag{7.5.5a}$$

or

$$\beta(g, \varepsilon) = \frac{-\varepsilon g}{1 + g\partial \ln Z/\partial g}. \tag{7.5.5b}$$

7.5.2 The calculation of $\beta(g, \varepsilon)$ and of $\gamma(g, \varepsilon)$ in the minimal scheme

By the definition of the minimal scheme, the renormalization constants contain nothing but poles in ε:

$$Z(g, \varepsilon) = 1 + \sum_{i=1}^{\infty} \frac{Z^{(i)}(g)}{\varepsilon^i}, \tag{7.5.6}$$

$$Z_3(g, \varepsilon) = 1 + \sum_{i=1}^{\infty} \frac{Z_3^{(i)}(g)}{\varepsilon^i} \tag{7.5.7}$$

(in this section i denotes the order of the pole and not the number of loops).

Start by examining equation (7.5.5b). In the minimal scheme we may write

$$\left(1 + g\frac{\partial \ln Z}{\partial g}\right)^{-1} = 1 + \sum_{i=1}^{\infty} \frac{z^{(i)}(g)}{\varepsilon^i}.$$

For $\beta(g, \varepsilon)$ to have a finite limit as $\varepsilon \to 0$, it is necessary that only $z^{(1)}$ should be nonzero, whence

$$\beta(g, \varepsilon) = -\varepsilon g + \beta_4(g), \tag{7.5.8}$$

where $\beta_4(g) = \beta(g)$ is the β function in 4 dimensions.

Note that the relation (7.5.8) is peculiar to the minimal scheme. We find further that

$$\beta(g) = -z^{(1)}(g),$$

but strictly speaking we have calculated Z rather than $\ln Z$. Let us therefore express $\beta(g)$ as a function of the $Z^{(i)}$, by rewriting (7.5.5a) as

$$\beta(g)\frac{\partial}{\partial g}(gZ(g, \varepsilon)) - \varepsilon g^2 \frac{\partial}{\partial g} Z(g, \varepsilon) = 0. \tag{7.5.9}$$

Substituting from (7.5.6) into (7.5.9) we find

$$\beta(g)\left(1 + \sum_i \frac{Z^{(i)}(g)}{\varepsilon^i}\right) - \varepsilon g^2 \sum_i \frac{Z'^{(i)}(g)}{\varepsilon^i} + g\beta(g)\sum_i \frac{Z'^{(i)}(g)}{\varepsilon^i} = 0.$$

At this point one need merely identify powers of $1/\varepsilon$ in order to write the expression for $\beta(g)$ as

$$\beta(g) = g^2 \frac{\partial Z^{(1)}(g)}{\partial g}, \tag{7.5.10}$$

and the recurrence relation as

$$g^2 \frac{\partial}{\partial g} Z^{(i+1)}(g) = \beta(g)\frac{\partial}{\partial g}(gZ^{(i)}(g)). \tag{7.5.11}$$

This last relation allows us for instance to calculate the poles proportional to ε^{-2} as functions of those proportional to ε^{-1}: we have already met an example of such a relation in the last chapter (Section 6.6).

The calculation of $\gamma(g, \varepsilon)$ starts from (7.5.4):

$$\left[\beta(g, \varepsilon)\frac{\partial}{\partial g} - \gamma(g, \varepsilon)\right] Z_3(g, \varepsilon) = 0.$$

Next we use (7.5.7):

$$\beta(g, \varepsilon)\sum_i \frac{\partial}{\partial g}\frac{Z_3^{(i)}(g)}{\varepsilon^i} = \gamma(g, \varepsilon)\left(1 + \sum_i \frac{Z_3^{(i)}(g)}{\varepsilon^i}\right).$$

By virtue of the form (7.5.8) of $\beta(g, \varepsilon)$ we find

$$\gamma(g, \varepsilon) = \gamma(g) = -g\frac{\partial}{\partial g} Z_3^{(1)}(g). \tag{7.5.12}$$

In the minimal scheme, $\gamma(g, \varepsilon)$ is independent of ε! We stress that it is only in the minimal scheme that $\beta(g, \varepsilon)$ and $\gamma(g, \varepsilon)$ assume the forms (7.5.8) and (7.5.12). For instance, appeal to the normalization conditions (7.1.25) would produce $\gamma(g, \varepsilon)$ in a form that does explicitly depend on ε. Note also that for the $Z_3^{(i)}$ one gets recurrence relations in i (Problem 7.5).

Evidently, the reasoning we have just given for the calculation of $\gamma(g, \varepsilon)$ applies also to the calculation of $\bar{\gamma}(g, \varepsilon)$:

$$\bar{\gamma}(g, \varepsilon) = \bar{\gamma}(g) = -g\frac{\partial}{\partial g}\bar{Z}^{(1)}(g). \tag{7.5.13}$$

7.5.3 The calculation of β, γ, and $\bar{\gamma}$ to two-loop order

In the last chapter we found Z_3 and Z to two-loop order:

$$Z_3(g, \varepsilon) = 1 - \frac{\alpha^2}{12\varepsilon}, \tag{7.5.14}$$

$$Z(g, \varepsilon) = 1 + \frac{3\alpha}{\varepsilon} + \frac{9\alpha^2}{\varepsilon^2} - \frac{17\alpha^2}{6\varepsilon}, \tag{7.5.15}$$

where $\alpha = g/(4\pi)^{D/2}$*. From (7.5.10) we have

$$\beta(g) = \frac{g^2}{(4\pi)^2}\frac{\mathrm{d}}{\mathrm{d}\alpha}\left(3\alpha - \frac{17}{6}\alpha^2\right) = g\left[3\alpha - \frac{17}{3}\alpha^2\right],$$

$$\beta(g) = \frac{3g^2}{(4\pi)^2} - \frac{17}{3}\frac{g^3}{(4\pi)^4} + O(g^4). \tag{7.5.16}$$

Since the coefficients β_0 and β_1 are independent of the renormalization scheme, the same result (7.5.16) would have emerged from using say the normalization conditions (7.1.25) and calculating directly in 4D with a cutoff (see Problem 7.3); $\gamma(g)$ is found from (7.5.12) and (7.5.14):

$$\gamma(g) = -\frac{g}{(4\pi)^2}\frac{\mathrm{d}}{\mathrm{d}\alpha}\left(-\frac{\alpha^2}{12}\right) = \frac{1}{6}\frac{g^2}{(4\pi)^4} = \frac{\alpha^2}{6}. \tag{7.5.17}$$

To calculate $\bar{\gamma}$, we must determine $\bar{Z}$; all the requisite integrals have been evaluated in the last chapter, since they are identical to those featured in the calculation of $\Gamma^{(4)}$. Only the symmetry factors differ. Let us evaluate the divergent part of $\Gamma^{(2,1)}$ to two-loop order:

$\Gamma^{(2,1)} =$

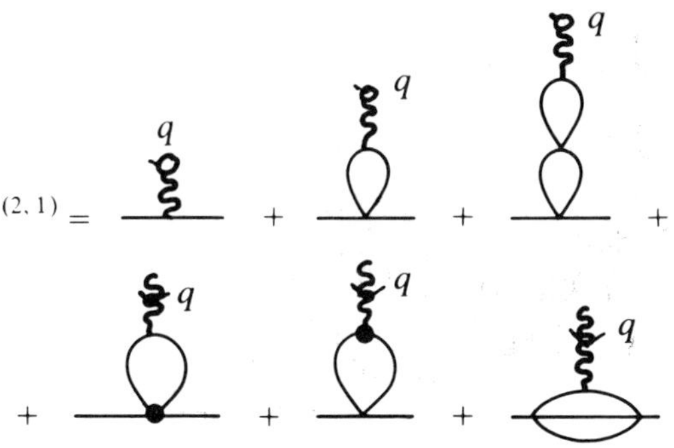

(7.5.18)

* Instead of g one can use a function of g, $\alpha = cgf(D)$, where $f(4) = 1$. It suffices to write all the equations ((7.5.6), (7.5.7), (7.5.8), . . .) as functions of α rather than of g.

The divergent part of the second graph in (7.5.18) is

$$\Gamma_{\mathrm{R}}^{(2,1)}|_{\mathrm{div}} = -\left(\frac{\alpha}{2}\right)\left(\frac{2}{\varepsilon}\right) = -\frac{\alpha}{\varepsilon},$$

whence

$$\bar{Z} = 1 + \frac{\alpha}{\varepsilon} + O(\alpha^2).$$

The divergent part of the last four graphs is

$$\frac{\alpha^2}{4}\frac{4}{\varepsilon^2}(1 + 2\varepsilon a_1) - \frac{1}{2}\frac{3\alpha^2}{\varepsilon}\frac{2}{\varepsilon}(1 + \varepsilon a_1)$$

$$-\frac{1}{2}\frac{\alpha^2}{\varepsilon}\frac{2}{\varepsilon}(1 + \varepsilon a_1) + \frac{\alpha^2}{2}\frac{2}{\varepsilon^2}\left(1 + 2\varepsilon a_1 + \frac{\varepsilon}{2}\right),$$

where $a_1 = 1 - \frac{1}{2}\gamma - \frac{1}{2}\ln(q^2/\mu^2)$ (see (6.6.3e)).

We observe once again the cancellation of the undesirable terms proportional to $\ln(q^2/\mu^2)$, and find for $\bar{Z}Z_3$ the result

$$\bar{Z}Z_3 = 1 + \frac{\alpha}{\varepsilon} + \frac{2\alpha^2}{\varepsilon^2} - \frac{\alpha^2}{2\varepsilon},$$

whence

$$\bar{Z} = 1 + \frac{\alpha}{\varepsilon} + \frac{2\alpha^2}{\varepsilon^2} - \frac{5\alpha^2}{12\varepsilon}. \qquad (7.5.19)$$

For $\bar{\gamma}$ this immediately yields

$$\bar{\gamma}(g) = -\alpha + \frac{5\alpha^2}{6}. \qquad (7.5.20)$$

7.5.4 The calculation of the critical exponents to order ε^2

The fixed point is given by $\beta(\alpha^*, \varepsilon) = 0$, whence

$$3\alpha^* - \frac{17}{3}\alpha^{*2} = \varepsilon,$$

namely

$$\alpha^* = \frac{\varepsilon}{3} + \frac{17}{81}\varepsilon^2 + O(\varepsilon^3). \qquad (7.5.21)$$

To order ε^2, η is given by

$$\eta = \gamma(\alpha^*) = \frac{\varepsilon^2}{54}, \qquad (7.5.22)$$

as we already know from Chapter 5 (equation (5.5.13)).

To derive ν, we must calculate $\bar{\gamma}(\alpha^*)$,

$$\bar{\gamma}(\alpha^*) = -\frac{\varepsilon}{3} - \frac{19}{162}\varepsilon^2,$$

and

$$\nu = \frac{1}{2 + \bar{\gamma}(\alpha^*)} = \frac{1}{2} + \frac{\varepsilon}{12} + \frac{7\varepsilon^2}{162} + O(\varepsilon^3). \tag{7.5.23}$$

The other critical exponents then follow from the scaling laws (3.2.16).

Problems

7.1 Solution of the CS equation

(a) Consider first an invariant charge obeying (7.1.10). Show that $\hat{\Gamma}$ is a function of the combination

$$\ln \mu - \int^{g} \frac{\mathrm{d}g'}{\beta(g')},$$

but not of $\ln \mu$ and g separately. (*Hint*: the solution of the partial differential equation $(\partial/\partial x + \partial/\partial y)\Phi(x, y) = 0$ is any arbitrary function $F(x - y)$.) Thence derive equation (7.1.20).

(b) For $N = 0$ obtain the solution of (7.2.7) in the form

$$\hat{\Gamma}_{\mathrm{R}}(k_i, t, g, \mu) = \hat{\Gamma}_{\mathrm{R}}(k_i, t(s), g(s), \mu(s) = s\mu),$$

and show that for $N = 0$ one has

$$\Gamma_{\mathrm{R}}(k_i, t, g, \mu) = \exp\left(-\frac{1}{2}N\int_0^{\ln s} \gamma(g(s'))\mathrm{d}\ln s'\right)\Gamma_{\mathrm{R}}(k_i, t(s), g(s), s\mu).$$

Use dimensional analysis to derive (7.2.8).

(c) Show that at the fixed point and for $T = T_{\mathrm{c}}$, the proper vertex $\Gamma_{\mathrm{R}}^{(N)}$ obeys

$$\left[\frac{\partial}{\partial \ln s} - \left(\frac{1}{2}N\eta - d_N\right)\right]\Gamma^{(N)}\left(\frac{k_i}{s}, g_0^*, \Lambda\right) = 0.$$

7.2 A different type of CS equation

(a) We start by deriving the equation originally established by Callan and Symanzik. Suppose that the renormalized theory is defined by the norming conditions (6.3.10), modified for $D < 4$ as follows:

$$\Gamma_{\mathrm{R}}^{(4)}(k_i = 0, m, g) = m^{\varepsilon} g.$$

By acting on

$$\Gamma^{(N)}(k_i, m_0, g_0, \Lambda) = Z_3^{-N/2}\left(\frac{\Lambda}{m}, g\right)\Gamma_{\mathrm{R}}^{(N)}(k_i, m, g)$$

with the operator $m(\mathrm{d}/\mathrm{d}m)_{g_0, \Lambda}$, show that one is led to the equation

$$\left[m\frac{\partial}{\partial m} + \beta(g, \varepsilon)\frac{\partial}{\partial g} - \frac{1}{2}N\gamma(g, \varepsilon)\right]\Gamma_{\mathrm{R}}^{(N)}(k_i, m, g)$$
$$= m^2(2 - \gamma(g, \varepsilon))\Gamma_{\mathrm{R}}^{(N, 1)}(k_i, q = 0, m, g),$$

where $\Gamma_R^{(N,1)}$ corresponds to the insertion of an operator φ^2 (see Section 6.5); and show that

$$\beta(g,\varepsilon) = m\frac{\mathrm{d}g}{\mathrm{d}m}\bigg|_{g_0,\Lambda}; \quad \gamma(g,\varepsilon) = m\frac{\mathrm{d}\ln Z_3}{\mathrm{d}m}\bigg|_{g_0,\Lambda}.$$

Hint: use the norming conditions (6.3.10b) and (6.3.10d).

(b) By using Weinberg's theorem one can show that the second term of the CS equation is negligible if $k_i/m \gg 1$:

$$\left[m\frac{\partial}{\partial m} + \beta(g,\varepsilon)\frac{\partial}{\partial g} - \frac{1}{2}N\gamma(g,\varepsilon)\right]\Gamma_R^{(N)}(k_i, g, m) \simeq 0.$$

Thence show, for $k_i/m \gg 1$ (ultraviolet regime) and for $g = g^*$ (why?), that

$$\Gamma_R^{(N)}(k_i, g^*, m) \sim k_i^{D - N(D/2-1) - 1/2N\gamma(g^*,\varepsilon)}.$$

(c) Setting $u_0 = m^\varepsilon f(g)$, show that $f(g)$ obeys

$$f(g) = g\exp\left[-\int_0^g \mathrm{d}g'\left(\frac{\varepsilon}{\beta(g')} + \frac{1}{g'}\right)\right]$$

(see (7.5.2)); and, in the vicinity of the fixed point, show that

$$u_0 \approx \text{constant} \times m^\varepsilon g|g^* - g|^{-\varepsilon/\omega},$$

where $\omega = \beta'(g^*, \varepsilon)$.

(d) Show that

$$Z_3 \sim m^{\gamma^*}, \quad \bar{Z} \sim m^{\bar{\gamma}^*},$$

where $\bar{\gamma}(g,\varepsilon) = m\dfrac{\mathrm{d}}{\mathrm{d}m}\ln \bar{Z}(g,\varepsilon)|_{g_0,\Lambda}$. Identify the exponents η and ν, by using the fact that $m_0^2 = a + bt_0$, where t_0 is the bare temperature difference $T - T_c$, and by taking into account the relation

$$2 - \gamma(g) = 2\bar{Z}^{-1}\frac{\partial m_0^2}{\partial m^2}\bigg|_{u_0,\Lambda}.$$

Compare with Section 3.5.3.

By using the results of Problem 6.3 and those of Section 7.4.1, calculate the function $\beta(g)$ to two-loop order. Verify that $\beta(g)$, normalized according to (6.3.10), is indeed identical to the result given in (7.5.16). (In other words the coefficients β_0 and β_1 are universal.)

7.4 The renormalization group for φ^3 theory

(a) By using the results of Problem 6.2, calculate the functions $\beta(g)$, $\gamma(g)$, and $\gamma_m(g)$ of φ^3 theory in 6 dimensions, to one-loop order. The function $\gamma_m(g)$ is defined by

$$\gamma_m(g) = -\,\mathrm{d}\ln m^2/\mathrm{d}\ln\mu|_{g_0,\Lambda}\,.$$

Use dimensional regularization, where $m_0^2 = Z_m m^2$, and the minimal scheme; then m is a mass parameter depending on the renormalization mass μ, and is analogous to the parameter t in φ^4 theory (see (7.2.3)).

The solution is

$$\beta(g) = -\frac{3}{4}\alpha g, \quad \gamma(g) = \frac{1}{6}\alpha, \quad \gamma_m(g) = \frac{5}{6}\alpha,$$

where $\alpha = g^2/(4\pi)^3$. Watch out for the replacement of ε by $\varepsilon/2$!

(b) Show that φ^3 theory is asymptotically free in the ultraviolet. Suppose first that the mass is zero. Show that as $s \to \infty$ the correlation functions behave according to

$$\Gamma_{\mathrm{R}}^{(N)}(sk_i, g, \mu) = s^{\,d_N - \frac{1}{2}N\int_0^{\ln s}\gamma(g(s'))\,\mathrm{d}\ln s'}\,\Gamma_{\mathrm{R}}^{(N)}(k_i, g(s), \mu)$$
$$\simeq s^{d_N}(\ln s)^{-N/18}\,\Gamma_{\mathrm{R}}^{(N)}(k_i, g(s), \mu).$$

In order to derive this equation one relies on the one-loop approximation to $\beta(g)$ and $\gamma(g)$, found in the preceding problem.

(c) Suppose now that the mass is nonzero. Show that the correlation functions obey the CS equation

$$\left[\frac{\partial}{\partial\ln\mu} + \beta(g)\frac{\partial}{\partial g} - \frac{1}{2}N\gamma(g) - \gamma_m(g)\frac{\partial}{\partial\ln m^2}\right]\Gamma_{\mathrm{R}}^{(N)}(k_i, g, m, \mu) = 0.$$

Discuss the influence of the mass on the asymptotic behaviour derived in the preceding problem.

7.5 Recurrence relations

(a) From the recurrence relations (7.5.11), show that the term $c'(\alpha^2/\varepsilon^2)$ can be calculated as a function of the term $c(\alpha/\varepsilon)$ in $Z(g, \varepsilon)$: $c' = c^2$.

(b) By writing the renormalization constant Z_3 in the minimal scheme in the form

$$Z_3(g, \varepsilon) = 1 + \sum_{i=1}^{\infty}\frac{Z_3^{(i)}(g)}{\varepsilon^i},$$

derive the recurrence relation

$$g\frac{\partial Z_3^{(i+1)}(g)}{\partial g} = \beta(g)\frac{\partial Z_3^{(i)}(g)}{\partial g} - \gamma(g)Z_3^{(i)}(g).$$

7.6 Ultraviolet-stable fixed point $g^* \neq 0$

Assume that in $D = 4$ the function β has the shape shown in Fig. 7.4, so that there is a nontrivial UV-stable fixed point.

(a) Make a qualitative sketch of the flow diagram (like Fig. 7.10).

(b) Supposing that there is no dimensional parameter in the bare theory other than Λ, show that the correlation length ξ takes the form

$$\xi \sim |g_0 - g^*|^{-1/\omega},$$

where $\omega = -\beta'(g^*)$.

7.7 The functions $\beta(g)$, $\gamma(g)$, . . . depend on the renormalization scheme, but some of their properties are universal, like the coefficients β_0 and β_1. Show that the following properties are independent of the renormalization scheme (Gross 1975, Section 4):

(i) the existence of a zero of $\beta(g)$: $\beta(g^*) = 0$;

(ii) the value of $\beta'(g^*)$ where $\beta(g^*) = 0$;

(iii) the value $\gamma(g^*)$ of $\gamma(g)$ at a fixed point;

(iv) the first coefficient γ_0 in $\gamma(g)$: $\gamma(g) = \gamma_0 g + \cdots$ or $\gamma(g) = \gamma_0 g^2 + \cdots$.

7.8 Corrections to scale invariance (Amit 1984, Chapter II.2)

We aim to calculate the corrections to scale-invariant behaviour. To ease the notation, take $T = T_c$, though the method generalizes readily to $T \neq T_c$. The correlation function at the fixed point is scale-invariant as $s \to 0$:

$$\Gamma_R^{(N)}(sk_i, g^*, \mu) = s^{d_N - \frac{1}{2}N\gamma(g^*)} \Gamma_R^{(N)}(k_i, g^*, \mu).$$

We define a function $C^{(N)}(k_i, g, \mu)$ by

$$\Gamma_R^{(N)}(k_i, g, \mu) = \exp\left(-\frac{1}{2}N \int_g^{g^*} \frac{\gamma(g') - \gamma(g^*)}{\beta(g')} dg'\right) \times C^{(N)}(k_i, g, \mu) \Gamma_R^{(N)}(k_i, g^*, \mu),$$

with $C^{(N)}(k_i, g^*, \mu) = 1$.

(a) Show that $C^{(N)}$ satisfies the equation

$$\left(\frac{\partial}{\partial \ln \mu} + \beta(g, \varepsilon)\frac{\partial}{\partial g}\right) C^{(N)}(k_i, g, \mu) = 0.$$

(b) By using

$$\beta(g, \varepsilon) = \omega(g - g^*) + O(g - g^*)^2$$

(where $\omega = \mathrm{d}\beta(g, \varepsilon)/\mathrm{d}g|_{g=g^*}$), show that $C^{(N)}$ takes the form

$$C^{(N)}(k_i, g, \mu) \approx 1 + (g - g^*)\mu^{-\omega}D(k_i).$$

Thence show that $\Gamma_{\mathrm{R}}^{(N)}$ behaves according to

$$\Gamma_{\mathrm{R}}^{(N)}(sk_i, g, \mu) \simeq s^{d - \frac{1}{2}N\gamma(g^*)}\Gamma(k_i, g^*, \mu)(1 + s^{\omega}(g - g^*)\mu^{-\omega}D(k_i)).$$

(c) Show that $\omega = \varepsilon + C_1\varepsilon^2$, and calculate C_1 explicitly. These corrections to scale invariance are precisely those already studied in Section 3.2.3; in the φ^4 model the exponent y_2 equals $-\varepsilon + O(\varepsilon^2)$ (see Problem 3.4(b)).

7.9 Renormalization-group invariance of the Coleman–Weinberg potential

Show that to order $g^2(\mu)$ the Coleman–Weinberg effective potential (Problem 6.9(c)) is invariant under changes in the renormalization point μ, provided $g(\mu)$ satisfies the evolution equation (7.3.2) in $D = 4$.

Prove generally that the effective potential satisfies the equation

$$\left[\frac{\partial}{\partial \ln \mu} + \beta(g)\frac{\partial}{\partial g} - \frac{1}{2}\gamma(g)\frac{\partial}{\partial \ln \bar{\varphi}}\right] V(\bar{\varphi}, g, \mu) = 0,$$

and that $V \sim \dfrac{2\pi^2}{9}\dfrac{\bar{\varphi}^4}{\ln \mu/\bar{\varphi}}$ when $\bar{\varphi}/\mu \ll 1$ and $D = 4$.

7.10 The critical exponent η to order ε^3

Calculate the contribution to the self-energy from the three-loop graph (which is unique). Show that

$$\frac{\partial \Sigma}{\partial k^2} = -\frac{\alpha^3}{3\varepsilon^2}\left[1 - \frac{3\varepsilon}{2}\ln\frac{k^2}{\mu^2} + \frac{\varepsilon}{2}(7 - 3\gamma)\right],$$

where $\alpha = g/(4\pi)^{D/2}$, and γ is Euler's constant. Add in the graphs with counterterms, and show that

$$Z_3 = 1 - \frac{\alpha^2}{12\varepsilon} - \frac{\alpha^3}{6\varepsilon^2} + \frac{\alpha^3}{24\varepsilon}.$$

Hence derive

$$\eta = \frac{\varepsilon^2}{54}\left[1 + \varepsilon\left(\frac{34}{27} - \frac{1}{4}\right) + O(\varepsilon^2)\right].$$

Further reading

The application of the renormalization group to critical phenomena is discussed in detail by Brézin *et al.* (1976, Sections V and VI); by Amit (1984, Chapters 8 and 9); by Itzykson and Drouffe (1989, Chapter 5); by Parisi (1988, Chapters 7–9) and by Zinn-Justin (1989, Chapters 22 to 25). There is a recent review article by Callaway (1988). See also two articles that are very clear and slanted specifically towards quantum field theory: Coleman (1973); and Gross (1975, Section 4). Further, see Collins (1983, Chapter 7). The role of irrelevant operators is discussed by Brézin *et al.* (1976, Section VIII.C), by Amit (1984, Chapter II.2), and by Itzykson and Drouffe (1989, Chapter 5).

III

The Quantum Theory of Scalar Fields

8 Path integrals in quantum and in statistical mechanics*

This chapter links Parts I and II, which were devoted to statistical mechanics, and Part III, which deals with quantum field theory. In Part III we shall revert to the tools developed for studying critical phenomena, namely functional integrals (often called path integrals in this new connection); Feynman diagrams; renormalization; and the renormalization group. A few adaptations are all that these tools need in their new context, and we shall use some elementary examples to illustrate the links between problems in statistical and in quantum mechanics.

Nevertheless, though the results of this chapter afford deeper insight into some later arguments, they will not be applied directly. Hence the present chapter may be skipped at a first reading.

In his lectures on quantum mechanics, Feynman (1965, Chapters 1 and 3) introduces a basic concept of the subject, that of *probability amplitudes*, by considering a Young-slits interference experiment performed with electrons. In this experiment, electrons emitted by a source A arrive at a (variable) target point B on a screen S′, having first passed through another screen S pierced by two slits (1) and (2). To the two possible trajectories there correspond respectively two probability amplitudes a_1 and a_2, given by the following rules (in a selfexplanatory notation: see Fig. 8.1):

$$a_1 = a_{\mathrm{B}1} a_{1\mathrm{A}}, \quad a_2 = a_{\mathrm{B}2} a_{2\mathrm{A}}.$$

The probability amplitude a for observing an electron at B is the sum of a_1 and a_2:

$$a = a_1 + a_2 = \sum_{j=1}^{2} a_{\mathrm{B}j} a_{j\mathrm{A}}.$$

Let us now complicate the experiment somewhat by introducing several intermediate screens J, K, L, each pierced by several slits numbered j, k, and l. Then the probability amplitude a reads

$$a = \sum_{j,k,l} a_{\mathrm{B}l} a_{lk} a_{kj} a_{j\mathrm{A}}.$$

It is a sum over all paths leading from A to B: in Fig. 8.2 for instance we have drawn the path A → J(1) → K(2) → L(1) → B.

* This chapter may be skipped at a first reading.

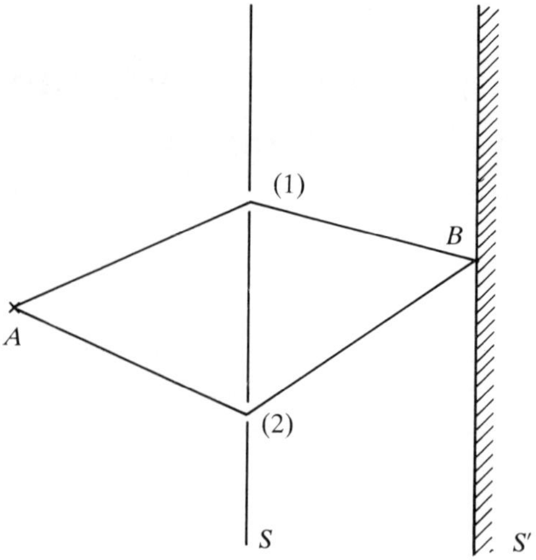

Fig. 8.1 The Young's slits experiment

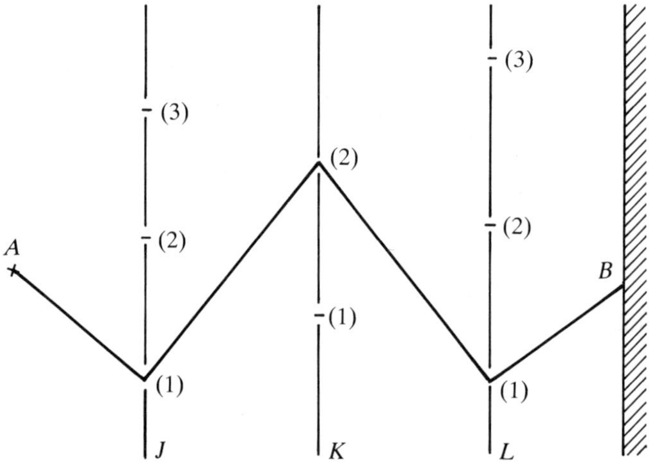

Fig. 8.2 A complicated variant of the Young's slits experiment

One can now replace the screens by a potential in which the electrons move, and associate with every path $[c]$ leading from A to B a probability amplitude $a[c]$, the total amplitude a being the sum (see Fig. 8.3)

$$a = \sum_{[c]} a[c]. \tag{8.0.1}$$

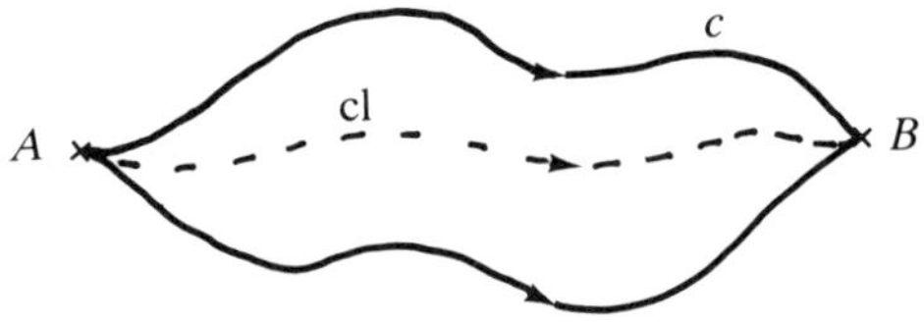

Fig. 8.3 Paths from A to B

It remains of course to give a prescription for $a[c]$, and for summing over the paths. This will be done in Section 8.2; there, starting from the properties of the evolution operator $\exp(-\mathrm{i}Ht)$, we shall show that the statistical weight of each path, namely $a[c]$, is given by

$$a[c] \sim \exp\left(\frac{\mathrm{i}}{\hbar} S(\mathrm{B}, \mathrm{A})\right), \tag{8.0.2}$$

where $\sim$ simply denotes proportionality, and where $S(\mathrm{B}, \mathrm{A})$ is the classical action evaluated along the path leading from A to B in a given time. Conversely, one can adopt (8.0.1) and (8.0.2) as fundamental postulates, and derive all the results of quantum mechanics from them: in other words one can adopt, as one's quantization postulate, equation (8.0.2) instead of the canonical commutation rule (CCR) $[Q, P] = \mathrm{i}\hbar$.

The probability amplitudes in (8.0.1) are matrix elements of the evolution operator in a representation where the position operator Q is diagonal. This matrix element can be continued analytically to complex values of the time $t = -\mathrm{i}\tau$. Then the probability amplitude becomes an element of the (unnormed) density matrix, still in the representation diagonalizing Q; the partition function of the *quantum* system is the trace of this density matrix, provided one chooses $\tau = 1/kT$. But the same density matrix can also be interpreted in a different way, as follows. To every path $[c]$ one assigns a configuration $[\gamma]$ of some *classical* system, the statistical weight of a configuration being given by

$$p(\gamma) \sim \exp(-H[\gamma]); \tag{8.0.3}$$

here H is nothing but the Hamiltonian of the classical system, which can be deduced from S. (H is also written as S_{E}, namely 'Euclidean action': see Section 8.3.3) The partition function of the classical system reads

$$Z = \sum_{[\gamma]} \exp(-H[\gamma]), \tag{8.0.4}$$

where the sum over configurations is the analogue of the sum over paths in (8.0.1).

Thus, starting from a quantum system with its probability amplitudes (8.0.2), one arrives at a classical system described by the probabilities (8.0.3).

In quantum mechanics, an important role is often played by the classical path from A to B, i.e. by the path which makes the action stationary. In the corresponding problem of classical statistical mechanics, the configuration analogous to the classical path is the 'Landau configuration', namely that configuration for which the Hamiltonian is stationary (see Section 2.2). To the quantum fluctuations around the classical path there correspond the statistical fluctuations around the Landau configuration. Such fluctuations are studied by means of perturbation theory, and it is not surprising that one should meet the same techniques in both types of problem.

Section 8.1 deals with an elementary example, the correspondence between the quantum dynamics of a spin 1/2 system and the dynamics of the one-dimensional Ising model. This example illustrates very clearly the passage from a quantum to a statistical problem, and allows one to pinpoint a number of analogies.

In Section 8.2 we establish the dictionary for translating the Schroedinger equation into the language of path integrals, by giving a precise meaning to the sum in (8.0.1). One important task of Section 8.2 is to study the boundary conditions on the path integral for the harmonic oscillator: a quantum field is in effect nothing more than a superposition of harmonic oscillators (mutually independent in the case of a free field, but coupled in general), and these boundary conditions prove crucial later, because they guarantee the unitarity and the causality of the theory.

Finally in Section 8.3 we shall spell out the connection between statistical mechanics and quantum mechanics, by introducing the 'Euclidean continuation' of the path integral defined in the previous section.

8.1 Quantum spin and Ising model

8.1.1 The path integral for spin 1/2

Consider in quantum mechanics a Hamiltonian describing the dynamics of a spin 1/2,

$$H = -K\sigma_1, \tag{8.1.1}$$

in units where $\hbar = 1$; $\sigma_1, \sigma_2, \sigma_3$ are the usual Pauli matrices (see equation (D.5)). In the representation diagonalizing σ_3 the eigenvectors of H are

$$|0\rangle = \frac{1}{\sqrt{2}}\begin{pmatrix}1\\1\end{pmatrix} \quad \text{and} \quad |1\rangle = \frac{1}{\sqrt{2}}\begin{pmatrix}1\\-1\end{pmatrix},$$

corresponding, respectively, to the eigenvalues $E_0 = -K$ (ground state), and to $E_1 = K$. Like all quantum systems, this system displays fluctuations: e.g.

$$\langle 0|\sigma_3^2|0\rangle \neq |\langle 0|\sigma_3|0\rangle|^2.$$

Consider now a matrix element of the evolution operator $\exp(-\mathrm{i}Ht)$ between two eigenstates of σ_3, $|S_a\rangle$ and $|S_b\rangle$ $(S_a, S_b = \pm 1)$:

$$F(t, S_b|0, S_a) = \langle S_b|\exp(-\mathrm{i}Ht)|S_a\rangle.$$

Here F is the probability amplitude for observing the eigenvalue S_b at time t, knowing that at time $t = 0$ the spin is in the state $|S_a\rangle$. The values of S_a and of S_b constitute the *boundary conditions* at the times $t = 0$ and t respectively. To ease the notation, we suppose that $t = N = \text{integer} \gg 1$, and divide the interval $[0, t]$ into N subintervals each of unit length:

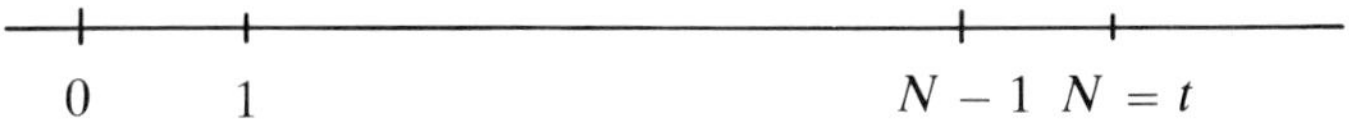

Now insert a complete set $|S_i\rangle$, $S_i = \pm 1$ of eigenstates of σ_3 at every point dividing one subinterval of $[0, t]$ from the next:

$$\langle S_b|\mathrm{e}^{-\mathrm{i}Ht}|S_a\rangle = \sum_{S_1=\pm 1} \cdots \sum_{S_{N-1}=\pm 1} \langle S_b|\mathrm{e}^{-\mathrm{i}H}|S_{N-1}\rangle \times \langle S_{N-1}| \cdots |S_1\rangle \langle S_1|\mathrm{e}^{-\mathrm{i}H}|S_a\rangle. \tag{8.1.2}$$

Further, define the (complex) function $V(S, S')$ by

$$\langle S|\mathrm{e}^{-\mathrm{i}H}|S'\rangle = \mathrm{e}^{-\mathrm{i}V(S,S')}. \tag{8.1.3}$$

Then equation (8.1.2) yields

$$\langle S_b|\mathrm{e}^{-\mathrm{i}H}|S_a\rangle = \sum_{[S_i]} \exp(-\mathrm{i}[V(S_b, S_{N-1}) + V(S_{N-1}, S_{N-2}) + \cdots + V(S_1, S_a)]), \tag{8.1.4}$$

where the sum runs over all intermediate configurations S_i:

$$\sum_{[S_i]} = \sum_{S_1=\pm 1} \cdots \sum_{S_{N-1}=\pm 1}.$$

No operators remain on the right of (8.1.4), which constitutes an elementary example of a *path integral*. Note that (8.1.4) indeed has the same form as (8.0.1).

We proceed by continuing (8.1.2) analytically to imaginary values of t: $t = -\mathrm{i}\tau$. The matrix element

$$F(-\mathrm{i}\tau, S_b; 0, S_a) = \langle S_b|\mathrm{e}^{-H\tau}|S_a\rangle$$

is given by an expression analogous to (8.1.4), and the corresponding matrix $V_{\mathrm{E}}(S, S')$ is easy to calculate on observing that, in a basis where σ_3 is diagonal, one has

$$\mathrm{e}^{-H} = \mathrm{e}^{K\sigma_1} = \begin{pmatrix} \cosh K & \sinh K \\ \sinh K & \cosh K \end{pmatrix}.$$

Writing $\exp(-V_E(S, S'))$ in the form

$$\exp(-V_E(S, S')) = \exp(A + BSS'),$$

we can identify

$$e^{A+B} = \cosh K, \quad e^{A-B} = \sinh K. \tag{8.1.5}$$

This identification supplies the link with a problem in *classical* statistical mechanics. Imagine that we have placed Ising spins S_i at the points $0, 1, \ldots, N$, each spin capable of assuming only the two values $S_i = +1$ and $S_i = -1$; and that the interaction Hamiltonian is

$$H[S_i] = -\sum_{i=0}^{N-1} (A + BS_iS_{i+1}), \tag{8.1.6a}$$

the probability of a configuration $[S_i]$ being given by

$$p[S_i] \sim \exp(-H[S_i]). \tag{8.1.6b}$$

The Hamiltonian (8.1.6a) is that of the one-dimensional Ising model (see Section 1.2.2). With $t = -i\tau$ the expression (8.1.4) becomes the partition function of the Ising model, when the values of the two end spins S_a and S_b are fixed. (The temperature is subsumed into the coefficients A and B; better still, one could identify it with $\hbar$.) Summing over S_a and S_b with periodic (cyclic) boundary conditions

$$S_a(=S_0) \equiv S_b(=S_N),$$

we obtain the following partition function

$$Z_N = \sum_{[S_i]} e^{\sum_i (A + BS_iS_{i+1})} = \sum_{[S_i]} e^{-H[S_i]}, \tag{8.1.7}$$

which equals

$$Z_N = \mathrm{Tr}\, e^{-H\tau}. \tag{8.1.8}$$

The partition function of a system of N *classical* spins is equal to the partition function of *one* quantum spin at temperature $kT = 1/N$. For this elementary example we have now implemented the programme announced in the introduction: starting from a probability amplitude having the form (8.1.4) (see (8.0.1)), we have constructed a classical system whose configurations have the probabilities given by (8.1.6) (see (8.0.3)). However, the quantum spin has no classical limit, and we have seen nothing analogous to (8.0.2). That is why the path integral (8.1.4) is useless in practice. Only the limit $N \to \infty$ yields interesting results, as we shall see below.

8.1.2 Analogues

1. Free energy and ground-state energy

In the limit $N \to \infty$, the free energy $\hat{F}$ per spin is given by

$$\hat{F} = \lim_{N\to\infty}\left(-\frac{1}{N}\ln Z_N\right) = \lim_{\tau\to\infty}\left(-\frac{1}{\tau}\ln \mathrm{Tr}\,\mathrm{e}^{-H\tau}\right). \tag{8.1.9}$$

But, as $\tau \to \infty$, $\mathrm{e}^{-H\tau}$ is dominated by the ground-state eigenvalue:

$$\mathrm{e}^{-H\tau} \approx |0\rangle \mathrm{e}^{-E_0\tau}\langle 0|,$$

and

$$\hat{F} = E_0.$$

The *free* energy per spin of the statistical system is therefore analogous to the energy of the ground state of the quantum system.

2. Correlation function and time-ordered product

Let us now examine the correlation function of two Ising spins,

$$\langle S_m S_l\rangle = \frac{1}{Z_N}\sum_{[S_i]} S_m S_l \mathrm{e}^{-H[S_i]}, \quad (m > l), \tag{8.1.10}$$

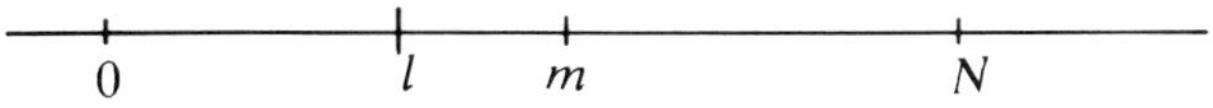

which by virtue of

$$\langle S_l|\sigma_3 \mathrm{e}^{-H}|S_{l-1}\rangle = S_l \mathrm{e}^{-V_E(S_l, S_{l-1})}$$

may be transformed into

$$\langle S_m S_l\rangle = \frac{1}{\mathrm{Tr}\,\mathrm{e}^{-H\tau}}\mathrm{Tr}[\mathrm{e}^{-(N-m)H}\sigma_3 \mathrm{e}^{-(m-l)H}\sigma_3 \mathrm{e}^{-lH}].$$

In the limit $N \to \infty$ with fixed $(m-l)$, one may write $\langle S_m S_l\rangle$ in the form

$$\langle S_m S_l\rangle = \mathrm{e}^{E_0(m-l)}\langle 0|\sigma_3 \mathrm{e}^{-(m-l)H}\sigma_3|0\rangle, \tag{8.1.11a}$$

or

$$\langle S_m S_l\rangle = |\langle 0|\sigma_3|0\rangle|^2 + |\langle 0|\sigma_3|1\rangle|^2 \mathrm{e}^{-(m-l)(E_1-E_0)}. \tag{8.1.12}$$

Before commenting on (8.1.12), we note that for $m < l$ we would have found

$$\langle S_m S_l\rangle = \mathrm{e}^{E_0(l-m)}\langle 0|\sigma_3 \mathrm{e}^{-(l-m)H}\sigma_3|0\rangle. \tag{8.1.11b}$$

Defining the operator $\sigma_3(t)$ in the Heisenberg picture by

$$\sigma_3(t) = \mathrm{e}^{\mathrm{i}Ht}\sigma_3 \mathrm{e}^{-\mathrm{i}Ht}, \tag{8.1.13}$$

and its analytic continuation to complex values of t by

$$\sigma_3(\tau) = e^{H\tau}\sigma_3 e^{-H\tau},$$

one sees that (8.1.11a) and (8.1.11b) are both covered by

$$\langle S_m S_l \rangle = \langle 0| T(\sigma_3(\tau_m)\sigma_3(\tau_l))|0\rangle . \tag{8.1.14}$$

Here we have introduced the *time-ordered product* (or T-product)

$$\begin{aligned} T(\sigma_3(t_m)\sigma_3(t_l)) &= \sigma_3(t_m)\sigma_3(t_l), \quad t_m > t_l, \\ &= \sigma_3(t_l)\sigma_3(t_m), \quad t_m < t_l, \end{aligned} \tag{8.1.15}$$

and $\tau_m > \tau_l$ if $t_m > t_l$.

Thus we see that the correlation function between two spins is the continuation, to complex t, of the ground-state expectation value of a T-product. To the quantum fluctuations of the operator σ_3 there correspond the statistical fluctuations of the classical variable S.

3. *Correlation length and energy gap*

We revert to equation (8.1.12); with our choice of Hamiltonian, we have $\langle 0|\sigma_3|0\rangle = 0$. If this condition is not satisfied (see e.g. Problem 8.1), then the correlation function must be defined by subtracting the product $\langle S_m\rangle\langle S_l\rangle$:

$$\langle S_m S_l\rangle \to \langle S_m S_l\rangle - \langle S_m\rangle\langle S_l\rangle$$

(see Section 1.4.1). Restricting ourselves for the moment to the Hamiltonian (8.1.1), we obtain

$$\langle S_m S_l\rangle = |\langle 0|\sigma_3|1\rangle|^2 e^{-|m-l|\Delta E},$$

where $\Delta E = E_1 - E_0$ is the energy difference between the ground and the excited state. The correlation length can be identified from

$$\langle S_m S_l\rangle \sim e^{-|m-l|/\xi}, \tag{8.1.16}$$

which yields $\xi = 1/\Delta E$: the correlation length is the inverse of the energy gap between the ground and the excited state. When there are several excited states, the behaviour of the correlation function is still governed by $\Delta E = E_1 - E_0$, where E_1 is the first-excited energy level; now however one must choose $(m-l)$ large enough to make the contributions of the other levels negligible.

Instead of dividing the interval $[0, \tau]$ into subintervals of unit length, we could have divided it into τ/ε subintervals each of length ε. Provided that $\tau/\varepsilon \gg 1$ (more precisely $\tau\Delta E/\varepsilon\hbar \gg 1$), the previous results are unaffected, the quantum problem remaining the same. This claim is readily verified through the explicit expression for the partition function (Problem 8.1). The result (8.1.9) however must be interpreted as follows: since the interval $[0, \tau]$ now contains τ/ε spins, $-(1/\tau)\ln Z$ is no longer the free energy per spin, but the free-energy *density* (per unit length). Notice that we have associated, with the quantum problem on a single site, a statistical-mechanics problem on a line, i.e. in a space of one dimension. A quantum field theory associates a quantum system with every point of space (the difficulties stemming from the fact that these systems interact with each other). *To a quantum field theory in a $(D-1)$-dimensional space (i.e. in a D-dimensional space–time) there corresponds a statistical system in a space of D dimensions.* Having observed this, we draw up a table which displays these analogies more generally;

free-energy density	~	ground-state energy density
correlation function	~	time-ordered product
reciprocal of the correlation length	~	energy gap

8.2 Particle in a potential

Envisage now a quantum system slightly more complicated than a spin 1/2, namely a (nonrelativistic) particle of mass m moving on a straight line in a potential $V(q)$. We denote the operators for position and momentum by Q and P respectively, and their eigenstates in the Schroedinger picture by $|q\rangle$ and $|p\rangle$. We choose the normalization $\langle p|p'\rangle = \delta(p-p')$, $\langle q|q'\rangle = \delta(q-q')$, and $\langle q|p\rangle = (2\pi)^{-1/2}\exp(\mathrm{i}q\cdot p)$. $Q(t)$ and $|q,t\rangle$ stand for the position operator and for its eigenvectors in the Heisenberg picture:

$$\begin{aligned} Q(t) &= \mathrm{e}^{\mathrm{i}Ht} Q \mathrm{e}^{-\mathrm{i}Ht}, \\ |q,t\rangle &= \mathrm{e}^{\mathrm{i}Ht}|q\rangle. \end{aligned} \tag{8.2.1}$$

8.2.1 The representation of probability amplitudes by path integrals

Let $F(q',t';q,t)$ be the probability amplitude that a particle initially at q at time t be found at q' at time t' (so that the boundary conditions are $q(t)=q$, $q(t')=q'$):

$$F(q',t';q,t) = \langle q',t'|q,t\rangle = \langle q'|\mathrm{e}^{-\mathrm{i}H(t'-t)}|q\rangle. \tag{8.2.2}$$

Divide the interval $[t,t']$ into $(n+1)$ subintervals each of length $\varepsilon = (t'-t)/(n+1)$, with $\varepsilon \to 0$:

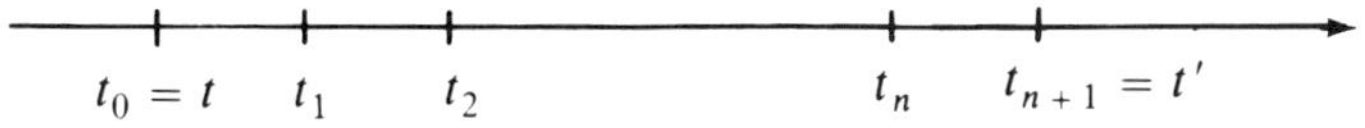

and let us write

$$\exp(-\mathrm{i}H(t'-t)) = \left(\exp\left[-\mathrm{i}\varepsilon\left(\frac{P^2}{2m} + V(Q)\right)\right]\right)^{n+1}.$$

We now use Trotter's (or the Lie product) theorem (Schulman 1981, Chapter 1; Glimm and Jaffe 1987, Chapter 3)

$$\lim_{n\to\infty} (\mathrm{e}^{A/n}\,\mathrm{e}^{B/n})^n = \mathrm{e}^{A+B}. \tag{8.2.3}$$

One can find in the two references just quoted heuristic and rigorous proofs of this formula, as well as the necessary conditions on the operators A and B. It is an instructive (and elementary) exercise to carry through the proof when $[A, B]$ is a c-number; it is also easy to prove (8.2.3) if A and B are bounded operators (see the references just quoted). Let us now insert at times $t_1, \ldots, t_n$ complete sets of eigenstates of the position operator Q,

$$F(q't'; q, t) = \int \prod_{l=1}^{n} \mathrm{d}q_l \prod_{l=0}^{n} \langle q_{l+1}|\exp\left(-\mathrm{i}\varepsilon\frac{P^2}{2m}\right)\exp(-\mathrm{i}\varepsilon\, V(Q))|q_l\rangle,$$

and evaluate the matrix element

$$\langle q_{l+1}|\exp\left(-\mathrm{i}\varepsilon\frac{P^2}{2m}\right)\exp(-\mathrm{i}\varepsilon\, V(Q))|q_l\rangle$$

$$= \exp(-\mathrm{i}\varepsilon\, V(q_l))\langle q_{l+1}|\exp\left(-\mathrm{i}\varepsilon\frac{P^2}{2m}\right)|q_l\rangle.$$

In order to compute the last matrix element we use

$$\langle q_{l+1}|P^2|q_l\rangle = \int \mathrm{d}p_l \langle q_{l+1}|P^2|p_l\rangle\,\langle p_l|q_l\rangle$$

$$= \int \frac{\mathrm{d}p_l}{2\pi}\, p_l^2\, \mathrm{e}^{\mathrm{i}(q_{l+1}-q_l)p_l}.$$

These results allow us to write $F(q', t'; q, t)$ in the form of a path integral:

$$F(q', t'; q, t) = \lim_{\varepsilon\to 0} \int \prod_{l=1}^{n} \mathrm{d}q_l \prod_{l=0}^{n} \left\{\frac{\mathrm{d}p_l}{2\pi}\exp(\mathrm{i}p_l(q_{l+1}-q_l))\right.$$
$$\left.\times\exp\left(-\mathrm{i}\varepsilon\left(\frac{p_l^2}{2m} + V\left(\frac{q_l+q_{l+1}}{2}\right)\right)\right)\right\}. \tag{8.2.4}$$

It is important to notice that in equation (8.2.4) q and p are *classical* variables (numbers!), as was the variable S in the previous section. We have taken $\frac{1}{2}(q_l + q_{l+1})$ as the argument of V for purely aesthetic reasons, since q_l or q_{l+1} would also be correct. However, for reasons to be discussed below, it is important to choose $\frac{1}{2}(q_l + q_{l+1})$ as the argument of the vector potential if one wants to write path integrals for propagation in a magnetic field. As V is a function only of q, it is possible to perform the p-integral in (8.2.4) (problem 8.2):

$$\int \frac{\mathrm{d}p}{2\pi} \exp\left(ipq - \mathrm{i}\varepsilon \frac{p^2}{2m}\right) = \left(\frac{m}{2\mathrm{i}\pi\varepsilon}\right)^{1/2} \exp\left(\frac{imq^2}{2\varepsilon}\right), \tag{8.2.5}$$

and equation (8.2.4) becomes

$$F(q', t'; q, t) = \lim_{\varepsilon \to 0} \left(\frac{m}{2\mathrm{i}\pi\varepsilon}\right)^{1/2} \int \prod_{l=1}^{n} \left[\left(\frac{m}{2\mathrm{i}\pi\varepsilon}\right)^{1/2} \mathrm{d}q_l\right]$$
$$\times \exp\left[\mathrm{i} \sum_{l=0}^{n} \frac{m(q_l - q_{l+1})^2}{2\varepsilon} - \mathrm{i}\varepsilon \sum_{l=0}^{n} V\left(\frac{q_l + q_{l+1}}{2}\right)\right]. \tag{8.2.6}$$

Here we introduce the compact symbol $\mathscr{D}q$ for integration over the q_i, and note that

$$\varepsilon \sum_{l=0}^{n} V\left(\frac{q_l + q_{l+1}}{2}\right) \to \int_t^{t'} V(q(t''))\,\mathrm{d}t'',$$
$$\varepsilon \sum_{l=0}^{n} \frac{m(q_l - q_{l+1})^2}{2\varepsilon^2} \to \int_t^{t'} m\left(\frac{\mathrm{d}q}{\mathrm{d}t''}\right)^2 \mathrm{d}t'';$$

then as the final form of the path integral one finds

$$F(q', t'; q, t) = \int \mathscr{D}q \exp\left(\frac{\mathrm{i}}{\hbar}\int_t^{t'} \left(\frac{1}{2}m\dot{q}^2 - V(q)\right)\mathrm{d}t''\right)$$
$$= \int \mathscr{D}q \exp\left(\frac{\mathrm{i}}{\hbar} S\right), \tag{8.2.7}$$

subject to the boundary conditions $q(t) = q$, $q(t') = q'$.

In equation (8.2.7) we have reinstated Planck's constant $\hbar$ and set $\dot{q} = \mathrm{d}q/\mathrm{d}t$;

$$L(q, \dot{q}) = \frac{1}{2}m\dot{q}^2 - V(q) \tag{8.2.8}$$

is the Lagrangean of the particle; and S is the corresponding *action*:

$$S = \int_t^{t'} L(q, \dot{q})\,\mathrm{d}t''. \tag{8.2.9}$$

In quantum mechanics, it is difficult to go further into the mathematical discussion of the paths, since it has not been possible to give a satisfactory mathematical meaning to the measure $\exp(\mathrm{i}S)\mathscr{D}q$. However one can write a path integral for matrix elements of the operator $\exp(-\tau H)$; as will be explained in Section 8.3., this corresponds to going to Euclidean space. It is then possible to give a rigorous definition of the measure $\exp(-S_{\mathrm{E}})\,\mathscr{D}q$, where S_{E} is the Euclidean action. If $V(q) = 0$, this measure is known by mathematicians under the name of 'conditional Wiener measure', and is closely related to sums over Brownian paths. From simple properties of Gaussian integrals one deduces that

$$\langle (q_{l+1} - q_l)^2 \rangle \sim \frac{\hbar\varepsilon}{m},$$

which corresponds to a well-known property of Brownian motions: the average distance travelled by the particle is of order $\varepsilon^{1/2} \sim (\Delta t)^{1/2}$. This means that most of the paths will be continuous, but not differentiable. Actually one can show that differentiable paths form a set of measure zero. This is the reason why one has to be careful in defining integral over paths, and in particular this is the reason why one has to choose the argument of the vector potential to be $\frac{1}{2}(q_l + q_{l+1})$ when studying a particle in a magnetic field (Schulman 1981, Chapters 4 and 5). In the above arguments, the important role is played by the kinetic term in the Lagrangean: in fact it is this term which determines the functional space over which one integrates.

Using the notations of (8.2.7) one can transform (8.2.4) and obtain the Hamiltonian form of the path integral, namely

$$F(q', t'; q, t) = \int \mathscr{D}p\, \mathscr{D}q \exp\left(\mathrm{i} \int_t^{t'} [p\dot{q} - H(p, q)]\, \mathrm{d}t'' \right). \tag{8.2.10}$$

This equation must be used when the Hamiltonian is not quadratic in p; however, it must be handled with care. Contrary to its appearance, it is not invariant under canonical transformations.

The interpretation of equation (8.2.7)

Consider a path from A: (q, t) to B: (q', t'); to it there corresponds a certain action S. Equation (8.2.7) can be interpreted as assigning to each path a statistical weight $\exp(\mathrm{i}S/\hbar)$, and asserting that the probability amplitude is obtained by summing over all such paths. In the procedure we have followed, the sum over paths has been defined thus. The paths are zigzags whose straight portions join the positions $q, q_1, \ldots, q'$ of the particle at times $t, t_1, \ldots, t'$ respectively (Fig. 8.4). Summing over the paths consists of integrating, with fixed q and q', over all the q_l corresponding to intermediate times t_l, with the integration measure

$$\left(\frac{m}{2\mathrm{i}\pi\varepsilon}\right)^{1/2} \prod_{l=1}^{n} \left(\frac{m}{2\mathrm{i}\pi\varepsilon}\right)^{1/2} \mathrm{d}q_l \to \mathscr{D}q.$$

Here, as already emphasized, the symbol $\mathscr{D}q$ means nothing more than is implied by (8.2.6).

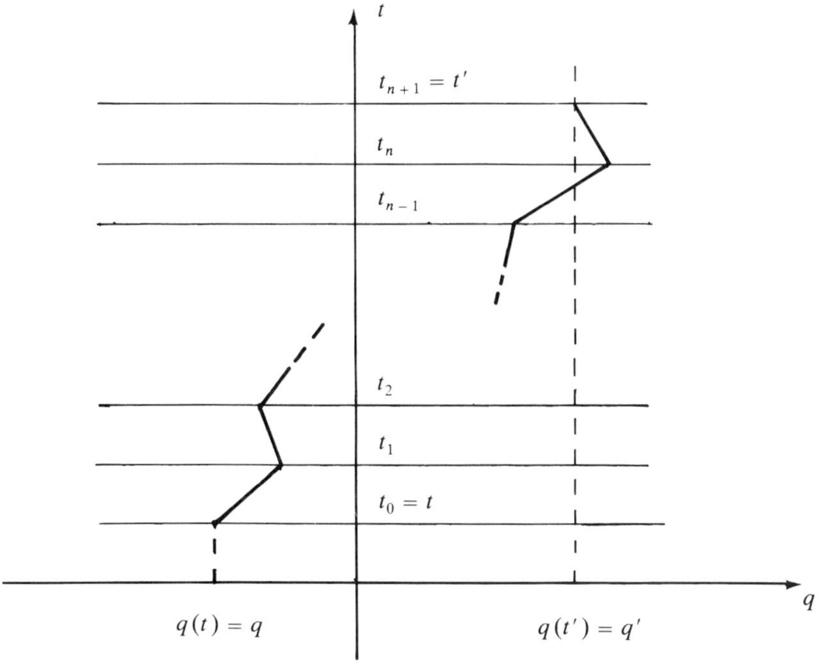

Fig. 8.4 Trajectories used in evaluating (8.2.6)

The action corresponding to a path element $q_l \to q_{l+1}$ is

$$\Delta S = \frac{1}{2} m \frac{(q_l - q_{l+1})^2}{\varepsilon} - \varepsilon V\left(\frac{q_l + q_{l+1}}{2}\right).$$

If $|q_l - q_{l+1}| \gg (\hbar\varepsilon/m)^{1/2}$, then the factor $\exp(\mathrm{i}S/\hbar)$ oscillates rapidly on switching from this path to a neighbouring one: this confirms that appreciable contributions arise only from trajectories that are 'sufficiently regular'.

Next we investigate the role of the classical trajectory $q_{cl}(t'')$. This path corresponds to a stationary action, i.e.

$$\left.\frac{\delta S}{\delta q(t'')}\right|_{q = q_{cl}(t'')} = 0 .$$

Consider a path $q(t'')$, and let S be the corresponding action (Fig. 8.5). If $|S - S_{cl}| \gg \hbar$, then the factor $\exp(\mathrm{i}S/\hbar)$ oscillates many times as q_{cl} is deformed into q. Hence the trajectories neighbouring q contribute only negligibly to the probability amplitude. The amplitude will be dominated by the trajectories near q_{cl}, whose action is such that $|S - S_{cl}| \lesssim \hbar$. In certain cases one can hope to take account of the quantum fluctuations around the classical trajectory by an expansion in powers of $\hbar$: in quantum mechanics, the corresponding approximation is known as the stationary phase approximation.

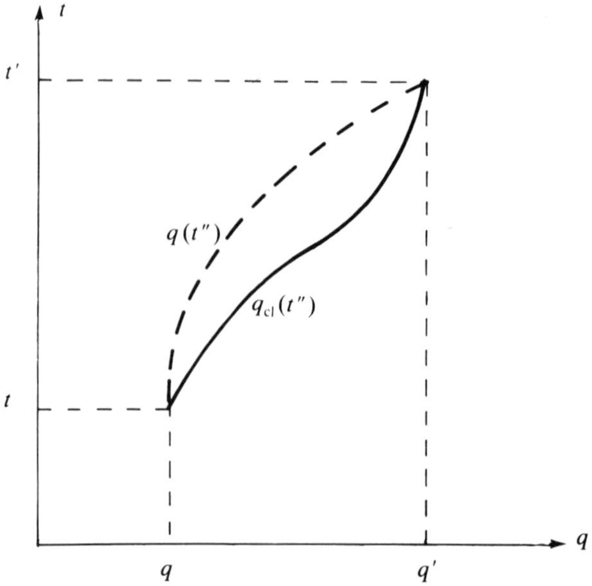

Fig. 8.5

8.2.2 The generating functional and the expressions for *T*-products

One very important construct in quantum field theory is the 'vacuum expectation value of *T*-products of operators', which has already been defined in the case of our toy model of section A. The 'vacuum state' is nothing but the ground state, or state of lowest energy, which is assumed to be nondegenerate, and will be denoted by $|0\rangle$. The *T*-product of $Q(t_1)$ and $Q(t_2)$ is defined by

$$T(Q(t_1)Q(t_2)) = Q(t_1)Q(t_2), \quad t_1 > t_2,$$
$$= Q(t_1)Q(t_2), \quad t_1 < t_2,$$

and the definition can be generalized obviously to the product of any number of operators: the *T*-product reorders the operators in the order of ascending times. Thus we shall be interested in objects like

$$\langle 0|T(Q(t_1)Q(t_2))|0\rangle .$$

From the path-integral representation of $F(q', t'; q, t)$ it is very easy to obtain the *T*-product matrix element between states $|q', t'\rangle$ and $|q, t\rangle$,

$$\langle q', t'|T(Q(t_1)Q(t_2))|q, t\rangle = \int \mathscr{D}q\, q(t_1)q(t_2)\,\mathrm{e}^{\mathrm{i}S} . \tag{8.2.11}$$

One has only to notice that the T-product ensures that times occur always in the right order when one divides the $[t, t']$ interval in order to establish the path integral formula. However, (8.2.11) is not exactly what we want, since we are interested in vacuum expectation values of T-products; hence we must now project (8.2.11) on the ground state.

The vacuum-to-vacuum amplitude and the generating functional

It will prove convenient to couple the quantum system to a 'source' $j(t)$. The values taken by such a function $j(t)$ are fixed in advance: it does not depend on the dynamics of the particle. The Lagrangean changes to

$$L = \frac{1}{2} m\dot{q}^2 - V(q) + j(t)q(t), \tag{8.2.12}$$

and one sees from the equations of motion that the particle is now subject to an extra force $j(t)$. The source $j(t)$ will be taken as zero outside an interval $[t, t']$ (Fig. 8.6):

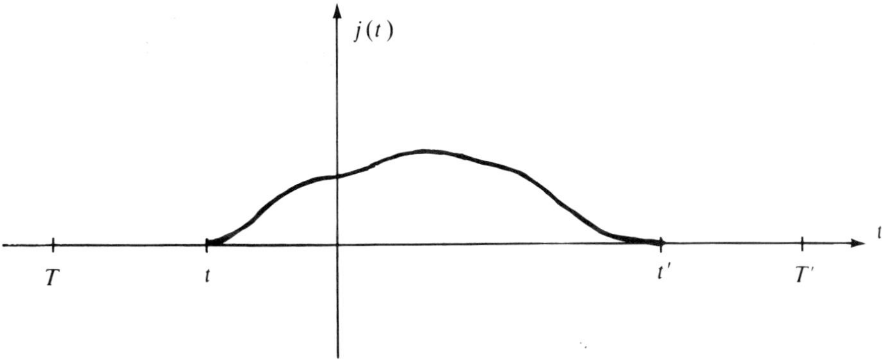

Fig. 8.6 The source $j(t)$

Inside this interval the evolution operator $U_j(t)$ obeys the differential equation

$$\mathrm{i}\frac{\mathrm{d}U_j}{\mathrm{d}t} = [H - j(t)Q]\, U_j(t), \tag{8.2.13}$$

where H is the Hamiltonian corresponding to the first two terms of (8.2.12).

Let T and T' be two times such that $T < t$ and $T' > t'$; then the probability amplitude $\langle Q', T' | Q, T \rangle$ in the presence of the source may be written as

$$\langle Q', T'|Q, T\rangle_j = \int \mathrm{d}q\, \mathrm{d}q' \langle Q', T'|q', t'\rangle \langle q', t'|q, t\rangle_j \langle q, t|Q, T\rangle .$$

We shall now pick out the ground state by taking appropriate limits of T and T'. Notice that

$$\langle q, t|Q, T\rangle = \langle q|\mathrm{e}^{-\mathrm{i}H(t-T)}|Q\rangle = \sum_n \varphi_n(q)\varphi_n^*(Q)\mathrm{e}^{-\mathrm{i}E_n(t-T)},$$

where $\varphi_n(q) = \langle q|n\rangle$ is the wavefunction of the state n, having energy E_n. The limit $T \to \mathrm{i}\infty$ picks out the ground state, as long as one imposes further conditions ensuring that the trajectories $Q(T)$ and $Q'(T')$ do not escape to infinity; e.g. one might require that $Q(T)$ and $Q'(T')$ tend to constants as T and $T' \to \infty$, which guarantees that the $\varphi_n(Q)$ tend to finite limits. In these circumstances we have

$$\lim_{T\to \mathrm{i}\infty} \mathrm{e}^{-\mathrm{i}E_0 T}\langle q, t|Q, T\rangle = \varphi_0(q, t)\varphi_0^*(Q),$$

where

$$\varphi_0(q, t) = \langle q, t|0\rangle = \mathrm{e}^{-\mathrm{i}E_0 t}\varphi_0(q).$$

Similarly one finds

$$\lim_{T'\to -\mathrm{i}\infty} \mathrm{e}^{\mathrm{i}E_0 T'}\langle Q', T'|q', t'\rangle = \varphi_0^*(q', t')\varphi_0(Q).$$

The *generating functional* $Z(j)$ for T-products is defined by

$$Z(j) = \lim_{\substack{T\to \mathrm{i}\infty \\ T'\to -\mathrm{i}\infty}} \frac{\langle Q', T'|Q, T\rangle_j}{\mathrm{e}^{-\mathrm{i}E_0(T'-T)}\varphi_0^*(Q)\varphi_0(Q')} \tag{8.2.14}$$

or

$$Z(j) = \int \mathrm{d}q\,\mathrm{d}q'\,\varphi_0^*(q', t')\langle q', t'|q, t\rangle_j\varphi_0(q, t). \tag{8.2.15}$$

Equation (8.2.14) supplies the 'projection onto the ground state' that we were looking for. The physical interpretation of $Z(j)$ is the following: if the system is in its ground state at time t, then $Z(j)$ is the probability amplitude for finding that at time t' it is again in the ground state (up to a phase factor); in other words $Z(j)$ is the vacuum-to-vacuum amplitude.

Equation (8.2.15) shows that $Z(j)$ can also be written as a matrix element,

$$Z(j) = \langle 0|\mathrm{e}^{\mathrm{i}Ht'}U_j(t', t)\mathrm{e}^{-\mathrm{i}Ht}|0\rangle, \tag{8.2.16}$$

which entails immediately that $Z(0) = 1$. This is not surprising, because without a source the system necessarily remains in its ground state. The derivation leading to (8.2.7) applies equally in the presence of a time-dependent potential $V(q, t)$; one need merely replace $\exp(-\mathrm{i}H(t' - t))$ by the evolution operator $U_j(t', t)$, and use the group composition rule

$$U_j(t', t) = U_j(t', t_n)U_j(t_n, t_{n-1}) \ldots U_j(t_1, t).$$

Accordingly, $\langle q', t'|q, t\rangle_j$ can be written as a path integral, and one has, up to a constant factor $\mathcal{N}$,

$$Z(j) = \mathcal{N} \lim_{\substack{T\to i\infty \\ T'\to -i\infty}} \int \mathcal{D}q \exp\left(i\int_T^{T'} (L(q, \dot{q}) + j(t)q(t))\,dt\right); \tag{8.2.17}$$

here the boundary conditions are $\lim_{T\to i\infty} q(T) = \text{constant}$ and $\lim_{T'\to -i\infty} q(T') = \text{constant}$. In general the multiplicative constant $\mathcal{N}$ is irrelevant, and the expression (8.2.17) is all that we shall ever need in practice.

The vacuum expectation-values of T-products are found by functional differentiation with respect to j: for instance

$$\langle 0|T(Q(t_1)Q(t_2))|0\rangle = \frac{(-i)^2}{Z(0)} \frac{\delta^2 Z(j)}{\delta j(t_1)\delta j(t_2)}\bigg|_{j=0}. \tag{8.2.18}$$

This follows at once from (8.2.17); one must divide by $Z(0)$ when $Z(0) \neq 1$.

8.2.3 The harmonic oscillator and Feynman boundary conditions

The results just found will now be applied to the harmonic oscillator. The properties of harmonic oscillators in quantum mechanics are well known, and rederiving them by functional methods is not, on the face of it, a very exciting exercise. Of course the point is that the formalism we are about to develop generalizes to quantum field theory.

Let us write down the generating functional for a harmonic oscillator of mass $m = 1$ and frequency ω, coupled to an external force (i.e. to a source) $j(t)$; the Lagrangean and the equation of motion are

$$\begin{cases} L(t) = \dfrac{1}{2}\dot{q}^2 - \dfrac{1}{2}\omega^2 q^2 + j(t)q(t), \\ \ddot{q} + \omega^2 q = j(t). \end{cases} \tag{8.2.19}$$

The generating functional reads

$$Z(j) = \int \mathcal{D}q \exp\left(i\int dt\left(\frac{1}{2}\dot{q}^2 - \frac{1}{2}\omega^2 q^2 + j(t)q(t)\right)\right), \tag{8.2.20}$$

where the limits $T\to i\infty$, $T'\to -i\infty$ as well as the multiplicative constant are understood but not written. Recall that $Z(j)$ is the vacuum-to-vacuum probability amplitude in the presence of the source j, and that it can be calculated by the standard techniques of quantum mechanics; the calculation is performed in Section 9.3. (The beginning of that section, up to equation (9.3.15), can be read independently of the rest of the chapter; equation (9.3.15) gives the expression for

$U_j(t', t)$.) If one is somewhat cavalier with the boundary conditions, then (8.2.20) can be integrated by parts:

$$Z(j) = \int \mathscr{D}q \exp\left(\mathrm{i} \int \mathrm{d}t \left(-\frac{1}{2} q \frac{\mathrm{d}^2 q}{\mathrm{d}t^2} - \frac{1}{2}\omega^2 q^2 + j(t)q(t) \right) \right). \tag{8.2.21}$$

In (8.2.21) the integral over q is Gaussian; up to a factor independent of j it can be written in the form

$$Z(j) = \exp\left(-\frac{1}{2} \int\int \mathrm{d}t\, \mathrm{d}t'\, j(t)\, D_{\mathrm{F}}(t-t')\, j(t') \right), \tag{8.2.22}$$

where D_{F} is a Green's function* for the harmonic oscillator:

$$\left(\frac{\mathrm{d}^2}{\mathrm{d}t^2} + \omega^2 \right) D_{\mathrm{F}}(t-t') = -\mathrm{i}\delta(t-t'). \tag{8.2.23}$$

The term with q^2 in (8.2.21) can be written

$$-\frac{1}{2} q \left[\mathrm{i} \left(\frac{\mathrm{d}^2}{\mathrm{d}t^2} + \omega^2 \right) \right] q,$$

and, formally, D_{F} is the inverse of the operator within the square brackets:

$$D_{\mathrm{F}} = \left[\mathrm{i} \left(\frac{\mathrm{d}^2}{\mathrm{d}t^2} + \omega^2 \right) \right]^{-1}.$$

One sees immediately that the boundary conditions must be treated with care, because (8.2.23) fails to define the function D_{F} uniquely: one can still add an arbitrary linear combination of the solutions $\exp(\pm \mathrm{i}\omega t)$ of the homogeneous equation. The same conclusion emerges in Fourier space: taking the Fourier transform of (8.2.23), and writing the variable conjugate to t as k_0, one finds

$$(k_0^2 - \omega^2) D_{\mathrm{F}}(k_0) = \mathrm{i}. \tag{8.2.24}$$

$D_{\mathrm{F}}(k_0)$ has poles at $k_0 = \pm\omega$, and we need a prescription for dealing with them. This prescription, which amounts to a correct treatment of the boundary conditions, will be derived from the explicit calculation of $Z(j)$. One could use the result (9.3.15), but it is interesting to show that everything follows from the path integral, and that no appeal to the operator formalism is necessary.

The calculation of $Z(j)$ is asked for in Problem 8.3, with enough hints to defuse any difficulties. With a change of notation $t \to t_i$, $t' \to t_f$, $T = t_f - t_i$, one

* If we wished to define a Green's function in the exact sense of the words, then in principle we should have no factor $(-\mathrm{i})$ on the right of (8.2.23); in fact D_{F} is a Green's function up to a factor i.

starts by determining $F(q_f, t_f; q_i, t_i) = \langle q_f t_f | q_i t_i \rangle_j$:

$$\langle q_f, t_f | q_i, t_i \rangle_j = \left(\frac{\omega}{2i\pi \sin \omega T} \right)^{1/2}$$
$$\times \exp\left\{ \frac{i\omega}{2\sin \omega T} [(q_i^2 + q_f^2)\cos \omega T - 2q_i q_f] \right.$$
$$+ \frac{iq_i}{\sin \omega T} \int_{t_i}^{t_f} \sin[\omega(t_f - t)] j(t)\, dt + \frac{iq_f}{\sin \omega T}$$
$$\times \int_{t_i}^{t_f} \sin[\omega(t - t_i)] j(t)\, dt - \frac{i}{\omega \sin \omega T}$$
$$\left. \times \iint dt\, dt' j(t) \theta(t - t') \sin[\omega(t_f - t) \sin[\omega(t' - t_i)] j(t') \right\}. \tag{8.2.25}$$

To find $Z(j)$ we need (8.2.15) with

$$\varphi_0(q, t) = \left(\frac{\omega}{\pi} \right)^{1/4} e^{-\frac{1}{2}\omega q^2} e^{-\frac{1}{2}\omega t}. \tag{8.2.26}$$

The integrals over q_i and q_f are Gaussian, and for $Z(j)$ we find (see Problem 8.3)

$$Z(j) = \exp\left\{ -\frac{1}{2} \iint_{-\infty}^{\infty} dt\, dt' j(t) D_F(t - t') j(t') \right\}, \tag{8.2.27}$$

where

$$D_F(t - t') = \frac{1}{2\omega} [\theta(t - t') e^{-i\omega(t - t')} + \theta(t' - t) e^{-i\omega(t' - t)}]. \tag{8.2.28}$$

One can verify at once that $D_F(t - t')$ indeed obeys the differential equation (8.2.23). The full calculation has enabled us to determine the Green's function D_F unambiguously. Next we establish a conclusion absolutely basic to what follows: *$D_F(t)$ is found from (8.2.24) by Feynman's prescription* $\omega^2 \to \omega^2 - i\varepsilon$, $(\varepsilon \to 0+)$, the convention for Fourier transforms being fixed by

$$D_F(t) = i \int_{-\infty}^{\infty} \frac{dk_0}{2\pi} \frac{e^{-ik_0 t}}{k_0^2 - \omega^2 + i\varepsilon}. \tag{8.2.29}$$

This prescription puts the poles at $k_0 = -\omega + i\varepsilon$ and $k_0 = \omega - i\varepsilon$. We can now evaluate $D_F(t)$ by the calculus of residues: for $t > 0$ the contour is closed by a large semicircle in the lower half-plane, and for $t < 0$ by a large semicircle in the upper half-plane (Fig. 8.7). In this way one does indeed reproduce (8.2.28).

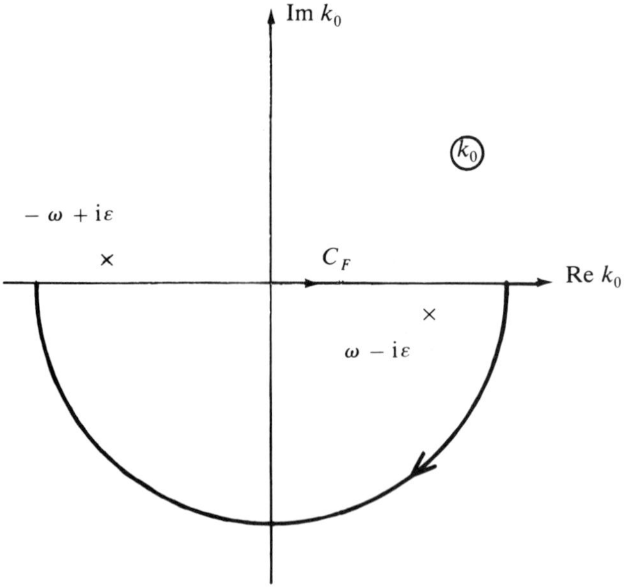

Fig. 8.7 The Feynman contour

For the harmonic oscillator it is thus convenient to replace equation (8.2.20) for $Z(j)$ by

$$Z(j) = \mathcal{N} \int \mathcal{D}q \exp\left(i \int_{-\infty}^{\infty} \left(\frac{1}{2}\dot{q}^2 - (\omega^2 - i\varepsilon)q^2 + jq \right) dt \right). \tag{8.2.30}$$

Note also that from (8.2.18) and (8.2.22) one can derive

$$\langle 0 | T(Q(t)Q(t')) | 0 \rangle = D_F(t - t').$$

This relation may be established equally well by direct calculation (see Problem 8.4), which will lead us to identify Green's functions with vacuum expectation-values of T-products of position operators.

The reader who has worked right through the calculations in Problem 8.3 will have noticed that, though not difficult, they are rather long. This suggests that the boundary conditions (q, t) and (q', t') on the path integral (8.2.7) are not the ones best suited to the harmonic-oscillator problem. In fact the most convenient formulation is of the Hamiltonian type, with boundary conditions on the variables $a(t)$ and $a^*(t)$ which are the classical analogues of annihilation and creation operators. But even with these variables one needs quite elaborate technical apparatus (Bargmann space), and for this aspect of path integrals we refer to the article by Faddeev (1975) or to the book by Itzykson and Zuber (1980, Chapter 9).

8.3 Euclidean continuation, and comments

8.3.1 The quantum partition-function

The limits $T \to \mathrm{i}\infty$, $T' \to -\mathrm{i}\infty$ may appear somewhat artificial in the first place. Much more natural limits are found by working with imaginary time $t = -\mathrm{i}\tau$ and with the evolution operator $\exp(-H\tau)$. The matrix element

$$\langle q'|\mathrm{e}^{-H(\tau'-\tau)}|q\rangle = F(q', -\mathrm{i}\tau'; q, -\mathrm{i}\tau)$$

is the analytic continuation of the matrix element (8.2.2) to $t = -\mathrm{i}\tau$, $t' = -\mathrm{i}\tau'$. The argument leading to (8.2.7) applies step by step, and allows the matrix element to be written as a path integral:

$$\langle q'|\mathrm{e}^{-H(\tau'-\tau)}|q\rangle = \int \mathscr{D}q \exp\left(-\int_{\tau}^{\tau'}\left(\frac{1}{2}m\dot{q}^2 + V(q)\right)\mathrm{d}\tau''\right). \tag{8.3.1}$$

The signs in the exponent are explained as follows. Analytic continuation leads from ε to $-\mathrm{i}\varepsilon'$:

$$-\mathrm{i}\varepsilon V(q) \to -\varepsilon' V(q),$$

$$\mathrm{i}\varepsilon\left(\frac{\mathrm{d}q}{\mathrm{d}t}\right)^2 \to \varepsilon'\left(\mathrm{i}\frac{\mathrm{d}q}{\mathrm{d}\tau}\right)^2 \to -\varepsilon'\left(\frac{\mathrm{d}q}{\mathrm{d}\tau}\right)^2.$$

Equation (8.3.1) features the 'Euclidean Lagrangean' (the name will be motivated in Section 10.3.3; one should note the + sign in front of $V(q)$)

$$L_{\mathrm{E}}(q, \dot{q}) = \frac{1}{2}m\dot{q}^2 + V(q), \tag{8.3.2}$$

whose integral with respect to τ'' is the 'Euclidean action' S_{E}.

The expression (8.3.1) is an element of an unnormed density matrix $\rho(q', q; \tau' - \tau)$. This density-matrix element obeys a diffusion equation, which is the Schroedinger equation with imaginary values of the time (Problem 8.2). Integration over q and q' with $q = q'$ gives the partition function of the quantum particle at temperature T ($\beta = 1/kT$),

$$Z(\beta) = \mathrm{Tr}\,\mathrm{e}^{-\beta H} = \int \mathrm{d}q \langle q|\mathrm{e}^{-\beta H}|q\rangle$$

$$= \int_{q(0)=q(\beta)} \mathscr{D}q \exp\left(-\int_0^{\beta}\left(\frac{1}{2}m\dot{q}^2 + V(q)\right)\mathrm{d}\tau''\right). \tag{8.3.3}$$

In the functional integral one must integrate over all trajectories subject to the periodic boundary conditions

$$q(0) = q(\beta).$$

The thermal averages of the 'T-products' (ordered with respect to τ) are defined by

$$\langle T(Q(\tau_1)Q(\tau_2))\rangle_\beta = \frac{1}{Z}\,\mathrm{Tr}(T(Q(\tau_1)Q(\tau_2)\mathrm{e}^{-\beta H}))$$
$$= \frac{1}{Z}\int_{q(0)=q(\beta)} \mathscr{D}q\, q(\tau_1)q(\tau_2)\exp\left(-\int_0^\beta L_\mathrm{E}(q,\dot{q})\,\mathrm{d}\tau''\right). \tag{8.3.4}$$

These T-products obey the periodicity condition

$$\langle T(Q(\beta)Q(\tau_2))\rangle_\beta = \langle T(Q(0)Q(\tau_2))\rangle_\beta, \tag{8.3.5}$$

which is evident from their functional form, and which it is also easy to derive by using the cyclic property of the trace (Problem 8.4).

To define a generating functional $Z_\mathrm{E}(j)$ for vacuum expectation-values of T-products, we must once again project onto the ground state. It is picked out by the limits $\tau' \to \infty$, $\tau \to -\infty$, and we find

$$Z_\mathrm{E}(j) = \mathcal{N}\int \mathscr{D}q \exp\left(-\int_{-\infty}^{\infty}\left(\frac{1}{2}m\dot{q}^2 + V(q) - j(\tau'')q(\tau'')\right)\mathrm{d}\tau''\right), \tag{8.3.6}$$

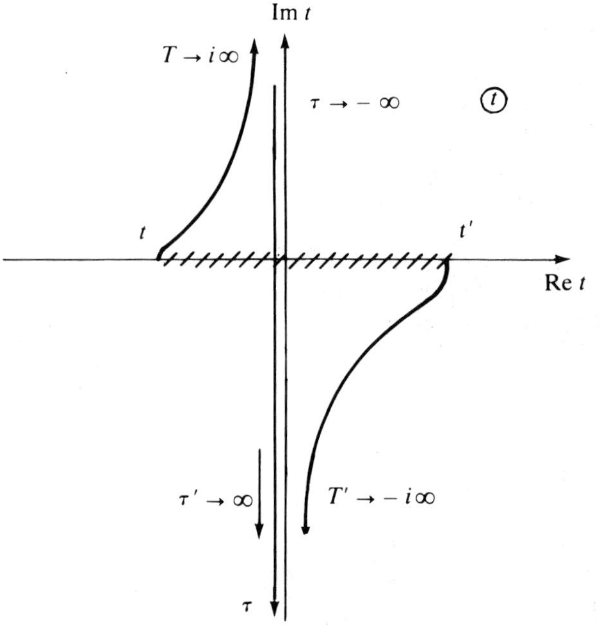

Fig. 8.8

with the boundary conditions

$$\lim_{\tau \to -\infty} q(\tau) = \text{constant}, \quad \lim_{\tau' \to \infty} q(\tau') = \text{constant}.$$

$Z_E(j)$ is called the Euclidean generating functional. Note that (8.3.6) is quite logical: to project onto the ground state one must let the temperature tend to zero, i.e. β to infinity. By differentiating with respect to j, we obtain the analytic continuations of the vacuum expectation values of ordinary T-products (see (8.2.18)):

$$\frac{(-\mathrm{i})^n}{Z(0)} \frac{\delta^n Z(j)}{\delta j(t_1)\ldots\delta j(t_n)}\bigg|_{\substack{t_l = -\mathrm{i}\tau_l \\ j=0}} = \frac{1}{Z_E(0)} \frac{\delta^n Z_E(j)}{\delta j(\tau_1)\ldots\delta j(\tau_n)}\bigg|_{j=0}. \tag{8.3.7}$$

Figure 8.8 makes it easier to visualize the boundary conditions in (8.2.17) and (8.3.6).

8.3.2 The classical analogue

Following the strategy outlined in the introduction to the present chapter, we proceed to find a problem in *classical* statistical mechanics which is analogous to the quantum problem. Changing the notation according to

$$t \to x, \quad \tau \to -L/2, \quad \tau' \to L/2$$

we consider a random function $q(x)$ (a 'field') on the interval $[-L/2, L/2]$, subject to the boundary conditions $q(-L/2) = q(L/2)$. Suppose that every configuration $q(x)$ of the field has the statistical weight

$$\exp\left[-\int_{-L/2}^{L/2}\left[\frac{1}{2}m\left(\frac{\mathrm{d}q}{\mathrm{d}x}\right)^2 + V(q) - j(x)q(x)\right]\mathrm{d}x\right].$$

Then the partition function (=sum over all configurations of the field $q(x)$) is given by a formula analogous to (8.3.3),

$$Z_E(j) = \int \mathscr{D}q \exp\left(-\int_{-L/2}^{L/2}\left[\frac{1}{2}m\left(\frac{\mathrm{d}q}{\mathrm{d}x}\right)^2 + V(q) - j(x)q(x)\right]\mathrm{d}x\right); \tag{8.3.8}$$

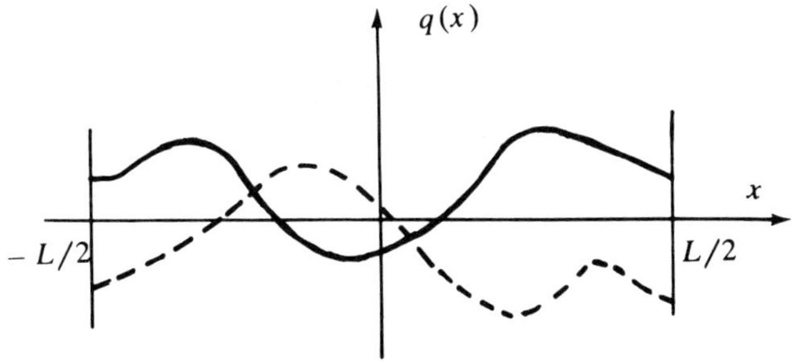

Fig. 8.9 Two different configurations of the field $q(x)$

here the integration measure $\mathscr{D}q$ is defined heuristically by a limiting procedure similar to the one we have just seen, but can be given a rigorous meaning. The change of notation $q(x) \to \varphi(x)$ evidently makes (8.3.8) recognizable as the partition function of the Ginzburg–Landau model in one dimension, with an interaction $V(\varphi)$. The correlations $\langle q(x_1)q(x_2)\rangle$ in the absence of the source are given by functional derivatives of (8.2.21),

$$\langle q(x_1)q(x_2)\rangle = \frac{1}{Z_{\mathrm{E}}(0)} \frac{\delta^2 Z_{\mathrm{E}}(j)}{\delta j(x_1)\,\delta j(x_2)}\bigg|_{j=0}; \tag{8.3.9}$$

they are related to the vacuum expectation-values of the T-products in the quantum problem by making L tend to infinity. Once again it is easy to see the correspondence between free energy per unit length and the ground-state energy, and between the reciprocal of the correlation length and the energy gap ΔE.

8.3.3 The Euclidean harmonic oscillator.

The Euclidean continuation of the generating functional (8.2.30) for the harmonic oscillator is

$$Z_{\mathrm{E}}(j) = \mathscr{N}\int \mathscr{D}q \exp\left(-\int_{-\infty}^{\infty}\left(\frac{1}{2}\dot{q}^2 + \frac{1}{2}\omega^2 q^2 - j(\tau'')q(\tau'')\right)\mathrm{d}\tau''\right). \tag{8.3.10}$$

The integral over q is Gaussian; in order to evaluate it we need the inverse $D_{\mathrm{E}}(\tau)$ of the operator

$$-\frac{\mathrm{d}^2}{\mathrm{d}\tau^2} + \omega^2,$$

or in other words the Green's function, satisfying

$$\left(-\frac{\mathrm{d}^2}{\mathrm{d}\tau^2} + \omega^2\right) D_{\mathrm{E}}(\tau - \tau') = \delta(\tau - \tau'). \tag{8.3.11}$$

Of course $D_{\mathrm{E}}(\tau)$ is nothing but the two-point correlation function of the one-dimensional Gaussian model. The solution of (8.3.11) takes the form

$$D_{\mathrm{E}}(\tau) = \frac{1}{2\omega}\mathrm{e}^{-\omega|\tau|} + A\,\mathrm{e}^{\omega\tau} + B\mathrm{e}^{-\omega\tau}, \tag{8.3.12}$$

and the Gaussian integration in (8.3.10) leads to

$$Z(j) = \mathscr{N}'\exp\left(\frac{1}{2}\iint \mathrm{d}\tau\,\mathrm{d}\tau' j(\tau) D_{\mathrm{E}}(\tau - \tau') j(\tau')\right). \tag{8.3.13}$$

However, if we require $\lim_{\tau\to\pm\infty} q(\tau) = \text{constant}$, then the solutions of the homogeneous equation must be eliminated from (8.3.12), and we are left with

$$D_{\mathrm{E}}(\tau) = \frac{1}{2\omega} \mathrm{e}^{-\omega|\tau|}. \tag{8.3.14}$$

In Fourier space, with ν the variable conjugate to τ, $D_{\mathrm{E}}(\nu)$ satisfies

$$(\nu^2 + \omega^2) D_{\mathrm{E}}(\nu) = 1, \tag{8.3.15}$$

while the inverse Fourier transform correctly reproduces (8.3.14). *The function $D_{\mathrm{E}}(\tau)$ is the Euclidean continuation of $D_{\mathrm{F}}(\tau)$*; from (8.2.28) we can verify that, accordingly,

$$D_{\mathrm{F}}(-\mathrm{i}\tau) = D_{\mathrm{E}}(\tau). \tag{8.3.16}$$

Finally, we note from (8.3.3) and (8.3.13) that

$$D_{\mathrm{E}}(\tau - \tau') = \langle 0 | T(Q(\tau) Q(\tau')) | 0 \rangle, \tag{8.3.17}$$

which can be checked by direct calculation (Problem 8.4).

All these results for the harmonic oscillator generalize to quantum field theory (Chapters 9 and 10). The only complication is that, instead of a single frequency ω, one must then integrate over a range of frequencies $\omega(\boldsymbol{k})$, which depend on the momenta $\boldsymbol{k}$.

Problems

8.1 (a) The partition function of the one-dimensional Ising model with the Hamiltonian

$$H = -\sum_{i=0}^{N-1} (A + BS_iS_{i+1}),$$

and with periodic boundary conditions, is given by

$$Z_N = \mathrm{e}^{NA} 2^N [(\cosh B)^N + (\sinh B)^N]$$

(see Section 1.2.2). For $N \to \infty$, the correlation function reads

$$\langle S_l S_m \rangle = (\tanh B)^{|l-m|}.$$

Use these results to verify the correspondence between the free energy per spin and the energy of the ground state. What happens if the interval $[0, \tau]$ is subdivided into $2N$ intervals instead of N?

(b) Investigate the case where $[0, \tau]$ is subdivided into τ/ε intervals, with $\varepsilon \to 0$.

(c) Replace the Hamiltonian (8.1.1) by

$$H = -K[\sigma_1 \cos\theta + \sigma_3 \sin\theta].$$

How does this modify the results in Section 8.1?

8.2 (a) Examine the convergence of the integral

$$\int \frac{\mathrm{d}p}{2\pi} \exp\left(ipq - \mathrm{i}\varepsilon \frac{p^2}{2m} \right),$$

and show that the sign of the square root in (8.2.5) has indeed been chosen correctly. (*Hint*: replace ε by $\varepsilon - i\eta$, and study the limit $\eta \to 0^+$.)

(b) Show that the wavefunction $\psi(q', t')$ at time t' can be expressed in terms of the wavefunction at time t through

$$\psi(q', t') = \int \mathrm{d}q\, F(q', t'; q, t) \psi(q, t).$$

Calculate $F(q', t'; q, t)$ when the potential vanishes ($V(q) = 0$).

(c) Show that $F(q', t'; q, t)$ obeys the Schroedinger equation

$$\left[\mathrm{i}\frac{\partial}{\partial t'} - H(q') \right] F(q', t', q, t) = 0,$$

with the initial condition

$$F(q', t; q, t) = \delta(q' - q).$$

(d) Show that the density-matrix element (8.3.1) obeys the diffusion equation

$$\left[\frac{\partial}{\partial \tau'} + H(q')\right]\rho(q', \tau'; q, \tau) = 0,$$

with the initial condition

$$\rho(q', \tau; q, \tau) = \delta(q' - q).$$

8.3 The aim is to derive equations (8.2.25) and (8.2.28).

(a) Let $\bar{q}(t)$ be a solution of the equation of motion for the forced oscillator,

$$\ddot{\bar{q}} + \omega^2 \bar{q} = j(t),$$

and let $\bar{q}_0(t)$ be a solution for the free oscillator. It will prove convenient to write, for $t_i \leqslant t \leqslant t_f$, $(T = t_f - t_i)$,

$$\bar{q}(t) = \bar{q}_0(t) + \int_{t_i}^{t_f} G(t, t') j(t')\,\mathrm{d}t',$$

where $G(t, t')$ is a Green's function for the harmonic oscillator, obeying

$$\left(\frac{\mathrm{d}^2}{\mathrm{d}t^2} + \omega^2\right) G(t, t') = \delta(t - t'); \quad G(t_i, t) = G(t_f, t) = 0.$$

Show that one may write

$$G(t,t') = \theta(t - t')u(t)v(t') + \theta(t' - t)u(t')v(t),$$

where u and v are solutions of the homogeneous equation, chosen as

$$u(t) = \lambda \sin \omega(t_f - t),$$

$$v(t) = \lambda' \sin \omega(t - t_i),$$

where

$$\lambda\lambda' = -[\omega \sin(\omega T)]^{-1}.$$

Why are there singularities at the points $\omega T = \pm\pi, \pm 2\pi$, etc.?

(b) Impose the boundary conditions

$$q(t_i) = q_i, \quad q(t_f) = q_f,$$

and let $\bar{q}(t)$ be a solution of the equation of motion. Write $q(t) = \bar{q}(t) + h(t)$, with $h(t_f) = h(t_i) = 0$. Show that the action $S(q, j)$ reads

$$S(q,j) = S(\bar{q}, j) + \int_{t_i}^{t_f} \mathrm{d}t \frac{1}{2}(\dot{h}^2 - \omega^2 h^2),$$

where the second term is independent of j.

(c) Evaluate $S(\bar{q}, j)$; note that $\bar{q}_{0i} = q_i$, $\bar{q}_{0f} = q_f$, and

$$\int_{t_i}^{t_f} L(\bar{q})\,\mathrm{d}t = \frac{1}{2}\bar{q}\dot{\bar{q}}\bigg|_{t_i}^{t_f} + \frac{1}{2}\int_{t_i}^{t_f} j(t)\bar{q}(t)\,\mathrm{d}t.$$

Hence derive $S(q_f, q_i; j)$, or in other words the exponent in (8.2.25) (up to a factor i).

(d) Carry out the integration over q_i and q_f, thus obtaining $Z(j)$ (up to a constant factor); use the ground-state wavefunction (8.2.26). (The integral is doubly Gaussian, with the A-matrix

$$A = \frac{-\mathrm{i}\omega}{\sin\omega T}\begin{pmatrix} \mathrm{e}^{\mathrm{i}\omega T} & -1 \\ -1 & \mathrm{e}^{\mathrm{i}\omega T} \end{pmatrix}).$$

Determine the exponent in (8.2.28).

(e) It remains to establish the prefactor of the exponential, i.e. in particular the factor $(\omega/2\mathrm{i}\pi\sin\omega T)^{1/2}$ in (8.2.25). For this one discretizes the integral

$$I = \int \mathscr{D}h \exp\left(\mathrm{i}\int_{t_i}^{t_f} \mathrm{d}t\frac{1}{2}(\dot{h}^2 - \omega^2 h^2)\right),$$

and one notes that, given an $n \times n$ matrix D_n of the form

$$D_n = \begin{pmatrix} 1 & a & 0 & 0 & \ldots & & \\ a & 1 & a & 0 & \ldots & & \\ 0 & a & 1 & a & 0 & \ldots & \\ \vdots & \vdots & & & & & \end{pmatrix},$$

its determinant Δ_n satisfies

$$\Delta_n = \Delta_{n-1} - a^2\Delta_{n-2}.$$

(f) Actually the calculation is warranted only if $|\omega T| < \pi$. To determine the prefactor in the general case, notice that

$$\int \mathrm{d}q'' F(q', t'; q'', t'')F(q'', t''; q, t) = F(q', t'; q, t),$$

and take $t = 0$, $t'' = \pi/2\omega$, $t' = \pi/\omega$, $j = 0$. From this, derive the prefactor

$$\mathrm{e}^{-\frac{\mathrm{i}\pi}{2}\left(\frac{1}{2} + \mathrm{Int}\left(\frac{\omega T}{\pi}\right)\right)}\left(\frac{\omega}{2\pi|\sin\omega T|}\right)^{1/2},$$

where $\mathrm{Int}(x)$ is the integer part of x.

8.4 (a) For the harmonic oscillator, show by direct calculation that

$$D_{\mathrm{F}}(t-t') = \langle 0|T(Q(t)Q(t'))|0\rangle .$$

Use the expression for Q in terms of the creation and annihilation operators,

$$Q = \frac{1}{\sqrt{2\omega}}(a + a^{\dagger}),$$

and show that

$$a(t) = a\mathrm{e}^{-\mathrm{i}\omega t}, \quad a^{\dagger}(t) = a^{\dagger}\mathrm{e}^{\mathrm{i}\omega t}.$$

The evaluation of the ground-state expectation value then becomes trivial.

(b) Do the same for $D_{\mathrm{E}}(\tau - \tau')$.

(c) Writing $\tau = \mathrm{i}t$, evaluate

$$D_{\beta}(t-t') = (\mathrm{Tr}\,\mathrm{e}^{\ \beta H})^{-1}\,\mathrm{Tr}(\mathrm{e}^{-\beta H}\,T(Q(t)Q(t'))),$$

$$\bar{D}_{\beta}(\tau-\tau') = (\mathrm{Tr}\,\mathrm{e}^{-\beta H})^{-1}\,\mathrm{Tr}(\mathrm{e}^{-\beta H}\,T(Q(\tau)Q(\tau'))).$$

Demonstrate the periodicity of $\bar{D}_{\beta}$:

$$\bar{D}_{\beta}(\tau = 0, \tau') = \bar{D}_{\beta}(\tau = \beta, \tau').$$

These results are easily generalized to quantum field theory at finite temperature (Dolan and Jackiw 1974); recent references are Landsman and van Weert (1987) and Kapusta (1989).

Further reading

The example in Section 8.1 is adapted from Shenker (1982, Section 1). See also Kogut (1979, Section III). For systematic discussions of path integrals in quantum mechanics, see Feynman and Hibbs (1965); and Schulman (1981). Some other useful references are Abers and Lee (1973, Section II); De Witt (1980); Parisi (1988, Chapter 13); Zinn-Justin (1989, Chapter 2); and Amit (1984, Chapter 3). For the path integral in statistical mechanics, see Feynman (1972, Chapter 3). Interesting comments on the treatment of the integration measure $\mathscr{D}q$ are made by Felsager (1981, Chapter 5). Detailed discussions of the Euclidean continuation and of the stationary phase approximation can be found in Parisi (1988).

9 Quantization of the Klein-Gordon field

Quantum field theory was born of the need to describe processes where particles are created or destroyed (annihilated). One elementary example is the radiative transition of an atom from an excited state A^* to the ground state A, emitting a photon (γ):

$$A^* \rightarrow A + \gamma.$$

The photon is created at the moment at which the transition takes place.

Another example is the β-decay of the neutron, where a neutron (n) produces a proton (p), an electron (e^-), and an antineutrino ($\bar{\nu}$):

$$n \rightarrow p + e^- + \bar{\nu}.$$

Here too the daughter particles are created at the moment of decay: a simple argument from the uncertainty principle shows that the electron emitted in the β-decay of the neutron cannot have existed within it beforehand. Reactions where particles are created or destroyed are the rule in elementary-particle physics; a π^0 meson for instance can be created in the collision between two protons:

$$p + p \rightarrow p + p + \pi^0.$$

Reactions where particles are created or destroyed are basically different from chemical reactions, which correspond to rearrangements of the atoms within the molecules; and different also from nuclear reactions, where the nucleons redistribute themselves to yield new nuclei. In both cases there is rearrangement rather than creation or destruction. However, the distinction may not be as sharp as it seems, because the process of rearrangement can be quite complex, with (virtual) destruction or creation of particles in its intermediate stages. The Schroedinger equation, which assumes that the number of particles is fixed, cannot describe creation or annihilation processes; but already in the 1930s the founders of quantum mechanics discovered how this could be done through the quantization of classical fields (improperly called 'second quantization', for historical reasons). Consider for example the electromagnetic field in a cavity; this field can be analysed into normal modes (eigenmodes), each mode in effect constituting a harmonic oscillator of frequency ω_k. The quantization of the field, which consists in the imposition of commutation rules between the field and its conjugate momentum (see equation (9.1.15)), allows one to demonstrate that every such harmonic oscillator becomes a quantum oscillator, having the energy levels $\frac{1}{2}\hbar\omega_k, \frac{3}{2}\hbar\omega_k, \ldots, (n+\frac{1}{2})\hbar\omega_k, \ldots$. The state with energy $(n+\frac{1}{2})\hbar\omega_k$ can be interpreted as a state with n photons having energy $\hbar\omega_k$ each. Thus the quantization of a classical field enables one to describe the creation and destruction of particles, the creation and destruction operators being simply the well-known operators $a^\dagger$ and a of the quantum harmonic oscillator.

Canonical quantization consists in postulating commutation relations between the field and its conjugate momentum at equal times. Except for the pioneering work of Feynman, this was the only quantization method available up to the late 1960s. Nevertheless it suffers from problems (even apart from questions of renormalization).

(i) It singles out a particular reference frame, leaving Lorentz invariance to be verified afterwards. In some cases (see (ii) below) it leads to explicitly non-covariant terms at intermediate stages.

(ii) It becomes very complicated when the interaction involves derivatives of the field: the definition of conjugate momenta then depends on the interaction.

(iii) It is ill-adapted (even) to Abelian gauge theories (electromagnetism); in these one needs to introduce an indefinite metric etc., while for non-Abelian gauge theories the canonical formalism becomes almost intractable.

In view of these difficulties, and of the growing interest in gauge theories, another quantization method has become very popular, exploiting path integrals. This method consists in using a path integral to represent the probability amplitude a for going from the initial to the final configuration of the field; thus it generalizes the notions explained in Section 8.2. Symbolically,

$$a = \int \mathscr{D}\varphi \exp(\mathrm{i}S/\hbar),$$

where φ is the field and S the action. This method has the advantage of being explicitly covariant, and incurs no special difficulties with derivative couplings. For gauge theories it is incomparably simpler than canonical quantization, and will be used in Chapter 11 for electrodynamics and in Chapter 13 for non-Abelian gauge theories.

However, it is obvious that not even this method can escape the problem of ultraviolet divergences, and the expression for a must be regularized in 4 dimensions. Moreover, it is not immediately clear whether the quantum theory defined by path integrals is unitary, or in other words whether it conserves probability. This property must be verified explicitly.

Be that as it may, as far as we can tell at present, canonical quantization and quantization through path integrals are equivalent. Hence we can always choose the method best adapted to each particular case.

The experience we have already acquired from critical phenomena would allow us to sidestep canonical quantization altogether. Indeed the reader already familiar with this formalism can go straight from here to Section 2 of the next Chapter. Nevertheless it would prove awkward to say nothing about canonical quantization at all, for the following reasons:

- the concept of a particle is rather un-intuitive in the path-integral formalism, while a creation operator is easy to visualize;
- much of the literature, and practically all of it before 1970, uses the canonical formalism;
- the case of fermions seems easier to understand if one tackles it first by canonical quantization.

The layout of this chapter is as follows. Field quantization is introduced in Section 9.1 through the example of the vibration field in a solid, and we show how particles (in this case phonons) are associated with a quantized field. Section 9.2 is devoted to quantizing the Klein–Gordon field, which describes particles of spin zero. In Section 9.3, the coupling of the Klein–Gordon field to a classical source allows us to derive Wick's theorem. By virtue of this theorem we shall be able to establish, in the next chapter, the equivalence of the perturbative expansions obtained through canonical quantization and through path integrals.

A final word about units: unless otherwise stated we shall always use units such that $\hbar$ (Planck's constant divided by 2π) $= c$ (the speed of light) $= 1$. In this system mass, momentum, and energy all have the same dimensions; length and time have the dimension of inverse mass.

9.1 The quantization of elastic vibrations

9.1.1 System with *N* degrees of freedom: Lagrangean, Hamiltonian, and quantization

We start by studying a mechanical system with N degrees of freedom. In classical mechanics, such a system is described by N generalized coordinates (i.e. dynamical variables) φ_i ($i = 0, 1, \ldots, N-1$), N generalized velocities $\dot{\varphi}_i = \mathrm{d}\varphi_i/\mathrm{d}t$, and a Lagrangean L that depends on the φ_i, the $\dot{\varphi}_i$, and possibly on the time t. The equations of motion follow from minimizing the action S,

$$S = \int_{t_1}^{t_2} L(\varphi_i, \dot{\varphi}_i, t)\,\mathrm{d}t, \tag{9.1.1}$$

subject to the boundary conditions

$$\delta\varphi_i(t_1) = \delta\varphi_i(t_2) = 0. \tag{9.1.2}$$

The condition $\delta S = 0$ plus the boundary conditions (9.1.2) yield the equations of motion, i.e. the Euler–Lagrange equations

$$\frac{\mathrm{d}}{\mathrm{d}t}\frac{\partial L}{\partial \dot{\varphi}_i} - \frac{\partial L}{\partial \varphi_i} = 0. \tag{9.1.3}$$

Recall that the equations of motion (9.1.3) remain unaffected if one augments the Lagrangean by the total time-derivative $(\mathrm{d}/\mathrm{d}t) f(\varphi_i, t)$ of a function of the coordinates and of the time.

This formalism can be applied at once to a one-dimensional model of elastic vibrations in solids. The atoms are represented by point masses m situated on a line, and their interactions by identical springs linking these masses; φ_i is the *displacement* of atom number i from its equilibrium position (Fig. 9.1):

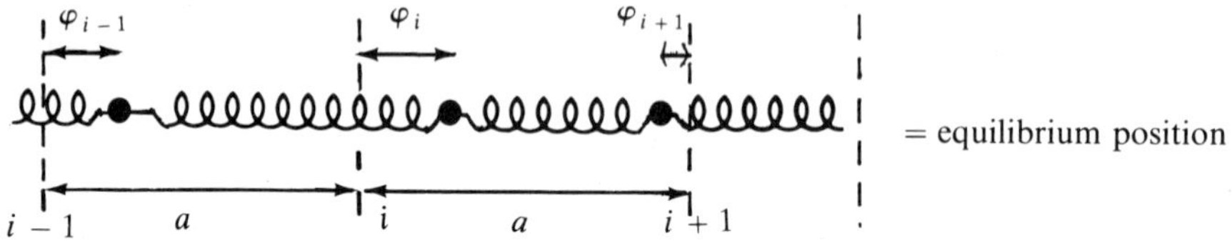

Fig. 9.1 A chain of springs

Since the Lagrangean is equal to the kinetic less the potential energy, one finds

$$L = \sum_{i=0}^{N-1} \left\{ \frac{1}{2} m \dot{\varphi}_i^2 - \frac{1}{2} K (\varphi_{i+1} - \varphi_i)^2 \right\}, \tag{9.1.4}$$

where K is the stiffness constant of the springs; as usual it proves convenient to impose periodic boundary conditions

$$\varphi_{i+N} = \varphi_i.$$

The equation of motion (9.1.3) now reads

$$m\ddot{\varphi}_i = K[(\varphi_{i+1} - \varphi_i) - (\varphi_i - \varphi_{i-1})]; \tag{9.1.5}$$

it is instructive to rederive it by arguing directly from the force experienced by atom number i.

The Hamiltonian follows from the Lagrangean through a Legendre transformation; one defines the conjugate momenta p_i by

$$p_i = \frac{\partial L}{\partial \dot{\varphi}_i}, \tag{9.1.6}$$

whereupon

$$H = \sum_i p_i \dot{\varphi}_i - L. \tag{9.1.7}$$

With the Lagrangean (9.1.4) equation (9.1.7) becomes

$$H = \sum_i \left\{ \frac{p_i^2}{2m} + \frac{1}{2} K (\varphi_{i+1} - \varphi_i)^2 \right\}. \tag{9.1.8}$$

The rules for quantizing a mechanical system with N degrees of freedom are classic: the coordinates φ_i and the momenta p_i become Hermitean operators acting in a Hilbert space, namely in the space of states, and they obey the canonical commutation rules (CCR)

$$\begin{cases} [\varphi_i(t), \varphi_j(t)] = [p_i(t), p_j(t)] = 0, \\ [\varphi_i(t), p_j(t)] = \mathrm{i}\delta_{ij}. \end{cases} \tag{9.1.9}$$

Here the operators φ_i and p_i have been written in the Heisenberg picture (Messiah 1961, Chapter 8), and the CCR (9.1.9) apply at *equal times.* Recall that we are using a system of units where $\hbar = 1$.

Once quantized, the Hamiltonian (9.1.8) is not difficult to diagonalize; but it is preferable to go directly to the continuum formulation, relegating this diagonalization to Problem 9.1.

9.1.2 Quantization of a continuous line

Given the mechanical system described by the Lagrangean (9.1.4), we go to its continuum limit following the strategy already employed in Part I. Let a be the equilibrium distance between successive atoms; we propose to let a tend to zero, keeping fixed the length $\bar{L} = Na$ of the system. Accordingly, the number N of degrees of freedom becomes infinite. In this limit one has

$$m/a \to \mu = \text{mass per unit length},$$

$$Ka \to Y = \text{Young's modulus}.$$

In fact, the definitions of the stiffness constant of a spring and of Young's modulus identify the fractional extension of a typical spring as

$$\frac{\varphi_{i+1} - \varphi_i}{a} = \frac{F}{Ka} = \frac{F}{Y}.$$

The version

$$L = a \sum_i \left\{ \frac{1}{2} \frac{m}{a} \dot{\varphi}_i^2 - \frac{1}{2} Ka \left(\frac{\varphi_{i+1} - \varphi_i}{a} \right)^2 \right\} = a \sum_i L_i$$

of (9.1.4) features a Riemann sum which can be transformed at once into the integral

$$L = \int_0^{\bar{L}} \mathrm{d}x \left\{ \frac{1}{2}\mu\left(\frac{\partial\varphi}{\partial t}\right)^2 - \frac{1}{2}Y\left(\frac{\partial\varphi}{\partial x}\right)^2 \right\}$$

$$= \int_0^{\bar{L}} \mathrm{d}x\, \mathscr{L}(\dot{\varphi}, \partial\varphi/\partial x). \tag{9.1.10}$$

The integrand $\mathscr{L}$ of (9.1.10) is the *Lagrangean density*. The Lagrangean (9.1.10) describes the longitudinal vibrations of a continuous medium, or along a string. It is important to realize that x is not a dynamical variable (generalized coordinate), but a *label* which identifies such a variable:

$$\varphi_i(t) \to \varphi(t, x).$$

In other words, the continuum formulation replaces $i \to x$; $\varphi(t, x)$ is called the (longitudinal) *displacement field* in the medium (or of the string), and it is a *classical field*. It is the displacement, from its equilibrium position, of the point of the string labelled x.

The Lagrangean density $\mathscr{L}$ in (9.1.10) depends neither on φ nor on t, but in general $\mathscr{L}$ does depend on both these quantities. The equations of motion follow from the principle of least action $\delta S = 0$, subject to the boundary conditions

$$\delta\varphi(t_1, x) = \delta\varphi(t_2, x) = 0,$$

$$\delta\varphi(t, 0) = \delta\varphi(t, \bar{L}) = 0.$$

In general x is allowed to range over the interval $]-\infty, +\infty[$, with the field required to vanish at infinity; then the second condition is satisfied automatically. It is equally possible to use periodic boundary conditions.

In the general case (still in one space dimension) the principle of least action allows one to write the equations of motion in the form

$$\frac{\partial}{\partial t}\frac{\partial \mathscr{L}}{\partial \dot{\varphi}} + \frac{\partial}{\partial x}\frac{\partial \mathscr{L}}{\partial(\partial\varphi/\partial x)} - \frac{\partial \mathscr{L}}{\partial \varphi} = 0; \tag{9.1.11}$$

for the particular Lagrangean (9.1.10) this yields the classical wave equation

$$\frac{\partial^2\varphi}{\partial t^2} - c_s^2\frac{\partial^2\varphi}{\partial x^2} = 0, \tag{9.1.12}$$

where $c_s = \sqrt{Y/\mu}$ is the speed of sound. One can check that (9.1.12) is indeed the continuum limit of (9.1.5).

To determine the Hamiltonian, we note that

$$p_i = \frac{\partial L}{\partial \dot{\varphi}_i} = a\frac{\partial L_i}{\partial \dot{\varphi}_i} \to a\frac{\partial \mathscr{L}}{\partial \dot{\varphi}},$$

whence

$$H = a \sum_i \left(\dot{\varphi}_i \frac{\partial L_i}{\partial \dot{\varphi}_i} - L_i \right) \to \int_0^{\bar{L}} \mathrm{d}x \left(\dot{\varphi} \frac{\partial \mathscr{L}}{\partial \dot{\varphi}} - \mathscr{L} \right). \tag{9.1.13}$$

Here $\partial \mathscr{L} / \partial \dot{\varphi}$ is the *momentum conjugate to the field* φ:

$$\pi(t, x) = \frac{\partial \mathscr{L}}{\partial \dot{\varphi}(t, x)}. \tag{9.1.14}$$

We are now in a position to write down the continuum version of the CCR; these follow from (9.1.9) on noting that

$$\lim_{a \to 0} \frac{1}{a} \delta_{ij} = \delta(x - x'),$$

and read

$$\begin{cases} [\varphi(t, x),\ \varphi(t, x')] = [\pi(t, x),\ \pi(t, x')] = 0, \\ [\varphi(t, x),\ \pi(t, x')] = \mathrm{i}\delta(x - x'). \end{cases} \tag{9.1.15}$$

It is essential to remember that the two operators involved in the CCR are taken *at equal times*. The equal-time CCR are independent of the dynamics. By contrast, the commutator

$$[\varphi(t, x), \varphi(t', x')] \quad \text{at} \quad t \neq t'$$

for instance does depend on the dynamics explicitly; moreover it cannot be calculated exactly except in especially simple cases.

9.1.3 Normal modes

There is a simple expression in terms of normal modes for the part of the Lagrangean (9.1.10) featuring $(\partial\varphi/\partial x)^2$, which is the limit of the interactions between nearest neighbours. Let us therefore Fourier-analyze the field (which is still classical),

$$\varphi(t, x) = \frac{1}{\sqrt{(\bar{L})}} \sum_k \varphi_k(t) \mathrm{e}^{\mathrm{i}kx}, \tag{9.1.16}$$

where k assumes the discrete values $k = 2\pi p/\bar{L}$, $p = 0, \pm 1, \pm 2, \cdots$. From the wave equation (9.1.12) we can obtain for the Fourier coefficients the differential equation

$$\frac{\mathrm{d}^2 \varphi_k}{\mathrm{d}t^2} + c_s^2 k^2 \varphi_k = 0. \tag{9.1.17}$$

As expected, the equations for the normal modes are decoupled. The solution of (9.1.17) is a linear combination of exponentials $\exp(\mathrm{i}\omega_k t)$ and $\exp(-\mathrm{i}\omega_k t)$, with

$\omega_k = c_s|k|$:

$$\varphi_k(t) = A_k e^{-i\omega_k t} + A^*_{-k} e^{i\omega_k t}. \tag{9.1.18}$$

This form of $\varphi_k(t)$ relies on the reality of $\varphi(t, x)$,

$$\varphi(t, x) = \varphi^*(t, x),$$

which imposes the condition

$$\varphi_k(t) = \varphi^*_{-k}(t).$$

Finally, the Fourier decompositions of the field and of its conjugate momentum are written as

$$\varphi(t, x) = \frac{1}{\sqrt{(\bar{L})}} \sum_k (A_k e^{-i\omega_k t + ikx} + A_k^* e^{i\omega_k t - ikx}), \tag{9.1.19a}$$

$$\pi(t, x) = \dot{\varphi}(t, x) = \frac{1}{\sqrt{(\bar{L})}} \sum_k (-i\omega_k A_k e^{-i\omega_k t + ikx} + i\omega_k A_k^* e^{i\omega_k t - ikx}). \tag{9.1.19b}$$

These relations can be inverted immediately to yield A_k and A_k^*:

$$A_k = \frac{i}{2\omega_k} \frac{1}{\sqrt{(\bar{L})}} \int_0^{\bar{L}} dx\, e^{i\omega_k t - ikx}(\pi(t, x) - i\omega_k \varphi(t, x)), \tag{9.1.20a}$$

$$A_k^* = \frac{-i}{2\omega_k} \frac{1}{\sqrt{(\bar{L})}} \int_0^{\bar{L}} dx\, e^{-i\omega_k t + ikx}(\pi(t, x) + i\omega_k \varphi(t, x)). \tag{9.1.20b}$$

For the time being, (9.1.19) and (9.1.20) are decompositions of a classical field: $\varphi(t, x)$ is a real number, namely the displacement field at the point x.

The vibrations are quantized by appeal to the CCR (9.1.15): φ and π become operators, as do the A_k and A_k^*. More precisely, the operator (9.1.20b) becomes the Hermitean conjugate of the operator (9.1.20a): $A_k^* \to A_k^\dagger$. From the CCR (9.1.15) it is easy to determine the commutators of the A_k; since these operators are time-independent, the calculation simplifies if in (9.1.20) one sets $t = 0$. One finds (Problem 9.2)

$$[A_k, A_{k'}] = 0; \ [A_k, A_{k'}^\dagger] = (2\omega_k)^{-1}\delta_{kk'}.$$

It proves convenient to rescale the A_k by writing

$a_k = \sqrt{2\omega_k} A_k$; this yields

$$[a_k, a_{k'}^\dagger] = \delta_{k,k'}, \tag{9.1.21}$$

while the Fourier decomposition of the field reads

$$\varphi(t, x) = \frac{1}{\sqrt{(\bar{L})}} \sum_k \frac{1}{\sqrt{2\omega_k}} [a_k e^{-i\omega_k t + ikx} + a_k^\dagger e^{i\omega_k t - ikx}]. \tag{9.1.22}$$

9.1.4 Phonons and Fock space

The expression we proceed to find for the Hamiltonian will lead to the basic physical interpretation of the operators a_k and $a_k^\dagger$. If we set $\mu = 1$, the classical expression for H reads

$$H = \frac{1}{2}\int_0^{\bar{L}} \mathrm{d}x\left(\pi^2(t, x) + c_s^2\left(\frac{\partial\varphi(t, x)}{\partial x}\right)^2\right). \tag{9.1.23}$$

The operator H is found by replacing φ and π by the corresponding operators; and its decomposition into normal modes by replacing φ and $\pi = \dot{\varphi}$ by their Fourier decompositions. Since the result is independent of the time, one can perform the calculation at $t = 0$. It proves convenient to use Parseval's theorem

$$\int_0^{\bar{L}} \varphi^2\, \mathrm{d}x = \sum_k \varphi_k \varphi_{-k},$$

and one finds

$$H = \frac{1}{2}\sum_k \hbar\omega_k(a_k^\dagger a_k + a_k a_k^\dagger) = \sum_k \hbar\omega_k\left(a_k^\dagger a_k + \frac{1}{2}\right) \tag{9.1.24}$$

(see Problem 9.3; we reinstate $\hbar$ for the rest of this section).

The normal-mode analysis shows that H *is a sum of independent harmonic-oscillator Hamiltonians having the frequencies* ω_k.

Let ω_k be the frequency of a normal mode (an eigenfrequency). Every normal mode is quantized as an independent harmonic oscillator. Recall that the eigenvectors of the simple-harmonic oscillator Hamiltonian take the form (Messiah 1961, Chapter 12; Cohen-Tannoudji *et al.* 1977, Chapter 5)

$$|n\rangle = \frac{1}{\sqrt{(n!)}}(a^\dagger)^n|0\rangle; \quad H|n\rangle = \hbar\omega\left(n + \frac{1}{2}\right)|n\rangle,$$

where $|0\rangle$ is the normed eigenvector obeying $a|0\rangle = 0$. The operators a and $a^\dagger$ are called the *annihilation and creation operators* (they turn $|n\rangle$ into $|n-1\rangle$ and $|n+1\rangle$ respectively); $|0\rangle$ is called the *vacuum state.* To construct the eigenstates of the Hamiltonian (9.1.24), we need merely take tensor products of the eigenstates for the individual normal modes; the vacuum state for instance reads

$$|0\rangle = |0_{k_1} 0_{k_2} \ldots 0_{k_i} \ldots\rangle,$$

and it is annihilated by any one of the a_{k_i}:

$$a_{k_i}|0\rangle = 0 \quad \forall k_i. \tag{9.1.25}$$

An eigenstate of H is characterized by the *occupation numbers* $n_{k_1} \ldots n_{k_i} \ldots$:

$$H|n_{k_1} \ldots n_{k_i} \ldots\rangle = \left[\sum_i \hbar\omega_{k_i}\left(n_{k_i} + \frac{1}{2}\right)\right]|n_{k_1} \ldots n_{k_i} \ldots\rangle, \tag{9.1.26}$$

where

$$|n_{k_1} \dots n_{k_i} \dots\rangle = \frac{1}{\sqrt{(n_{k_1}! \dots n_{k_i}!)}} (a_{k_1}^\dagger)^{n_{k_1}} \dots (a_{k_i}^\dagger)^{n_{k_i}} |0\rangle. \tag{9.1.27}$$

Accordingly, the operator $a_k^\dagger$ applied to an eigenstate of H belonging to eigenvalue E produces a state with energy $E + \hbar\omega_k$: one may interpret this by saying that the operator $a_k^\dagger$ creates a particle having energy $\hbar\omega_k$, called a *phonon*, or in other words a quantum of the acoustic vibrations. The state $|n_{k_1} \dots n_{k_i} \dots\rangle$ contains n_{k_1} phonons of energy $\hbar\omega_{k_1}, \dots, n_{k_i}$ phonons of energy $\hbar\omega_{k_i}$, etc.

We see that by quantizing a classical field, namely the field of longitudinal vibrations along a string, we have been enabled to describe the creation and annihilation of particles. This interpretation hinges on the fact that the energy of a quantized wave with frequency ω_k cannot assume arbitrary values, but only the values $\frac{1}{2}\hbar\omega_k, \frac{3}{2}\hbar\omega_k, \dots, (n + \frac{1}{2})\hbar\omega_k \dots$.

What we have just done for acoustic vibrations could be repeated (with a few complications) for the electromagnetic field. For instance, the quantization of the normal modes of the electromagnetic field in a cavity leads to the concept of photons: the energy of the electromagnetic field associated with an oscillation having frequency ω_k can assume only the values $\frac{1}{2}\hbar\omega_k$, $\frac{3}{2}\hbar\omega_k, \dots$, $(n + \frac{1}{2})\hbar\omega_k, \dots$, corresponding to states containing $0, 1, \dots, n, \dots$ photons.

The careful reader will have noticed that we have skated rather lightly over a serious problem: all the states defined in (9.1.27) have infinite energy. For the vacuum state for instance

$$\sum_k \tfrac{1}{2}\hbar\omega_k = +\infty,$$

since $\omega_k = 2\pi k/L$, $k = 1, 2, \dots$. However, short of destroying the crystal or the vibrating string, all that we can observe are energy *differences* with respect to the ground state $|0\rangle$. We could, by convention, take the ground-state energy as zero, which amounts to redefining H:

$$H = \sum_k \omega_k a_k^\dagger a_k, \quad H|0\rangle = 0. \tag{9.1.28}$$

To ensure that the vacuum energy vanish automatically, we agree to write all operator products in their *normal form*, by placing the creation operators on the left and the annihilation operators on the right; in this reordering process, their commutators are ignored. The normal product AB is written as :AB:; thus we have, for example,

$$\tfrac{1}{2}{:}a_k^\dagger a_k + a_k a_k^\dagger{:} = a_k^\dagger a_k.$$

The construction of the eigenstates of H as above has enabled us to identify a Hilbert space, and explicit expressions for the operators a_k and $a_k^\dagger$; mathematically speaking, we have found a *representation* of the CCR (9.1.21). In ordinary quantum mechanics, the customary representation of the CCR $[Q, P] = \mathrm{i}$ is the following: the Hilbert space is the space L^2 of functions square-integrable over the interval $]-\infty, \infty[$, and

$$Q \to x, \quad P \to -\mathrm{i}\frac{\partial}{\partial x}.$$

A theorem due to von Neumann affirms that up to unitary transformations this representation is unique. The same theorem applies to any system with a finite number of degrees of freedom. But if there are infinitely many degrees of freedom, then there exist representations of the CCR different from the one we have just constructed. The latter is called the *Fock-space representation of the CCR*, and its identifying characteristic is that it has a vacuum state $a_k|0\rangle = 0$, under the supplementary condition that the vacuum be a cyclic vector (see Streater and Wightman 1964). Other representations will not be used in this book.

9.2 Quantization of the Klein–Gordon field

9.2.1 Wave equation and Lagrangean

Consider a classical field which is a Lorentz scalar, $\varphi'(x') = \varphi(x)$, where $x' = \Lambda x$, with Λ a Lorentz transformation (see Appendix D). Such a field must obey a partial differential equation analogous to (9.1.12). The only Lorentz-invariant second-order operator one can construct from the derivatives $\partial_\mu = \partial/\partial x^\mu$ is the d'Alembertian

$$\Box = \partial_\mu \partial^\mu = \frac{\partial^2}{\partial t^2} - \nabla^2 \tag{9.2.1}$$

(recall that in our units $c = 1$). The simplest possible Lorentz-invariant second-order equation reads

$$(\Box + m^2)\varphi(x) = 0. \tag{9.2.2}$$

Equation (9.2.2) is called the equation for the *free Klein–Gordon field*. We stress that until further notice $\varphi(x)$ must be interpreted as a classical field analogous to the classical electromagnetic field, and not as the wavefunction in some relativistically generalized Schroedinger equation. The parameter m has the dimensions of inverse length (or of time, seeing that $c = 1$).

Let $j(x)$ be a scalar function assigned in advance, i.e. a *classical source*; then the Klein–Gordon equation coupled to a classical source reads

$$(\Box + m^2)\varphi(x) = j(x).$$

It will be studied in Section 9.3. Finally, the right-hand side of (9.2.2) could be replaced by a function of φ (and even of its derivatives), yielding

$$(\Box + m^2)\varphi(x) = -V'(\varphi).$$

In that case one has to do with an interacting Klein–Gordon field, which will be treated in the next chapter.

Equation (9.2.2) can be obtained by minimizing the action

$$S = \int d^4x \left(\frac{1}{2} (\partial_\mu \varphi)(\partial^\mu \varphi) - \frac{1}{2} m^2 \varphi^2 \right)$$
$$= \int d^4x \left(\frac{1}{2} \left(\frac{\partial \varphi}{dt} \right)^2 - \frac{1}{2} (\nabla \varphi)^2 - \frac{1}{2} m^2 \varphi^2 \right)$$
$$= \int d^4x \left(-\frac{1}{2} \varphi (\Box + m^2) \varphi \right). \tag{9.2.3a}$$

The last expression follows from an integration by parts. This integration is justified along the space axes, because we assume as always that the fields vanish fast enough at infinity. Along the time axis it is of a formal nature, and serves to identify the propagator: see Section 8.2.3. (One should not conclude from (9.2.3a) that $\mathscr{L} = 0$ if φ obeys (9.2.2).) The equations of motion, generalizations of (9.1.11), read

$$\partial_\mu \frac{\partial \mathscr{L}}{\partial (\partial_\mu \varphi)} - \frac{\partial \mathscr{L}}{\partial \varphi} = 0. \tag{9.2.3b}$$

9.2.2 Fourier decomposition

As with vibrations in a solid, one naturally looks for the normal modes. We define Fourier transforms by integrals over all of space, so as to preserve formal Lorentz invariance, i.e. the *covariance* of the equations; accordingly,

$$\varphi(x) = \int \frac{d^4k}{(2\pi)^4} e^{-ikx} \varphi(k), \tag{9.2.4a}$$

$$\varphi(k) = \int d^4x \, e^{ikx} \varphi(x), \tag{9.2.4b}$$

where $kx = k^0x^0 - \boldsymbol{k}\cdot\boldsymbol{x}$. The reality of φ entails $\varphi(k) = \varphi^*(-k)$. Lastly, the wave equation (9.2.2) applied to (9.2.4a) entails

$$(k^2 - m^2)\varphi(k) = 0.$$

The Fourier coefficient $\varphi(k)$ is nonzero only on the *mass hyperboloid* ('*mass shell*') $k^2 - m^2 = 0$, whence $\varphi(k)$ must be proportional to $\delta(k^2 - m^2)$. Further, the mass shell $k^2 = m^2$ splits into two sheets, one with $k^0 > 0$ and the other with $k^0 < 0$; these two sheets cannot be connected by transformations belonging to the proper Lorentz group. The relation

$$\delta(k^2 - m^2) = \frac{1}{2\omega_{\boldsymbol{k}}}(\delta(k^0 - \omega_{\boldsymbol{k}}) + \delta(k^0 + \omega_{\boldsymbol{k}})),$$
$$\omega_{\boldsymbol{k}} = \sqrt{(\boldsymbol{k}^2 + m^2)}\ (>0) \tag{9.2.5}$$

allows us to rewrite $\varphi(k)$ as

$$\varphi(k) = 2\pi\delta(k^2 - m^2)[\theta(k^0)\varphi^{(+)}(\boldsymbol{k}) + \theta(-k^0)\varphi^{(-)}(\boldsymbol{k})], \tag{9.2.6}$$

where $\theta(k^0)$ is the Heaviside step function: $\theta(k^0) = 1$ if $k^0 > 0$, and $\theta(k^0) = 0$ if $k^0 < 0$. The factor (2π) is conventional. Substituting (9.2.6) into (9.2.4a) we obtain

$$\varphi(x) = \int \frac{\mathrm{d}^3 k}{(2\pi)^3 2\omega_{\boldsymbol{k}}}(\varphi^{(+)}(\boldsymbol{k})\mathrm{e}^{-\mathrm{i}kx} + \varphi^{(-)}(-\boldsymbol{k})\mathrm{e}^{\mathrm{i}kx}), \tag{9.2.7}$$

where

$$kx = \omega_{\boldsymbol{k}} t - \boldsymbol{k}\cdot\boldsymbol{x}.$$

In fact the notation $k \cdot x$ is ambiguous. For a four-dimensional Fourier transform (like (9.2.4) or (9.2.20)), $kx = k^0 x^0 - \boldsymbol{k}\cdot\boldsymbol{x}$, where k^0 can have either sign. But for a three-dimensional Fourier transform (like (9.2.12) or (9.2.22)), $kx = \omega_{\boldsymbol{k}} x^0 - \boldsymbol{k}\cdot\boldsymbol{x}$ where $\omega_{\boldsymbol{k}} = \sqrt{(\boldsymbol{k}^2 + m^2)} > 0$. The expression (9.2.7) is very similar to (9.1.19a), but with three differences:

- the space has three dimensions and not only one;
- $\omega_{\boldsymbol{k}} = (\boldsymbol{k}^2 + m^2)^{1/2}$ instead of $\omega_k = |k|$ (i.e. $m = 0$ in (9.1.19a));
- the wavevector assumes continuous rather than discrete values; we have integrated over all of space rather than only over a finite interval. In a self-explanatory notation, (9.2.7) is written

$$\varphi(x) = \varphi^{(+)}(x) + \varphi^{(-)}(x),$$

where $\varphi^{(+)}$ ($\varphi^{(-)}$) is the positive- (negative-) frequency part: in quantum mechanics, a system with energy E evolves in time, by convention, according to $\exp(-\mathrm{i}Et/\hbar) = \exp(-\mathrm{i}\omega t)$.

The integration measure in (9.2.7), namely

$$\mathrm{d}\tilde{k} = \frac{\mathrm{d}^3 k}{(2\pi)^3 2\omega_{\boldsymbol{k}}}, \tag{9.2.8}$$

is Lorentz-invariant, since it can be written in the alternative and manifestly invariant form

$$\mathrm{d}\tilde{k} = \frac{\mathrm{d}^4 k}{(2\pi)^4} 2\pi\delta(k^2 - m^2)\theta(k^0).$$

It is also possible to verify its invariance directly (Problem 9.4).

One can construct wavepackets from positive- (or negative-) energy solutions; for instance,

$$f(x) = \int \frac{d^3k}{(2\pi)^3 2\omega_{\boldsymbol{k}}} e^{-ikx} f(\boldsymbol{k}). \tag{9.2.9}$$

The scalar product of two wavepackets is given by

$$(g, f) = i \int d^3x\, g^* \overleftrightarrow{\partial}_0 f = \int \frac{d^3k}{(2\pi)^3 2\omega_{\boldsymbol{k}}} g^*(\boldsymbol{k}) f(\boldsymbol{k}), \tag{9.2.10}$$

where $\overleftrightarrow{\partial}_0$ is defined by

$$f \overleftrightarrow{\partial}_0 g = f \frac{\partial g}{\partial x_0} - \frac{\partial f}{\partial x^0} g. \tag{9.2.11}$$

This scalar product is, by construction, positive definite for positive-energy solutions, and is constant in time: $\partial_0(g, f) = 0$ (Problem 9.5). Under the scalar product (9.2.10), a wavepacket with negative energy is orthogonal to any packet with positive energy.

9.2.3 Canonical quantization

We shall obtain the quantized Klein–Gordon field, analogously to Section 9.1, by replacing the Fourier coefficients in (9.2.7) with operators; we need merely take care over the normalization. Equation (9.2.7) becomes

$$\varphi(x) = \int \frac{d^3k}{(2\pi)^3 2\omega_{\boldsymbol{k}}} (a(\boldsymbol{k}) e^{-ikx} + a^\dagger(\boldsymbol{k}) e^{ikx}), \tag{9.2.12}$$

where

$$[a(\boldsymbol{k}), a^\dagger(\boldsymbol{k}')] = (2\pi)^3 2\omega_{\boldsymbol{k}} \delta^{(3)}(\boldsymbol{k} - \boldsymbol{k}'). \tag{9.2.13}$$

It is easy to check (Problem 9.6) that (9.2.12) and (9.2.13) lead to the canonical commutation relations

$$\begin{aligned} &[\varphi(t, \boldsymbol{x}), \varphi(t, \boldsymbol{x}')] = [\pi(t, \boldsymbol{x}), \pi(t, \boldsymbol{x}')] = 0, \\ &[\varphi(t, \boldsymbol{x}), \pi(t, \boldsymbol{x}')] = i\delta^{(3)}(\boldsymbol{x} - \boldsymbol{x}'). \end{aligned} \tag{9.2.14}$$

Occasionally it proves convenient to quantize in a box of finite volume V, generally a cube of edge-length $\bar{L}$ ($V = \bar{L}^3$), so as to work with normable exponentials. The immediate generalization of (9.1.22) reads

$$\varphi(x) = \frac{1}{\sqrt{(V)}} \sum_{\boldsymbol{k}} \frac{1}{\sqrt{(2\omega_{\boldsymbol{k}})}} (a_{\boldsymbol{k}} e^{-ikx} + a_{\boldsymbol{k}}^\dagger e^{ikx}), \tag{9.2.15}$$

$$[a_{\boldsymbol{k}}, a^{\dagger}_{\boldsymbol{k}}] = \delta_{\boldsymbol{k},\boldsymbol{k}'}. \tag{9.2.16}$$

Notice the difference between the normalizations in (9.2.12), (9.2.13), and (9.2.15), (9.2.16). The expression

$$H = \sum_{\boldsymbol{k}} \omega_{\boldsymbol{k}} a^{\dagger}_{\boldsymbol{k}} a_{\boldsymbol{k}}$$

for the Hamiltonian, found for instance from (9.2.15), (9.2.16), shows that the operator $a^{\dagger}_{\boldsymbol{k}}$ ($a_{\boldsymbol{k}}$) creates (destroys) a particle having wave-vector $\boldsymbol{k}$ (hence momentum $\boldsymbol{k}$) and energy $\omega_{\boldsymbol{k}} = (\boldsymbol{k}^2 + m^2)^{1/2}$. These particles have no other degrees of freedom; accordingly they are scalar particles, of zero spin. The Klein–Gordon field will be used to describe particles like π mesons. Such particles obey Bose statistics by construction: indeed our construction of the Fock space ensures that the wavefunction is symmetric under the exchange of any pair of particles.

Since we are using units with $\hbar = 1$, $\boldsymbol{k}$ is not only a wave-vector but also a momentum ($\boldsymbol{p} = \hbar\boldsymbol{k}$), and $\omega_{\boldsymbol{k}}$ is not only a frequency but also an energy ($E_{\boldsymbol{k}} = \hbar\omega_{\boldsymbol{k}}$). In units with $\hbar = c = 1$, the parameter m has the dimensions of mass ($\hbar/mc$ is a length); m is the mass of the particles described by the quantized Klein–Gordon field:

$$E_{\boldsymbol{k}} = \sqrt{(\boldsymbol{k}^2 + m^2)}\ (= \sqrt{(\hbar^2\boldsymbol{k}^2c^2 + m^2c^4)}).$$

In quantizing we have chosen a special reference frame, since the CCR are written at equal times. In principle one needs to check that quantization in a different reference frame leads to an equivalent theory. To this end, we would need to construct the infinitesimal generators of translations and of Lorentz transformations, namely the energy-momentum, the angular momentum and the boost operators. This construction is somewhat laborious (it involves many indices), and can be found in all the classic texts; see also Appendix C. We can dispense with the details, because later on we shall use a method that is manifestly covariant.

9.2.4 The commutator at $t \neq t'$

For the *free* Klein–Gordon field (i.e. for the field obeying (9.2.2)) the commutators can be calculated even at $t \neq t'$. We have stressed already that in general this calculation is impossible, since it depends on the dynamics, while the CCR (9.2.14) at *equal times* do not.

Let us therefore evaluate $[\varphi(x), \varphi(x')]$ by using the Fourier representation (9.2.12):

$$[\varphi(x), \varphi(x')] = \int \mathrm{d}\tilde{k}\,\mathrm{d}\tilde{k}'\{[a(\boldsymbol{k}), a^{\dagger}(\boldsymbol{k}')]\mathrm{e}^{-\mathrm{i}kx + \mathrm{i}k'x'} + [a^{\dagger}(\boldsymbol{k}), a(\boldsymbol{k}')]\mathrm{e}^{\mathrm{i}kx - \mathrm{i}k'x'}\}$$

$$= \int \mathrm{d}\tilde{k}(\mathrm{e}^{-\mathrm{i}k(x-x')} - \mathrm{e}^{\mathrm{i}k(x-x')}) \underset{\mathrm{def}}{=} \mathrm{i}\Delta(x - x').$$

One can write $\Delta(x - x')$ in the explicitly covariant form

$$\Delta(x - x') = -\mathrm{i}[\varphi(x), \varphi(x')] = -\mathrm{i}\int \frac{\mathrm{d}^4 k}{(2\pi)^4} \mathrm{e}^{-\mathrm{i}k(x-x')} \times 2\pi\varepsilon(k^0)\delta(k^2 - m^2), \tag{9.2.17}$$

where $\varepsilon(k^0) = \theta(k^0) - \theta(-k^0)$. Several properties of Δ follow from (9.2.17):

(i) $(\Box + m^2)\Delta(x) = 0$,

(ii) $\Delta(x) = \Delta(x^2, \varepsilon(x^0))$,

(ii′) $\Delta(\Lambda x) = \Delta(x)$ if $\Lambda \in$ orthochronous Lorentz group,

(iii) $\Delta(x) = -\Delta(-x)$,

(iv) $\dfrac{\partial}{\partial x^0}\Delta(x - x') = -\delta^{(3)}(\boldsymbol{x} - \boldsymbol{x}')$,

(v) $\Delta(x) = 0$ if $x^2 < 0$.

Property (v) reflects what is called the *locality* of the theory: observables at space-like-separated points must commute. No light signal can connect any pair of such points, whence observations made there must be mutually independent. Being Hermitean, the Klein–Gordon field is indeed an observable, and the commutator $[\varphi(x), \varphi(x')]$ must vanish when $(x - x')^2 < 0$.

9.2.5 The propagator

We proceed to study the Green's functions of the Klein–Gordon equation, i.e. the solutions of

$$(\Box + m^2)G(x) = \delta^{(4)}(x); \tag{9.2.18}$$

or, going over to Fourier space, of

$$(-k^2 + m^2)G(k) = 1. \tag{9.2.19}$$

Thus

$$G(x) = -\int \frac{\mathrm{d}^4 k}{(2\pi)^4} \frac{\mathrm{e}^{-\mathrm{i}kx}}{(k^0)^2 - (\boldsymbol{k}^2 + m^2)}. \tag{9.2.20}$$

The k^0-integrand has poles at $k^0 = \pm\omega_{\boldsymbol{k}}$, which enable one to define several Green's functions, differing in the contours used to avoid these poles. We shall restrict ourselves to the Green's function G_{F}, obtained by using the Feynman contour C_{F} of Fig. 9.2; the study of certain other Green's functions is relegated to Problem 9.7. Note that alternatively we may write

$$G_{\mathrm{F}} = -\int_{C_{\mathrm{F}}} \frac{\mathrm{d}^4 k}{(2\pi)^4} \frac{\mathrm{e}^{-\mathrm{i}kx}}{k^2 - m^2} = -\int \frac{\mathrm{d}^4 k}{(2\pi)^4} \frac{\mathrm{e}^{-\mathrm{i}kx}}{k^2 - m^2 + \mathrm{i}\varepsilon}, \tag{9.2.21}$$

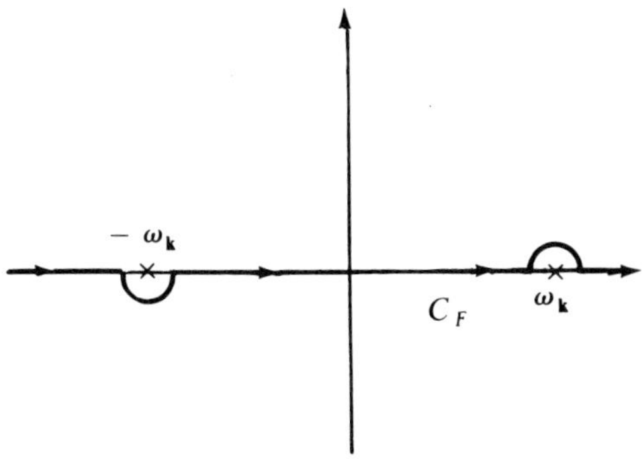

Fig. 9.2 The Feynman contour

where the replacement $m^2 \to m^2 - i\varepsilon$, $\varepsilon \to 0+$ ensures that the pole at $k^0 = \omega_k$ $(-\omega_k)$ is shifted into the half-plane $\operatorname{Im} k^0 < 0$ $(\operatorname{Im} k^0 > 0)$. G_F can now be evaluated by the calculus of residues. For $x^0 > 0$, the contour may be closed by a semicircle in the lower half-plane, and only the pole at $k^0 = \omega_k - i\varepsilon$ contributes:

$$G_F(x)|_{x^0>0} = i\int \frac{d^3k}{(2\pi)^3 2\omega_k} e^{-ikx} = i\langle 0|\,\varphi(x)\,\varphi(0)\,|0\rangle. \tag{9.2.22a}$$

For $x^0 < 0$, the contour is closed by a semicircle in the upper half-plane, and only the pole at $k^0 = -\omega_k + i\varepsilon$ contributes:

$$G_F(x)|_{x^0<0} = i\int \frac{d^3k}{(2\pi)^3 2\omega_k} e^{ikx} = i\langle 0|\,\varphi(0)\,\varphi(x)\,|0\rangle. \tag{9.2.22b}$$

Both the equations (9.2.22) are subsumed into

$$G_F(x - x') = i\langle 0|\, T(\varphi(x)\varphi(x'))\,|0\rangle, \tag{9.2.23}$$

where the *T-product* of two fields is defined by

$$\begin{aligned} T(\varphi(x)\varphi(x')) &= \varphi(x)\varphi(x') \qquad x^0 > x'^0, \\ T(\varphi(x)\varphi(x')) &= \varphi(x')\varphi(x) \qquad x^0 < x'^0. \end{aligned} \tag{9.2.24}$$

One could reasonably worry about the covariance of such a definition, since the order in time is not always invariant under Lorentz transformations. However, it can be reversed only if the separation between x and x' is space-like, $(x - x')^2 < 0$, in which case the fields commute by virtue of locality.

In order to avoid certain factors i in our formulae, we define the *Feynman propagator* $\Delta_F(x - x')$ by

$$\Delta_F(x-x') = \langle 0|\, T(\varphi(x)\varphi(x'))\,|0\rangle, \tag{9.2.25a}$$

$$\Delta_F(x-x') = \int \frac{d^4k}{(2\pi)^4}\, e^{-ik(x-x')} \frac{i}{k^2-m^2+i\varepsilon}. \tag{9.2.25b}$$

The convention (9.2.25) is not the usual one. Most writers adopt the definition $\Delta_F = -i\langle 0|\, T(\varphi\varphi)\,|0\rangle$. The propagator Δ_F equals $-iG_F$, whence it obeys

$$(\Box + m^2)\Delta_F = -i\delta^{(4)}(x). \tag{9.2.26}$$

It is instructive to verify this equation directly, by starting from the definition (9.2.25a) of Δ_F (Problem 9.8).

9.2.6 The singularities on the light cone

Strictly speaking the 'functions' $\Delta(x)$, $\Delta_F(x)$, etc. are not functions at all, but distributions, with singularities on the light cone $x^2 = 0$. We shall evaluate a simple case, and to this end restrict ourselves to $m = 0$; actually this will allow us to determine the strongest singularities. For $m = 0$ the standard notation is

$$D_i = \Delta_i(m = 0).$$

Start by evaluating $D_+(x)$:

$$D_+(x) = \langle 0|\, \varphi(x)\varphi(0)\,|0\rangle = \int \frac{d^3k}{(2\pi)^3\, 2k}\, e^{-i(kx^0 - \boldsymbol{k}\cdot\boldsymbol{x})},$$

where

$$k = \|\boldsymbol{k}\|\ (=\omega_{\boldsymbol{k}}), \quad \boldsymbol{k}\cdot\boldsymbol{x} = kr\cos\theta.$$

Integration over $\cos\theta$ yields

$$D_+(x) = \frac{-i}{2(2\pi)^2 r}\int_0^\infty dk\,(e^{-ik(x^0-r)} - e^{-ik(x^0+r)}).$$

The integral over k must be interpreted as a distribution, being in fact the Fourier transform of $\theta(k)$; to make the integral converge, we replace $(x^0 - r)$ by $(x^0 - r - i\varepsilon)$:

$$\int_0^\infty dk\, e^{-ik(x^0-r-i\varepsilon)} = \frac{-i}{x_0 - r - i\varepsilon} = -iP\frac{1}{x^0 - r} + \pi\delta(x^0 - r),$$

where P denotes the Cauchy principal value. This relation allows us to complete the evaluation of $D_+(x)$:

$$D_+(x) = \frac{-1}{4\pi^2} P\left(\frac{1}{x^2}\right) - \frac{i}{4\pi}\delta(x^2)\varepsilon(x^0). \tag{9.2.27}$$

(where $x^2 = (x^0)^2 - r^2$). Defining $D_-(x)$ by

$$D_-(x) = \langle 0|\varphi(0)\varphi(x)|0\rangle = \langle 0|\varphi(-x)\varphi(0)|0\rangle = D_+(-x),$$

we immediately find the functions $D(x)$ and $D_F(x)$:

$$\begin{aligned} D(x) &= -i(D_+(x) - D_-(x)) = \frac{-1}{2\pi}\,\varepsilon(x^0)\delta(x^2), \\ D_F(x) &= \theta(x^0)D_+(x) + \theta(-x^0)D_-(x). \end{aligned} \tag{9.2.28}$$

The explicit expression of $D_F(x)$ then reads

$$D_F(x) = \frac{-1}{4\pi^2}\,P\frac{1}{x^2} - \frac{i}{4\pi}\,\delta(x^2) = \frac{-1}{4\pi^2(x^2 - i\varepsilon)}. \tag{9.2.29}$$

When $m \neq 0$, one can derive the following expression for $\Delta_F(x)$ (Bogoliubov and Shirkov 1959, Appendix 1):

$$\begin{aligned} \Delta_F(x) = {} & \frac{-1}{4\pi^2}\,P\left(\frac{1}{x^2}\right) - \frac{i}{4\pi}\,\delta(x^2) + \frac{im^2}{16\pi}\,\theta(x^2) \\ & + \frac{m^2}{8\pi^2}\ln\frac{m\sqrt{(|x^2|)}}{2} + O(\sqrt{(|x^2|)}\ln|x^2|). \end{aligned} \tag{9.2.30}$$

The presence of singularities on the light-cone shows that one cannot without special precautions multiply together two fields at one and the same point. In fact, the field operators must be considered as 'operator-valued distributions'; the vacuum expectation-values of products of fields are ordinary distributions. A programme, called axiomatic field theory, has been developed on this basis, mainly by Wightman and his collaborators (Streater and Wightman 1964). The axiomatic theory supplies rigorous proofs of the PCT and of the spin-statistics theorems, and (granted some supplementary assumptions) establishes certain asymptotic bounds on scattering amplitudes. But further elaboration of the programme has been hampered by the fact that it is not well adapted to gauge theories.

9.3 Coupling to a classical source, and Wick's theorem

9.3.1 The evolution operator, and Dyson's equation

In quantum mechanics, the evolution operator $U(t, t_0)$ relates the Schroedinger-picture state vector at time t to the state vector at time t_0:

$$|\psi_s(t)\rangle = U(t, t_0)|\psi_s(t_0)\rangle. \tag{9.3.1}$$

It obeys the differential equation

$$i\frac{d}{dt}\,U(t, t_0) = H(t)U(t, t_0), \tag{9.3.2}$$

equivalent to the Schroedinger equation; the group composition rule

$$U(t, t_1)U(t_1, t_0) = U(t, t_0); \tag{9.3.3}$$

and the unitarity condition

$$U^\dagger(t, t_0) = U^{-1}(t, t_0) = U(t_0, t). \tag{9.3.4}$$

As a rule one cannot calculate the evolution operator exactly; but often it happens that one can subdivide

$$H = H_0 + H_1 \tag{9.3.5}$$

in a way that does allow one to calculate the evolution operator $U_0(t, t_0)$ corresponding to H_0. In practice this means that one can diagonalize H_0. Then one treats H_1 as a perturbation; H_0 is generally called the 'free Hamiltonian', and H_1 the 'interaction Hamiltonian'. In the perturbative approach it proves convenient to write

$$U(t, t_0) = U_0(t, t_0)U_\mathrm{I}(t, t_0); \tag{9.3.6}$$

then it is easy to show that U_I satisfies the differential equation

$$\mathrm{i}\frac{\mathrm{d}U_\mathrm{I}}{\mathrm{d}t} = H_\mathrm{I}(t)U_\mathrm{I}(t, t_0), \tag{9.3.7a}$$

$$H_\mathrm{I}(t) = U_0^{-1}(t, t_0)H_1(t)U_0(t, t_0). \tag{9.3.7b}$$

H_I is the Hamiltonian H_1 in the '*interaction picture*'. The differential equation (9.3.7) can be rewritten as an integral equation,

$$U_\mathrm{I}(t, t_0) = 1 - \mathrm{i}\int_{t_0}^{t} H_\mathrm{I}(t')U_\mathrm{I}(t', t_0)\,\mathrm{d}t', \tag{9.3.8}$$

which automatically incorporates the initial condition $U_\mathrm{I}(t_0, t_0) = 1$. Note that in general one *cannot* write

$$U_\mathrm{I}(t, t_0) = \mathrm{e}^{-\mathrm{i}\int_{t_0}^{t} H_\mathrm{I}(t')\,\mathrm{d}t'}, \tag{9.3.9}$$

because in general the commutator $[H_\mathrm{I}(t'), H_\mathrm{I}(t'')]$ is nonzero, and because (except in some special cases) $\exp(A + B) \neq \exp(A)\exp(B)$ when $[A, B] \neq 0$. Nevertheless (9.3.8) can be solved by iteration, which yields

$$U_\mathrm{I}(t, t_0) = 1 + \sum_{n=1}^{\infty} \frac{(-\mathrm{i})^n}{n!}\int_{t_0}^{t} \mathrm{d}t_1 \ldots \mathrm{d}t_n\, T(H_\mathrm{I}(t_1) \ldots H_\mathrm{I}(t_n)). \tag{9.3.10}$$

This, called Dyson's equation, can be expressed formally as

$$U_\mathrm{I}(t, t_0) = T\left(\mathrm{e}^{-\mathrm{i}\int_{t_0}^{t} H_\mathrm{I}(t')\,\mathrm{d}t'}\right), \tag{9.3.11}$$

where the T-product instructs one to use (9.3.10) in expanding the exponential.

Equation (9.3.10) is evidently a perturbation expansion by powers of H_1: if H_1 has a coupling constant g as a factor, then the nth-order term in (9.3.10) is proportional to g^n. In (9.3.11) the T-product corrects the erroneous expression given by (9.3.9); it ensures that $U_{\text{I}}(t, t_0)$ obeys the group composition rule (9.3.3), and that it satisfies the unitarity relation (9.3.4).

9.3.2 The harmonic oscillator coupled to a classical source

Before we try to couple the Klein–Gordon field to a classical source, let us study the coupling of a simple-harmonic oscillator to such a source. This will allow us to introduce, in an elementary context, the concepts of incoming, or 'in', states; of outgoing, or 'out', states; and of the S matrix. We choose (in the Schroedinger picture) the Hamiltonian

$$H = \omega a^\dagger a - aj^*(t) - a^\dagger j(t) = H_0 + H_1, \tag{9.3.12}$$

where $H_0 = \omega a^\dagger a$ is the Hamiltonian of a simple-harmonic oscillator. If the source $j(t)$ is real, then $H_1 = -\sqrt{(2\omega)}jx$, and $\sqrt{(2\omega)}j(t)$ simply represents an externally applied force; thus we are faced with the problem of a forced quantum oscillator. Complex j entails an additional coupling linear in the velocity. In order to simplify the equations, we choose the time origin so that $t_0 = 0$, and set $U(t, 0) = U(t)$. To calculate $a_{\text{I}}(t)$ (see (9.3.7b)),

$$a_{\text{I}}(t) = \mathrm{e}^{\mathrm{i}H_0 t} a \mathrm{e}^{-\mathrm{i}H_0 t},$$

we use

$$\frac{\mathrm{d}a_{\text{I}}}{\mathrm{d}t} = \mathrm{i}[H_0, a_{\text{I}}(t)] = -\mathrm{i}\omega a_{\text{I}}(t).$$

Hence

$$a_{\text{I}}(t) = \mathrm{e}^{-\mathrm{i}\omega t} a, \quad a_{\text{I}}^\dagger(t) = \mathrm{e}^{\mathrm{i}\omega t} a^\dagger, \tag{9.3.13}$$

and the Hamiltonian $H_{\text{I}}(t)$ reads

$$H_{\text{I}}(t) = -[a\mathrm{e}^{-\mathrm{i}\omega t} j^*(t) + a^\dagger \mathrm{e}^{\mathrm{i}\omega t} j(t)]. \tag{9.3.14}$$

The evolution equation (9.3.7a) is 'almost' solved by (9.3.9), because even though $H_{\text{I}}(t)$ and $H_{\text{I}}(t')$ fail to commute, nevertheless their commutator is just a number:

$$[H_{\text{I}}(t), H_{\text{I}}(t')] = \mathrm{e}^{-\mathrm{i}\omega(t-t')} j^*(t) j(t') - \mathrm{e}^{-\mathrm{i}\omega(t'-t)} j(t) j^*(t').$$

This property allows us to guess the solution of the evolution equation (see Problem 9.10):

$$U_{\text{I}}(t) = \exp\left(\mathrm{i}a^\dagger \int_0^t j(t')\mathrm{e}^{\mathrm{i}\omega t'}\,\mathrm{d}t'\right) \exp\left(\mathrm{i}a \int_0^t j^*(t')\mathrm{e}^{-\mathrm{i}\omega t'}\,\mathrm{d}t'\right)$$

$$\times \exp\left(-\iint_0^t \mathrm{d}t'\,\mathrm{d}t''\,\theta(t'-t'') j^*(t') \mathrm{e}^{-\mathrm{i}\omega(t'-t'')} j(t'')\right). \tag{9.3.15}$$

One checks by direct calculation that (9.3.15) indeed satisfies the evolution equation (9.3.7a). The solution can be written in other forms (see Problem 9.10); note that in (9.3.15) $U_I(t)$ is expressed in normal form, all the $a^\dagger$ being to the left of all the a.

In order to simplify the argument, we suppose that the source $j(t)$ vanishes outside some interval $[T_1, T_2]$, with $T_1 > 0$ (Fig. 9.3), and study the annihilation operator in the Heisenberg picture,

$$a_H(t) = U^{-1}(t)aU(t) = e^{-i\omega t}U_I^{-1}(t)aU_I(t).$$

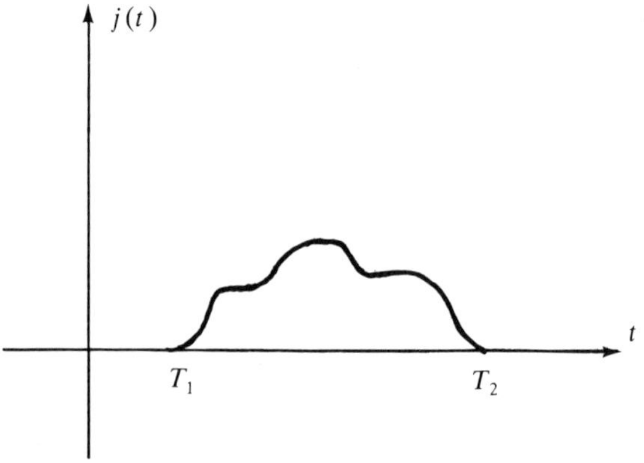

Fig. 9.3 The source $j(t)$

By virtue of (9.3.15) and of the identity

$$e^{-i\alpha a^\dagger} a\, e^{i\alpha a^\dagger} = a + i\alpha$$

we find

$$a_H(t) = e^{-i\omega t}\left[a + i\int_0^t j(t')e^{i\omega t'}\,dt'\right]. \tag{9.3.16}$$

For $t < T_1$, $a_H(t)$ coincides with $a(t) = a\exp(-i\omega t)$, which we shall call $a_{in}(t)$:

$$\lim_{t\to-\infty} a_H(t) = a_{in}(t);$$

on the other hand, for $t \to +\infty$, we have

$$\lim_{t\to+\infty} a_H(t) = e^{-i\omega t}\left(a + i\int_{-\infty}^{\infty} j(t')e^{i\omega t'}\,dt'\right),$$

whence

$$\lim_{t\to+\infty} a_H(t) = e^{-i\omega t}(a + i\tilde{j}(\omega)) = a_{out}(t).$$

The operators $a_{\rm in}(t)$ and $a_{\rm out}(t)$ have factors $\exp(-{\rm i}\omega t)$, and we can define time-independent operators $a_{\rm in}$ and $a_{\rm out}$ by

$$a_{\rm in} = a, \quad a_{\rm out} = a + {\rm i}j(\omega). \tag{9.3.17}$$

The relation between $a_{\rm in}$ and $a_{\rm out}$ can also be written as

$$a_{\rm out} = S^{-1} a_{\rm in} S; \tag{9.3.18}$$

here, the unitary operator S (generally called the *S matrix*, and not to be confused with the action) is given by

$$S = U_{\rm I}(\infty, -\infty), \quad S^\dagger = S^{-1}. \tag{9.3.19}$$

Note that the limit $t \to \infty$ features only one Fourier component of the source $j(t)$, namely $j(\omega)$; the limit 'projects onto the energy shell'. We investigate the transition probabilities rather briefly; starting with a state $|n_{\rm in}\rangle$ at $t = t_1$ ($t_1 < T_1$),

$$|n_{\rm in}\rangle = \frac{1}{\sqrt{n!}} (a_{\rm in}^\dagger)^n |0_{\rm in}\rangle,$$

we aim to calculate the probability that at time $t = t_2$ ($t_2 > T_2$) one observes a state $|m_{\rm in}\rangle$. The probability amplitude is

$$\begin{aligned}\langle m_{\rm in}|U(t_2, t_1)|n_{\rm in}\rangle &= \langle m_{\rm in}|U_0(t_2, t_1)U_{\rm I}(t_2, t_1)|n_{\rm in}\rangle \\ &= {\rm e}^{-{\rm i}E_m(t_2-t_1)}\langle m_{\rm in}|S|n_{\rm in}\rangle \\ &= {\rm e}^{-{\rm i}E_m(t_2-t_1)}\langle m_{\rm out}|n_{\rm in}\rangle. \end{aligned} \tag{9.3.20}$$

The phase factor here is evidently irrelevant to the transition probability. It is easy to calculate $\langle m_{\rm out}|n_{\rm in}\rangle$ when $n = 0$; in fact (suppressing the label 'in' to ease the notation) we can evaluate

$$\langle m|{\rm e}^{{\rm i}\alpha a^\dagger}|0\rangle = \frac{{\rm i}^m\alpha^m}{\sqrt{m!}},$$

$$\int {\rm d}t'\,{\rm d}t''\,\theta(t'-t'')j^*(t'){\rm e}^{-{\rm i}\omega(t'-t'')}j(t'')$$

$$= -\frac{1}{2{\rm i}\pi} P\int \frac{{\rm d}E}{E} j^*(\omega+E)j(\omega+E) + \frac{1}{2}|j(\omega)|^2 = {\rm i}\varphi + \frac{1}{2}|j(\omega)|^2.$$

The last line is derived simply by using the integral representation of the θ function. This yields the probability amplitude

$$\langle m|U(t_2, t_1)|0\rangle = {\rm e}^{-{\rm i}E_m(t_2-t_1)}{\rm e}^{-{\rm i}\varphi}\frac{{\rm i}^m(j(\omega))^m}{\sqrt{m!}}{\rm e}^{-\frac{1}{2}|j(\omega)|^2},$$

and the transition probability

$$P_m = |\langle m|U(t_2, t_1)|0\rangle|^2 = \frac{(|j(\omega)|^2)^m {\rm e}^{-|j(\omega)|^2}}{m!}. \tag{9.3.21}$$

The probability distribution for the occupation number m is the *Poisson distribution*, with an average $\bar{m} = |j(\omega)|^2$. This is a general feature of radiation from classical sources, provided one starts from the ground state at $t = -\infty$. The final state is a *coherent state* (Problem 9.11), given, up to a phase, by

$$e^{-\frac{1}{2}|j(\omega)|^2} e^{ia^\dagger j(\omega)}|0\rangle. \tag{9.3.22}$$

One can check that it is normed to unity ($\sum_m P_m = 1$): probability is conserved, as a direct consequence of the unitarity of the S matrix.

9.3.3 The Klein–Gordon field coupled to a classical source

Next we tackle the problem of a Klein–Gordon field coupled to a classical source $j(x)$. The Lagrangean density is

$$\mathscr{L} = \frac{1}{2}(\partial_\mu \varphi)(\partial^\mu \varphi) - \frac{1}{2} m^2 \varphi^2 + j\varphi,$$

leading to the equation of motion

$$(\Box + m^2)\varphi(x) = j(x), \tag{9.3.23}$$

and to the interaction Hamiltonian

$$H_1 = -\int d^3x\, j(x)\varphi(x),$$

where the field $\varphi(x)$ is in the Heisenberg picture. The source $j(x)$ is supposed to vanish for $t < T_1$, at which times $\varphi(t, \boldsymbol{x})$ is a free field, to be written accordingly as $\varphi_{\text{in}}(x)$:

$$\varphi(t, \boldsymbol{x}) = \varphi_{\text{in}}(t, \boldsymbol{x}), \quad t < T_1.$$

Start at some time $t_1 < T_1$, and consider the time-evolution of $\varphi(t, \boldsymbol{x})$:

$$\begin{aligned}\varphi(t, \boldsymbol{x}) &= U^{-1}(t, t_1)\varphi(t_1, \boldsymbol{x})U(t, t_1)\\ &= U_{\text{I}}^{-1}(t, t_1)U_0^{-1}(t, t_1)\varphi_{\text{in}}(t_1, \boldsymbol{x})U_0(t, t_1)U_{\text{I}}(t, t_1);\end{aligned}$$

given that the evolution operator of the free field φ_{in} is $U_0(t, t_1)$, this entails

$$\varphi(t, \boldsymbol{x}) = U_{\text{I}}^{-1}(t, t_1)\varphi_{\text{in}}(t, \boldsymbol{x})U_{\text{I}}(t, t_1). \tag{9.3.24}$$

The Hamiltonian H_1 in the interaction picture becomes

$$H_{\text{I}}(t) = -\int d^3x\, j(x)\varphi_{\text{in}}(x). \tag{9.3.25}$$

Since $\varphi_{\text{in}}(x)$ is a free field, we can use the Fourier decomposition (9.2.12). Defining the three-dimensional Fourier transform of $j(t, \boldsymbol{x})$ by

$$j(t, \boldsymbol{k}) = j^*(t, -\boldsymbol{k}) = \int d^3x\, e^{-i\boldsymbol{k}\cdot\boldsymbol{x}} j(t, \boldsymbol{x}),$$

we rewrite (9.2.25) as a sum of Fourier components,

$$H_{\mathrm{I}}(t) = -\int \frac{\mathrm{d}^3 k}{(2\pi)^3 2\omega_k}\,(a_{\mathrm{in}}(\boldsymbol{k}) j^*(t,\boldsymbol{k})\mathrm{e}^{-\mathrm{i}\omega_k t} + a^\dagger_{\mathrm{in}}(\boldsymbol{k}) j(t,\boldsymbol{k})\mathrm{e}^{\mathrm{i}\omega_k t}). \tag{9.3.26}$$

This is a sum of Hamiltonians of the type (9.3.14), whence the evolution operator $U_{\mathrm{I}}(t)$ can be expressed analogously to (9.3.15). All we need is the matrix $S = U_{\mathrm{I}}(\infty, -\infty)$, though we could equally well compute $U_{\mathrm{I}}(t_2, t_1)$; note that

$$\int \mathrm{d}\tilde{k}\, a_{\mathrm{in}}(\boldsymbol{k}) j^*(t,\boldsymbol{k})\mathrm{e}^{-\mathrm{i}\omega_k t} = \int \mathrm{d}^3 x\, \varphi^{(+)}_{\mathrm{in}}(x) j(x),$$

and that the double integral in (9.3.15) reduces to

$$\begin{aligned}
&\int \mathrm{d}t\,\mathrm{d}t'\,\mathrm{d}\tilde{k}\,\theta(t-t') j^*(t,\boldsymbol{k})\mathrm{e}^{-\mathrm{i}\omega_k (t-t')} j(t',\boldsymbol{k}) \\
&= \int \mathrm{d}^4 x\,\mathrm{d}^4 x'\, j(x) j(x') \int \mathrm{d}\tilde{k}\,\theta(t-t')\mathrm{e}^{-\mathrm{i}\omega_k(t-t') + \mathrm{i}\boldsymbol{k}\cdot(\boldsymbol{x}-\boldsymbol{x}')} \\
&= \frac{1}{2}\int \mathrm{d}^4 x\,\mathrm{d}^4 x'\, j(x) j(x') \int \mathrm{d}\tilde{k} \\
&\quad \times [\theta(t-t')\mathrm{e}^{-\mathrm{i}\omega_k(t-t') + \mathrm{i}\boldsymbol{k}\cdot(\boldsymbol{x}-\boldsymbol{x}')} + \theta(t'-t)\mathrm{e}^{-\mathrm{i}\omega_k(t'-t) + \mathrm{i}\boldsymbol{k}\cdot(\boldsymbol{x}'-\boldsymbol{x})}] \\
&= \frac{1}{2}\int \mathrm{d}^4 x\,\mathrm{d}^4 x'\, j(x) \Delta_{\mathrm{F}}(x-x') j(x').
\end{aligned}$$

Collecting these results, we may write the S matrix in the form

$$\begin{aligned}
S &= T\left(\exp\left(\mathrm{i}\int \mathrm{d}^4 x\, j(x)\varphi_{\mathrm{in}}(x)\right)\right) \\
&= :\exp\left(\mathrm{i}\int \mathrm{d}^4 x\, j(x)\varphi_{\mathrm{in}}(x)\right): \\
&\quad \times \exp\left(-\frac{1}{2}\int \mathrm{d}^4 x\,\mathrm{d}^4 x'\, j(x)\Delta_{\mathrm{F}}(x-x') j(x')\right).
\end{aligned} \tag{9.3.27}$$

This expression is important partly because it prescribes the S matrix for coupling to a classical source, which is an interesting result in itself. But the chief reason for our concern with (9.3.27) is that it leads to a very simple proof of Wick's theorem.

9.3.4 Wick's theorem

Wick's theorem is derived simply by expanding the exponentials on both sides of (9.3.27), and by equating coefficients of powers of $j(x)$. Some care is needed, however: the terms encountered in the expansion of

$$T\left(e^{i\int d^4x\, j(x)\varphi_{in}(x)}\right),$$

for instance the term

$$\frac{(i)^2}{2!}\int d^4x_1\, d^4x_2\, j(x_1)j(x_2)T(\varphi_{in}(x_1)\varphi_{in}(x_2)),$$

are symmetric under the exchange of the arguments of φ_{in} and of j. Therefore the expansion on the right of (9.3.27) should also be symmetrized. In fact, from

$$\int dx_1\, dx_2\, j(x_1)j(x_2)T(x_1, x_2) = \int dx_1\, dx_2\, j(x_1)j(x_2)F(x_1, x_2),$$

with $T(x_1, x_2) = T(x_2, x_1)$, we readily conclude that

$$T(x_1, x_2) = F_S(x_1, x_2),$$

where $F_S(x_1, x_2)$ is that part of F which is symmetric under interchanging (x_1, x_2); the integral over the antisymmetric part $F_A(x_1, x_2)$ vanishes automatically.

Consider now a typical term in the expansion of the right-hand side of (9.3.27), featuring $(j(x))^p(j(x))^{2m}$; it is prefaced by the coefficient

$$i^p\left(-\frac{1}{2}\right)^m\frac{1}{p!\,m!}.$$

The coefficient of $(j(x))^p(j(x))^{2m}$ must be symmetrized in all its indices. The number of independent terms is

$$\frac{(2m+p)!}{p!\,m!\,2^m},$$

and the coefficient of $(j(x))^{2m+p}$ on the right of (9.3.27) is

$$i^p\left(-\frac{1}{2}\right)^m\frac{1}{m!p!}\frac{2^m p!\,m!}{(2m+p)!} = \frac{(i)^{2m+p}}{(2m+p)!}.$$

This is precisely the coefficient found by expanding the T-product. When $(2m+p) = 2n$ is even, equating the coefficients (and dropping the index 'in') yields

$$\begin{aligned} T(\varphi(x_1)\ldots\varphi(x_{2n})) = {} & :\varphi(x_1)\ldots\varphi(x_{2n}): \\ & + \{:\overbrace{\varphi(x_1)\varphi}(x_2)\varphi(x_3)\ldots\varphi(x_{2n}): + \text{perm.}\} \\ & + :\overbrace{\varphi(x_1)\varphi}(x_2)\overbrace{\varphi(x_3)\varphi}(x_4)\varphi(x_5)\ldots\varphi(x_{2n}): + \text{perm.}\} \\ & + \{\cdots\} + \cdots + \{\cdots\} \\ & + \{\overbrace{\varphi(x_1)\varphi}(x_2)\ldots\overbrace{\varphi(x_{2n-1})\varphi}(x_{2n}) + \text{perm.}\}, \end{aligned} \tag{9.3.28}$$

where the 'contraction' $\wick{\c\varphi(x_1)\c\varphi(x_2)}$ is defined by

$$\wick{\c\varphi(x_1)\c\varphi(x_2)} = \langle 0|\,T(\varphi_{\text{in}}(x_1)\varphi_{\text{in}}(x_2))|0\rangle = \Delta_{\text{F}}(x_1 - x_2).$$

When $2n = 4$ for example, (9.3.28) reads

$$\begin{aligned} T(\varphi_1\varphi_2\varphi_3\varphi_4) = {} & :\varphi_1\varphi_2\varphi_3\varphi_4: + \{:\wick{\c\varphi_1\c\varphi_2}\varphi_3\varphi_4: \\ & + :\wick{\c\varphi_1\varphi_2\c\varphi_3}\varphi_4: + :\wick{\c\varphi_1\varphi_2\varphi_3\c\varphi_4}: + :\varphi_1\wick{\c\varphi_2\c\varphi_3}\varphi_4: \\ & + :\varphi_1\wick{\c\varphi_2\varphi_3\c\varphi_4}: + :\varphi_1\varphi_2\wick{\c\varphi_3\c\varphi_4}:\} + \{\wick{\c\varphi_1\c\varphi_2}\wick{\c\varphi_3\c\varphi_4} \\ & + \wick{\c1\varphi_1\c2\varphi_2\c1\varphi_3\c2\varphi_4} + \wick{\c1\varphi_1\c2\varphi_2\c2\varphi_3\c1\varphi_4}\}. \end{aligned} \tag{9.3.29}$$

Note that $\wick{\c\varphi_1\c\varphi_2}$ for instance can be taken outside the normal product, because it is just a number.

Wick's theorem applies equally when some products are already in normal form, provided we drop all contractions between terms that are inside any given normal product from the outset. A simple illustration reads

$$\begin{aligned} T(:\varphi_1\varphi_2::\varphi_3\varphi_4:) = {} & :\varphi_1\varphi_2\varphi_3\varphi_4: + \{:\wick{\c\varphi_1\varphi_2\c\varphi_3}\varphi_4: \\ & + :\wick{\c\varphi_1\varphi_2\varphi_3\c\varphi_4}: + :\varphi_1\wick{\c\varphi_2\c\varphi_3}\varphi_4: + :\varphi_1\wick{\c\varphi_2\varphi_3\c\varphi_4}:\} \\ & + \{\wick{\c1\varphi_1\c2\varphi_2\c1\varphi_3\c2\varphi_4} + \wick{\c1\varphi_1\c2\varphi_2\c2\varphi_3\c1\varphi_4}\}. \end{aligned} \tag{9.3.30}$$

Finally we evaluate the vacuum expectation-value of (9.3.28). Because of the normal products only the last term on the right is nonzero:

$$\begin{aligned} & \langle 0|\,T(\varphi(x_1)\ldots\varphi(x_{2n}))|0\rangle \\ & \qquad = \Delta_{\text{F}}(x_1 - x_2)\ldots\Delta_{\text{F}}(x_{2n-1} - x_{2n}) + \text{perm.} \end{aligned} \tag{9.3.31}$$

Notice the striking similarity to equation (5.1.13) which we derived earlier for Gaussian integrals. The next chapter will exploit this similarity to prove the equivalence between canonical and path-integral quantization.

Problems

9.1 Quantization on the 'discrete line'

We aim to quantize the mechanical system described by the Lagrangean L in equation (9.1.4), subject to periodic boundary conditions.

(a) Show that the normal modes are

$$\varphi_k = \frac{1}{\sqrt{N}} \sum_s \varphi_s(t) \mathrm{e}^{-iks},$$

where $k = 2\pi p/N$, $p = 0, 1, 2, \ldots, N-1$, and that $\varphi_k(t)$ obeys the differential equation

$$\ddot{\varphi}_k + \omega_k^2 \varphi_k = 0; \quad \omega_k = \sqrt{\left(\frac{4K}{m}\right)} \left|\sin\frac{k}{2}\right|.$$

(b) Use the Fourier decomposition

$$\varphi_s = \frac{1}{\sqrt{N}} \sum_k \frac{1}{\sqrt{(2m\omega_k)}} (a_k \mathrm{e}^{-\mathrm{i}\omega_k t + iks} + a_k^\dagger \mathrm{e}^{\mathrm{i}\omega_k t - iks})$$

to verify the CCR, and to show that

$$H = \sum_k \frac{1}{2} \omega_k (a_k^\dagger a_k + a_k a_k^\dagger).$$

9.2 Use the relations (9.1.20) at $t = 0$, and the CCR (9.1.15), to show that $[A_k, A_{k'}^\dagger] = (2\omega_k)^{-1} \delta_{kk'}$.

9.3 Derive the expression (9.1.24) for the Hamiltonian. *Hint*: determine the Fourier coefficients π_k and $(\partial\varphi/\partial x)_k$ of π and $\partial\varphi/\partial x$ respectively, and use Parseval's theorem.

9.4 Consider a Lorentz transformation in the z-direction:

$$k_0' = k_0 \cosh\varphi - k_z \sinh\varphi,$$

$$k_z' = -k_0 \sinh\varphi + k_z \cosh\varphi, \quad k_x' = k_x, \quad k_y' = k_y.$$

Verify that the integration measure $\mathrm{d}\tilde{k}$ in (9.2.8) is invariant under this transformation.

9.5 Derive equation (9.2.10); check that the solutions with positive energy are orthogonal to those with negative energy and show that the scalar product (g, f) is constant in time:

$$\frac{\partial}{\partial t}(g, f) = 0.$$

(Use the Klein–Gordon equation and Green's theorem.)

9.6 Verify the CCR between φ and $\pi = \dot{\varphi}$, starting from (9.2.12) and (9.2.13).

9.7 Green's functions of the Klein–Gordon equation

Evaluate (9.2.20) using the contours C_R and C_A from Fig. 9.4. Show that $G_R(x) = -\theta(x^0)\Delta(x)$, and that $G_A(x) = \theta(-x^0)\Delta(x)$. Hence derive

$$G_R(x - y) = i\langle 0|\,\theta(x^0 - y^0)[\varphi(x), \varphi(y)]|0\rangle,$$

$$G_A(x - y) = -i\langle 0|\,\theta(-x^0 + y^0)[\varphi(x), \varphi(y)]|0\rangle.$$

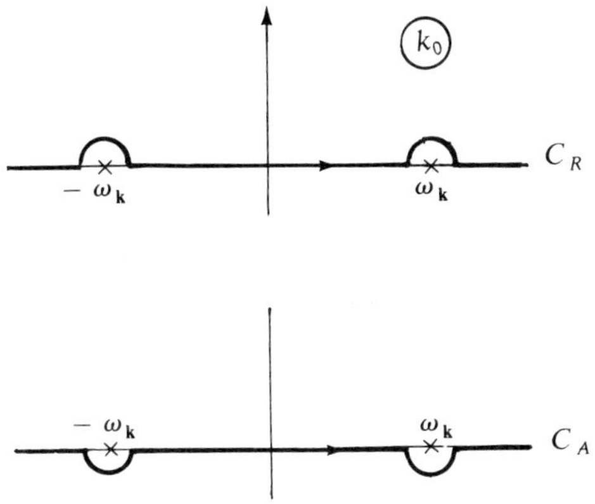

Fig. 9.4 The contours C_A and C_R

9.8 By calculating the derivatives explicitly, check that

$$(\Box_x + m^2)\langle 0|\,T(\varphi(x)\varphi(0))|0\rangle = -i\delta^{(4)}(x).$$

Hint: use the Klein–Gordon equation and note that $(\partial/\partial x_0)\theta(x^0) = \delta(x^0)$.

9.9 Some useful operator identities

(a) A and B are arbitrary operators. Show that

$$e^A B e^{-A} = B + [A, B] + \frac{1}{2!}[A, [A, B]] + \cdots.$$

Hint: consider the Taylor expansion of $F(t) = \exp(tA)B\exp(-tA)$ around $t = 0$.

(b) Let A and B be two operators such that both commute with $[A, B]$. Prove that

$$\exp(A + B) = \exp(A)\exp(B)\exp(-[A, B]/2).$$

Hint: Find a differential equation for $F(t) = \exp(At)\exp(Bt)$. Thence derive $\exp(A)\exp(B) = \exp(B)\exp(A)\exp([A, B])$.

(*Caution*: the two identities just above are not general truths, i.e. they are not satisfied by arbitrary A and B.)

(c) Application: $e^{\alpha a^\dagger} e^{\beta a} e^{-\alpha a^\dagger} = e^{-\alpha\beta} e^{\beta a}$.

9.10 Other forms of equation (9.3.15)

Equation (9.3.15) may be written

$$U_{\mathrm{I}}(t) = e^{i\alpha a^\dagger} e^{i\alpha^* a} e^{-X}; \quad \alpha = \int_0^t j(t')e^{i\omega t'}\,dt',$$

where X is a double integral.

(a) Let $A_1, \ldots, A_n$ be n operators such that $[A_i, A_j]$ is just a number. Prove that

$$e^{A_n} e^{A_{n-1}} \ldots e^{A_1} = e^{A_n + \cdots + A_1} e^{\frac{1}{2}\sum_{j>i}[A_j, A_i]}.$$

(b) By subdividing the interval $[0, t]$ into n short subintervals each of length Δt, show that

$$U_{\mathrm{I}}(t) \simeq \exp\left(-i\sum_{j=1}^{n} H_{\mathrm{I}}(t_j)\Delta t\right)\exp\left(-\frac{1}{2}(\Delta t)^2 \sum_{j>i}[H_{\mathrm{I}}(t_j), H_{\mathrm{I}}(t_i)]\right);$$

thence derive

$$U_{\mathrm{I}}(t) = e^{i(\alpha a^\dagger + \alpha^* a)} e^{-X'},$$

$$X' = \frac{1}{2}\iint dt'\,dt''\,\varepsilon(t' - t'')j^*(t')e^{-i\omega(t'-t'')}j(t'').$$

(c) Show that this expression for $U_{\mathrm{I}}(t)$ coincides with (9.3.15) as it should.

9.11 Field coupled to a classical source

Consider a Hermitean scalar field $\varphi(x)$ coupled to a classical field:

$$(\Box + m^2)\varphi(x) = j(x).$$

(a) Show that the 'in' and 'out' fields are related through

$$\varphi_{\mathrm{out}}(x) = \varphi_{\mathrm{in}}(x) - \int \Delta(x - x')j(x')\,d^4x'.$$

(See Problem 9.7; and assume that $j(x)$ vanishes outside some interval $[T_1, T_2]$.)

(b) Use the Fourier decomposition

$$j(x) = \int \frac{d^4k}{(2\pi)^4}\, e^{-ikx} j(k)$$

of $j(x)$ to show that the operators a_{in} and a_{out} are related through

$$a_{out}(\boldsymbol{k}) = a_{in}(\boldsymbol{k}) + i j(\omega_{\boldsymbol{k}}, \boldsymbol{k}).$$

(c) Defining the S matrix by

$$S^{-1} a_{in}(\boldsymbol{k}) S = a_{out}(\boldsymbol{k}),$$

show that, up to a phase factor,

$$S = \exp\left\{ i \int \frac{d^3k}{(2\pi)^3 2\omega_{\boldsymbol{k}}} (a_{in}(\boldsymbol{k}) j^*(\omega_{\boldsymbol{k}}, \boldsymbol{k}) + a_{in}^\dagger(\boldsymbol{k}) j(\omega_{\boldsymbol{k}}, \boldsymbol{k})) \right\}.$$

Show that this expression differs from $U_I(\infty, -\infty)$ (equation (9.3.27)) by a phase factor, and determine this factor.

9.12 Coherent states (Cohen–Tannoudji *et al.* 1977, Chapter 5)

Let $|z\rangle$ be the 'coherent state' $\exp(a^\dagger z)|0\rangle$, where z is a complex number.

(a) Show that $|z\rangle$ is an eigenvector of a: $a|z\rangle = z|z\rangle$, and that

$$\langle z|z\rangle = \exp(|z|^2),$$

where $z = x + iy$.

(b) Show that $\int \frac{dx\, dy}{\pi} e^{-|z|^2} |z\rangle\langle z| = \mathbb{1}$. (This is the closure relation.)

(c) Let $D(z) = \exp(za^\dagger - z^* a)$. Show that $D(z)|0\rangle$ is a coherent state normed to unity. Calculate $\langle x|D(z)|0\rangle$, where $|x\rangle$ is an eigenvector of the position operator.

(d) Show that $y^{a^\dagger a}|z\rangle = |yz\rangle$; thence derive

$$x^{a^\dagger a} = :e^{(x-1)a^\dagger a}:.$$

9.13 Short-distance behaviour of a product of currents

Let $\varphi(x)$ be a scalar massless free field and set $j(x) = :\varphi^2(x):$ ($j(x)$ is the prototype of a current, an operator bilinear in the fields).

(a) Show, using Wick's theorem, that

$$T(j(x)j(0)) = \frac{1}{8\pi^4(x^2 - i\varepsilon)^2} + \frac{:\varphi(x)\varphi(0):}{\pi^2(x^2 - i\varepsilon)} + :\varphi^2(x)\varphi^2(0):$$

(b) Compute $[j(x), j(0)]$. *Hint*: use the identity

$$\frac{1}{(x^2 - i\varepsilon)^n} - \frac{1}{(x^2 + i\varepsilon)^n} = 2i\pi \frac{(-1)^{n-1}}{(n-1)!} \delta^{(n-1)}(x^2)$$

Further reading

The mechanical system featured in Section 9.1 is described in detail by Goldstein (1980). An introduction to the quantization of elastic vibrations and of the electromagnetic field is given by C. Cohen-Tannoudji *et al.* (1977, Chapter 5). The quantization of the Klein–Gordon field is discussed in all the classic texts, e.g. by Bjorken and Drell (1965, Chapter 12), or by Itzykson and Zuber (1980, Chapter 3). For the Lagrangean formalism for classical fields, and the construction of the energy–momentum and of the angular momentum operators, see e.g. Bjorken and Drell (1965, Chapter 11). Wightman's programme (axiomatic field theory) is set out by Streater and Wightman (1964). For the various 'pictures' and their evolution operators, see e.g. Messiah, (1961, Chapter 8). The coupling of a quantized field (the electromagnetic field) to a classical source is discussed by Bjorken and Drell (1965, Chapter 17), and by Itzykson and Zuber (1980, Chapter 4).

10 Green's functions and the S matrix

In this chapter we study the Klein–Gordon field interacting with itself. This model has very limited physical applications, but it allows us to introduce several important ideas without having to use the more complicated formalism for particles of nonzero spin. The Lagrangean density $\mathscr{L}$ reads

$$\mathscr{L} = \frac{1}{2}(\partial_\mu \varphi)(\partial^\mu \varphi) - \frac{1}{2} m^2 \varphi^2 - V(\varphi), \tag{10.0.1}$$

where $V(\varphi)$ is a polynomial in φ and its derivatives. (By an abuse of language, $\mathscr{L}$ is often referred to as the Lagrangean; the notation $\mathscr{L}$ instead of L suffices to resolve any ambiguity.) When V involves the derivatives $\partial_\mu \varphi$ one is faced with *derivative couplings*; these lead to complications in the canonical formalism, because the conjugate momentum $\pi = \partial \mathscr{L} / \partial \dot{\varphi}$ then depends on V. Until further notice we assume that V contains no derivative couplings, which will be studied later in Section 10.5. The first two terms in (10.0.1), denoted by $\mathscr{L}_0$, constitute the *free Lagrangean*, while $-V(\varphi)$, sometimes denoted by $\mathscr{L}_1$, is the *interaction Lagrangean.* The Hamiltonian is subdivided similarly: $H = H_0 + H_1$. The equations of motion derived from (9.2.3b) read

$$(\Box + m^2)\varphi(x) = -V'(\varphi(x)). \tag{10.0.2}$$

Our favourite example will be 'φ^4 theory', where

$$\mathscr{L}_1 = -\frac{g}{4!}\varphi^4(x), \tag{10.0.3a}$$

very reminiscent, evidently, of Ginzburg–Landau theory. The connection between the two theories will be examined in Section 10.2. Note that g must be positive if one requires the Hamiltonian to be positive definite. Occasionally one uses instead of (10.0.3a) the normal-ordered interaction

$$\mathscr{L}'_1 = -\frac{g}{4!}:\varphi^4(x):. \tag{10.0.3b}$$

Using (10.0.3b) instead of (10.0.3a) is a matter of taste and of opportunity, the differences being absorbed by the mass renormalization. As a rule we shall not use the normal product if we need to subject φ to inhomogeneous transformations (e.g. to translations) while maintaining certain symmetries. Instead of 'φ^4

theory' we shall on occasion use the example of 'φ^3 theory', where

$$\mathscr{L}_1 = -\frac{g}{3!}\varphi^3(x). \tag{10.0.4}$$

Under this theory in the number of dimensions where it is renormalizable, namely $D = 6$ (see Problem 6.2), field renormalization (related to Z_3) is non-trivial already to one-loop order. By contrast, in φ^4 theory one must wait until two-loop order for Z_3 to differ from 1. Of course φ^3 theory is less realistic even than φ^4 theory; moreover it is pathological in that its Hamiltonian is not positive definite. Nevertheless it is perfectly well defined in perturbation theory.

Our first aim (in Sections 10.1 and 10.2) is to determine the perturbative expansions of the Green's functions, defined as vacuum expectation-values (vacuum = ground state) of T-products of the fields:

$$\langle 0|\, T(\varphi(x_1)\, .\,.\,.\, \varphi(x_N))|0\rangle.$$

The Feynman rules for these perturbation expansions, obtained from Wick's theorem, will enable us to relate the Green's functions to the correlation functions from Part II, and to set up the path-integral formalism for quantization.

Our second aim is to relate the Green's functions to observables. Collisions are described by means of the S matrix, and we shall need to establish a relation between Green's functions and S-matrix elements (Section 10.3). The important unitarity property of the S matrix is studied in Section 10.4, and some generalizations (complex scalar field, vector field, and derivative couplings) in Section 10.5.

10.1 Perturbation expansion of the Green's functions

10.1.1 The *S*-matrix in the interaction picture

In the last chapter we used the formula (9.3.24), which relates the field $\varphi(t, \boldsymbol{x})$ interacting with a classical source to the incoming field $\varphi_{\text{in}}(t, \boldsymbol{x})$. This formula was easy to derive provided the source vanishes outside some finite-time interval. Similarly we had no difficulty in finding the S-matrix element between an initial and a final state, each containing a given number of particles.

Now we wish to deal with a self-interacting field, and this is a far more complicated problem because the interaction $V(\varphi)$ never vanishes. In the last chapter, it was natural (but by no means necessary) to identify the field $\varphi(t, \boldsymbol{x})$ with a free field $\varphi_{\text{in}}(t, \boldsymbol{x})$ as $t \to -\infty$. Now there is no reason to privilege $t \to -\infty$, and we shall prefer to make the identification at some finite time. If we were dealing with a system having a finite number of degrees of freedom, $\varphi(x)$ and $\varphi_{\text{in}}(x)$ would be connected by a unitary transformation

$$\varphi(t, \boldsymbol{x}) = U^{-1}(t)\varphi_{\text{in}}(t, \boldsymbol{x})U(t), \tag{10.1.1}$$

by virtue of von Neumann's theorem. Unfortunately, in a relativistic quantum field theory, according to a theorem due to Haag (see Streater and Wightman 1964, Chapter 4), the existence of such a unitary transformation implies that $\varphi(x)$ must be a free field. Here we shall ignore this difficulty and accept that, formally at least, one can proceed from equation (10.1.1), even though we must bear in mind that the 'derivations' which follow are merely heuristic.

We choose a particular time, say $t = 0$, and assume that at $t = 0$ the fields $\varphi(0, \boldsymbol{x})$ and $\varphi_{\text{in}}(0, \boldsymbol{x})$ coincide, as well as their conjugate momenta. In other words, we choose boundary conditions on the free field $\varphi_{\text{in}}(x)$ as follows:

$$\varphi_{\text{in}}(0, \boldsymbol{x}) = \varphi(0, \boldsymbol{x}), \tag{10.1.2a}$$

$$\partial_0 \varphi_{\text{in}}(0, \boldsymbol{x}) = \partial_0 \varphi(0, \boldsymbol{x}). \tag{10.1.2b}$$

For simplicity we have excluded derivative interactions, so that the momentum conjugate to φ is simply its time-derivative. Because φ and φ_{in} both obey the CCR at $t = 0$, the choice (10.1.2) for φ_{in} is always possible. The Heisenberg equations of motion read

$$\varphi(t, \boldsymbol{x}) = \mathrm{e}^{\mathrm{i}Ht} \varphi(0, \boldsymbol{x}) \mathrm{e}^{-\mathrm{i}Ht}, \tag{10.1.3a}$$

$$\varphi_{\text{in}}(t, \boldsymbol{x}) = \mathrm{e}^{\mathrm{i}H_0 t} \varphi_{\text{in}}(0, \boldsymbol{x}) \mathrm{e}^{-\mathrm{i}H_0 t}; \tag{10.1.3b}$$

here H is the full Hamiltonian and H_0 the Hamiltonian for a free field with mass m, where m is the *physical* mass of the particles associated with our neutral scalar field. This means that the interaction Hamiltonian H_1 must contain mass counterterms. From (10.1.2) and (10.1.3) we obtain

$$\varphi(x) = \mathrm{e}^{\mathrm{i}Ht} \mathrm{e}^{-\mathrm{i}H_0 t} \varphi_{\text{in}}(x) \mathrm{e}^{\mathrm{i}H_0 t} \mathrm{e}^{-\mathrm{i}Ht},$$

and an analogous equation for $\pi(x)$. We see that $\varphi(x)$ and $\varphi_{\text{in}}(x)$ are indeed connected by a unitary transformation which we denote by $U(t, 0)$,

$$U(t, 0) = \mathrm{e}^{\mathrm{i}H_0 t} \mathrm{e}^{-\mathrm{i}Ht}, \tag{10.1.4}$$

$$\varphi(x) = U^{-1}(t, 0) \varphi_{\text{in}}(x) U(t, 0). \tag{10.1.5}$$

As usual we have the unitary and group composition properties:

$$U^{-1}(t, 0) = U^{\dagger}(t, 0) = U(0, t), \tag{10.1.6a}$$

$$U(t_1, t_2) = U(t_1, 0) U(0, t_2). \tag{10.1.6b}$$

A simple calculation shows that

$$\mathrm{i}\frac{\mathrm{d}}{\mathrm{d}t} U(t, 0) = \mathrm{e}^{\mathrm{i}H_0 t}[H(\varphi) - H_0(\varphi_{\text{in}})] \mathrm{e}^{-\mathrm{i}H_0 t} U(t, 0).$$

However, $H(\varphi)$ and $H_0(\varphi_{\text{in}})$ are time-independent, and we can make the subtraction in the square bracket at $t = 0$, which leaves us with

$$\mathrm{i}\frac{\mathrm{d}}{\mathrm{d}t} U(t, 0) = H_1(\varphi_{\text{in}}(x)) U(t, 0). \tag{10.1.7}$$

Thus $U(t, 0)$ is nothing but the evolution operator in the interaction picture (see (9.3.6)), which was denoted by U_{I} in the last chapter. The solution of (10.1.7) has already been written in (9.3.11),

$$U(t, 0) = T\left(\exp\left(-\mathrm{i} \int_0^t H_1(\varphi_{\mathrm{in}}(x'))\,\mathrm{d}t' \right)\right). \tag{10.1.8}$$

Since we have excluded derivative couplings, $H_1 = -L_1$ and (10.1.8) can be rewritten

$$U(t, 0) = T\left(\exp\left(\mathrm{i} \int_0^t \mathscr{L}_1(\varphi_{\mathrm{in}}(x'))\mathrm{d}^4x' \right)\right).$$

This result generalizes those of Section 9.3.3. We can also write the S-matrix, which is formally equal to $U(\infty, -\infty)$,

$$S = \lim_{t_2 \to \infty} \lim_{t_1 \to -\infty} \mathrm{e}^{\mathrm{i}H_0 t_2}\,\mathrm{e}^{-\mathrm{i}H(t_2 - t_1)}\,\mathrm{e}^{-\mathrm{i}H_0 t_1},$$

or

$$S = T\left(\exp\left(\mathrm{i} \int_{-\infty}^{\infty} \mathscr{L}_1(\varphi_{\mathrm{in}}(x))\,\mathrm{d}^4x \right)\right). \tag{10.1.9}$$

For example, in the framework of the φ^4 theory (10.0.3b), equation (10.1.9) reads

$$S = T\left(\exp\left(\frac{-\mathrm{i}g}{4!} \int \mathrm{d}^4x : \varphi_{\mathrm{in}}(x) : \mathrm{d}^4x \right)\right). \tag{10.1.10}$$

Let us emphasize that it is the *free* field $\varphi_{\mathrm{in}}(x)$ which appears in (10.1.9) and (10.1.10).

When the interaction $\mathscr{L}_1(\varphi)$ depends on derivatives of φ, the momentum conjugate to φ is not simply $\dot{\varphi}$, and H_1 does not equal $-L_1$. One example of such a situation is the electrodynamics of scalar particles (Itzykson and Zuber 1980, Chapter 6). Nevertheless, after several intermediate complications which demonstrate the unsuitability of canonical quantization to this kind of problem, one does eventually recover equation (10.1.9). By contrast, as we shall see in Section 10.5, the path-integral method leads to this result without any difficulty.

10.1.2 The formula of Gell-Mann and Low

We now derive a formula which makes possible the perturbative calculation of Green's functions defined by

$$G^{(N)}(x_1, \ldots, x_N) = \langle 0 | T(\varphi(x_1) \ldots \varphi(x_N)) | 0 \rangle, \tag{10.1.11}$$

where the φ are interacting fields, and $|0\rangle$ the ground state, or physical vacuum, of the interacting theory. We set $x_i^0 = t_i$ and start by studying the case with the times ordered in the sequence

$$t_1 > t_2 > \ldots > t_N,$$

expressing the $\varphi(x_i)$ as functions of the $\varphi_{\text{in}}(x_i)$ through (10.1.5) and using the group composition rule (10.1.6b),

$$\langle 0|T(\varphi(x_1)\ldots\varphi(x_N))|0\rangle$$
$$= \langle 0|U^{-1}(t_1,0)\varphi_{\text{in}}(x_1)U(t_1,t_2)\ldots\varphi_{\text{in}}(x_N)U(t_N,0)|0\rangle. \tag{10.1.12}$$

Equation (10.1.12) is written in terms of the free field $\varphi_{\text{in}}(x)$, but the expectation value is taken with the physical vacuum $|0\rangle$. Let us instead try to express the expectation value in terms of the *perturbative* vacuum $|0_{\text{in}}\rangle$, namely the ground state for the free field. In order to obtain the necessary relation, consider the following matrix element of $\exp(\mathrm{i}Ht)$,

$$\lim_{t\to\pm\infty}\langle\Psi|\mathrm{e}^{\mathrm{i}Ht}|\chi\rangle$$

where $|\Psi\rangle$ and $|\chi\rangle$ are two normalizable states chosen in the Fock space of the free field. Insert a complete set of eigenstates $|n\rangle$ of the full Hamiltonian H:

$$\lim_{t\to\pm\infty}\langle\Psi|\mathrm{e}^{\mathrm{i}Ht}|\chi\rangle = \lim_{t\to\pm\infty}\left\{\langle\Psi|0\rangle\langle 0|\chi\rangle + \sum_{n\neq 0}\mathrm{e}^{\mathrm{i}E_n t}\langle\Psi|n\rangle\langle n|\chi\rangle\right\} \tag{10.1.13}$$

If there are no massless particles in the theory, the second term in (10.1.13) oscillates rapidly when $t\to\pm\infty$. Thus it is reasonable, although not proven—but remember that our discussion is purely heuristic—to assume that we may drop this second term and write

$$\lim_{t\to\pm\infty}\langle\Psi|\mathrm{e}^{\mathrm{i}Ht}|\chi\rangle = \langle\Psi|0\rangle\langle 0|\chi\rangle,$$

for all states $|\Psi\rangle, |\chi\rangle$ belonging to the Fock space of the free field. We may write, in the sense of weak convergence, i.e. for matrix elements,

$$\lim_{t\to\pm\infty}\mathrm{e}^{\mathrm{i}Ht}|\chi\rangle = |0\rangle\langle 0|\chi\rangle.$$

In particular we have

$$|0\rangle = \lim_{t\to-\infty}\frac{\mathrm{e}^{\mathrm{i}Ht}|0_{\text{in}}\rangle}{\langle 0|0_{\text{in}}\rangle} = \lim_{t\to-\infty}\frac{U^\dagger(t,0)|0_{\text{in}}\rangle}{\langle 0|U^\dagger(t,0)|0_{\text{in}}\rangle}, \tag{10.1.14a}$$

$$|0\rangle = \lim_{t\to\infty}\frac{U^\dagger(t,0)|0_{\text{in}}\rangle}{\langle 0|U^\dagger(t,0)|0_{\text{in}}\rangle}. \tag{10.1.14b}$$

We may thus cast (10.1.12) into

$$\langle 0|\varphi(x_1)\ldots\varphi(x_N)|0\rangle$$
$$= \lim_{t\to\infty}\frac{\langle 0_{\text{in}}|U(t,t_1)\varphi_{\text{in}}(x_1)U(t_1,t_2)\ldots\varphi_{\text{in}}(x_N)U(t_N,-t)|0_{\text{in}}\rangle}{\langle 0_{\text{in}}|U(t,0)|0\rangle\langle 0|U^\dagger(-t,0)|0_{\text{in}}\rangle}.$$

The denominator may be transformed by using again the argument following (10.1.13):

$$\lim_{t\to\infty}\sum_n \langle 0_{\text{in}}|\mathrm{e}^{\mathrm{i}H_0t}\,\mathrm{e}^{-\mathrm{i}Ht}|n\rangle\langle n|\mathrm{e}^{-\mathrm{i}Ht}\,\mathrm{e}^{\mathrm{i}H_0t}|0_{\text{in}}\rangle$$
$$=\lim_{t\to\infty}\langle 0_{\text{in}}|U(t,0)U^{\dagger}(-t,0)|0_{\text{in}}\rangle=\lim_{t\to\infty}\langle 0_{\text{in}}|U(t,\,-t)|0_{\text{in}}\rangle.$$

One recognizes the (perturbative) vacuum to vacuum element of the S-matrix in the interaction picture. Now the numerator can be written as

$$\lim_{t\to\infty}\langle 0_{\text{in}}|T\left(\varphi_{\text{in}}(x_1)\ldots\varphi_{\text{in}}(x_N)\mathrm{e}^{-\mathrm{i}\int_{-t}^{t}\mathrm{d}t'H_1(\varphi_{\text{in}}(x'))}\right)|0_{\text{in}}\rangle.$$

In order to obtain this formula, we have used the explicit solution (10.1.8) for $U(t,0)$ as well as

$$T\left(A(t)\mathrm{e}^{-\mathrm{i}\int_{t_1}^{t_2}H_1(t')\,\mathrm{d}t'}\right)=T\left(\mathrm{e}^{-\mathrm{i}\int_{t}^{t_2}H_1(t')\,\mathrm{d}t'}\right)A(t)\,T\left(\mathrm{e}^{-\mathrm{i}\int_{t_1}^{t}H_1(t')\,\mathrm{d}t'}\right),$$

valid for any operator $A(t)$ when $t_1\leqslant t\leqslant t_2$, from the very definition of the T-product. One notes that if $A(t)=1$, this equation is nothing but the group law (9.3.3). Thanks to the T-product, one can now take any ordering of the times $(t_1,\ldots,t_N)$, and one writes the final result, namely the formula of Gell-Mann and Low, as

$$G^{(N)}(x_1,\ldots,x_N)=\frac{\langle 0_{\text{in}}|T\left(\varphi_{\text{in}}(x_1)\ldots\varphi_{\text{in}}(x_N)\mathrm{e}^{\mathrm{i}\int\mathrm{d}^4x\,\mathscr{L}_1(\varphi_{\text{in}}(x))}\right)|0_{\text{in}}\rangle}{\langle 0_{\text{in}}|T\left(\mathrm{e}^{\mathrm{i}\int\mathrm{d}^4x\,\mathscr{L}_1(\varphi_{\text{in}}(x))}\right)|0_{\text{in}}\rangle}.\tag{10.1.15}$$

Starting from (10.1.15), it is easy to construct a generating functional for the Green's functions. Define the functional

$$Z(j)=\langle 0_{\text{in}}|T\left(\exp\left(\mathrm{i}\int\mathrm{d}^4x(\mathscr{L}_1(\varphi_{\text{in}}(x))+j(x)\varphi_{\text{in}}(x))\right)\right)|0_{\text{in}}\rangle.\tag{10.1.16}$$

It is obvious that $G^{(N)}$ then follows by functional differentiation,

$$G^{(N)}(x_1,\ldots,x_N)=\frac{(-\mathrm{i})^N}{Z(0)}\left.\frac{\delta^{(N)}Z(j)}{\delta j(x_1)\ldots\delta j(x_N)}\right|_{j=0},\tag{10.1.17}$$

and that $Z(j)/Z(0)$ can also be written as

$$Z(j)/Z(0)=\langle 0|T\left(\exp\left(\mathrm{i}\int\mathrm{d}^4x\,j(x)\varphi(x)\right)\right)|0\rangle.\tag{10.1.18}$$

Notice that (10.1.16) features the free field φ_{in}, while (10.1.18) features the interacting field φ.

10.1.3 The perturbation expansion

The importance of the formulae (10.1.15) and (10.1.16) stems from the fact that they allow the Green's functions to be expanded perturbatively. Thus, expanding the exponential in the numerator of (10.1.15), we find

$$\langle 0_{\text{in}}|T\left(\varphi_{\text{in}}(x_1)\dots\varphi_{\text{in}}(x_N)\mathrm{e}^{\mathrm{i}\int \mathrm{d}^4y\,\mathscr{L}_1(\varphi_{\text{in}}(y))}\right)|0_{\text{in}}\rangle$$
$$= \sum_{p=0}^{\infty}\frac{\mathrm{i}^p}{p!}\int \mathrm{d}^4y_1\dots\mathrm{d}^4y_p\langle 0_{\text{in}}|T(\varphi_{\text{in}}(x_1)\dots\varphi_{\text{in}}(x_N)$$
$$\times\mathscr{L}_1(\varphi_{\text{in}}(y_1))\dots\mathscr{L}_1(\varphi_{\text{in}}(y_p)))|0_{\text{in}}\rangle. \tag{10.1.19}$$

The vacuum expectation-values in (10.1.15) can be evaluated with the help of Wick's theorem: to such expectation values only the last term of (9.3.28) gives a nonzero contribution. Moreover, this contribution has exactly the structure (5.1.13) of an average over a Gaussian distribution. Accordingly we can apply the Feynman rules established in Chapter 5, provided we make two changes:

(i) Replace the contraction in (5.1.13) by that in (9.3.31), i.e. by $\Delta_{\text{F}}(y_i - y_j)$.

(ii) For the φ^4 model from Chapter 5, the perturbation expansion stems from the expansion of the exponential

$$\mathrm{e}^{-\frac{g}{4!}\int \mathrm{d}^4y\varphi^4(y)} = \sum_p\left(\frac{-g}{4!}\right)^p\frac{1}{p!}\int \mathrm{d}^4y_1\dots\mathrm{d}^4y_p\varphi^4(y_1)\dots\varphi^4(y_p),$$

where p is the order of perturbation theory. To each vertex we assign a factor $-g$. In our present case, if the Lagrangean is given by (10.0.3a), then the pth-order term of the perturbation series features a factor $(-\mathrm{i}g)^p$, and to each vertex we must assign a factor $-\mathrm{i}g$.

The perturbative expansion of the denominator in (10.1.15) yields vacuum-to-vacuum diagrams, whose vertices are not linked to any external point. They play the same role as did the terms in the expansion of $Z(0)$ in (5.1.16) (whence the terminology we introduced in Chapter 5). To take the denominator into account, we need merely eliminate (as in Chapter 5) all diagrams containing any vacuum-to-vacuum parts. Thus we can formulate the x-space Feynman rules for the $G^{(N)}$ to pth order of perturbation theory:

(i) Draw all topologically inequivalent diagrams having N external points $x_1, \dots, x_N$, and p vertices $y_1, \dots, y_p$, and having no vacuum-to-vacuum parts. If the interaction is written as a normal product, then certain diagrams are missing (see (9.3.30)). For instance, the diagram in Fig. 10.1 does not contribute to $G^{(2)}$ if the Lagrangean is given by (10.0.3b):

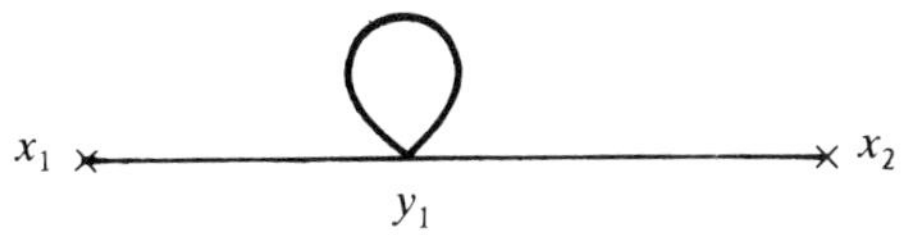

Fig. 10.1 A (tadpole) graph eliminated by the normal product

(ii) To every line of the diagram assign a factor $\Delta_F(x-y)$ $(=\Delta_F(y-x))$.

(iii) To every vertex assign a factor $-ig$.

(iv) Integrate over all internal points y_j.

(v) Multiply by the symmetry factor of the diagram.

The arguments of Chapter 5 about connected correlation functions and about proper vertices apply equally to Green's functions. Connected Green's functions $G_c^{(N)}$ are defined via the logarithm of $Z(j)$ $(Z(j) = \exp(W(j)))$, and proper vertices via the generating functional $\Gamma(\bar{\varphi})$:

$$i\Gamma(\bar{\varphi}) = W(j) - i\int d^4x\, j(x)\bar{\varphi}(x),$$

$$\bar{\varphi}(x) = -i\delta W(j)/\delta j(x).$$

The proper vertex $\Gamma^{(2)}$ is given, up to a factor i, by the inverse of the propagator:

$$\Gamma^{(2)}(k) = i[G^{(2)}(k)]^{-1} = k^2 - m^2 - \Sigma(k), \tag{10.1.20a}$$

where $\Sigma(k)$ is the self-energy. The proper vertices $\Gamma^{(N)}$, $N > 2$, are related to the 1-particle irreducible connected Green's functions, shorn of their *full* external propagators $\bar{G}_c^{(N)}|_{1\text{-PI}}$. Note that the Green's functions $\bar{G}_c^{(N)}$ need not be 1-PI, and that in general they differ from proper vertices. Thus we have

$$\bar{G}_c^{(N)}(k_1, \ldots, k_N) = \prod_{i=1}^{N} [G^{(2)}(k_i)]^{-1} G_c^{(N)}(k_1, \ldots, k_N), \tag{10.1.20b}$$

$$\Gamma^{(N)}(k_1, \ldots, k_N) = -i\bar{G}_c^{(N)}(k_1, \ldots, k_N)|_{1\text{-PI}}, \tag{10.1.20c}$$

with Fourier transforms defined by

$$(2\pi)^4 \delta^{(4)}\left(\sum_{i=1}^{N} k_i\right) G_c^{(N)}(k_1, \ldots, k_N)$$
$$= \int d^4x_1 \ldots d^4x_N\, e^{-i\sum_{j=1}^{N} k_j\cdot x_j} G_c^{(N)}(x_1, \ldots, x_N) \tag{10.1.21}$$

(see (5.2.15)). Under this definition all the momenta k_i run *inwards* into the diagram (Section 10.3.3). For instance, the Feynman rules for the proper vertices $\Gamma^{(N)}(k_i)$ read as follows.

(i) Draw all 1-particle irreducible diagrams of order p that have N external 4-momenta $k_1, \ldots, k_N$ entering the diagram.

(ii) To every internal line of the diagram assign a factor $\mathrm{i}/(k^2 - m^2 + \mathrm{i}\varepsilon)$.

(iii) To every vertex assign a factor $-\mathrm{i}g$.

(iv) Write the internal momenta so as to ensure 4-momentum conservation at every vertex, and integrate over all the independent variables q, or in other words over all closed loops, with the integration measure $\mathrm{d}^4q/(2\pi)^4$.

(v) Multiply by the symmetry factor of the graph, and by an overall factor $(-\mathrm{i})$ (see (10.1.20c)).

10.1.4 Renormalization and the norming conditions

In general the Green's functions defined by the Feynman rules above diverge in 4 dimensions. As in Part II, the theory must be renormalized. Hence we must settle on norming conditions analogous to (6.3.10) or to (6.4.6). In quantum field theory one often renormalizes *on the mass shell* (from choice rather than from necessity); the renormalized propagator must have a pole at $k^2 = m^2$, where m is the physical mass of the particle*, and one requires that the residue at this pole be i. Then, in the vicinity of the pole at $k^2 = m^2$, the propagator approximates as closely as possible to the free propagator $\mathrm{i}/(k^2 - m^2 + \mathrm{i}\varepsilon)$. The mass-shell norming conditions read

$$\Gamma_{\mathrm{R}}^{(2)}(k^2 = m^2) = 0, \tag{10.1.22a}$$

$$\frac{\mathrm{d}}{\mathrm{d}k^2}\Gamma_{\mathrm{R}}^{(2)}(k^2 = m^2) = 1, \tag{10.1.22b}$$

$$\Gamma_{\mathrm{R}}^{(4)}(k_i = 0) = -g. \tag{10.1.22c}$$

If the mass is zero, then the presence of singularities at $k_i^2 \geqslant 0$ leads one to choose a spacelike subtraction point (see equation (6.4.7)) such that

$$k_i \cdot k_j|_{\mathrm{SP}} = -\frac{1}{4}\mu^2(4\delta_{ij} - 1), \quad k_i^2 = -\frac{3}{4}\mu^2, \tag{10.1.23}$$

* In general the renormalized mass m_{R} does not equal the physical mass m, but differs from it by a finite renormalization (say if one uses the norming condition $\Gamma_{\mathrm{R}}^{(2)}(k^2 = 0) = m_{\mathrm{R}}^2$). Of course $\Gamma_{\mathrm{R}}^{(2)}(k^2 = m^2) = 0$ is obligatory in every case. With renormalization on the mass shell one does have $m^2 = m_{\mathrm{R}}^2$.

and the conditions (6.4.6) read

$$\Gamma_{\mathrm{R}}^{(2)}(k^2 = 0) = 0, \tag{10.1.24a}$$

$$\frac{\mathrm{d}}{\mathrm{d}k^2}\Gamma_{\mathrm{R}}^{(2)}(k^2 = -\mu^2) = 1, \tag{10.1.24b}$$

$$\Gamma_{\mathrm{R}}^{(4)}(k_i|_{\mathrm{SP}}) = -g. \tag{10.1.24c}$$

10.2 Path integrals and the Euclidean theory

10.2.1 The path integral for $Z(j)$

In the perturbative expansion of the Green's functions, Wick's theorem has enabled us to identify the outcome of a Gaussian integration. Therefore it must be possible to express the generating functional (10.1.16) of Green's functions as a path integral. Start by noting that, as long as one refrains from using the normal product, $Z(j)$ may be written as

$$Z(j) = \mathcal{N} \exp\left(\mathrm{i}\int \mathrm{d}^4x\, \mathcal{L}_1\left(-\mathrm{i}\frac{\delta}{\delta j(x)}\right)\right)$$
$$\times \exp\left(-\frac{1}{2}\int \mathrm{d}^4x\, \mathrm{d}^4x'\, j(x)\Delta_{\mathrm{F}}(x - x')j(x')\right) \tag{10.2.1}$$

(see (5.1.22)), where $\mathcal{N}$ is a multiplicative constant irrelevant to calculating Green's functions. (It is instructive to check the factors i, say in the case of the Lagrangean (10.0.3a).) But the second exponential in (10.2.1) can be expressed as a Gaussian integral:

$$\exp\left(-\frac{1}{2}\int \mathrm{d}^4x\, \mathrm{d}^4x'\, j(x)\Delta_{\mathrm{F}}(x - x')j(x')\right)$$
$$= \mathcal{N}'\int \mathcal{D}\varphi \exp\left(-\int \mathrm{d}^4x\left(\frac{1}{2}\varphi[\mathrm{i}(\Box + m^2 - \mathrm{i}\varepsilon)]\varphi - \mathrm{i}j(x)\varphi(x)\right)\right)$$
$$= \mathcal{N}'\int \mathcal{D}\varphi \exp\left(\mathrm{i}\int \mathrm{d}^4x(\mathcal{L}_0(x) + j(x)\varphi(x))\right). \tag{10.2.2}$$

An integration by parts allows one to identify the expressions $\mathcal{L}_0(x)$ occurring in (10.2.2) and in (10.0.1) (see the comments following equation (9.2.3a). In (10.2.2), the integral over φ is Gaussian, featuring the inverse of the operator

$$\mathrm{i}(\Box + m^2 - \mathrm{i}\varepsilon),$$

which we may symbolize by $[\mathrm{i}(\Box + m^2 - \mathrm{i}\varepsilon)]^{-1}$. This inverse however is nothing other than Δ_F, given that

$$[i(\Box_x + m^2 - i\varepsilon)]\Delta_F(x - x') = \delta^{(4)}(x - x')$$

(see (9.2.26)). One could reason equally well in Fourier space:

$$\Delta_F(k) = i/(k^2 + m^2 - i\varepsilon); \quad i(\Box + m^2 - i\varepsilon) \to -i(k^2 - m^2 - i\varepsilon).$$

Note that the factor $(-i\varepsilon)$ ensures that the Gaussian integral (10.2.2) converges: without this factor the integrand would oscillate as $\varphi \to \pm\infty$. Finally, (10.2.1) may be expressed in the form

$$Z(j) = \mathcal{N}'' \int \mathcal{D}\varphi \exp\left(i \int d^4x(\mathcal{L}_0(\varphi) + \mathcal{L}_1(\varphi) + j(x)\varphi(x))\right)$$
$$= \mathcal{N}'' \int \mathcal{D}\varphi \exp(i(\text{action}/\hbar)), \tag{10.2.3}$$

where we recognize the Gaussian component $\mathcal{L}_0(\varphi)$ as the free Lagrangean. We must draw attention to a manifest abuse of notation: in (10.2.2) and (10.2.3) $\varphi(x)$ stands for a *classical* field configuration, and not for an operator. Strictly speaking one should use a different symbol, e.g. $A(x)$, in (10.2.2) and (10.2.3). However, observing common usage we shall continue with the incorrect notation, which is endemic in the literature. The expressions (10.1.16), (10.2.1), and (10.2.3) for $Z(j)$ contain different multiplicative constants, which may be disregarded because they cancel in the process of calculating the Green's functions (see equation (10.1.17)).

Evidently it is not by mere coincidence that (10.1.16) yields $Z(j)$ in the form of a path integral. Rather, we have already shown in Section 8.2 that probability amplitudes in quantum mechanics can be written as such integrals, i.e. as sums over all configurations $q(t)$ that satisfy certain initial and final conditions; whence we derived a generating functional for the vacuum expectation-values of T-products (equation (8.2.17)). In our present case the dynamical variable is the field φ rather than a position-coordinate q, and it is not surprising that the generating functional for the vacuum expectation-values of T-products should be expressible as a sum over configurations of the field. The only ambiguity in the functional formulation stems from the choice of an inverse of the operator $(\Box + m^2)$. The prescription $m^2 \to m^2 - i\varepsilon$ can be justified in the same way as in Chapter 8. In the approach we have followed here, it is obviously a consequence of the canonical formulation.

From what has just been said, we see that if we are given a classical Lagrangean $\mathcal{L}(x)$, then we have the *option* of quantizing it through the generating functional (10.2.3), rather than through the canonical formalism. In view of the uncertainties afflicting the latter (because of the interaction picture), the former option seems no worse a priori. It needs saying however that (10.2.3) assumes that one regularizes at some intermediate stage, and that one verifies locality and the unitarity of the S matrix produced by this method.

The functional formulations (10.2.1) and (10.2.3) suggest that there must exist a very precise relation between the Green's functions of the present chapter and the correlation functions of Chapter 5; indeed the latter are quite commonly referred to as the Green's functions of the Euclidean theory, for reasons that will become clear presently. The relation in question is independent of the perturbative expansion. However, to begin with we do use this expansion, which will permit a digression to examine how Green's functions are calculated in practice. We start with a simple example, namely with $\Gamma^{(4)}$ in the context of the φ^4 model (10.0.3), to the first nontrivial order of perturbation theory.

10.2.2 $\Gamma^{(4)}$ to second order in g. Wick rotation

Let us calculate the contribution of the graph in Fig. 10.2 to $\Gamma^{(4)}$, calling it $\bar{\Gamma}^{(4)}$:

$$\bar{\Gamma}^{(4)} = -\mathrm{i}\frac{(-\mathrm{i}g)^2}{2} I(k_i)$$

$$I(k_i) = \int \frac{\mathrm{d}^D q}{(2\pi)^D} \frac{(\mathrm{i})^2}{(q^2 - m^2 + \mathrm{i}\varepsilon)((k-q)^2 - m^2 + \mathrm{i}\varepsilon)}, \tag{10.2.4}$$

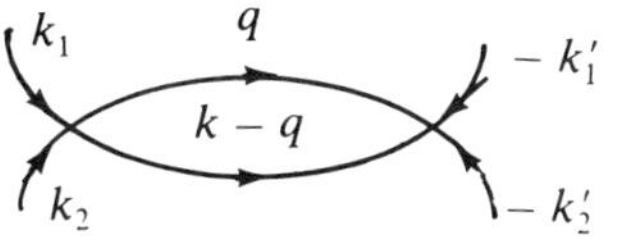

Fig. 10.2 A contribution of order g^2 to $\Gamma^{(4)}$

where $k = k_1 + k_2$. We have regularized this expression (dimensionally), because the integral diverges in $4D$ (D always denotes the dimensionality of *space–time*: the ordinary part of the space is $(D-1)$-dimensional.) In order to evaluate (10.2.4) we again use the Feynman-parameter method, while indicating how the present calculation differs from that of Chapter 5. First we find

$$I(k^2) = \int_0^1 \mathrm{d}x \int \frac{\mathrm{d}^D q}{(2\pi)^D} \frac{(\mathrm{i})^2}{[q^2 + x(1-x)k^2 - m^2 + \mathrm{i}\varepsilon]^2}. \tag{10.2.5}$$

Now suppose that $k^2 < 0 : k^2 = -\boldsymbol{k}_{\mathrm{E}}^2$. This is what would happen say for $k_0 = 0$:

$$k^2 = -\boldsymbol{k}^2 = -k_1^2 - \cdots - k_{D-1}^2 = -\boldsymbol{k}_{\mathrm{E}}^2.$$

Then the q_0-integrand has poles at

$$q_0 = \pm(m^2 + \boldsymbol{q}^2 + x(1-x)\boldsymbol{k}_{\mathrm{E}}^2)^{1/2},$$

and the prescription for avoiding these is perfectly well-defined, because in actual fact m^2 stands for $m^2 - \mathrm{i}\varepsilon$. By setting $q_0 = \mathrm{i}q_4$ we can deform the contour (still avoiding the poles) so that the integration runs along the imaginary axis

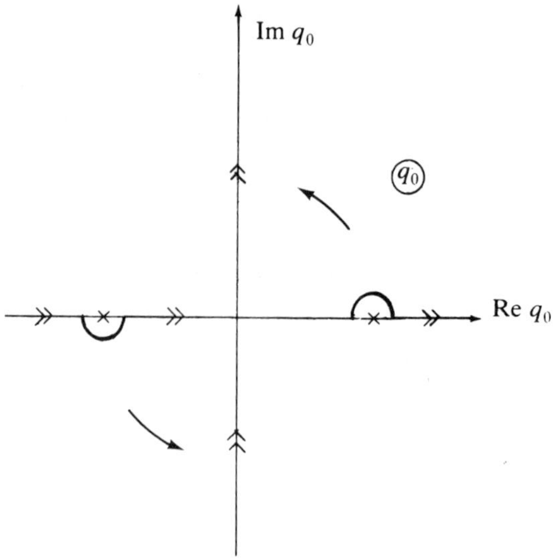

Fig. 10.3 The Wick rotation

$-\infty < \operatorname{Im} q_0 < +\infty$ (Fig. 10.3); thus

$$I(k^2) = -\mathrm{i}\int_0^1 \mathrm{d}x \int \frac{\mathrm{d}^D q_\mathrm{E}}{(2\pi)^D} \frac{1}{[\boldsymbol{q}_\mathrm{E}^2 + x(1-x)\boldsymbol{k}_\mathrm{E}^2 + m^2]^2}, \tag{10.2.6}$$

where $\mathrm{d}^D q_\mathrm{E} = \mathrm{d}q_4 \mathrm{d}q_1 \cdots \mathrm{d}q_{D-1}$. Up to a factor $(-\mathrm{i})$ this coincides with the expression we found in Chapter 5. The deformation of the contour, which amounts to a rotation through $+\pi/2$, is called a *Wick rotation*. Lastly the index E on q_E is an abbreviation for 'Euclidean'; for $D = 4$

$$\boldsymbol{q}_\mathrm{E}^2 = q_1^2 + q_2^2 + q_3^2 + q_4^2$$

is indeed a Euclidean metric, while the original metric was Minkowskian:

$$q^2 = q_0^2 - q_1^2 - q_2^2 - q_3^2.$$

When $k^2 = -\boldsymbol{k}_\mathrm{E}^2 < 0$, one says that the Green's function is being evaluated in the Euclidean region. Collecting all the factors i, one sees that

$$\bar{\Gamma}^{(4)}(k^2 = -\boldsymbol{k}_\mathrm{E}^2 < 0) = -\bar{\Gamma}_\mathrm{E}^{(4)}(\boldsymbol{k}_\mathrm{E}^2), \tag{10.2.7}$$

where $\bar{\Gamma}_\mathrm{E}^{(4)}(\boldsymbol{k}_\mathrm{E}^2)$ is the proper vertex calculated in Chapter 5; from now on we call it the proper vertex of the Euclidean theory.

We can now clinch the calculation of $\bar{\Gamma}^{(4)}$:

$$\bar{\Gamma}^{(4)}(k^2) = \frac{g^2\Gamma(2-D/2)}{2(4\pi)^{D/2}} \int_0^1 \mathrm{d}x[m^2 - x(1-x)k^2]^{(D/2)-2}. \tag{10.2.8}$$

As we already know, this expression diverges for $D = 4$, i.e. for $\varepsilon = 4 - D = 0$*. In view of $x^{-\varepsilon/2} \approx 1 - \frac{1}{2}\varepsilon \ln x$, equation (10.2.8) yields

$$\bar{\Gamma}^{(4)}(k^2) = \frac{g^2}{(4\pi)^2\varepsilon}\left[1 - \frac{\varepsilon}{2}\int_0^1 \mathrm{d}x \ln(m^2 - x(1-x)k^2) + \text{constant} + O(\varepsilon^2)\right], \tag{10.2.9}$$

and the term diverging like $1/\varepsilon$ is eliminated by the renormalization. For instance, if we adopt the norming condition (10.1.22c), then we find that the renormalized Green's function is given by

$$\bar{\Gamma}^{(4)}_{\mathrm{R}} = \frac{-g^2}{2(4\pi)^2}\int_0^1 \mathrm{d}x \ln\left(1 - \frac{x(1-x)k^2}{m^2}\right). \tag{10.2.10}$$

The logarithm is well-defined for $k^2 < 4m^2$; in particular, it is well-defined in the Euclidean region. On the other hand, $\bar{\Gamma}^{(4)}_{\mathrm{R}}$ has a branch cut along $k^2 \geqslant 4m^2$, whose physical origin we shall examine in Section 10.4.

10.2.3 The connection with the Euclidean theory

To discuss the general case of a proper vertex $\Gamma^{(N)}$ having N external lines, to order V of perturbation theory, we follow the method introduced in Chapter 5. We start from the Schwinger representation of the propagator,

$$\frac{\mathrm{i}}{p^2 - m^2 + \mathrm{i}\varepsilon} = \int_0^\infty \mathrm{d}\alpha\, \mathrm{e}^{\mathrm{i}\alpha(p^2 - m^2 + \mathrm{i}\varepsilon)}, \tag{10.2.11}$$

where the iε in the exponent ensures that the integral converges. It would be pointless to repeat the calculation performed in Section 5.5.3, and it will suffice if we keep track of the factors i. These factors stem (see Problem 10.1) from the two integrals

$$\int \frac{\mathrm{d}^D k}{(2\pi)^D}\, \mathrm{e}^{\mathrm{i}ak^2} = \frac{\mathrm{i}\,\mathrm{e}^{-\mathrm{i}\pi D/4}}{(4\pi a)^{D/2}} \quad (a > 0) \tag{10.2.12}$$

and

$$\int_0^\infty \mathrm{d}\lambda\, \lambda^{\alpha - 1}\, \mathrm{e}^{-\mathrm{i}\lambda(m^2 - \mathrm{i}\varepsilon)} = \mathrm{e}^{-\mathrm{i}\pi\alpha/2}(m^2)^{-\alpha}\Gamma(\alpha). \tag{10.2.13}$$

While following the calculation in Chapter 5 one notes

- a factor $(\mathrm{i}\mathrm{e}^{-\mathrm{i}\pi D/4})^I$ stemming from the I integrals over the p_i;
- a factor $(\mathrm{i}\mathrm{e}^{-\mathrm{i}\pi D/4})^{V-1}$ stemming from the integral over the $(V-1)$ variables z;
- a factor $\mathrm{e}^{-\mathrm{i}(\pi/2)(I - LD/2)}$ stemming from the integral over the homogeneity variable λ (see the step from (5.5.9) to (5.5.10)).

We thus find an overall factor $(-\mathrm{i})^{V-1}$.

* Of course $\varepsilon = 4 - D$ should not be confused with the ε in the propagator.

Accordingly, for the integral J we obtain

$$J = \frac{(-\mathrm{i})^{V-1}\Gamma(I - LD/2)}{(4\pi)^{LD/2}} \int \frac{\prod \mathrm{d}x_i \delta(1 - \sum x_i)}{[P(x_i)]^{D/2}} (m^2 - k^{\mathrm{T}} A^{-1} k)^{(LD/2)-I}, \tag{10.2.14}$$

where the matrix A is identically the same as the matrix in (5.5.8). Next we define *the Euclidean region*: a set of momenta k_i are in this region if the inequality

$$(\sum_i \lambda_i k_i)^2 < 0$$

holds for arbitrary real values of the λ_i. All linear combinations of vectors k_i in this region are orthogonal to some time-like vector, whose components in a suitable reference frame can be written as (1, 0, 0, 0). In such a frame the time-components of the k_i vanish ($k_{i0} = 0$), and we have

$$k \cdot k' = -\boldsymbol{k} \cdot \boldsymbol{k}' = -(k_1 k_1' + k_2 k_2' + k_3 k_3').$$

With every k_i we can associate a vector $\boldsymbol{k}_{i\mathrm{E}}$ in a 4-dimensional Euclidean space, such that $k_{i4} = 0$ with respect to some special reference frame, where

$$\boldsymbol{k}_{\mathrm{E}} \cdot \boldsymbol{k}'_{\mathrm{E}} = \boldsymbol{k} \cdot \boldsymbol{k}'.$$

The term $k^{\mathrm{T}} A^{-1} k$ in (10.2.14) becomes $-\boldsymbol{k}_{\mathrm{E}}^{\mathrm{T}} A^{-1} \boldsymbol{k}_{\mathrm{E}}$ (recall that $\boldsymbol{k}_{\mathrm{E}}^{\mathrm{T}} A^{-1} \boldsymbol{k}_{\mathrm{E}} \geqslant 0$ because A is a positive matrix). We must now keep track of the factor i stemming from the substitution $-g \to -\mathrm{i}g$; of the factor $-\mathrm{i}$ in the definition of $\Gamma^{(N)}$ (see (10.1.20c)); and of the factor -1 in the definition of $\Gamma_{\mathrm{E}}^{(N)}$; in the Euclidean region these lead one to identify

$$\Gamma^{(N)}(k_i)|_{\mathrm{E}} = -\Gamma_{\mathrm{E}}^{(N)}(\boldsymbol{k}_{i\mathrm{E}}). \tag{10.2.15}$$

(Note that our definition of the $\Gamma^{(N)}$ coincides with the definition adopted by Itzykson and Zuber, but that the $\Gamma_{\mathrm{E}}^{(N)}$ differ by a factor -1.) To go from the $\Gamma^{(N)}$ to the connected Green's functions $G_{\mathrm{c}}^{(N)}$, one needs certain products of external propagators, namely $\prod_{j=1}^{N} G^{(2)}(k_j)$ in the Minkowskian case, and $\prod_{j=1}^{N} G_{\mathrm{E}}^{(2)}(\boldsymbol{k}_{j\mathrm{E}})$ in the Euclidean case; since $G^{(2)}(k)|_{\mathrm{E}} = -\mathrm{i}G_{\mathrm{E}}^{(2)}(\boldsymbol{k}_{\mathrm{E}})$, and in view of the factors $-\mathrm{i}$ in (10.1.20c) and -1 in the definition of $\Gamma_{\mathrm{E}}^{(N)}$, one finds that the connected Green's functions are given by

$$G_{\mathrm{c}}^{(N)}(k_i)|_{\mathrm{E}} = \mathrm{i}(-\mathrm{i})^N G_{\mathrm{c,E}}^{(N)}(\boldsymbol{k}_{i,\mathrm{E}}). \tag{10.2.16}$$

As we saw in detail in Parts I and II, the correlation functions $G_{\mathrm{c,E}}^{(N)}(\boldsymbol{k}_{\mathrm{E}})$ can be calculated from the functional integral

$$Z(j) = \int \mathscr{D}\varphi \exp\left(-\int \mathrm{d}^D x(\mathscr{H}(\varphi) - j(x)\varphi(x))\right); \tag{10.2.17}$$

therefore it would have been possible to derive (10.2.15) and (10.2.16) directly from a comparison between (10.2.3) and (10.2.17), without appealing to the perturbation expansion at all. Indeed $\mathscr{H}(x)$ is nothing but the analytic continuation of $\mathscr{L}(x)$ under $t \to -\mathrm{i}\tau$:

$$\square + m^2 \to -\sum_{i=1}^{4} \frac{\partial^2}{\partial x_i^2} + m^2.$$

This generalizes what we saw in Chapter 8 for the case of just one degree of freedom: there, the Euclidean continuation $t \to -\mathrm{i}\tau$ took us from a quantum-mechanics problem to a problem of statistical mechanics in one dimension. In our present case we observe that analytic continuation turns a problem from quantum field theory into a problem of statistical mechanics in a Euclidean D-dimensional space. Accordingly, the 'Hamiltonian' $\mathscr{H}(x)$ in Parts I and II of this book must be considered as the Euclidean continuation $\mathscr{L}_{\mathrm{E}}(x)$ of the *Lagrangean* $\mathscr{L}(x)$ of a quantum field theory. This is why one often meets the symbol $\mathscr{L}_{\mathrm{E}}(x)$ instead of $\mathscr{H}(x)$. Relations like (10.2.15) and (10.2.16) can be derived from formulae that generalize (8.3.7). It suffices to consider the time coordinate, because it alone is affected by Euclidean continuation.

Finally we note that the prescription $m^2 \to m^2 - \mathrm{i}\varepsilon$ can be derived from the following so-called 'Euclidean postulate': by definition, the Green's functions of the quantal theory are given by the analytic continuation $\tau \to it$ of the correlation functions of the Euclidean theory.

This connection between Euclidean and Minkowski space has been put on a rigorous basis in the framework of the Osterwalder–Schrader axioms (Glimm and Jaffe 1987, Chapter 6; Parisi 1988, Chapter 15). These axioms give the conditions under which a Euclidean field theory can be continued into a quantum field theory. It is then possible to prove that the Euclidean Green's functions can be continued analytically to Minkowski space, and that their continuations obey the Wightman axioms.

10.2.4 The equations of motion

We illustrate the functional formulation by establishing equations of motion for the Green's functions. In the generating functional (10.2.3) we change variables

$$\varphi(x) \to \varphi(x) + \varepsilon f(x),$$

where $\varepsilon \to 0$, and where $f(x)$ is an arbitrary function. A change of variables does not affect $Z(j)$; as $\mathscr{D}\varphi$ is invariant, we can write $\delta Z = 0$ by expanding to order ε,

$$\int \mathrm{d}^4 y\, f(y) \int \mathscr{D}\varphi \exp\left(\mathrm{i}S[\varphi] + \mathrm{i}\int \mathrm{d}^4 x\, j(x)\varphi(x)\right)\left[\frac{\delta S}{\delta\varphi(y)} + j(y)\right] = 0.$$

Because the function f is arbitrary, this entails

$$\int \mathscr{D}\varphi \exp\left(\mathrm{i}S[\varphi] + \mathrm{i}\int \mathrm{d}^4 x\, j(x)\varphi(x)\right)\left[\frac{\delta S}{\delta\varphi(y)} + j(y)\right] = 0. \qquad (10.2.18)$$

The equations of motion we want are derived by differentiating with respect to j and then setting $j = 0$; differentiating once, we find

$$[Z(0)]^{-1} \int \mathscr{D}\varphi\, \mathrm{e}^{\mathrm{i}S[\varphi]}\left[\mathrm{i}\varphi(x)\frac{\delta S}{\delta\varphi(y)} + \delta^{(4)}(x-y)\right] = 0.$$

With the Lagrangean (10.0.1) for example, provided $V(\varphi)$ does not contain any derivatives of φ, this equation reads

$$[Z(0)]^{-1}\int \mathscr{D}\varphi\, e^{iS[\varphi]}\{\varphi(x)[(\Box_y + m^2)\varphi(y) + V'(\varphi(y))]\} + i\delta^{(4)}(x-y) = 0.$$

We can also express the result as a T-product, on noticing that the differentiations with respect to y may be taken outside the functional integral:

$$(\Box_y + m^2)\langle 0|T(\varphi(x)\varphi(y))|0\rangle + \langle 0|T(\varphi(x)V'(\varphi(y)))|0\rangle = -i\delta^{(4)}(x-y). \tag{10.2.19}$$

Although $\delta S/\delta\varphi(y) = 0$ at the classical level, we note that

$$\langle 0|T\left(\varphi(x)\frac{\delta S}{\delta\varphi(y)}\right)|0\rangle,$$

which is equivalent to the right-hand side of (10.2.19), is nonzero: the insertion of $\delta S/\delta\varphi(y)$ into a Green's function gives a nonzero result, because one integrates over field configurations that do not obey the classical equations of motion. In operator language, the delta function $\delta^{(4)}(x-y)$ evidently stems from the fact that the Klein–Gordon operator $(\Box_y + m^2)$ fails to commute with the T-product (see Problem 9.8). From this observation one could of course derive (10.2.19) directly. The important point in this discussion that we wish to register is the following. Suppose we have derived a certain identity by means of the functional formalism; *when it is expressed in terms of vacuum expectation-values of T-products, all differential operators must be moved outside these products.*

Differentiating (10.2.18) N times we obtain the general equation of motion,

$$(\Box_y + m^2)G^{(N+1)}(y, x_1, \ldots, x_N) + \langle 0|T(V'(\varphi(y))\varphi(x_1)\ldots\varphi(x_N))|0\rangle$$
$$= -i\sum_{j=1}^{N}\delta^{(4)}(y-x_j)G^{(N-1)}(x_1, \ldots, x_{j-1}, x_{j+1}, \ldots, x_N). \tag{10.2.20}$$

One observes once again that the insertion of $\delta S/\delta\varphi(y)$ leads to a nonzero result, though only at the points $y = x_1, \ldots, y = x_N$.

10.3 Cross-sections and the S-matrix

The dynamics of elementary particles is accessible to observation partly through their decays, but mainly through their collisions in accelerators. The quantities measured in collision processes are the cross-sections, and our first task is to relate these to elements of the S-matrix. Afterwards we shall have to relate the S-matrix elements to the Green's functions, which we have learned how to calculate at least in perturbation theory. The relation we need is established via the 'reduction formulae'; however, the machinery required to derive these formulae is quite elaborate (especially for particles with nonzero spin), and we shall start with a less rigorous but more intuitive method. As a rule one defines a T-matrix, containing the nontrivial part of the S-matrix, through the following expression for the matrix elements $i \to f$:

$$S_{fi} = \delta_{fi} + \mathrm{i}(2\pi)^4 \delta^{(4)}(K_f - K_i) T_{fi}; \tag{10.3.1}$$

here K_i (K_f) is the four-momentum of the initial (final) state.

10.3.1 Cross-sections

In calculating these it is important to be specific about the normalization of the state vectors. If $|\boldsymbol{k}\rangle = a^\dagger(\boldsymbol{k})|0\rangle$, then according to (9.2.13) the orthonormality relation reads

$$\langle \boldsymbol{k}'|\boldsymbol{k}\rangle = (2\pi)^3 2\omega_{\boldsymbol{k}} \delta^{(3)}(\boldsymbol{k} - \boldsymbol{k}'),$$

while the closure relation, in the one-particle subspace of Fock-space, becomes

$$\int \frac{\mathrm{d}^3 k}{(2\pi)^3 2\omega_{\boldsymbol{k}}} |\boldsymbol{k}\rangle\langle \boldsymbol{k}| = \mathbb{1}_1.$$

Let us adopt an initial state formed from two wave-packets,

$$|i, \mathrm{in}\rangle = \int \mathrm{d}\tilde{k}_1 \, \mathrm{d}\tilde{k}_2 \, f_1(\boldsymbol{k}_1) f_2(\boldsymbol{k}_2) |\boldsymbol{k}_1, \boldsymbol{k}_2, \mathrm{in}\rangle,$$

where

$$\mathrm{d}\tilde{k} = \mathrm{d}^3 k / [(2\pi)^3 2\omega_{\boldsymbol{k}}].$$

To each wave-packet there corresponds a positive-energy solution of the Klein–Gordon equation,

$$f(x) = \int \mathrm{d}\tilde{k} \, \mathrm{e}^{-\mathrm{i}kx} f(\boldsymbol{k}), \tag{10.3.2}$$

and the scalar product is defined by

$$(f, g) = \mathrm{i} \int \mathrm{d}^3 x \, f^* \overleftrightarrow{\partial}_0 g = \int \mathrm{d}\tilde{k} \, f^*(\boldsymbol{k}) g(\boldsymbol{k})$$

(see (9.2.10)). The transition probability

$$w_{fi} = |\langle f, \mathrm{out}|i, \mathrm{in}\rangle|^2$$

is given in terms of the T-matrix element $\langle f|T|\boldsymbol{k}_1, \boldsymbol{k}_2\rangle$ by

$$\begin{aligned} w_{fi} = \int & \mathrm{d}\tilde{k}_1 \, \mathrm{d}\tilde{k}_2 \, \mathrm{d}\tilde{k}'_1 \, \mathrm{d}\tilde{k}'_2 \, f_1^*(\boldsymbol{k}_1) f_2^*(\boldsymbol{k}_2) f_1(\boldsymbol{k}'_1) f_2(\boldsymbol{k}'_2) \\ & \times (2\pi)^4 \delta^{(4)}(K_f - k_1 - k_2)(2\pi)^4 \delta^{(4)}(k_1 + k_2 - k'_1 - k'_2) \\ & \times \langle f|T|\boldsymbol{k}_1 \boldsymbol{k}_2\rangle^* \langle f|T|\boldsymbol{k}'_1, \boldsymbol{k}'_2\rangle \end{aligned} \tag{10.3.3}$$

(see (10.3.1); the index 'in' has been dropped in order to ease the notation). If the momentum of the wave-packet is well defined, then $f(\boldsymbol{k}_i)$ peaks near $\boldsymbol{k}_i = \underline{\boldsymbol{k}}_i$ We shall assume that the matrix element varies little in the vicinity of $(\underline{\boldsymbol{k}}_1, \underline{\boldsymbol{k}}_2)$:

$$\langle f|T|\boldsymbol{k}_1, \boldsymbol{k}_2\rangle \approx \langle f|T|\underline{\boldsymbol{k}}_1, \underline{\boldsymbol{k}}_2\rangle. \tag{10.3.4}$$

Now write one of the delta functions as a Fourier transform,

$$(2\pi)^4 \delta^{(4)}(k_1 + k_2 - k_1' - k_2') = \int \mathrm{d}^4 x\, \mathrm{e}^{\mathrm{i}(k_1 + k_2 - k_1' - k_2')x},$$

and integrate over the $\boldsymbol{k}_i$ and the $\boldsymbol{k}_i'$ using the approximation (10.3.4):

$$w_{fi} = \int \mathrm{d}^4 x |f_1(x)|^2 |f_2(x)|^2 (2\pi)^4 \delta^{(4)}(K_f - \underline{k}_1 - \underline{k}_2) |\langle f|T|\underline{\boldsymbol{k}}_1, \underline{\boldsymbol{k}}_2\rangle|^2. \tag{10.3.5}$$

Here, w_{fi} is a transition probability integrated over space and time; it is in fact the integral of a transition probability per unit volume and per unit time:

$$\frac{\mathrm{d}w_{fi}}{\mathrm{d}V\,\mathrm{d}t} = |f_1(x)|^2 |f_2(x)|^2 (2\pi)^4 \delta^{(4)}(K_f - \underline{k}_1 - \underline{k}_2) |\langle f|T|\underline{\boldsymbol{k}}_1, \underline{\boldsymbol{k}}_2\rangle|^2. \tag{10.3.6}$$

On the other hand, for the packet momentum to be well defined one must have

$$f(x) \approx \mathrm{e}^{-\mathrm{i}\underline{k}\cdot x} F(x),$$

where the function $F(x)$ is slowly-varying because the wave-packet is wide. But in that case

$$\mathrm{i} f^*(x) \overleftrightarrow{\partial}_\mu f(x) \simeq 2k_\mu |f(x)|^2.$$

Now consider particle 1 as the projectile, and particle 2 as the target at rest in the laboratory frame. The number of target particles per unit volume is

$$\frac{\mathrm{d}n_2}{\mathrm{d}V} \approx 2\underline{\omega}_2 |f_2(x)|^2 \tag{10.3.7}$$

(where we use the shorthand notation $\omega_2 = \omega_{k_2}$), and we have $\omega_2 = m_2$ because the target is at rest. The incident particle flux is

$$\frac{\|\boldsymbol{k}_1\|}{\omega_1} 2\omega_1 |f_1(x)|^2 = 2\|\boldsymbol{k}_1\| |f_1(x)|^2. \tag{10.3.8}$$

It remains only to introduce the definition

$$\mathrm{d}\sigma = \frac{\mathrm{d}w_{fi}}{\mathrm{d}V\,\mathrm{d}t} \frac{\mathrm{d}\Phi}{(\text{flux})(\text{target density})}$$

of the cross-section, where $\mathrm{d}\Phi$ is an element of phase space; equations (10.3.6),

(10.3.7), and (10.3.8) then yield

$$\mathrm{d}\sigma = \frac{(2\pi)^4 \delta^{(4)}(K_f - k_1 - k_2)}{4 m_2 \| \boldsymbol{k}_1 \|} |\langle f | T | \boldsymbol{k}_1, \boldsymbol{k}_2 \rangle|^2 \, \mathrm{d}\Phi.$$

The 'flux factor' $m_2 \| \boldsymbol{k}_1 \| = F$ can be expressed in a Lorentz invariant way:

$$F = [(k_1 \cdot k_2)^2 - m_1^2 m_2^2]^{1/2}. \tag{10.3.9}$$

If we wish to calculate the cross-section for observing N final-state particles with momenta $(\boldsymbol{k}'_1, \ldots, \boldsymbol{k}'_N)$ in a certain region of phase space, then we must integrate with the integration measure

$$\mathrm{d}\Phi^{(N)} = \mathrm{d}\tilde{k}'_1 \cdots \mathrm{d}\tilde{k}'_N = \frac{\mathrm{d}^3 k'_1}{(2\pi)^3 2\omega'_1} \cdots \frac{\mathrm{d}^3 k'_N}{(2\pi)^3 2\omega'_N}. \tag{10.3.10}$$

Accordingly, we write the end-result for $\mathrm{d}\sigma$ as

$$\mathrm{d}\sigma = \frac{1}{4F} (2\pi)^4 \delta^{(4)}(K_f - k_1 - k_2) |\langle f | T | \boldsymbol{k}_1, \boldsymbol{k}_2 \rangle|^2$$
$$\times \frac{\mathrm{d}^3 k'_1}{(2\pi)^3 2\omega'_1} \cdots \frac{\mathrm{d}^3 k'_N}{(2\pi)^3 2\omega'_N} \mathscr{S}, \tag{10.3.11}$$

where $K_f = \sum_{i=1}^{N} k'_i$, and $\mathscr{S}$ is a statistical factor, equal to $\prod_i (1/m_i!)$ if the final state contains m_i identical particles of type (i).

A similar calculation (Problem 10.3) shows that the rate for a particle of energy ω to disintegrate into N daughter particles with momenta $k'_1, \ldots, k'_N$ is

$$\mathrm{d}\Gamma = \frac{1}{2\omega} (2\pi)^4 \delta^{(4)}\left(k - \sum_1^N k'_i \right) |\langle f | T | \boldsymbol{k} \rangle|^2$$
$$\times \frac{\mathrm{d}^3 k'_1}{(2\pi)^3 2\omega'_1} \cdots \frac{\mathrm{d}^3 k'_N}{(2\pi)^3 2\omega'_N} \mathscr{S}. \tag{10.3.12}$$

10.3.2 Application to 2 particle → 2 particle processes

We illustrate (10.3.11) through the 2 particle → 2 particle reaction

$$k_1 + k_2 \to k'_1 + k'_2,$$

by calculating the differential cross-section $\mathrm{d}\sigma/\mathrm{d}\Omega$ in the centre-of-mass frame. By invariance under rotations around $\boldsymbol{k}_1$, the cross-section obviously cannot depend on the azimuthal angle φ, and depends only on the polar angle θ (Fig. 10.4).

We go to the centre-of-mass frame, where

$$k_1 = (\omega_1, \boldsymbol{k}), \quad k_2 = (\omega_2, -\boldsymbol{k}), \quad s = E^2 = (k_1 + k_2)^2.$$

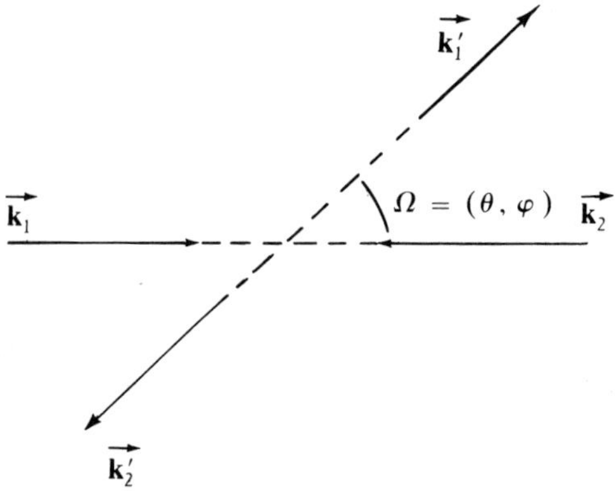

Fig. 10.4 Two-particle kinematics

Simple kinematics (Problem 10.2) yields

$$F = \sqrt{s}\,\|\boldsymbol{k}\|. \tag{10.3.13}$$

Next we calculate the 2-particle phase-space element

$$\mathrm{d}\Phi^{(2)} = \int \frac{\mathrm{d}^3 k'_1}{(2\pi)^3 2\omega'_1} \frac{\mathrm{d}^3 k'_2}{(2\pi)^3 2\omega'_2} (2\pi)^4 \delta^{(4)}(K_i - k'_1 - k'_2).$$

The integration over $\mathrm{d}^3 k'_2$ is immediate; it leaves us with one delta function, whence

$$\int \mathrm{d}^3 k'\, \delta(E - \omega'_1 - \omega'_2) = \int k'^2\, \mathrm{d}k'\, \mathrm{d}\Omega\, \delta(E - \omega'_1 - \omega'_2)$$
$$= \frac{\omega'_1 \omega'_2 k'}{E}\, \mathrm{d}\Omega,$$

where Ω stands for solid angle, and where we have used

$$\frac{\mathrm{d}}{\mathrm{d}k'}(\omega'_1 + \omega'_2) = \frac{k'}{\omega'_1} + \frac{k'}{\omega'_2} = \frac{k'E}{\omega'_1 \omega'_2} \quad \text{since} \quad \omega'_i = \sqrt{(\boldsymbol{k}'^2 + m_i'^2)}.$$

Thus the phase-space element is

$$\mathrm{d}\Phi^{(2)} = \frac{1}{(2\pi)^2} \frac{k'}{4E}\, \mathrm{d}\Omega = \frac{k'\, \mathrm{d}\Omega}{16\pi^2 \sqrt{s}}. \tag{10.3.14}$$

Substituting from (10.3.13) and (10.3.14) into (10.3.11) we find the end-result for the two-body differential cross-section,

$$\frac{\mathrm{d}\sigma}{\mathrm{d}\Omega} = \frac{1}{64\pi^2 s} \frac{\|\boldsymbol{k}'\|}{\|\boldsymbol{k}\|} |\langle \boldsymbol{k}'_1, \boldsymbol{k}'_2 | T | \boldsymbol{k}_1, \boldsymbol{k}_2 \rangle|^2. \tag{10.3.15}$$

10.3.3 Evaluation of an S-matrix element

Before establishing the general relation between S (or T) matrix elements and Green's functions, we use a simple example in order to explain how such matrix elements can be calculated perturbatively. We shall require Wick's theorem for the operator products $(\varphi(x_1)\ldots\varphi(x_N))$. The result is very simple: we need merely return to equation (9.3.28), and on its right replace the contractions $\overbrace{\varphi(x_1)\varphi(x_2)}$ by

$$\underbrace{\varphi(x_1)\varphi(x_2)} = \langle 0|\varphi(x_1)\varphi(x_2)|0\rangle.$$

Here we have dropped the index 'in': $\varphi_{\text{in}}(x) \to \varphi(x)$. For the product of say two fields one finds

$$\varphi(x_1)\varphi(x_2) = :\varphi(x_1)\varphi(x_2): + \underbrace{\varphi(x_1)\varphi(x_2)}.$$

The proof of this theorem is elementary, and is set as Problem 10.4. Within the framework described by the Lagrangean (10.0.3b), let us try to evaluate, to second order of perturbation theory, the S-matrix element for the process

$$k_1 + k_2 \to k_1' + k_2'.$$

The expansion of (10.1.9) to second order reads

$$S_2(x', x) = \left(\frac{-ig}{4!}\right)^2 T(:\varphi^4(x')::\varphi^4(x):). \tag{10.3.16}$$

One's first reaction to the problem of evaluating S_{fi} is to write the matrix element as

$$\begin{aligned}&\langle \boldsymbol{k}_1', \boldsymbol{k}_2'|S_2(x', x)|\boldsymbol{k}_1, \boldsymbol{k}_2\rangle\\ &\quad = \langle 0|a(\boldsymbol{k}_1')\,a(\boldsymbol{k}_2')S_2(x', x)a^\dagger(\boldsymbol{k}_1)a^\dagger(\boldsymbol{k}_2)|0\rangle.\end{aligned} \tag{10.3.17}$$

This expression is correct to this order of perturbation theory; but in general one must multiply by a further factor which we shall determine later but which we ignore for the moment. In order to evaluate S_2 one expands (10.3.16) by appeal to Wick's theorem,

$$\begin{aligned}\frac{1}{(4!)^2}T(:\varphi^4(x)::\varphi^4(x'):) &= \frac{1}{(4!)^2}:\varphi^4(x')\varphi^4(x): + \frac{1}{(3!)^2}\overbrace{\varphi(x')\varphi(x)}\\ &\quad\times:\varphi^3(x')\varphi^3(x): + \frac{1}{8}(\overbrace{\varphi(x')\varphi(x)})^2:\varphi^2(x')\varphi^2(x):\\ &\quad + \frac{1}{3!}(\overbrace{\varphi(x')\varphi(x)})^3:\varphi(x')\varphi(x): + \frac{1}{4!}(\overbrace{\varphi(x')\varphi(x)})^4,\end{aligned} \tag{10.3.18}$$

where factors like 1/8 are of course related to symmetry factors. Wick's theorem can now be applied to the operator products so as to evaluate the vacuum

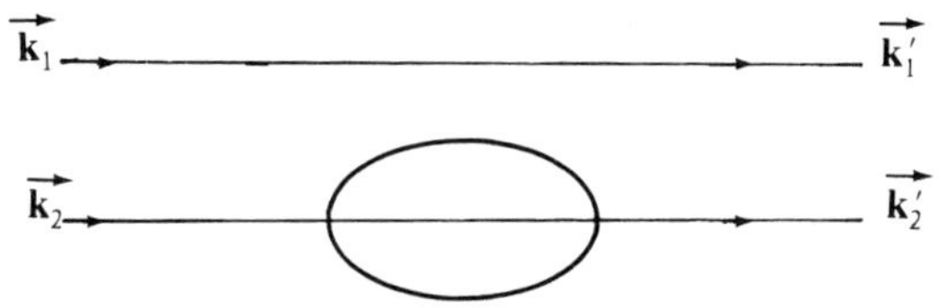

Fig. 10.5 Disconnected diagram for 2 particle → 2 particle scattering

expectation-value in (10.1.17). The only nonzero terms are those that contain nothing but contractions. Moreover, contractions like $\underbrace{a(\boldsymbol{k}_1')\,a^\dagger(\boldsymbol{k}_1)}$ yield factors $\delta^{(3)}(\boldsymbol{k}_1 - \boldsymbol{k}_1')$, corresponding to particles that propagate without interacting (Fig. 10.5).

Such terms correspond to disconnected diagrams of the S-matrix. Here however we are seeking to calculate the connected terms, and observe that all 'external' operators a and $a^\dagger$ must then be contracted with 'interior' operators $\varphi(x')$, $\varphi(x)$. Only the third term of (10.3.18) contributes to connected S-matrix elements, by virtue of contractions like

$$\underbrace{a(\boldsymbol{k}_1')\varphi(x')}\underbrace{a(\boldsymbol{k}_2')\varphi(x')}(\overbrace{\varphi(x')\varphi(x)})^2\underbrace{\varphi(x)a^\dagger(\boldsymbol{k}_1)}\underbrace{\varphi(x)a^\dagger(\boldsymbol{k}_2)},$$

which can be represented by the graph in Fig. 10.6:

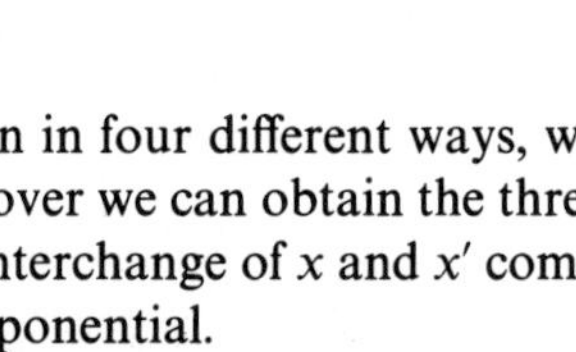

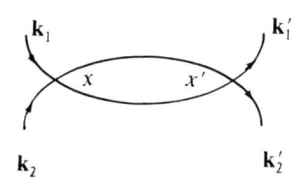

Fig. 10.6 A contribution to 2 particle → 2 particle scattering

This graph can be drawn in four different ways, whence the usual symmetry factor $4 \times 1/8 = 1/2$. Moreover we can obtain the three other graphs by permuting the external lines; the interchange of x and x' compensates for the factor $1/2!$ in the expansion of the exponential.

The contractions $\underbrace{a(\boldsymbol{k})\varphi(x)}$ can be evaluated immediately from the Fourier representation (9.2.12) of $\varphi(x)$, and from the commutation relation (9.2.13), which entails

$$\langle 0|a(\boldsymbol{k})a^\dagger(\boldsymbol{k}')|0\rangle = (2\pi)^3 2\omega_{\boldsymbol{k}}\delta^{(3)}(\boldsymbol{k} - \boldsymbol{k}').$$

Hence we find

$$\underbrace{a(\boldsymbol{k})\varphi(x)} = e^{ikx}, \quad \underbrace{\varphi(x)a^\dagger(\boldsymbol{k})} = e^{-ikx}. \tag{10.3.19}$$

The S-matrix element to second-order perturbation theory reads

$$\langle \boldsymbol{k}'_1, \boldsymbol{k}'_2 | S | \boldsymbol{k}_1 \boldsymbol{k}_2 \rangle = \frac{(-ig)^2}{2} \int \mathrm{d}^4x\, \mathrm{d}^4x'\, \mathrm{e}^{\mathrm{i}(k'_1 + k'_2)x'}\, \mathrm{e}^{-\mathrm{i}(k_1 + k_2)\cdot x} [\Delta_\mathrm{F}(x' - x)]^2$$

$$+ \text{ permutations.} \tag{10.3.20}$$

If we compare this with the result of calculating a connected four-point Green's function $G_\mathrm{c}^{(4)}(x_1, x_2, x_3, x_4)$, we see that the external propagators like $\Delta_\mathrm{F}(x_1 - x)$ have been replaced by exponentials, namely by $\exp(-ikx)$ *for an incoming and* $\exp(ikx)$ *for an outgoing particle*: thus, S-matrix elements are Fourier transforms of Green's functions* G_c, *shorn of their external propagators* $\Delta_\mathrm{F}(x_1 - y_1)$. Accordingly, in Fourier space they are Green's functions shorn of the factors $\mathrm{i}/(k^2 - m^2)$. Moreover, the external lines are 'on the mass shell', i.e. $k_i^2 = k_i'^2 = m^2$, $k_{i0}, k'_{i0} > 0$. One must be careful of the fact that such a Green's function is not a proper vertex, since in general it is not 1-particle irreducible; in particular it can contain self-energy insertions on external lines (see Fig. 10.7). We now proceed to a more comprehensive derivation, by establishing the reduction formulae.

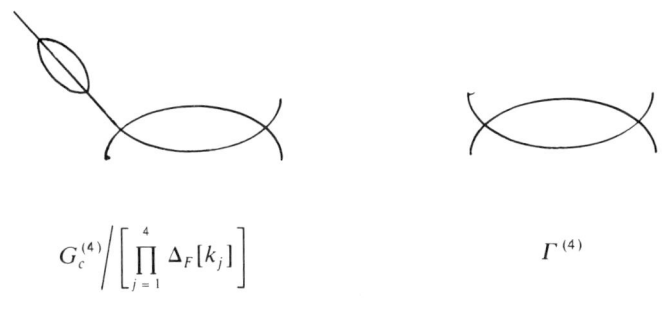

$G_c^{(4)} \Big/ \left[\prod_{j=1}^{4} \Delta_F[k_j] \right]$ $\qquad\qquad$ $\Gamma^{(4)}$

Fig. 10.7

10.3.4 The *S*-matrix in the Heisenberg picture

In the previous derivation of S-matrix elements, we have dropped a multiplicative factor; the reason is that states like $|\boldsymbol{k}_1, \boldsymbol{k}_2\rangle$ are interaction-picture states, and not proper asymptotic states. We must therefore analyse more thoroughly the relationship between the asymptotic states of a collision between elementary particles and their field-theoretical description.

A collision experiment can be analysed as follows. Long before the collision ($t \to -\infty$), the experimentalist has prepared widely separated wave-packets (usually two) which evolve freely and independently. The wave-packets collide at

* In (10.3.19) and (10.3.20), $\exp(ikx) = \exp(\mathrm{i}\omega_k t - \mathrm{i}\boldsymbol{k}\cdot\boldsymbol{x})$; it is understood that the exponential is multiplied by a factor $\theta(k^0)$, and that $k^2 = m^2$.

$t \simeq 0$, and, during the interaction time ($\approx 10^{-23}$ s), it is impossible to follow the fate of individual particles: by contrast to nonrelativistic quantum mechanics, the number of particles is not conserved. After the collision, wave-packets leave the interaction region and, as $t \to \infty$, they evolve again freely and independently, although the number and type of particles are in general different from those of the initial state. The main point of this discussion is to show that the notion of particle has a meaning only in the case of free particles, for $t \to \pm\infty$; in other words, *the notion of particle is only asymptotic*. It does not make sense during the interaction.

Let us now turn to the field theoretic description. We work in the Heisenberg picture, where the state vector is time-independent, while the observables evolve with time. The observables for $t \to -\infty$ will be denoted by O_{in}, those for $t \to +\infty$ by O_{out}. The 'in' and 'out' states will be defined by complete sets of commuting 'in' and 'out' observables, the Hilbert spaces spanned by these states being denoted by $\mathscr{H}_{\text{in}}$ and $\mathscr{H}_{\text{out}}$ respectively. These Hilbert spaces can be identified with the Fock space of *free* fields $\varphi_{\text{in}}(x)$ and $\varphi_{\text{out}}(x)$ of mass m; note that m is the physical mass of the particles, with the self-interactions of the field already accounted for (for simplicity we limit ourselves to the case of a self-interacting neutral scalar field). The fields $\varphi_{\text{in}}(x)$ and $\varphi_{\text{out}}(x)$ will be used to construct the initial and final states of the collision. Although we use the same notations, the field $\varphi_{\text{in}}(x)$ should not be confused with the field φ_{in} of Section 10.1.

We also assume asymptotic completeness: the Hilbert spaces $\mathscr{H}_{\text{in}}$ and $\mathscr{H}_{\text{out}}$ are identical to the Hilbert space $\mathscr{H}$ of the interacting theory, i.e.

$$\mathscr{H}_{\text{in}} = \mathscr{H}_{\text{out}} = \mathscr{H}.$$

The S-matrix describes the totality of all possible measurements at $t \to +\infty$, once the state at $t = -\infty$ has been given. Let $\{|\varphi_\alpha, \text{out}\rangle\}$ and $\{|\varphi_\alpha, \text{in}\rangle\}$ be complete sets of states (α denotes the eigenvalues of a complete set of commuting observables); one can then expand a given state $|\varphi_\alpha, \text{in}\rangle$ with respect to the basis $\{|\varphi_\alpha, \text{out}\rangle\}$, with coefficients $S_{\beta\alpha}$:

$$|\varphi_\alpha, \text{in}\rangle = \sum_\beta S_{\beta\alpha} |\varphi_\beta, \text{out}\rangle,$$

where $|S_{\beta\alpha}|^2$ gives the probability of finding the system in the state $|\varphi_\beta, \text{out}\rangle$ when an 'out'-measurement is performed. It follows that

$$S_{\beta\alpha} = \langle \varphi_\beta, \text{out} | \varphi_\alpha, \text{in} \rangle = \langle \varphi_\beta, \text{in} | S | \varphi_\alpha, \text{in} \rangle. \tag{10.3.21}$$

Since both the sets $\{|\varphi_\alpha, \text{in}\rangle\}$ and $\{|\varphi_\alpha, \text{out}\rangle\}$ are complete sets of states of $\mathscr{H}$, the S-matrix must be unitary,

$$S^\dagger = S^{-1},$$

or in other words

$$\sum_\beta S_{\alpha\beta} S^*_{\beta\alpha'} = \delta_{\alpha\alpha'}.$$

This last equation makes the conservation of probability perfectly explicit. One must always keep in mind that the states $|\varphi_\alpha, \text{in}\rangle$ and $|\varphi_\alpha, \text{out}\rangle$ refer to *different* bases. Assume for example that there are two kinds of particles, say a and b, to which correspond particle-number operators N_a and N_b. The operators $N^{\text{in}}_{a(b)}$ and $N^{\text{out}}_{a(b)}$ will be different in an interacting theory. If $|\varphi_a, \text{in}\rangle$ describes a two-particle state of type (a) we shall have

$$N_a^{\text{in}}|\varphi_a, \text{in}\rangle = 2|\varphi_a, \text{in}\rangle,$$

$$N_b^{\text{in}}|\varphi_a, \text{in}\rangle = 0,$$

but in general $|\varphi_a, \text{in}\rangle$ will not be an eigenstate of N_a^{out}, and the scalar product

$$\langle\varphi_b, \text{out}|\varphi_a, \text{in}\rangle$$

will in general be nonzero.

The second form of $S_{\beta\alpha}$ in (10.3.21) is formally similar to that found in the interaction picture:

$$S_{\beta\alpha} = \langle\varphi_\beta|S|\varphi_\alpha\rangle.$$

However, in the interaction picture the state vectors evolve with time, and the state $|\varphi_\alpha\rangle$ is (loosely speaking) the state vector at $t = -\infty$, while $S|\varphi_\alpha\rangle$ is that state which has developed at $t = +\infty$ out of $|\varphi_\alpha\rangle$. In that case, the states $|\varphi_\alpha\rangle$ and $|\varphi_\beta\rangle$ refer to the *same* basis of $\mathscr{H}$.

The problem is now the following: given an interacting field $\varphi(x)$, construct the asymptotic 'in' and 'out' states and the corresponding fields $\varphi_{\text{in}}(x)$ and $\varphi_{\text{out}}(x)$. This can be done thanks to the (rather involved) Haag–Ruelle construction, which is entirely independent of perturbation theory, and relies on the following axioms.

(1) There exists a unique vacuum $|0\rangle$, and this vacuum is a cyclic vector: repeated applications of $\varphi(x)$ to $|0\rangle$ generate a set of states which is dense in $\mathscr{H}$.

(2) There exists an energy-momentum operator P_μ which acts as an infinitesimal generator of space-time translations:

$$e^{ia_\mu P^\mu}\varphi(x)e^{-ia_\mu P^\mu} = \varphi(x+a). \tag{10.3.22}$$

(3) The spectrum of $P^2 = P^\mu P_\mu$ contains two isolated points $P^2 = 0$ and $P^2 = m^2$; the corresponding eigenstates are the vacuum and one-particle states $|\boldsymbol{k}\rangle$. The continuum (including possible bound states) begins at $P^2 = M^2$, where $m^2 < M^2 \leqslant 4m^2$. These assumptions exclude the case of massless particles, which lead to long-range interactions.

(4) The field $\varphi(x)$ obeys the causality (or locality) axiom:

$$[\varphi(x), \varphi(y)] = 0 \quad \text{if} \quad (x-y)^2 < 0.$$

(5) If $|\boldsymbol{k}\rangle$ is a one-particle state, then

$$\langle 0|\varphi(x)|\boldsymbol{k}\rangle \neq 0.$$

One then constructs the asymptotic states in the following way. Let $\varphi(f, t)$ be the operator (smeared field) defined by

$$\varphi(f, t) = \int \frac{d^4k}{(2\pi)^4}\,\varphi(-k)f(k)e^{i(k_0-\omega_k)t},$$

where $\varphi(k)$ is the Fourier transform of $\varphi(x)$ and $f(k)$ that of a function $f(t, \boldsymbol{x})$, $f(k)$ being nonzero only if $k^0 > 0$ and if

$$(m - \mu)^2 \leqslant k^2 \leqslant (m + \mu)^2,$$

where $(m + \mu) < M$. Let us define the state

$$|\varphi_n(f_1, \ldots, f_n; t)\rangle = \varphi(f_1, t) \ldots \varphi(f_n, t)|0\rangle.$$

One can understand the physical meaning of this state by looking at the free field case, where it is time independent, as

$$\varphi^{\text{in}}_{\text{out}}(f, t) = \int \frac{d^3k}{(2\pi)^3 2\,\omega_k} f(\omega_k, \boldsymbol{k})\, a^{\dagger\text{in}}_{\text{out}}(\boldsymbol{k}).$$

One can then demonstrate the asymptotic condition

$$\lim_{t \to \mp\infty} |\varphi_n(f_1, \ldots, f_n; t)\rangle = |\varphi^{\text{in}}_{n\,\text{out}}(\hat{f}_1, \ldots, \hat{f}_n)\rangle, \tag{10.3.23a}$$

where the $\hat{f}_i$'s are the mass-shell restrictions of the f_i's: $\hat{f}_i(k) = f_i(\omega_k, \boldsymbol{k})$, and the convergence in (10.3.23a) is strong convergence. The limit vectors defined in (10.3.23a) allow the construction of two Fock spaces $\mathscr{H}_{\text{in}}$ and $\mathscr{H}_{\text{out}}$ and of the corresponding free fields φ_{in} and φ_{out}. One must finally add the axiom of asymptotic completeness,

$$\mathscr{H} = \mathscr{H}_{\text{in}} = \mathscr{H}_{\text{out}}.$$

One can relax the condition on the support of $f(k)$ and smear $\varphi(x)$ with a smooth function $f(x)$; then the asymptotic condition (10.3.23a), which can be written schematically as

$$\lim_{t \to \mp\infty} \varphi(f, t) = \varphi^{\text{in}}_{\text{out}}(\hat{f}), \tag{10.3.23b}$$

can be shown to hold, but only in the sense of weak convergence: the above equation has to be sandwiched between 'in' or 'out' states.

Among all observables, the energy-momentum operator P_μ plays a special role, since it is time independent. Since the vacuum state is the only state which obeys $P_\mu|0\rangle = 0$, there is no distinction between 'in' and 'out' vacua, and we have

$$|0\rangle = |0, \text{in}\rangle = |0, \text{out}\rangle.$$

In other words the vacuum is stable. Similarly the one-particle states are stable, and (apart from a possible phase factor which can be taken to be unity) we have

$$|\boldsymbol{k}\rangle = |\boldsymbol{k}, \text{in}\rangle = |\boldsymbol{k}, \text{out}\rangle.$$

Before going further, let us mention that the correspondence between the interacting fields and the free fields describing the asymptotic particles need not be one-to-one, and may even be very complicated. Also, for simplicity, we have assumed the matrix element $\langle 0|\varphi(x)|\boldsymbol{k}\rangle$ to be nonzero, but we could have weakened this assumption by taking a polynomial in $\varphi(x)$, instead of $\varphi(x)$: the asymptotic states may have quantum numbers different from those of the fields!

We now wish to make the connection with perturbation theory, by normalizing the field $\varphi(x)$ through the canonical commutation relations (9.2.14). Let us point out that requiring CCR for an interacting field $\varphi(x)$ is not possible within an axiomatic framework, so that from now on we give up any claim to rigour. Since the normalization implied by (10.3.23b) and that implied by the CCR are in general incompatible, we must introduce in the asymptotic condition a multiplicative factor $Z_3^{1/2}$:

$$\lim_{t\to-\infty} \varphi(x) = Z_3^{1/2}\varphi_{\text{in}}(x);\ \lim_{t\to\infty} \varphi(x) = Z_3^{1/2}\varphi_{\text{out}}(x). \tag{10.3.24}$$

Again this relation must be understood in the sense of weak convergence for the smeared field (otherwise we would soon discover that $Z_3 = 1$ and that $\varphi(x)$ is a free field (Problem 10.5)). We shall show later on that the constant Z_3 is indeed the same for 'in' and 'out' fields, and that it is related to the field renormalization. The field $\varphi(x)$ is often called the interpolating field, and need not be unique: it is not completely determined by the S-matrix.

10.3.5 Reduction formulae

The reduction formulae allow one to relate S-matrix elements to Green's functions, which in turn can be evaluated in perturbation theory. It would be possible to start from (10.3.23a) (see Bros 1971 or Hepp 1965), but it is simpler to follow the original derivation of Lehmann, Symanzik and Zimmermann (LSZ) by using the asymptotic condition in the form (10.3.24).

We shall study the simplest case, 2 particle → 2 particle scattering. For the S-matrix element, (10.3.21) gives

$$S_{fi} = \langle \boldsymbol{k}'_1, \boldsymbol{k}'_2\,\text{out}|\boldsymbol{k}_1, \boldsymbol{k}_2\,\text{in}\rangle = \langle \boldsymbol{k}'_1\boldsymbol{k}'_2, \text{out}|a^\dagger_{\text{in}}(\boldsymbol{k}_1)|\boldsymbol{k}_2\rangle. \tag{10.3.25}$$

As pointed out previously, we have $|\boldsymbol{k}, \text{out}\rangle = |\boldsymbol{k}, \text{in}\rangle = |\boldsymbol{k}\rangle$. We exploit the expression

$$a^\dagger_{\text{in}}(\boldsymbol{k}_1) = -\mathrm{i}\int \mathrm{d}^3x\, \mathrm{e}^{-\mathrm{i}k_1x}\overleftrightarrow{\partial}_0\, \varphi_{\text{in}}(x), \tag{10.3.26}$$

easily obtained by comparison with (9.1.20b): we also use (10.3.24), and find

$$S_{fi} = \lim_{t\to-\infty} -\mathrm{i}Z_3^{-1/2}\int_t \mathrm{d}^3x\, \mathrm{e}^{-\mathrm{i}k_1x}\overleftrightarrow{\partial}_0\, \langle \boldsymbol{k}'_1\boldsymbol{k}'_2, \text{out}|\varphi(x)|\boldsymbol{k}_2\rangle.$$

In view of

$$\left(\lim_{t\to\infty} - \lim_{t\to-\infty}\right)\int_t \mathrm{d}^3x\,\psi(t, \boldsymbol{x}) = \int_{-\infty}^{\infty} \mathrm{d}t\,\frac{\mathrm{d}}{\mathrm{d}t}\int \mathrm{d}^3x\,\psi(t, \boldsymbol{x})$$

this yields

$$S_{fi} = \langle \boldsymbol{k}'_1, \boldsymbol{k}'_2, \text{out}|a^\dagger_{\text{out}}(\boldsymbol{k}_1)|\boldsymbol{k}_2\rangle$$
$$+ \mathrm{i}Z_3^{-1/2}\int \mathrm{d}^4x\,\partial_0[\mathrm{e}^{-\mathrm{i}k_1x}\overleftrightarrow{\partial}_0\langle \boldsymbol{k}'_1\boldsymbol{k}'_2\,\text{out}|\varphi(x)|\boldsymbol{k}_2\rangle]. \tag{10.3.27}$$

The first term is disconnected, containing for example a factor $\delta^{(3)}(\boldsymbol{k}_1 - \boldsymbol{k}'_1)$, and we disregard it. The second term we integrate by parts (strictly speaking one should work with wave-packets rather than plane waves, which would justify dropping the terms integrated over space):

$$\partial_0^2 e^{-ik_1x} = -\omega_1^2 e^{-ik_1x} = -(\boldsymbol{k}_1^2 + m^2)e^{-ik_1x} \to (\nabla^2 - m^2)e^{-ik_1x}$$

$$\int d^4x[(\nabla^2 - m^2)e^{-ik_1x}]\langle \ldots |\varphi(x)| \ldots \rangle$$
$$= \int d^4x\, e^{-ik_1x}(\nabla^2 - m^2)\langle \ldots |\varphi(x)| \ldots \rangle;$$

then (10.3.27) becomes

$$S_{fi} = (\text{disconnected term}) + iZ_3^{-1/2}\int d^4x\, e^{-ik_1x}(\Box_x + m^2)\langle \boldsymbol{k}_1', \boldsymbol{k}_2' \text{ out}|\varphi(x)|\boldsymbol{k}_2\rangle. \tag{10.3.28}$$

We proceed by rewriting the last matrix element as

$$\langle \boldsymbol{k}_1', \boldsymbol{k}_2' \text{ out}|\varphi(x)|\boldsymbol{k}_2\rangle = \langle \boldsymbol{k}_2'|a_{\text{out}}(\boldsymbol{k}_1')\varphi(x)|\boldsymbol{k}_2\rangle$$
$$= \lim_{t'\to\infty} iZ_3^{-1/2}\int d^3x'\, e^{ik_1'x'}\overleftrightarrow{\partial_0}\langle \boldsymbol{k}_2'|\varphi(x')\varphi(x)|\boldsymbol{k}_2\rangle.$$

Since $t' = x_0' \to \infty$, we can replace $\varphi(x')\varphi(x)$ by the T-product $T(\varphi(x')\varphi(x))$. Merely by repeating the foregoing manipulations one obtains

$$\langle \boldsymbol{k}_1', \boldsymbol{k}_2' \text{ out}|\varphi(x)|\boldsymbol{k}_2\rangle = \langle \boldsymbol{k}_2'|\varphi(x)a_{\text{in}}(\boldsymbol{k}_2')|\boldsymbol{k}_2\rangle$$
$$+ iZ_3^{-1/2}\int d^4x'\, e^{ik_1'x'}(\Box_{x'} + m^2)\langle \boldsymbol{k}_2'|T(\varphi(x')\varphi(x))|\boldsymbol{k}_2\rangle, \tag{10.3.29}$$

where the first term is disconnected. The operation can obviously be continued for particles $\boldsymbol{k}_2$ and $\boldsymbol{k}_2'$, yielding the connected S-matrix element (since there are only two initial particles) as a vacuum expectation-value:

$$S_{fi|\text{connected}} = (i)^4(Z_3^{-1/2})^4\int d^4x_1\, d^4x_2\, d^4x_1'\, d^4x_2'$$
$$\times e^{-i(k_1x_1 + k_2x_2)}e^{i(k_1'x_1' + k_2'x_2')}(\Box_{x_1'} + m^2)(\Box_{x_2'} + m^2)$$
$$\times(\Box_{x_1} + m^2)(\Box_{x_2} + m^2)\langle 0|T(\varphi(x_1)\varphi(x_2)\varphi(x_1')\varphi(x_2'))|0\rangle. \tag{10.3.30}$$

Clearly this formula generalizes to arbitrary S-matrix elements according to the following rules:

- to every incoming particle assign a factor $iZ_3^{-1/2}e^{-ikx}(\Box_x + m^2)$;
- to every outgoing particle assign a factor $iZ_3^{-1/2}e^{ik'x'}(\Box_{x'} + m^2)$.

The vacuum expectation-value in (10.3.30) is a Green's function, and the factor $i(\Box_x + m^2)$ amputates an external propagator Δ_F; in Fourier space this factor amounts to $-i(k^2 - m^2)$. The end-result for the connected S-matrix element $\langle \boldsymbol{k}_1' \ldots \boldsymbol{k}_M' \text{ out}|\boldsymbol{k}_1 \ldots \boldsymbol{k}_N, \text{in}\rangle$ is the general rule

$$\langle \boldsymbol{k}'_1 \ldots \boldsymbol{k}'_M \text{ out} | \boldsymbol{k}_1 \ldots \boldsymbol{k}_N \text{ in} \rangle_c$$
$$= \lim_{k_j^2 \to m^2} \prod_{j=1}^{N} [-iZ_3^{-1/2}(k_j^2 - m^2)\theta(k_j^0)] \lim_{k_l'^2 \to m^2} \prod_{l=1}^{M} [-iZ_3^{-1/2}$$
$$\times (k_l'^2 - m^2)\theta(k_l'^0)] \, G_c^{(N+M)}(k_1, \ldots, k_N; -k'_1, \ldots, -k'_M) \tag{10.3.31}$$

To eliminate all ambiguity, the factors $\theta(k_0)$, tacitly understood in the equations that went before, have now been written out explicitly. The $(-)$ sign for final-state particles allows for the fact that under our conventions (see (10.1.21)) all momenta are taken as running towards (into) the diagram, while the k' for outgoing particles run away from (out of) the diagram.

10.3.6 The *S*-matrix and the renormalized Green's functions

It remains to establish the connection between the factors Z_3 appearing in the reduction formula (10.3.31), and the renormalization constant Z_3 introduced in Chapter 6. Start by considering the matrix element of the field $\varphi(x)$ between the vacuum and the single-particle state $|\boldsymbol{k}\rangle$ (recall that with suitably chosen phases $|\boldsymbol{k}_{\text{in}}\rangle = |\boldsymbol{k}_{\text{out}}\rangle = |\boldsymbol{k}\rangle$); using the energy-momentum operator P_μ which is also the space–time translation operator, we obtain

$$\langle 0|\varphi(x)|\boldsymbol{k}\rangle = \langle 0|e^{iPx}\varphi(0)e^{-iPx}|\boldsymbol{k}\rangle = (\text{constant})e^{-ikx}, \tag{10.3.32}$$

while

$$\langle 0|\varphi_{\text{in}}(x)|\boldsymbol{k}\rangle = e^{-ikx}.$$

By taking the limit $t \to -\infty$, one sees in the light of the asymptotic condition (10.3.24) that the constant in (10.3.32) must equal $Z_3^{1/2}$; and taking the limit $t \to +\infty$, one can show that $\lim_{t\to\infty} \varphi(x) = Z_3^{1/2}\varphi_{\text{out}}(x)$, which we have used already, though without proof. Now suppose that in trying to calculate the full propagator $G^{(2)}(x-y)$, we isolate the contribution from single-particle states,

$$\langle 0|T(\varphi(x)\varphi(y))|0\rangle = \theta(x^0 - y^0)\langle 0|\varphi(x)\varphi(y)|0\rangle$$
$$+ \theta(y^0 - x^0)\langle 0|\varphi(y)\varphi(x)|0\rangle = \Big\{ \theta(x^0 - y^0)\sum_{\boldsymbol{k}} \langle 0|\varphi(x)|\boldsymbol{k}\rangle$$
$$\times \langle \boldsymbol{k}|\varphi(y)|0\rangle + \theta(y^0 - x^0)\sum_{\boldsymbol{k}} \langle 0|\varphi(y)|\boldsymbol{k}\rangle\langle \boldsymbol{k}|\varphi(x)|0\rangle \Big\}$$
$$+ \Big\{ \theta(x^0 - y^0)\sum_{n} \langle 0|\varphi(x)|n\rangle\langle n|\varphi(y)|0\rangle$$
$$+ \theta(y^0 - x^0)\sum_{n} \langle 0|\varphi(y)|n\rangle\langle n|\varphi(x)|0\rangle \Big\},$$

where $\sum_n$ sums over a complete set of eigenstates of the full Hamiltonian, excluding single-particle states (and the vacuum state).

Inside the first pair of curly brackets we can evaluate the matrix elements by using (10.3.32), with constant $= Z_3^{1/2}$. These brackets contain simply $Z_3 \Delta_F(x-y; m^2)$; if we had used a free field, then only the single-particle states could contribute. Accordingly we find

$$G^{(2)}(x-y) = Z_3 \Delta_F(x-y; m^2) + \cdots,$$

or, in Fourier space (see Problem 10.5),

$$G^{(2)}(k) = \frac{iZ_3}{k^2 - m^2 + i\varepsilon} + \int_{4m^2}^{\infty} dm'^2 \rho(m'^2) \Delta_F(k; m'^2), \tag{10.3.33}$$

where the 'spectral function' $\rho(m^2)$ is positive. Equation (10.3.33) is the Källen–Lehmann representation of the propagator. So far we have not renormalized anything (except the mass: m in (10.3.33) is the physical mass), and for the moment we must think of $\varphi(x)$ as the bare field; in other words $\varphi(x) \to \varphi_0(x)$, whose normalization is fixed by the CCR. After renormalization we have

$$\varphi_0(x) = \hat{Z}_3^{1/2} \varphi_R(x),$$

where $\varphi_R(x)$ is the renormalized field and $\hat{Z}_3$ the field-renormalization constant (see (6.4.3))*. Start by studying the behaviour of the renormalized propagator in the vicinity of $k^2 = m^2$:

$$G_R^{(2)}(k) \underset{k^2 \to m^2}{\simeq} \frac{iZ_3 \hat{Z}_3^{-1}}{k^2 - m^2 + i\varepsilon} = \frac{iz_3}{k^2 - m^2 + i\varepsilon}. \tag{10.3.34}$$

This defines the constant z_3: iz_3 is the residue at the pole of the renormalized propagator at $k^2 = m^2$. Next we express the cross-section in terms of the Green's functions shorn of their full propagators $\bar{G}_c^{(N)}(k_i)$ (see (10.1.20b)); the index 0 identifies a bare correlation function. Note that

$$\lim_{k^2 \to m^2} (k^2 - m^2) G_0^{(2)}(k) = iZ_3, \tag{10.3.35}$$

whence, given that the $\bar{G}^{(N)}$ transform under renormalization like the $\Gamma^{(N)}$, we have

$$\begin{aligned} \lim_{k_i^2 \to m^2} \left\{ \prod_{i=1}^{N+M} -i(k_i^2 - m^2) \right\} Z_3^{-(N+M)/2} G_{c,0}^{(N+M)}(k_i) \\ = Z_3^{(N+M)/2} \bar{G}_{c,0}^{(N+M)}(k_i) = (Z_3 \hat{Z}_3^{-1})^{(N+M)/2} \bar{G}_{c,R}^{(N+M)}(k_i) \\ = z_3^{(N+M)/2} \bar{G}_{c,R}^{(N+M)}(k_i); \end{aligned} \tag{10.3.36}$$

* In view of the divergences in δm^2 and in Z_3, the argument leading to (10.3.37) cannot claim to be more than heuristic. We note only two crucial features. (1) The S-matrix elements are invariants of the renormalization group; (2) the S-matrix is unitary, at least to order by order of perturbation theory (Section 10.4.2).

here we ease the notation by not troubling to distinguish between incoming and outgoing particles. The end-result for the S-matrix in terms of renormalized Green's functions reads

$$\langle \boldsymbol{k}'_1, \ldots, \boldsymbol{k}'_M \text{ out} | \boldsymbol{k}_1, \ldots, \boldsymbol{k}_N \text{ in} \rangle_c$$
$$= \lim_{k_i^2 \to m^2} \lim_{k_j'^2 \to m^2} z_3^{(N+M)/2} \bar{G}_{c,R}^{(N+M)}(k_1, \ldots, k_N; -k'_1, \ldots, -k'_M). \quad (10.3.37)$$

When we renormalize on the mass shell, the factor z_3 is unity by definition (one of the advantages of this scheme); but in any other scheme (MS, etc.) it is essential to remember the presence of this constant. The argument that led us to (10.3.37) shows that the S-matrix element is independent of the renormalization scheme; in other words it is, as it must be, an invariant of the renormalization group.

To calculate S-matrix elements in practice, one might use the technique of Section 10.3.3, which for particles with spin is easier than the reduction formulae. One need merely multiply the result by the appropriate power of z_3.

It is far from obvious that the unitarity of the S-matrix and the locality properties of the theory survive under renormalization. The conclusion that everything comes out all right in the end follows from technically quite complex work by authors referred to collectively as BPHZ, and by Epstein and Glaser (1973).

10.4 The unitarity of the *S*-matrix

Unitarity ($S^\dagger = S^{-1}$) is a crucial property of the S-matrix, since it is the guarantee of the conservation of probability. Without such a guarantee the theory becomes totally meaningless. The present section studies two problems.

- Assuming $S^\dagger = S^{-1}$, we use this relation to derive for a simple case an important property of the elements of the T-matrix.
- Quite generally we establish rules for 'cutting' diagrams, called the Cutkosky rules. These rules enable one to verify the (perturbative) unitarity of the S-matrix; and they prove very useful when one constructs the theory from a path integral, from which the unitarity property is not in the least selfevident.

10.4.1 Unitarity and the dispersion relations

Revert once more to the favourite example of the present chapter, namely to a second-order contribution to the scattering process $k_1 + k_2 \to k'_1 + k'_2$ in φ^4 theory, corresponding to the graph in Fig. 10.6. This graph was calculated in Section 10.2, equation (10.2.13), where the factor $-\mathrm{i}$ stemming from the definition (10.3.4) of the T-matrix is identically the same as in $\Gamma^{(N)}$ (see (10.1.20c)). On the other hand, instead of normalizing at $k_i = 0$, one can normalize at some point $s_0 < 4m^2$, where $s = (k_1 + k_2)^2$. Under these conditions the expression for $T^{(2)}$ becomes

$$T^{(2)}(s, s_0) = \frac{-g^2}{2(4\pi)^2} \int_0^1 \mathrm{d}x \ln\left[\frac{m^2 - sx(1-x)}{m^2 - s_0 x(1-x)}\right]. \tag{10.4.1}$$

It proves convenient to change variables,

$$x = \frac{1}{2} + \frac{1}{2}y, \quad x(1-x) = \frac{1}{4}(1 - y^2),$$

$$T^{(2)}(s, s_0) = \frac{-g^2}{16(2\pi)^2} \int_{-1}^1 \mathrm{d}y \ln\left[\frac{4m^2 - s(1-y^2)}{4m^2 - s_0(1-y^2)}\right]. \tag{10.4.2}$$

The logarithm is well-defined when $(1 - y^2) < 4m^2/s$, and has a branch point at $s = 4m^2(1 - y^2)$. Since the branch point remains on the real axis as y varies, we define $T^{(2)}(s, s_0)$ in the complex s-plane cut from $4m^2$ to infinity (Fig. 10.8); we choose the logarithm to be real when $s < 4m^2$, while for $s(1 - y^2) > 4m^2$

$$\ln(4m^2 - s(1-y^2)) = \ln|4m^2 - s(1-y^2)| + \mathrm{i}(\theta - \pi). \tag{10.4.3}$$

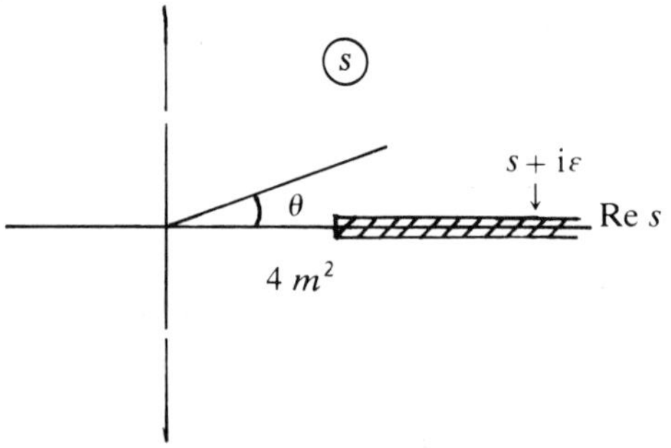

Fig. 10.8 The cut complex s plane

Then the function $T^{(2)}(s)$ is analytic in the cut s-plane, because the integral over y is uniformly convergent. The prescription $m^2 \to m^2 - \mathrm{i}\varepsilon$ entails that for $s \geqslant 4m^2$ we must define $T^{(2)}(s, s_0)$ by

$$\lim_{\varepsilon \to 0^+} T^{(2)}(s + \mathrm{i}\varepsilon, s_0). \tag{10.4.4}$$

In other words the cut must be approached from the upper half-plane ($\mathrm{Im}\, s > 0$). This determines $\mathrm{Im}\, T^{(2)}(s, s_0)$:

$$\mathrm{Im}\, T^{(2)}(s, s_0) = \frac{\pi g^2}{16(2\pi)^2} \int_{-(1-4m^2/s)^{1/2}}^{(1-4m^2/s)^{1/2}} \mathrm{d}y = \frac{\pi g^2}{8(2\pi)^2} \sqrt{\left(\frac{s - 4m^2}{s}\right)}. \tag{10.4.5}$$

Quite generally, we consider a function $f(s)$ of the complex variable s, analytic in the complex plane cut from $4m^2$ to infinity, and required moreover to be *real analytic*, i.e. to satisfy the condition

$$f^*(s) = f(s^*). \tag{10.4.6}$$

Applying Cauchy's theorem to the contour C in Fig. 10.9, and assuming that the integral over the large circle tends to zero, one obtains

$$f(s) = \frac{1}{2i\pi}\int_{4m^2}^{\infty}\frac{ds'}{s'-s}\,[f(s'+i\varepsilon) - f(s'-i\varepsilon)],$$

whence

$$f(s) = \frac{1}{\pi}\int_{4m^2}^{\infty}\frac{\operatorname{Im} f(s')\,ds'}{s'-s}. \tag{10.4.7}$$

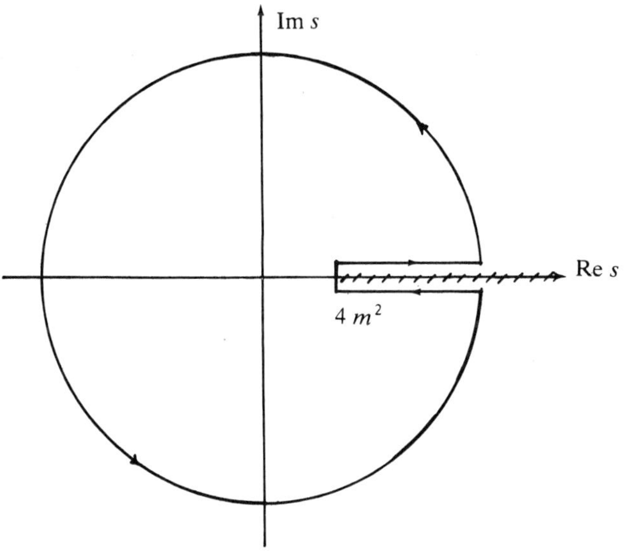

Fig. 10.9 Contour for Cauchy's theorem

Here the second line follows from the first by virtue of the property (10.4.6):

$$f(s'+i\varepsilon) - f(s'-i\varepsilon) = 2i\operatorname{Im} f(s'+i\varepsilon) = 2i\operatorname{Im} f(s')$$

(when no confusion can result we shall by convention write $f(s)$ instead of $f(s+i\varepsilon)$). If the integral along the large circle does not vanish, then we make one (or more) subtractions:

$$f(s) - f(s_0) = \frac{s-s_0}{\pi}\int_{4m^2}^{\infty}\frac{\operatorname{Im} f(s')\,ds'}{(s'-s)(s'-s_0)}. \tag{10.4.8}$$

Equations (10.4.7) and (10.4.8) are called *dispersion relations*. In cases with one subtraction, knowledge of $\operatorname{Im} f(s)$ determines $f(s)$ up to an arbitrary constant $f(s_0)$. If N subtractions are needed, then one must introduce N arbitrary constants. In the case of $T^{(2)}$ a single subtraction suffices, and (10.4.5) yields

$$T^{(2)}(s, s_0) = T^{(2)}(s_0, s_0) + \frac{g^2(s - s_0)}{8(2\pi)^2} \int_{4m^2}^{\infty} \mathrm{d}s' \sqrt{\left(\frac{s' - 4m^2}{s'}\right)} \times \frac{1}{(s' - (s + \mathrm{i}\varepsilon))(s' - s_0)}. \tag{10.4.9}$$

The imaginary part (10.4.5) can be found from the unitarity condition. It happens that the two other second-order graphs are real for $s \geqslant 4m^2$; they might have cuts in the variables $t = (k_1 - k_3)^2$ and $u = (k_1 - k_4)^2$, but t and u are negative when $s \geqslant 4m^2$. In terms of the T-matrix defined by (10.3.4) the relation $SS^\dagger = 1$ reads

$$T_{fi} - T^\dagger_{fi} = \mathrm{i} \sum_n (2\pi)^4 \delta^{(4)}(K_f - K_i) T_{fn} T^\dagger_{ni}. \tag{10.4.10}$$

Invariance under time-reversal entails

$$\langle \boldsymbol{k}'_1 \boldsymbol{k}'_2 | T | \boldsymbol{k}_1 \boldsymbol{k}_2 \rangle = \langle -\boldsymbol{k}_1 - \boldsymbol{k}_2 | T | -\boldsymbol{k}'_1 - \boldsymbol{k}'_2 \rangle = \langle \boldsymbol{k}_1 \boldsymbol{k}_2 | T | \boldsymbol{k}'_1 \boldsymbol{k}'_2 \rangle,$$

and to second-order perturbation theory (10.4.10) yields

$$\operatorname{Im} T^{(2)} = \frac{1}{4} \int \frac{\mathrm{d}^3k \, \mathrm{d}^3k'}{(2\pi)^3 2\omega (2\pi)^3 2\omega'} (2\pi)^4 \delta^{(4)}(k + k' - k_1 - k_2) |T^{(1)}|^2. \tag{10.4.11}$$

Here we recognize the phase-space element $\mathrm{d}\Phi^{(2)}$ from (10.3.14); integrated over $\mathrm{d}\Omega$ it becomes

$$\frac{\|\boldsymbol{k}\|}{4\pi\sqrt{s}} = \frac{1}{8\pi} \sqrt{\left(\frac{s - 4m^2}{s}\right)}.$$

In (10.4.11) we have taken into account too a factor 1/2 due to the indistinguishability of the particles; the matrix element $T^{(1)}$ is of the simplest,

$$T^{(1)} = -g,$$

while

$$\operatorname{Im} T^{(2)} = \frac{\pi g^2}{8(2\pi)^2} \sqrt{\left(\frac{s - 4m^2}{s}\right)},$$

in exact agreement with (10.4.5). One sees from this example that the prescription $m^2 \to m^2 - \mathrm{i}\varepsilon$ is essential to guarantee the unitarity of the S-matrix. Moreover, the existence of the dispersion relations is closely linked with the causality property. To this property too the prescription $m^2 \to m^2 - \mathrm{i}\varepsilon$ is essential.

Our example furnishes an elementary illustration of several properties that we expect in T-matrix elements generally, and which have been much studied.

(i) T-matrix elements can be continued analytically into the complex plane(s) of their kinematic variable(s).

(ii) Admissible intermediate states give rise to cuts in these complex planes. Bound states or elementary particles correspond to poles; in φ^3 theory for example the $2 \to 2$ scattering amplitude has a pole at $s = m^2$ and a cut along $s \geqslant 4m^2$ (Fig. 10.10).

Pole at $s = m^2$

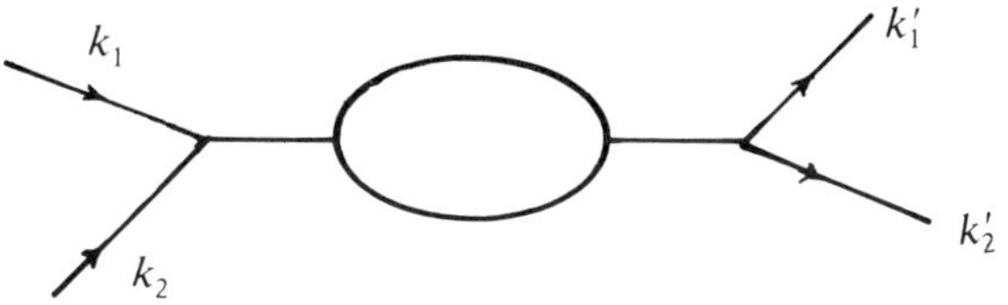

Pole at $s = m^2$ + cut along $s \geqslant 4m^2$

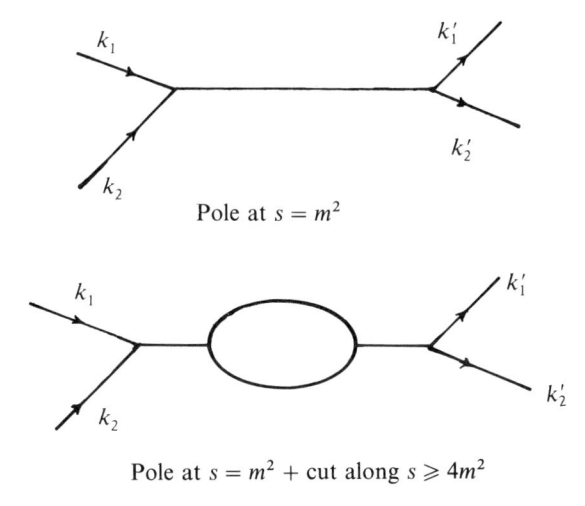

Fig. 10.10 Diagrams responsible for poles and cuts in s

(iii) By summing over the intermediate states one can calculate the imaginary part of the amplitude in terms of expressions to lower orders of perturbation theory. This observation leads to one possible method for calculating certain diagrams.

(iv) The real part is calculated from a dispersion relation, which may need subtractions. These introduce unknown constants, reflecting the necessity of renormalization.

10.4.2 The Cutkosky rules

Following 't Hooft and Veltman (1974) we shall now establish the general rules for computing the imaginary parts of Green's functions: they are called the *Cutkosky* (or *cutting*) *rules*. We use the functions

$$\Delta^{\pm}(x) = \int \frac{d^4k}{(2\pi)^4} 2\pi\delta(k^2 - m^2)\theta(\pm k^0)e^{-ikx}, \tag{10.4.12}$$

which satisfy

$$\begin{aligned} &\Delta^{\pm}(x) = (\Delta^{\mp}(x))^*, \quad \Delta^{\pm}(-x) = \Delta^{\mp}(x), \\ &\Delta^{+}(x) = \langle 0|\varphi(x)\varphi(0)|0\rangle, \quad \Delta^{-}(x) = \langle 0|\varphi(0)\varphi(x)|0\rangle. \end{aligned} \tag{10.4.13}$$

Recall that

$$\Delta_F(x) = \theta(x^0)\Delta^+(x) + \theta(-x^0)\Delta^-(x). \tag{10.4.14}$$

It will prove convenient to use the symbols

$$\Delta_F(x_i - x_j) = \Delta_{ij}, \quad \Delta^{\pm}(x_i - x_j) = \Delta^{\pm}_{ij}, \tag{10.4.15}$$

and to note that complex conjugation interchanges positive and negative frequencies:

$$\Delta^*_{ij} = \theta(x_i^0 - x_j^0)\Delta^-_{ij} + \theta(x_j^0 - x_i^0)\Delta^+_{ij}.$$

Green's functions shorn of their full external propagators will be written $F(x_1, \ldots, x_n)$; they consist of vertices and propagators, e.g.

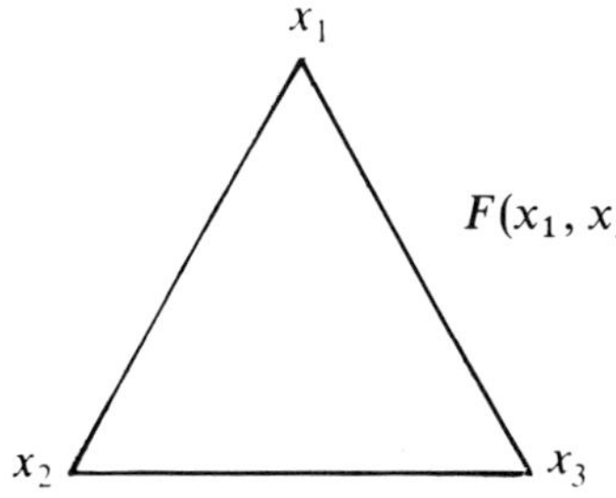

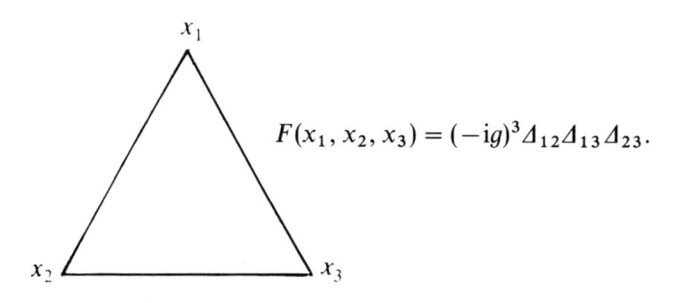

$$F(x_1, x_2, x_3) = (-ig)^3\Delta_{12}\Delta_{13}\Delta_{23}.$$

To be definite, the discussion will be illustrated by φ^3 theory; moreover we shall disregard problems of renormalization, which is justifiable if we confine ourselves to a space–time of appropriate dimensionality.

Starting from $F(x_1, \ldots, x_i, \ldots, x_j, \ldots, x_n)$ we define a function $F(x_1, \ldots, \underline{x}_i, \ldots, x_j, \ldots, x_n)$, in which some of the x_i are underlined, according to the following conventions:

(i) $\Delta_{kl} \to \Delta_{kl}$ if neither x_k nor x_l is underlined;

(ii) $\Delta_{kl} \to \Delta^+_{kl}$ if x_k is underlined but x_l is not;

(iii) $\Delta_{kl} \to \Delta^-_{kl}$ if x_l is underlined but x_k is not;

(iv) $\Delta_{kl} \to \Delta^*_{kl}$ if both x_k and x_l are underlined;

(v) $-ig \to ig$, where $(+ig)$ corresponds to an underlined vertex.

In brief,

$$\Delta_{\underline{k}l} = \Delta^+_{kl}; \quad \Delta_{k\underline{l}} = \Delta^-_{kl}; \quad \Delta_{\underline{k}\,\underline{l}} = \Delta^*_{kl}.$$

Note that these rules are selfconsistent. Indeed $\Delta_{kl} = \Delta_{lk}$, $\Delta_{\underline{k}l} = \Delta^{+}_{kl}$, and $\Delta_{l\underline{k}} = \Delta^{-}_{lk}$; but, from (10.4.13), $\Delta^{+}_{kl} = \Delta^{-}_{lk}$. If x_i is underlined, then on the graph we mark the corresponding vertex with a circle. For example:

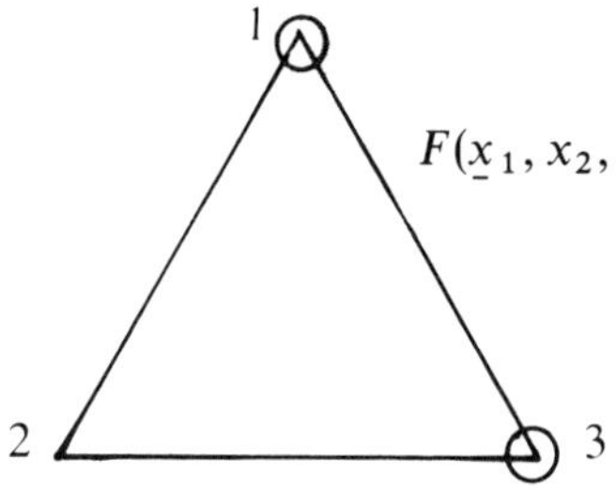

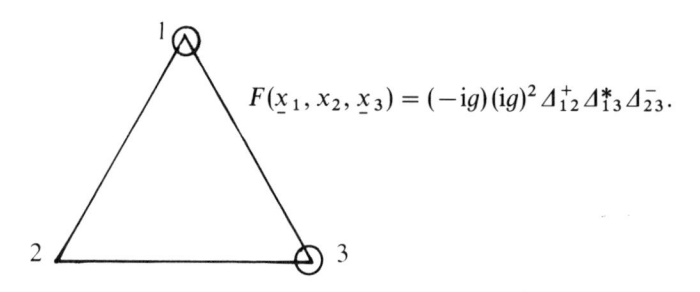

$$F(\underline{x}_1, x_2, \underline{x}_3) = (-ig)(ig)^2 \Delta^{+}_{12} \Delta^{*}_{13} \Delta^{-}_{23}.$$

On the other hand, if $x_i^0 > x_j^0$, then by virtue of (10.4.14) we have $\Delta_{\underline{i}j} = \Delta_{ij}$ and $\Delta_{i\underline{j}} = \Delta_{\underline{i}\,\underline{j}}$; underlining $\underline{x}_i$ leaves the propagator unaffected. We thus derive the following theorem:

Theorem 10.1: Suppose that x_i^0 exceeds x_j^0 for all $j \neq i$. Then

$$F(x_1, \ldots, \underline{x}_i, \ldots, \underline{x}_j, \ldots, x_n) = -F(x_1, \ldots, x_i, \ldots, \underline{x}_j, \ldots, x_n), \quad (10.4.16)$$

where the two sides differ only in that x_i is underlined (not underlined) on the left (right), the $-$ sign stemming from the replacement $-ig \to ig$ at the vertex x_i. Now note that in calculating Δ_{ij} or $\Delta^{\pm}_{ij}$ from (10.4.12) and (10.4.15), the factor $\exp[-ik(x_i - x_j)]$ corresponds to a momentum k *entering* the vertex (i). This shows that if two vertices are linked by a factor Δ^{+} or Δ^{-}, then the energy flow is always directed towards the circled vertex:

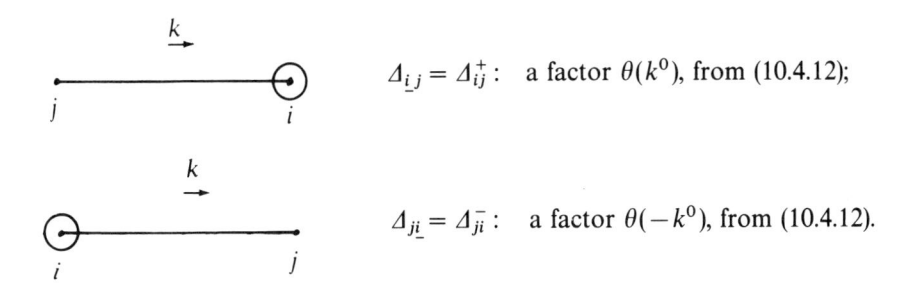

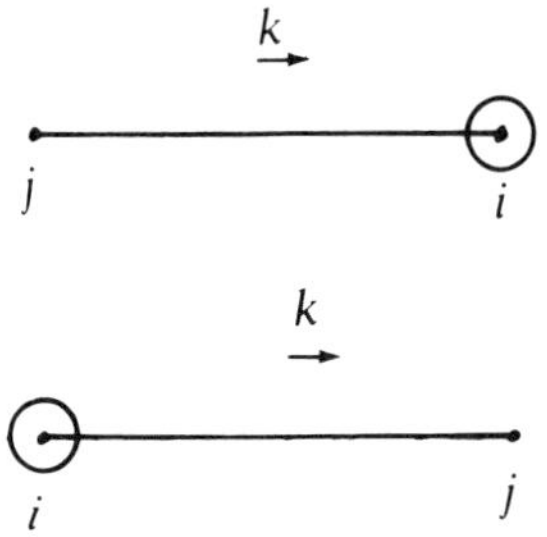

$\Delta_{\underline{i}j} = \Delta^{+}_{ij}$: a factor $\theta(k^0)$, from (10.4.12);

$\Delta_{j\underline{i}} = \Delta^{-}_{ji}$: a factor $\theta(-k^0)$, from (10.4.12).

By contrast, if the internal line corresponds to a factor Δ_{ij} or to a factor Δ^{*}_{ij}, then there is energy flow in both directions.

Next we calculate an S-matrix element; by convention, incoming paticles are drawn on the left of the diagram and outgoing ones on the right. The graph in Fig. 10.11 serves as an example:

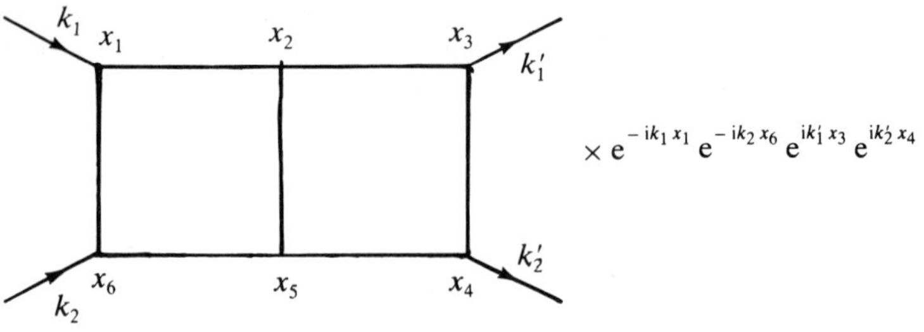

384 | GREEN'S FUNCTIONS AND THE S MATRIX

Fig. 10.11

$$\times e^{-ik_1 x_1} e^{-ik_2 x_6} e^{ik_1' x_3} e^{ik_2' x_4}$$

The function $F(x_1, \ldots, x_6)$ must be multiplied by a product of exponentials given by the rules established in Section 10.3:

incoming particle: e^{-ikx}, outgoing particle: $e^{ik'x}$,

followed by integration over all the x_i. Theorem 10.1 entails the corollary

$$\sum F(x_1, \ldots, \underline{x}_i, \ldots, x_j, \ldots, \underline{x}_n) = 0, \tag{10.4.17}$$

where $\sum$ indicates that one has summed over all possible ways of underlining. Indeed, let x_i be the point having the greatest time-component; all the terms of (10.4.17) can be grouped by pairs so that the two terms in each pair are identical except for just one point, x_i being underlined in one term but not in the other. By virtue of Theorem 10.1, the sum of the two terms vanishes.

At this stage one sees that many diagrams will give zero because they are subject to mutually conflicting requirements. The diagram in Fig. 10.12 for instance vanishes because at the vertex i momentum conservation imposes $k_{01} + k_{02} + k_{03} = 0$, while k_{01}, k_{02}, and k_{03} must all be positive since momentum always flows towards the circled vertex.

A diagram can be nonzero only if it satisfies both of the following conditions:

(i) the circled vertices must constitute a connected region, linked to one or more outgoing lines;

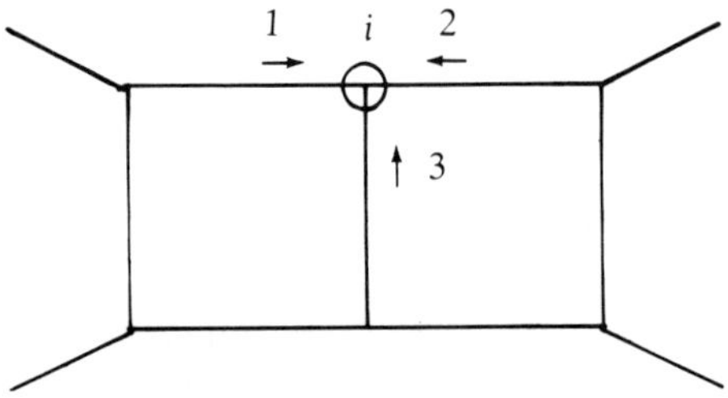

Fig. 10.12

(ii) the vertices that are not circled must constitute a connected region linked to one or more incoming lines.

For suppose that (i) is not satisfied. Then one can cut the diagram along a line bounding a subdiagram such that all the momenta entering the subdiagram are positive (Fig. 10.13), a condition that conflicts with energy conservation. Similarly, the diagram in Fig. 10.14 vanishes, because only positive energies leave

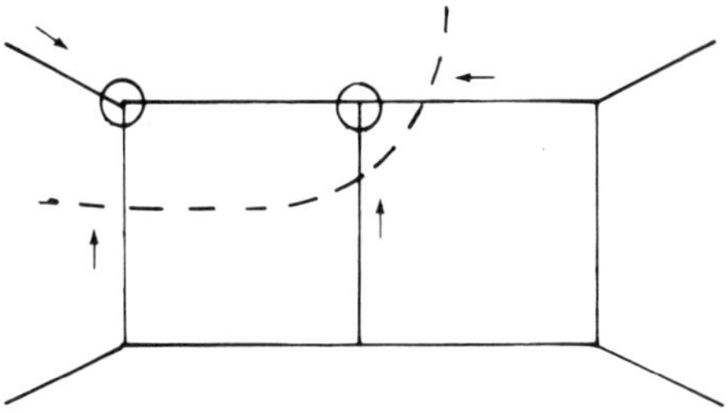

Fig. 10.13

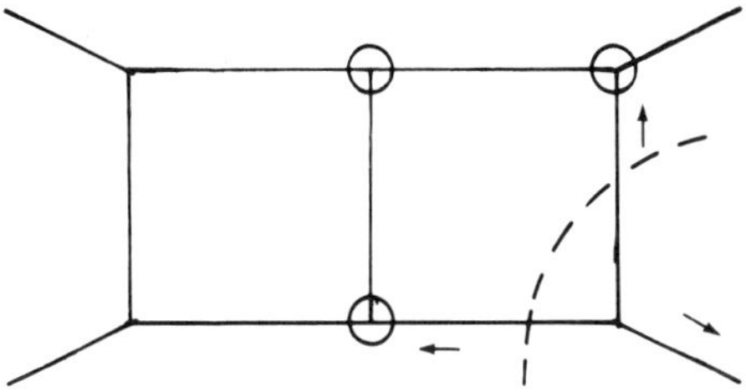

Fig. 10.14

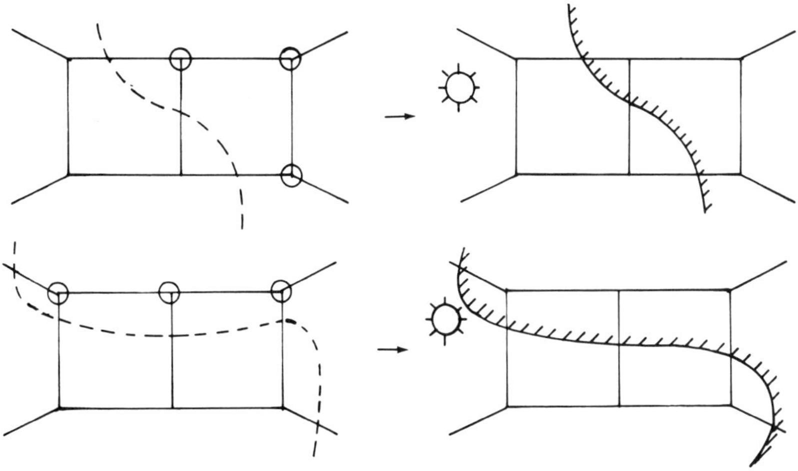

Fig. 10.15 A diagram cut into two regions

the subdiagram bounded by the demarcation line. This observation allows one to subdivide every nonzero diagram into two regions: one on the 'sunny' side, containing only non-circled vertices, and the other on the 'shaded' side, containing the circled vertices (Fig. 10.15).

No particular physical significance attaches to the cutting of external lines by the demarcation line. The internal lines are given by the following rules, which rely on the first version of the Feynman rules given in Section 5.2.5.

sunny-side propagator: 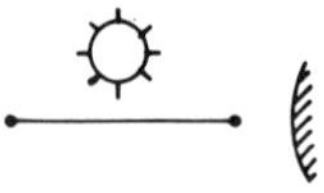 $$\frac{\mathrm{i}}{(2\pi)^4}\frac{1}{k^2 - m^2 + \mathrm{i}\varepsilon}, \qquad (10.4.18a)$$

shaded-side propagator:  $$\frac{-\mathrm{i}}{(2\pi)^4}\frac{1}{k^2 - m^2 - \mathrm{i}\varepsilon}, \qquad (10.4.18b)$$

cut propagator: 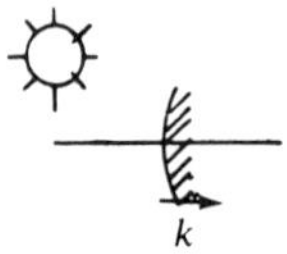

$$\frac{1}{(2\pi)^4} 2\pi\delta(k^2 - m^2)\theta(k^0). \qquad (10.4.18c)$$

The vertices are given by

$$\begin{aligned} &\text{sunny-side vertex:} \quad (-\mathrm{i}g)(2\pi)^4\delta^{(4)}(\textstyle\sum k_i), \\ &\text{shaded-side vertex:} \quad (\mathrm{i}g)(2\pi)^4\delta^{(4)}(\textstyle\sum k_i) \end{aligned} \qquad (10.4.18d)$$

Then (10.4.17) entails

$$F(k_1, \ldots, k_n) + \bar{F}(k_1, \ldots, k_n) = -\sum_{\text{cuts}} F_c(k_1, \ldots, k_n). \qquad (10.4.19)$$

Equation (10.4.19) gives the Cutkosky rules. On the left of (10.4.19), F is evaluated using the sunny-side rules, $\bar{F}$ using the shaded-side rules, and, on the right, F_c according to (10.4.18). The sum runs over all possible cuts, with energy flowing from sun to shade. Certain cuts can give zero on account of the kinematics.

It remains to make the connection with the unitarity condition. Define the Lagrangean $\mathscr{L}^*$ as the complex-conjugate of $\mathscr{L}$; and define also the S-matrix $\bar{S}$ calculated from $\mathscr{L}^*$ by using $-\mathrm{i}/(k^2 - m^2 - \mathrm{i}\varepsilon)$ for the propagators and by replacing $(-\mathrm{i}g)$ by $(\mathrm{i}g)$ (in other words, using the shaded-side rules).

It is easy to convince oneself that

$$\langle f|S^\dagger(\mathscr{L})|\mathrm{i}\rangle = \langle f|\bar{S}(\mathscr{L}^*)|\mathrm{i}\rangle.$$

If $\mathscr{L} = \mathscr{L}^*$, then the unitarity condition reads

$$\bar{T}_{fi} + T_{fi} = -\mathrm{i} \sum_p \int \mathrm{d}\Phi^{(p)}\, \bar{T}_{fp} T_{pi}, \tag{10.4.20}$$

where $\mathrm{d}\Phi^{(p)}$ is the phase-space element appropriate to p intermediate particles. This however is nothing but (10.4.19) (with $T = -\mathrm{i}F$), because $\bar{T}_{fp}$ is evaluated using the shaded-side rules, while

$$\mathrm{d}\Phi^{(p)} = (2\pi)^4 \delta(K_f - K_p) \prod_{j=1}^{p} \frac{(2\pi)}{(2\pi)^4} \delta(k_j^2 - m^2)\theta(k_j^0).$$

Thus, if the Lagrangean $\mathscr{L}$ is real, equation (10.4.19) allows one to compute the imaginary part of T-matrix elements.

Finally we must examine the factors Z_3, because the T-matrix elements are related to the Green's functions by a factor $Z_3^{-(N+M)/2}$, where N is the number of incoming and M the number of outgoing particles (see (10.3.31)). For a state with p intermediate particles this entails a factor Z_3^{-p}. But every propagator cut on the mass shell contributes a factor $Z_3 \delta(k^2 - m^2)$, and the factors Z_3 cancel exactly (see Problem 10.6).

It follows that *if $\mathscr{L}$ is real* then every diagram satisfies the unitarity condition; we say that we have *perturbative unitarity*. This property entails that the sum over all diagrams is unitary, or in other words it entails the unitarity of the S-matrix. The question of causality can be treated similarly (see t'Hooft and Veltman 1974).

10.5 Generalizations

As yet we have considered only the case of a neutral scalar field. Now we proceed to generalize the arguments to a charged scalar field and to a finite-mass field with spin 1; we shall also consider derivative couplings, which have been ignored so far.

10.5.1 Charged scalar field

We wish to describe the interactions of charged particles, having charges ± 1* (when measured in some convenient units), and having spin 0; π^+ and π^- mesons furnish an example. It would be possible to use two Hermitean fields φ_1 and φ_2, but in order to construct eigenstates of the charge it proves more convenient to introduce the fields $\varphi(x)$ and $\varphi^\dagger(x)$:

$$\varphi = \frac{1}{\sqrt{2}}(\varphi_1 + \mathrm{i}\varphi_2), \quad \varphi^\dagger = \frac{1}{\sqrt{2}}(\varphi_1 - \mathrm{i}\varphi_2). \tag{10.5.1}$$

* The charge need not be electrical charge.

The Fourier decompositions of the free fields $\varphi(x)$ and $\varphi^\dagger(x)$ read

$$\varphi(x) = \int \mathrm{d}\tilde{k}[a(\boldsymbol{k})\mathrm{e}^{-\mathrm{i}kx} + b^\dagger(\boldsymbol{k})\mathrm{e}^{\mathrm{i}kx}], \tag{10.5.2a}$$

$$\varphi^\dagger(x) = \int \mathrm{d}\tilde{k}[a^\dagger(\boldsymbol{k})\mathrm{e}^{\mathrm{i}kx} + b(\boldsymbol{k})\mathrm{e}^{-\mathrm{i}kx}], \tag{10.5.2b}$$

where a (b) destroys particles of charge $+1$ (-1) (see (10.5.6)). The commutation rules are

$$[a(\boldsymbol{k}), a^\dagger(\boldsymbol{k}')] = (2\pi)^3 2\omega_k \delta^{(3)}(\boldsymbol{k} - \boldsymbol{k}'), \tag{10.5.3a}$$

$$[b(\boldsymbol{k}), b^\dagger(\boldsymbol{k}')] = (2\pi)^3 2\omega_k \delta^{(3)}(\boldsymbol{k} - \boldsymbol{k}'), \tag{10.5.3b}$$

with all other commutators vanishing. The free-field Lagrangean is

$$\mathscr{L} = (\partial_\mu \varphi^\dagger)(\partial^\mu \varphi) + m^2 \varphi^\dagger \varphi; \tag{10.5.4}$$

it is invariant under the phase transformation

$$\varphi \to \mathrm{e}^{-\mathrm{i}\Lambda}\varphi, \quad \varphi^\dagger \to \mathrm{e}^{\mathrm{i}\Lambda}\varphi^\dagger.$$

Noether's theorem (see Section 11.3.3 and Appendix C) allows us to infer the existence of a conserved current

$$j_\mu = \mathrm{i}:\varphi^\dagger \overleftrightarrow{\partial}_\mu \varphi:, \quad \partial^\mu j_\mu = 0, \tag{10.5.5}$$

which in turn entails the conservation of the charge

$$Q = \int \mathrm{d}^3x\, j^0(x) = \int \mathrm{d}\tilde{k}[a^\dagger(\boldsymbol{k})a(\boldsymbol{k}) - b^\dagger(\boldsymbol{k})b(\boldsymbol{k})]. \tag{10.5.6}$$

This equation confirms that a (b) indeed destroys particles of charge $+1$ (-1). The φ^4 coupling can be generalized by adopting for instance the interaction Lagrangean

$$\mathscr{L}_1(x) = -\frac{g}{(2!)^2}:\varphi^\dagger(x)^2\varphi(x)^2:. \tag{10.5.7}$$

In order to establish the Feynman rules it will prove convenient *to assign a definite direction to the propagator, namely the direction in which positive charge flows*; in $\langle 0|T|\varphi_{\text{in}}(y)\varphi^\dagger_{\text{in}}(x)|0\rangle$, $\varphi^\dagger_{\text{in}}(x)$ creates a π^+ (or annihilates a π^-) at x, while $\varphi_{\text{in}}(y)$ annihilates a π^+ (or creates a π^-) at y; in either case charge $+1$ is created at x and annihilated at y. Note that

$$\langle 0|T(\varphi_{\text{in}}(y)\varphi_{\text{in}}(x))|0\rangle = \langle 0|T(\varphi^\dagger_{\text{in}}(y)\varphi^\dagger_{\text{in}}(x))|0\rangle = 0.$$

If (by convention) one decides to follow the positive charge, then the propagator is directed from x to y:

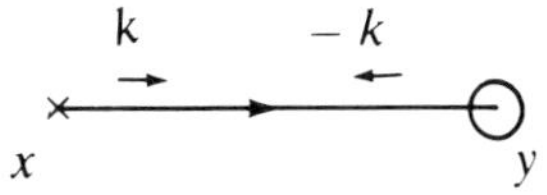

$$: \Delta_F(y-x) = \overline{\varphi(y)\varphi^\dagger(x)}$$

$$= \int \frac{d^4k}{(2\pi)^4} \frac{i}{k^2 - m^2 + i\varepsilon} e^{ik(x-y)}. \qquad (10.5.8)$$

It is most important not to confuse the direction of the propagator with the directions of the momenta. To make the symmetry factors easy to determine, we indicate, provisionally, the starting point of the propagator by a cross (×), and its destination point by a circle (○)*: in the perturbative expansion, every propagator must join a cross to a circle.

As an exercise let us calculate the contribution of order g^2 to the four-point correlation function

$$G^{(4)} = \langle 0 | T(\varphi(x_3)\,\varphi(x_4)\varphi^\dagger(x_1)\varphi^\dagger(x_2)) | 0 \rangle,$$

using the interaction Lagrangean (10.5.7). Every vertex will be split into two crosses (×) and two circles (○). We obtain two kinds of graphs (Fig. 10.16), to which we must of course add their permutations:

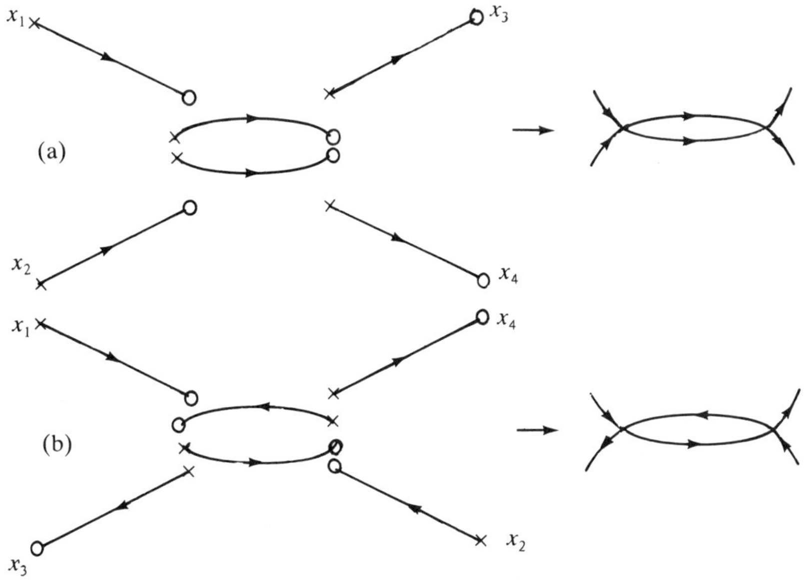

Fig. 10.16 Contributions to $G^{(4)}$

* Obviously, the vertices circled here have nothing to do with the vertices circled in Section 10.4.2.

The symmetry factor of graph (a) is

$$\left[\frac{1}{(2!)^2}\right]^2 \times (2!)^3 = \frac{1}{2}.$$

This graph corresponds to the scattering of identical particles (π^+ on π^+), and the symmetry factor is just the usual one. By contrast, the graph in Fig. 10.16(b) corresponds to (π^+, π^-) scatter, and its symmetry factor is 1.

The functional formalism generalizes straightforwardly, provided one defines $Z(j)$ by

$$Z(j) = \int \mathscr{D}(\varphi, \varphi^*) \exp\left(\mathrm{i}S(\varphi, \varphi^*) + \mathrm{i} \int \mathrm{d}^4x (j(x)\varphi^*(x) + j^*(x)\varphi(x)) \right), \tag{10.5.9}$$

and provided one uses the identity (A.2.4), which yields

$$\int \mathscr{D}(\varphi, \varphi^*) \exp\left(\mathrm{i}S_0(\varphi, \varphi^*) + \mathrm{i} \int \mathrm{d}^4x (j(x)\varphi^*(x) + j^*(x)\varphi(x)) \right)$$
$$= \mathscr{N} \exp\left(- \int \mathrm{d}^4x\, \mathrm{d}^4x'\, j^*(x) \Delta_{\mathrm{F}}(x - x') j(x') \right). \tag{10.5.10}$$

Notice that the exponent in (10.5.10) differs by a factor of 2 from that in (10.2.1). The results found by applying (10.5.9) and (10.5.10) are of course the same as those that follow from the application of Wick's theorem.

To calculate S-matrix elements we use the following contractions, with their diagrammatic representations (see (10.3.19)):

$\underbracket{\varphi(x)a^\dagger(\boldsymbol{k})} = \mathrm{e}^{-ikx}$: incoming π^+ meson:  (10.5.11a)

$\underbracket{a(\boldsymbol{k})\varphi^\dagger(x)} = \mathrm{e}^{ikx}$: outgoing π^+ meson: 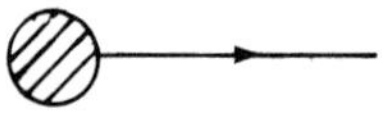 (10.5.11b)

$\underbracket{\varphi^\dagger(x)b^\dagger(\boldsymbol{k})} = \mathrm{e}^{-ikx}$: incoming π^- meson:  (10.5.11c)

$\underbracket{b(k)\varphi(x)} = \mathrm{e}^{ikx}$: outgoing π^- meson:  (10.5.11d)

10.5.2 Massive vector field

To avoid any loss of generality (beyond the restriction to Bose fields), one can consider real fields φ_i provided with indices (i) for spin, or for internal symmetries, or for both; and also complex fields ψ_i. In order to simplify the notation, let us agree that (*i*) indicates not only the spin and the internal symmetries, but also location in space–time. Under these conditions we can write down a very general Lagrangean in the form

$$\mathscr{L} = \frac{1}{2}\varphi_i W_{ij}\varphi_j + (\psi_i^* \bar{W}_{ij}\psi_j + \text{h.c.}) + \text{interactions}$$

where W is a symmetric matrix. The propagators are the inverses of the W_{ij} and of the $\bar{W}_{ij}$, up to a factor i:

$$\underset{j \qquad\qquad i}{\Delta^{\mathrm{F}}_{ij} = \mathrm{i}W^{-1}_{ij}}, \qquad \underset{j \qquad\qquad i}{\bar{\Delta}^{\mathrm{F}}_{ij} = \mathrm{i}\bar{W}^{-1}_{ij}}. \tag{10.5.12}$$

We assign no direction to the propagator Δ^{F}_{ij}, because the matrix W_{ij} is symmetric; $\bar{\Delta}^{\mathrm{F}}_{ij}$ however will as a rule be directional.

As an example we take the Lagrangean for a particle with spin 1 and nonzero mass, coupled to a conserved current j_μ ($\partial^{\mu}_{j_\mu} = 0$):

$$\mathscr{L} = A^\mu[(\Box + m^2)g_{\mu\nu} - \partial_\mu\partial_\nu]A^\nu - j_\mu A^\mu. \tag{10.5.13}$$

The equations of motion read

$$(\Box + m^2)A^\mu - \partial^\mu(\partial_\nu A^\nu) = j^\mu, \tag{10.5.14}$$

while the conservation of j^μ imposes the condition

$$m^2(\partial^\mu A_\mu) = 0 \;\Rightarrow\; \partial^\mu A_\mu = 0 \quad (m \neq 0). \tag{10.5.15}$$

The outcome is a Klein–Gordon equation for each component,

$$(\Box + m^2)A^\mu = 0.$$

However, on account of (10.5.15) only three out of the four components are independent, conformably with the number of degrees of freedom for spin 1. To clinch the assignment of spin 1 would call for the construction of the angular-momentum tensor, and for a proof that under rotations its components transform like a vector.

The propagator can be read directly off (10.5.13):

$$\Delta^{\mu\nu}_{\mathrm{F}}(k) = \mathrm{i}\,\frac{-g^{\mu\nu} + k^\mu k^\nu/m^2}{k^2 - m^2 + \mathrm{i}\varepsilon}. \tag{10.5.16}$$

Its longitudinal part has no pole at $k^2 = m^2$,

$$k_\mu \Delta^{\mu\nu}_{\mathrm{F}}(k) = m^{-2}k^\nu, \tag{10.5.17}$$

which confirms (10.5.15): namely that the longitudinal part does not constitute a dynamical degree of freedom. The cut propagator reads

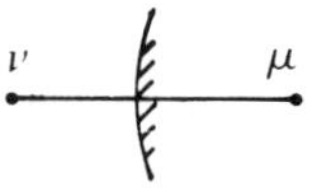

$$= \left(- g_{\mu\nu} + \frac{k_\mu k_\nu}{m^2} \right) \frac{2\pi}{(2\pi)^4} \theta(k^0)\delta(k^2 - m^2). \qquad (10.5.18)$$

On the other hand, the spin state of a particle having spin 1 can be described by means of its three possible polarization directions. One could for instance choose these directions along the three coordinate axes (x, y, z) in the rest-frame of the particle; then, if the particle momentum $\boldsymbol{k}$ is parallel to $0z$, they transform into

$$\varepsilon_\mu^{(x)} = (0, 1, 0, 0), \qquad (10.5.19a)$$

$$\varepsilon_\mu^{(y)} = (0, 0, 1, 0), \qquad (10.5.19b)$$

$$\varepsilon_\mu^{(0)} = (k/m, 0, 0, k^0/m). \qquad (10.5.19c)$$

Here, $\varepsilon_\mu^{(x)}$ and $\varepsilon_\mu^{(y)}$ are transverse polarizations; the combinations $\varepsilon_\mu^{(\pm)} = \mp (\varepsilon_\mu^{(x)} \pm \mathrm{i}\varepsilon_\mu^{(y)})/\sqrt{2}$ correspond to values ± 1 of the spin projection along $0z$; and $\varepsilon_\mu^{(0)}$ corresponds to spin-projection zero, which is also called longitudinal polarization (Messiah 1961, Chapter XIII).

One can readily derive the identity

$$\sum_{\lambda = \pm, 0} \varepsilon_\mu^{(\lambda)} \varepsilon_\nu^{*(\lambda)} = - g_{\mu\nu} + \frac{k_\mu k_\nu}{m^2}. \qquad (10.5.20)$$

It enters into the unitarity relation, because the sum over intermediate states includes a sum over polarizations. Actually the sum (10.5.20) is identical to the result found by cutting the propagator, whence the theory is indeed unitary. The crucial point is that only the physical degrees of freedom enter when the propagator is cut; the polarization state parallel to k_μ contributes nothing, because of (10.5.17). By contrast, renormalizability is, a priori, at risk, because the factor m^{-2} of the propagator behaves like a coupling constant of negative dimensionality. Chapter 12 will show that, as regards physical quantities, renormalizability can be preserved nevertheless.

10.5.3 Derivative couplings

So far we have been careful to avoid derivative couplings. Their treatment is particularly delicate in the canonical formalism, because noncovariant quantities enter at intermediate stages. By contrast, in the functional formalism such couplings present no difficulties. Consider for example a vector field A_μ coupled to a charged scalar field through the interaction Lagrangean

$$\mathscr{L}_1 = \mathrm{i}gA^\mu(x)[\varphi^\dagger \overleftrightarrow{\partial}_\mu \varphi(x)], \qquad (10.5.21)$$

where the factor i ensures that $\mathscr{L}_1$ is Hermitean. This interaction induces terms

like

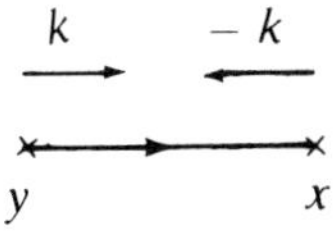

$$\frac{\partial}{\partial x^{\mu}} \overbrace{\varphi(x)\varphi^{\dagger}(y)} = \frac{\partial}{\partial x^{\mu}} \int \frac{\mathrm{d}^4 k}{(2\pi)^4} \frac{\mathrm{i}}{k^2 - m^2 + \mathrm{i}\varepsilon} \mathrm{e}^{-\mathrm{i}k(x-y)}$$

$$= -\mathrm{i}k_{\mu} \int \frac{\mathrm{d}^4 k}{(2\pi)^4} \frac{\mathrm{i}}{k^2 - m^2 + \mathrm{i}\varepsilon} \mathrm{e}^{-\mathrm{i}k(x-y)}.$$

Accordingly, the term of $\mathscr{L}_1$ featuring $\partial_{\mu}\varphi$ leads to a factor $-\mathrm{i}k_{\mu}$, where k_{μ} is the momentum running into the vertex x. The simplicity of the functional formalism stems from the fact that one need not commute derivatives with T-products (see the discussion just after equation (10.2.19)). By contrast, in the canonical formalism such commutators do occur, and they lead to noncovariant terms, consequences of the fact that the momentum conjugate to φ is not simply $\dot{\varphi}$; but the terms compensating for this emerge only after a complicated calculation. Again there are no problems with unitarity or with causality; but derivatives tend to introduce coupling constants having negative dimensionality, which do endanger renormalizability.

Problems

10.1 (a) For $a > 0$ derive the identity

$$\int_{-\infty}^{\infty} e^{iak^2}\, dk = e^{i\pi/4}\sqrt{(\pi/a)}\, e^{-ib^2/4a}.$$

(*Hint*: consider the limit of $\int_{-\infty}^{\infty} e^{iak^2 - \varepsilon k^2} dk$ as $\varepsilon \to 0^+$.) Thence derive (10.2.12).

(b) Prove equation (10.2.13), starting from the identity

$$\int d\lambda\, \lambda^{\alpha-1} e^{-\lambda} = \Gamma(\alpha).$$

10.2 Show that the flux factor (10.3.9) reduces to $\sqrt{s}\,\|\boldsymbol{k}\|$ for a two-particle collision in the centre-of-mass frame.

10.3 By adapting the argument in Section 10.3.1, derive (10.3.12) for the decay rate of an unstable particle.

10.4 Wick's theorem for ordinary products

(a) Derive the identity

$$\exp\left(i\int d^4x\, j(x)\varphi_{\text{in}}(x)\right)$$

$$= :\exp\left(i\int d^4x\, j(x)\varphi_{\text{in}}(x)\right):$$

$$\times \exp\left(-\frac{1}{2}\int d^4x\, d^4x'\, j(x)\, \underbrace{\varphi(x)\varphi(x')}\, j(x')\right),$$

where

$$\underbrace{\varphi(x)\varphi(x')} = \langle 0|\varphi_{\text{in}}(x)\varphi_{\text{in}}(x')|0\rangle.$$

(*Hint*: $\varphi_{\text{in}}(x) = \varphi_{\text{in}}^{+}(x) + \varphi_{\text{in}}^{-}(x)$, and Problem 9.9.)

(b) Thence derive Wick's theorem for

$\varphi_{\text{in}}(x_1)\varphi_{\text{in}}(x_2) \ldots \varphi_{\text{in}}(x_{2n})$.

10.5 The Källen–Lehmann representation (Itzykson and Zuber 1980, Chapter 5; or Bjorken and Drell 1965, Chapter 16)

(a) Let $F(x-y)$ be the vacuum expectation-value of the commutator $[\varphi(x), \varphi(y)]$:

$$F(x-y) = \langle 0|[\varphi(x), \varphi(y)]|0\rangle.$$

By introducing a complete set of intermediate states $|n\rangle$ together with the identity

$$1 = \int d^4q\,\delta^{(4)}(q - p_n),$$

show that

$$F(x - y) = i\int_0^\infty dm'^2\,\rho(m'^2)\Delta(x - y; m'^2),$$

where

$$\rho(q) = (2\pi)^3\sum_n \delta^{(4)}(p_n - q)|\langle 0|\varphi(0)|n\rangle|^2.$$

The argument should exploit Lorentz invariance and the fact that the energy is positive, to establish that

$$\rho(q) = \rho(q^2)\theta(q^0), \quad \text{with } q^2 > 0.$$

(b) By isolating the 1-particle contributions and using the CCR, prove the relation

$$1 = Z_3 + \int_{4m^2}^\infty dm'^2\,\rho(m'^2).$$

Thence derive the inequality $0 \leqslant Z_3 \leqslant 1$. ($Z_3 = 1$ is the hallmark of a free field.)

(c) Obtain for the full propagator $G^{(2)}(x - y)$ a representation analogous to the one in (a), with the same spectral function ρ (see (10.3.33)).

10.6 Write down the perturbative unitarity condition for the graph of φ^3 theory, of order g^6, shown in Fig. 10.17. Pay special attention to the factors Z_3. The following identity is useful:

$$\delta(x)\left(\frac{1}{x + i\varepsilon} + \frac{1}{x - i\varepsilon}\right) = -\,\delta'(x),$$

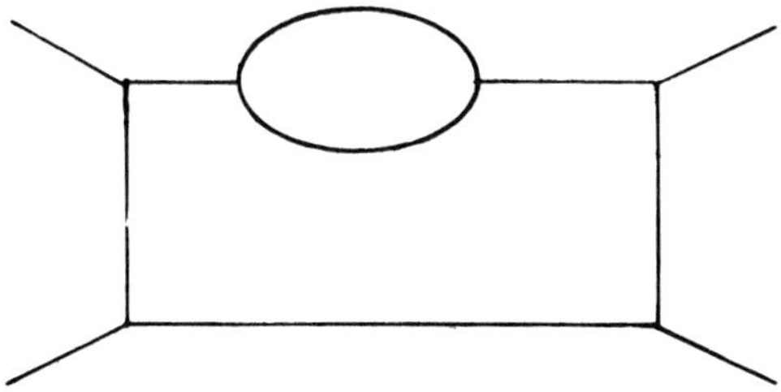

Fig. 10.17

with

$$\delta(x) = \lim_{\varepsilon \to 0} \frac{\varepsilon}{\pi(x^2 + \varepsilon^2)}.$$

10.7 The ρ^0 meson is a particle of mass $M = 765$ MeV; it decays into a $\pi^+\pi^-$ pair ($m_\pi = 140$ MeV) with a width of $\Gamma = 125$ MeV. The decay is described by the phenomenological Lagrangean (10.5.21). What is the value of g?

Further reading

The Haag–Ruelle construction is described by Jost (1965), and by Hepp (1965). An elementary discussion can be found in Bros (1971). Our derivation of the formula of Gell-Mann and Low follows Gasiorowicz (1966, Chapter 8). The reduction formulae are established, for instance, by Itzykson and Zuber (1980, Chapter 5), Bjorken ad Drell (1965, Chapter 16), and by Gasiorowicz (1966, Chapter 6). Our proof of equation (10.3.11) for cross-sections follows Itzykson and Zuber (1980, Chapter 5). The analytic properties of scattering amplitudes are studied by Itzykson and Zuber (1980, Chapter 6), Bjorken and Drell (1965, Chapter 18), and in the book by Eden *et al.* (1966). The quantization of the charged scalar field is considered in detail by all the classic texts (Itzykson and Zuber 1980, Bjorken and Drell 1965, Gasiorowicz 1966, etc.). Finally, Sections 1 to 9 of t'Hooft and Veltman (1974) give a general survey of all the problems tackled in this chapter.

IV Gauge Theories

11 Quantization of the Dirac field and of the electromagnetic field

Modern elementary-particle physics is dominated by gauge theories, whose prototype is quantum electrodynamics (QED). The fundamental particles in such theories are

(i) spin 1/2 particles, described by fields called 'matter fields';

(ii) spin 1 particles, described by fields called 'gauge fields' (an evidently unfortunate nomenclature, because a gauge particle like the photon is just as 'material' as a spin 1/2 particle like the neutrino).

These particles are assumed to be elementary, *in the sense that their interactions are local* (*pointlike*): *such interactions are described by products of fields all at the same space–time point*. In electrodynamics, the spin 1/2 particles are electrons (and their antiparticles, i.e. positrons), while the gauge particles are photons. In this case the interactions between spin 1/2 particles are mediated by photons: Fig. 11.1 shows a Feynman graph contributing to electron–electron scattering.

The electromagnetic and the weak interactions have been unified into the electroweak interaction, whose gauge particles are the photon and the $W^{\pm}$ and Z^0 bosons (predicted by theory in 1967 and found experimentally in 1983). Figure 11.2 shows a Feynman graph contributing to electron–neutrino scattering. Further, electroweak theory needs to introduce spin 0 particles, called Higgs bosons, which so far (up to 1990) have escaped all attempts at detection (see Section 13.3.3).

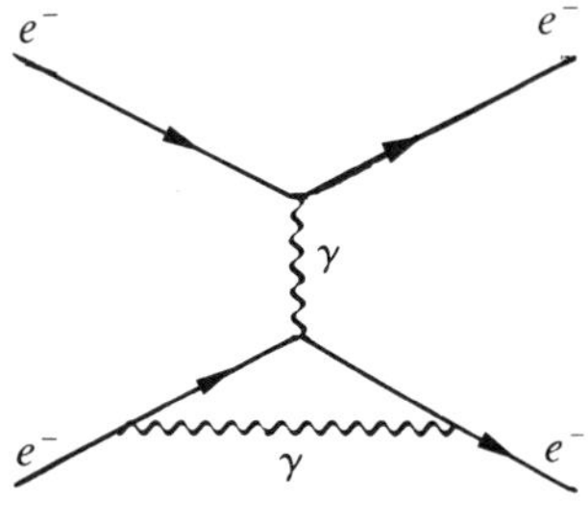

Fig. 11.1 A Feynman diagram for electron–electron scattering

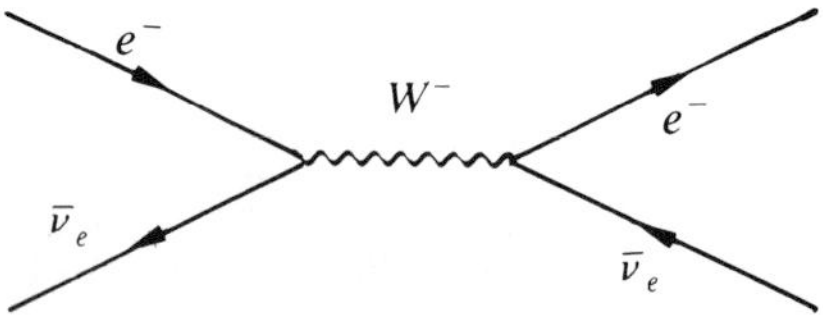

Fig. 11.2 A diagram contributing to $\bar{\nu}_e e^-$ scattering

The strongly interacting particles (proton, neutron, mesons, etc.) are not pointlike, and in this sense they are not elementary; rather they are made of quarks and antiquarks. The gauge theory of the strong interactions is quantum chromodynamics (QCD), whose spin 1/2 particles are quarks and antiquarks, and whose gauge particles are called gluons (see Section 13.4). Quarks and antiquarks have electroweak interactions as well.

So far, the theory we have developed in the two preceding chapters allows us to deal only with spin 0 particles; in other words, only with the very limited and somewhat academic case of interactions between Higgs bosons. We have yet to learn how to quantize fields that describe particles having spin 1/2, i.e. Dirac fields, and also how to quantize gauge fields.

The canonical quantization of the Dirac field is dealt with in Section 11.1. To avoid too long a tale, we shall treat the Dirac equation rather succinctly; readers unfamiliar with it will probably wish to turn to other discussions, in particular as regards its transformation properties under the Poincaré group and under the discrete symmetry operations. But we do introduce explicitly all the ideas essential to the applications treated in this book. Section 11.2 is devoted to Wick's theorem for fermions, and to the functional formulation.

The theory of the electromagnetic field is the simplest example of a gauge theory: the gauge group is Abelian, by contrast to the non-Abelian gauge groups of the electroweak interactions and of quantum chromodynamics. But canonical quantization is complex even in the Abelian case, and after recalling some basics in Section 11.3, it will prove more convenient to adopt quantization by path integrals, as we do in Section 11.4. This is, moreover, a good introduction to the more complicated case of non-Abelian theories.

11.1 Quantization of the Dirac field

The Klein–Gordon field, studied in Chapter 9, describes particles of spin 0. Further, in Chapter 10 we saw how to describe particles of spin 1 with finite mass. Particles with spin 0 and spin 1 have in common that they are *bosons*, i.e. that they obey Bose–Einstein statistics. By contrast, it is well known that spin 1/2 particles like electrons and protons obey Fermi–Dirac statistics; they are *fermions*. The formalism of earlier chapters is manifestly ill-adapted to fermions, because a state vector like

$$a^\dagger(\boldsymbol{k}_1)a^\dagger(\boldsymbol{k}_2)|0\rangle$$

is symmetric under the interchange of the two particles, by virtue of the commutation rules for the $a^\dagger(\boldsymbol{k})$. Hence we must modify Fock space so that antisymmetry (rather than symmetry) is ensured automatically. The other ingredient we need is a field equation transforming appropriately to spin 1/2; this is the Dirac equation. The Dirac field will be quantized along the same lines as the Klein–Gordon field, but in a Fock space adapted to fermions.

11.1.1 Fock space for fermions

We start from a simple problem in nonrelativistic quantum mechanics, namely from a system of N mutually noninteracting fermions in a potential $V(\boldsymbol{x})$. To avoid complicated notations we even assume that they have spin 0. (We shall see presently that in *relativistic* quantum field theory this assumption is not selfconsistent; but there is nothing selfcontradictory about it in nonrelativistic quantum mechanics.) Write the eigenfunctions of the single-particle Hamiltonian

$$H = \frac{\boldsymbol{p}^2}{2m} + V(\boldsymbol{x})$$

as $u_{\alpha_i}(\boldsymbol{x})$. Label the different energy levels E_{α_i}, with indices α_i arranged in some definite order $\alpha_1, \alpha_2, \ldots, \alpha_N, \ldots$:

$$Hu_{\alpha_i}(\boldsymbol{x}) = E_{\alpha_i}u_{\alpha_i}(\boldsymbol{x})$$

(if the levels are degenerate, we evidently need some additional quantum numbers). The wavefunction of a set of N fermions occupying the levels $\alpha_{i_1} \ldots \alpha_{i_N}$, correctly antisymmetrized and normed, is a Slater determinant:

$$\Psi(\boldsymbol{x}_1, \ldots, \boldsymbol{x}_N) = \langle \boldsymbol{x}_1, \ldots, \boldsymbol{x}_N \,|\, \Psi\rangle = \frac{1}{\sqrt{N!}}\det(u_{\alpha_i}(\boldsymbol{x}_j)). \tag{11.1.1}$$

For instance, in the case of two particles occupying levels 3 and 5 the wavefunction reads

$$\begin{aligned}\Psi(\boldsymbol{x}_1, \boldsymbol{x}_2) &= \frac{1}{\sqrt{2!}}(u_3(\boldsymbol{x}_1)u_5(\boldsymbol{x}_2) - u_3(\boldsymbol{x}_2)u_5(\boldsymbol{x}_1)) \\ &= \frac{1}{\sqrt{2!}}\begin{vmatrix} u_3(\boldsymbol{x}_1) & u_3(\boldsymbol{x}_2) \\ u_5(\boldsymbol{x}_1) & u_5(\boldsymbol{x}_2)\end{vmatrix}.\end{aligned} \tag{11.1.2}$$

Information equivalent to (11.1.1) is carried by the *occupation numbers* n_{α_i} of the levels α_i; in the case of the wavefunction (11.1.2) for instance these occupation numbers are

$$n_3 = 1; \quad n_5 = 1; \quad n_i = 0, \quad (i \neq 3, 5).$$

For bosons, the n_α can take the values 0, 1, 2, . . . ; but for fermions, $n_\alpha = 0$ or 1, because a (nondegenerate) energy level cannot hold more than one fermion. Now consider just a single level α, and try to construct creation and annihilation operators by analogy to those for bosons. The state space is two-dimensional, and for our basis we can choose the vectors $|0\rangle_\alpha$ and $|1\rangle_\alpha$ corresponding to $n_\alpha = 0$ and to $n_\alpha = 1$ respectively:

$$\text{unoccupied state } |0\rangle_\alpha = \begin{pmatrix} 0 \\ 1 \end{pmatrix}_\alpha; \quad \text{occupied state } |1\rangle_\alpha = \begin{pmatrix} 1 \\ 0 \end{pmatrix}_\alpha.$$
$$(n_\alpha = 0) \qquad\qquad (n_\alpha = 1)$$

The operators a_α and $a_\alpha^\dagger$ must satisfy the following conditions:

(i) a_α acting on the vacuum gives zero: $a_\alpha|0\rangle_\alpha = 0$;

(ii) $a_\alpha^\dagger$ acting on the vacuum creates an occupied state having $n_\alpha = 1$: $a_\alpha^\dagger|0\rangle_\alpha = |1\rangle_\alpha$;

(iii) because we cannot put more than one fermion into the state α, we must have $a_\alpha^\dagger|1\rangle_\alpha = 0$.

These three conditions fully determine the matrix representations of a_α and of $a_\alpha^\dagger$; they read

$$a_\alpha = \begin{pmatrix} 0 & 0 \\ 1 & 0 \end{pmatrix}, \quad a_\alpha^\dagger = \begin{pmatrix} 0 & 1 \\ 0 & 0 \end{pmatrix},$$

and entail the *anticommutation* relation

$$\{a_\alpha, a_\alpha^\dagger\} = a_\alpha a_\alpha^\dagger + a_\alpha^\dagger a_\alpha = \mathbb{1}_\alpha. \tag{11.1.3}$$

Just as for bosons, the operator for the number of particles in state (α) is $N_\alpha = a_\alpha^\dagger a_\alpha$:

$$N_\alpha = a_\alpha^\dagger a_\alpha = \begin{pmatrix} 1 & 0 \\ 0 & 0 \end{pmatrix},$$

$$N_\alpha|0\rangle_\alpha = 0, \quad N_\alpha|1\rangle_\alpha = |1\rangle_\alpha.$$

In order to construct the state space, one need merely take the tensor product of the two-dimensional spaces corresponding to the individual levels. The vacuum for instance is constructed as the tensor product of the vectors $|0\rangle_\alpha$:

$$|0\rangle = \prod_\alpha |0\rangle_\alpha = \prod_\alpha \begin{pmatrix} 0 \\ 1 \end{pmatrix}_\alpha.$$

However, all this still fails to ensure antisymmetry under the interchange of two fermions, because the operators a_{α_i} and a_{α_j} commute, acting as they do on different spaces. To construct anticommuting operators we introduce the operator

$$\eta_{\alpha_i} = \prod_{\alpha_1}^{\alpha_i - 1} (1 - 2N_\alpha) = \prod_{\alpha_1}^{\alpha_i - 1} \begin{pmatrix} -1 & 0 \\ 0 & 1 \end{pmatrix}_\alpha, \tag{11.1.4}$$

and define another operator $b_{\alpha_i}^\dagger$ by

$$b_{\alpha_i}^\dagger = \eta_{\alpha_i} a_{\alpha_i}^\dagger = a_{\alpha_i}^\dagger \eta_{\alpha_i}. \tag{11.1.5}$$

Notice that if $\alpha_i < \alpha_j$, then

$$a_{\alpha_i}^\dagger \eta_{\alpha_j} = -\eta_{\alpha_j} a_{\alpha_i}^\dagger, \quad a_{\alpha_j}^\dagger \eta_{\alpha_i} = \eta_{\alpha_i} a_{\alpha_j}^\dagger;$$

it follows that

$$\{b_{\alpha_i}^\dagger, b_{\alpha_j}^\dagger\} = 0.$$

This anticommutation rule does ensure that the state vector is antisymmetric:

$$|\Psi\rangle = b_{\alpha_i}^\dagger b_{\alpha_j}^\dagger|0\rangle = -b_{\alpha_j}^\dagger b_{\alpha_i}^\dagger|0\rangle.$$

The other anticommutation relations follow readily:

$$\{b_{\alpha_i}, b_{\alpha_j}\} = \{b^\dagger_{\alpha_i}, b^\dagger_{\alpha_j}\} = 0, \quad \{b_{\alpha_i}, b^\dagger_{\alpha_j}\} = \delta_{ij}. \tag{11.1.6}$$

Now one can construct a field operator $\psi(t, \boldsymbol{x})$, given in the Heisenberg picture by

$$\psi(t, \boldsymbol{x}) = \sum_\alpha b_\alpha u_\alpha(t, \boldsymbol{x}), \tag{11.1.7}$$

and satisfying the equal-time anticommutation rules

$$\begin{aligned} &\{\psi(t, \boldsymbol{x}), \psi(t, \boldsymbol{x}')\} = 0, \\ &\{\psi(t, \boldsymbol{x}), \psi^\dagger(t, \boldsymbol{x}')\} = \delta^{(3)}(\boldsymbol{x} - \boldsymbol{x}'). \end{aligned} \tag{11.1.8}$$

The second of equations (11.1.8) follows by virtue of the closure relation

$$\sum_\alpha u_\alpha(t, \boldsymbol{x}) u^*_\alpha(t, \boldsymbol{x}') = \delta^{(3)}(\boldsymbol{x} - \boldsymbol{x}').$$

It can be shown (Problem 11.1) that the wavefunction $\Psi(\boldsymbol{x}_1, \ldots, \boldsymbol{x}_N)$ in (11.1.1) is

$$\Psi(\boldsymbol{x}_1, \ldots, \boldsymbol{x}_N) = \frac{1}{\sqrt{N!}} \langle 0 | \psi(\boldsymbol{x}_1) \ldots \psi(\boldsymbol{x}_N) | \Psi \rangle, \tag{11.1.9}$$

(where $\psi(\boldsymbol{x}) = \psi(t = 0, \boldsymbol{x})$). The formalism just presented proves very useful in nonrelativistic quantum mechanics for studying the dynamics of systems of N identical fermions (as well as systems of N identical bosons, for which one naturally switches from anticommutators to commutators). This is what is often called the 'many-body problem', with important applications to solid-state and to nuclear physics. Nevertheless the physics of the equations that emerge from this approach, called 'second quantization', is strictly the same as the physics of the Schroedinger equation; the formalism is simply a technical convenience.

11.1.2 The Dirac equation

Originally the Dirac equation was introduced as a relativistic generalization of the Schroedinger equation, designed to overcome the difficulties in the way of a quantum-relativistic interpretation of the Klein–Gordon equation. We must stress that in both cases one was dealing with a *single-particle* equation, meant to describe for example the behaviour of a relativistic particle in a potential. Actually the Dirac equation does not (any more than the Klein–Gordon equation) admit of a *completely* selfconsistent single-particle interpretation. Nevertheless such an interpretation is physically acceptable, and very useful in practice, provided the potential varies little over distances of the order of the Compton wavelength ($\hbar/mc$) of the particle in question. For example, it allows one to calculate, in a first approximation, the relativistic corrections to the spectrum of the hydrogen atom. We shall ignore this aspect of the Dirac equation altogether, since there are many books that treat it in fine detail. As with the Klein–Gordon equation, we shall consider the Dirac equation initially *as governing a classical field*, which we shall then have to turn into a quantized field.

Let us start by writing down the Dirac equation directly, relying on other discussions to motivate or to 'derive' it. It reads

$$(\mathrm{i}\gamma^{\mu}_{\alpha\beta}\partial_{\mu} - m\delta_{\alpha\beta})\psi_{\beta}(t, \boldsymbol{x}) = 0. \tag{11.1.10a}$$

Here, $\psi_{\beta}(t, \boldsymbol{x})$ is an object (a spinor) with four components, $\beta = 1, 2, 3, 4$; and the $\gamma^{\mu}_{\alpha\beta}$ are a set of 4×4 matrices, called *the Dirac matrices*, obeying the anticommutation rules

$$\{\gamma^{\mu}, \gamma^{\nu}\} = \gamma^{\mu}\gamma^{\nu} + \gamma^{\nu}\gamma^{\mu} = 2g^{\mu\nu}\mathbb{1}. \tag{11.1.11}$$

The indices α and β, by contrast to μ, are *not* Lorentz indices. But μ *is* a Lorentz index, $\mu = 0, 1, 2, 3$, while m is the mass of the particle. Often one uses the notation

$$\partial\!\!\!/ = \gamma^{\mu}\partial_{\mu}, \tag{11.1.12}$$

or more generally $a\!\!\!/ = \gamma^{\mu}a_{\mu}$ for an arbitrary four-vector a_{μ}; then (11.1.10a) can be rewritten in matrix form as

$$(\mathrm{i}\partial\!\!\!/ - m)\psi = 0. \tag{11.1.10b}$$

The irreducible representations of the anticommutation relations (11.1.11) are unique up to a similarity transformation. The standard representation of the matrices γ^{μ} reads

$$\gamma^0 = \gamma_0 = \begin{pmatrix} \mathbb{1} & 0 \\ 0 & -\mathbb{1} \end{pmatrix}; \quad \gamma^i = -\gamma_i = \begin{pmatrix} 0 & \sigma_i \\ -\sigma_i & 0 \end{pmatrix}, \tag{11.1.13}$$

where $\mathbb{1}$ is the unit 2×2 matrix, and $\sigma_1, \sigma_2, \sigma_3$ are the Pauli matrices. The Hermitean-conjugation relation

$$\gamma^{\mu\dagger} = \gamma^0\gamma^{\mu}\gamma^0 \tag{11.1.14}$$

is often useful. In terms of the matrices γ_{μ}, one defines other matrices

$$\sigma_{\mu\nu} = \frac{\mathrm{i}}{2}[\gamma^{\mu}, \gamma^{\nu}], \tag{11.1.15}$$

and also the matrix

$$\gamma_5 = \mathrm{i}\gamma^0\gamma^1\gamma^2\gamma^3 = \gamma_5^{\dagger}, \tag{11.1.16}$$

which satisfies

$$\{\gamma_5, \gamma^{\mu}\} = 0, \quad \gamma_5^2 = 1, \quad \gamma_5^{\dagger} = -\gamma_0\gamma_5\gamma_0. \tag{11.1.17}$$

Equation (11.1.10b) implies that $\psi(t, x)$ obeys the Klein–Gordon equation: on multiplying (11.1.10b) from the left by $(\mathrm{i}\partial\!\!\!/ + m)$ and using (11.1.11), one finds

$$(\Box + m^2)\psi(t, \boldsymbol{x}) = 0,$$

which confirms our interpretation of m as the particle mass.

The Dirac equation is Lorentz-covariant: if Λ is a Lorentz transformation ($x' = \Lambda x$), then there exists a matrix $S(\Lambda)$ transforming $\psi(x)$ according to

$$\psi'(x') = S(\Lambda)\psi(x); \tag{11.1.18}$$

further, S satisfies

$$S^{-1}(\Lambda)\gamma^\mu S(\Lambda) = \Lambda^\mu_{\ \nu}\gamma^\nu, \quad S(\Lambda_1\Lambda_2) = \pm S(\Lambda_1)S(\Lambda_2),$$
$$S^{-1}(\Lambda) = \gamma^0 S^\dagger(\Lambda)\gamma^0 \tag{11.1.19}$$

(the ambiguity in sign stems from the half-integer nature of the spin). Accordingly, $\psi'(x')$ obeys the Dirac equation in the transformed (primed) inertial frame:

$$(\mathrm{i}\gamma^\mu \partial'_\mu - m)\psi'(x') = 0.$$

The transformation rule (11.1.18) is more complicated than the rule for scalar fields, which reads simply $\varphi'(x') = \varphi(x)$. We refer to the classic texts for the derivation of (11.1.19), for the explicit form of $S(\Lambda)$, and for proof of the fact that the field described by the Dirac equation indeed has spin 1/2.

The Dirac equation can be derived from the Lagrangean density

$$\begin{aligned}\mathscr{L} &= \bar{\psi}(x)(\mathrm{i}\overrightarrow{\not{\partial}} - m)\psi(x) = -\bar{\psi}(x)(\mathrm{i}\overleftarrow{\not{\partial}} + m)\psi(x) \\ &= \bar{\psi}(x)\left(\frac{\mathrm{i}}{2}\overleftrightarrow{\not{\partial}} - m\right)\psi(x),\end{aligned} \tag{11.1.20}$$

where the conjugate spinor $\bar{\psi}(x)$ is defined by

$$\bar{\psi}_\alpha(x) = \psi^*_\beta(x)\gamma^0_{\beta\alpha}. \tag{11.1.21}$$

The three different expressions in (11.1.20) are related through integrations by parts. When the variational principle is applied, ψ and $\bar{\psi}$ must be treated as mutually independent variables. Note that $\bar{\psi}(x)$ obeys the equation

$$\bar{\psi}(x)(\mathrm{i}\overleftarrow{\not{\partial}} + m) = 0, \tag{11.1.22}$$

which one can derive from (11.1.20), or by Hermitean conjugation from (11.1.10b). The momentum π_α canonically conjugate to ψ_α is

$$\pi_\alpha = \frac{\partial\mathscr{L}}{\partial(\partial_0\psi_\alpha)} = \mathrm{i}\bar{\psi}_\beta\gamma^0_{\beta\alpha} = \mathrm{i}\psi^\dagger_\alpha; \tag{11.1.23}$$

for the Hamiltonian density this yields

$$\mathscr{H} = \pi_\alpha\dot{\psi}_\alpha - \mathscr{L} = \mathrm{i}\psi^\dagger(\partial_0\psi) - \mathscr{L}. \tag{11.1.24}$$

In the calculations below, it will be useful to remark that $\mathscr{L} = 0$ if ψ obeys the Dirac equation. Coupling to the electromagnetic field $A_\mu(x)$ is implemented by the minimal-coupling substitution

$$\partial_\mu \to \partial_\mu + \mathrm{i}eA_\mu,$$

where e is the charge of the particle in question (for details see e.g. Section 11.3);

then the interaction Lagrangean reads

$$\mathscr{L}_1 = -e\bar{\psi}(x)A\!\!\!/\psi(x) = -j^\mu(x)A_\mu(x). \tag{11.1.25}$$

Accordingly, the electromagnetic current density is given by

$$j_\mu(x) = e\bar{\psi}(x)\gamma_\mu\psi(x). \tag{11.1.26}$$

One can easily satisfy oneself (Problem 11.2) that this current is indeed conserved, i.e. that

$$\partial^\mu j_\mu(x) = 0, \tag{11.1.27}$$

which entails conservation of the charge Q:

$$Q = \int_t \mathrm{d}^3x\, j^0(x), \quad \frac{\mathrm{d}Q}{\mathrm{d}t} = 0. \tag{11.1.28}$$

Thus we can check, by using (11.1.27) and Gauss's theorem, that

$$\frac{\mathrm{d}}{\mathrm{d}t}\int_t \mathrm{d}^3x\, j^0(x) = -\int \mathrm{d}^3x(\nabla\cdot\boldsymbol{j}) = -\int \boldsymbol{j}\cdot\mathrm{d}\boldsymbol{S} = 0,$$

provided we assume, as usual, that the fields vanish fast enough at infinity.

11.1.3 Solutions of the Dirac equation

Since the Dirac field obeys the Klein–Gordon equation, its solutions are linear superpositions of plane waves of the form

$$\psi_n(x) = w_n(\boldsymbol{p})\mathrm{e}^{-\mathrm{i}\varepsilon_n(p\cdot x)}, \tag{11.1.29}$$

where

$$p\cdot x = E_p t - \boldsymbol{p}\cdot\boldsymbol{x}, \quad E_p = \sqrt{(\boldsymbol{p}^2 + m^2)} > 0,$$

and where $\varepsilon_n = \pm 1$; $\varepsilon_n = +1$ corresponds to the positive-energy solutions, and $\varepsilon_n = -1$ to the negative-energy solutions. Substitution from (11.1.29) into (11.1.10b) yields the following equations for the $w_n(\boldsymbol{p})$:

$$(p\!\!\!/ - m)w_n(\boldsymbol{p}) = 0: \quad \varepsilon_n = +1 \tag{11.1.30a}$$

$$(p\!\!\!/ + m)w_n(\boldsymbol{p}) = 0: \quad \varepsilon_n = -1. \tag{11.1.30b}$$

If $\boldsymbol{p} = 0$, then $p\!\!\!/ = m\gamma^0$, and one finds four linearly independent solutions which may be chosen as follows:

$$w_1 = \hat{u}^{(1)}(\mathbf{0}) = \begin{pmatrix}1\\0\\0\\0\end{pmatrix}, \quad w_2 = \hat{u}^{(2)}(\mathbf{0}) = \begin{pmatrix}0\\1\\0\\0\end{pmatrix},$$

$$w_3 = \hat{v}^{(1)}(\mathbf{0}) = \begin{pmatrix} 0 \\ 0 \\ 1 \\ 0 \end{pmatrix}, \quad w_4 = \hat{v}^{(2)}(\mathbf{0}) = \begin{pmatrix} 0 \\ 0 \\ 0 \\ 1 \end{pmatrix}. \tag{11.1.31}$$

The $\hat{u}^{(r)}$ ($\hat{v}^{(r)}$) correspond to $\varepsilon_n = +1$ (-1); they are positive-energy (negative-energy) spinors. For arbitrary $\boldsymbol{p}$ we notice that

$$(\not{p} - m)(\not{p} + m) = (p^2 - m^2) = 0.$$

This allows one to write the general solution of (11.1.30) as

$$u^{(r)}(\boldsymbol{p}) = C(\not{p} + m)\hat{u}^{(r)}(\mathbf{0}), \tag{11.1.32a}$$

$$\hat{v}^{(r)}(\boldsymbol{p}) = -\,C'(\not{p} - m)\hat{v}^{(r)}(\mathbf{0}), \tag{11.1.32b}$$

where C and C' are norming constants. In order to display these solutions explicitly, one can appeal to the representation (11.1.13) of the Dirac matrices, which yields

$$\not{p} = \begin{pmatrix} E_p & -\boldsymbol{\sigma}\cdot\boldsymbol{p} \\ \boldsymbol{\sigma}\cdot\boldsymbol{p} & -E_p \end{pmatrix}. \tag{11.1.33}$$

The spinors $u^{(r)}(\boldsymbol{p})$ and $v^{(r)}(\boldsymbol{p})$ can be expressed in terms of two-component spinors $\chi^{(r)}$,

$$\chi^{(1)} = \begin{pmatrix} 1 \\ 0 \end{pmatrix}, \quad \chi^{(2)} = \begin{pmatrix} 0 \\ 1 \end{pmatrix},$$

in the form

$$\text{energy} > 0: \quad u^{(r)}(\boldsymbol{p}) = \frac{1}{\sqrt{E_p + m}} \begin{pmatrix} (E_p + m)\chi^{(r)} \\ (\boldsymbol{\sigma}\cdot\boldsymbol{p})\chi^{(r)} \end{pmatrix}, \tag{11.1.34a}$$

$$\text{energy} < 0: \quad v^{(r)}(\boldsymbol{p}) = \frac{1}{\sqrt{E_p + m}} \begin{pmatrix} (\boldsymbol{\sigma}\cdot\boldsymbol{p})\chi^{(r)} \\ (E_p + m)\chi^{(r)} \end{pmatrix}. \tag{11.1.34b}$$

In (11.1.32), the norming constants $C = C' = (E_p + m)^{-1/2}$ are chosen so that*

$$\begin{aligned} \bar{u}^{(r)}(\boldsymbol{p})u^{(s)}(\boldsymbol{p}) &= 2m\delta_{rs}, \\ \bar{v}^{(r)}(\boldsymbol{p})v^{(s)}(\boldsymbol{p}) &= -\,2m\delta_{rs}, \\ \bar{u}^{(r)}(\boldsymbol{p})v^{(s)}(\boldsymbol{p}) &= 0. \end{aligned} \tag{11.1.35}$$

* Most writers adopt the normalization

$$\bar{u}^{(r)}u^{(s)} = \delta_{rs}, \quad \bar{v}^{(r)}v^{(s)} = -\,\delta_{rs}.$$

The alternative normalization (11.1.35), though in some cases it proves rather inelegant (see (11.1.37)), allows one to use the same formulae for boson and for fermion cross-sections.

Note the minus sign in the orthonormality relation for negative-energy spinors. The closure relations too are useful, e.g. in calculating cross-sections. Let us for instance evaluate the matrix $\Lambda_+(\boldsymbol{p})$:

$$
\begin{aligned}
(\Lambda_+(\boldsymbol{p}))_{\alpha\beta} &= \sum_{r=1}^{2} u_\alpha^{(r)}(\boldsymbol{p})\bar{u}_\beta^{(r)}(\boldsymbol{p}) \\
&= \sum_r \frac{1}{(E_p + m)} \hat{u}_\lambda^{(r)}(0)(\not{p} + m)_{\alpha\lambda}(\not{p} + m)_{\gamma\beta}\hat{\bar{u}}_\gamma^{(r)}(0) \\
&= \frac{1}{2(E_p + m)}(\not{p} + m)_{\alpha\lambda}(1 + \gamma^0)_{\lambda\gamma}(\not{p} + m)_{\gamma\beta} \\
&= (\not{p} + m)_{\alpha\beta}.
\end{aligned}
$$

A similar calculation yields

$$
(\Lambda_-(\boldsymbol{p}))_{\alpha\beta} = -\sum_{r=1}^{2} v_\alpha^{(r)}(\boldsymbol{p})\bar{v}_\beta^{(r)}(\boldsymbol{p}) = -(\not{p} - m)_{\alpha\beta}.
$$

Accordingly we can define two 'projectors' $\Lambda_+(\boldsymbol{p})$ and $\Lambda_-(\boldsymbol{p})$, projecting onto positive and negative energy states respectively:

$$
\Lambda_+(\boldsymbol{p}) = \sum_r u^{(r)}(\boldsymbol{p})\bar{u}^{(r)}(\boldsymbol{p}) = \not{p} + m, \tag{11.1.36a}
$$

$$
\Lambda_-(\boldsymbol{p}) = -\sum_r v^{(r)}(\boldsymbol{p})\bar{v}^{(r)}(\boldsymbol{p}) = -(\not{p} - m). \tag{11.1.36b}
$$

These operators satisfy the relations

$$
\Lambda_+ + \Lambda_- = 2m, \quad \Lambda_+\Lambda_- = \Lambda_-\Lambda_+ = 0, \quad \Lambda_\pm^2 = 2m\Lambda_\pm. \tag{11.1.37}
$$

Because of the normalization adopted in (11.1.35), Λ_+ and Λ_- are not projection operators in the strict sense. In fact they are non-Hermitean even if the spinors are normed to unity, unless $\boldsymbol{p} = 0$: $\Lambda_\pm^\dagger = \gamma^0\Lambda_\pm\gamma^0$. The superscript (r) is a spin index, and one can readily define projection operators onto the different spin states. The four degrees of freedom of the Dirac field correspond to two spin degrees of freedom for each sign of the energy.

11.1.4 Quantization of the Dirac field

As in the Klein–Gordon case, we start from the Fourier decomposition of the classical Dirac field. For each value of $\boldsymbol{p}$ the spinors $u^{(r)}(\boldsymbol{p})$ and $v^{(r)}(\boldsymbol{p})$ constitute a basis for the four-dimensional space of the Dirac spinors, and we may write

$$
\psi(t, \boldsymbol{x}) = \sum_{r=1}^{2} \int \frac{d^3p}{(2\pi)^3 2E_p} [b_r(\boldsymbol{p})u^{(r)}(\boldsymbol{p})e^{-ipx} + d_r^*(\boldsymbol{p})v^{(r)}(\boldsymbol{p})e^{ipx}], \tag{11.1.38a}
$$

$$
\bar{\psi}(t, \boldsymbol{x}) = \sum_{r=1}^{2} \int \frac{d^3p}{(2\pi)^3 2E_p} [b_r^*(\boldsymbol{p})\bar{u}^{(r)}(\boldsymbol{p})e^{ipx} + d_r(\boldsymbol{p})\bar{v}^{(r)}(\boldsymbol{p})e^{-ip\cdot x}]. \tag{11.1.38b}
$$

Because ψ is a complex field, one must introduce two independent coefficients $b_r(\boldsymbol{p})$ and $d_r^*(\boldsymbol{p})$, complex conjugation being associated with negative energies.

In view of our investigations at the start of this section, it is logical to *postulate* that the Fourier coefficients $b_r(\boldsymbol{p})$, $d_r(\boldsymbol{p})$ and their complex conjugates be replaced by operators obeying the anticommutation rules

$$\begin{aligned}\{b_r(\boldsymbol{p}), b_{r'}^\dagger(\boldsymbol{p}')\} &= (2\pi)^3 2E_p \delta_{rr'} \delta^{(3)}(\boldsymbol{p}-\boldsymbol{p}'),\\ \{d_r(\boldsymbol{p}), d_{r'}^\dagger(\boldsymbol{p}')\} &= (2\pi)^3 2E_p \delta_{rr'} \delta^{(3)}(\boldsymbol{p}-\boldsymbol{p}'),\end{aligned} \tag{11.1.39}$$

with all other anticommutators vanishing. We shall show presently that this quantization postulate is perfectly selfconsistent, while postulating commutation rules would prove unacceptable.

The quantized fields $\psi(x)$ and $\bar{\psi}(x)$ are obtained from (11.1.38) by replacing the Fourier coefficients $b_r(\boldsymbol{p})$, $d_r(\boldsymbol{p})$ by (identically-designated) operators:

$$\psi(x) = \sum_{r=1}^{2} \int \frac{d^3p}{(2\pi)^3 2E_p} [b_r(\boldsymbol{p}) u^{(r)}(\boldsymbol{p}) e^{-ipx} + d_r^\dagger(\boldsymbol{p}) v^{(r)}(\boldsymbol{p}) e^{ipx}], \tag{11.1.40a}$$

$$\bar{\psi}(x) = \sum_{r=1}^{2} \int \frac{d^3p}{(2\pi)^3 2E_p} [b_r^\dagger(\boldsymbol{p}) \bar{u}^{(r)}(\boldsymbol{p}) e^{ipx} + d_r(\boldsymbol{p}) \bar{v}^{(r)}(\boldsymbol{p}) e^{-ipx}]. \tag{11.1.40b}$$

Just as for the Klein–Gordon field, the annihilation (creation) operators are linked with positive (negative) energies. It is instructive to compare equations (11.1.40) with (10.5.2), i.e. with those found for a charged scalar field.

One can check (Problem 11.6) that for the fields the rules (11.1.39) entail the anticommutation relations

$$\{\psi_\alpha(t, \boldsymbol{x}), \psi_\beta^\dagger(t, \boldsymbol{x}')\} = \delta_{\alpha\beta} \delta^{(3)}(\boldsymbol{x}-\boldsymbol{x}'). \tag{11.1.41}$$

But according to (11.1.23) the momentum conjugate to ψ_β is just $i\psi_\beta^\dagger$, which allows us to rewrite (11.1.41) as

$$\{\psi_\alpha(t, \boldsymbol{x}), \pi_\beta(t, \boldsymbol{x}')\} = i\delta_{\alpha\beta} \delta^{(3)}(\boldsymbol{x}-\boldsymbol{x}'). \tag{11.1.42}$$

In fact the assumptions (11.1.39) amount to postulating anticommutation rules between the field and its conjugate momentum, which is of course the most natural assumption once one realizes that commutators fail.

Next we consider the Hamiltonian and the charge. Equation (11.1.24) yields the Hamiltonian as

$$H = i \int d^3x \, \bar{\psi}(x) \gamma^0 \frac{\partial \psi(x)}{\partial t}.$$

To calculate H we use Parseval's theorem: if

$$\psi(\boldsymbol{x}) = \int \frac{d^3p}{(2\pi)^3} e^{i\boldsymbol{p}\cdot\boldsymbol{x}} \psi(\boldsymbol{p}),$$

then

$$\int d^3x\, \psi(\boldsymbol{x})\varphi(\boldsymbol{x}) = \int \frac{d^3p}{(2\pi)^3}\psi(\boldsymbol{p})\varphi(-\boldsymbol{p}).$$

For the case under study, the Fourier decomposition (11.1.40) yields the coefficients

$$\bar{\psi}(\boldsymbol{p}) = \sum_r \frac{1}{2E_p}[d_r(\boldsymbol{p})\bar{v}^{(r)}(\boldsymbol{p})e^{-iE_p t} + b_r^\dagger(-\boldsymbol{p})\bar{u}^{(r)}(-\boldsymbol{p})e^{iE_p t}],$$

$$\frac{\partial}{\partial t}\psi(-\boldsymbol{p}) = \frac{-i}{2}\sum_s [b_s(-\boldsymbol{p})u^{(s)}(-\boldsymbol{p})e^{-iE_p t} - d_s^\dagger(\boldsymbol{p})v^{(s)}(\boldsymbol{p})e^{iE_p t}].$$

Here we must take note of the orthonormality conditions

$$\begin{aligned} &\bar{v}^{(r)}(\boldsymbol{p})\gamma^0 u^{(s)}(-\boldsymbol{p}) = 0, \\ &\bar{v}^{(r)}(\boldsymbol{p})\gamma^0 v^{(s)}(\boldsymbol{p}) = \bar{u}^{(r)}(\boldsymbol{p})\gamma^0 u^{(s)}(\boldsymbol{p}) = 2E_p\delta_{rs}, \end{aligned} \tag{11.1.43}$$

(Problem 11.5), in order to express the Hamiltonian in terms of the Fourier coefficients:

$$H = \sum_{r=1}^{2}\int \frac{d^3p}{(2\pi)^3 2E_p} E_p[b_r^\dagger(\boldsymbol{p})b_r(\boldsymbol{p}) - d_r(\boldsymbol{p})d_r^\dagger(\boldsymbol{p})]. \tag{11.1.44}$$

The term prefaced by the minus sign can be transformed by virtue of the anticommutation rules (11.1.39). For clarity we adopt discrete normalization as in (9.2.16):

$$-d_{r,\boldsymbol{p}}d_{r,\boldsymbol{p}}^\dagger = -1 + d_{r,\boldsymbol{p}}^\dagger d_{r,\boldsymbol{p}}.$$

By redefining the zero of energy, the Hamiltonian can be put into the form*

$$H = \sum_{r=1}^{2}\int \frac{d^3p}{(2\pi)^3 2E_p} E_p(b_r^\dagger(\boldsymbol{p})b_r(\boldsymbol{p}) + d_r^\dagger(\boldsymbol{p})d_r(\boldsymbol{p})). \tag{11.1.45}$$

Alternatively, (11.1.45) can be derived from (11.1.44) by introducing normal products, but with one important difference from the case of bosons: now, when the creation operators are shifted towards the left, it is (all the) *anti*commutators that are dropped, so that

$$:bb^\dagger: = -b^\dagger b.$$

The crucial point in this calculation is that the anticommutation rules are essential for making the Hamiltonian positive definite. With commutation rules, the second term in (11.1.44) could not be made positive, not even by redefining

* Equation (11.1.45) is readily generalized to any component of the four-momentum:

$$P_\mu = \sum_{r=1}^{2}\int \frac{d^3p}{(2\pi)^3 2E_p} p_\mu[b_r^\dagger(\boldsymbol{p})b_r(\boldsymbol{p}) + d_r^\dagger(\boldsymbol{p})d_r(\boldsymbol{p})].$$

the energy zero. This is a special case of the *spin-statistics theorem*: fields with half-integer spin (1/2, 3/2, . . .) must be quantized through *anticommutation* rules. Conversely, fields with integer spin (0, 1, . . .) must be quantized through *commutation* rules; one can show that, otherwise, the locality condition breaks down. The connection between spin and statistics is a deep (and fascinating) consequence of marrying quantum mechanics with relativity. As far as the writer knows, there is no simple and intuitive argument for the theorem.

To complete the physical interpretation of our quantization process, it remains to calculate the charge,

$$Q = e \int \mathrm{d}^3 x : \bar{\psi}(x) \gamma^0 \psi(x) : .$$

The calculation proceeds exactly like the one above, and yields

$$Q = e \sum_{r=1}^{2} \int \frac{\mathrm{d}^3 p}{(2\pi)^3 2E_p} (b_r^\dagger(\boldsymbol{p}) b_r(\boldsymbol{p}) - d_r^\dagger(\boldsymbol{p}) d_r(\boldsymbol{p})). \tag{11.1.46}$$

Note the similarity to equation (10.5.6) for the charged scalar field.

Bearing in mind that $b_r^\dagger(\boldsymbol{p}) b_r(\boldsymbol{p})$ and $d_r^\dagger(\boldsymbol{p}) d_r(\boldsymbol{p})$ are the number operators for particles of momentum $\boldsymbol{p}$ and spin r, we see from (11.1.45) and (11.1.46) that $b_r^\dagger(\boldsymbol{p})$ creates a particle of charge e and momentum $\boldsymbol{p}$, while $d_r^\dagger(\boldsymbol{p})$ creates a particle with charge $-e$ and momentum $\boldsymbol{p}$. The field $\psi(x)$ creates a charge $-e$ (through $d_r^\dagger(\boldsymbol{p})$), or destroys a charge e (through $b_r^\dagger(\boldsymbol{p})$): conversely, $\bar{\psi}(x)$ changes the charge by $+e$. This property is formulated more generally in Problem 11.9.

If by convention we call the particle with charge e the 'particle', then the particle with charge $-e$ is the 'antiparticle'; the classic example is furnished by the electron (the particle) and the positron (the antiparticle). Note that one can equally well define charges other than the electric charge; for instance the baryonic charge, which distinguishes neutrons from antineutrons.

To sum up, the four degrees of freedom of a Dirac field enable it to describe a particle plus an antiparticle, each having two spin degrees of freedom.

By choosing a special representation of the Dirac matrices, namely the Majorana representation, one can conveniently describe spin 1/2 particles that are their own antiparticles; such objects are called 'Majorana particles'. In this case the Dirac field has only two degrees of freedom (see Problem 12.13).

11.1.5 The propagator of the Dirac field

The two preceding chapters have registered the crucial role played by the vacuum expectation-value of the T-product of two fields. Thus it is natural to study the T-product of two Dirac fields. But, because of the anticommutation rules, one must change a sign in the definition of such a T-product, and set

$$\begin{aligned} T(\psi_\alpha(x)\bar{\psi}_\beta(x')) = {} & \theta(x^0 - x'^0)\psi_\alpha(x)\bar{\psi}_\beta(x') \\ & - \theta(x'^0 - x^0)\bar{\psi}_\beta(x')\psi_\alpha(x). \end{aligned} \tag{11.1.47}$$

As always, the permutation of two Dirac fields introduces a minus sign. Now evaluate the vacuum expectation-value of this T-product, using the Fourier representation (11.1.40) of the free Dirac field:

$$\langle 0|\, T(\psi_\alpha(x)\bar{\psi}_\beta(x'))|0\rangle = \int \frac{\mathrm{d}^3 p}{(2\pi)^3 2E_p}[\theta(t-t')(\not{p}+m)_{\alpha\beta}\mathrm{e}^{-\mathrm{i}p(x-x')}$$

$$-\,\theta(t'-t)(\not{p}-m)_{\alpha\beta}\mathrm{e}^{\mathrm{i}p(x-x')}]. \tag{11.1.48}$$

To derive this equation we have used the closure relation (11.1.36). Just as for the Klein–Gordon field, the result can be expressed by means of the Feynman contour (Problem 11.7):

$$\langle 0|\, T(\psi_\alpha(x)\bar{\psi}_\beta(x'))|0\rangle = [S_{\mathrm{F}}(x-x')]_{\alpha\beta}$$

$$= \int \frac{\mathrm{d}^4 p}{(2\pi)^4}\mathrm{e}^{-\mathrm{i}p(x-x')}\frac{\mathrm{i}(\not{p}+m)_{\alpha\beta}}{p^2-m^2+\mathrm{i}\varepsilon}. \tag{11.1.49}$$

In Fourier space the Feynman propagator $S_{\mathrm{F}}(p)$ for the Dirac field is given by

$$S_{\mathrm{F}}(p) = \frac{\mathrm{i}(\not{p}+m)}{p^2-m^2+\mathrm{i}\varepsilon} = \frac{\mathrm{i}}{\not{p}-m+\mathrm{i}\varepsilon}. \tag{11.1.50}$$

It is easy to check that this propagator is indeed a Green's function for the Dirac equation:

$$(\mathrm{i}\not{\partial}-m)S_{\mathrm{F}}(x) = \mathrm{i}\delta^{(4)}(x). \tag{11.1.51}$$

Further, one can express $S_{\mathrm{F}}(x-x')$ in terms of $\Delta_{\mathrm{F}}(x-x')$; in view of (11.1.49) one has

$$S_{\mathrm{F}}(x-x') = (\mathrm{i}\not{\partial}_x + m)\Delta_{\mathrm{F}}(x-x'). \tag{11.1.52}$$

11.2 Wick's theorem for fermions

11.2.1 'Fermion oscillator' coupled to an external source

Rather than discuss the general case, we shall elucidate the method we intend to follow by investigating an elementary example. In fact our example is just the fermionic analogue of the harmonic oscillator coupled to a classical source (see Section 9.3.2). Consider two operators ψ and $\bar{\psi}$ such that

$$\psi^2 = \bar{\psi}^2 = 0, \quad \{\psi, \bar{\psi}\} = 1; \tag{11.2.1}$$

and consider also the 'Hamiltonian' of a 'fermion oscillator',

$$H_0 = E\bar{\psi}\psi.$$

In order to couple ψ and $\bar{\psi}$ to an external source, one introduces the elements

$\eta(t)$ and $\bar{\eta}(t)$ of a Grassmann algebra*; these are 'anticommuting numbers' in the sense that $\eta(t)$ and $\bar{\eta}(t)$ anticommute with each other and also with the ψ and $\bar{\psi}$:

$$\{\eta(t), \eta(t')\} = \{\eta(t), \bar{\eta}(t')\} = \{\eta(t), \psi\} = \cdots = 0. \tag{11.2.2}$$

The full 'Hamiltonian' is

$$H = E\bar{\psi}\psi - \bar{\eta}(t)\psi - \bar{\psi}\eta(t). \tag{11.2.3}$$

We stress that we are talking about a purely mathematical problem; H need not be Hermitean, and ψ and $\bar{\psi}$ need not be Hermitean conjugates. Only the relations (11.2.1) and (11.2.2) matter. We can now employ the method of Section 9.3.2, by introducing the interaction-picture operators $\psi_{\mathrm{I}}(t)$ and $\bar{\psi}_{\mathrm{I}}(t)$:

$$\psi_{\mathrm{I}}(t) = \mathrm{e}^{\mathrm{i}H_0 t}\psi\mathrm{e}^{-\mathrm{i}H_0 t}, \quad \bar{\psi}_{\mathrm{I}}(t) = \mathrm{e}^{\mathrm{i}H_0 t}\bar{\psi}\mathrm{e}^{-\mathrm{i}H_0 t}.$$

Since $[H_0, \psi] = -E\psi$ and $[H_0, \bar{\psi}] = E\bar{\psi}$, we have

$$\psi_{\mathrm{I}}(t) = \mathrm{e}^{-\mathrm{i}Et}\psi, \quad \bar{\psi}_{\mathrm{I}}(t) = \mathrm{e}^{\mathrm{i}Et}\bar{\psi},$$

and the interaction-picture Hamiltonian becomes

$$H_{\mathrm{I}}(t) = -[\bar{\eta}(t)\mathrm{e}^{-\mathrm{i}Et}\psi + \bar{\psi}\mathrm{e}^{\mathrm{i}Et}\eta(t)]. \tag{11.2.4}$$

The commutator $[H_{\mathrm{I}}(t), H_{\mathrm{I}}(t')]$ is independent of the operators ψ and $\bar{\psi}$:

$$[H_{\mathrm{I}}(t), H_{\mathrm{I}}(t')] = \bar{\eta}(t)\eta(t')\mathrm{e}^{-\mathrm{i}E(t-t')} - \bar{\eta}(t')\eta(t)\mathrm{e}^{-\mathrm{i}E(t'-t)}; \tag{11.2.5}$$

moreover, because $(\bar{\eta}\eta)$ commutes with ψ and η, we can check at once that

$$[[H_{\mathrm{I}}(t), H_{\mathrm{I}}(t')], H_{\mathrm{I}}(t'')] = 0.$$

This allows us to use the identity from Problem 9.10(b) in order to rewrite $U_{\mathrm{I}}(t)$ as

$$U_{\mathrm{I}}(t) = \exp\left(-\mathrm{i}\int^t H_{\mathrm{I}}(t')\mathrm{d}t'\right)$$

$$\times \exp\left(-\frac{1}{2}\iint^t \mathrm{d}t'\,\mathrm{d}t''\,\theta(t'-t'')[H_{\mathrm{I}}(t'), H_{\mathrm{I}}(t'')]\right). \tag{11.2.6}$$

The normal-ordered version of $U_{\mathrm{I}}(t)$ is found by appealing once again to the identity

$$\mathrm{e}^{A+B} = \mathrm{e}^A\mathrm{e}^B\mathrm{e}^{-\frac{1}{2}[A,B]},$$

* A Grassmann algebra $\mathscr{A}$ is defined as follows. Let there be N variables (generators) $\eta_1, \ldots, \eta_N$ such that

(i) $\{\eta_i, \eta_j\} = 0$,

(ii) if $\eta_i \in \mathscr{A} : \lambda\eta_i \in \mathscr{A}$ $(\lambda \in \mathbb{C})$,

(iii) $\eta_i, \eta_j \in \mathscr{A} : \lambda\eta_i + \mu\eta_j \in \mathscr{A}$,

(iv) $\eta_i, \eta_j \in \mathscr{A} : \eta_i\eta_j \in \mathscr{A}$;

Then the η_i generate a Grassmann algebra.

and the end-result reads

$$U_{\mathrm{I}}(t) = :\exp\left(-\mathrm{i}\int^{t} H_{\mathrm{I}}(t')\,\mathrm{d}t'\right): \\ \times\exp\left(-\iint^{t} \mathrm{d}t'\,\mathrm{d}t''\bar{\eta}(t')\theta(t'-t'')\mathrm{e}^{-\mathrm{i}E(t'-t'')}\eta(t'')\right). \quad (11.2.7)$$

Note the similarity between this and (9.3.15). If we now identify ψ and $\bar{\psi}$ with creation and annihilation operators a and $a^{\dagger}$ (see (11.1.3)), then

$$\langle 0|T(\psi_{\mathrm{I}}(t')\bar{\psi}_{\mathrm{I}}(t''))|0\rangle = \theta(t'-t'')\mathrm{e}^{-\mathrm{i}E(t'-t'')},$$

and the double integral in (11.2.7) is equivalent to

$$\iint \mathrm{d}t'\,\mathrm{d}t''\,\bar{\eta}(t')\langle 0|T(\psi_{\mathrm{I}}(t')\bar{\psi}_{\mathrm{I}}(t''))|0\rangle\eta(t).$$

Given our experience with boson fields, it is not hard to guess the form of Wick's theorem for fermions:

$$T\left(\exp\left(\mathrm{i}\int \mathrm{d}^4x(\bar{\eta}(x)\psi(x)+\bar{\psi}(x)\eta(x))\right)\right) \\ = :\exp\left(\mathrm{i}\int \mathrm{d}^4x(\bar{\eta}(x)\psi(x)+\bar{\psi}(x)\eta(x))\right): \\ \times\exp\left(-\iint \mathrm{d}^4x\,\mathrm{d}^4x'\,\bar{\eta}(x')\langle 0|T(\psi(x)\bar{\psi}(x'))|0\rangle\eta(x')\right). \quad (11.2.8)$$

The proof follows exactly the same scenario as before, and is indicated in Problem 11.8. In equation (11.2.8), $\psi(x)$ and $\bar{\psi}(x)$ are *free* fields, just as $\varphi(x)$ was in (9.3.27). Note that the coefficient of the double integral in (11.2.8) is (-1), while in (9.3.27) it was $(-1/2)$; the difference reflects simply the presence of two different charges, i.e. of two types of particle. We noticed exactly the same phenomenon with charged bosons (see (10.5.10)).

Expanding the exponentials in (11.2.8), and identifying the coefficients of η and of $\bar{\eta}$, one derives identities like

$$T(\psi(x)\bar{\psi}(x')) = :\psi(x)\bar{\psi}(x'): + \langle 0|T(\psi(x)\bar{\psi}(x'))|0\rangle.$$

The correct signs follow automatically from the anticommutation rules for the external sources $\eta(x)$ and $\bar{\eta}(x)$. In practice one needs to watch the order of the factors; thus, in the fermionic equivalent of (9.3.28), every term must be multiplied by the signature of the permutation from the initial to the final ordering of the ψ and the $\bar{\psi}$.

11.2.2 The functional formulation: integration over Grassmann variables

It is perfectly possible to write down a generating functional for fermion Green's functions. However, instead of integrating with respect to numbers as for bosons, one must integrate with respect to the elements of a Grassmann algebra. Start with a single degree of freedom, by introducing two Grassmann variables ψ and $\bar{\psi}$:

$$\psi^2 = \bar{\psi}^2 = 0, \quad \{\psi, \bar{\psi}\} = 0. \tag{11.2.9}$$

Note carefully the difference between (11.2.1) and (11.2.9): the former features a (field) operator, while the latter features a classical field, or rather the fermionic generalization of a classical field. The most general polynomial one can construct from ψ and $\bar{\psi}$ in view of (11.2.9) is*

$$P(\psi) = a_0 + a_1\psi + \tilde{a}_1\bar{\psi} + a_{12}\bar{\psi}\psi. \tag{11.2.10}$$

Now we *define* integrals with respect to Grassmann variables by

$$\int \mathrm{d}\psi = 0, \quad \int \mathrm{d}\psi\, \psi = 1,$$

$$\int \mathrm{d}\bar{\psi} = 0, \quad \int \mathrm{d}\bar{\psi}\, \bar{\psi} = 1,$$

whence

$$\int \mathrm{d}\bar{\psi}\, \mathrm{d}\psi\, P(\psi) = \int \mathrm{d}\bar{\psi}(a_1 - a_{12}\bar{\psi}) = -a_{12}. \tag{11.2.11}$$

(Indeed $\int \mathrm{d}\psi\, \bar{\psi}\psi = -\int \mathrm{d}\psi\, \psi\bar{\psi} = -\bar{\psi}$.) As an immediate application of (11.2.11) we can evaluate the 'Gaussian' integral

$$\int \mathrm{d}\bar{\psi}\, \mathrm{d}\psi\, \mathrm{e}^{-a\bar{\psi}\psi} = a.$$

Next we try to generalize the Gaussian integral to several variables, by examining the case of two degrees of freedom:

$$\int \mathrm{d}\bar{\psi}_1\, \mathrm{d}\psi_1\, \mathrm{d}\bar{\psi}_2\, \mathrm{d}\psi_2\, \mathrm{e}^{-\sum_{i,j=1}^{2} \bar{\psi}_i A_{ij} \psi_j}. \tag{11.2.12}$$

When the exponential is expanded, only the term containing the factor $\bar{\psi}_1\psi_1\bar{\psi}_2\psi_2$ gives a nonzero contribution; by virtue of the anticommutation rules, the integral (11.2.12) equals

$$A_{11}A_{22} - A_{12}A_{21} = \det A.$$

* If a_0 and a_{12} are numbers, then a_1 ($\tilde{a}_1$) is a Grassmann constant with respect to ψ ($\bar{\psi}$). Note also the relation $\mathrm{d}(\lambda\psi) = \lambda^{-1}\,\mathrm{d}\psi$.

It is not hard to convince oneself that the general result reads

$$\int \prod_{i=1}^{N} \mathrm{d}\bar{\psi}_i \mathrm{d}\psi_i \mathrm{e}^{-\sum_{i,j=1}^{N} \bar{\psi}_i A_{ij} \psi_j} = \det A. \tag{11.2.13}$$

This should be compared with the result in (A.2.4) (for $j = 0$; $\mathrm{d}x\,\mathrm{d}y = (\mathrm{d}z\,\mathrm{d}z^*)/2\mathrm{i}$), namely with

$$\int \prod_{i=1}^{N} \frac{\mathrm{d}z_i \, \mathrm{d}z_i^*}{2\mathrm{i}\pi} \exp\left(- \sum_{i,j=1}^{N} z_i^* A_{ij} z_j \right) = (\det A)^{-1}. \tag{11.2.14}$$

Notice that through our using Grassmann variables the determinant A has risen from the denominator to the numerator. This property proves very convenient for expanding Jacobians in diagrammatic form, for instance in non-Abelian gauge theories. An application of this kind is given in Problem 11.14. As with ordinary variables, one can add linear terms to the exponent in (11.2.13), and define an integral $I(\eta, \bar{\eta})$ depending on the (Grassmann) sources η and $\bar{\eta}$:

$$I(\bar{\eta}, \eta) = \int \prod_{i=1}^{N} \mathrm{d}\bar{\psi}_i \mathrm{d}\psi_i \exp(-\bar{\psi}_i A_{ij} \psi_j + \bar{\eta}_i \psi_i + \bar{\psi}_i \eta_i).$$

Equation (A.2.4) can be transcribed into Grassmann variables by the change of variables

$$\psi_j = \psi_j' + A_{jk}^{-1} \eta_k; \quad \bar{\psi}_j = \bar{\psi}_j' + \bar{\eta}_k A_{kj}^{-1},$$

which yields

$$I(\bar{\eta}, \eta) = (\det A) \exp(\bar{\eta}_i A_{ij}^{-1} \eta_j). \tag{11.2.15}$$

This expression enables us to write down a generating functional $Z(\bar{\eta}, \eta)$ for Green's functions, namely

$$Z(\bar{\eta}, \eta) = \int \mathscr{D}(\bar{\psi}, \psi) \times \exp\left(\mathrm{i} \int \mathrm{d}^4 x [\bar{\psi}(\mathrm{i}\overrightarrow{\not\partial} - m)\psi + V(\bar{\psi}, \psi) + \bar{\eta}\psi + \bar{\psi}\eta] \right), \tag{11.2.16}$$

where the prescription $m \to m - \mathrm{i}\varepsilon$ guarantees that the ensuing propagator is indeed $S_{\mathrm{F}}(p)$. As an example, from $Z_0(\bar{\eta}, \eta)$ we can reconstruct the equivalent of Wick's theorem,

$$Z_0(\bar{\eta}, \eta) = \int \mathscr{D}(\bar{\psi}, \psi) \exp\left(\mathrm{i} \int \mathrm{d}^4 x [\bar{\psi}(\mathrm{i}\overrightarrow{\not\partial} - m)\psi + \bar{\eta}\psi + \bar{\psi}\eta] \right)$$

$$= \mathscr{N} \exp\left(- \int \mathrm{d}^4 x \, \mathrm{d}^4 x' \bar{\eta}(x) \mathrm{i} (\mathrm{i}\not\partial - m)^{-1} \eta(x') \right). \tag{11.2.17}$$

From equation (11.1.51), it follows that $i(i\not{\partial} - m)^{-1}$ is, as it should be, equal to the propagator $S_F(x)$. By following up the indications given in the preceding chapter, one can show that the Green's functions calculated by functional differentiation of $Z(\bar{\eta}, \eta)$ are the same as those found through Wick's theorem.

11.3 The Lagrangean formalism for the classical electromagnetic field

11.3.1 Maxwell's equations and the electromagnetic potentials

In the elementary formulation of Maxwell's equations one introduces an electric field $\boldsymbol{E}$ and a magnetic field $\boldsymbol{B}$, coupled to a charge density ρ and to a current density $\boldsymbol{j}$, which are also called the *sources* of the electromagnetic field:

$$\nabla \cdot \boldsymbol{E} = \rho, \qquad \nabla \times \boldsymbol{B} - \frac{\partial \boldsymbol{E}}{\partial t} = \boldsymbol{j}, \tag{11.3.1a, b}$$

$$\nabla \times \boldsymbol{E} = -\frac{\partial \boldsymbol{B}}{\partial t}, \qquad \nabla \cdot \boldsymbol{B} = 0. \tag{11.3.2a, b}$$

(Here $c = 1$; under the conventions governing (11.3.1), Coulomb's law reads $\|\boldsymbol{F}\| = |q_1 q_2|/4\pi r^2$.) The equations (11.3.1) imply the continuity equation for $(\rho, \boldsymbol{j})$,

$$\frac{\partial \rho}{\partial t} + \nabla \cdot \boldsymbol{j} = 0, \tag{11.3.3}$$

which expresses the local conservation of charge.

Equations (11.3.2), which are independent of $(\rho, \boldsymbol{j})$, imply the existence of a scalar potential φ and of a vector potential $\boldsymbol{A}$ such that

$$\boldsymbol{E} = -\nabla \varphi - \frac{\partial \boldsymbol{A}}{\partial t}, \qquad \boldsymbol{B} = \nabla \times \boldsymbol{A}. \tag{11.3.4a, b}$$

The fields $\boldsymbol{E}$ and $\boldsymbol{B}$ remain unchanged by *gauge transformations*

$$\varphi \to \varphi' = \varphi + \frac{\partial \Lambda}{\partial t}, \qquad \boldsymbol{A} \to \boldsymbol{A}' = \boldsymbol{A} - \nabla \Lambda, \tag{11.3.5}$$

where $\Lambda(t, \boldsymbol{x})$ is an arbitrary function of time and of position.

11.3.2 Covariant formulation

The potentials $(\varphi, \boldsymbol{A})$ are collected into a four-vector $A^\mu(x)$:

$$A^0(x) = \varphi(t, \boldsymbol{x}), \quad A^i(x) = (\boldsymbol{A})_i(t, \boldsymbol{x})$$

(at least, $A^\mu(x)$ behaves as a four-vector up to the arbitrariness allowed by gauge transformations); and similarly for $(\rho, \boldsymbol{j})$:

$$j^0(x) = \rho(t, \boldsymbol{x}), \quad j^i(x) = (\boldsymbol{j})_i(t, \boldsymbol{x}).$$

$j^\mu(x)$ is the electromagnetic current four-vector, and satisfies the covariant version of the continuity equation (11.3.3), also called the *conservation law for the current* j_μ:

$$\partial^\mu j_\mu(x) = 0. \tag{11.3.6}$$

In this covariant notation the gauge transformation (11.3.5) reads

$$A_\mu(x) \to A'_\mu(x) = A_\mu(x) + \partial_\mu \Lambda(x),$$

where $\Lambda(x)$ is an arbitrary function of x.

The electromagnetic field tensor $F^{\mu\nu}$ is defined by

$$F^{\mu\nu} = -F^{\nu\mu} = \partial^\mu A^\nu - \partial^\nu A^\mu; \tag{11.3.7}$$

in matrix form it reads

$$F^{\mu\nu} = \begin{matrix} \\ \mu \\ \downarrow \end{matrix} \overset{\nu \to}{\begin{pmatrix} 0 & -E_1 & -E_2 & -E_3 \\ E_1 & 0 & -B_3 & B_2 \\ E_2 & B_3 & 0 & -B_1 \\ E_3 & -B_2 & B_1 & 0 \end{pmatrix}}.$$

The dual tensor $\tilde{F}^{\mu\nu}$ is related to $F^{\mu\nu}$ by

$$\tilde{F}^{\mu\nu} = \frac{1}{2}\varepsilon^{\mu\nu\rho\sigma} F_{\rho\sigma} \tag{11.3.8}$$

(see (D.1.1)), and arises through the substitutions $\boldsymbol{E} \to \boldsymbol{B}$, $\boldsymbol{B} \to -\boldsymbol{E}$, which leave Maxwell's equations invariant in the absence of sources $(\rho, \boldsymbol{j})$.

In this covariant formulation, Maxwell's equations become

$$\partial^\mu F_{\mu\nu} = j_\nu, \tag{11.3.9}$$

$$\partial^\mu \tilde{F}_{\mu\nu} = 0. \tag{11.3.10}$$

They follow from the Lagrangean density

$$\mathscr{L} = -\frac{1}{4} F^{\mu\nu} F_{\mu\nu} - j^\mu A_\mu, \tag{11.3.11}$$

which in terms of $\boldsymbol{E}$ and $\boldsymbol{B}$ reads

$$\mathscr{L} = \frac{1}{2}(\boldsymbol{E}^2 - \boldsymbol{B}^2) - \rho\varphi + \boldsymbol{j} \cdot \boldsymbol{A}. \tag{11.3.12}$$

The form of the Lagrangean in (11.3.11) can be found by writing down the most general expression that is Lorentz-invariant, gauge-invariant, and quadratic in A_μ and $\partial_\nu A_\mu$. A form equivalent to (11.3.11) reads

$$\mathscr{L} = \frac{1}{2} A^\mu(\Box g_{\mu\nu} - \partial_\mu \partial_\nu) A^\nu - j^\mu A_\mu \tag{11.3.13}$$

(see the comments following equation (9.2.3a)). Expressed in terms of the potentials, (11.3.9) becomes

$$(\Box g_{\mu\nu} - \partial_\mu \partial_\nu) A^\nu = j_\mu. \tag{11.3.14}$$

The *Lorentz gauge* is defined by $\partial_\mu A^\mu = 0$, in which case (11.3.14) simplifies to

$$\Box A^\mu = j^\mu. \tag{11.3.15}$$

It is of interest to determine the conjugate momenta:

$$\pi^0 = \frac{\partial \mathscr{L}}{\partial(\partial_0 A_0)} = 0,$$

$$\pi^k = \frac{\partial \mathscr{L}}{\partial(\partial_0 A_k)} = F^{k0} = E_k.$$

Crucially, there is no momentum conjugate to A_0!

11.3.3 Gauge invariance and current conservation

The current-conservation law (11.3.6) is most remarkable, and deserves careful study.

A general technique for identifying conserved currents is supplied by *Noether's theorem*, which links current conservation to invariances of the Lagrangean. (For a general discussion see Appendix C.) Suppose for instance that a Lagrangean density $\mathscr{L}$ depending on N fields is invariant under certain transformations

$$\varphi_r(x) \to \varphi'_r(x) = (e^{-i\Lambda T})_{rs} \varphi_s(x) \tag{11.3.16}$$

specified by a parameter Λ, where T is a Hermitean matrix. Now take Λ as infinitesimal, and expand to first order in Λ:

$$\varphi'_r(x) = \varphi_r(x) - i\Lambda T_{rs} \varphi_s(x). \tag{11.3.17}$$

The Lagrangean changes by

$$\delta\mathscr{L} = \frac{\partial \mathscr{L}}{\partial(\partial_\mu \varphi_r)} \delta(\partial_\mu \varphi_r) + \frac{\partial \mathscr{L}}{\partial \varphi_r} \delta\varphi_r$$

$$= \partial_\mu \left(\frac{\partial \mathscr{L}}{\partial(\partial_\mu \varphi_r)} \delta\varphi_r \right) - \left(\partial_\mu \frac{\partial \mathscr{L}}{\partial(\partial_\mu \varphi_r)} \right) \delta\varphi_r + \frac{\partial \mathscr{L}}{\partial \varphi_r} \delta\varphi_r.$$

In the rightmost expression the last two terms cancel by virtue of the equations of motion for the *classical* field; since $\delta\mathscr{L} = 0$, it follows that

$$\partial_\mu \left[\frac{\partial \mathscr{L}}{\partial(\partial_\mu \varphi_r)} \delta\varphi_r \right] = 0. \tag{11.3.18}$$

This is Noether's theorem. Substituting from (11.3.17) into (11.3.18) one obtains the conserved current:

$$j_\mu(x) = -\mathrm{i} \frac{\partial \mathscr{L}}{\partial(\partial_\mu \varphi_r)} T_{rs} \varphi_s, \quad \partial^\mu j_\mu(x) = 0. \tag{11.3.19}$$

Let us now apply this result to the Dirac Lagrangean (11.1.20),

$$\mathscr{L} = \bar{\psi}(x)(\mathrm{i}\overrightarrow{\not{\partial}} - m)\psi.$$

This Lagrangean is invariant under the *global gauge transformations*

$$\psi \to \mathrm{e}^{-\mathrm{i}q\Lambda}\psi, \quad \bar{\psi} \to \bar{\psi}\mathrm{e}^{\mathrm{i}q\Lambda}. \tag{11.3.20}$$

(To avoid any confusion between $\mathrm{e} = 2.71\ldots$ and the charge of the electron, we write the latter as q in the rest of this chapter.) Because $\partial\mathscr{L}/\partial(\partial_\mu\psi) = \mathrm{i}\bar{\psi}\gamma^\mu$, we recover our former expression for the conserved current,

$$j_\mu(x) = q\bar{\psi}(x)\gamma_\mu\psi(x).$$

Local gauge invariance

Suppose now that Λ, instead of being a constant, depends on x; the transformation

$$\psi \to \mathrm{e}^{-\mathrm{i}q\Lambda(x)}\psi(x), \quad \bar{\psi} \to \bar{\psi}(x)\mathrm{e}^{\mathrm{i}q\Lambda(x)} \tag{11.3.21}$$

is called a *local gauge transformation.* It is said to be Abelian, because the product of two transformations is commutative. The Lagrangean (11.1.20) is not invariant under such transformations, because

$$\begin{aligned}\partial_\mu\psi'(x) &= \partial_\mu(\mathrm{e}^{-\mathrm{i}q\Lambda(x)}\psi(x)) \\ &= -\mathrm{i}q(\partial_\mu\Lambda)\psi'(x) + \mathrm{e}^{-\mathrm{i}q\Lambda(x)}(\partial_\mu\psi(x)) \neq \mathrm{e}^{-\mathrm{i}q\Lambda(x)}(\partial_\mu\psi(x)).\end{aligned}$$

In order to obtain a Lagrangean that *is* invariant under (11.3.21), we write down the full Lagrangean of the electromagnetic field coupled to the Dirac field; this is nothing but the *Lagrangean of quantum electrodynamics* (QED) (though for the moment we are still dealing with classical fields). It reads

$$\mathscr{L}_{\mathrm{QED}} = -\frac{1}{4}F_{\mu\nu}F^{\mu\nu} + \bar{\psi}(\mathrm{i}\overrightarrow{\not{\partial}} - m)\psi - q\bar{\psi}(x)\gamma_\mu\psi(x)A^\mu(x). \tag{11.3.22}$$

The first two terms in (11.3.22) correspond to the free Lagrangeans of the electromagnetic and of the Dirac fields respectively, while the last is an interaction coupling the two fields to each other. We see that the first term in $\partial_\mu\psi'$ is cancelled if, simultaneously with (11.3.21), the electromagnetic field is subjected to the gauge transformation

$$A_\mu \to A'_\mu = A_\mu(x) + \partial_\mu \Lambda(x). \tag{11.3.23}$$

Then the *full* Lagrangean (11.3.22) remains invariant under local gauge transformations. Conversely, one can look for a derivative D_μ, called the *covariant derivative*, such that

$$(D_\mu \psi)'(x) = \mathrm{e}^{-\mathrm{i}q\Lambda(x)} D_\mu \psi(x).$$

Evidently this covariant derivative reads

$$D_\mu = \partial_\mu + \mathrm{i}qA_\mu(x), \tag{11.3.24}$$

and we note that, indeed,

$$(\partial_\mu + \mathrm{i}qA_\mu + \mathrm{i}q(\partial_\mu \Lambda))\mathrm{e}^{-\mathrm{i}q\Lambda(x)}\psi(x) = \mathrm{e}^{-\mathrm{i}q\Lambda(x)}(\partial_\mu + \mathrm{i}qA_\mu)\psi(x).$$

In other words, under gauge transformations $D_\mu \psi(x)$ behaves in the same way as $\psi(x)$ does, which explains the name 'covariant' for this derivative. Whenever a Lagrangean is invariant under a global gauge transformation (11.3.20), one can make it invariant under the local transformation (11.3.21), provided one introduces a *vector* field $A_\mu(x)$ (sometimes called the 'compensating field') that transforms according to (11.3.23), and couples to the original fields according to the minimal-coupling prescription

$$\partial_\mu \to \partial_\mu + \mathrm{i}qA_\mu. \tag{11.3.25}$$

This amounts to replacing the derivatives ∂_μ in the original Lagrangean by the covariant derivatives D_μ given by (11.3.24). The same method (just marginally more sophisticated) will enable us to transform global into local gauge invariance in non-Abelian cases (see Section 13.1.2).

11.4 Quantization of the electromagnetic field

11.4.1 Problems with quantizing the electromagnetic field

In the examples studied so far, the degrees of freedom of the field at a point x equalled in number the physical degrees of freedom: one for a neutral Klein–Gordon field, two for a charged field, and four for the Dirac field (in the case of a vector field, the extra component is easily eliminated: see Section 10.5.2). The problem with quantizing the electromagnetic field stems from the fact that one must use more than the number of physical degrees of freedom; of physical degrees of freedom there are only *two*, because a photon has only two polarization states, e.g. right- and left-circularly polarized. The photon has spin 1; but the general theory of the representations of the Poincaré group shows that a zero-mass spin-j particle can have only two polarization states, provided its interactions conserve parity. The projection of the spin onto the propagation direction, called the *helicity*, equals $\pm j$; for the photon, right (left) circular polarization corresponds to helicity $+1$ (-1). The polarization of a photon having momentum $k_\mu = (k_0, 0, 0, k_0)$ and helicity $+1$ (-1) can be described by a four-vector $\varepsilon_\mu^{(+)}$ $(\varepsilon_\mu^{(-)})$:

$$\varepsilon_\mu^{(+)} = \left(0, \frac{-1}{\sqrt{2}}, \frac{-\mathrm{i}}{\sqrt{2}}, 0\right), \quad \varepsilon_\mu^{(-)} = \left(0, \frac{1}{\sqrt{2}}, \frac{-\mathrm{i}}{\sqrt{2}}, 0\right). \tag{11.4.1}$$

The phase conventions in (11.4.1) have been chosen so as to respect the standard conventions of angular-momentum theory (Messiah 1961, Chapter 13). Note that $k^\mu \varepsilon_\mu^{(\pm)} = \boldsymbol{k} \cdot \boldsymbol{\varepsilon}^{(\pm)} = 0$: the photon is *transversely* polarized.

The electromagnetic field $F^{\mu\nu}$ $(= \boldsymbol{E}, \boldsymbol{B})$ has six components; but, by virtue of the constraints (11.3.2) or (11.3.10), only two of the six are independent. Thus one might contemplate quantizing the fields $\boldsymbol{E}$ and $\boldsymbol{B}$ subject to these constraints. However, such attempts run into an immediate and serious difficulty: in quantum mechanics, the electromagnetic interaction when expressed in terms of $\boldsymbol{E}$ and $\boldsymbol{B}$ is nonlocal. One very explicit illustration of this fact is the Bohm–Aharanov effect. The effect is observed in experiments of the Young's-slits type on electron interference; between the two slits and just behind the screen E_1 one puts a long thin solenoid, so designed that the B-field outside the solenoid is zero to a very good approximation (Fig. 11.3).

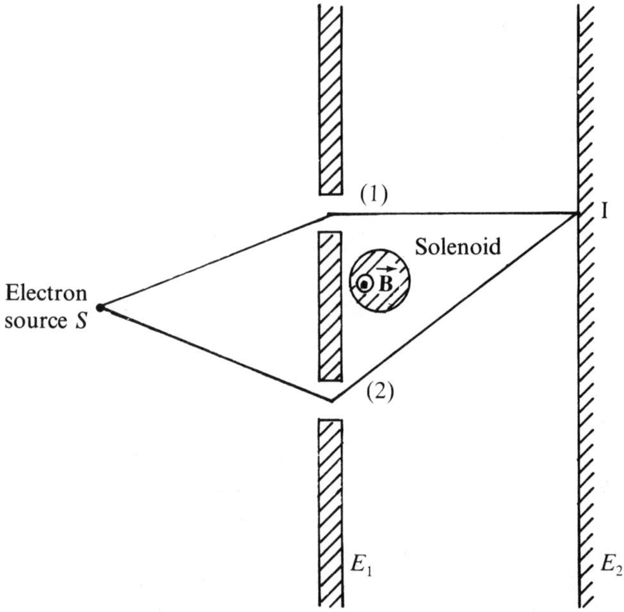

Fig. 11.3 The Bohm–Aharonov experiment

Without the solenoid one observes a certain interference pattern on the screen E_2. Let a_1 (a_2) be the probability amplitude for an electron emitted by the source S to arrive at the detection point I after following the trajectory 1 (2). These two probability amplitudes have different phases δ_1 and δ_2; and the phase difference $\Delta = \delta_1 - \delta_2$ governs the interference pattern (we assume $|a_1| \sim |a_2|$). In the presence of the field $\boldsymbol{B} = \nabla \times \boldsymbol{A}$ the Lagrangean for the electrons is modified,

$$L \to L' = L + q\boldsymbol{v} \cdot \boldsymbol{A} = L + \delta L$$

(see (11.3.12)), and so are the phases associated with the two trajectories; for trajectory 1, say, we have

$$\delta_1 \to \delta_1' = \delta_1 + \delta S = \delta_1 + \frac{q}{\hbar} \int_{t_S(1)}^{t_I} \boldsymbol{v} \cdot \boldsymbol{A} \, \mathrm{d}t = \delta_1 + \frac{q}{\hbar} \int_{S(1)}^{I} \boldsymbol{A} \cdot \mathrm{d}\boldsymbol{l}. \tag{11.4.2}$$

In fact the probability amplitude for a trajectory is proportional to $\exp(\mathrm{i}S/\hbar)$: see (8.0.2).

In presence of the solenoid the two trajectories differ in phase by

$$\begin{aligned} \Delta' = \delta_1' - \delta_2' &= \Delta + \frac{q}{\hbar}\left[\int_{S(1)}^{I} \boldsymbol{A} \cdot \mathrm{d}\boldsymbol{l} - \int_{S(2)}^{I} \boldsymbol{A} \cdot \mathrm{d}\boldsymbol{l}\right] \\ &= \Delta + \frac{q}{\hbar} \iint_{\Gamma} \boldsymbol{B} \cdot \mathrm{d}\boldsymbol{S}, \end{aligned} \tag{11.4.3}$$

where Γ is a surface bounded by the trajectories. This expression is gauge-invariant, but we observe that the B-field affects the propagation of the electron, even though $\boldsymbol{B}$ vanishes throughout the region where the probability of finding the electron is nonzero. Therefore any interaction between the electron and the B-field would have to be nonlocal: by contrast, the interaction between the electron and the vector potential $\boldsymbol{A}$ is perfectly local, in that $\int \boldsymbol{A} \cdot \mathrm{d}\boldsymbol{l}$ is evaluated along the actual trajectory*.

There is no way to sidestep quantizing the potential (though we shall continue to speak about quantizing the electromagnetic *field*), whence we must tackle gauge-invariance head-on; then, because potentials are not unique, it is not surprising that there should be more than the physical number of degrees of freedom. These difficulties affect canonical quantization as much as path integrals. Thus:

- canonical quantization: there is no momentum conjugate to A_0: $\pi_0 = 0$;
- path integrals: try to find a naïve generalization to electromagnetism of the generating functional for scalar fields. The action depends on the free

* It should not surprise one that in quantum mechanics the potentials play such an important part. The elementary formulation of classical mechanics (Newton's laws) features forces, but the elementary formulation of quantum mechanics (the Schroedinger equation) features the potential energy.

Lagrangean $\mathscr{L}_0$, and we assume that the electromagnetic field couples to a *classical* conserved current $j_\mu(x)$; then

$$Z(j) = \int \mathscr{D}A_\mu \exp\left(\mathrm{i}\int \mathrm{d}^4x(\mathscr{L}_0(A_\mu) - j_\mu A^\mu)\right)$$

$$= \int \mathscr{D}A_\mu \exp(\mathrm{i}S[A]), \tag{11.4.4}$$

where (see (11.3.13))

$$\mathscr{L}_0 = \frac{1}{2}A^\mu[\Box g_{\mu\nu} - \partial_\mu\partial_\nu]A^\nu. \tag{11.4.5}$$

Notice the signs in (11.4.5): as regards the time component, the d'Alembertian has a sign opposite to the sign appropriate to a Klein–Gordon field, while these signs are the same as regards the space components. A good way to remember this is to realize that it is the space components of A_μ that are the physical ones, and that for them the signs must be the same as for scalar fields. The experience acquired in Chapter 9 tempts one to write the photon propagator as

$$D^{\mathrm{F}}_{\mu\nu} = -\mathrm{i}[\Box g_{\mu\nu} - \partial_\mu\partial_\nu]^{-1}.$$

Unfortunately the operator $[\Box g_{\mu\nu} - \partial_\mu\partial_\nu]$ has no inverse, since it gives zero when applied to $\partial^\nu\Lambda$; equivalently, in Fourier space $(g_{\mu\nu} - k_\mu k_\nu/k^2)$ projects onto the subspace which is Minkowski-orthogonal to k_μ. By contrast, for the vector field (Section 10.5.2), $(g_{\mu\nu} - k_\mu k_\nu/m^2)$ can be inverted unless $k^2 = m^2$.

To quantize A_μ, one must choose ('fix') a gauge. The standard choices are

(i) $\partial^\mu A_\mu(x) = 0$: Lorentz gauge; (11.4.6)

(ii) $\nabla\cdot\boldsymbol{A}(x) = 0$: Coulomb gauge; (11.4.7)

(iii) $n^\mu A_\mu(x) = 0$: axial gauge $(n^2 < 0)$. (11.4.8)

The Lorentz gauge fails to fix the potential uniquely, since one can still perform the gauge transformation

$$A_\mu \to A_\mu + \partial_\mu\theta(x),$$

provided $\theta(x)$ obeys

$$\Box\theta(x) = 0.$$

By contrast, the choices (11.4.7) and (11.4.8) do fix the potential uniquely, provided we require it to vanish at infinity. However, quantization in the Coulomb gauge (11.4.7) or in the axial gauge (11.4.8) violates the formal covariance of the theory, and we shall restrict ourselves to gauge-fixing equations that are explicitly covariant, though in fact the method used below is perfectly general. With covariant gauges, calculations are much simpler than with the gauges (11.4.7) or (11.4.8); but they have the drawback of introducing unphysical degrees of freedom. In cases where it is essential to keep only the physical degrees of freedom, the choice of an axial gauge for instance may prove unavoidable.

11.4.2 Quantization in the Lorentz gauge: the generating functional

The problem with the functional integral (11.4.4) stems from the fact that one is integrating over far too many configurations, seeing that, physically, $A_\mu(x)$ and $A_\mu + \partial_\mu \Lambda(x)$ are equivalent. Instead of integrating over all configurations, we should find a way of integrating only over equivalence-classes, two configurations being equivalent if one can be turned into the other by a gauge transformation.

The method to be described is purely heuristic; we shall need to check (and will in the next chapter) that the Feynman rules it gives define a theory that is unitary, local, and renormalizable. Here we remark only that, as compared to canonical quantization, our method is far quicker, and has the advantage of preparing the way for quantizing non-Abelian gauge theories.

In order to integrate over equivalence-classes of configurations one first fixes the gauge by a condition of the type

$$f(A_\mu) = 0, \tag{11.4.9}$$

needing to ensure only that this condition really does determine a unique representative of the class in question. Write gauge transformations as

$$A_\mu^\Lambda(x) = A_\mu(x) + \partial_\mu \Lambda(x). \tag{11.4.10}$$

Since this amounts simply to a change of origin for A_μ, it leaves the integration measure $\mathscr{D}A_\mu$ unaffected:

$$\mathscr{D}A_\mu^\Lambda = \mathscr{D}A_\mu\left(= \prod_{x,\,\mu} \mathrm{d}A_\mu(x)\right).$$

Next we define a gauge-invariant quantity $\Delta_f(A)$ by

$$\Delta_f(A)\int \prod_x \delta(f(A^\Lambda(x)))\mathrm{d}\Lambda(x) = 1. \tag{11.4.11}$$

This is indeed gauge-invariant, because, for a given function $\Lambda_0(x)$, we have

$$\int \prod_x \delta(f(A^{\Lambda+\Lambda_0}(x)))\mathrm{d}\Lambda(x) = \int \prod_x \delta(f(A^\Lambda(x)))\mathrm{d}\Lambda(x),$$

whence $\Delta_f(A) = \Delta_f(A^{\Lambda_0})$. We can now substitute from (11.4.11) into the path-integral (11.4.4), using current-conservation to show that $S[A] = S[A^\Lambda]$:

$$\begin{aligned} Z(j) &= \int \prod_{x,\mu} \mathrm{d}A_\mu(x) \mathrm{e}^{\mathrm{i}S[A]} \Delta_f(A) \prod_x \delta(f(A^\Lambda(x)))\mathrm{d}\Lambda(x) \\ &= \int \prod_x \mathrm{d}\Lambda(x) \int \prod_{x,\mu} \mathrm{d}A_\mu^{-\Lambda}(x) \mathrm{e}^{\mathrm{i}S[A^{-\Lambda}]} \Delta_f(A^{-\Lambda}) \prod_x \delta(f(A(x))) \\ &= \int \prod_x \mathrm{d}\Lambda(x) \int \prod_{x,\mu} \mathrm{d}A_\mu(x) \mathrm{e}^{\mathrm{i}S[A]} \Delta_f(A) \prod_x \delta(f(A(x))). \end{aligned} \tag{11.4.12}$$

In the last line we have factored out a 'volume' $\prod_x \mathrm{d}\Lambda(x)$; this is independent of the fields, and as such can be ignored because it yields only a multiplicative constant. Accordingly, we find for $Z(j)$ the result

$$Z(j) = \int \mathscr{D}A_\mu \mathrm{e}^{\mathrm{i}S[A]} \Delta_f(A) \prod_x \delta(f(A)). \tag{11.4.13}$$

Now $\Delta_f(A)$ is given by

$$\Delta_f^{-1}(A) = \int \prod_x \mathrm{d}\Lambda(x) \prod_x \delta(f(A_\mu^\Lambda)),$$

and in principle this expression cannot be evaluated unless the condition $f(A_\mu^\Lambda) = 0$ fixes Λ uniquely; but unfortunately this is not the case for the Lorentz condition in Minkowski space. However, by appeal to the Euclidean postulate, we can proceed in Euclidean space which is immune to such problems. Let us therefore evaluate $\Delta_f(A)$ in the gauge fixed by the Lorentz condition (11.4.6). By virtue of the delta function, we need consider only configurations A_μ^Λ close to a configuration having $\partial^\mu A_\mu = 0$:

$$\partial^\mu A_\mu^\Lambda \approx \partial^\mu(\partial_\mu \Lambda) = \Box \Lambda.$$

This shows that $\Delta_f(A)$ is actually independent of A. With more complicated choices of gauge this property would fail (see Problem 11.14), as it would in non-Abelian theories; but in the present case $\Delta_f(A)$ is just a multiplicative constant which we can ignore altogether. Hence $Z(j)$ is given by

$$Z(j) = \int \mathscr{D}A_\mu \mathrm{e}^{\mathrm{i}S[A]} \prod_x \delta(\partial^\mu A_\mu). \tag{11.4.14}$$

This form of $Z(j)$ would yield the photon propagator in transverse form; but it proves convenient to generalize it slightly by adopting the gauge condition

$$\partial^\mu A_\mu - c(x) = 0.$$

The consequent changes in (11.4.14) for $Z(j)$ are trivial. We can then average over $c(x)$ with a Gaussian distribution featuring an arbitrary parameter α, according to

$$Z(j) = \int \mathscr{D}A\, \mathscr{D}c(x) \exp\left(-\frac{\mathrm{i}}{2\alpha}\int \mathrm{d}^4x\, c^2(x) + \mathrm{i}S[A]\right) \times \prod_x \delta(\partial^\mu A_\mu - c(x)),$$

whence

$$Z(j) = \int \mathscr{D}A \exp\left(\mathrm{i}S[A] - \frac{\mathrm{i}}{2\alpha}\int (\partial_\mu A^\mu)^2 \mathrm{d}^4x\right). \tag{11.4.15}$$

Equation (11.4.15) shows that the initial Lagrangean $\mathscr{L}_0$, equation (11.4.5), may be replaced by the *Stueckelberg Lagrangean*

$$\mathscr{L}_S = \frac{1}{2} A^\mu [\Box g_{\mu\nu} - \partial_\mu \partial_\nu] A^\nu - \frac{1}{2\alpha}(\partial_\mu A^\mu)^2. \tag{11.4.16}$$

The non-gauge-invariant part $(1/2\alpha)(\partial^\mu A_\mu)^2$ of (11.4.16) is called the '*gauge-fixing term*'. One can now determine the propagator by inspection, since we may rewrite $\mathscr{L}_S$ as

$$\mathscr{L}_S = \frac{1}{2} A^\mu \left[\Box g_{\mu\nu} - \left(1 - \frac{1}{\alpha}\right)\partial_\mu \partial_\nu\right] A^\nu = \frac{1}{2} A^\mu K_{\mu\nu} A^\nu.$$

The operator $K_{\mu\nu}$ is invertible, and its inverse is easily found in Fourier space (Problem 11.12):

$$D_F^{\mu\nu} = \frac{\mathrm{i}}{k^2 + \mathrm{i}\varepsilon}\left(-g^{\mu\nu} + (1-\alpha)\frac{k^\mu k^\nu}{k^2 + \mathrm{i}\varepsilon}\right). \tag{11.4.17}$$

In order to establish the factor $(k^2 + \mathrm{i}\varepsilon)^{-1}$ in (11.4.17), $\mathscr{L}$ has been augmented by an addend $-\mathrm{i}\varepsilon A^\mu A_\mu$. To justify the latter, we observe that for a finite-mass vector meson the Lagrangean contains the mass term $m^2 A^\mu A_\mu$, to which the prescription $m^2 \to m^2 - \mathrm{i}\varepsilon$ adds $-\mathrm{i}\varepsilon A^\mu A_\mu$. Alternatively one can appeal to the Euclidean postulate (Section 10.2.2).

The special cases $\alpha = 1$ and $\alpha = 0$ yield the propagator in the 'Feynman gauge' and the 'Landau gauge' respectively:

$$\alpha = 1 \text{ (Feynman):} \quad D_F^{\mu\nu} = \frac{-\mathrm{i}g^{\mu\nu}}{k^2 + \mathrm{i}\varepsilon}, \tag{11.4.18}$$

$$\alpha = 0 \text{ (Landau):} \quad D_F^{\mu\nu} = \frac{\mathrm{i}}{k^2 + \mathrm{i}\varepsilon}\left(-g^{\mu\nu} + \frac{k^\mu k^\nu}{k^2 + \mathrm{i}\varepsilon}\right). \tag{11.4.19}$$

In the Landau gauge the propagator is transverse, $k_\mu D_F^{\mu\nu} = 0$. The Lagrangean for electrodynamics given by the manipulations above is renormalizable (as the next chapter will show), but this is not enough to make it into a physically satisfactory theory. It still remains to verify that the S-matrix is unitary, and that physical quantities are independent of the choice of gauge condition (in particular of the parameter $\mathfrak{a}$); in other words we should verify that the expressions for physical quantities are *gauge-independent*. It is not obvious that they are, because the Green's functions do depend on $\mathfrak{a}$, and on the gauge-condition more generally. Indeed, in order to construct a generating functional for the Green's functions, we must now couple the electromagnetic potential A_μ to an arbitrary (non-conserved!) current density $J_\mu(x)$; and, after gauge fixing, the generating functional $Z(J)$ itself is $\mathfrak{a}$- (and more generally gauge-) dependent.

Let us revert briefly to the equations of motion derived from the Stueckelberg Lagrangean (11.4.16) coupled to a conserved current j^μ; they read

$$\Box A^\mu - \left(1 - \frac{1}{\mathfrak{a}}\right)\partial^\mu(\partial_\nu A^\nu) = j^\mu. \tag{11.4.20}$$

Current-conservation entails that $(\partial_\mu A^\mu)$ is a *free field*:

$$\Box(\partial_\mu A^\mu) = 0.$$

Classically, $\partial_\mu A^\mu = 0$ could be enforced as an initial condition at $t = -\infty$, and $\partial_\mu A^\mu$ would then remain zero everywhere: the term $(-1/2\mathfrak{a})(\partial_\mu A^\mu)^2$ has no physical relevance at all. By contrast, in a quantum theory $\partial_\mu A^\mu = 0$ cannot be enforced as a constraint on the operators; in fact the momentum operator conjugate to A^0, call it π^0, is given by

$$\pi^0 = -\frac{1}{\mathfrak{a}}(\partial_\mu A^\mu).$$

Instead, we can demand (only) that the positive-frequency part $\partial^\mu A_\mu^{(+)}$ yield zero when applied to any physical state $|\psi\rangle$, namely that

$$\partial^\mu A_\mu^{(+)}|\psi\rangle = 0,$$

which guarantees $\langle\psi|\partial_\mu A^\mu|\psi\rangle = 0$. Because A^0 does now possess a conjugate momentum, canonical quantization can proceed unimpeded, provided we introduce an indefinite metric.

Problems

11.1 Derive the identity

$$\langle \boldsymbol{x}_1 \ldots \boldsymbol{x}_N | \Psi \rangle = \frac{1}{\sqrt{N!}} \langle 0 | \psi(\boldsymbol{x}_1) \ldots \psi(\boldsymbol{x}_N) | \Psi \rangle.$$

(*Hint*: consider the case where $|\Psi\rangle = |n_{\alpha_1} \ldots n_{\alpha_i} \ldots \rangle$.)

11.2 Given $j_\mu(x) = \bar{\psi}(x)\gamma_\mu\psi(x)$, show that $\partial^\mu j_\mu(x) = 0$ when (a) $A_\mu = 0$; (b) $A_\mu \neq 0$.

11.3 Verify the orthonormality relations (11.1.35).

11.4 Using (11.1.19), show that under Lorentz transformations $\bar{\psi}(x)\gamma^\mu\psi(x)$ and $\bar{\psi}(x)\sigma^{\mu\nu}\psi(x)$ transform, respectively, as a four-vector and as an antisymmetric tensor.

11.5 Prove the Gordon identity

$$\bar{u}(\boldsymbol{p}')\gamma^\mu u(\boldsymbol{p}) = \frac{1}{2m}\bar{u}(\boldsymbol{p}')[p + p')^\mu + \mathrm{i}\sigma^{\mu\nu}(p' - p)_\nu]u(\boldsymbol{p}).$$

Find the corresponding expressions for $\bar{v}(\boldsymbol{p}')\gamma^\mu v(\boldsymbol{p})$ and for $\bar{v}(\boldsymbol{p}')\gamma^\mu u(\boldsymbol{p})$. Thence derive the orthonormality relations (11.1.43). Check the results by using the explicit expressions (11.1.34) for u and v.

11.6 Verify the anticommutation rule (11.1.41). (Use the closure relation (11.1.36).)

11.7 Check in detail the steps from (11.1.47) to (11.1.48) and from (11.1.48) to (11.1.49). Use the anticommutation rule (11.1.41) to show that $S_{\mathrm{F}}(x)$ is a Green's function of the Dirac equation, i.e. that $(\mathrm{i}\not{\partial} - m)S_{\mathrm{F}}(x) = \mathrm{i}\delta^{(4)}(x)$.

11.8 Complete the proof of Wick's theorem for fermions. One can either proceed directly (see Itzykson and Zuber 1980), or use the method of Chapter 9 and sum over $\boldsymbol{p}$ and the spin index r:

$$\bar{\eta}(t, \boldsymbol{p}) = \int \mathrm{d}^3 x\, \mathrm{e}^{\mathrm{i}\boldsymbol{p}\cdot\boldsymbol{x}}\bar{\eta}(t, \boldsymbol{x}), \quad \eta(t, \boldsymbol{p}) = \int \mathrm{d}^3 x\, \mathrm{e}^{-\mathrm{i}\boldsymbol{p}\cdot\boldsymbol{x}}\eta(t, \boldsymbol{x}).$$

Note that $\bar{\eta}(t, \boldsymbol{p})$ is coupled to $d_r^\dagger(\boldsymbol{p})$, and $\eta(t, -\boldsymbol{p})$ to $d_r(\boldsymbol{p})$, whence η and $\bar{\eta}$ exchange roles in this case.

11.9 Derive the commutators

$$[Q, \psi(x)] = -e\psi(x), \quad [Q, \bar{\psi}(x)] = e\bar{\psi}(x).$$

How do you interpret them physically?

11.10 Derive Maxwell's equations from the Lagrangean (11.3.11).

11.11 Show that the Schroedinger equation for a single particle of charge q in an electromagnetic potential $(\varphi, \boldsymbol{A})$ is invariant under gauge transformations in the following sense: if $\varphi \to \varphi + \partial_0 \Lambda$, $\boldsymbol{A} \to \boldsymbol{A} - \nabla\Lambda$, then the transformed wave-function $\psi'(t, \boldsymbol{x}) = \exp(-\mathrm{i}q\Lambda)\psi(t, \boldsymbol{x})$ obeys the Schroedinger equation with the potential $(\varphi', \boldsymbol{A}')$. Hence show that the transition probabilities are invariant (Cohen-Tannoudji *et al.* 1977, Chapter 3).

11.12 Writing $D^{\mathrm{F}}_{\mu\nu} = \alpha g_{\mu\nu} + \beta k_\mu k_\nu$, determine α and β, and establish the form (11.4.17) of the propagator.

11.13 (a) Starting with the Lagrangean

$$\mathscr{L} = \frac{1}{2} A^\alpha [(\Box + \lambda^2) g_{\alpha\beta} - \partial_\alpha \partial_\beta] A^\beta - j_\alpha A^\alpha$$

for a vector meson of mass λ coupled to a current j^α, we add to it a term $-(1/2\mathfrak{a})(\partial_\alpha A^\alpha)^2$. What are the equations of motion? Show that if the current j^μ is conserved, then $\partial_\alpha A^\alpha$ is a free field.

(b) For the propagator, derive the expression

$$D^{\mu\nu}_{\mathrm{F}} = \frac{\mathrm{i}}{q^2 - \lambda^2 + \mathrm{i}\varepsilon} \left[-g^{\mu\nu} + (1 - \mathfrak{a}) \frac{q^\mu q^\nu}{q^2 - \mathfrak{a}\lambda^2 + \mathrm{i}\varepsilon} \right]$$

(which is nonsingular as $\lambda \to 0$).

(c) Show that the new Lagrangean is invariant under the transformation

$$A_\mu \to A_\mu + \partial_\mu \theta,$$

provided $(\Box + \mathfrak{a}\lambda^2)\theta(x) = 0$. Show that A_μ can be made transverse $(\partial^\mu A_\mu = 0)$ by a gauge transformation of this kind, and that an addend $(-1/2\mathfrak{a})(\partial_\mu A^\mu)^2$ makes no difference to the physics of $\mathscr{L}$ at the classical level.

11.14 Quantize the electromagnetic field using the gauge condition $f(A_\mu) = \partial_\mu A^\mu + \frac{1}{2} g A_\mu A^\mu = 0$. (This gauge is of no interest in electrodynamics, but provides good practice for quantizing non-Abelian gauge theories: see 't Hooft and Veltman (1974, Section 11).)

(a) Evaluate

$$\Delta_f^{-1}(A) = \int \prod_x \mathrm{d}\Lambda(x) \prod_x \delta(f(A^\Lambda(x))).$$

In view of the form (11.4.13) of $Z(j)$, we need calculate Δ_f only in the vicinity of a configuration $[A_\mu]$ such that $f(A_\mu) = 0$.

$$f(A_\mu^\Lambda) \simeq f(A_\mu) + \frac{\partial f}{\partial A^\mu}\delta A^\mu = \frac{\partial f}{\partial A^\mu}\partial^\mu \Lambda.$$

Thence derive

$$\Delta_f = \det[(\Box_x + gA_\mu(x)\partial_x^\mu)\delta^{(4)}(x-y)].$$

(*Hint*: $\int \prod_i \mathrm{d}\Lambda_i \delta(M_{ij}\Lambda_j) = (\det M_{ij})^{-1}$.)

(b) Average over a function $c(x)$, as in going from (11.4.14) to (11.4.15), and derive $Z(j)$ in the form

$$Z(j) = \int \mathscr{D}A_\mu \mathscr{D}(\bar\eta, \eta) \exp\left(\mathrm{i}\left[S[A] - \frac{1}{2\alpha}\int \mathrm{d}^4x(f(A))^2\right]\right)$$

$$\times \exp\left(-\mathrm{i}\int \mathrm{d}^4x[\bar\eta(\Box + gA_\mu\partial^\mu)\eta]\right).$$

(*Hint*: use (11.2.13). The fermionic fields η and $\bar\eta$ should not be confused with the sources featured in Section 11.2.)

(c) The effective Lagrangean contains fictitious scalar particles (Fadeev–Popov ghosts) described by the fields η and $\bar\eta$. Derive the Feynman rules for electrodynamics in this gauge (see Section 10.5.3).

(d) Show that (without electrons) photon–photon scattering remains trivial, in spite of the presence of 3-photon and 4-photon vertices:

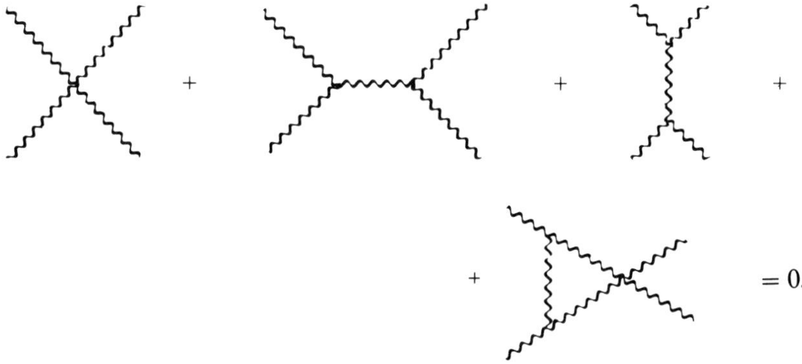

$= 0.$

Show further that the corrections to the photon propagator remain zero at the one-loop level:

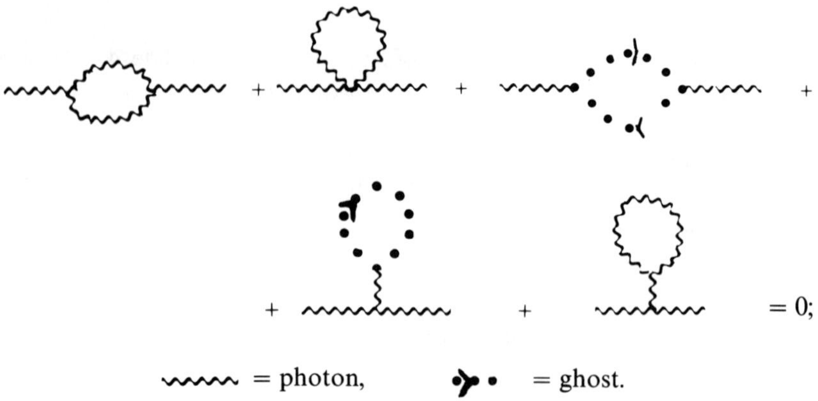

(One must use a regularization that is gauge-invariant, like dimensional regularization.)

Further reading

One classic reference for the many-body problem is Fetter and Walecka (1971). A recent one is Negele and Orland (1988). For full information about the Dirac equation see Messiah (1961, Chapter 20), Bjorken and Drell (1965, Chapter 2), Itzykson and Zuber (1980, Chapter 2), or Ramond (1980, Chapter I). Invariance under the Poincaré group is discussed by Wightman (1960), and by Gasiorowicz (1966, Chapter 4). For integration over Grassmann variables see for instance Berezin (1966); Zinn-Justin (1989, Chapter 1); Itzykson and Drouffe (1989, Chapter 2); or Ramond (1980, Chapter V). The Bohm–Aharonov effect is discussed by Feynman (1964, Chapter 15). See also the article by 't Hooft (1980). The canonical quantization of the electromagnetic field is discussed by Itzykson and Zuber (1980, Chapter 3), and by Gasiorowicz (1966, Chapter 3). Quantization through path integrals is adapted from Lee (1975). See also Popov (1983, Chapter 3), Itzykson and Zuber (1980, Chapter 9), or Ramond (1980, Chapter VII).

12 Quantum electrodynamics

Quantum electrodynamics, i.e. the relativistic quantum theory of the interactions between electrons and photons, is the physical theory with incomparably the most precise confirmation from experiment. We quote just one example, the anomalous magnetic moment of the electron. Every particle with charge e and mass m has a magnetic moment $\boldsymbol{\mu}$, whose value can be displayed, in terms of the particle spin $\boldsymbol{S}$, in the form

$$\boldsymbol{\mu} = g\frac{e}{2m}\boldsymbol{S},$$

where g is the gyromagnetic ratio ('g-factor'). Classically one has $g = 1$, a result established simply by calculating the magnetic moment of an arbitrary rotating charge distribution. For the electron the Dirac equation predicts $g = 2$, in excellent agreement with observation (though it is known that the prediction is not peculiar to the Dirac equation: Galilean wave equations for spin 1/2 likewise lead to $g = 2$: Lévy-Leblond 1967). *Radiative corrections*, or in other words terms from perturbation theory featuring diagrams with closed loops, predict a departure from the Dirac value; to first order in α (defined by equations (12.1.2) below) one finds

$$g = 2\left(1 + \frac{\alpha}{2\pi}\right).$$

We shall derive this in Section 12.3. In fact the perturbative expansion has been taken much further: to third order in α (72 graphs) it has been evaluated in full, and to fourth order (891 graphs!) in part (Kinoshita 1983, 1988). If we write

$$g = 2(1 + a_e),$$

then the theoretical value of a_e is

$$a_e^{\text{th}} = \frac{1}{2}\frac{\alpha}{\pi} - 0.328478445\left(\frac{\alpha}{\pi}\right)^2 + 1.1765(13)\left(\frac{\alpha}{\pi}\right)^3 + O\left(\frac{\alpha}{\pi}\right)^4,$$

where the digits '(13)' reflect uncertainty about the last two significant figures in the coefficient of $(\alpha/\pi)^3$. This yields

$$a_e^{\text{th}} = 1159652133(29).10^{-12},$$

while the measured value is (Schwinberg *et al.* 1981; Dehmelt 1988)

$$a_e^{\exp} = 1159652188(4).10^{-12},$$

in truly spectacular agreement. There is agreement to high accuracy also about the anomalous magnetic moment of the muon, the Lamb shift, the fine structure of positronium, etc. Such agreement demonstrates that radiative corrections, and the corresponding renormalization program, are not mere speculative fancies of the theorists. The present chapter can of course give only a very limited view of the ramifications of quantum electrodynamics; nevertheless it will enable us, for the first time in this book, to use a quantum field theory to calculate observable effects.

Section 12.1 is devoted to the Feynman rules for electrodynamics; the only difficulty in practice, due to the presence of fermions, relates to a question of signs. Section 12.2 applies the rules to two simple cases featuring only tree diagrams. One-loop diagrams are tackled in Section 12.3, which calculates radiative corrections to the photon and electron propagators, and to the electron–photon vertex. Finally, Section 12.4 studies the important topic of the Ward identities, and gives a very schematic outline of the renormalization programme. Section 12.4.4 contains a qualitative discussion of renormalization and of the renormalization group in electrodynamics, and can be read independently of the rest of the Chapter.

12.1 The Feynman rules for quantum electrodynamics

We start from the Lagrangean identified in the last chapter:

$$\mathscr{L}_{\mathrm{QED}} = \left[-\frac{1}{4}F_{\mu\nu}F^{\mu\nu} + \frac{1}{2}\lambda^2 A_\mu A^\mu - \frac{1}{2a}(\partial_\mu A^\mu)^2\right] + \left[\bar{\psi}\left(\frac{i}{2}\overleftrightarrow{\partial\!\!\!/} - m\right)\psi\right] + [-e\bar{\psi}\gamma_\mu\psi A^\mu]. \tag{12.1.1}$$

The first pair of brackets contain the Lagrangean $\mathscr{L}_S$ for the free electromagnetic field, as modified by Stueckelberg. In some intermediate calculations one needs to attribute to the photon a nonzero mass λ. If the problem is well formulated, the physical predictions must remain finite in the limit $\lambda \to 0$. Problem (11.13) shows how to modify the propagator when introducing such a mass. The second pair of brackets in (12.1.1) contain the Lagrangean $\mathscr{L}_D$ of the free Dirac field, equation (11.1.20), and the third pair the interaction $\mathscr{L}_I$ between the Dirac and the electromagnetic fields, determined by the minimal-coupling prescription (11.3.25).

The charge e in (12.1.1) is the electron charge ($e < 0$); physical results are generally expressed in terms of *the fine-structure constant* α, which is a dimensionless quantity:

$$\alpha = \frac{e^2}{4\pi} \left(= \frac{e^2}{4\pi\hbar c} \right) \simeq \frac{1}{137}. \tag{12.1.2}$$

12.1.1 The Green's functions in configuration space

Using the results of the last chapter and our experience with the scalar field, we can write down in short order the Feynman rules for the Green's functions, found as functional derivatives of the generating functional

$$Z(J, \bar{\eta}, \eta) = \int \mathscr{D}(A_\mu, \bar{\psi}, \psi) \times \exp\left(\mathrm{i} \int \mathrm{d}^4 x (\mathscr{L}_{\text{QED}} + J^\mu A_\mu + \bar{\eta}\psi + \bar{\psi}\eta) \right). \tag{12.1.3}$$

The Green's functions $G^{(2n,m)}$ are given by

$$G^{(2n,m)}_{\mu_1 \ldots \mu_m}(y_1 \ldots y_n; x_1 \ldots x_n; z_1 \ldots z_m) = \frac{(-\mathrm{i})^{2n+m}}{Z(0)} \times \left. \frac{\delta^{(2n+m)} Z(J, \bar{\eta}, \eta)}{\delta\bar{\eta}(y_1) \ldots \delta\bar{\eta}(y_n) \delta\eta(x_1) \ldots \delta\eta(x_n) \delta J^{\mu_1}(z_1) \ldots \delta J^{\mu_m}(z_m)} \right|_{J=\bar{\eta}=\eta=0}. \tag{12.1.4}$$

Just like η and $\bar{\eta}$, the source J_μ is an auxiliary mathematical source serving to identify the Green's functions; there is no reason why it should satisfy $\partial^\mu J_\mu = 0$. There are equal numbers of derivatives with respect to η and $\bar{\eta}$, because otherwise integration over the anticommuting variables would give zero, seeing that the Lagrangean features only the combination $\bar{\psi}\psi$. For convenience in writing we shall on occasion adopt the operator notation

$$G^{(2n,m)}_{\mu_1 \ldots \mu_m}(y_1, \ldots, y_n; x_1, \ldots, x_n; z_1, \ldots, z_m) = \langle 0 | T(\psi(y_1) \ldots \psi(y_n) \bar{\psi}(x_1) \ldots \bar{\psi}(x_n) A_{\mu_1}(z_1) \ldots A_{\mu_m}(z_m)) | 0 \rangle. \tag{12.1.5}$$

Here one can recognize the conclusion we have just reached: charge conservation implies that there are equal numbers of ψ and $\bar{\psi}$.

The electron propagator is drawn as an unbroken line *directed along the flow of electronic charge*:

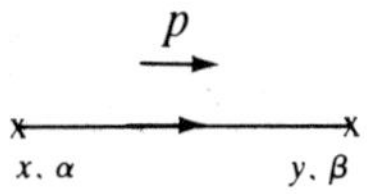

$$: S^{\mathrm{F}}_{\beta\alpha}(y - x) = \overbracket{\psi_\beta(y)\bar{\psi}_\alpha}(x). \tag{12.1.6}$$

In writing (12.1.6) one meets a familiar difficulty: reactions are read from left to right (the electron enters the diagram from the left); but operator multiplication proceeds from right to left. The operator $\bar{\psi}_\alpha(x)$ creates an electron (or destroys a positron) at x; this electron propagates from x to y, and is destroyed at y by $\psi_\beta(y)$; thus charge flows from x to y. The propagator can likewise represent the creation of a positron at y followed by its destruction at x: the Feynman rules ignore order in time.

We recall the Fourier transforms defined by

$$S^{\mathrm{F}}_{\beta\alpha}(y-x) = \int \frac{\mathrm{d}^4 p}{(2\pi)^4}\,\mathrm{e}^{-\mathrm{i}p(y-x)}\left(\frac{\mathrm{i}}{\not{p} - m + \mathrm{i}\varepsilon}\right)_{\beta\alpha}, \tag{12.1.7a}$$

$$\left(\frac{\mathrm{i}}{\not{p} - m + \mathrm{i}\varepsilon}\right)_{\beta\alpha} = \int \mathrm{d}^4(y-x)\,\mathrm{e}^{\mathrm{i}p(y-x)} S^{\mathrm{F}}_{\beta\alpha}(y-x). \tag{12.1.7b}$$

Note that these conventions for Fourier transforms are perfectly compatible with the conventions defining the Green's functions: $\exp(-\mathrm{i}px)$ in (12.1.7b) corresponds to a momentum p entering the diagram from the left, while $\exp(\mathrm{i}py)$ corresponds to a momentum $(-p)$ entering from the right. If we define the full propagator $S(y-x)$ by

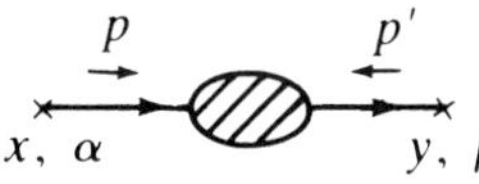

$$: S_{\beta\alpha}(y-x) = \langle 0|\, T(\psi_\beta(y)\bar{\psi}_\alpha(x))\,|0\rangle, \tag{12.1.8}$$

then, according to our conventions, the Fourier transform of a Green's function reads

$$(2\pi)^4\delta^{(4)}(p+p')S_{\beta\alpha}(p',p) = \int \mathrm{d}^4x\,\mathrm{d}^4y\,\mathrm{e}^{-\mathrm{i}(px+p'y)}S_{\beta\alpha}(y-x). \tag{12.1.9}$$

In order to avoid all risk of confusion, one should distinguish carefully between the arrow on the propagator, which indicates the direction in which the charge flows, and the arrow indicating the momentum flow.

The photon propagator presents fewer problems, because it has no direction; it is represented by a wavy line (undirected),

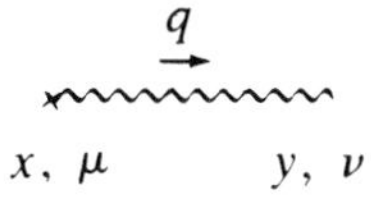

$$D^{\mathrm{F}}_{\nu\mu}(y-x) = D^{\mathrm{F}}_{\mu\nu}(x-y) = \overline{A_\nu(y)A_\mu}(x), \tag{12.1.10}$$

$$D^{\mathrm{F}}_{\nu\mu}(y-x) = \int \frac{\mathrm{d}^4 q}{(2\pi)^4}\,\frac{\mathrm{i}\,\mathrm{e}^{-\mathrm{i}q(y-x)}}{q^2-\lambda^2+\mathrm{i}\varepsilon}\left[-g_{\nu\mu} + \frac{(1-\mathrm{a})q_\nu q_\mu}{q^2 - \mathrm{a}\lambda^2 + \mathrm{i}\varepsilon}\right]. \tag{12.1.11}$$

The full photon propagator $D_{\nu\mu}(y-x)$ is

$$D_{\nu\mu}(y-x) = \langle 0|\, T(A_\nu(y)A_\mu(x))|0\rangle. \tag{12.1.12}$$

The electron–photon vertex can be identified directly from the Lagrangean (12.1.1): to every vertex one assigns a factor $-ie\gamma^\mu$:

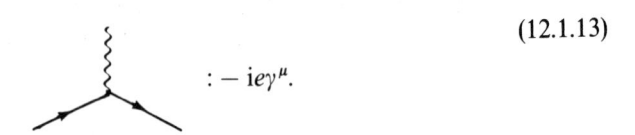

$$: -ie\gamma^\mu. \tag{12.1.13}$$

The only delicate point relates to the signs, which require very careful attention whenever fermions are involved. Because of charge conservation, an electron line has only two options:

(i) it can bite its own tail, as in closed electron loops;

(ii) it can traverse the entire diagram, or in other words it can enter at some point x_i and leave at another point y_j.

First we study a case with one closed loop, for instance Fig. 12.1; expanding the exponential $\exp(iS_1)$, where S_1 is the action corresponding to the electron–photon coupling in (12.1.1), we find

$$\bar{\psi}(1)\psi(1)\bar{\psi}(2)\psi(2)\bar{\psi}(3)\psi(3)\bar{\psi}(4)\psi(4), \tag{12.1.14a}$$

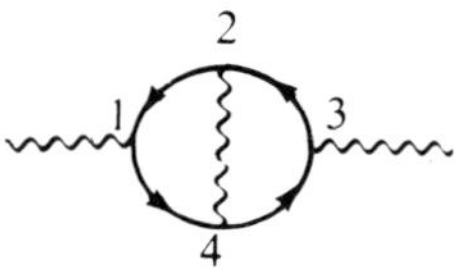

Fig. 12.1 A fermion loop

where we have dropped all factors irrelevant to the sign (like (γ^μ, A_μ, . . .). Note that the order of the factors $\bar{\psi}\psi$ is irrelevant too, because the $(\bar{\psi}\psi)$ commute with each other. But in order to shape (12.1.14a) into a product of two propagators, $\psi(4)$ for instance must be moved to the left; according to Wick's theorem this yields a contribution

$$-\,\psi(4)\bar{\psi}(1)\psi(1)\bar{\psi}(2)\psi(2)\bar{\psi}(3)\psi(3)\bar{\psi}(4), \tag{12.1.14b}$$

where the $-$ sign stems from the fact that $\psi(4)$ has been anticommuted with an odd number of fields. The result is evidently the same for any arbitrary number of factors $\bar{\psi}\psi$, and we conclude that *to every closed fermion loop one must assign a factor* (-1).

As regards lines that traverse the diagram, consider for instance the graph in Fig. 12.2: the corresponding factor is

$$\psi(y_1)\bar{\psi}(t_4)\psi(t_4)\bar{\psi}(t_1)\psi(t_1)\bar{\psi}(x_1), \tag{12.1.15a}$$

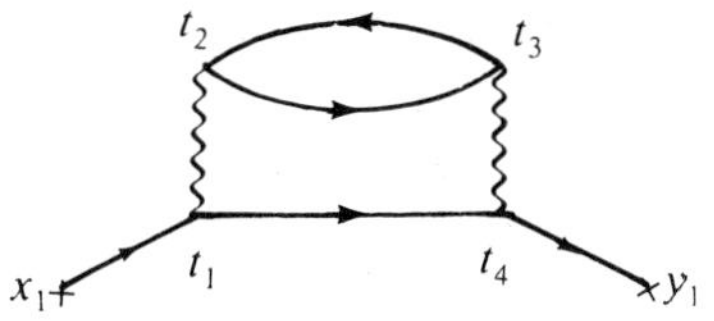

Fig. 12.2

while at the start one had

$$\psi(y_1)\bar{\psi}(x_1)\bar{\psi}(t_1)\psi(t_1)\bar{\psi}(t_4)\psi(t_4). \tag{12.1.15b}$$

In this diagram we must keep track of two sign factors:

- a factor (-1) assigned to the electron loop:
- a factor equal to the signature of the permutation that takes the arrangement (12.1.15b) into (12.1.15a). In the present example this signature is $+1$.

Finally we must determine the symmetry factor. To this end we divide every vertex into a circle, corresponding to a factor ψ, and into a cross, corresponding to a factor $\bar{\psi}$, in such a way that every fermion line goes from a cross to a circle:

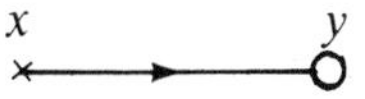

$$= \quad \psi(y)\bar{\psi}(x).$$

Every contraction necessarily joins a cross to a circle. Thus, in the diagram of Fig. 12.3, there is only one way to contract, and the application of Wick's

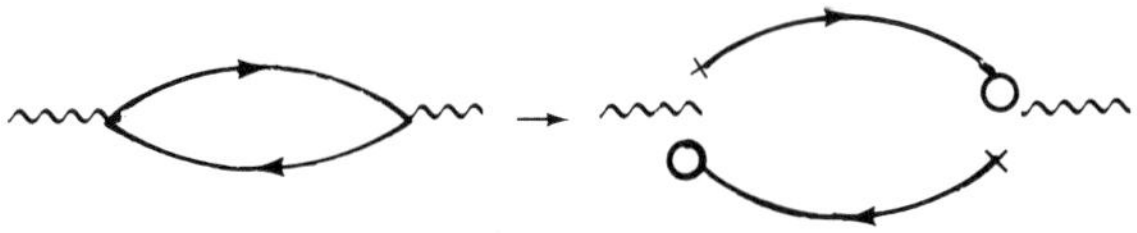

Fig. 12.3

theorem yields just one term. Since the Lagrangean $\mathscr{L}_1$ contains no factor (1/factorial), *all topologically inequivalent diagrams have a symmetry factor equal to 1.*

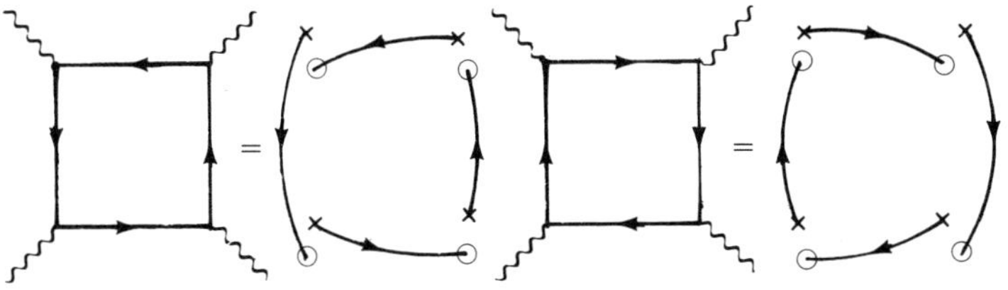

Fig. 12.4 Diagrams for photon–photon scattering

Note that the two diagrams in Fig. 12.4, contributing to photon–photon scattering, are topologically inequivalent.

We can now summarize the Feynman rules in configuration space:

(1) Electron propagator: 

$: S^{\mathrm{F}}_{\beta\alpha}(y-x).$

(2) Photon propagator: 

$: D^{\mathrm{F}}_{\nu\mu}(y-x).$

(3) Vertex: 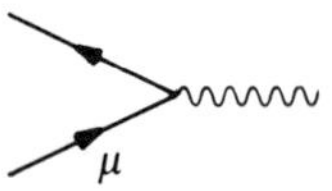

$: -\mathrm{i}e\gamma^{\mu}.$

(4) A factor (-1) for every closed fermion loop.

(5) An overall sign associated with the configuration of the external lines.

(6) A symmetry factor $+1$ for all topologically inequivalent diagrams.

(7) Integrate over all internal points of the diagram.

12.1.2 *S*-matrix elements

We proceed to the Feynman rules for *S*-matrix elements. On adapting the method explained in Section 10.3.3, we see that the following contractions are needed:

$\underbrace{\psi(x)b_r^{\dagger}(\boldsymbol{p})} = u^{(r)}(\boldsymbol{p})\mathrm{e}^{-\mathrm{i}px}$ 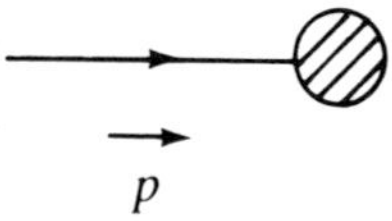

incoming electron;

$\underbrace{b_r(\boldsymbol{p})\bar{\psi}(x)} = \bar{u}^{(r)}(\boldsymbol{p})\mathrm{e}^{\mathrm{i}px}$ 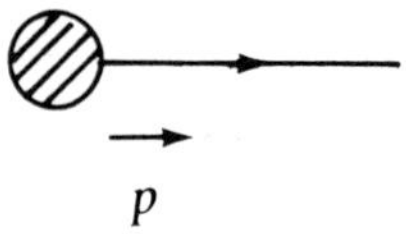

outgoing electron;

$\underbrace{\bar{\psi}(x)\mathrm{d}_r^{\dagger}(\boldsymbol{p})} = \bar{v}^{(r)}(\boldsymbol{p})\mathrm{e}^{-\mathrm{i}px}$ 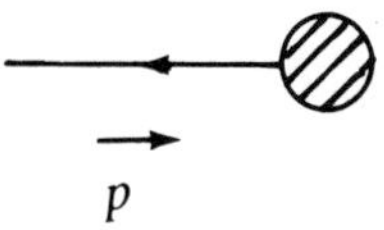

incoming positron;

$\underbrace{\mathrm{d}_r(\boldsymbol{p})\psi(x)} = v^{(r)}(\boldsymbol{p})\mathrm{e}^{\mathrm{i}px}$ 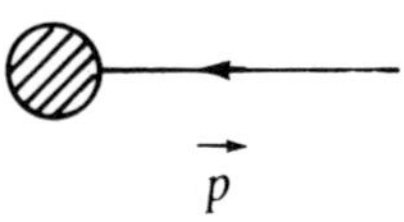

outgoing positron;

$\underbrace{A_{\mu}(x)a_s^{\dagger}(\boldsymbol{k})} = \varepsilon^{(s)}_{\mu}(\boldsymbol{k})\mathrm{e}^{-\mathrm{i}kx}$ 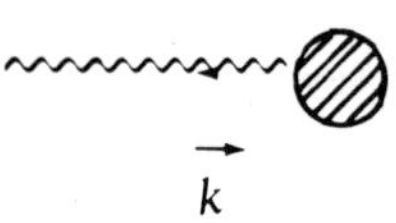

incoming photon;

$$\underbrace{a_s(\boldsymbol{k})A_\mu(x)} = \varepsilon_\mu^{(s)*}(\boldsymbol{k})\,\mathrm{e}^{\mathrm{i}kx}$$

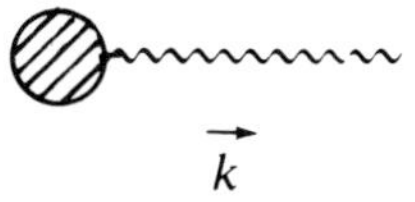

outgoing photon.

In the last two equations, $\varepsilon_\mu^{(s)}(\boldsymbol{k})$ is a polarization four-vector (see (11.4.1)), and s a polarization index (specifying circular, linear, or some other polarization). Since the incoming and outgoing photons are physical, contractions can be calculated from a Fourier decomposition of the electromagnetic field with only two transverse (physical) degrees of freedom,

$$A_\mu(x) = \sum_{s=1,2} \int \frac{\mathrm{d}^3k}{(2\pi)^3 2\omega_k} [\varepsilon_\mu^{(s)}(\boldsymbol{k})a_s(\boldsymbol{k})\mathrm{e}^{-\mathrm{i}kx} + \varepsilon_\mu^{(s)*}(\boldsymbol{k})a_s^\dagger(\boldsymbol{k})\mathrm{e}^{\mathrm{i}kx}], \tag{12.1.16}$$

where $k^\mu \varepsilon_\mu = \boldsymbol{k}\cdot\boldsymbol{\varepsilon} = 0$.

Let us apply these rules to calculate say the S-matrix element for the Compton effect,

$$\gamma(\boldsymbol{k}, s) + e(\boldsymbol{p}, r) \to \gamma(\boldsymbol{k}', s') + e(\boldsymbol{p}', r'),$$

to first order in e^2. We need to evaluate

$$S_{fi}^{(2)} = \frac{(-\mathrm{i}e)^2}{2!}\int \mathrm{d}^4x\,\mathrm{d}^4y \langle \boldsymbol{k}', s'; \boldsymbol{p}', r' | T(\bar{\psi}(x)\gamma^\mu\psi(x)A_\mu(x)$$
$$\times \bar{\psi}(y)\gamma^\nu\psi(y)A_\nu(y))|\boldsymbol{k}, s; \boldsymbol{p}, r\rangle, \tag{12.1.17}$$

where the factor 1/2! stems from the expansion of $\exp(\mathrm{i}S_1)$. However, the assignment ($x \leftrightarrow$ electron $p : y \leftrightarrow$ electron p') contributes exactly the same as the inverse assignment; as usual therefore we can omit the 1/2! provided we ignore permutations of the vertices x and y. We have two diagrams, given, up to a factor, by Fig. 12.5.

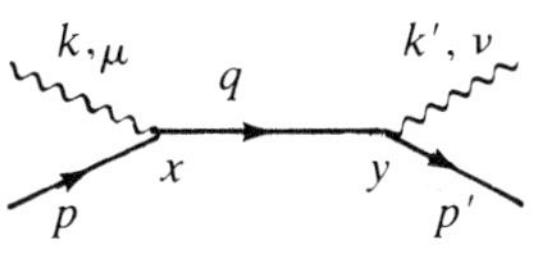

$$\begin{cases} \varepsilon_\nu^{(s')*}(\mathbf{k}')\,\varepsilon_\mu^{(s)}(\mathbf{k})\,\bar{u}^{(r')}(\mathbf{p}')\,\gamma^\nu S_F(y-x) \\ \times\, \gamma^\mu u^{(r)}(\mathbf{p})\,\mathrm{e}^{-\mathrm{i}px}\,\mathrm{e}^{-\mathrm{i}kx}\,\mathrm{e}^{\mathrm{i}p'y}\,\mathrm{e}^{\mathrm{i}k'y} \end{cases}$$

(a)

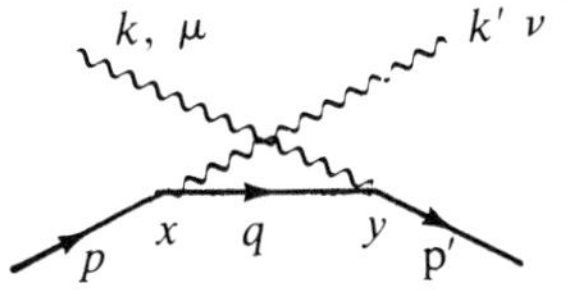

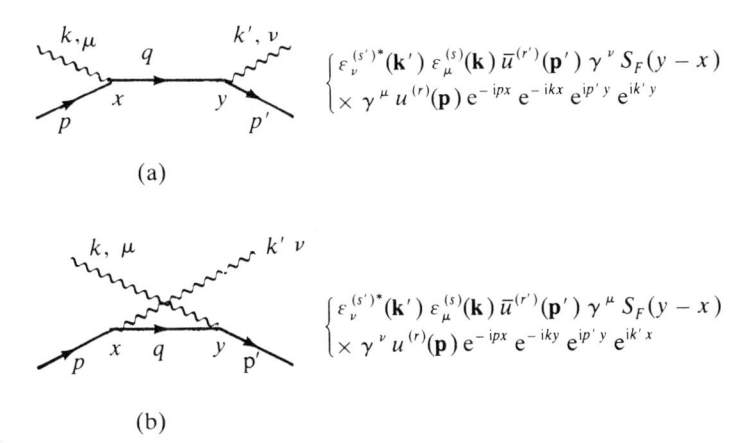

$$\begin{cases} \varepsilon_\nu^{(s')*}(\mathbf{k}')\,\varepsilon_\mu^{(s)}(\mathbf{k})\,\bar{u}^{(r')}(\mathbf{p}')\,\gamma^\mu S_F(y-x) \\ \times\, \gamma^\nu u^{(r)}(\mathbf{p})\,\mathrm{e}^{-\mathrm{i}px}\,\mathrm{e}^{-\mathrm{i}ky}\,\mathrm{e}^{\mathrm{i}p'y}\,\mathrm{e}^{\mathrm{i}k'x} \end{cases}$$

(b)

Fig. 12.5 Diagrams for the Compton effect to order e^2

In view of the expression (12.1.7a) for S_F as a Fourier transform, we can integrate over x and y, and obtain $S_{fi}^{(2)}$ in the form

$$S_{fi}^{(2)} = \varepsilon_\nu^{(s')*}(\boldsymbol{k}')\varepsilon_\mu^{(s)}(\boldsymbol{k})\bar{u}^{(r)}(\boldsymbol{p}')\Big\{(2\pi)^4\,\delta^{(4)}(p+k-q)$$

$$\times(2\pi)^4\,\delta^{(4)}(p'+k'-q)(-\mathrm{i}e\gamma^\nu)\frac{\mathrm{i}}{(2\pi)^4}\frac{\not{q}+m}{q^2-m^2+\mathrm{i}\varepsilon}(-\mathrm{i}e\gamma^\mu)$$

$$+(2\pi)^4\,\delta^{(4)}(p-k'-q)(2\pi)^4\,\delta^{(4)}(p'-k-q)(-\mathrm{i}e\gamma^\mu)$$

$$\times\frac{\mathrm{i}}{(2\pi)^4}\frac{\not{q}+m}{q^2-m^2+\mathrm{i}\varepsilon}(-\mathrm{i}e\gamma^\nu)\Big\}u^{(r)}(\boldsymbol{p}). \qquad (12.1.18)$$

The Feynman rules assign a factor $(2\pi)^4\delta^{(4)}(\ldots)$ to every vertex, and a factor $1/(2\pi)^4$ to every propagator. In view of

$$S_{fi} = \delta_{fi} + \mathrm{i}(2\pi)^4\,\delta^{(4)}(P_f - P_i)T_{fi}, \qquad (12.1.19)$$

we extract a factor $(2\pi)^4\delta^{(4)}(p'+k'-p-k)$ in order to calculate the T-matrix element. As noted in Section 5.2.5, it is easier to keep track of the factors $(2\pi)^4$ if we associate them with the loop integrals through factors $\int \mathrm{d}^4q/(2\pi)^4$.

We can now summarize the Feynman rules for the T-matrix elements in Fourier space.

(1) Draw all connected and topologically inequivalent diagrams free of self-energy insertions on external lines.

(2) To every *internal* electron line assign a factor

$$\frac{\mathrm{i}}{\not{p}-m+\mathrm{i}\varepsilon} = \frac{\mathrm{i}(\not{p}+m)}{p^2-m^2+\mathrm{i}\varepsilon}.$$

(3) To every *internal* photon line assign a factor

$$\mathrm{i}(-g_{\mu\nu}+(1-a)k_\mu k_\nu/(k^2+\mathrm{i}\varepsilon))/(k^2+\mathrm{i}\varepsilon).$$

If necessary, perform the intermediate calculations with a finite photon mass λ (see (12.1.11)).

(4) To every vertex assign a factor $-\mathrm{i}e\gamma^\mu$.

(5) Put all external lines on their mass shells ($p^2 = m^2$, $k^2 = 0$), and assign to them the following factors:
incoming electron: $u^{(r)}(\boldsymbol{p})$; outgoing electron: $\bar{u}^{(r)}(\boldsymbol{p})$;
incoming positron: $\bar{v}^{(r)}(\boldsymbol{p})$; outgoing position: $v^{(r)}(\boldsymbol{p})$;
incoming photon: $\varepsilon_\mu^{(s)}(\boldsymbol{k})$; outgoing photon: $\varepsilon_\mu^{(s)*}(\boldsymbol{k})$.

(6) Multiply every external electron line by a factor $z_2^{1/2}$, and every external photon line by $z_3^{1/2}$; z_2 and z_3 are related to the renormalization constants of the electron and photon fields (see Section 12.3 and (10.3.37)).

(7) Assign a factor (-1) to every closed electron loop, and determine the overall sign factor of the diagram, which depends on the configuration of the external lines.

(8) Multiply by $-\mathrm{i}$ (see (12.1.19)).

(9) Integrate over all closed loops, with measure $\mathrm{d}^4q/(2\pi)^4$.

Since the creation and annihilation operators for fermions and for bosons are normed in the same way, all cross-sections are given by (10.3.11). These cross-sections depend on the polarization indices (s) of the photons, and on the spin indices (r) of the electrons. In many cases the final spins are not observed, and the initial particles are unpolarized; then one must sum over the final spins and average over the initial spins.

12.2 Applications

We give two simple applications of these Feynman rules, calculating to lowest order of perturbation theory,

- the scattering of an electron by a Coulomb field;
- the cross-section for $e^+e^- \to \mu^+\mu^-$.

These two examples feature only tree diagrams, so that questions of divergence and of renormalization do not arise.

12.2.1 Electron scattering by a Coulomb field

Electron scattering by an external potential $A_\mu^{(e)}(x)$ is described by the Lagrangean

$$\mathcal{L}_1 = -e\bar{\psi}(x)\gamma^\mu\psi(x)A_\mu^{(e)}(x). \tag{12.2.1}$$

To first-order perturbation theory, the S-matrix element reads

$$S_{fi} = -\mathrm{i}e\int \mathrm{d}^4x \langle \boldsymbol{p}', r'|\bar{\psi}(x)\gamma^\mu\psi(x)|\boldsymbol{p}, r\rangle A_\mu^{(e)}(x), \tag{12.2.2}$$

where $\boldsymbol{p}(\boldsymbol{p}')$ and $r(r')$ are the momentum and the spin index of the initial (final) electron. The Feynman graph corresponding to (12.2.2) is shown in Fig. 12.6. We calculate only the scattering by a Coulomb potential,

$$A_0^{(e)}(x) = \frac{-Ze}{4\pi\|\boldsymbol{x}\|}, \quad \boldsymbol{A}^{(e)}(x) = 0. \tag{12.2.3}$$

Equations (12.2.1) and (12.2.3) describe to a good approximation the scattering of an electron by a heavy nucleus of charge Z, provided the electron energy is

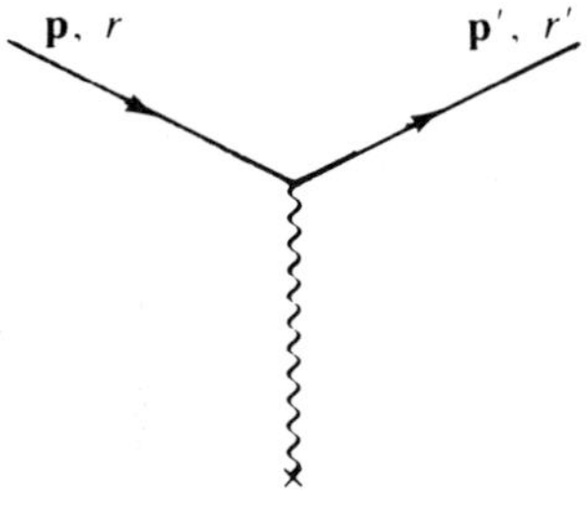

Fig. 12.6 Scattering by an external field

small compared with the mass of the nucleus; then the latter can be considered as a static source of the potential $A_\mu^{(e)}$. The momentum $(\boldsymbol{p} - \boldsymbol{p}')$ lost by the electron is absorbed by the nucleus; the momentum of the electron is not conserved, but its energy is. Clearly this is all very similar to potential scattering. When $Z\alpha \ll 1$, the first-order approximation (12.2.7) is adequate; but even when $Z\alpha$ fails to satisfy this condition, the result (12.2.7) remains valid by virtue of certain properties peculiar to the Coulomb field. From (12.2.2), S_{fi} can be evaluated immediately:

$$S_{fi} = \frac{\mathrm{i}Ze^2}{4\pi}\,\bar{u}^{(r')}(\boldsymbol{p}')\gamma^0 u^{(r)}(\boldsymbol{p})\int \mathrm{d}^4x\,\mathrm{e}^{\mathrm{i}(p'-p)x}\frac{1}{\|\boldsymbol{x}\|}$$

$$= \mathrm{i}Z\alpha\,2\pi\delta(E' - E)\bar{u}^{(r')}(\boldsymbol{p}')\gamma^0 u^{(r)}(\boldsymbol{p})\frac{4\pi}{(\boldsymbol{p}' - \boldsymbol{p})^2}. \tag{12.2.4}$$

To determine the cross-section we need a slight modification of the method used in Section 10.3.1. Because only energy is now conserved, we define, as in potential scattering,

$$S_{fi} = \delta_{fi} + \mathrm{i}(2\pi)\delta(E_f - E_i)T_{fi}. \tag{12.2.5}$$

The cross-section is given in terms of T_{fi} by

$$\mathrm{d}\sigma = \frac{1}{2\|\boldsymbol{p}\|}\left(\overline{\sum}|T_{fi}|^2\right)2\pi\delta(E' - E)\frac{\mathrm{d}^3p'}{(2\pi)^3\,2E'}. \tag{12.2.6}$$

The detailed proof is relegated to Problem 12.1. The symbol $\overline{\Sigma} = \frac{1}{2}\Sigma$ prescribes summation over the final and averaging over the initial spin, because we assume that the incoming electron is unpolarized and that the polarization of the outgoing electron is not observed. Performing these sums we find

$$\overline{\sum}|\bar{u}^{(r)}(\boldsymbol{p}')\gamma^0 u^{(s)}(\boldsymbol{p})|^2 = \frac{1}{2}\mathrm{Tr}\,[\gamma^0(\not{p} + m)\gamma^0(\not{p}' + m)]$$

$$= 2(EE' + \boldsymbol{p}\cdot\boldsymbol{p}' + m^2).$$

On the other hand, the phase-space factor in (12.2.6) can be written as

$$\frac{2\pi\delta(E-E')\,\mathrm{d}^3p'}{(2\pi)^3 2E'} = \frac{p\,\mathrm{d}\Omega}{2(2\pi)^2},$$

where $\Omega = (\theta, \varphi)$ is the scattering angle of the outgoing electron. Collecting all these factors we obtain the differential cross-section

$$\frac{\mathrm{d}\sigma}{\mathrm{d}\Omega} = \frac{(Z\alpha)^2}{8p^4 \sin^4(\theta/2)}(E^2 + m^2 + p^2\cos\theta). \tag{12.2.7}$$

This expression, called the Mott formula, is the relativistic generalization of the Rutherford cross-section; it could have been derived equally well from the single-particle Dirac equation. In the nonrelativistic limit ($\boldsymbol{p}^2/m^2 \to 0$) equation (12.2.7) reduces to

$$\frac{\mathrm{d}\sigma}{\mathrm{d}\Omega} = \frac{(Z\alpha)^2}{4m^2 v^4 \sin^4(\theta/2)}$$

where $v = p/E$ is the velocity of the electron. This of course is just the classic Rutherford formula.

12.2.2 Calculation of $e^+e^- \to \mu^+\mu^-$

The lepton called μ (or the muon) has exactly the same properties as the electron, apart from its mass: $m_\mu \approx 200\, m_e$.

The reaction

$$e^+e^- \to \mu^+\mu^-$$

occurs many times over in e^+e^- colliders (at Stanford, DESY, and now at LEP), and is easier to calculate than $e^+e^- \to e^+e^-$ (or $e^-e^- \to e^-e^-$), because in lowest-order perturbation theory only one graph contributes. This is the 'annihilation diagram' drawn in Fig. 12.7, which also defines the kinematics: the electron and the positron 'annihilate each other' to give rise to a (virtual) photon γ, which then disintegrates to produce the final $\mu^+\mu^-$ pair. To simplify the kinematics, it is convenient to suppose that, in the reference frame where the calculation is done, the energies (k_0, p_0, k_0', p_0') are much larger than the electron and muon masses m_e and m_μ. The T-matrix element corresponding to the

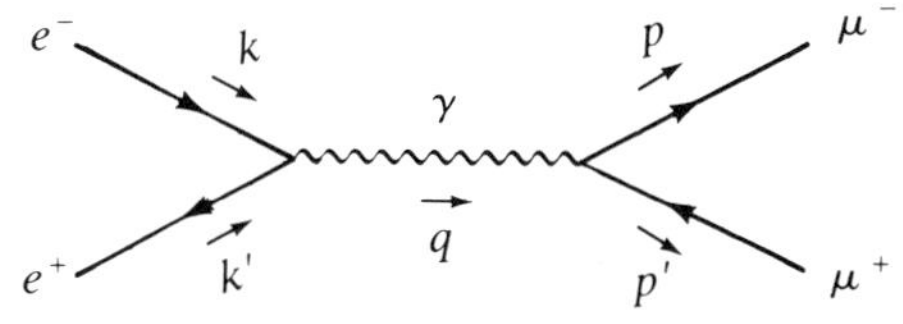

Fig. 12.7 The reaction $e^+e^- \to \mu^+\mu^-$ with one intermediate photon

diagram is

$$T_{fi} = -\frac{(-\mathrm{i}e)^2}{q^2}\bar{v}(\boldsymbol{k}')\gamma^\mu u(\boldsymbol{k})\bar{u}(\boldsymbol{p})\gamma_\mu v(\boldsymbol{p}'). \tag{12.2.8}$$

Note that, fortunately, the contribution from the term in the propagator featuring $q_\mu q_\nu$ vanishes because of current conservation. Since the polarizations are not observed, one must sum over the spin indices:

$$\begin{aligned}\sum|\bar{u}(\boldsymbol{p})\gamma_\mu v(\boldsymbol{p}')|^2 &= \mathrm{Tr}[(\not{p} + m_\mu)\gamma_\mu(\not{p}' - m_\mu)\gamma_\nu] \\ &\simeq \mathrm{Tr}(\not{p}\gamma_\mu\not{p}'\gamma_\nu) = 4(p_\mu p'_\nu + p'_\mu p_\nu) - 4(p\cdot p')g_{\mu\nu} = L_{\mu\nu}.\end{aligned} \tag{12.2.9}$$

For the electron one finds similarly

$$\sum|\bar{v}(\boldsymbol{k}')\gamma_\mu u(\boldsymbol{k})|^2 = l_{\mu\nu} \simeq 4(k_\mu k'_\nu + k'_\mu k_\nu) - 4(k\cdot k')g_{\mu\nu}.$$

Since $q^\mu L_{\mu\nu} = q^\nu L_{\mu\nu} = 0$, we can in calculating $L_{\mu\nu}l^{\mu\nu}$ replace the factor $l_{\mu\nu}$ by

$$l'_{\mu\nu} = -8k_\mu k_\nu - 2q^2 g_{\mu\nu};$$

this yields

$$\frac{1}{4}l_{\mu\nu}L^{\mu\nu} = 16(p\cdot k)^2 - 16(p\cdot k)(q\cdot k) - 4q^2(p\cdot q) + 4q^4$$

(the factor 1/4 stems from the average over the initial spins).

Now go to the centre-of-mass frame, which in the case of e^+e^- collider experiments is also the laboratory frame; then

$$q = (\sqrt{q^2}, 0, 0, 0), \quad k = \left(\frac{1}{2}\sqrt{q^2}, 0, 0, \frac{1}{2}\sqrt{q^2}\right),$$

$$k' = \left(\frac{1}{2}\sqrt{q^2}, 0, 0, -\frac{1}{2}\sqrt{q^2}\right),$$

$$p = \left(\frac{1}{2}\sqrt{q^2}, \frac{1}{2}\sqrt{q^2}\sin\theta, 0, \frac{1}{2}\sqrt{q^2}\cos\theta\right),$$

where θ is the angle between the direction of the outgoing μ^- and the incident electron beam. In this frame one finds

$$\frac{1}{4}l_{\mu\nu}L^{\mu\nu} = q^4(1 + \cos^2\theta). \tag{12.2.10}$$

To obtain the differential cross-section we refer to (10.3.15):

$$\frac{\mathrm{d}\sigma}{\mathrm{d}\Omega} = \frac{\alpha^2}{4q^2}(1 + \cos^2\theta), \tag{12.2.11}$$

whence the total cross-section reads

$$\sigma_{\mathrm{tot}} = \frac{4\pi\alpha^2}{3q^2}. \tag{12.2.12}$$

12.2.3 Application: calculation of the ratio *R*

In quantum chromodynamics one can calculate the cross-section for the process $e^+e^- \to$ hadrons; according to QCD, hadrons consist of point quarks q_i, and this cross-section is given simply as

$$\sum_i \sigma(e^+e^- \to \bar{q}_i q_i),$$

where the sum runs over all types of quark that can be 'produced' at the energy in question. Actually the theoretical prediction applies in the limit $q^2 \to -\infty$, i.e. in the unphysical region. It can be shown that in the region $q^2 > 0$ the prediction applies to averaged values (see Section 13.4.4). The quarks known to date, with their charges e_i measured in units of the proton charge, are

quark u (up) : $e = 2/3$, quark d (down) : $e = -1/3$,
quark s (strange) : $e = -1/3$, quark c (charmed) : $e = 2/3$,
quark b (beauty) : $e = -1/3$.

Moreover every quark comes in three versions differing in 'colour' (nothing to do of course with colour in the ordinary sense!).

The first three quarks are light ($m <$ proton mass). The c and b quarks have masses ≈ 1.5 GeV and ≈ 5 GeV respectively; hence one needs energies of at least 3 and 10 GeV respectively in order to produce them in a collider. Quarks are coupled to the photon by their charges, and diagrammatically one is led to write the equation in Fig. 12.8:

$$R = \frac{\sigma(e^+e^- \to \text{hadrons})}{\sigma(e^+e^- \to \mu^+\mu^-)} =$$

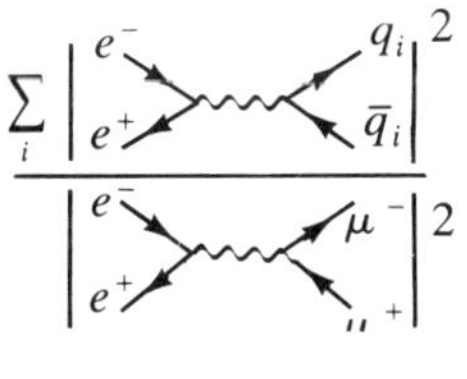

Fig. 12.8 Diagrammatic representation of the ratio *R*

For $q \gg 10$ GeV, the ratio R should be given by

$$R = 3\left(\left(\frac{2}{3}\right)^2 + \left(\frac{1}{3}\right)^2 + \left(\frac{1}{3}\right)^2 + \left(\frac{2}{3}\right)^2 + \left(\frac{1}{3}\right)^2\right) = \frac{11}{3}.$$

The experimental results are in good agreement with this prediction (Fig. 12.9). In fact quantum chromodynamics predicts that R should be slightly larger than 11/3, on account of radiative corrections (see Section 13.4.4).

There are good theoretical reasons for predicting the existence of a sixth quark, called t (for 'top'), with charge 2/3. At the threshold for producing this quark, the ratio R should jump from 11/3 to 15/3.

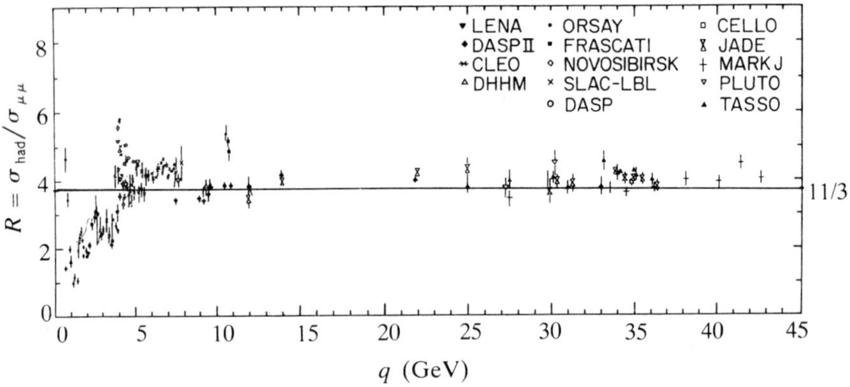

Fig. 12.9 A compilation of the ratio R (Duke and Roberts 1985)

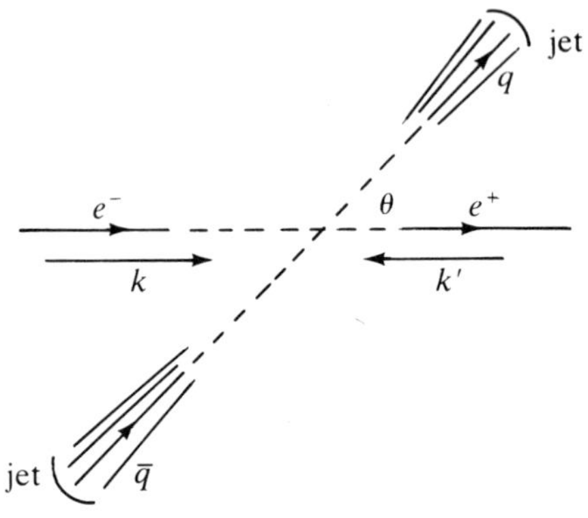

Fig. 12.10 Jet production

The hadrons produced by the decay of the (unobservable) quarks are strongly collimated around the direction of the quark and antiquark which are formed initially; one is in fact dealing with 'jets' (Fig. 12.10). From these jets one can reconstruct (approximately) the quark–antiquark direction; it follows a $(1 + \cos^2 \theta)$ distribution, just as predicted by (12.2.11). This confirms that, as other arguments suggest, the quarks do indeed have spin 1/2. Quarks of spin 0 for instance would have an angular distribution proportional to $\sin^2 \theta$ (Problem 12.8).

12.3 One-loop diagrams in electrodynamics

This section embarks on our study of renormalization in quantum electrodynamics, by calculating diagrams with one closed loop. But it proves worth while to preface this by a look at power-counting, in order to determine which diagrams diverge. On account of the particle spins, power-counting is now more complicated than it was for the scalar case in Chapter 5. On the other hand, certain symmetry properties (like gauge invariance) lower the degree of divergence of some diagrams, and power-counting proves too pessimistic. A spectacular example of such reduced divergence is the 'supersymmetric' Lagrangean of Problem 12.13.

12.3.1 Power-counting in electrodynamics

Because the Dirac action

$$\int d^D x \bar{\psi}\left(\frac{i}{2}\overleftrightarrow{\not{\partial}} - m\right)\psi$$

is dimensionless, the dimensions $[\psi]$ of the Dirac field are

$$[\psi] = \frac{D}{2} - \frac{1}{2}, \; [\psi] = \frac{3}{2} \text{ if } D = 4. \tag{12.3.1}$$

Since the electromagnetic field has its usual dimensionality $[A] = D/2 - 1$, the dimensions of e are

$$[e] = 2 - D/2. \tag{12.3.2}$$

In 4D the coupling constant is dimensionless, as we know already (see (12.1.2)), and this suggests that in the physical 4D space-time electrodynamics is renormalizable*.

The electron propagator has dimension (-1), the photon propagator dimension (-2), and the superficial degree of divergence $\omega(G)$ of a proper vertex G is given by

$$\omega(G) = 4L - I_F - 2I_B \tag{12.3.3}$$

(see Section 5.6.1), where L is the number of loops, I_F the number of internal electron (fermion) lines, and I_B the number of internal photon (boson) lines. If V is the number of vertices, and E_F (E_B) the number of external fermion (boson) lines, then we can also write

$$2V = 2I_F + E_F; \quad V = 2I_B + E_B; \quad L = I_F + I_B - V + 1, \tag{12.3.4}$$

* Recent theoretical ideas suggest that the physical dimensionality of space–time might be higher than 4, say 10, or 26, or some other number. Fortunately the extra dimensions are unobservable short of energies of the order of $\approx 10^{19}$ GeV.

whence

$$\omega(G) = 4 - \frac{3}{2}E_{\mathrm{F}} - E_{\mathrm{B}}. \tag{12.3.5}$$

This formula is the electrodynamic analogue of (5.6.2). A priori, the superficially divergent Green's functions are the following:

(1) photon propagator:  $\omega(G) = 2$;

(2) electron propagator:  $\omega(G) = 1$;

(3) electron-photon vertex:  $\omega(G) = 0$;

(4) three-photon vertex:  $\omega(G) = 1$;

(5) four-photon Green's function: 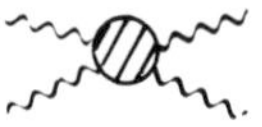 $\omega(G) = 0$.

However, such power-counting is over-pessimistic. Indeed, let us examine each case more thoroughly.

(1) From the Ward identity for the photon propagator (see 12.4.16) one can show that the corrections to the free propagator are proportional to

$$- g_{\mu\nu}k^2 + k_\mu k_\nu,$$

where k is the photon momentum. Hence we can extract two powers of k, and one has $\omega(G) = 0$ instead of 2. Since this property relies on gauge invariance, one must of course regularize in a way that respects the invariance, on pain of reverting to $\omega(G) = 2$.

(2) Let us write the full inverse electron propagator $S^{-1}(p)$ in the form

$$\mathrm{i}S^{-1}(p) = \not{p} - m - \Sigma(p). \tag{12.3.6}$$

The self-energy $\Sigma(p)$ is a matrix in Dirac space, expressible in terms of the unit matrix $\mathbb{1}$ and of $\not{p}$; the one-loop calculation (Section 12.3.3) shows that the self-energy $\Sigma(p)$ takes the form

$$\Sigma(p) = m\bar{A}(p^2)\mathbb{1} + (\not{p} - m)B(p^2), \tag{12.3.7}$$

where $\bar{A}$ and B have superficial degree of divergence $\omega(G) = 0$. In the general case, the coefficient $A(p^2)$ of the unit matrix $\mathbb{1}$ is still of the form $m\bar{A}(p^2)$. This stems from the invariance of the Lagrangean under the transformation

$$\psi \to \mathrm{e}^{\mathrm{i}\alpha\gamma_5}\psi$$

when the electron mass is zero; the mass counterterm δm is always proportional to m, and $\omega(G) = 0$.

(3) The original conclusion $\omega(G) = 0$ remains valid.

(4) This vertex is excluded by charge-conjugation, which induces $A_\mu(x) \to -A_\mu(x)$. An explicit check is furnished by Furry's theorem (Problem 12.2): diagrams with an odd number of photon lines attached to a closed electron loop contribute zero.

(5) The proper vertex $\Gamma^{(4)}$ depends on four Lorentz indices (μ, ν, ρ, σ), corresponding to the four momenta k_1, k_2, k_3, k_4; but one of the Ward identities (see Problem 12.10) implies

$$k_1^\mu \Gamma^{(4)}_{\mu\nu\rho\sigma}(k_1, k_2, k_3, k_4) = 0.$$

This allows us to extract four powers of k, and the superficial degree of divergence is -4 rather than 0.

To summarize, in studying divergences we need consider only the following Green's functions:

- the photon propagator;
- the electron propagator;
- the electron–photon vertex.

12.3.2 The photon propagator and vacuum polarization

The first-order correction to the photon propagator $D_{\mu\nu}$ is given by the one-loop diagram of Fig. 12.11; expressed analytically it reads

$$D_F^{\mu\alpha}(q)\Pi^{(1)}_{\alpha\beta}(q)D_F^{\beta\nu}(q),$$

where

$$\Pi^{(1)}_{\alpha\beta} = -e^2\mu^\varepsilon \int \frac{d^D p}{(2\pi)^D} \frac{\mathrm{Tr}(\gamma_\alpha(\not p + m)\gamma_\beta(\not p - \not q + m))}{(p^2 - m^2 + i\varepsilon)((p-q)^2 - m^2 + i\varepsilon)}. \qquad (12.3.8)$$

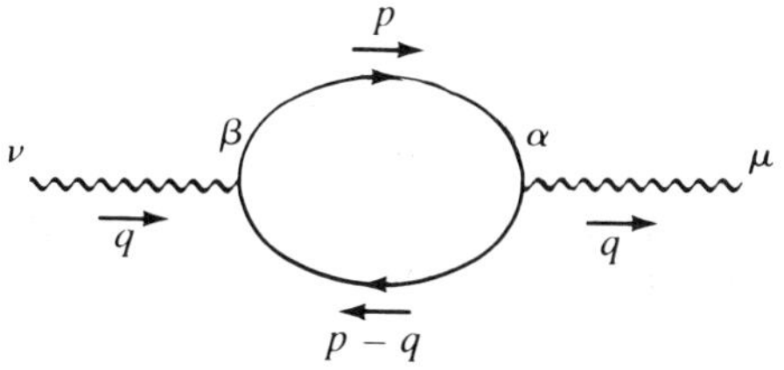

Fig. 12.11 One-loop correction to the photon propagator

In applying the Feynman rules one must not forget the factor (-1) due to the closed electron loop. We have adopted dimensional regularization, writing

$$\varepsilon = 4 - D$$

as usual (which must not be confused with the ε in the propagators); the factor μ^{ε} compensates for the dimensionality of the coupling constant when $D \neq 4$. The denominators in (12.3.8) are combined by appeal to the Feynman identity

$$\frac{1}{ab} = \int_0^1 \frac{\mathrm{d}x}{[ax + b(1-x)]^2};$$

then we change variables $p \to p + xq$ so as to eliminate from the denominator terms linear in p, which turns (12.3.8) into

$$\Pi^{(1)}_{\alpha\beta} = -e^2\mu^{\varepsilon}\int_0^1 \mathrm{d}x \int \frac{\mathrm{d}^D p}{(2\pi)^D} \frac{N_{\alpha\beta}(p + xq, q)}{(p^2 + x(1-x)q^2 - m^2 + \mathrm{i}\varepsilon)^2}. \tag{12.3.9}$$

The numerator $N_{\alpha\beta}$ is found by evaluating the trace in (12.3.8). Terms linear in p can be dropped because they cancel under the integration; noting that

$$\int \mathrm{d}^D p p_\mu p_\nu f(p^2) = \frac{1}{D} g_{\mu\nu} \int \mathrm{d}^D p p^2 f(p^2),$$

we find the following expression for $N_{\alpha\beta}$:

$$N_{\alpha\beta} = -8x(1-x)[q_\alpha q_\beta - q^2 g_{\alpha\beta}] + 4g_{\alpha\beta}\left[\left(\frac{2}{D} - 1\right)p^2 - x(1-x)q^2 + m^2\right]. \tag{12.3.10}$$

The expression (12.3.10) contains a transverse component, proportional to

$$d_{\alpha\beta} = -g_{\alpha\beta} + \frac{q_\alpha q_\beta}{q^2}, \tag{12.3.11}$$

plus a term proportional to $g_{\alpha\beta}$. In order to carry out the integration over p in (12.3.9) we shall need two integrals (see equations (B.4) and (B.5)),

$$I_0 = \int \frac{\mathrm{d}^D p}{(2\pi)^D} \frac{1}{(p^2 - \mathscr{M}^2 + \mathrm{i}\varepsilon)^2} = \frac{\mathrm{i}}{(4\pi)^{D/2}} \frac{\Gamma(2 - D/2)}{[\mathscr{M}^2 - \mathrm{i}\varepsilon]^{2-D/2}}, \tag{12.3.12a}$$

$$I_1 = \int \frac{\mathrm{d}^D p}{(2\pi)^D} \frac{p^2}{(p^2 - \mathscr{M}^2 + \mathrm{i}\varepsilon)^2} = \frac{-\mathrm{i}}{2(4\pi)^{D/2}} \frac{D\Gamma(1 - D/2)}{[\mathscr{M}^2 - \mathrm{i}\varepsilon]^{1-D/2}}, \tag{12.3.12b}$$

where the factors $\pm\mathrm{i}$ stem from the Wick rotation.

Let us show that in (12.3.10) the term proportional to $g_{\alpha\beta}$ vanishes on integration over p. On isolating the factor

$$\frac{4\mathrm{i}}{(4\pi)^{D/2}}[m^2 - x(1-x)q^2]^{-(1-D/2)}$$

the coefficient of $g_{\alpha\beta}$ is found to be

$$\left(\frac{2}{D}-1\right)\left(\frac{-1}{2}\right)D\Gamma\left(1-\frac{D}{2}\right)+\Gamma\left(2-\frac{D}{2}\right)=0.$$

The correction $\Pi^{(1)}_{\alpha\beta}$ is purely transverse, in agreement with the Ward identity

$$q^{\alpha}\Pi_{\alpha\beta}=0$$

which we shall meet in the next section. But $\Pi_{\alpha\beta}$ has this property only if one uses a gauge-invariant regularization, like dimensional regularization; gauge invariance is independent of the dimensionality of space, because the requisite algebraic identities (D.2.7.11) and (D.2.14) can be generalized consistently to arbitrary dimensionality D (problems arise when one needs γ_5, which fortunately does not feature in electrodynamics). Another gauge-invariant regularization is the method of Pauli and Villars. By contrast, the Schwinger regularization (5.5.6) violates gauge invariance; it incurs quadratically divergent terms proportional to $g_{\alpha\beta}$, which yield a correction to the photon mass (Bogoliubov and Shirkov 1959, Chapter 4).

Collecting all the factors, we find to one-loop order that

$$\Pi^{(1)}_{\alpha\beta}(q)=\frac{8e^2\mu^{\varepsilon}}{(4\pi)^{D/2}}\mathrm{i}\Gamma\left(2-\frac{D}{2}\right)(-q^2g_{\alpha\beta}+q_{\alpha}q_{\beta})$$
$$\times\int_0^1\mathrm{d}x\,\frac{x(1-x)}{(m^2-\mathrm{i}\varepsilon-x(1-x)q^2)^{2-D/2}}. \qquad (12.3.13)$$

Before exploiting this equation, let us repeat the manipulations which in Chapter 5 led to the expression for the proper vertex $\Gamma^{(2)}$,

$$D^{\mu\nu}=D_{\mathrm{F}}^{\mu\nu}+D_{\mathrm{F}}^{\mu\alpha}\Pi_{\alpha\beta}D_{\mathrm{F}}^{\beta\nu}+D_{\mathrm{F}}^{\mu\alpha}\Pi_{\alpha\beta}D_{\mathrm{F}}^{\beta\delta}\Pi_{\delta\rho}D_{\mathrm{F}}^{\rho\nu}+\dots. \qquad (12.3.14)$$

Diagrammatically, this reads as Fig. 12.12, where the unshaded bubbles represent 1-particle irreducible graphs. Writing

$$D_{\mathrm{F}}^{\mu\nu}=\frac{\mathrm{i}d^{\mu\nu}}{q^2+\mathrm{i}\varepsilon}-\frac{\mathrm{i}a\,q^{\mu}q^{\nu}}{(q^2+\mathrm{i}\varepsilon)^2},$$
$$\Pi_{\alpha\beta}=\mathrm{i}q^2d_{\alpha\beta}\bar{\omega}(q^2), \qquad (12.3.15)$$

and summing the geometric series in (12.3.14), one finds

$$D^{\mu\nu}=\frac{\mathrm{i}d^{\mu\nu}}{(q^2+\mathrm{i}\varepsilon)(1+\bar{\omega}(q^2))}-\frac{\mathrm{i}aq^{\mu}q^{\nu}}{(q^2+\mathrm{i}\varepsilon)^2}. \qquad (12.3.16)$$

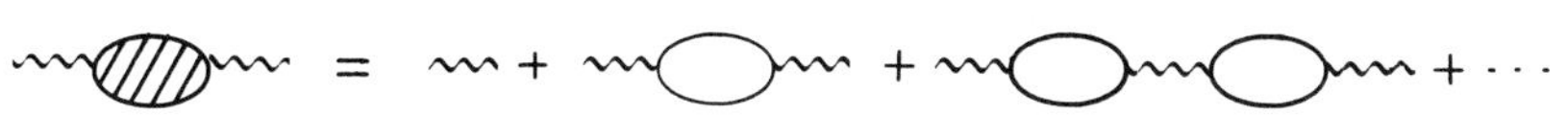

Fig. 12.12 Summation for the propagator

The denominator here can serve to define the proper vertex $\Gamma^{(2)}$:

$$\Gamma^{(2)} = q^2(1 + \bar{\omega}(q^2)). \tag{12.3.17}$$

This proper vertex satisfies the relation $\Gamma^{(2)}(q^2 = 0) = 0$; the photon mass remains zero if it is zero initially. The term $\bar{\omega}(q^2)$ will however lead to a renormalization of the electromagnetic potential A_μ, which we shall implement by introducing a counterterm depending on the renormalization constant Z_3.

Next we study $\bar{\omega}(q^2)$ in one-loop order, namely

$$\bar{\omega}^{(1)}(q^2) = \frac{8e^2\mu^\varepsilon \Gamma(\varepsilon/2)}{(4\pi)^{D/2}} \int_0^1 \frac{\mathrm{d}x\, x(1-x)}{(m^2 - \mathrm{i}\varepsilon - x(1-x)q^2)^{\varepsilon/2}}. \tag{12.3.18}$$

As was to be expected, this expression diverges when $D = 4$. Using

$$\Gamma\left(\frac{\varepsilon}{2}\right) \approx \frac{2}{\varepsilon} - \gamma$$

we find, to order ε^0, that

$$\bar{\omega}^{(1)}(q^2) = \frac{2\alpha}{3\pi}\left(\frac{1}{\varepsilon} + \frac{1}{2}\ln C\right) - \frac{2\alpha}{\pi}\int_0^1 \mathrm{d}x\, x(1-x)$$

$$\times \ln\left(\frac{m^2 - \mathrm{i}\varepsilon - x(1-x)q^2}{\mu^2}\right), \tag{12.3.19}$$

where $C = 4\pi e^{-\gamma}$.

The divergent term is $(2\alpha/3\pi)(1/\varepsilon)$. Pauli–Villars regularization with a cutoff Λ would yield

$$\int \frac{\mathrm{d}^4 k}{(2\pi)^4}\frac{1}{k^4} \simeq \frac{\mathrm{i}}{16\pi^2}\ln\frac{\Lambda^2}{m^2},$$

while

$$\int \frac{\mathrm{d}^D k}{(2\pi)^D}\frac{1}{k^4} \simeq \frac{\mathrm{i}}{16\pi^2}\frac{2}{\varepsilon}.$$

In order to compare the divergent parts in these two regularizations, we need merely replace $2/\varepsilon$ by $\ln(\Lambda^2/m^2)$. Of course the finite parts then differ. Similarly, this substitution would not yield correct results in a two-loop calculation; it merely allows one to compare the most divergent components, namely those featuring $1/\varepsilon^2$ and $\ln^2(\Lambda^2/m^2)$ respectively. To renormalize the divergence in (12.3.18) one augments the Lagrangean (12.1.1) by a counterterm

$$\delta\mathscr{L} = -\frac{1}{4}Z_3^{(1)}F_{\mu\nu}F^{\mu\nu} = \frac{1}{2}Z_3^{(1)}A^\mu(\Box g_{\mu\nu} - \partial_\mu\partial_\nu)A^\nu. \tag{12.3.20}$$

This implies that, in (12.1.1), no counterterms are associated with the mass term $(A_\mu A^\mu)$, nor with the gauge-fixing term $(\partial_\mu A^\mu)^2$. To one-loop order,

$Z_3 - 1 = Z_3^{(1)}$, where $Z_3^{(l)}$ is the contribution at l-loop order:

$$Z_3 = 1 + Z_3^{(1)} + Z_3^{(2)} + \cdots + Z_3^{(l)} + \cdots .$$

Treated perturbatively, the counterterm (12.3.20) leads to a vertex

$$= \mathrm{i}Z_3^{(1)}(-q^2 g_{\mu\nu} + q_\mu q_\nu), \tag{12.3.21}$$

which transforms $\Gamma^{(2)}$ into $\Gamma_{\mathrm{R}}^{(2)}$:

$$\Gamma_{\mathrm{R}}^{(2)} = q^2[(1 + \bar{\omega}(q^2)) + Z_3^{(1)}] \tag{12.3.22}$$

(compare (12.3.15) and (12.3.21)). At this point one can choose between several different ways of renormalizing; for instance, to one-loop order the minimal scheme (MS) yields*

$$Z_3^{(1)} = -\frac{2\alpha_{\mathrm{MS}}}{3\pi\varepsilon}, \tag{12.3.23a}$$

while the scheme $(\overline{\mathrm{MS}})$ leads to

$$Z_3^{(1)} = -\frac{2\alpha_{\overline{\mathrm{MS}}}}{3\pi\varepsilon}\left(1 + \frac{\varepsilon}{2}\ln C\right). \tag{12.3.23b}$$

However, schemes like (MS) and $(\overline{\mathrm{MS}})$, though perfectly consistent, are not customary in electrodynamics, because this theory has a classical zero-frequency limit. Consider for example two sources $j_\mu^{(1)}$ and $j_\mu^{(2)}$ of the electromagnetic field; these will interact by exchanging photons, or in other words through a coupling term

$$-\mathrm{i}j_\mu^{(1)}(q)D^{\mu\nu}(q)j_\nu^{(2)}(-q) = -j_\mu^{(1)}(q)\frac{1}{q^2(1 + \bar{\omega}(q^2))}j^{\mu(2)}(-q), \tag{12.3.24}$$

where we have used current conservation $q_\mu j^\mu(q) = 0$. In the static limit $q_0 = 0$, (12.3.24) becomes

$$j_0(\boldsymbol{q})\frac{1}{\boldsymbol{q}^2(1 + \bar{\omega}(-\boldsymbol{q}^2))}j_0(-\boldsymbol{q}). \tag{12.3.25}$$

Here $1/\boldsymbol{q}^2$ is the Fourier transform of $1/4\pi r$ (see (12.2.4)). Considering (12.3.25) as $\boldsymbol{q}^2 \to 0$ we see that, if we are to recover Coulomb's law at large distances together with the usual definition of electric charge, then $\bar{\omega}(q^2)$ must be renormalized subject to the condition $\bar{\omega}_{\mathrm{R}}(q^2 = 0) = 0$, which fails in the MS and $\overline{\mathrm{MS}}$ schemes. Traditionally one uses the on-the-mass-shell scheme, where

$$\begin{aligned}\bar{\omega}_{\mathrm{R}}^{(1)}(q^2) &= \bar{\omega}^{(1)}(q^2) - \bar{\omega}^{(1)}(0)\\ &= -\frac{2\alpha}{\pi}\int_0^1 \mathrm{d}x\, x(1-x)\ln\left(1 - \frac{q^2 x(1-x)}{m^2}\right).\end{aligned} \tag{12.3.26}$$

* From now on, α denotes the value of the fine structure constant in the on-shell renormalization scheme.

With this choice, the residue of the full photon propagator at its pole at $q^2 = 0$ is $\mathrm{i}d_{\mu\nu}$; in other words the constant z_3 (see (10.3.34)) equals 1. In the on-shell renormalization scheme, the full renormalized propagator finally becomes

$$D_{\mu\nu}^{(\mathrm{R})} = \frac{\mathrm{i}d_{\mu\nu}}{(q^2 + \mathrm{i}\varepsilon)(1 + \bar{\omega}^{(1)}(q^2) - \bar{\omega}^{(1)}(0))} - \frac{\mathrm{i}a q_\mu q_\nu}{(q^2 + \mathrm{i}\varepsilon)^2}. \tag{12.3.27}$$

The integral over the Feynman parameter x in (12.3.26) is elementary (Problem 12.4). We confine ourselves to just two comments.

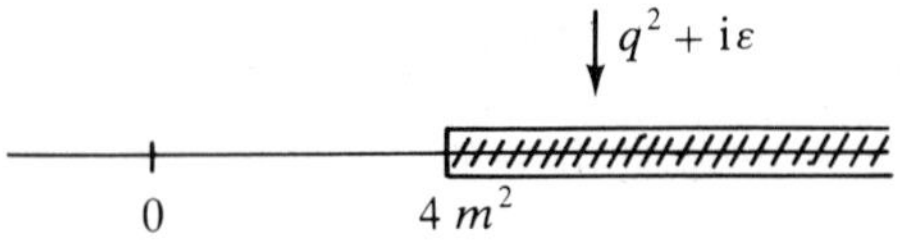

Fig. 12.13 The definition of $\bar{\omega}_{\mathrm{R}}^{(1)}(q^2)$ in the cut q^2 plane

(1) The definition of the integral in (12.3.26) is unproblematic when $q^2 < 4m^2$, because then the argument of the logarithm is positive. Like the example considered in Section 10.4.1. $\bar{\omega}_{\mathrm{R}}^{(1)}(q^2)$ has a cut for $q^2 \geqslant 4m^2$, which is the threshold for e^+e^- pair-production; in view of the prescription $m^2 \to m^2 - \mathrm{i}\varepsilon$, the physically requisite value is the limit of the integral with $q^2 = q^2 + \mathrm{i}\varepsilon$. The Cutkosky rules relate the imaginary part of $\bar{\omega}_{\mathrm{R}}^{(1)}(q^2)$ to the disintegration rate of a virtual photon with squared mass $\geqslant 4m^2$ into two electrons (Problem 12.4):

$$\mathrm{Im}\,\bar{\omega}_{\mathrm{R}}^{(1)}(q^2) = \sum \left| \quad \right|^2.$$

In fact the dispersion relation satisfied by $\bar{\omega}_{\mathrm{R}}^{(1)}(q^2)$ requires one subtraction, reflecting the divergence of the perturbative calculation and the need to renormalize it.

(2) In the static limit $q^2 = -\boldsymbol{q}^2$, and for $\boldsymbol{q}^2 \ll m^2$, one can expand (12.3.26) by powers of $\boldsymbol{q}^2$:

$$1 + \bar{\omega}_{\mathrm{R}}^{(1)}(-\boldsymbol{q}^2) = 1 - \frac{2\alpha}{\pi}\frac{\boldsymbol{q}^2}{m^2}\int_0^1 \mathrm{d}x[x(1-x)]^2 = 1 - \frac{\alpha}{15\pi}\frac{\boldsymbol{q}^2}{m^2}.$$

Coulomb's law in Fourier space is modified as follows:

$$\frac{1}{\boldsymbol{q}^2} \to \frac{1}{\boldsymbol{q}^2\left(1 - \dfrac{\alpha}{15\pi}\dfrac{\boldsymbol{q}^2}{m^2}\right)} \simeq \frac{1}{\boldsymbol{q}^2} + \frac{\alpha}{15\,\pi m^2},$$

whence, in ordinary space,

$$\frac{1}{4\pi r} \to \frac{1}{4\pi r} + \frac{\alpha}{15\,\pi m^2}\,\delta(\boldsymbol{r}). \tag{12.3.28}$$

This modification of Coulomb's law affects the s-states of the hydrogen atom; for example it shifts the $2s_{1/2}$ level by -27 MHz, and splits the levels $2s_{1/2}$ and $2p_{1/2}$ which according to the Dirac equation should be degenerate. The calculation above gives only a small part of the $2p_{1/2}$–$2s_{1/2}$ splitting, which amounts to roughly $+1000$ MHz. But the calculation does show that *radiative corrections*, i.e. perturbative corrections from diagrams with closed loops, are perfectly real effects.

Radiative corrections to the photon propagator are often called vacuum-polarization corrections; the production of (virtual) e^+e^- pairs induces a true screening effect, and the consequent departure from Coulomb's law can be ascribed to the dielectric constant of the vacuum. Physically, screening means that the charge is weaker seen far off than nearby. This is confirmed by a more complete calculation (Problem 12.5) of the deviations from Coulomb's law:

$$\frac{e}{4\pi r} \to V(r) = \frac{e}{4\pi r} Q(r),$$

$$Q(r) \simeq 1 + \frac{\alpha}{3\pi}\left(\ln\frac{1}{(mr)^2} - 2\gamma - \frac{5}{3}\right): \quad mr \ll 1, \tag{12.3.29a}$$

$$Q(r) \simeq 1 + \frac{\alpha}{\sqrt{(4\pi)}\,(mr)^{3/2}}\,\mathrm{e}^{-2mr} \quad : \quad mr \gg 1. \tag{12.3.29b}$$

For $mr \gg 1$ one notes the factor $\exp(-2mr)$ in (12.3.29b); since the effect is due to virtual e^+e^- pairs with mass $\geqslant 2m$, it must at large distances fall exponentially with a characteristic length $1/(2m)$. To see this one need merely recall the classic reasoning of Yukawa: in the static limit the Klein–Gordon equation coupled to a point source $\delta(\boldsymbol{r})$ reads

$$(\nabla^2 - m^2)\varphi(\boldsymbol{r}) = \delta(\boldsymbol{r}),$$

whose solution is

$$\varphi(\boldsymbol{r}) = -\frac{\mathrm{e}^{-mr}}{4\pi r}.$$

The field falls exponentially with a characteristic length $1/m$.

As $r \to 0$, or more precisely when $mr \ll 1$, one might think that the electron mass becomes negligible and that electrodynamics becomes scale-invariant. Since the potential has dimension 1, scale invariance naïvely understood would suggest that at short distances the potential should be proportional to $1/r$. But we know that this is by no means the case, for reasons discussed in Chapter 7: scale invariance is broken by the introduction of a mass scale necessary to allow renormalization. On the other hand, the presence specifically of the electron mass in (12.3.29a) stems from on-shell renormalization; more generally m will be replaced by some subtraction mass μ. All this will be discussed presently, when we come to study the renormalization group of electrodynamics.

12.3.3 The electron propagator

The radiative corrections to the electron propagator depend explicitly on the gauge, and in particular on the parameter $\mathfrak{a}$ that fixes the gauge in the covariant form of the (photon) propagator. We shall retain this full form of the propagator, by calculating first in the Feynman gauge, and then adding the contribution from $q_\mu q_\nu$. Moreover we shall need to attribute a nonzero mass to the photon, in order to take care of infrared divergences. Using the technique explained earlier for the photon propagator, we write the full electron propagator $S(p)$ in terms of the self-energy $\Sigma(p)$:

$$S(p) = \frac{\mathrm{i}}{\not p - m - \Sigma(p)}, \tag{12.3.30}$$

$\Sigma(p) = \mathrm{i}$

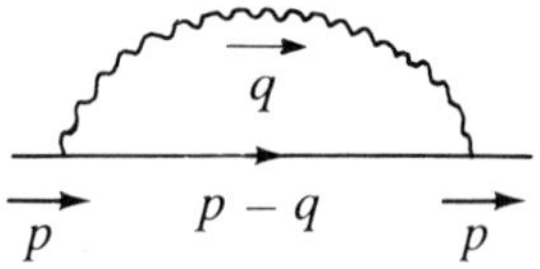

In the Feynman gauge $\mathfrak{a} = 1$, $\Sigma(p)$ reads

$$\Sigma(p) = -\mathrm{i}e^2\mu^\varepsilon \int \frac{\mathrm{d}^D q}{(2\pi)^D} \frac{\gamma_\mu(\not p - \not q + m)\gamma^\mu}{(q^2 - \lambda^2 + \mathrm{i}\varepsilon)((p-q)^2 - m^2 + \mathrm{i}\varepsilon)}. \tag{12.3.31}$$

We calculate using exactly the same techniques as for the photon propagator; the identities (D.2.7) and(D.2.8) yield

$$\Sigma(p) = \frac{e^2\mu^\varepsilon \Gamma(\varepsilon/2)}{(4\pi)^{D/2}} \int_0^1 \mathrm{d}x \frac{mD + (2-D)(1-x)\not p}{[xm^2 - \mathrm{i}\varepsilon + (1-x)\lambda^2 - x(1-x)p^2]^{\varepsilon/2}}.$$

One expands to order ε^0:

$$\begin{aligned}\Sigma(p) = {} & \frac{\alpha}{2\pi\varepsilon}(3m - (\not p - m))\left(1 + \frac{\varepsilon}{2}\ln C\right) \\ & + \frac{\alpha}{4\pi}\{-m + (\not p - m) - \int_0^1 \mathrm{d}x[2m(1+x) \\ & - 2(1-x)(\not p - m)]\ln[f(x, p^2)/\mu^2]\}\end{aligned} \tag{12.3.32}$$

where

$$f(x, p^2) = xm^2 + (1-x)\lambda^2 - x(1-x)p^2 - \mathrm{i}\varepsilon.$$

The self-energy takes the form

$$\Sigma(p) = A(p^2) + (\not p - m)B(p^2), \tag{12.3.33}$$

with divergent coefficients $A(p^2)$ and $B(p^2)$. We renormalize by introducing a mass counterterm and a field-renormalization counterterm, expressed in terms of the electron-field renormalization constant Z_2:

$$\delta\mathscr{L} = \delta m^{(1)}\bar{\psi}\psi + Z_2^{(1)}\left(\frac{\mathrm{i}}{2}\bar{\psi}\overleftrightarrow{\not{\partial}}\psi - m\bar{\psi}\psi\right). \tag{12.3.34}$$

These counterterms introduce two extra vertices which are treated perturbatively,

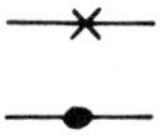

$: \mathrm{i}\delta m^{(1)}$

$: \mathrm{i}Z_2^{(1)}(\not{p} - m);$

they lead to the renormalized self-energy

$$\Sigma_{\mathrm{R}}(p) = A(p^2) - \delta m^{(1)} + (\not{p} - m)[B(p^2) - Z_2^{(1)}]. \tag{12.3.35}$$

Again it proves easy to determine $\delta m^{(1)}$ and $Z_2^{(1)}$ simply by inspecting (12.3.33) in the (MS) or in the $(\overline{\mathrm{MS}})$ schemes; in the MS scheme for instance

$$\delta m^{(1)} = \frac{3\alpha_{\mathrm{MS}} m}{2\pi\varepsilon},$$

$$Z_2^{(1)} = -\frac{\alpha_{\mathrm{MS}}}{2\pi\varepsilon}.$$

In general one prefers to renormalize on the mass shell, by requiring that near the pole at $\not{p} = m$ the renormalized propagator behave like $\mathrm{i}(\not{p} - m)^{-1}$. Hence one must choose

$$\delta m^{(1)} = A(m^2), \tag{12.3.36a}$$

$$Z_2^{(1)} = 2mA'(m^2) + B(m^2) = \left.\frac{\partial\Sigma}{\partial\not{p}}\right|_{\not{p}=m}. \tag{12.3.36b}$$

The integral giving $A(m^2)$ is infrared-finite, but $Z_2^{(1)}$ presents a problem: we are in effect trying to renormalize precisely at the threshold for electron–photon states, and it is not surprising that we run into an infrared divergence. Here one must keep a finite photon mass, which shifts the threshold for producing electron–photon states to $p^2 = (m + \lambda)^2$, higher than the renormalization point. We can now calculate $Z_2^{(1)}$ from (12.3.33):

$$\begin{aligned} Z_2^{(1)} = &-\frac{\alpha}{2\pi\varepsilon}\left(1 + \frac{\varepsilon}{2}\ln C\right) \\ &+ \frac{\alpha}{4\pi}\left\{1 + 4m^2\int_0^1 \mathrm{d}x\,\frac{x(1 - x^2)}{x^2m^2 + (1 - x)\lambda^2}\right. \\ &\left. + 2\int_0^1 \mathrm{d}x(1 - x)\ln(x^2m^2/\mu^2)\right\}. \end{aligned} \tag{12.3.37}$$

In the integrals that are infrared-finite we have set $\lambda = 0$. The infrared-divergent integral reads

$$I = \int_0^1 \mathrm{d}x \frac{x(1 - x^2)}{x^2 m^2 + (1 - x)\lambda^2};$$

if $\lambda = 0$, then at $x = 0$ this plainly diverges like $\mathrm{d}x/x$. Since the dangerous point is $x = 0$, we can within an error of order $(\lambda^2/m^2)\ln(m^2/\lambda^2)$ replace I by

$$I' = \int_0^1 \mathrm{d}x \frac{x}{x^2 m^2 + \lambda^2} - \int_0^1 \mathrm{d}x \frac{x}{m^2} = \frac{1}{2m^2}\ln\frac{m^2}{\lambda^2} - \frac{1}{2m^2}.$$

The end-result for $Z_2^{(1)}$ is

$$Z_2^{(1)} = -\frac{\alpha}{2\pi\varepsilon}\left(1 + \frac{\varepsilon}{2}\ln C\right) + \frac{\alpha}{4\pi}\left(2\ln\frac{m^2}{\lambda^2} - 4 + \ln\frac{m^2}{\mu^2}\right). \tag{12.3.38}$$

Next we calculate the contribution Σ_a due to the terms featuring $q_\mu q_\nu$ in the propagator:

$$\begin{aligned}\Sigma_a(p) = \mathrm{i}e^2(1 - a)\mu^\varepsilon \int \frac{\mathrm{d}^D q}{(2\pi)^D}\bigg[&\frac{\not{q}(p^2 - m^2)}{(p - q)^2 - m^2 + \mathrm{i}\varepsilon} \\ &- \frac{q^2(\not{p} - m)}{(p - q)^2 - m^2 + \mathrm{i}\varepsilon} - \not{q}\bigg]\frac{1}{(q^2 - \lambda^2 + \mathrm{i}\varepsilon)(q^2 - a\lambda^2 + \mathrm{i}\varepsilon)} \\ = \mathrm{i}e^2(1 - a)(I_1 - I_2 - I_3).&\end{aligned} \tag{12.3.39}$$

The integral I_3 vanishes by symmetry. The integral I_1 is ultraviolet-finite by power-counting, while I_2 is ultraviolet-divergent. When calculating $\partial\Sigma_a/\partial\not{p}|_{\not{p}=m}$, the opposite applies as regards convergence in the infrared: we see that I_2 acquires one extra power of q as $q \to 0$. On account of the factors $(p^2 - m^2)$ in I_1 and $(\not{p} - m)$ in I_2, Σ_a does not contribute to mass renormalization. Moreover, this conclusion applies to all orders of perturbation theory; δm is gauge-independent (as a consequence of the gauge-invariance of $\bar{\psi}(x)\psi(x)$: see Problem 12.12(a)).

In order to evaluate Σ_a we must combine three denominators; adapting (B.3) to the calculation of I_1 we find

$$\left.\frac{\partial I_1}{\partial \not{p}}\right|_{\not{p}=m} = \frac{-2\mathrm{i}m^2}{(4\pi)^2}\int \frac{x_1\,\mathrm{d}x_1\,\mathrm{d}x_2\,\theta(1 - x_1 - x_2)}{x_1^2 m^2 - a\lambda^2 x_1 + a\lambda^2 + \lambda^2(1 - a)x_2}, \tag{12.3.40}$$

which is indeed infrared-divergent. Evaluation yields

$$\left.\frac{\partial I_1}{\partial \not{p}}\right|_{\not{p}=m} = \frac{-\mathrm{i}}{(4\pi)^2}\left[\ln\frac{m^2}{\lambda^2} + \frac{a}{1 - a}\ln a - 1\right]. \tag{12.3.41}$$

The integral I_2 becomes

$$\left.\frac{\partial I_2}{\partial \not p}\right|_{\not p=m} = \frac{iD\Gamma(\varepsilon/2)\mu^\varepsilon}{2(4\pi)^{D/2}} \int dx_1\, dx_2\, \theta(1-x_1-x_2)$$

$$\times [x_1^2 m^2 + \mathfrak{a}\lambda^2(1-x_1) + \lambda^2(1-\mathfrak{a})x_2]^{-\varepsilon/2}$$

$$-\frac{i\Gamma(3-D/2)}{(4\pi)^{D/2}} \int dx_1\, dx_2\, \theta(1-x_1-x_2)$$

$$\times \frac{x_1^2 m^2}{[x_1^2 m^2 + \mathfrak{a}\lambda^2(1-x_1) + \lambda^2(1-\mathfrak{a})x_2]^{1+\varepsilon/2}}.$$

The end-result for the gauge-dependent part $\partial\Sigma_\mathfrak{a}/\partial\not p|_{\not p=m}$ reads

$$\left.\frac{\partial \Sigma_\mathfrak{a}}{\partial \not p}\right|_{\not p=m} = \frac{\alpha(1-\mathfrak{a})}{4\pi}\left[\frac{2}{\varepsilon}\left(1+\frac{\varepsilon}{2}\ln C\right) + \ln\frac{\mu^2}{\lambda^2} + \frac{\mathfrak{a}}{1-\mathfrak{a}}\ln \mathfrak{a} + 1\right]. \tag{12.3.42}$$

Collecting results from (12.3.38) and (12.3.42) we find

$$Z_2^{(1)} = -\frac{\alpha\mathfrak{a}}{2\pi\varepsilon}\left(1+\frac{\varepsilon}{2}\ln C\right)$$

$$+\frac{\alpha}{4\pi}\left[(3-\mathfrak{a})\ln\frac{m^2}{\lambda^2} + \mathfrak{a}\ln\frac{\mathfrak{a}m^2}{\mu^2} - 3 - \mathfrak{a}\right]. \tag{12.3.43}$$

The ultraviolet-divergent or infrared-divergent terms both depend on the gauge; *to this order* of perturbation theory, the ultraviolet divergences vanish in the Landau gauge ($\mathfrak{a}=0$), while the infrared divergences vanish in the Yennie–Fried gauge $\mathfrak{a}=3$.

12.3.4 The electron–photon vertex

Our normalization is such that to zero order of perturbation theory the proper vertex Γ_μ equals γ_μ. We shall not calculate the vertex in full, but confine ourselves to the case where both the external electrons are on their mass shells. This allows us to use

$$\overrightarrow{\not p} = m, \quad \overleftarrow{\not p}' = m,$$

as well as the Gordon identity (Problem 11.5)

$$\gamma^\mu = \frac{1}{2m}(p'+p)^\mu + \frac{i}{2m}\sigma^{\mu\nu}(p'-p)_\nu. \tag{12.3.44}$$

Start by calculating, in the Feynman gauge, the one-loop contribution $\Lambda^{(1)\mu}$ to

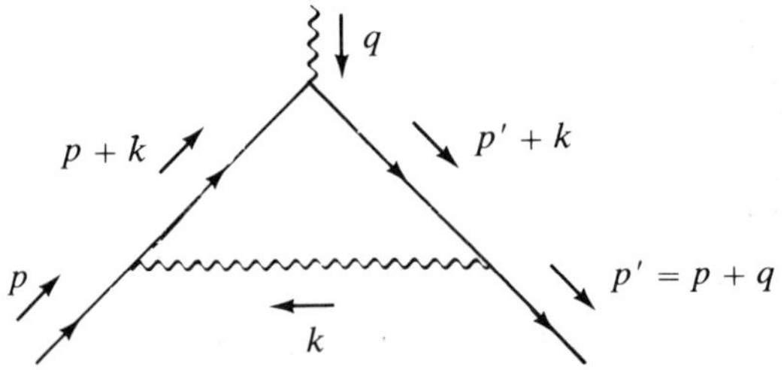

Fig. 12.14 The electron–photon vertex

Γ_μ (Fig. 12.14 defines the kinematics):

$$\Lambda^{(1)\mu} = -\mathrm{i}e^2\mu^\varepsilon \int \frac{\mathrm{d}^D k}{(2\pi)^D} \gamma_\alpha \frac{\not p' + \not k + m}{(p' + k)^2 - m^2 + \mathrm{i}\varepsilon} \cdot$$

$$\times \gamma^\mu \frac{\not p + \not k + m}{(p + k)^2 - m^2 + \mathrm{i}\varepsilon} \gamma^\alpha \frac{1}{k^2 - \lambda^2 + \mathrm{i}\varepsilon}. \tag{12.3.45}$$

The numerator $N_\mu(p, p')$ simplifies on appeal to the Dirac equation and to certain identities from Appendix D.2:

$$N^\mu(p', p, k) = \gamma^\mu[4k\cdot(p + p') + 4p\cdot p' - (2 - D)k^2] + [2(2 - D)k^\mu - 4(p + p')^\mu]\not k + 4mk^\mu.$$

We now adapt the identity (B.3) to the calculation in hand, and note that we have in effect changed variables

$$k \to k - x'p' - xp.$$

After this change, terms linear in k may be dropped from the numerator; moreover the denominator

$$D(x, x', q^2) = m^2(x + x')^2 + \lambda^2(1 - x - x') - xx'q^2 - \mathrm{i}\varepsilon \tag{12.3.46}$$

is symmetric in x and x', whence from the numerator we may also drop all terms antisymmetric under the interchange $x \leftrightarrow x'$. After such simplification, the numerator reduces to

$$N^\mu = \gamma^\mu \left\{ \frac{(2 - D)^2}{D} k^2 - q^2[2(1 - x)(1 - x') - (4 - D)xx'] + m^2[4(1 - x - x') + (2 - D)(x + x')^2] \right\} - \mathrm{i}m\sigma^{\mu\nu} q_\nu (x + x')(2 + (2 - D)(x + x')). \tag{12.3.47}$$

Having performed these somewhat tedious calculations, we are left with the vertex

$$\Lambda_\mu^{(1)} = -2\mathrm{i}e^2\mu^\varepsilon \int \mathrm{d}x\,\mathrm{d}x'\,\theta(1-x-x') \int \frac{\mathrm{d}^D k}{(2\pi)^D} \frac{N_\mu(p',p,k)}{(k^2 - D(x,x',q^2))^3}. \qquad (12.3.48)$$

Only the term with k^2 in (12.3.47) yields a divergent contribution. In all the other terms we can set $D = 4$. On the other hand, it would prove convenient to express Λ_μ in terms of two *form factors* $F_1(q^2)$ and $F_2(q^2)$,

$$\Lambda_\mu = \gamma_\mu F_1(q^2) + \frac{\mathrm{i}}{2m}\sigma_{\mu\nu}q^\nu F_2(q^2). \qquad (12.3.49)$$

After intergrating over k and expanding to order ε^0, one can identify

$$F_1^{(1)}(q^2) = \frac{\alpha}{2\pi\varepsilon}\left(1 + \frac{\varepsilon}{2}\ln C\right) - \frac{\alpha}{2\pi}\left\{1 + \int \mathrm{d}x\,\mathrm{d}x'\,\theta(1-x-x')\left\{\ln\frac{D(x,x',q^2)}{\mu^2} - \frac{q^2(1-x)(1-x')}{D(x,x',q^2)} + \frac{m^2[2(1-x-x')-(x+x')^2]}{D(x,x',q^2)}\right\}\right\}, \qquad (12.3.50a)$$

$$F_2^{(1)}(q^2) = \frac{\alpha m^2}{\pi}\int \frac{\mathrm{d}x\,\mathrm{d}x'\,\theta(1-x-x')(x+x')(1-x-x')}{D(x,x',q^2)}. \qquad (12.3.50b)$$

Before commenting on these expressions, let us calculate the gauge-dependent part of the vertex; application of the Dirac equation reduces it to

$$\Lambda_\mathrm{a}^\mu(p',p) = \mathrm{i}e^2\mu^\varepsilon(1-\mathrm{a})\gamma^\mu \int \frac{\mathrm{d}^D k}{(2\pi)^D}\frac{1}{(k^2-\lambda^2+\mathrm{i}\varepsilon)(k^2-\mathrm{a}\lambda^2+\mathrm{i}\varepsilon)} \qquad (12.3.51)$$

(Problem 12.6). This integral is both infrared and ultraviolet divergent; if $\lambda = 0$, it becomes $\int \mathrm{d}^D k/[(2\pi)^D k^4]$. Hence it is proportional to $\int \mathrm{d}k/k$, which diverges logarithmically when $k \to 0$ and also when $k \to \infty$. The end-result reads

$$\Lambda_\mathrm{a}^\mu = \frac{-\alpha(1-\mathrm{a})}{2\pi}\left[\left(\frac{1}{\varepsilon} + \frac{1}{2}\ln C\right) + \frac{1}{2}\left(\ln\frac{\mu^2}{\lambda^2} + \frac{\mathrm{a}}{1-\mathrm{a}}\ln \mathrm{a} + 1\right)\right]. \qquad (12.3.52)$$

This expression is independent of q, and contributes only to the form factor F_1, as could have been foreseen directly from the initial expression (12.3.51).

While the form factor F_1 is simultaneously infrared and ultraviolet divergent, F_2 is neither; notice that infrared divergences stem from the region $x, x' \to 0$, and that the factor $(x + x')$ in the numerator of (12.3.50b) suffices to eliminate them. To eliminate the ultraviolet divergences, one renormalizes by adding to the

Lagrangean a counterterm

$$\delta\mathscr{L} = -eZ_1^{(1)}\bar{\psi}\gamma_\mu\psi A^\mu, \tag{12.3.53}$$

which leads to the extra vertex

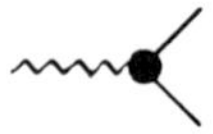 $-ieZ_1^{(1)}\gamma_\mu$.

As an illustration we calculate $Z_1^{(1)}$ in the MS scheme. In view of (12.3.50a) and (12.3.52), elimination of the pole at $\varepsilon = 0$ yields

$$Z_1^{(1)} = -\frac{\alpha_{\text{MS}}}{2\pi\varepsilon}(1-(1-a)) = -\frac{\alpha_{\text{MS}}a}{2\pi\varepsilon}. \tag{12.3.54}$$

Comparing with (12.3.43) one notes that the terms proportional to $1/\varepsilon$ are the same in $Z_2^{(1)}$ and in $Z_1^{(1)}$. Of course this is not a coincidence, but a consequence of the Ward identity for the vertex (see Section 12.4.1). When one renormalizes on the mass shell, Z_1 contains infrared divergences just as Z_2 does; on-shell renormalization consists of imposing on the renormalized vertex $F_{1\mathbf{R}}(q^2)$ the condition

$$F_{1\mathbf{R}}(q^2 = 0) = 0. \tag{12.3.55}$$

This amounts to defining the charge e of the electron so as to ensure that the probability amplitude for absorbing (or emitting) a zero-frequency photon equals $e\gamma^\mu$. As regards $Z_1^{(1)}$ this condition yields

$$Z_1^{(1)} = -F_1(q^2 = 0); \tag{12.3.56}$$

in the Feynman gauge $a = 1$, $F_1(q^2 = 0)$ is (in view of (12.3.50a)) given by

$$\begin{aligned} F_1(q^2 = 0) = {}& \frac{\alpha}{2\pi\varepsilon}\left(1 + \frac{\varepsilon}{2}\ln C\right) \\ & - \frac{\alpha}{2\pi}\left\{1 + \int \mathrm{d}x\,\mathrm{d}x'\,\theta(1-x-x')\left[\ln(m^2(x+x')^2/\mu^2)\right.\right. \\ & \left.\left. + \frac{m^2[2(1-x-x')-(x+x')^2]}{m^2(x+x')^2+\lambda^2(1-x-x')}\right]\right\}. \end{aligned}$$

The integrals over x and x' are evaluated by setting

$$u = x + x', \quad v = x - x', \quad \int \mathrm{d}x\,\mathrm{d}x'\,\theta(1-x-x')f(u) = \int_0^1 uf(u)\,\mathrm{d}u, \tag{12.3.57}$$

and one finds

$$F^{(1)}(q^2 = 0) = \frac{\alpha}{2\pi\varepsilon}\left(1 + \frac{\varepsilon}{2}\ln C\right) - \frac{\alpha}{2\pi}\left[\ln\frac{m^2}{\lambda^2} + \frac{1}{2}\ln\frac{m^2}{\mu^2} - 2\right]. \tag{12.3.58}$$

Adding the gauge-dependent part (12.3.52), we obtain

$$Z_1^{(1)} = -\frac{\alpha \mathfrak{a}}{2\pi\varepsilon}\left(1 + \frac{\varepsilon}{2}\ln C\right) + \frac{\alpha}{4\pi}\left[(3 - \mathfrak{a})\ln\frac{m^2}{\lambda^2} + \mathfrak{a}\ln\frac{\mathfrak{a}m^2}{\mu^2} - 3 - \mathfrak{a}\right]. \tag{12.3.59}$$

Note the equality $Z_1^{(1)} = Z_2^{(1)}$ when $Z_2^{(1)}$ is calculated by on-shell renormalization (see (12.3.43)). This is a consequence of the Ward identity (12.4.21). In actual fact that identity *imposes* the condition $F_1^{\mathrm{R}}(0) = 0$, provided one chooses $\partial\Sigma_{\mathrm{R}}/\partial\not{p}|_{\not{p}=m} = 0$ (Problem 12.11).

The infrared divergences constitute very reasonable grounds for concern. In fact however they cancel from the answer to any question that is physically well-posed: with finite experimental energy-resolution ΔE it is impossible to distinguish an isolated electron from an electron accompanied by low-energy photons whose total energy is less than ΔE. If we take the experimental energy-resolution into account, then in the calculation of cross-sections the $\ln \lambda^2$ are replaced by $\ln \Delta E$, and the photon mass introduced at intermediate stages of the calculation disappears from the physically verifiable predictions. Another example of the elimination of infrared divergences is given in Section 13.4.

12.3.5 The anomalous magnetic moment of the electron

No such problems beset the form factor $F_2(q^2)$. We calculate it at $q^2 = 0$:

$$F_2^{(1)}(0) = \frac{\alpha m^2}{\pi}\int \frac{\mathrm{d}x\,\mathrm{d}x'\,\theta(1 - x - x')(x + x')(1 - x - x')}{m^2(x + x')^2}.$$

The change of variables (12.3.57) yields the result at once:

$$F_2^{(1)}(0) = \frac{\alpha}{\pi}\int_0^1 \mathrm{d}u(1 - u) = \frac{\alpha}{2\pi},$$

$$F_2^{(1)}(0) = \frac{\alpha}{2\pi}. \tag{12.3.60}$$

This term can be interpreted as a *radiative correction to the gyromagnetic ratio of the electron*; it is often called the 'anomalous magnetic moment'.

To see why, consider the scattering of an electron by an external field, taking into account the one-loop corrections calculated above. The diagrams we need to consider to first order in $Z\alpha$ are those shown in Fig. 12.15 (to which one must of course add the counterterms); but if we renormalize on the mass shell, then diagrams (d) and (e) do not contribute, the constant z_2 being equal to 1 (see (10.3.37)). The contribution from diagrams (a) + (b) + (c) is

$$-\,\mathrm{i}Ze^2 A_\mu^{(e)}(\gamma^\mu + \Lambda_{\mathrm{R}}^{(1)\mu} + \Pi_{\mathrm{R}}^{(1)\mu\nu} D_{\nu\sigma}^{\mathrm{F}}\gamma^\sigma).$$

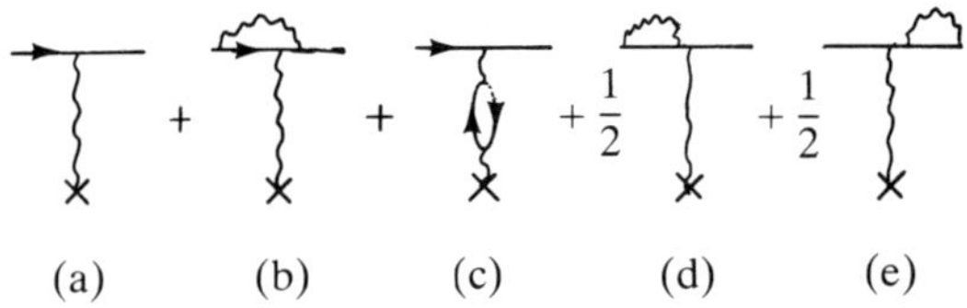

Fig. 12.15 Radiative corrections to scattering by an external field

We confine attention to small momentum transfers $q = p' - p$, with $q^2 \ll m^2$. Then $F^{(1)}_{1\mathrm{R}}(q^2)$ may be replaced by $q^2 F'^{(1)}_{1\mathrm{R}}(q^2 = 0)$; this reduces the expression just given to

$$-\mathrm{i}Ze^2 A^{(e)}_\mu \left[\gamma^\mu \left(1 + q^2 \left(F'^{(1)}_{1\mathrm{R}}(0) - \frac{\alpha}{15\,\pi m^2} \right) \right) + \frac{\mathrm{i}}{2m} \sigma^{\mu\nu} q_\nu F^{(1)}_2(0) \right],$$

where the factor $-\alpha/(15\pi m^2)$ stems from diagram (c) (see (12.3.28)). Going to configuration space and using the Gordon identity, we find for an electron in a slowly-varying external field the Hamiltonian

$$H = e \int \mathrm{d}^3 x \left\{ \frac{\mathrm{i}}{2m} \bar{\psi}(x) \overleftrightarrow{\partial}_\mu \psi(x) (1 - (F'^{(1)}_{1\mathrm{R}}(0) - \alpha/(15\,\pi m^2)\Box) \right.$$
$$\left. \times A^{\mu(e)}(x) + (1 + F^{(1)}_2(0)) \frac{1}{4m} \bar{\psi}(x) \sigma_{\mu\nu} \psi(x) F^{\mu\nu(e)} \right\}, \tag{12.3.61}$$

where (see (11.3.7))

$$F^{12} = -B_3, \quad F^{23} = -B_1, \quad F^{31} = -B_2.$$

In view of the Dirac representation of the matrices $\sigma_{\mu\nu}$ (Appendix D), the second term of (12.3.61) can be written as

$$-\boldsymbol{B}\left[\frac{e}{2m} \left(1 + \frac{\alpha}{2\pi} \right) 2 \int \mathrm{d}^3 x \bar{\psi}(x) \frac{1}{2} \boldsymbol{\sigma} \psi(x) \right]. \tag{12.3.62}$$

Disregarding the radiative corrections one recognizes the gyromagnetic ratio $2(e/2m)$ of the Dirac electron. The correction relative to this factor is just $\alpha/2\pi \sim 10^{-3}$.

To summarize, we focus on the two important points of this section:

(i) Adding to the Lagrangean (12.1.1) the counterterms (12.3.20), (12.3.34), and (12.3.53) has enabled us to obtain finite results from one-loop calculations.

(ii) Radiative corrections have observable physical consequences, and agree with the results of experiment.

12.4 Ward identities, unitarity, and renormalization

12.4.1 Ward identities

As explained in Appendix C, Ward identities are the quantum equivalent of current conservation $\partial^\mu j_\mu(x) = 0$ when this relation holds in the classical theory as a consequence of a continuous symmetry. A proof along these lines is set in Problem 12.9; it relies on the *global* gauge invariance of the QED Lagrangean (12.1.1). One can now ask the following question: what is the role of local gauge invariance, which proved so crucial in establishing the form of the QED Lagrangean? Once we have fixed a gauge, local gauge invariance of course is lost; it is broken explicitly in (12.1.1) by the term $-(\partial^\mu A_\mu)^2/2\mathfrak{a}$, and also by $\frac{1}{2}\lambda^2 A_\mu A^\mu$. However, it was discovered by Becchi, Rouet, and Stora (BRS) that a remnant of local gauge invariance does survive, namely a global symmetry of the gauge-fixed Lagrangean, called BRS invariance. In general it involves the so-called Fadeev–Popov ghost fields (see Problem 11.14 or Section 13.2). In an Abelian theory like QED these ghost fields decouple in the Lorentz gauge (and more generally under any linear gauge condition). But one can reintroduce them for the purpose of deriving Ward identities; in a manner that at this point may seem somewhat artificial, one adds to $\mathscr{L}_{\text{QED}}$ a piece involving a scalar field φ with $(\text{mass})^2 = \mathfrak{a}\lambda^2$, decoupled from photons and electrons, namely

$$\mathscr{L} = \mathscr{L}_{\text{QED}} + \mathscr{L}_\varphi, \quad \mathscr{L}_\varphi = -\frac{1}{2}\varphi(\Box + \mathfrak{a}\lambda^2)\varphi. \tag{12.4.1}$$

In an Abelian gauge theory, φ may be taken to be a Bose field, though in general ghost fields have a fermionic character. The field φ will be used to perform local gauge transformations, depending on an infinitesimal parameter ω:

$$A_\mu(x) \to A'_\mu(x) = A_\mu(x) + \omega\partial_\mu\varphi(x), \tag{12.4.2a}$$

$$\psi(x) \to \psi'(x) = \psi(x) - ie\omega\psi(x)\varphi(x), \tag{12.4.2b}$$

$$\bar{\psi}(x) \to \bar{\psi}'(x) = \bar{\psi}(x) + ie\omega\varphi(x)\bar{\psi}(x). \tag{12.4.2c}$$

It would also be possible to make $\mathscr{L}$ invariant by transforming the field φ appropriately.

Let $S[A, \psi, \bar{\psi}]$ be the action arising from the Lagrangean (12.1.1), and let $\omega\Delta S$ be its variation arising from (12.4.2) to first order in ω,

$$S[A', \psi', \bar{\psi}'] \simeq S[A, \psi, \bar{\psi}] + \omega\Delta S; \tag{12.4.3}$$

further, let $X(A, \psi, \bar{\psi})$ be a product of fields taken at the points $(y_1, \ldots, y_n)$, denoted collectively by y:

$$X'(y) = X(A', \psi', \bar{\psi}') \simeq X(A, \psi, \bar{\psi}) + \omega\Delta X(y). \tag{12.4.4}$$

Since φ is a free field, we have the identity

$$\langle 0|T(X(y)\varphi(z))|0\rangle = \langle 0|T(X'(y)\varphi(z))|0\rangle \tag{12.4.5}$$

$$= 0.$$

Using the invariance of the integration measure under (12.4.2), we may write the second Green's function in (12.4.5) as

$$\langle 0|T(X'(y)\varphi(x))|0\rangle = \mathcal{N}\int \mathscr{D}(A,\psi,\bar{\psi},\varphi)\exp\{\mathrm{i}S[A,\psi,\bar{\psi}] + \mathrm{i}\omega\Delta S + \mathrm{i}S_\varphi\}[X(y)\varphi(z) + \omega\Delta X(y)\varphi(z)].$$

Expanding to first order in ω (with $\Delta\mathscr{L}(x)$ the variation of the Lagrangean) we find

$$\mathcal{N}\int \mathscr{D}(A,\psi,\bar{\psi},\varphi)\left[\Delta X(y)\varphi(z) + \mathrm{i}\int \mathrm{d}^4x\,\Delta\mathscr{L}(x)X(y)\varphi(z)\right]\mathrm{e}^{\mathrm{i}S[A,\psi,\bar{\psi}]}\mathrm{e}^{\mathrm{i}S_\varphi},$$

or in other words

$$\langle 0|T(\Delta X(y)\varphi(z))|0\rangle = -\mathrm{i}\int \mathrm{d}^4x\langle 0|T(\Delta\mathscr{L}(x)X(y)\varphi(z))|0\rangle. \tag{12.4.6}$$

Only the gauge-dependent part

$$-\frac{1}{2\mathfrak{a}}(\partial_\mu A^\mu)^2 + \frac{1}{2}\lambda^2 A_\mu A^\mu$$

of $\mathscr{L}$ is affected by the transformation (12.4.2), whence an integration by parts immediately turns $\Delta\mathscr{L}$ into

$$\Delta\mathscr{L}(x) = -[\partial_\mu A^\mu(x)]\left[\left(\frac{1}{\mathfrak{a}}\Box + \lambda^2\right)\varphi(x)\right]. \tag{12.4.7}$$

This equation shows clearly that *the form of the Ward identities depends on one's choice of the gauge-fixing term in the Lagrangean.* The Ward identities found from the gauge in Problem 11.14 would differ from those we shall now write in the Lorentz gauge, and would in fact be much more complicated (see Problem 13.5).

In the arguments that follow, *all differential operators must be considered as positioned outside the T-product*, even if we write them inside in order to make the equations more easily legible (see the discussion following (10.2.19)). Substituting from (12.4.7) into (12.4.6), we find the identity

$$\int \langle 0|T(\partial_\mu A^\mu(x)X(y)(\Box_x + \mathfrak{a}\lambda^2)\varphi(x)\varphi(z))|0\rangle\,\mathrm{d}^4x = -\mathrm{i}\mathfrak{a}\langle 0|T(\Delta X(y)\varphi(z))|0\rangle. \tag{12.4.8}$$

If we take into account the fact that $\varphi(x)$ is a free field with mass $\mathfrak{a}\lambda^2$,

$$(\Box_x + \mathfrak{a}\lambda^2)\langle 0|T(\varphi(x)\varphi(z))|0\rangle = -\mathrm{i}\delta^{(4)}(x-z),$$

then we can derive from (12.4.8) the generic Ward identity

$$\langle 0|T(\partial^\mu A_\mu(x)X(y))|0\rangle = \mathfrak{a}\langle 0|T(\varphi(x)\Delta X(y))|0\rangle. \tag{12.4.9}$$

Next we illustrate this result in two special cases, namely for the photon propagator

$$D_{\mu\nu}(x-y) = \langle 0|T(A_\mu(x)A_\nu(y))|0\rangle$$

(see 12.1.12), and for the electron–photon vertex

$$V_\mu(x, y, z) = \langle 0|T(A_\mu(x)\psi(y)\bar{\psi}(z))|0\rangle. \tag{12.4.10}$$

In the first example we choose

$$X(y) = A_\nu(y), \quad \Delta X(y) = \frac{\partial}{\partial y^\nu}\varphi(y);$$

then (12.4.9) yields

$$\langle 0|T(\partial^\mu A_\mu(x)A_\nu(y))|0\rangle = \mathfrak{a}\langle 0|T\left(\varphi(x)\frac{\partial}{\partial y^\nu}\varphi(y)\right)|0\rangle,$$

or

$$\frac{\partial}{\partial x_\mu}D_{\mu\nu}(x-y) = -\mathfrak{a}\frac{\partial}{\partial x^\nu}\Delta_{\mathrm{F}}(x-y). \tag{12.4.11}$$

The Ward identity can also be derived by acting with the Klein–Gordon operator $(\Box_x + \mathfrak{a}\lambda^2)$ on both sides of (12.4.11):

$$(\Box_x + \mathfrak{a}\lambda^2)\frac{\partial}{\partial x^\mu}\langle 0|T(A^\mu(x)A^\nu(y))|0\rangle = i\mathfrak{a}\frac{\partial}{\partial x_\nu}\delta^{(4)}(x-y). \tag{12.4.12}$$

For the vertex one chooses

$$X(y, z) = \psi(y)\bar{\psi}(z)$$
$$\Delta X = -ie\psi(y)\varphi(y)\bar{\psi}(z) + ie\psi(y)\varphi(z)\bar{\psi}(z);$$

then (12.4.9) yields

$$\frac{\partial}{\partial x_\mu}V_\mu(x, y, z) = -ie\mathfrak{a}[\Delta_{\mathrm{F}}(x-y) - \Delta_{\mathrm{F}}(x-z)]S(y-z), \tag{12.4.13}$$

where $S(y-z)$ is the full electron propagator (12.1.8). Again we act with the Klein–Gordon operator on both sides of (12.4.13):

$$(\Box_x + \mathfrak{a}\lambda^2)\frac{\partial}{\partial x_\mu}V_\mu(x, y, z) = -e\mathfrak{a}[\delta(x-y) - \delta(x-z)]S(y-z). \tag{12.4.14}$$

Equations (12.4.12) and (12.4.14) are the equations of motion for the operator

$\partial^\mu A_\mu(x)$; they express the fact that it is a free field of $(\text{mass})^2 = \mathfrak{a}\lambda^2$:

$$(\Box_x + \mathfrak{a}\lambda^2)\partial_\mu A^\mu(x) = 0,$$

an identity that follows from the *classical* equations of motion (see Problem 11.13). Note that in fact the right-hand sides of (12.4.12) and (12.4.14) have the same form as those of (10.2.19) and (10.2.20), provided there is no interaction.

In order to write down the Ward identities explicitly, it proves convenient to go to Fourier space. Consider a Green's function $G^{(2n,m)}$ (see (12.1.4)). In Fourier space, $\partial/\partial x^\mu$ yields a factor iq_μ, where q is the variable conjugate to x; the momentum q is directed away from the vertex x. If we now vary an operator $A_\nu(y)$ on the right of (12.4.9), then we find in Fourier space a factor

$$\frac{\mathfrak{a}q_\nu}{q^2 - \mathfrak{a}\lambda^2}(2\pi)^4\,\delta^{(4)}(q + k),$$

where k is the variable conjugate to y.

Now consider an incoming electron line entering the diagram with momentum p, corresponding to a field operator $\bar{\psi}(z)$. Then the right-hand side of (12.4.9) yields a factor

$$ie\mathfrak{a}\,\Delta_F(x - z)\langle 0|\,T(\bar{\psi}(z)\ldots)|0\rangle,$$

representing the insertion of the composite operator $\varphi(z)\bar{\psi}(z)$; in Fourier space (see Fig. 12.16; p is conjugate to z) this corresponds to

$$\frac{-e\mathfrak{a}}{q^2 - \mathfrak{a}\lambda^2}\,G^{(2n,m-1)}(\ldots;p + q,\ldots).$$

Finally, for an outgoing electron line leaving the diagram with momentum p', we

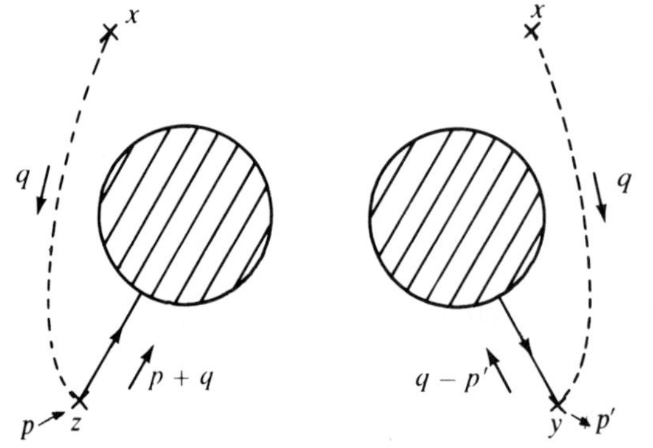

Fig. 12.16 Diagrammatic representation of the right-hand side of (12.4.9) for incoming and outgoing electron lines

need merely change the sign of the last result:

$$\frac{e\mathfrak{a}}{q^2 - \mathfrak{a}\lambda^2} G^{(2n, m-1)}(\ldots; -p' + q, \ldots).$$

(Recall the convention that all momenta run inwards into the diagram.) For an arbitrary Green's function $G^{(2n,m)}$ (into whose definition it proves convenient as in (10.1.21) to include the factor $(2\pi)^4\delta^{(4)}$) we find accordingly the general form of the Ward identity,

$$\begin{aligned} q^\mu G^{(2n, m+1)}_{\mu, \nu_1, \ldots, \nu_m}&(q, k_1, \ldots, k_m; p_1, \ldots, p_n; \ -p'_1, \ldots, -p'_n) \\ &= \frac{-i\mathfrak{a}}{q^2 - \mathfrak{a}\lambda^2} \{(2\pi)^4 \delta^{(4)}(q + k_1) \\ &\quad \times q_{\nu_1} G^{(2n, m-1)}_{\nu_2, \ldots, \nu_m}(k_2, \ldots; \ldots; \ldots) + \text{perm.} \\ &\quad - eG^{(2n, m)}_{\nu_1, \ldots, \nu_m}(\ldots; p_1 + q, \ldots; \ldots) + \text{perm.} \\ &\quad + eG^{(2n, m)}_{\nu_1, \ldots \nu_m}(\ldots; \ldots; -p'_1 + q, \ldots) + \text{perm.}\}. \end{aligned} \tag{12.4.15}$$

This equation applies also to connected Green's functions, provided we drop all but the last two lines on the right.

Let us now apply this result to two special cases.

(a) The photon propagator ($n = 0$, $m = 1$):

$$q^\mu D_{\mu\nu}(q) = -\frac{i\mathfrak{a}q_\nu}{q^2 - \mathfrak{a}\lambda^2}. \tag{12.4.16}$$

This is evidently the Fourier transform of (12.4.11) (notice that $G^{(0,2)}_{\mu\nu}$ differs from $D_{\mu\nu}$ by a factor $(2\pi)^4\delta^{(4)}$). But the free propagator $D^{\mathrm{F}}_{\mu\nu}$, equation (12.1.11), satisfies

$$q^\mu D^{\mathrm{F}}_{\mu\nu}(q) = \frac{-i\mathfrak{a}q_\nu}{q^2 - \mathfrak{a}\lambda^2}. \tag{12.4.17}$$

Comparison between (12.4.16) and (12.4.17) shows that *the radiative corrections to the photon propagator are purely transverse*, i.e.

$$q^\mu \Pi_{\mu\nu}(q) = 0, \tag{12.4.18}$$

a conclusion we have already checked at one-loop order by explicit calculation.

(b) The electron–photon vertex ($n = 1$, $m = 0$).

In this case equation (12.4.15) yields the Fourier transform of (12.4.13),

$$q^\mu V_\mu(q, p, p') = \frac{ie\mathfrak{a}}{q^2 - \mathfrak{a}\lambda^2} [S(p + q) - S(p' - q)], \tag{12.4.19}$$

where the notation is the same as in Section 12.3.4 (Fig. 12.14). It proves convenient to transform this relation by using the proper vertex

$$V_\mu = D_{\mu\nu}(q)S(p')(-\mathrm{i}e\Gamma^\nu)S(p);$$

then the identity (12.4.16) puts the result into its final form*

$$q^\mu \Gamma_\mu(p, p') = \mathrm{i}[S^{-1}(p') - S^{-1}(p)]. \tag{12.4.20}$$

It is instructive to check that the signs are correct to zero-loop order, where

$$\Gamma_\nu = \gamma_\nu, \quad \mathrm{i}S_F^{-1}(p) = (\not{p} - m).$$

The limit $q \to 0$ also yields the useful identity

$$\Gamma_\mu(p, p) = \mathrm{i}\frac{\partial S^{-1}}{\partial p^\mu} = \gamma_\mu - \frac{\partial \Sigma}{\partial p^\mu}. \tag{12.4.21}$$

12.4.2 Unitarity

The basic function of the Ward identities is to preserve the unitarity of electrodynamics†. In fact unitarity depends on the validity of the Cutkosky rules; the intermediate states entering these rules must give identically the same contribution as one gets from physical states. As an example we consider photon–electron scattering with an intermediate state containing one electron and two photons (Fig. 12.17); this has all the ingredients of the general argument. To fix our ideas and to ease the notation, take the zero-mass case and the Feynman gauge. On cutting the propagator of the momentum-k photon in the intermediate state we get

$$-g_{\rho\nu}\frac{2\pi}{(2\pi)^4}\theta(k^0)\delta(k^2), \tag{12.4.22}$$

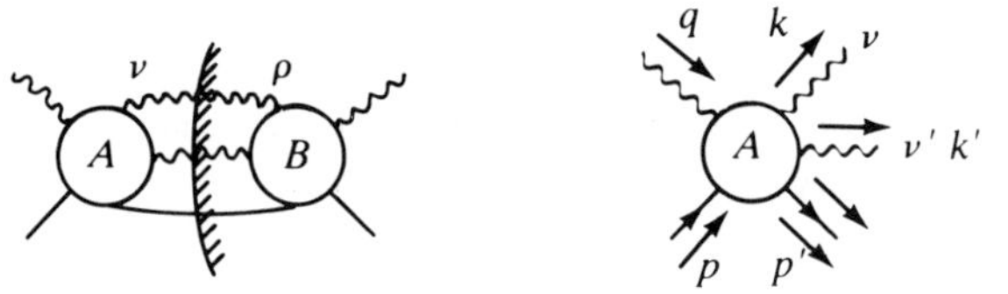

Fig. 12.17 Intermediate state with one electron and two photons

* The factor i on the right of (12.4.20) stems from the definition of the propagator. Like the factors (4π) in electrostatics, the factors i ejected from certain formulae always surface somewhere else instead.

† In non-Abelian gauge theories it can happen that a current corresponding to some local gauge invariance, and therefore conserved at the classical level, is incapable of being conserved after renormalization. This is the problem of 'anomalies'; for the same reason as in electrodynamics, it spells mortal danger to unitarity.

while summation over the polarization states of a physical photon yields not $-g_{\rho\nu}$ but

$$\sum_s \varepsilon_\rho^{(s)}(k)\varepsilon_\nu^{(s)*}(k) = -g_{\rho\nu} + \frac{k_\rho n_\nu + k_\nu n_\rho}{k \cdot n}. \tag{12.4.23}$$

Here n_μ is the vector $(k_0, -\boldsymbol{k})$, given $k_\mu = (k_0, \boldsymbol{k})$; the right-hand side of (12.4.23) projects onto the subspace orthogonal to k_μ and to n_μ. Perturbative unitarity will be satisfied only if the contribution from (12.4.22) is the same as the contribution from (12.4.23). This follows in consequence of a Ward identity, which we shall write diagrammatically for the Green's function corresponding to the scattering amplitude A; to do so we calculate $k^\nu G^{(2;3)}_{\nu\nu';\mu}$ (Fig. 12.18):

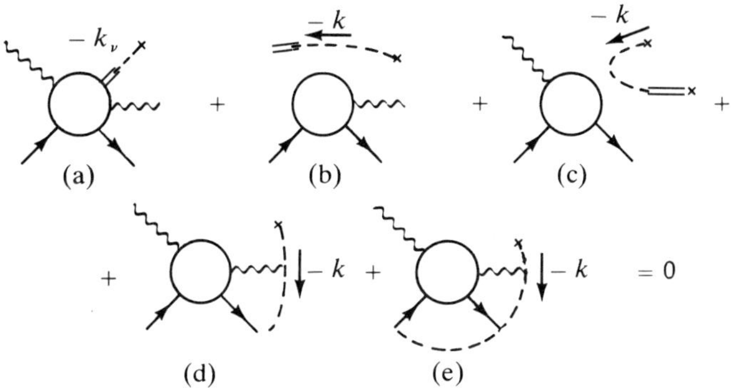

Fig. 12.18 The Ward identity for $G^{(2,3)}$

The contribution (b) vanishes because the initial photon is physical: $q^\mu \varepsilon_\mu^{(s)}(q) = 0$. We must show that the contribution of (c) also vanishes, even if the polarization of the photon k' is unphysical; the desired result follows because the condition $k + k' = 0$ is incompatible with the step functions $\theta(k^0)$ and $\theta(k'^0)$ in the cut propagator. The result also follows immediately from the remarks after equation (12.4.15), if one considers only connected S-matrix elements.

The electrons present more of a problem. Recall that one calculates the S-matrix element by multiplying the Green's function by the full inverse propagators, this being the operation that yields the Green's function with the full propagators amputated:

$$\bar{G}^{(2,3)}_{\nu,\nu';\mu}(k, k', p'; q, p).$$

The effect of the operation is to turn the term (e) of Fig. 12.18 into

$$\bar{G}^{(2;2)}_{\nu';\mu}(k', p'; q, p-k)S(p-k)S^{-1}(p).$$

On taking the mass-shell limit $\not{p} = m$, $S^{-1}(p)$ vanishes, and (e) vanishes likewise; a similar argument shows that the term (d) vanishes if the electron p' is on its mass shell. In more sophisticated language, the composite operators $\varphi(x)\psi(x)$ and $\varphi(x)\bar{\psi}(x)$ have no single-particle poles.

Summarizing, with external electrons on their mass shells one gets the Ward identity for the connected Green's function shorn of its full propagators:

$$k^\nu \bar{G}^{(2,3)}_{\nu,\nu';\mu}(k, k', p'; q, p) = 0. \tag{12.4.24}$$

Clearly this is a general result, applicable to all Green's functions $\bar{G}^{(2n,m)}_{\mu,\ldots}$.

The identity (12.4.24) shows that the projection (12.4.23) onto physical states may be replaced by $-g_{\rho\nu}$; by virtue of the Ward identity, only physical states contribute to the Cutkosky rules.

12.4.3 Renormalization

Up to this point our reasoning has been somewhat formal, in that it has ignored divergences. Now however we must tackle the problems of renormalization. As in Chapter 6, we aim not at systematic proofs, but only to make renormalizability as plausible for electrodynamics as we made it for φ^4 theory. For definiteness we use the minimal scheme, defining the renormalization constant Z_m through $m_0 = Z_m m$. To one-loop order, and collecting the counterterms (12.3.20), (12.3.34), and (12.3.53), we find

$$\begin{aligned}\mathscr{L} = &-\frac{1}{4}F_{\mu\nu}F^{\mu\nu} + \frac{1}{2}\lambda^2 A_\mu A^\mu - \frac{1}{2\alpha}(\partial_\mu A^\mu)^2 \\ &+ \bar{\psi}\left(\mathrm{i}\frac{\overleftrightarrow{\partial\!\!\!/}}{2} - m\right)\psi - e\bar{\psi}\gamma_\mu\psi A^\mu - \frac{1}{4}(Z_3 - 1)F_{\mu\nu}F^{\mu\nu} \\ &- (Z_2 Z_m - 1)m\bar{\psi}\psi + (Z_2 - 1)\bar{\psi}\,\mathrm{i}\frac{\overleftrightarrow{\partial\!\!\!/}}{2}\psi \\ &- e(Z_1 - 1)\bar{\psi}\gamma_\mu\psi A^\mu. \end{aligned} \tag{12.4.25}$$

If we want the renormalized Green's functions to satisfy the Ward identities, then we must validate the renormalized versions of (12.4.16) and (12.4.21). As regards (12.4.16) this implies that there are no counterterms featuring $A_\mu A^\mu$ or $(\partial_\mu A^\mu)^2$. Next we consider (12.4.21). Under renormalization

$$\Gamma_\mu(p, p') \to \Gamma_{\mu,\mathrm{R}}(p, p') = \Gamma_{\mu,\,\mathrm{reg}}(p, p') + Z_1^{(1)}\gamma_\mu,$$

$$\mathrm{i}S^{-1}(p) \to \mathrm{i}S_\mathrm{R}^{-1}(p) = \mathrm{i}S_\mathrm{reg}^{-1}(p) + Z_2^{(1)}(p\!\!\!/ - m),$$

where $\Gamma_{\mu,\,\mathrm{reg}}(p, p')$ and $S_\mathrm{reg}(p)$ have been calculated with dimensional regularization. Because such regularization preserves gauge invariance (this being its chief virtue), one has

$$\Gamma_{\mu,\,\mathrm{reg}}(p, p) = \mathrm{i}\frac{\partial S_\mathrm{reg}^{-1}(p)}{\partial p^\mu}.$$

Hence, to validate the same identity for the renormalized functions, we must ensure that

$$Z_1^{(1)} = Z_2^{(1)}. \tag{12.4.26a}$$

In the minimal scheme this identity is automatic, given that the divergent parts of the vertex and of the propagator are necessarily identical. By manipulating the finite parts of the counterterms one could make $Z_1 \neq Z_2$, but such choices would be fatal to unitarity.

Next we introduce the bare fields and the bare constants by writing

$$A_0^\mu = Z_3^{1/2} A^\mu \tag{12.4.27a}$$

$$e_0 = Z_1 Z_2^{-1} Z_3^{-1/2} e = Z_3^{-1/2} e \tag{12.4.27b}$$

$$\lambda_0^2 = Z_3^{-1} \lambda^2 \tag{12.4.27c}$$

$$\psi_0 = Z_2^{1/2} \psi \tag{12.4.27d}$$

$$m_0 = Z_m m \tag{12.4.27e}$$

$$\alpha_0 = Z_3 \alpha. \tag{12.4.27f}$$

As a function of these bare quantities the Lagrangean (12.4.25) reads

$$\mathscr{L} = -\frac{1}{4} F_0^{\mu\nu} F_{0_{\mu\nu}} + \frac{1}{2} \lambda_0^2 A_{0_\mu} A_0^\mu - \frac{1}{2\alpha_0} (\partial_\mu A_0^\mu)^2$$
$$+ \bar{\psi}_0 \left(\frac{\mathrm{i}}{2} \overleftrightarrow{\not\partial} - m_0 \right) \psi_0 - e_0 \bar{\psi}_0 \gamma_\mu \psi_0 A_0^\mu, \tag{12.4.28}$$

which shows that its structure is preserved under renormalization. The role of the identity $Z_1 = Z_2$ is crucial to this end; since

$$\partial_\mu + \mathrm{i} e A^\mu \to \partial_\mu + \mathrm{i} Z_1^{-1} Z_2 e_0 A_0^\mu = \partial_\mu + \mathrm{i} e_0 A^\mu, \tag{12.4.29}$$

minimal coupling is indeed maintained. In the general case one proceeds recursively. Suppose we have calculated the renormalized Green's functions and the counterterms to l-loop order. The regularized Green's functions are found from the Lagrangean (12.4.28), where $e_0, m_0, \ldots$ are functions of ε having poles proportional to $\varepsilon^{-1}, \ldots, \varepsilon^{-l}$. Because of the structure of the Lagrangean (12.4.28), the regularized Green's functions obey the Ward identities. The procedure we have just considered in one-loop order can be extended to define the counterterms to $(l+1)$-loop order, respecting all the while the identity (12.4.21) for the renormalized Green's functions, provided one chooses $Z_1^{(l+1)} = Z_2^{(l+1)}$. With perfect generality one finds the relation

$$Z_1 = Z_2. \tag{12.4.26b}$$

The bare Green's functions to $(l + 1)$-loop order will satisfy the general Ward identity (12.4.15); this follows because they are calculated from the Lagrangean (12.4.19), whose parameters are evaluated to l-loop order. (Recall that, apart from $G^{(0,2)}$, $G^{(2,0)}$, and $G^{(2,1)}$ which we have studied above, the Green's functions are superficially convergent.) The Ward identities for the renormalized Green's functions follow from the multiplicative structure of renormalization, i.e. from

$$G_{\mathrm{R}}^{(2n,\,m)} = Z_2^{-n} Z_3^{-m/2} G_0^{(2n,\,m)}$$

and from the identities (see (12.4.27))

$$\mathfrak{a}_0 \lambda_0^2 = \mathfrak{a}\lambda^2, \quad Z_3^{-1}\mathfrak{a}_0 = \mathfrak{a}, \quad Z_3^{-1/2}\mathfrak{a}_0 e_0 = \mathfrak{a}e,$$

where $\mathfrak{a}$, λ, and e are the renormalized quantities.

To summarize, renormalized electrodynamics is indeed unitary and can also be shown to be local. It remains to show that physical quantities (say S-matrix elements) are gauge-independent (see Problem 12.13). For instance, it can be shown that S-matrix elements are independent of $\mathfrak{a}$, though this is not evident a priori, resulting as it does from a cancellation between the factor $z_2(\mathfrak{a})$ (see (10.3.37)) and the $\mathfrak{a}$-dependence of certain Green's functions. In view of the fact that physical quantities are indeed independent of $\mathfrak{a}$, one can in the propagator (12.1.11) take the limit $\mathfrak{a} \to \infty$; this returns one to the naïve form (10.5.16) of the propagator for a massive vector field, which yields a manifestly unitary theory. The Lagrangean of a massive vector field coupled to a conserved vector current is physically equivalent to a renormalizable Lagrangean, in spite of the severe divergences stemming from the factors $1/m^2$ in the Green's functions; it is in fact physically equivalent to the Lagrangean as modified by Stueckelberg.

12.4.4 The renormalization group for electrodynamics

Before we proceed to the formalism it is worth describing intuitively the variation of the electric charge (or of α) with distance, which affords an elementary account of renormalization and of the renormalization group. Consider two static (infinitely massive) charges $+e$ and $-e$ separated by a distance $r \ll 1/m$; to zero-loop order their mutual potential energy is $-\alpha_0/r$. It suffers corrections from screening by virtual e^+e^- pairs (Fig. 12.19), or in other words from vacuum polarization. To determine the effects of this polarization we integrate over the (wavelength λ) fluctuations of the electromagnetic field due to

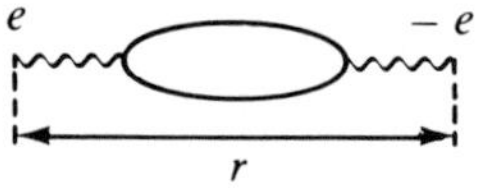

Fig. 12.19

such e^+e^- pairs, according to

$$\alpha(r) = \alpha_0 - d\alpha_0^2 \int_{\Lambda^{-1}}^{r} \frac{d\lambda}{\lambda}. \tag{12.4.30}$$

Here, the combination $d\lambda/\lambda$ is imposed by dimensional considerations; the constant d is positive (since screening reduces the potential energy). Λ is an ultraviolet and r an infrared cutoff; if $\lambda \gtrsim r$, then the fluctuations cannot distinguish between the two charges. The integration in (12.4.30) yields

$$\alpha(r) = \alpha_0 - d\alpha_0^2 \ln(\Lambda r). \tag{12.4.31}$$

Now let R be some reference distance, defining the charge $\alpha(R)$ at distance R. We express α_0 in terms of $\alpha(R)$,

$$\alpha_0 = \alpha(R) + d\alpha^2(R)\ln(\Lambda R),$$

and substitute into (12.4.31):

$$\alpha(r) = \alpha(R)\left(1 - d\alpha(R)\ln\frac{r}{R}\right). \tag{12.4.32}$$

This operation eliminates the bare charge α_0 in favour of the reference charge at distance R, and simultaneously eliminates the cutoff Λ; we have in effect renormalized the charge. The renormalization has introduced a length scale R which breaks the naïve scale-invariance of electrodynamics; for $mr \ll 1$ such invariance would predict that $\alpha(r) \to$ constant, which is manifestly not the case. Dimensional analysis shows that $\alpha(r)$ takes the form

$$\alpha(r) = F\left(\alpha(R), \frac{r}{R}\right),$$

whence differentiation with respect to r yields

$$\left.\frac{d\alpha(r)}{d\ln r}\right|_{r=R} = F'(\alpha(r), 1) = -\beta(\alpha(r))$$
$$= -d\alpha^2(r) + O(\alpha^3(r)). \tag{12.4.33}$$

This equation defines the variation of the charge as a function of distance, or, which amounts to the same thing, as a function of a mass $\mu \sim 1/r$. We recognize in (12.4.33) the definition of the Callan–Symanzik function β. For α small enough, equation (12.4.33) shows that the charge increases as the distance shrinks; as $r \to 0$, one approaches the 'true' (i.e. infinite) value of α.

Revert now to more formal calculations exploiting the results of Chapter 7. Since the fine-structure constant has the same dimensionality as the coupling constant g of scalar theory, we can use the results of Chapter 7 provided we replace g by α. (If we were to work with e instead of α, we should need to replace ε by $\varepsilon/2$.) The relation between α_0 and α reads

$$\alpha_0 = Z_3^{-1}\alpha. \tag{12.4.34}$$

The renormalization constant Z_3 has been calculated in Section 12.3.2; to one-loop order (see (12.3.23)) its divergent part is

$$Z_3^{(1)} = -\frac{2\alpha}{3\pi\varepsilon}.$$

By virtue of equation (7.5.10), and remembering that g_0 and g are related by $g_0 = Zg$, we immediately obtain the function $\beta(\alpha)$ to first order in α:

$$\beta(\alpha) = \mu\frac{d\alpha(\mu)}{d\mu} = \frac{2\alpha^2}{3\pi} + O(\alpha^3). \tag{12.4.35}$$

The value of the constant d in (12.3.33) is $2/3\pi$.

Still to one-loop order one can calculate $\alpha(q^2)$ as a function of $\alpha(\mu^2)$:

$$\alpha(q^2) = \frac{\alpha(\mu^2)}{1 - \dfrac{2\alpha(\mu^2)}{3\pi}\ln\left(\dfrac{q^2}{\mu^2}\right)}. \tag{12.4.36}$$

The coupling constant increases with q^2 or, equivalently, it increases with decreasing distance, conformably to the heuristic arguments given above.

The sign of the term proportional to α^2 shows that electrodynamics is not asymptotically free. It seems probable that, like φ^4 theory, electrodynamics cannot in 4 dimensions be defined nontrivially as a limit of a theory with cutoff. Nevertheless, the value of q^2 where $\alpha(q^2)$ becomes of order 1 is astronomically high. Since it is generally believed at present that electrodynamics should be subsumed into some larger non-Abelian gauge theory, its lack of asymptotic freedom is probably quite irrelevant, and casts no shadow on the success of the renormalization programme.

A comment finally about the definition of electric charge. In electrodynamics a natural definition is supplied by the existence of a classical zero-frequency limit. When one renormalizes on the mass shell, the low-energy limit of Compton scattering is given, to all orders of perturbation theory, by the classical result for the scattering of an electromagnetic wave by a point charge, namely by Thomson's formula

$$\sigma = \frac{8\pi}{3}\left(\frac{\alpha}{m}\right)^2. \tag{12.4.37}$$

This result follows from the Ward identity for the Compton effect (Problem 12.10). If α denotes the value of the fine-structure constant in the on-shell scheme, and α_{MS} its value in say the MS scheme, then it is easy to find the connection between these two definitions of α:

$$\alpha_{\text{MS}} = \frac{Z_3^{\text{MS}}}{Z_3}\alpha = \alpha + \frac{\alpha^2}{3\pi}\ln\left(\frac{C\mu^2}{m^2}\right) + O(\alpha^3). \tag{12.4.38}$$

Clearly this relation depends on the mass scale used in the minimal scheme. The advantage of such a scheme is that it admits the $m \to 0$ limit of electrodynamics without difficulty. In the on-shell scheme this limit is sabotaged by infrared divergences. One good example is supplied by the formula (12.3.29a) for the short-distance potential $V(r)$; one cannot set $m = 0$ even when $mr \to 0$, in spite of the fact that in this regime the mass of the electron is quite irrelevant. The difficulty is wholly peculiar to this particular renormalization scheme; nothing like it occurs say in the MS scheme.

Problems

12.1 Derive equation (12.2.6) by adapting the calculation in Section 10.3.1 to the scattering of a single (relativistic) particle by a potential.

12.2 Furry's theorem

(a) Using the uniqueness of the irreducible representations of the anticommutation rules (11.1.11) for the γ_μ matrices, show that there exists a matrix C (called the charge-conjugation matrix) such that

$$C\gamma_\mu C^{-1} = -\gamma_\mu^{\mathrm{T}}.$$

(b) Thence derive $CS_{\mathrm{F}}(x-y)C^{-1} = [S_{\mathrm{F}}(y-x)]^{\mathrm{T}}$.

(c) Consider a closed electron loop with an odd number of photon lines attached. Show that summing over both directions of the loop gives zero.

12.3 Photoproduction of π^0 mesons

We aim to calculate the cross-section for the photoproduction of π^0 mesons in the reaction

$$\gamma(k) + \text{proton}(p) \rightarrow \pi^0(q) + \text{proton}(p'),$$

starting from the phenomenological interaction Lagrangean

$$\mathscr{L}_1(x) = -e\bar{\psi}\gamma_\mu\psi A^\mu - \mathrm{i}G\bar{\psi}\gamma_5\psi\varphi.$$

(This Lagrangean must not be taken literally: it yields very poor results except at the production threshold. The main object of the exercise is to illustrate, without excessive algebra, how cross-sections are calculated.)

In $\mathscr{L}_1(x)$, ψ is the proton field, φ the π^0 field, e the proton charge, and G a coupling constant such that $G^2/4\pi \approx 15$.

(a) Draw the Feynman graphs contributing to this reaction to order eG, and write down their analytic expressions. Check that the sum $\mathscr{M}_\mu$ of these two graphs indeed satisfies the condition $k^\mu\mathscr{M}_\mu = 0$ when the protons p and p' are on the mass shell ($p^2 = p'^2 = m^2$).

(b) Calculate the differential photoproduction cross-section $\mathrm{d}\sigma/\mathrm{d}\Omega$ for unpolarized photons and protons, to order $(eG)^2$. Evaluate $\mathrm{d}\sigma/\mathrm{d}\Omega$ numerically, and plot it as a function of θ in the centre-of-mass frame, for a laboratory energy $E_\gamma = 300$ MeV. Compare with the experimental results, and comment.

Hint: use $k^\mu\mathscr{M}_\mu = 0$ to replace the photon polarization sum by $-g_{\mu\nu}$.

The solution reads

$$\frac{1}{4}\sum_{\text{spins}} |\varepsilon_\mu^{(s)}\mathscr{M}^\mu|^2 = e^2G^2\left[2m^2\mu^2\left(\frac{1}{(s-m^2)^2}+\frac{1}{(u-m^2)^2}\right) - \frac{(t-\mu^2)^2+2\mu^2(t-2m^2)}{(s-m^2)(u-m^2)}\right],$$

where μ is the π^0 mass, m the proton mass, $s=(p+k)^2$, $u=(p'-k)^2$, and $t=(q-k)^2$.

12.4 A study of the photon propagator (de Rafael 1976, Chapter 7)

(a) Carry out the integration over x in the expression (12.3.26) for $\bar{\omega}_R^{(1)}(q^2)$, and derive the end-result (for $q^2 < 4m^2$) in the form

$$\bar{\omega}_R^{(1)}(q^2) = \frac{\alpha}{3\pi}\left[\frac{8}{3} - X^2 - \frac{1}{2}(3-X^2)X\ln\left|\frac{1+X}{1-X}\right|\right],$$

where $X=(1-4m^2/q^2)^{1/2}$. A change of variables to $y=1-2x$ may help.

(b) When $q^2 > 4m^2$, $\bar{\omega}_R^{(1)}$ acquires an imaginary part. Integrating by parts and changing variables to $s=4m^2/(1-y^2)$, put $\bar{\omega}_R^{(1)}$ into the form

$$\bar{\omega}_R^{(1)}(q^2) = \frac{\alpha q^2}{3\pi}\int_{4m^2}^{\infty}\frac{\mathrm{d}s}{s}\frac{1}{s-q^2+\mathrm{i}\varepsilon}\left(1+\frac{2m^2}{s}\right)\sqrt{\left(1-\frac{4m^2}{s}\right)}.$$

Thence show that $\bar{\omega}_R^{(1)}(q^2)$ satisfies a dispersion relation with one subtraction, and that its imaginary part is given by

$$\frac{1}{\pi}\operatorname{Im}\bar{\omega}_R^{(1)}(q^2) = \frac{\alpha}{3\pi}\left(1+\frac{2m^2}{q^2}\right)\sqrt{\left(1-\frac{4m^2}{q^2}\right)}\theta(q^2-4m^2).$$

(c) Relate $\operatorname{Im}\bar{\omega}_R^{(1)}(q^2)$ to the 'decay rate' $\Gamma(q^2)$ of a virtual photon with mass q^2 into an electron–positron pair (see (10.3.12)):

$$\sqrt{q^2}\operatorname{Im}\bar{\omega}_R^{(1)}(q^2) = \Gamma(q^2).$$

12.5 Vacuum-polarization corrections to the Coulomb potential (Berestetskii *et al.* 1980, Section 112)

(a) We aim to calculate the potential $V(r)$, taking into account the vacuum-polarization correction to one-loop order, in the limits $mr \ll 1$ and $mr \gg 1$. Accordingly we must calculate the Fourier transform of $\bar{\omega}_R^{(1)}(q^2)$ in the static limit $q^2 = -\boldsymbol{q}^2$. By deforming the integration contour to run along the positive imaginary axis, show that the correction $\delta V(r)$ to the Coulomb potential can be written

$$\delta V(r) = \frac{e}{4\pi r}\frac{2\alpha}{3\pi}\int_1^\infty \mathrm{d}x\, \mathrm{e}^{-2mrx}\left(1+\frac{1}{2x^2}\right)\frac{\sqrt{(x^2-1)}}{x^2}.$$

(b) Derive the expressions (12.3.29a) and (12.3.29b) in the limits $mr \ll 1$ and $mr \gg 1$ respectively. For $mr \ll 1$, introduce a parameter x_1 such that $1/mr \gg x_1 \gg 1$, and split the integral into $\int_1^{x_1} + \int_{x_1}^\infty$.

12.6 (a) Derive the much-employed identity

$$\frac{1}{\not p + \not k - m}\not k \frac{1}{\not p - m} = \frac{1}{\not p - m} - \frac{1}{\not p + \not k - m}.$$

(b) By using this identity in the form

$$\frac{1}{\not p + \not k - m}\not k = 1 - \frac{1}{\not p + \not k - m}(\not p - m),$$

together with the Dirac equation, derive equation (12.3.51) for the gauge-dependent part of the electron–photon vertex.

12.7 Behaviour of the form factor $F_1(q^2)$ at large q^2

We aim to determine the behaviour of the form factor $F_{1\mathrm{R}}(q^2)$ as $q^2 \to \pm\infty$. In order to simplify the calculation we consider the case $m = 0$. (For $m \neq 0$, see Itzykson and Zuber 1980, Chapter 7.) We start by studying the limit $q^2 \to -\infty$, where $q^2 = -Q^2$, and renormalize on the mass shell: $F_{1\mathrm{R}}(q^2 = 0) = 0$.

(a) Show that in (12.3.50a) one can take the limit $m = 0$, which simplifies the formula giving $F_{1\mathrm{R}}(q^2)$ to

$$F_{1\mathrm{R}}(q^2) = \frac{\alpha}{2\pi}\int \mathrm{d}x\,\mathrm{d}x' \left\{ -\ln\frac{xx'Q^2 + \lambda^2(1-x-x')}{\lambda^2(1-x-x')} - \frac{Q^2(1-x)(1-x')}{xx'Q^2 + \lambda^2(1-x-x')}\right\}.$$

(b) Thence derive

$$F_{1\mathrm{R}}(Q^2) = \frac{2\alpha}{3\pi}\left[-\frac{1}{2}\ln^2\left(\frac{Q^2}{\lambda^2}\right) + \frac{3}{2}\ln\frac{Q^2}{\lambda^2} - \frac{7}{4} - \frac{\pi^2}{3} + O\left(\frac{\lambda^2}{Q^2}\right)\right].$$

How must this result be modified for $q^2 \to +\infty$?

12.8 The electrodynamics of scalar particles

(a) Starting from the Lagrangean of the free charged scalar field,

$$\mathscr{L}(x) = (\partial_\mu\varphi^\dagger)(\partial^\mu\varphi) - m^2\varphi^\dagger\varphi,$$

find the Lagrangean of this field coupled to the electromagnetic field, and also the Feynman rules, namely

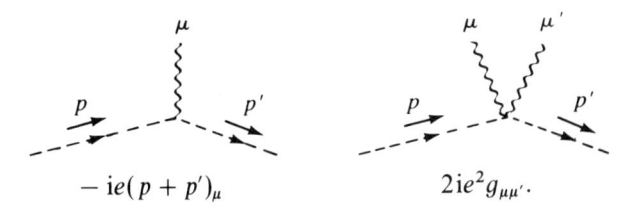

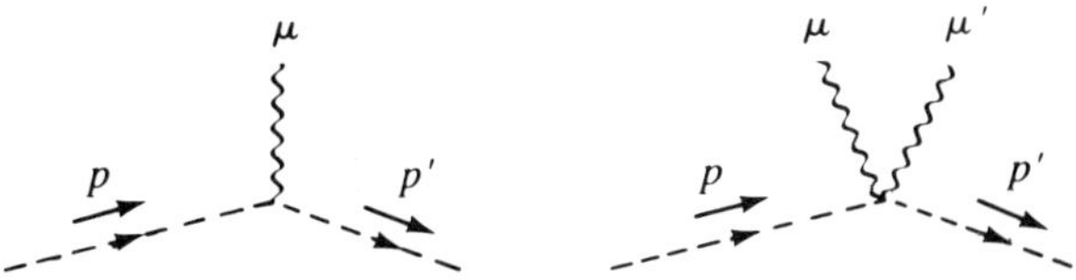

$$-\mathrm{i}e(p+p')_\mu \qquad\qquad 2\mathrm{i}e^2 g_{\mu\mu'}.$$

(b) Calculate in the laboratory frame (π-meson initially at rest), the differential Compton cross-section for a π^+ meson to order e^4, assuming that the initial photons are unpolarized:

$$\gamma(k) + \pi^+(p) \to \gamma(k') + \pi^+(p').$$

Pretend (though it is false) that photons couple to π^+ mesons as if the latter were point particles.

Hint: choose $\varepsilon_\mu^{(s)}$ so that $\varepsilon\cdot p = \varepsilon'\cdot p = 0$.

(c) Calculate the differential cross-section for the reaction

$$e^+e^- \to \pi^+\pi^-$$

to order e^4, assuming that the initial electrons are unpolarized. Show that

$$\frac{\mathrm{d}\sigma}{\mathrm{d}\Omega} = \frac{\alpha^2}{8q^2}\sin^2\theta.$$

What would be the value of the ratio R in Section 12.2.3 if quarks had spin zero?

(d) Determine from dimensional analysis all graphs that are superficially divergent. Show that the superficial degree of divergence is reduced in the following cases:

photon propagator;
γ-π-π vertex;
3 and 4-photon vertices.

(e) Draw all divergent diagrams at one loop order and determine their divergent part using dimensional regularization. Check that the photon propagator is transverse. The renormalization constants Z_1, Z_2, Z_3, Z_4 and $\bar{Z}_4$ are defined through

$$\delta\mathscr{L} = -\frac{1}{4}(Z_3-1)F_{\mu\nu}F^{\mu\nu} - \delta m^2\varphi^\dagger\varphi + (Z_2-1)(\partial_\mu\varphi^\dagger)(\partial^\mu\varphi)$$

$$-\mathrm{i}e(Z_1-1)\varphi^\dagger\overleftrightarrow{\partial}_\mu\varphi + e^2(Z_4-1)(A_\mu A^\mu)(\varphi^\dagger\varphi) - \frac{1}{(2!)^2}\bar{Z}_4(\varphi^\dagger\varphi)^2.$$

Compute $Z_1, Z_2, Z_3, Z_4, \bar{Z}_4$ at the one-loop order in the Feynman gauge:

$$Z_1 = Z_2 = Z_4 = 1 + \frac{\alpha}{2\pi}\frac{2}{\varepsilon}; \quad Z_3 = 1 - \frac{\alpha}{12\pi}\frac{2}{\varepsilon}; \quad \bar{Z}_4 = -\frac{24\alpha^2}{\varepsilon}$$

Since $\bar{Z}_4$ is nonzero, one must add a term $(\varphi^\dagger\varphi)^2$ to the Lagrangean, and a normalization condition for the 4-π proper vertex.

(f) Show that $Z_4 = Z_1$ is required if one wants to preserve the structure of the Lagrangean while renormalizing the theory. This relation also follows from a Ward identity. Let $C_{\alpha\beta} = -\mathrm{i}\bar{G}^{(4)}_{\alpha\beta}$, where $\bar{G}^{(4)}_{\alpha\beta}$ is the connected γ-γ-π-π Green's function shorn of its full external propagators. Show that $C_{\alpha\beta}$ obeys the Ward identity

$$k^\alpha C_{\alpha\beta}(k, k'; p, p') = e^2\Gamma_\beta(p + k, p')S(p + k)S^{-1}(p) - e^2S^{-1}(p') \\ \times S(p' - k)\Gamma_\beta(p, p' - k),$$

where the momenta are defined in (b), Γ_β is the γ-π-π proper vertex normalized by

$$\Gamma_\beta(p, p') = -\mathrm{i}e(p + p')_\beta \Lambda(p, p'),$$

and $\Lambda(p, p') = 1$ in the tree approximation. $S(p)$ is the π-propagator. Show that

$$k^\beta\Gamma_\beta(p, p') = \mathrm{i}S^{-1}(p') - \mathrm{i}S^{-1}(p); \quad k = p' - p.$$

Subdivide $C_{\alpha\beta}$ into 1-PR and 1-PI parts, and show that the proper vertex $\Gamma^{(4)}_{\alpha\beta}$ obeys

$$k^\alpha\Gamma^{(4)}_{\alpha\beta} = e^2\Gamma_\beta(p + k, p') - e^2\Gamma_\beta(p, p' - k).$$

Check this equation, as well as the previous Ward identities, at the tree level, and deduce from it the relation $Z_1 = Z_4$.

(g) Low-energy theorem: use the previous Ward identity for $\Gamma^{(4)}_{\alpha\beta}$ and the gauge $\varepsilon \cdot p = \varepsilon' \cdot p = 0$ to show that

$$\lim_{q, q' \to 0} \Gamma^{(4)}_{\alpha\beta} = 2e^2 g_{\alpha\beta}.$$

Thus, in the mass-shell renormalization scheme, the low-energy limit of Compton scattering is given to all orders by the tree approximation.

12.9 Another derivation of Ward identities

(a) Use the global gauge invariance of the QED Lagrangean and a slight modification of equation (C.3.5) in order to prove the Ward identities. One has only to note that, since A_μ is not affected by a (global!) gauge transformation, one can insert products of A_μ's in the Green's functions on both sides of equation (C.3.5), to obtain

$$\partial_x^\mu \langle 0 | T(j_\mu(x) A_{\nu_1}(x_1) \ldots A_{\nu_m}(x_m) \psi(y_1) \ldots \psi(y_n) \bar{\psi}(z_1) \ldots \bar{\psi}(z_n) | 0 \rangle$$
$$= e \sum [\delta(x - z_i) - \delta(x - y_i)]$$
$$\times \langle 0 | T(A_{\nu_1}(x_1) \ldots A_{\nu_m}(x_m) \psi(y_1) \ldots \psi(y_n) \bar{\psi}(z_1) \ldots \bar{\psi}(z_n) | 0 \rangle,$$

where $j_\mu = e\bar{\psi}\gamma_\mu\psi$.

(b) In the case of the photon propagator, use

$$D_{\mu\nu}(z - x) = D^{\mathrm{F}}_{\mu\nu}(z - x) - \mathrm{i} \int \mathrm{d}^4 y D^{\mathrm{F}}_{\mu\alpha}(z - y) \hat{\Pi}^\alpha_\nu(y - x)$$

where

$$\hat{\Pi}^\alpha_\nu(y - x) = \langle 0 | T(j^\alpha(y) A_\nu(x)) | 0 \rangle$$

to recover equation (12.4.16).

(c) Prove equation (12.4.15) by using the results of (a) and (b).

12.10 The Ward identities for the 4-photon vertex and for the Compton effect (de Rafael 1976, Chapter 6)

(a) Apply the identity (12.4.15) to the case $n = 0$, $m = 3$, and use (12.4.16), to derive the Ward identity for the proper four-photon vertex,

$$k_1^{\mu_1} \Gamma^{(0,4)}_{\mu_1\mu_2\mu_3\mu_4}(k_1, k_2, k_3, k_4) = 0.$$

(Do not forget the disconnected parts of $G^{(0,4)}$.)

(b) Next, apply the identity (12.4.15) to the case $n = 1, m = 2$, corresponding to the Compton effect (or to the production of an e^+e^- pair by two photons). Let $G^{(2,2)}_{\mathrm{c},\mu\mu'} = \mathbb{C}_{\mu\mu'}$ be the connected Green's function

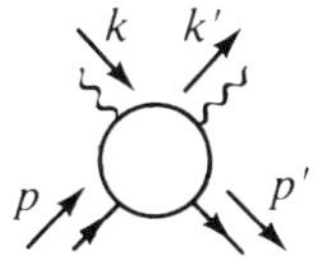

Fig. 12.20 The Compton amplitude

shorn of its external propagators (Fig. 12.20). Derive the relation

$$k^\mu \mathbb{C}_{\mu\mu'}(k, -k'; p, -p') = \mathrm{i}e^2 \Gamma_{\mu'}(p + k, p') S(p + k) S^{-1}(p)$$
$$- \mathrm{i}e^2 S^{-1}(p') S(p' - k) \Gamma_{\mu'}(p, p' - k),$$

where Γ_μ is the proper electron–photon vertex (see (12.4.20)). Verify this identity in the tree approximation.

(c) For $k \to 0$, show that the forward ($k = k'$) Compton amplitude can be expressed in terms of the self-energy $\Sigma(p)$ alone.

12.11 For zero momentum transfer and on the mass shell ($\not{p} = m$), the inverse propagator and the proper vertex can be written in the form

$$\mathrm{i}S^{-1}(p) = \not{p} - m - A(p^2) - (\not{p} - m)B(p^2),$$

$$\Gamma_\mu(p, p)|_{\not{p}=m} = \gamma_\mu(1 + F_1(0) + F_2(0)) - \frac{p_\mu}{m} F_2(0).$$

Use the identity (12.4.21) to relate $F_1(0)$ and $F_2(0)$ to $A'(m^2)$ and $B(m^2)$. Thence show that on-shell renormalization and the condition $Z_1 = Z_2$ entail $F_{1\mathrm{R}}(0) = 0$

12.12 Gauge dependence of Z_2 (Collins 1983, Chapter 12; Lautrup 1976)

(a) Preliminaries.

(i) Show that the Green's function $\langle 0|T(F^0_{\mu\nu}(x)F_0^{\mu\nu}(y))|0\rangle$ is independent of the gauge parameter α_0. (The reasoning in Section 11.4.2 may help.) Hence show that the same is true for the transverse part of the propagator $D^0_{\mu\nu}$, and use this to prove that Z_3 is independent of α.

(ii) Suppose that the Lagrangean depends on a parameter α. Show that the α-dependence of the Green's functions is governed by

$$\frac{\partial}{\partial\alpha}\langle 0|T(X)|0\rangle = \mathrm{i}\left[\langle 0|T\left(\frac{\partial S}{\partial\alpha}X\right)|0\rangle - \langle 0|T(X)|0\rangle \times \langle 0|\frac{\partial S}{\partial\alpha}|0\rangle\right] = \mathrm{i}\langle 0|T\left(:\frac{\partial S}{\partial\alpha}:X\right)|0\rangle,$$

where the symbol : : indicates that the vacuum expectation-value is to be subtracted.

(iii) Show that the insertion of $\int \mathrm{d}^4x : \varphi(x)\,\delta S/\delta\varphi(x):$ counts the number of external lines of the Green's function. More precisely,

$$[Z(0)]^{-1}\int \mathrm{d}^4x \int \mathscr{D}\varphi : \varphi(x)\frac{\delta S}{\delta\varphi(x)} : \varphi(x_1)\ldots\varphi(x_N)\,\mathrm{e}^{\mathrm{i}S} = -\mathrm{i}N\langle 0|T(\varphi(x_1)\ldots\varphi(x_N))|0\rangle.$$

(b) For simplicity we consider the case where the electron mass vanishes ($m = 0$), and use the minimal subtraction scheme. Derive the identity

$$\frac{\partial}{\partial \mathrm{a}}\langle 0|T(X)|0\rangle = \frac{\mathrm{i}}{2}\int \mathrm{d}^4x\langle 0|T(:\varphi^2(x):\Delta^2 X)|0\rangle$$
$$- N_\psi\langle 0|T(X)|0\rangle\frac{\partial \ln Z_2}{\partial \mathrm{a}},$$

where ΔX is the variation of X under a gauge transformation (see (12.4.9)), and N_ψ is the number of ψ fields in X.

(c) By taking $X = \psi(y)\bar{\psi}(z)$, derive the identity written diagrammatically in Fig. 12.21.

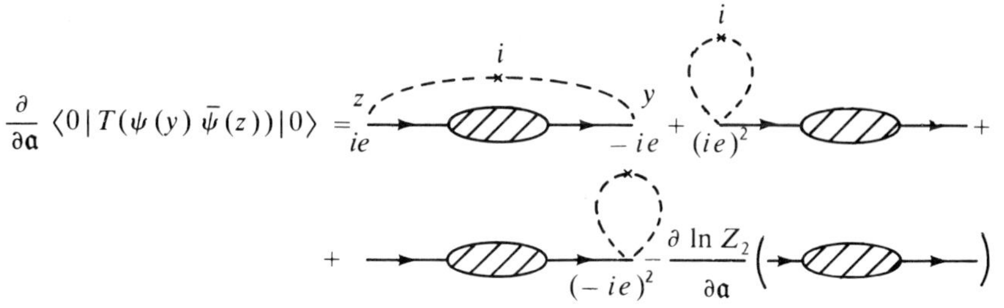

Fig. 12.21 The equation for $\partial \ln Z_2/\partial \mathrm{a}$

Noting that the integral for the first graph converges, show that in the MS scheme

$$Z_2(\mathrm{a}) = Z_2(\mathrm{a}=0)\exp\left[\frac{-\alpha \mathrm{a}}{2\pi\varepsilon}\right].$$

Check this result to one-loop order (see (12.3.43)).

12.13 Gauge independence of *S*-matrix elements (Lee 1975)

(a) Let us consider first a φ^4-theory and define a generating function $Z(j)$ through

$$Z(j) = \int \mathscr{D}\varphi \exp\left(\mathrm{i}S + \mathrm{i}\int \mathrm{d}^4x\, j(x) f(\varphi(x))\right).$$

To be specific, one can think of $f(\varphi)$ as $\varphi + \varphi^3$, for example. Draw the first terms of the diagrammatic expansion of the Green's functions $G_{\mathrm{f}}^{(2)}$ and $G_{\mathrm{f}}^{(4)}$, defined by functional differentiation of $Z(j)$. Define

$$Z_3 = \lim_{p^2\to m^2}[-\mathrm{i}(p^2-m^2)G^{(2)}(p)],$$

$$Z_{3f} = \lim_{p^2\to m^2}[-\mathrm{i}(p^2-m^2)G_{\mathrm{f}}^{(2)}(p)],$$

where $G^{(2)}(p)$ corresponds to the usual case $f(\varphi) = \varphi$. Show that one can write

$$Z_{3f} = \sigma^2 Z_3,$$

and draw the first terms of the diagrammatic expansion of σ.

Show that σ disappears from the expressions for S-matrix elements. (*Hint*: only those diagrams contributing to $G_f^{(N)}$ that have one-particle poles in all p_i's have a nonzero limit when $p_i^2 \to m^2$.)

(b) Reverting to QED, let $Z_\omega(j)$ be the generating functional ($\partial^\mu j_\mu \neq 0!$)

$$Z_\omega(j) = \int \mathscr{D}A \exp\left(iS + i \int d^4x \left(-\frac{1}{2\alpha} (\partial_\mu A^\mu - \omega(x))^2 + j_\mu A^\mu \right)\right)$$

$$S = \int d^4x \left(-\frac{1}{4} F_{\mu\nu} F^{\mu\nu} \right).$$

Show that

$$Z_\omega(j) = \int \mathscr{D}A \exp\left(iS + i \int d^4x \left(-\frac{1}{2\alpha} (\partial_\mu A^\mu) + j^\mu (A_\mu + \partial_\mu \Box^{-1} \omega) \right)\right)$$

Generalize this result to the full QED Lagrangean. Show that a variation of $\alpha : \alpha \to \alpha + \delta\alpha$ corresponds to

$$\omega(x) = \frac{\delta\alpha}{2\alpha} \partial_\mu A^\mu.$$

These observations afford an alternative proof of the results derived in Problem 12.12.

(c) Use (a) and (b) to show that S-matrix elements are α-independent. Do not forget that for a physical photon $\varepsilon \cdot k = 0$.

(d) Try to generalize to any gauge condition $f(A_\mu) = 0$ (see Problem 11.14).

12.14 Majorana particles and supersymmetric Lagrangeans (see e.g. Haber and Kane 1985)

(a) The charge-conjugation matrix C defined in Problem 12.2 relates positive- and negative-energy solutions of the Dirac equation,

$$u(\boldsymbol{p}, s) = C\bar{v}^{\mathrm{T}}(\boldsymbol{p}, s), \quad v(\boldsymbol{p}, s) = C\bar{u}^{\mathrm{T}}(\boldsymbol{p}, s).$$

Check this in the Dirac representation, where $C = \begin{pmatrix} 0 & -\mathrm{i}\sigma_2 \\ -\mathrm{i}\sigma_2 & 0 \end{pmatrix}$.
Note the phase conventions for the spinors when $\boldsymbol{p} = 0$.

(b) The Fourier decomposition of a Majorana field $\lambda(x)$ reads

$$\lambda(x) = \sum_s \int \mathrm{d}\tilde{p}[b_s(\boldsymbol{p})u(\boldsymbol{p}, s)\,\mathrm{e}^{-\mathrm{i}px} + b_s^\dagger(\boldsymbol{p})v(\boldsymbol{p}, s)\,\mathrm{e}^{\mathrm{i}px}].$$

Verify that $\lambda(x)$ is equal to its charge conjugate $\lambda^{\mathrm{c}}(x)$,

$$\lambda(x) = \lambda^{\mathrm{c}}(x) = C\bar{\lambda}^{\mathrm{T}}(x).$$

The field $\lambda(x)$ has only two degrees of freedom instead of four: Majorana particles are their own antiparticles.

(c) We aim to establish the Feynman rules for a Majorana field coupled to a scalar field,

$$\mathscr{L} = \frac{1}{2}\bar{\lambda}(\mathrm{i}\overleftrightarrow{\partial\!\!\!/} - m)\lambda - \frac{1}{2}\varphi(\Box + m^2)\varphi + g(\bar{\lambda}\Gamma\lambda)\varphi,$$

where $\Gamma = 1$ or $\mathrm{i}\gamma_5$. Establish the following rules for contractions:

$$\lambda(x)\bar{\lambda}(y) = S_{\mathrm{F}}(x - y),$$

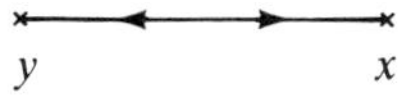

$$\lambda(x)\lambda(y) = -\,S_{\mathrm{F}}(x - y)C,$$

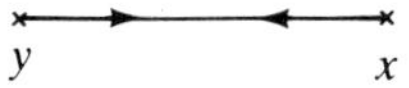

$$\bar{\lambda}(x)\bar{\lambda}(y) = C^{-1}S_{\mathrm{F}}(x - y).$$

Calculate the probability amplitude for the decay of a virtual meson into two fermions,

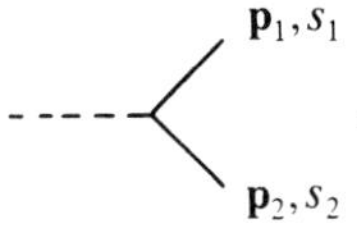

$$= -\,2\mathrm{i}g\bar{u}(\boldsymbol{p}_1, s_1)\Gamma v(\boldsymbol{p}_2, s_2) = -\,2\mathrm{i}g\bar{u}(\boldsymbol{p}_1, s_1)\Gamma C\bar{u}^{\mathrm{T}}(\boldsymbol{p}_2)$$

(up to a sign ambiguity).

Show that in calculating the closed loop in Fig. 12.22, one may assign to the vertex a factor $(-\,2\mathrm{i}g)$, provided one also takes account of a symmetry factor equal to 1/2.

Fig. 12.22 Self-energy correction to the meson propagator

(d) Consider the supersymmetric Lagrangean*

$$\mathscr{L} = \frac{1}{2}(\partial_\mu A)^2 + \frac{1}{2}(\partial_\mu B)^2 - \frac{1}{2}m^2(A^2 + B^2) + \frac{1}{2}\bar{\lambda}(\mathrm{i}\overleftrightarrow{\partial\!\!\!/} - m)\lambda$$

$$- gmA(A^2 + B^2) - \frac{1}{2}g^2(A^2 + B^2)^2 - g\bar{\lambda}(A - \mathrm{i}\gamma_5 B)\lambda,$$

where A is a scalar and B a pseudoscalar field. Establish the following Feynman rules:

$= -12\,\mathrm{i}g^2,$

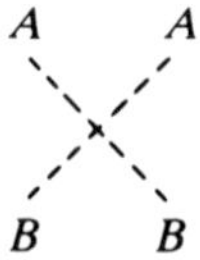

$= -4\,\mathrm{i}g^2,$

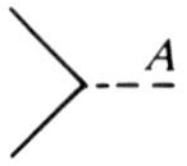

$= -2\,\mathrm{i}g,$

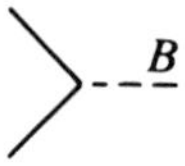

$= -2g\gamma_5.$

Show that the correction δm to the meson propagator diverges only logarithmically. (Use Pauli–Villars regularization; why?) Show that a renormalization of the field suffices to make the theory finite at one-loop order. These conclusions generalize to an arbitrary number of loops.

* This Lagrangean is known under the names of Wess and Zumino (1974); see also Haber and Kane (1985, Appendix E), and Iliopoulos and Zumino (1974).

Further reading

The very first thing to read is Feynman (1985). The Feynman rules and their classic application to the Compton effect are discussed by Bjorken and Drell (1965, Chapter 7), and by Itzykson and Zuber (1980, Chapters 5 and 6). One could well compare our one-loop calculations with those of Bjorken and Drell (1965, Chapter 8) and of Itzykson and Zuber (1980, Chapter 7), who regularize by using a cutoff. The treatment of infrared divergences is illustrated by V. Berestetskii *et al.* (1980, Section 118). The classic paper on infrared divergences is that by Yennie *et al.* (1961). The renormalizability of electrodynamics is studied by Itzykson and Zuber (1980, Chapters 7 and 8), and by Collins (1983, Chapter 12). Our discussion of Ward identities has been adapted from 't Hooft and Veltman (1974, Section 11). A compact proof can be found in Zinn-Justin (1989, Chapter 17). Another useful reference is de Rafael (1976).

13 Non-Abelian gauge theories

Non-Abelian gauge theories are today the basis of the 'standard model' in elementary-particle physics. This model has two props:

(i) the Glashow–Salam–Weinberg (GSW) model, which unifies the electromagnetic and the weak interactions into a single theory of electroweak interactions;

(ii) quantum chromodynamics (QCD), which is currently supposed to be the theory of the strong interactions.

There have been attempts to merge the GSW model and QCD into unified theories embracing the strong as well as the weak and the electromagnetic interactions; these attempts at 'grand unification' also work with non-Abelian gauge theories as one of their ingredients. But at the time of writing there is no consensus over such grand unification, because the simplest model (SU(5)) disagrees with experiment over the stability of the proton.

Non-Abelian gauge theories are at least an order of magnitude more complicated than quantum electrodynamics; from their discovery by Yang and Mills (1954) it took a good twenty years before they were brought under control in perturbation theory, and there are still many problems that remain open in the non-perturbative domain. Moreover their field of application is enormous. Thus there can be no question in a single chapter of giving more than a very succinct introduction; any discussion aiming to be at all thorough would certainly be at least as long as this entire book. In fact the object of the present chapter is twofold:

(i) to give an elementary first introduction, which should allow the reader easier access to the more specialized texts cited in the references;

(ii) to describe certain basic calculations in some detail (the Higgs phenomenon, the function $\beta(g)$, the Altarelli–Parisi equation), thus supplying an introduction to papers devoted more specifically to particle physics.

The chapter starts by describing the classical theory (Section 13.1), and follows this by discussing quantization (Section 13.2). However, gauge invariance and renormalization will be treated very sketchily. Section 13.3 considers the GSW model restricted to leptons; the chief aim here is to show how, through the Higgs phenomenon, a broken symmetry allows one to construct a renormalizable theory of massive vector bosons. Section 13.4 deals with quantum chromodynamics; it describes the calculation of the function $\beta(g)$ to one-loop order, which allows one to demonstrate the crucial property of asymptotic freedom. The example of e^+e^- annihilation to order α_s is also studied in detail. Finally Section 13.5 sketches current attempts to enter the non-perturbative domain through lattice calculations. Sections 13.3 and 13.5 are largely independent of Section 13.2, and can be tackled directly after reading Section 13.1.

13.1 Non-Abelian gauge fields: the classical theory

As their basic ingredient, non-Abelian gauge theories use compact Lie groups. To introduce these in an elementary way we briefly review the properties of the simplest nontrivial Lie group, namely the group SU(2). This will allow us to introduce the ideas we shall need later (infinitesimal generators etc.). Readers familiar with Lie groups should proceed directly to Section 13.1.2, after just a brief look at our notations.

13.1.1 The group SU(2)

Consider the group of 2×2 unitary matrices having determinant 1; we can write these in terms of four complex numbers *a, b, c, d* (eight parameters in all) as

$$U = \begin{pmatrix} a & b \\ c & d \end{pmatrix}. \tag{13.1.1}$$

The numbers *a, b, c, d* satisfy the constraints

$$ad - bc = 1 \tag{13.1.2}$$

and

$$a = d^*, \quad b^* = -c. \tag{13.1.3}$$

This leaves three independent parameters. The number of such parameters is called the *dimension* of the Lie group: accordingly SU(2) has dimension 3. Combining (13.1.2) and (13.1.3) one obtains the identity

$$|a|^2 + |b|^2 = 1,$$

which implies that the region over which a and b vary is finite and closed; thus the parameters a and b assume values from a compact domain, whence the designation as a compact Lie group.

Now consider such a matrix U in the neighbourhood of the identity, writing it as

$$U = \mathbb{1} - \mathrm{i}\xi, \quad U^\dagger = \mathbb{1} + \mathrm{i}\xi^\dagger. \tag{13.1.4}$$

The conditions $UU^{\dagger} = \mathbb{1}$ and $\det U = 1$ show that the matrix ξ is Hermitean and has zero trace,

$$\xi = \xi^{\dagger}, \quad \operatorname{Tr} \xi = 0. \tag{13.1.5}$$

Any matrix that satisfies (13.1.5) can be expressed in terms of the Pauli matrices τ_a, $a = 1, 2, 3$,

$$\tau_1 = \begin{pmatrix} 0 & 1 \\ 1 & 0 \end{pmatrix}, \quad \tau_2 = \begin{pmatrix} 0 & -\mathrm{i} \\ \mathrm{i} & 0 \end{pmatrix}, \quad \tau_3 = \begin{pmatrix} 1 & 0 \\ 0 & -1 \end{pmatrix}, \tag{13.1.6}$$

by introducing three real parameters Λ_a through

$$\xi = \Lambda_a(\tau_a/2), \quad \Lambda_a = \operatorname{Tr}(\xi\tau_a); \tag{13.1.7}$$

these are infinitesimal for the moment, but will be allowed to assume finite values in due course. Recall that the Pauli matrices satisfy the identity

$$\tau_a \tau_b = \delta_{ab} + \mathrm{i}\varepsilon_{abc}\tau_c. \tag{13.1.8}$$

The set of parameters Λ_a allows one to define a unit vector $\hat{\boldsymbol{n}}$ by

$$\Lambda_a = \varepsilon \hat{\boldsymbol{n}}_a, \quad \varepsilon = (\Lambda_a \Lambda_a)^{1/2},$$

where $\varepsilon \ll 1$. An element of the group SU(2) can be written as a product of N infinitesimal elements in the limit $N \to \infty$,

$$U_{\hat{\boldsymbol{n}}}(\theta) = \lim_{N \to \infty} \left[U_{\hat{\boldsymbol{n}}}\left(\frac{\theta}{N}\right) \right]^N$$

$$= \lim_{N \to \infty} \left(1 - \frac{\mathrm{i}}{2}\frac{\theta}{N} \hat{\boldsymbol{n}} \cdot \boldsymbol{\tau} \right)^N = \mathrm{e}^{-\mathrm{i}\theta(\hat{\boldsymbol{n}} \cdot \boldsymbol{\tau})/2}; \tag{13.1.9}$$

conversely it is easy to show that any element of the group can be expressed in the form (13.1.9). The matrix $\frac{1}{2}\boldsymbol{\tau} \cdot \hat{\boldsymbol{n}}$ is called the *infinitesimal generator* of the transformations around the direction $\hat{\boldsymbol{n}}$. Obviously there are as many independent infinitesimal generators as there are parameters; for SU(2) one has the three infinitesimal generators $\frac{1}{2}\tau_a$. They satisfy the commutation rules

$$\left[\frac{1}{2}\tau_a, \frac{1}{2}\tau_b \right] = \mathrm{i}\varepsilon_{abc}\frac{1}{2}\tau_c, \tag{13.1.10}$$

which constitute the *Lie algebra of the group*. At this point it proves convenient to make the terminology somewhat more precise. The group SU(2) exists independently of its realization (representation) by matrices; it is defined by a certain composition rule prescribing the product of any two elements, and the inverse of every element. To every element g of SU(2) one can assign a matrix $D(g)$, acting on an n-dimensional space, and such that

$$D(g_1 g_2) = D(g_1)\, D(g_2). \tag{13.1.11}$$

The set of all such matrices constitutes an n-dimensional *representation* of the

Lie group; if the correspondence is one-to-one, then the representation is faithful, and if the matrices D are unitary, the representation is unitary. What we have described above (in equations (13.1.1–10)) is called the *fundamental* representation. Clearly there exists a trivial one-dimensional representation, assigning the number 1 to every element g. Another well-known representation (not a faithful one) is furnished by the rotation matrices in three-dimensional space, which constitute the group SO(3). The groups SU(2) and SO(3) are homomorphic; to every element of SO(3) there correspond two elements of SU(2).

We give a quick demonstration of these classic properties. Let $\boldsymbol{x}$, having components x_a, be a vector in three-dimensional space, and construct the 2×2 Hermitean and traceless matrix

$$X = \tfrac{1}{2} x_a \tau_a. \tag{13.1.12}$$

Define a matrix $X_a(\theta)$ and a vector $\boldsymbol{x}_{\hat{\boldsymbol{n}}}(\theta)$ by

$$X_{\hat{\boldsymbol{n}}}(\theta) = U_{\hat{\boldsymbol{n}}}(\theta) X U_{\hat{\boldsymbol{n}}}^{\dagger}(\theta) = \frac{1}{2} x_{\hat{\boldsymbol{n}}a}(\theta) \tau_a. \tag{13.1.13}$$

Using (13.1.10) one can show that $\boldsymbol{x}_{\hat{\boldsymbol{n}}}(\theta)$ satisfies

$$\frac{\mathrm{d}\boldsymbol{x}_{\hat{\boldsymbol{n}}}(\theta)}{\mathrm{d}\theta} = \hat{\boldsymbol{n}} \times \boldsymbol{x}_{\hat{\boldsymbol{n}}}(\theta), \tag{13.1.14}$$

which implies that $\boldsymbol{x}_{\hat{\boldsymbol{n}}}(\theta)$ can be derived from $\boldsymbol{x}$ by a rotation through an angle θ around the axis $\hat{\boldsymbol{n}}$. This exhibits the correspondence between the elements of SU(2) and rotations in three dimensions; moreover it yields the infinitesimal generators of the 3-dimensional representation:

$$T_1 = \begin{pmatrix} 0 & 0 & 0 \\ 0 & 0 & -\mathrm{i} \\ 0 & \mathrm{i} & 0 \end{pmatrix}, \quad T_2 = \begin{pmatrix} 0 & 0 & \mathrm{i} \\ 0 & 0 & 0 \\ -\mathrm{i} & 0 & 0 \end{pmatrix}, \quad T_3 = \begin{pmatrix} 0 & -\mathrm{i} & 0 \\ \mathrm{i} & 0 & 0 \\ 0 & 0 & 0 \end{pmatrix}. \tag{13.1.15}$$

The matrices T_a obey the same commutation rules as the matrices $\frac{1}{2}\tau_a$:

$$[T_a, T_b] = \mathrm{i}\varepsilon_{abc} T_c. \tag{13.1.16}$$

In fact the matrix elements $(T_a)_{bc}$ are given explicitly by

$$(T_a)_{bc} = -\mathrm{i}\varepsilon_{abc}. \tag{13.1.17}$$

Finally, on taking $\theta = 2\pi$, one can see that this rotation, which amounts to the identity, corresponds to two distinct matrices of SU(2), namely to $+\mathbb{1}$ and to $-\mathbb{1}$.

We have just constructed, explicitly, three representations of SU(2), having dimensions 1, 2, 3 respectively; from the standard theory of angular momentum we know that in fact there exist infinitely many representations, with dimensions 1, 2, 3, 4, . . . , $(2j + 1)$, . . . , describing angular momenta 0, 1/2, 1, 3/2, . . . , j, ... In a $(2j + 1)$-dimensional space the generators of the Lie algebra will be represented by $(2j+1)\times(2j+1)$ matrices, satisfying the commutation rules (13.1.16). By an abuse of language, a basis vector of such a vector space is often called an 'element of the $(2j+1)$-dimensional representation', though strictly speaking it is the matrices that constitute the elements of the representation.

An r-dimensional Lie group G has r infinitesimal generators, obeying an algebra

$$[t_a, t_b] = \mathrm{i} f_{abc} t_c, \tag{13.1.18a}$$

where the f_{abc} (which generalize the ε_{abc} of SU(2)) are called the *structure constants* of the Lie group (or of the Lie algebra). The structure constants are real, and antisymmetric under any odd permutation of pairs of indices:

$$f_{abc} = -f_{bac} = -f_{acb} = -f_{cba} = f_{bca} = f_{cab}. \tag{13.1.19}$$

We shall use only unitary representations $U(g)$ of the group G (in fact for compact Lie groups all representations are equivalent to a unitary representation); in that case the infinitesimal generators are represented by Hermitean matrices obeying the commutation rules

$$[T_a, T_b] = \mathrm{i} f_{abc} T_c. \tag{13.1.18b}$$

Unless otherwise stated, it will always be understood that the representation is irreducible (for a definition see Pontryagin (1966) or Georgi (1982). In the *adjoint representation* of the Lie group, the infinitesimal generators are represented by $r\times r$ matrices given by

$$(T_a)_{bc} = -\mathrm{i} f_{abc} \tag{13.1.20}$$

(see (13.1.17)). Accordingly, in the adjoint representation the infinitesimal generators are represented by pure imaginary matrices, and the matrices $U(g)$ are real and orthogonal. For SU(2), the matrices of the 3-dimensional adjoint representation are simply the rotation matrices, which are indeed real and orthogonal.

The groups most widely used in elementary-particle physics are the groups $\mathrm{SU}(N)$, i.e. the groups of unitary $N\times N$ matrices with determinant 1. The number of parameters is (N^2-1); except for $N=2$, there are *two* fundamental N-dimensional representations, called N and $\bar{N}$. Their infinitesimal generators are generally normed through the relation

$$\mathrm{Tr}(T_a T_b) = \tfrac{1}{2}\delta_{ab}. \tag{13.1.21}$$

13.1.2 Parallel transport and covariant derivative

Reverting to electromagnetism as discussed in Chapter 11, consider an example from elementary quantum mechanics. Let $\varphi(\boldsymbol{x})$ be the wave-function of a charged particle, say of an electron; the phase of this wave-function can be modified globally, i.e. in the same way at all points of space, without any change in physical predictions. But it would seem that it cannot be modified locally; for instance, if the phase of the wave-function is changed by 180° near one of the two slits in a Young's slits experiment, then the interference pattern shifts radically. However, we have seen that such local modification does become possible if the particles are coupled to an electromagnetic field. In the presence of such a field, the original field-free propagator

$$F_0(2, 1) = \langle \boldsymbol{x}_2 | U(t_2, t_1,) | \boldsymbol{x}_1 \rangle$$

of the Schroedinger equation (see Section 8.2.1) is altered; when the propagator is calculated as a path integral, the statistical weight of every trajectory between the points (1) and (2) must be multiplied by the factor

$$R(C; \boldsymbol{A}) = \mathrm{e}^{\mathrm{i}\int_1^2 \boldsymbol{A}\cdot \mathrm{d}\boldsymbol{x}}. \tag{13.1.22}$$

(We take the charge as unity in order to ease the notation.) A gauge transformation

$$\boldsymbol{A} \to \boldsymbol{A}' = \boldsymbol{A} - \nabla \Lambda$$

transforms $R(C; \boldsymbol{A})$ into $R(C; \boldsymbol{A}')$,

$$R(C; \boldsymbol{A}') = \mathrm{e}^{-\mathrm{i}\Lambda(\boldsymbol{x}_2)} R(C; \boldsymbol{A}) \mathrm{e}^{\mathrm{i}\Lambda(\boldsymbol{x}_1)}; \tag{13.1.23}$$

this absorbs the local phase change

$$\varphi(\boldsymbol{x}) \to \varphi'(\boldsymbol{x}) = \mathrm{e}^{-\mathrm{i}\Lambda(\boldsymbol{x})} \varphi(\boldsymbol{x}), \tag{13.1.24}$$

and leaves invariant the probability amplitudes

$$\int \mathrm{d}x_1\, \mathrm{d}x_2\, \psi^*(\boldsymbol{x}_2) F(2, 1)\, \varphi(\boldsymbol{x}_1).$$

In the language of differential geometry* one says that the factor $R(C; \boldsymbol{A})$ effects a 'parallel transport' of the wave-function between the points x_1 and x_2, thus allowing the phases at these two points to be compared. Let us now restrict ourselves to an infinitesimal displacement,

$$\boldsymbol{x}_1 = \boldsymbol{x}, \quad \boldsymbol{x}_2 = \boldsymbol{x} + \mathrm{d}\boldsymbol{x},$$

$$R(C; \boldsymbol{A}) = 1 + \mathrm{i}\boldsymbol{A}\cdot \mathrm{d}\boldsymbol{x}. \tag{13.1.25}$$

Parallel transport of the wave-function $\varphi(\boldsymbol{x})$ to $\boldsymbol{x} + \mathrm{d}\boldsymbol{x}$ yields

$$\varphi^t(\boldsymbol{x}) = (1 + \mathrm{i}\boldsymbol{A}\cdot \mathrm{d}\boldsymbol{x}) \varphi(\boldsymbol{x}). \tag{13.1.26}$$

The covariant derivative $\boldsymbol{D}\varphi$ is defined not through $[\varphi(\boldsymbol{x} + \mathrm{d}\boldsymbol{x}) - \varphi(\boldsymbol{x})]$, but through $[\varphi(\boldsymbol{x} + \mathrm{d}\boldsymbol{x}) - \varphi^t(\boldsymbol{x})]$:

$$\varphi(\boldsymbol{x} + \mathrm{d}\boldsymbol{x}) - \varphi^t(\boldsymbol{x}) = \mathrm{d}\boldsymbol{x}\cdot \boldsymbol{D}\varphi. \tag{13.1.27}$$

* Readers familiar with differential geometry will recognize that $\boldsymbol{A}$ is a connection, and that $F_{\mu\nu}$ (see Section 13.1.3) is a curvature tensor.

Under gauge transformations, $\varphi'(\boldsymbol{x})$ transforms, by construction, like $\varphi(\boldsymbol{x} + \mathrm{d}\boldsymbol{x})$; this ensures that $\boldsymbol{D}\varphi$, by contrast to $\boldsymbol{\nabla}\varphi$, transforms like φ:

$$[\boldsymbol{D}\varphi(\boldsymbol{x})]' = (\boldsymbol{\nabla} - \mathrm{i}\boldsymbol{A}')\varphi'(\boldsymbol{x}) = \mathrm{e}^{-\mathrm{i}\Lambda(\boldsymbol{x})}\boldsymbol{D}\varphi(\boldsymbol{x}). \tag{13.1.28}$$

The gauge transformations (13.1.24) are said to be Abelian, because the product of two such transformations is commutative. The gauge group is the Lie group $U(1)$, parametrized by the real numbers Λ in the interval $[0, 2\pi]$.

We shall now generalize all this to the non-Abelian case. Let G be a compact, semi-simple Lie group, whose Lie algebra is defined by structure constants f_{abc}; we call G the *gauge group* of the non-Abelian theory. We shall use only unitary representations of G, where the infinitesimal generators are represented by Hermitean matrices T_a obeying

$$[T_a, T_b] = \mathrm{i}f_{abc}T_c. \tag{13.1.18b}$$

Any arbitrary element of the representation can be expressed in terms of real parameters Λ_a in the form

$$U(g) = \mathrm{e}^{-\mathrm{i}\Lambda_a T_a}. \tag{13.1.29}$$

Now consider (reverting to Minkowski space) a set of classical fields $\{\varphi_i(x)\}$, $i = 1, \ldots, n$, transforming under an n-dimensional representation of the group G,

$$[U(g)\varphi]_i(x) = [\mathrm{e}^{-\mathrm{i}\Lambda_a T_a}]_{ij}\varphi_j(x). \tag{13.1.30}$$

The label i is an internal symmetry index, whose physical significance need not be specified for the moment. Equation (13.1.30) defines a global non-Abelian gauge transformation of the fields φ_i, which generalizes the phase transformation (11.3.20) of the Abelian case. One can easily write down interactions invariant under such a transformation. For example, if $G = \mathrm{SU}(2)$, if φ_i belongs to the 2-dimensional representation, and A_i^μ to the 3-dimensional representation, then the interactions

$$\varphi_i^\dagger(x)\Box\varphi_i(x) \quad \text{or} \quad \varphi_j^\dagger(x)(\tau_i)_{jk}(\partial_\mu\varphi_k(x))A_i^\mu(x)$$

are manifestly invariant under (13.1.30). (Here we suppose that the field φ_i is a Lorentz scalar, though it would be equally easy to construct examples with a spinor field.) But the situation changes radically if we allow the gauge transformation (13.1.30) to become *local*, with parameters Λ depending on x:

$$[U(g(x))\varphi]_i(x) = [\mathrm{e}^{-\mathrm{i}\Lambda_a(x)T_a}]_{ij}\varphi_j(x). \tag{13.1.31}$$

In order to write down interactions that are invariant, we must as in the Abelian case introduce a *gauge field* $A_a^\mu(x)$, depending on an internal symmetry index a; for a given representation one defines $\mathscr{A}^\mu(x)$ by

$$\mathscr{A}^\mu(x) = T_a A_a^\mu(x). \tag{13.1.32}$$

Parallel transport requires the field $\mathscr{A}^\mu(x)$, which takes its values in the Lie algebra of the group. For an infinitesimal path we generalize (13.1.25) to

$$R(x+\mathrm{d}x, x; A) = 1 - \mathrm{i}\mathrm{d}x_\mu \mathscr{A}^\mu(x), \tag{13.1.33}$$

but for a finite path one must be careful of the fact that the $\mathscr{A}^\mu(x)$ do not commute with each other. The problem here is exactly the same as for the evolution operator (see Chapter 9); instead of splitting the interval $[0, T]$ into N subintervals T/N, with $N \to \infty$, we split the path C into N infinitesimal paths $\mathrm{d}x^\mu(l)$, (Fig. 13.1), and define $R(C; A)$ through the limit

$$R(C; A) = \lim_{N\to\infty} \prod_{l=1}^{N} (1 - \mathrm{i}\mathrm{d}x^\mu(l)\mathscr{A}_\mu(l))$$

$$= \lim_{N\to\infty} \prod_{l=1}^{N} \mathrm{e}^{-\mathrm{i}\mathrm{d}x^\mu(l)\mathscr{A}_\mu(l)}.$$

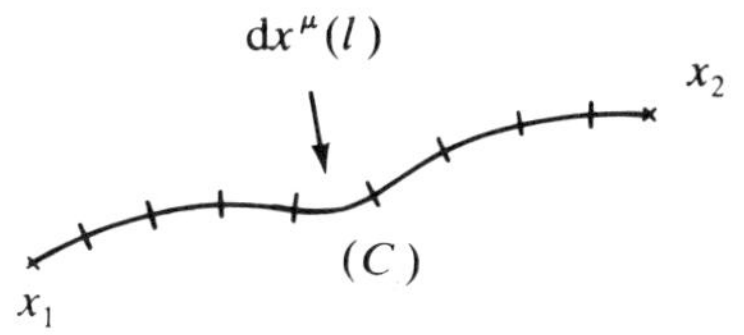

Fig. 13.1

The result can be expressed formally as

$$R(C; A) = P\left(\mathrm{e}^{-\mathrm{i}\int_{x_1}^{x_2} \mathscr{A}_\mu(x)\cdot \mathrm{d}x^\mu}\right), \tag{13.1.34}$$

where the symbol P plays the same role as the symbol T plays in (9.3.11). As in the Abelian case, the role of the connection $\mathscr{A}_\mu$ is to compensate for local phase changes,

$$R(C; A') = U(g(x_2))R(C; A)U^{-1}(g(x_1)). \tag{13.1.35}$$

In order to derive the transformation law for $\mathscr{A}^\mu$ from (13.1.35), we need merely take C as an infinitesimal path:

$$1 - \mathrm{i}\mathrm{d}x_\mu \mathscr{A}'^\mu = U(g(x+\mathrm{d}x))(1 - \mathrm{i}\mathrm{d}x_\mu \mathscr{A}^\mu)U^{-1}(g(x))$$

$$= 1 - \mathrm{i}\mathrm{d}x_\mu[\mathrm{i}(\partial^\mu U)U^{-1} + U\mathscr{A}^\mu U^{-1}],$$

whence

$$\mathscr{A}'_\mu = U(g)\mathscr{A}_\mu U^{-1}(g) + \mathrm{i}[\partial_\mu U(g)]U^{-1}(g). \tag{13.1.36}$$

This generalizes the gauge transformation (11.3.23) of the electromagnetic field; from now on we shall often write $\mathscr{A}^g_\mu$ instead of $\mathscr{A}'_\mu$.

The transformation law (13.1.36) features the matrices $U(g)$ of some particular representation, and it seems at first that the law depends on the representation being considered; this would be a serious nuisance, because one would need to introduce a new field A_a^μ for each new representation. Fortunately it is not so, since the transformation law depends only on the Lie algebra of the group; this is readily seen from the infinitesimal form of (13.1.36):

$$U(g) = \mathbb{1} - \mathrm{i}\Lambda_a T_a,$$

$$T_a \delta A_a^\mu = -\mathrm{i}\Lambda_a [T_a, T_b] A_b^\mu + (\partial_\mu \Lambda_a) T_a.$$

Since the generators T_a are linearly independent, one finds

$$\delta A_a^\mu = f_{abc} \Lambda_b A_c^\mu + \partial^\mu \Lambda_a. \tag{13.1.37a}$$

When Λ is independent of x (as in a global gauge transformation) this equation shows that A_a^μ *transforms under the adjoint representation of the group G*, because in that representation one has $(T_a)_{bc} = -\mathrm{i}f_{abc}$. By contrast to the Abelian case, the field A_a^μ is not neutral with respect to the internal symmetry in question; the electromagnetic field carries no charge, but the field A_a^μ does carry the quantum numbers associated with the symmetry. This is the property that makes non-Abelian gauge theories nonlinear even at the classical level: the superposition principle fails from the outset.

The discussion above might seem rather plodding, but has the advantage that it supplies parallel transport and the covariant derivative directly; under parallel transport one has

$$\varphi(x) \to \varphi'(x) = R(x + \mathrm{d}x, x; A)\varphi(x)$$
$$= \varphi(x) - \mathrm{i}(\mathrm{d}x_\mu \mathscr{A}^\mu)\varphi(x).$$

By construction (see (13.1.35)) $\varphi'(x)$ transforms like $\varphi(x + \mathrm{d}x)$ in (13.1.31) (and not like $\varphi(x)$!), and one can legitimately compare $\varphi(x + \mathrm{d}x)$ with $\varphi'(x)$; in particular one can calculate the covariant derivative at once:

$$\varphi(x + \mathrm{d}x) - \varphi'(x) = \mathrm{d}x^\mu[\partial_\mu \varphi + \mathrm{i}\mathscr{A}_\mu \varphi] = \mathrm{d}x^\mu(D_\mu \varphi),$$

whence

$$D^\mu = \partial^\mu \mathbb{1} + \mathrm{i}\mathscr{A}^\mu = \partial^\mu \mathbb{1} + \mathrm{i}T_a A_a^\mu. \tag{13.1.38}$$

On restoring indices this becomes

$$D_{ij}^\mu = \partial^\mu \delta_{ij} + \mathrm{i}(T_a)_{ij} A_a^\mu. \tag{13.1.39a}$$

By construction, the covariant derivative possesses the basic property

$$[D_\mu \varphi(x)]' = U(g(x)) D_\mu \varphi(x), \tag{13.1.40}$$

which allows gauge-invariant interactions to be written down immediately. Notice also that in the adjoint representation the covariant derivative reads

$$D^\mu_{ab} = \partial^\mu \delta_{ab} + f_{abc} A^\mu_c . \tag{13.1.39b}$$

It is useful to notice that the infinitesimal transformation law of A_μ can be written

$$\delta A^\mu_a = D^\mu_{ab} \Lambda_b . \tag{13.1.37b}$$

13.1.3 The tensor $F^{\mu\nu}$ and the Lagrangean

To determine the form of the (field strength) tensor $F^{\mu\nu}_a$, which is the non-Abelian generalization of the electromagnetic field tensor $F^{\mu\nu}$, we try to generalize a classic formula derived from Stokes's theorem in electromagnetism, namely

$$\int_{\bar{C}} \boldsymbol{A} \cdot \mathrm{d}\boldsymbol{x} = \iint \boldsymbol{B} \cdot \mathrm{d}\boldsymbol{S},$$

where $\bar{C}$ is a closed curve. Let us consider an infinitesimal rectangular contour $\bar{C}$, centred on a point x, with sides of lengths δa_μ and δb_μ. The points 1, 2, 3, 4 are chosen at the centres of the four sides of the rectangle (Fig. 13.2). We calculate $R(\bar{C}; A)$ for this contour,

$$R(\bar{C}; A) = \mathrm{e}^{\mathrm{i}\delta b_\mu \mathscr{A}^\mu(4)} \mathrm{e}^{\mathrm{i}\delta a_\mu \mathscr{A}^\mu(3)} \mathrm{e}^{-\mathrm{i}\delta b_\mu \mathscr{A}^\mu(2)} \mathrm{e}^{-\mathrm{i}\delta a_\mu \mathscr{A}^\mu(1)}.$$

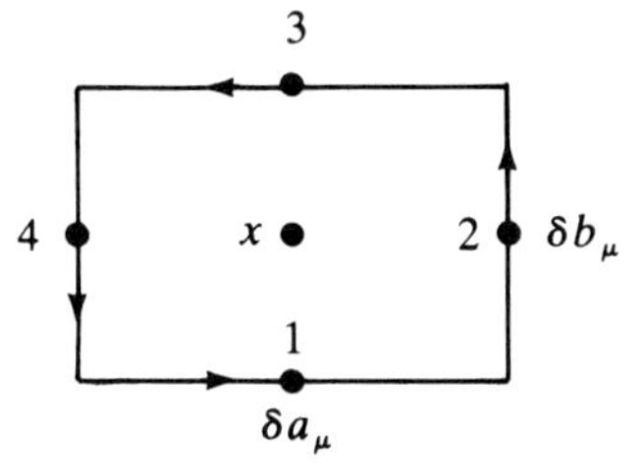

Fig. 13.2

The values of the $\mathscr{A}^\mu(i)$ are found from an expansion restricted to the vicinity of x; for example,

$$\mathscr{A}^\mu(1) = \mathscr{A}^\mu(x) - \tfrac{1}{2}\delta b_\nu \partial^\nu \mathscr{A}^\mu(x).$$

If the $\mathscr{A}^\mu$ were numbers, as in electromagnetism, one would have simply

$$R(\bar{C}; A) = \mathrm{e}^{-\mathrm{i}\int_{\bar{C}} A^\mu \mathrm{d}x_\mu} \simeq \mathrm{e}^{-\mathrm{i}\delta a_\mu \delta b_\nu (\partial^\mu A^\nu - \partial^\nu A^\mu)}$$

$$= \mathrm{e}^{-\mathrm{i}\int \mathrm{d}\sigma_{\mu\nu} F^{\mu\nu}}, \tag{13.1.41}$$

where $d\sigma^{\mu\nu} = \delta a^\mu \wedge \delta b^\nu$ is an oriented surface element. But when $\mathscr{A}^\mu$ is a matrix, one must add a term stemming from non-commutativity, given by

$$e^{i\delta b_\mu \mathscr{A}^\mu(x)} e^{i\delta a_\mu \mathscr{A}^\mu(x)} e^{-i\delta b_\mu \mathscr{A}^\mu(x)} e^{-i\delta a_\mu \mathscr{A}^\mu(x)}. \tag{13.1.42}$$

From the classic formula

$$e^{i\varepsilon h} e^{i\varepsilon' g} e^{-i\varepsilon h} e^{-i\varepsilon' g} \simeq 1 - \varepsilon\varepsilon'[h, g] \simeq e^{-\varepsilon\varepsilon'[h, g]}$$

of the theory of Lie groups we can evaluate (13.1.42) as

$$e^{\delta a_\mu \delta b_\nu [\mathscr{A}^\mu, \mathscr{A}^\nu]}.$$

Combining this with (13.1.41), we notice that $R(\bar{C}; A)$ can be put into the form

$$R(\bar{C}; A) = \exp\left(-i \int d\sigma_{\mu\nu} \{\partial^\mu \mathscr{A}^\nu - \partial^\nu \mathscr{A}^\mu + i[\mathscr{A}^\mu, \mathscr{A}^\nu]\} \right). \tag{13.1.43}$$

This equation does not generalize to a finite contour. Actually, one can observe that for a finite contour $R(\bar{C}; A)$ depends on the origin. Equation (13.1.43) leads us to define a field strength tensor $\mathscr{F}^{\mu\nu}$ by

$$\mathscr{F}^{\mu\nu} = \partial^\mu \mathscr{A}^\nu - \partial^\nu \mathscr{A}^\mu + i[\mathscr{A}^\mu, \mathscr{A}^\nu]; \tag{13.1.44}$$

on restoring indices, this reads

$$F_a^{\mu\nu} = \partial^\mu A_a^\nu - \partial^\nu A_a^\mu - f_{abc} A_b^\mu A_c^\nu. \tag{13.1.45}$$

Now consider how to interpret the calculation we have just performed. The meaning of the quantity $R(\bar{C}; A)$ is quite specific: it gives the 'rotation' of the phase of the field around a closed trajectory. Under a gauge transformation, $R(\bar{C}; A)$ transforms according to

$$R(\bar{C}; A') = U(g(x)) R(\bar{C}; A) U^{-1}(g(x)),$$

whence it is not gauge invariant, by contrast to the Abelian case. It does however have a gauge-invariant trace.

Since $d\sigma^{\mu\nu}$ and $\mathscr{F}^{\mu\nu}$ are both antisymmetric under the interchange $\mu \leftrightarrow \nu$, $\mathscr{F}'^{\mu\nu}$ must be given by the same transformation rule, namely by

$$\mathscr{F}'^{\mu\nu} = U(g(x))\, \mathscr{F}^{\mu\nu} U^{-1}(g(x)); \tag{13.1.46}$$

for an infinitesimal transformation this yields

$$\delta F_a^{\mu\nu} = f_{abc} \Lambda_b F_c^{\mu\nu}. \tag{13.1.47}$$

We see therefore that on the one hand $F_a^{\mu\nu}$ transforms under the adjoint representation, and that on the other hand $\partial_\mu \Lambda_a$ does not feature in the transformation law; nevertheless, and by contrast to electromagnetism, this is not enough to make $F^{\mu\nu}$ gauge invariant. However, equation (13.1.46) does suggest a generalization of the action for the electromagnetic field that is Lorentz invariant, gauge invariant, and reduces to (11.3.11) for an Abelian group: namely

$$S = -\frac{1}{2\bar{g}^2}\int d^4x\,\mathrm{Tr}(\mathscr{F}^{\mu\nu}\mathscr{F}_{\mu\nu}) = \frac{-1}{4\bar{g}^2}\int d^4x\,F_a^{\mu\nu}F_{\mu\nu a}, \tag{13.1.48}$$

where the second step relies on (13.1.21), assuming that the infinitesimal generators are taken in the fundamental representation, and where $\bar{g}$ is a coupling constant. Later we shall rescale the fields according to $A_\mu \to \bar{g}A_\mu$, which eliminates the factor $1/\bar{g}^2$ from the action. The expression (13.1.48) leads to interactions involving three and four gauge particles, since $F^{\mu\nu}$ is quadratic in A^μ. This is the property that makes non-Abelian gauge fields wholly nontrivial even in the absence of any other particles, by contrast to the electromagnetic field which without electrons is a free field.

To determine the equations of motion one couples A^μ to a current j_a^μ belonging to the adjoint representation of the gauge group,

$$S = -\int d^4x\left[\frac{1}{2\bar{g}^2}\mathrm{Tr}(\mathscr{F}^{\mu\nu}\mathscr{F}_{\mu\nu}) + \frac{1}{\bar{g}}j_a^\mu A_{\mu a}\right]. \tag{13.1.49}$$

Before generalizing Maxwell's equations, we note the identity

$$[D_\mu, D_\nu] = i\mathscr{F}_{\mu\nu}, \tag{13.1.50}$$

whence in particular

$$D_\mu D_\nu F^{\mu\nu}(=(D_\mu D_\nu)_{ab}F_b^{\mu\nu}) = 0. \tag{13.1.51}$$

The variational principle applied to the action S, equation (13.1.49), yields the equations of motion

$$D_\mu F^{\mu\nu} = \bar{g}j^\nu \tag{13.1.52}$$

(Problem 13.1); and, in view of (13.1.51), the equation of continuity for j^ν reads

$$D_\nu j^\nu = 0. \tag{13.1.53}$$

Notice that this equation features the covariant derivative, and thereby the field A_μ. This should have been foreseen, because the field A_μ carries the quantum numbers related to the internal symmetry, thus featuring as its own source. Similarly (Problem 13.2) one can derive analogues of the Bianchi identities of differential geometry,

$$D_\sigma F_{\mu\nu} + D_\mu F_{\nu\sigma} + D_\nu F_{\sigma\mu} = 0. \tag{13.1.54}$$

Finally (Problem 13.3) we note that the condition $\mathscr{F}_{\mu\nu} = 0$ in the vicinity of some point implies that, in this vicinity, $\mathscr{A}_\mu$ takes the form

$$\mathscr{A}_\mu(x) = i[\partial_\mu U(g)]U^{-1}(g); \tag{13.1.55}$$

such an $\mathscr{A}_\mu$ is called a 'pure gauge'.

We end this section by summarizing the essential differences between Abelian and non-Abelian gauge fields at the classical level. In the non-Abelian case

(i) the field strength tensor $F^{\mu\nu}$ is not gauge invariant;

(ii) it is impossible even at the classical level to formulate the theory without appeal to the potentials (see (13.1.52–54));

(iii) finally, non-Abelian gauge fields have nontrivial topological properties. In the Abelian case, $F^{\mu\nu} = 0$ is equivalent to $A^\mu = \partial^\mu \Lambda$, and in the absence of singularities one can pass continuously from this value to $A^\mu = 0$. In the non-Abelian case one cannot. If for instance one chooses the gauge $A^0 = 0$, then the pure gauge field

$$\boldsymbol{A} = i(\boldsymbol{\nabla} U(g))U^{-1}(g), \qquad \lim_{\|\boldsymbol{x}\| \to \infty} U(g(\boldsymbol{x})) = \mathbb{1}$$

is characterized by an integer stemming from the topology. The properties of the classical solutions of the Yang–Mills fields are fascinating, but beyond the scope of the present discussion.

13.2 The quantization of non-Abelian gauge theories

13.2.1 The generating functional

We proceed to adapt the method of Section 11.4 to a more complex situation, following the same heuristic approach as before. The object is to integrate only over equivalence classes of configurations $[A_\mu]$, instead of integrating over all configurations. Two configurations belong to the same equivalence class if they are related by a gauge transformation (13.1.36). One must fix the gauge through some condition like

$$f_a(A_b^\mu) = 0.$$

At a given point x, the gauge transformation (13.1.36) amounts to a unitary transformation followed by a change of origin; it leaves invariant the integration measure

$$\mathscr{D}A_\mu \left(= \prod_{x,a,\mu} \mathrm{d}A_a^\mu(x) \right) = \mathscr{D}A_\mu^g.$$

The action S (13.1.48) too is invariant under gauge transformations. Assuming that the equation $f_a(A_\mu^g) = 0$ defines g uniquely, we introduce a quantity $\Delta_f(A)$ by

$$\Delta_f^{-1}(A) = \int \prod_x \mathrm{d}g(x) \prod_{x,a} \delta(f_a(A^g(x))). \tag{13.2.1}$$

To forestall a proliferation of indices, we shall as a rule omit the Lorentz index, and sometimes the group label as well. It is easy to check that $\Delta_f(A)$ is gauge invariant, by appeal to the invariance of the Haar measure (Pontryagin 1966)

$$dg = d(g_0 g),$$

where g_0 is a fixed group element. (The Haar measure is the invariant measure for integrating over the group parameters; one can show that up to a multiplicative factor it is unique.) Now write the 'partition function'* Z in the form

$$Z = \int \prod_{x,a} dA_a(x) e^{iS[A]} \Delta_f(A) \prod_{x,a} \delta(f_a(A^g(x))) \prod_x dg(x)$$

$$= \int \prod_x dg(x) \int \prod_{x,a} dA_a^{g^{-1}}(x) e^{iS[A^{g^{-1}}]} \Delta_f(A^{g^{-1}}) \prod_{x,a} \delta(f_a(A(x))).$$

The second integral is independent of g, which allows one to factorize the product of the integrals over the group volume at each point. Dividing by this factor (infinite but just a constant) we obtain

$$Z = \int \mathscr{D}A e^{iS[A]} \Delta_f(A) \prod_{x,a} \delta(f_a(A(x))). \tag{13.2.2}$$

At this point we must evaluate $\Delta_f(A)$ from (13.2.1); by virtue of the delta function we need integrate only over configurations $[A]$ close to those that satisfy the gauge condition. Starting from a configuration $[A]$ satisfying $f_a(A) = 0$, we perform an infinitesimal gauge transformation (13.1.37). In order to simplify the discussion we consider only the Lorentz gauge $\partial_\mu A_a^\mu = 0$, though the argument is perfectly general (see also Problem 11.14):

$$\partial_\mu A_a^{\mu g} = \partial_\mu A_a^\mu + \partial_\mu[\partial^\mu \Lambda_a + f_{abc}\Lambda_b A_c^\mu]$$

$$= \Box \Lambda_a + f_{abc} A_c^\mu \partial^\mu \Lambda_b.$$

Let us recast this result into the form

$$\partial_\mu A_a^{\mu g} = \int d^4y [M_f(x, y)]_{ab} \Lambda_b(y),$$

where

$$[M_f(x, y)]_{ab} = (\Box \delta_{ab} + f_{abc} A_c^\mu \partial_\mu) \delta^{(4)}(x - y). \tag{13.2.3}$$

In this notation, and in view of

$$\prod_x dg(x) = \prod_{x,a} d\Lambda_a(x),$$

$\Delta_f(A)$ becomes†

$$\Delta_f(A) = \det M_f$$

* It would also be possible to consider the expectation values of gauge-invariant operators.

† For a general gauge condition $f(A) = 0$, it is easy to prove from (13.1.37b) that $M_f = \det(f^\mu D_\mu) = \det(D^\mu f_\mu)$, where $f^\mu = \delta f(x, A(x))/\delta A_\mu(x)$.

(see Problem 11.14(a)). To make the determinant manageable it must be transformed into a perturbation series, by rewriting it, in virtue of (11.2.13), as an integral over Grassmann variables $\eta(x)$ and $\bar{\eta}(x)$,

$$\det M_f = \int \mathscr{D}(\eta, \bar{\eta}) \exp\left(-\mathrm{i}\int \mathrm{d}^4x\,\mathrm{d}^4y\,\bar{\eta}_a(x)[M_f(x, y)]_{ab}\eta_b(y)\right).$$

The variables η and $\bar{\eta}$ must not be confused with the sources η and $\bar{\eta}$ from Section 11.2.2, even though the symbols are the same. This allows us to write the partition function, disregarding multiplicative constants, in the form

$$Z = \int \mathscr{D}(A, \eta, \bar{\eta}) \exp\left(\mathrm{i}\int \mathrm{d}^4x\,\mathscr{L}_{\mathrm{eff}}(x)\right) \prod_{x,a} \delta(\partial_\mu A_a^\mu(x)), \tag{13.2.4a}$$

where

$$\mathscr{L}_{\mathrm{eff}}(x) = -\frac{1}{4\bar{g}^2} F_a^{\mu\nu} F_{\mu\nu a} - \bar{\eta}_a(x)[\Box\,\delta_{ac} - f_{abc} A_b^\mu \partial_\mu]\eta_c(x). \tag{13.2.4b}$$

The Lagrangean (13.2.4b) features fictitious fermionic fields $\bar{\eta}_a(x)$ and $\eta_a(x)$, called Faddeev–Popov ghosts, which transform under the adjoint representation of the gauge group. Obviously these fields appear only as internal lines, and never as external particles in S-matrix elements. Note that these 'fermions' are particles of spin 0; if they did appear as external lines, they would violate the spin-statistics theorem. Since they are fermion fields, one must not forget the factor -1 whenever they form a closed loop in a diagram. It remains only to repeat, without any changes, the steps that led from (11.4.14) to (11.4.15). The Lagrangean $\mathscr{L}_{\mathrm{eff}}(x)$ that is eventually used in (13.2.4a) includes the Lagrangean $\mathscr{L}_{\mathrm{G}}$ for the gauge field, a term $\mathscr{L}_{\mathrm{GF}}$ fixing the gauge, and a term $\mathscr{L}_{\mathrm{FP}}$ corresponding to the Faddeev–Popov ghosts. One could also add an interaction with fermions, taken in the representation with infinitesimal generators T_a; unless we say otherwise, we shall restrict ourselves to cases where the T_a belong to the fundamental representations. Lastly it proves convenient to rescale the gauge fields,

$$A_a^\mu \to g A_a^\mu,$$

writing g (instead of $\bar{g}$) for the coupling constant. The (gauge-dependent) generating functional $Z(J, \rho, \bar{\rho})$ is obtained by coupling the gauge and the fermion fields to sources J, ρ, and $\bar{\rho}$. One can also define generating functionals for the Green's functions of the ghost fields.

It remains only to summarize our results (after integrating $\mathscr{L}_{\mathrm{FP}}$ by parts):

$$\mathscr{L} = \mathscr{L}_{\mathrm{G}} + \mathscr{L}_{\mathrm{GF}} + \mathscr{L}_{\mathrm{FP}} + \mathscr{L}_{\mathrm{F}}, \tag{13.2.5a}$$

$$\mathscr{L}_{\mathrm{G}} = -\frac{1}{4}(\partial^\mu A_a^\nu - \partial^\nu A_a^\mu - g f_{abc} A_b^\mu A_c^\nu)$$
$$\times(\partial_\mu A_{\nu a} - \partial_\nu A_{\mu a} - g f_{ade} A_{\mu d} A_{\nu e}), \tag{13.2.5b}$$

$$\mathscr{L}_{\mathrm{GF}} = -\frac{1}{2\alpha}(\partial_\mu A_a^\mu)^2, \tag{13.2.5c}$$

$$\begin{aligned}\mathscr{L}_{\mathrm{FP}} &= (\partial^\mu \bar{\eta}_a)(\partial_\mu \delta_{ac} - g f_{abc} A_{\mu b})\eta_c \\ &= (\partial^\mu \bar{\eta}_a)(D_\mu \eta)_a,\end{aligned} \tag{13.2.5d}$$

$$\mathscr{L}_{\mathrm{F}} = \bar{\psi}_i[\mathrm{i}\gamma_\mu(\partial^\mu \delta_{ij} + \mathrm{i}g A_a^\mu (T_a)_{ij}) - m\delta_{ij}]\psi_j. \tag{13.2.5e}$$

13.2.2 The Feynman rules

The Feynman rules are visible from (13.2.5) on inspection, once one remembers that a term $\partial_\mu \varphi(x)$ in the Lagrangean corresponds in Fourier space to a factor $-\mathrm{i}q_\mu$, where q_μ is the momentum entering the vertex. Start with the expressions for the propagators.

(i) Gauge field:

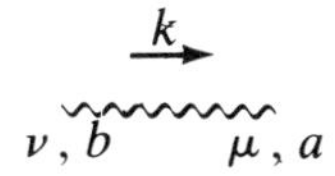

$$\frac{\mathrm{i}\delta_{ab}}{k^2 + \mathrm{i}\varepsilon}\left(-g_{\mu\nu} + (1-\alpha)\frac{k_\mu k_\nu}{k^2 + \mathrm{i}\varepsilon}\right). \tag{13.2.6a}$$

(ii) Faddeev–Popov ghost:

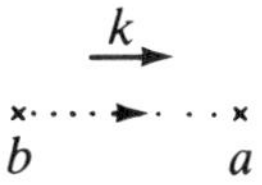

$$\frac{\mathrm{i}\delta_{ab}}{k^2 + \mathrm{i}\varepsilon}. \tag{13.2.6b}$$

(iii) Fermion:

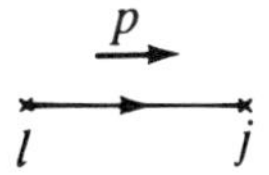

$$\frac{\mathrm{i}\delta_{jl}}{\not{p} - m + \mathrm{i}\varepsilon} = \frac{\mathrm{i}\delta_{jl}(\not{p} + m)}{p^2 - m^2 + \mathrm{i}\varepsilon}. \tag{13.2.6c}$$

We shall accept without proof the prescription $k^2 \to k^2 + \mathrm{i}\varepsilon$ for the propagator (13.2.6b); it can be shown to be the only prescription compatible with unitarity and causality. (The Faddeev–Popov ghosts are essential to guaranteeing unitarity through the Cutkosky rules.)

Next we study the vertex with three gauge particles. One term of the Lagrangean (13.2.5b) contributing to this vertex reads

$$\frac{g}{4} f_{abc} \partial^\nu A_a^\mu A_{\nu b} g_{\mu\rho} A_c^\rho.$$

On application of Wick's theorem this term yields a contribution of the type

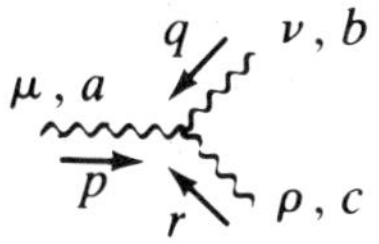

$$\frac{g}{4} f_{abc} p^\nu g_{\mu\rho}.$$

Actually there are 3! = 6 contributions to Wick's theorem, and four terms in the Lagrangean (13.2.5b). The full vertex reads

$$-gf_{abc}[g_{\mu\nu}(p-q)_\rho + g_{\nu\rho}(q-r)_\mu + g_{\rho\mu}(r-p)_\nu]. \tag{13.2.7a}$$

Similarly one finds the expression for the vertex with four gauge particles,

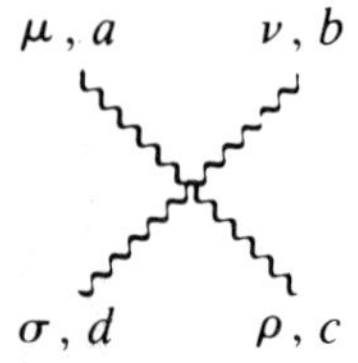

$$\begin{aligned}&-ig^2[f_{eab}f_{ecd}(g_{\mu\rho}\,g_{\nu\sigma}-g_{\mu\sigma}\,g_{\nu\rho})\\&+f_{eac}f_{edb}(g_{\mu\sigma}\,g_{\rho\nu}-g_{\mu\nu}\,g_{\rho\sigma})\\&+f_{ead}f_{ebc}(g_{\mu\nu}\,g_{\sigma\rho}-g_{\mu\rho}\,g_{\sigma\nu})].\end{aligned} \tag{13.2.7b}$$

The gluon–ghost coupling is given by the Lagrangean (13.2.5d). Let us evaluate the vertex

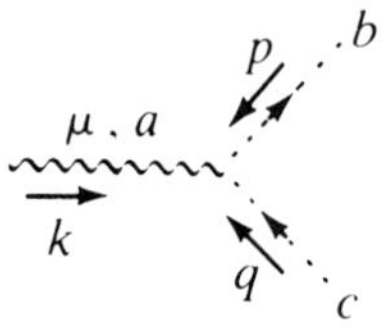

;

the contribution to it from (13.2.5d) reads

$$g\,f_{abc}(\partial_\mu \bar{\eta}_b)\,A_a^\mu \eta_c,$$

and yields

$$gf_{abc}p^\mu. \tag{13.2.7c}$$

This vertex can be symmetrized only in the Landau gauge ($\alpha = 0$). Finally, the gluon–fermion vertex is determined by (13.2.5e),

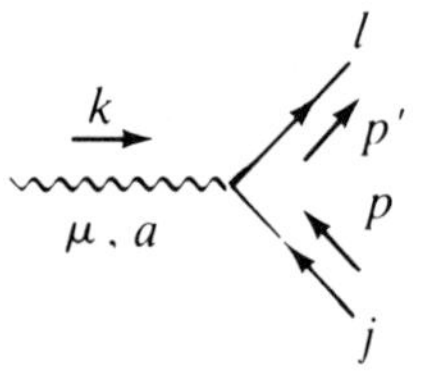

$$-ig\gamma^\mu(T_a)_{lj}. \tag{13.2.7d}$$

Obviously these rules are more complicated than the rules for quantum electrodynamics. When evaluating a Feynman graph, it is recommended that in general one start by calculating the 'group-theory factor' stemming from the f_{abc} and from the $(T_a)_{ij}$. The general rules for calculating this factor are given by Cvitanovic (1976). Here we merely quote the group-theory factors appropriate to the simplest one-loop calculations. One uses the following identities, applicable to a given representation R (Problem 13.4):

$$\mathrm{Tr}(T_a T_b) = T(R)\delta_{ab}, \tag{13.2.8a}$$

$$\sum_a (T_a)^2 = C(R)\mathbb{1}. \tag{13.2.8b}$$

Let r be the dimension of the group, and n the dimension of the representation for the fermions (for $SU(N)$ this is generally N); from (13.2.8) one then derives the relation

$$nC(R) = rT(R). \tag{13.2.9}$$

As a rule we use the symbol C_F for the value of $C(R)$ for the fermion representation; for the fundamental representation of $SU(N)$ one finds

$$C(R) = C_F = \frac{N^2 - 1}{2N} \qquad (SU(N)). \tag{13.2.10}$$

For the adjoint representation one has $n = r$ and $C(R) = T(R) = C_A$ ($C_A = N$ for $SU(N)$). Consider some important one-loop diagrams.

(i) The self-energy of a gauge particle:

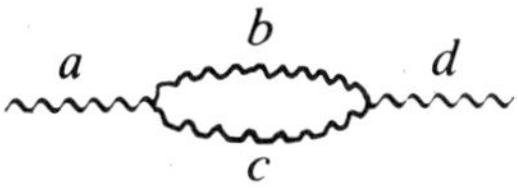

$$f_{abc} f_{dbc} = (T_a)_{bc}(T_d)_{cb} = C_A \delta_{ad}$$
$$= N\delta_{ad} \qquad (SU(N)). \tag{13.2.11a}$$

(ii) The self-energy of a fermion belonging to the fundamental representation of $SU(N)$:

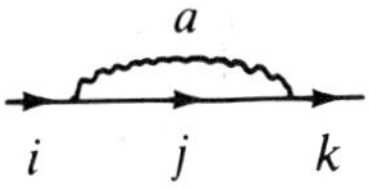

$$: (T_a)_{kj}(T_a)_{ji} = C_F \delta_{ik}$$
$$= \frac{N^2 - 1}{2N}\delta_{ik} \qquad (SU(N)). \tag{13.2.11b}$$

(iii) A fermion loop contributing to the self-energy of the gauge particles (see (13.1.21)):

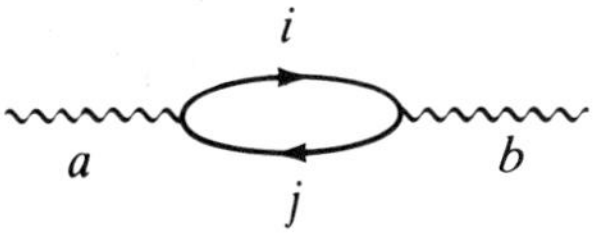

$$(T_b)_{ij}(T_a)_{ji} = T_F \delta_{ab}$$
$$= \frac{1}{2}\delta_{ab}. \tag{13.2.11c}$$

For the group SU(2) the reader can check these results straightforwardly. Other examples are set in Problem 13.4.

From this point one proceeds as in any field theory, integrating over closed loops, and so on. One must be careful not to forget the symmetry factors,

 : symmetry factor $\frac{1}{2}$,

nor the factors (-1) associated with the Fadeev–Popov loops,

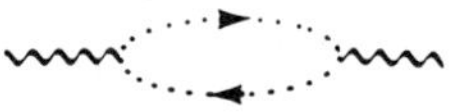 : factor (-1).

13.2.3 Renormalization and the Ward identities

The Feynman rules formulated above naturally lead to divergent integrals, and it becomes necessary to renormalize. One good exercise (though it demands some perseverance) is to calculate all the divergent diagrams in one-loop order, and from them to derive the renormalization constants. In order to maintain gauge invariance throughout, it is recommended that one regularize dimensionally, and that the renormalization constants be determined in the minimal scheme. Readers with the courage to embark on this calculation will find some further hints in Section 13.4.

The primitively divergent diagrams are identified by power-counting. Since one is dealing with Green's functions it is quite legitimate for ghosts to appear as external particles, seeing that it will for instance be necessary to renormalize their propagators. On the other hand, in view of the transverse nature of the gauge-field propagator, and of the fact that p_μ factors from the self-energy of the ghosts, there are neither mass counterterms, nor counterterms with $(\partial_\mu A^\mu)^2$. By the light of these remarks it is easy to list the diagrams we need to evaluate (the renormalization constants are given in brackets):

(a) gauge-particle self-energy:  (Z_3);

(b) ghost self-energy:  $(\tilde{Z}_3)$;

(c) fermion self-energy:  (Z_2);

(d) vertex with three gauge particles:  (Z_1);

(e) gauge-particle-and-ghost vertex:  $(\tilde{Z}_1)$;

(f) gauge-particle-and-fermion vertex:  (Z_{1F});

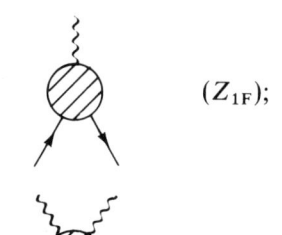

(g) four-gauge-particle vertex:  (Z_4).

Proceeding as in Chapter 6 we add the counterterms to the Lagrangean $\mathscr{L}_{\text{eff}}$, equation (13.2.5), and find

$$\begin{aligned}\mathscr{L} + \delta\mathscr{L} = &-\frac{1}{4}Z_3(\partial^\mu A^\nu - \partial^\nu A^\mu)(\partial_\mu A_\nu - \partial_\nu A_\mu)\\ &+\frac{g}{2}Z_1(\partial^\mu A^\nu - \partial^\nu A^\mu)A_\mu A_\nu - \frac{1}{4}g^2 Z_4 A_\mu A_\nu A^\mu A^\nu\\ &-\frac{1}{2\alpha}(\partial_\mu A^\mu)^2 + \tilde{Z}_3(\partial_\mu \bar{\eta})(\partial^\mu \eta) - g\tilde{Z}_1(\partial^\mu\bar{\eta})A_\mu \eta\\ &+Z_2\bar{\psi}\mathrm{i}\gamma_\mu\partial^\mu\psi - Z_{1F}g\bar{\psi}\gamma_\mu\psi A^\mu. \end{aligned} \tag{13.2.12}$$

(The group indices and group-theory factors are not written explicitly, and the fermion mass has been taken as zero.) The bare fields and the bare parameters are defined by

$$\begin{aligned} &A_0 = Z_3^{1/2}A; \quad \eta_0(\bar{\eta}_0) = \tilde{Z}_3^{1/2}\eta(\bar{\eta}); \quad \psi_0 = Z_2^{1/2}\psi;\\ &g_0 = Z_1 Z_3^{-3/2}g; \quad \alpha_0 = Z_3\alpha; \qquad g_{0F} = Z_{1F}Z_3^{-1/2}Z_2^{-1}g. \end{aligned} \tag{13.2.13}$$

The Lagrangean $\mathscr{L} + \delta\mathscr{L}$ may be considered as our initial Lagrangean $\mathscr{L}(A_0, \eta_0, \psi_0, \alpha_0, g_0, g_{0F} = g_0)$, provided the identities

$$\frac{Z_4}{Z_1} = \frac{Z_1}{Z_3} = \frac{\tilde{Z}_1}{\tilde{Z}_3} = \frac{Z_{1F}}{Z_2} \tag{13.2.14}$$

are satisfied. They generalize the relation $Z_1 = Z_2$ in quantum electrodynamics, and are satisfied explicitly by calculations to one-loop order. They can be proved more generally (as they can in electrodynamics) by appeal to generalizations of the Ward identities to non-Abelian theories. These generalizations are called the Slavnov–Taylor identities; unfortunately they are far more complicated than their Abelian counterparts. The most elegant and compact method for deriving them uses the Becchi–Rouet–Stora (BRS) transformation, described for example by Itzykson and Zuber (1980) (see also Problem 13.5). Here we shall merely indicate where the difficulty resides.

Recall that in order to quantize the gauge field we had to 'fix the gauge' by imposing some condition $f(A) = 0$. This introduces into the effective Lagrangean two terms which are not gauge invariant, namely

$$\mathscr{L}_{\text{GF}} = -\frac{1}{2a}(f(A))^2 \quad \text{and} \quad \mathscr{L}_{\text{FP}}.$$

Suppose now that one tries to repeat the reasoning of Section 12.4.1, adding a field φ_a belonging to the adjoint representation of the gauge group, and serving to induce the infinitesimal ($\omega \ll 1$) gauge transformation

$$\delta A_a^\mu = \omega f_{abc} \varphi_b A_c^\mu + \omega(\partial_\mu \varphi_a) \tag{13.2.15}$$

that generalizes (12.4.5). Then the terms $\mathscr{L}_{\text{GF}}$ and $\mathscr{L}_{\text{FP}}$ will engender interactions of the type φAA, $\varphi A\eta$, $\varphi\bar{\eta}A\eta$ in addition to the term (12.4.7), which make it much more complicated to write down identities like (12.4.9). In electrodynamics, the simplicity of the Lorentz gauge $\partial^\mu A_\mu = 0$ stems from the fact that the Faddeev–Popov ghosts, though present in principle, are decoupled, and can be disregarded. But the situation in the gauge $\partial_\mu A^\mu + \frac{1}{2}g A_\mu A^\mu = 0$ of Problem 11.4 is quite different; in this gauge one needs identities of the Slavnov–Taylor type (Problem 13.5). What one should remember from this rather sketchy discussion are the interrelations between the gauge condition, the Lagrangean for the Faddeev–Popov ghosts, and the Ward–Slavnov–Taylor identities.

The role of the Faddeev–Popov 'particles' in ensuring unitarity can be seen from a simple calculation. Suppose we compute the contribution of two intermediate gauge particles to the reaction $f\bar{f} \to f\bar{f}$, where f and $\bar{f}$ denote fermions coupled to the gauge fields. One can check that the Cutkosky rules yield contributions that depend on unphysical polarization states of the gauge particles. These unwanted contributions are however cancelled by contributions from the intermediate state with two ghosts. The cancellation is a consequence of the Ward identity for the 4-point Green's function.

13.3 The Glashow–Salam–Weinberg model of electroweak interactions

In an Abelian gauge theory with coupling to a conserved current one can introduce a mass term for the gauge boson without impairing renormalizability. What happens is that the dangerous component of the propagator with mass, namely

$$k_\mu k_\nu / m^2,$$

which introduces a 'coupling constant' m^{-2} having dimension -2, is controlled by current conservation, at least as far physical quantities are concerned. But in non-Abelian theories the situation is quite different, and introducing a mass term by brute force destroys their renormalizability. If one wishes to make the gauge particles massive, as one must in order to cater for short-range forces like the weak interactions, then one must do so by a construction involving a (spontaneously) broken symmetry.

13.3.1 Goldstone bosons and the Higgs phenomenon

When a continuous symmetry is broken spontaneously, there appear zero-mass particles called *Goldstone bosons.* A very simple model for explaining this phenomenon is supplied by Ginzburg–Landau theory with an order parameter of dimension $n = 2$. We shall write down the effective action $\Gamma(\varphi)$ for a Euclidean theory, with $\mu = 1, 2, 3, 4$, and $(\partial_\mu \varphi)^2 = \sum_{\mu=1}^{4} (\partial_\mu \varphi)^2$. The position of the indices ($A_\mu A_\mu$ instead of $A_\mu A^\mu$) will distinguish between the theories in Euclidean and in Minkowski space. In the tree approximation the effective action reads

$$\Gamma(\varphi) = \int \mathrm{d}^4 x \left[\frac{1}{2}(\partial_\mu \varphi_1)^2 + \frac{1}{2}(\partial_\mu \varphi_2)^2 - \frac{1}{2}\mu^2(\varphi_1^2 + \varphi_2^2) + \frac{\lambda}{4}(\varphi_1^2 + \varphi_2^2)^2 \right]. \tag{13.3.1}$$

The coefficient of $(\varphi_1^2 + \varphi_2^2)$ is negative, corresponding to temperatures below the critical, $T < T_c$. The coupling constant λ must be positive. The effective potential $V(\varphi_1, \varphi_2)$, which we may write as

$$V(\varphi_1, \varphi_2) = \frac{\lambda}{4}\left[(\varphi_1^2 + \varphi_2^2) - \frac{\mu^2}{\lambda} \right]^2 + \text{constant}, \tag{13.3.2}$$

has a minimum at

$$\varphi_1^2 + \varphi_2^2 = \frac{\mu^2}{\lambda}.$$

Accordingly, in the (φ_1, φ_2) plane this minimum must lie on a circle (Fig. 13.3). The general form of the potential is sketched in Fig. 13.4. Conformably with the usual mechanism of spontaneous symmetry breaking, we choose a definite ground state (i.e. a field-theory vacuum); in the language of magnetism, we choose a specific direction for the magnetization. One possible choice is

$$\varphi_1 = v = \sqrt{\frac{\mu^2}{\lambda}}, \quad \varphi_2 = 0, \tag{13.3.3}$$

where v is the 'vacuum expectation-value' of the field φ_1.

The fluctuations of (φ_1, φ_2) around the ground state (13.3.3) are described by φ_2 and by a field η (the notation η is standard; of course the field η should not be confused with an F–P ghost),

$$\eta = \varphi_1 - v. \tag{13.3.4}$$

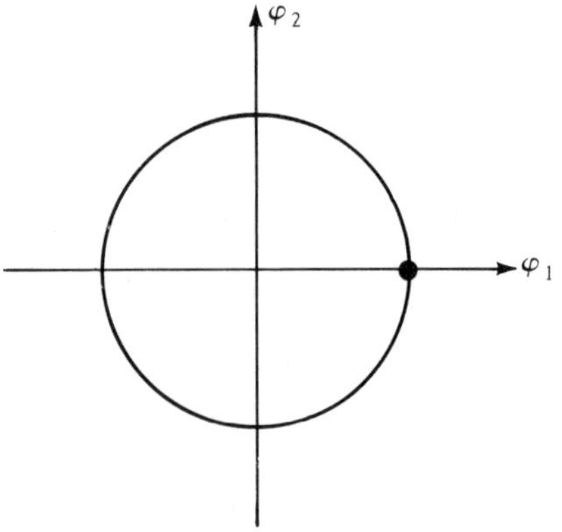

Fig. 13.3

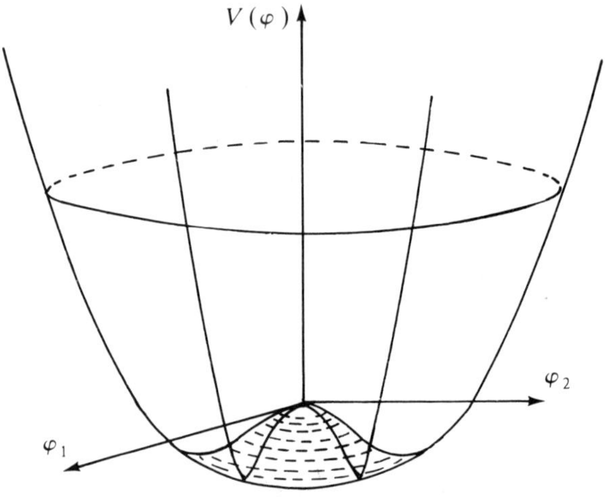

Fig. 13.4

Substituting for φ_1 in the effective potential (13.3.2) we find

$$V(\eta, \varphi_2) = \mu^2\eta^2 + \lambda v\eta(\eta^2 + \varphi_2^2) + \frac{\lambda}{4}(\eta^2 + \varphi_2^2)^2. \tag{13.3.5}$$

This form of the effective potential shows that we are in the presence of a massive particle (η) with mass $\sqrt{2}\mu$, and of a zero-mass particle (φ_2) which is the Goldstone boson. These particles interact amongst themselves through cubic and quartic couplings. The existence of Goldstone bosons is easily interpreted: fluctuations around the ground state in a direction normal to the φ_1 axis cost no energy, while fluctuations parallel to the φ_1 axis experience a harmonic potential.

It is easy to prove the following result (Problem 13.6). Let H be that subgroup of the symmetry group G of the effective action which leaves the ground state invariant. If G has N independent generators, while H has M, then there exist $(N–M)$ Goldstone bosons. The example above $(G = U(1))$ has $N = 1$ and $M = 0$.

The introduction of a gauge field coupled to the field φ entails the disappearance of the Goldstone boson, which resurfaces as the longitudinal component of the massive gauge field. This is the Higgs phenomenon. Let us revert to the effective action (13.3.1), and add to it an (Abelian) gauge field A_μ. It proves convenient to introduce the charged fields φ and φ^*,

$$\varphi = \frac{1}{\sqrt{2}}(\varphi_1 + \mathrm{i}\varphi_2), \quad \varphi^* = \frac{1}{\sqrt{2}}(\varphi_1 - \mathrm{i}\varphi_2), \tag{13.3.6}$$

and the covariant derivative

$$D_\mu \varphi = (\partial_\mu - \mathrm{i}qA_\mu)\varphi.$$

The effective action becomes*

$$\Gamma = \int \mathrm{d}^4x \left[\frac{1}{4} F_{\mu\nu}F_{\mu\nu} + (D_\mu\varphi)^*(D_\mu\varphi) + \lambda\left(\varphi^*\varphi - \frac{\mu^2}{2\lambda}\right)^2 \right]. \tag{13.3.7}$$

The ground state is again chosen as

$$\varphi_1 = v, \quad \varphi_2 = 0, \quad \varphi = v/\sqrt{2}.$$

Instead of the fields (φ_1, φ_2) we now use the real fields (ξ, η), writing

$$\varphi(x) = \mathrm{e}^{\mathrm{i}\xi(x)/v}\frac{v + \eta}{\sqrt{2}}.$$

Without the field A_μ, and restricted to terms quadratic in η, ξ, the effective action (13.3.7) would read

$$\Gamma(\varphi) = \int \mathrm{d}^4x \left[\frac{1}{2}(\partial_\mu\eta)^2 + \frac{1}{2}(\partial_\mu\xi)^2 + \frac{1}{2}(2\mu^2)\eta^2 + O(\eta(\partial_\mu\xi)^2, \eta^3, \ldots) \right].$$

We see that the field ξ has zero mass; it is just the Goldstone boson. On the other hand, operating as we are in the framework of a gauge-field theory, we are at

* In three dimensions, Γ is the free energy of the Ginzburg–Landau theory of superconductivity; μ^{-1} is the coherence length and $(qv)^{-1}$ the penetration depth of the magnetic field (see e.g. Fetter and Walecka 1971, p. 430).

liberty to perform gauge transformations, and choose the special transformation

$$\varphi(x) \to \varphi'(x) = \mathrm{e}^{-\mathrm{i}\xi(x)/v}\varphi(x), \tag{13.3.8a}$$

$$A_\mu(x) \to A'_\mu(x) = A_\mu - \frac{1}{qv}\partial_\mu\xi(x), \tag{13.3.8b}$$

which leaves the effective action (13.3.7) invariant. Since

$$\varphi'(x) = \frac{v+\eta}{\sqrt{2}},$$

we can substitute this into (13.3.7), writing (A_μ, φ) instead of (A'_μ, φ'):

$$\begin{aligned}\Gamma = \int \mathrm{d}^4x \Bigg[& \frac{1}{4}F_{\mu\nu}F_{\mu\nu} + \frac{1}{2}(\partial_\mu\eta + \mathrm{i}qA_\mu(v+\eta)) \\ & \times(\partial_\mu\eta - \mathrm{i}qA_\mu(v+\eta)) + \frac{\lambda}{4}\left[(v+\eta)^2 - \frac{\mu^2}{\lambda}\right]^2\Bigg] \\ = \int \mathrm{d}^4x \Bigg[& \frac{1}{4}F_{\mu\nu}F_{\mu\nu} + \frac{1}{2}(\partial_\mu\eta)^2 \\ & + \frac{1}{2}q^2v^2A_\mu A_\mu + q^2\eta vA_\mu A_\mu \\ & + \frac{1}{2}q^2\eta^2A_\mu A_\mu + \mu^2\eta^2 + \lambda v\eta^3 + \frac{\lambda}{4}\eta^4\Bigg].\end{aligned} \tag{13.3.9}$$

The physical content is apparent on inspection; one has

- a massive vector field with squared mass $m^2 = q^2v^2 = q^2\mu^2/\lambda$;
- a scalar field η with squared mass $2\mu^2$.

These two fields interact amongst themselves through cubic and quartic couplings. The gauge transformation has eliminated the zero-mass field ξ, which reappears, in effect, as the longitudinal component of a massive vector field A_μ. The number of degrees of freedom is of course conserved. At the start we had two degrees of freedom for the gauge field and two for the scalar fields; in the end we have three degrees of freedom for the massive vector field, plus one for the scalar field.

What we have just described is an illustration of the Higgs phenomenon: a zero-mass gauge field acquires mass by propagation through the vacuum of the scalar fields, which are accordingly called 'Higgs fields'. The same phenomenon occurs in non-Abelian gauge theories. By means of the Higgs phenomenon, gauge bosons can acquire mass while still preserving renormalizability; indeed, since the couplings of the Higgs fields are all renormalizable, the initial Lagrangean itself is likewise renormalizable, and one can show that this renormalizability survives the symmetry breaking. The gauge transformation (13.3.8) delivers a theory that is manifestly unitary, because on cutting the propagator of a massive vector boson one encounters only physical states.

Hence we conclude (and a full analysis of the problem confirms) that one can in this way construct a unitary and renormalizable theory of massive bosons within the framework of a non-Abelian gauge theory (which cannot in fact be done in any other way!).

13.3.2 The Fermi theory of weak interactions

What is known as the 'Fermi theory' is actually a modification, in the late 1950s, of the theory proposed by Fermi twenty years before. Its characteristics are that

(i) it is a current–current theory;

(ii) it leads to a non-renormalizable effective four-fermion interaction;

(iii) it exhibits maximal parity violation.

Let us consider these three characteristics briefly, referring for supporting detail to the book by Gasiorowicz (1966). We restrict ourselves to the weak interactions of the leptons $e^{\pm}$, ν_e, $\bar{\nu}_e$ (electron neutrino and antineutrino), $\mu^{\pm}$, ν_μ, $\bar{\nu}_\mu$ (muon neutrino and antineutrino). Recall that neutrinos are electrically neutral (whence their name), have spin 1/2, and a mass-value compatible with zero. In terms of the weak current $J_\lambda(x)$ and of a coupling constant G_{F} (the Fermi constant), the Fermi interaction reads

$$\mathscr{L}(x) = -\frac{G_{\mathrm{F}}}{\sqrt{2}} J_\lambda(x) J^{\lambda\dagger}(x), \tag{13.3.10}$$

where J_λ contains an electronic part $J_\lambda^{(e)}$, plus a muonic part $J_\lambda^{(\mu)}$,

$$J_\lambda(x) = J_\lambda^{(e)}(x) + J_\lambda^{(\mu)}(x), \tag{13.3.11}$$

with

$$J_\lambda^{(e)}(x) = \bar{\psi}_e(x)\gamma_\lambda(1-\gamma_5)\psi_{\nu_e}(x) \tag{13.3.12a}$$

and

$$J_\lambda^{(\mu)}(x) = \bar{\psi}_\mu(x)\gamma_\lambda(1-\gamma_5)\psi_{\nu_\mu}(x). \tag{13.3.12b}$$

(J_λ also contains parts associated, respectively, with the lepton τ, and with hadrons, which will not be considered here.) In (13.3.12), ψ_e, ψ_{ν_e}, ψ_μ, and ψ_{ν_μ} are Dirac fields associated with the four types of particles. For instance, the Lagrangean (13.3.12) describes the decay

$$\mu^- \to e^- + \bar{\nu}_e + \nu_\mu$$

of the muon (Problem 13.7), and the elastic scattering of $\nu_e e^-$, $\bar{\nu}_e e^-$ (Problem 13.8).

Since Fermi fields have dimension 3/2, the Fermi constant G_F has dimension -2. Its value is

$$G_F \approx 1.16 \times 10^{-5}\ (\text{GeV})^{-2}.$$

The Fermi theory is non-renormalizable, because it has no special symmetry capable to reducing the degree of divergence.

Finally, the factor $(1-\gamma_5)$ in (13.3.12) entails maximal parity violation; neutrinos, if their mass is zero, always have helicity -1 (i.e. left-circular polarization), while antineutrinos always have helicity $+1$ (i.e. right-circular polarization). It proves convenient to introduce projection operators onto the states with helicities ± 1 respectively, namely

$$P_L = \frac{1}{2}(1-\gamma_5), \quad P_R = \frac{1}{2}(1+\gamma_5). \tag{13.3.13}$$

(For zero mass, the *chirality*, defined as the eigenvalue of γ_5, coincides with the helicity.) Finally (though purely formally for the moment) we can assign the neutrino and the left-handed electron to a doublet (i.e. to a 2-dimensional representation) of a group SU(2), called $SU(2)_L$, and we can proceed similarly with the muon and its neutrino:

$$\chi_L(e) = \begin{pmatrix} \nu_e \\ e^- \end{pmatrix}_L, \quad \chi_L(\mu) = \begin{pmatrix} \nu_\mu \\ \mu^- \end{pmatrix}_L, \quad e_L = \frac{1}{2}(1-\gamma_5)\psi_e. \tag{13.3.14}$$

The current $J_\lambda^{(e)}$ for example is then written in the form

$$J_\lambda^{(e)} = 2\bar{\chi}_L(e)\gamma_\lambda \tau^- \chi_L(e), \tag{13.3.15}$$

where

$$\tau^- = \frac{1}{2}(\tau_1 - i\tau_2) = \begin{pmatrix} 0 & 0 \\ 1 & 0 \end{pmatrix}.$$

This notation suggests an internal symmetry described by a group $SU(2)_L$, called weak isospin (not to be confused with isospin in the strong interactions). However, this is not yet a true symmetry, because the weak neutral current does not correspond to the matrix τ_3; in other words the neutral current cannot be obtained simply by replacing τ^- in (13.3.15) with τ_3.

On the other hand, what we are trying to do is to replace the Fermi theory, severely tarnished through its non-renormalizability, by a theory of weak interactions mediated by a charged massive vector boson W (Fig. 13.5).

If one requires renormalizability, then these bosons must be the gauge particles of some non-Abelian gauge theory, made massive through the Higgs phenomenon. We cannot adopt $SU(2)_L$ as our gauge group, for the reasons explained above. Hence it is natural to introduce another neutral current into the theory, namely the electromagnetic current, a step that leads to the unification of the weak and the electromagnetic interactions. At the time (1967) when

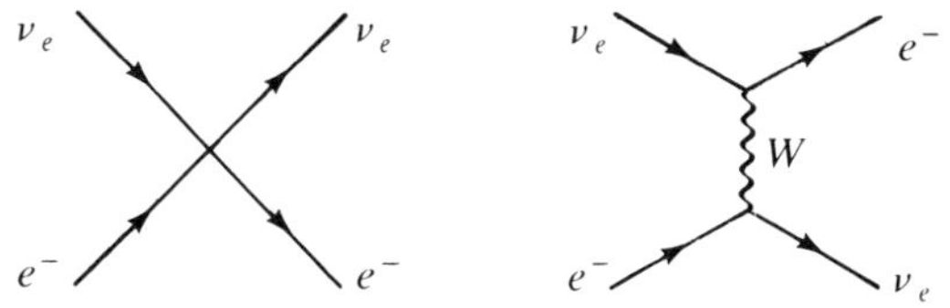

Fig. 13.5 $\nu_e e^-$ scattering, (a) in Fermi theory, (b) in a theory with exchange of a massive vector boson W

this model was being formulated, very little was known about neutral currents, and two possibilities were open.

(i) One could identify the neutral current with the electromagnetic current, and settle for three gauge bosons, namely two massive bosons W^+ and W^- mediating the weak interactions, plus one photon (γ). This is possible at the price of extra leptons. In this model, due to Georgi and Glashow, there is no weak neutral current.

(ii) One could introduce *two* neutral currents, one electromagnetic and one weak. Then one needs four gauge bosons, $W^\pm$, Z^0, and γ, and four infinitesimal generators for the gauge group. This is the GSW model.

Mathematically, the Georgi–Glashow model is more elegant than the GSW model, because it relies on a simple Lie group (SO(3)), which has the advantage that it explains the quantization of electric charge. It is not however the model chosen by Nature.

13.3.3 The Glashow–Salam–Weinberg (GSW) model

We proceed to some technicalities of the GSW model, which at the time of writing has wide experimental support. The gauge group G is the product of an SU(2) group and of a group $U(1)$, which indeed produce four generators:

$$G = \mathrm{SU}(2)_\mathrm{L} \times U(1)_Y.$$

The index L refers to the weak isospin of the left-handed leptons, and the index Y to the weak hypercharge defined below. One needs two coupling constants, g for $\mathrm{SU}(2)_\mathrm{L}$ and g' for $U(1)_Y$.

The representations of $\mathrm{SU}(2)_\mathrm{L}$ are labelled by the weak isospin I, which can assume the values 0, 1/2, 1, 3/2, . . . (though only the first three occur in practice). The elements of a given representation are labeled by the third component of the isospin, $I_3 = -I, -I+1, \ldots I$. The coupling constant of a particle to a gauge boson of $U(1)_Y$ is $g'\,Y/2$, where the factor 1/2 is conventional. Since this group is Abelian, the value of Y is arbitrary, and must be determined a posteriori. The left-handed leptons are arranged into two $\mathrm{SU}(2)_\mathrm{L}$ doublets (i.e. into two representations with isospin $I = 1/2$),

$$\chi_\mathrm{L}(e) = \begin{pmatrix} \nu_e \\ e^- \end{pmatrix}_\mathrm{L}, \quad \chi_\mathrm{L}(\mu) = \begin{pmatrix} \nu_\mu \\ \mu^- \end{pmatrix}_\mathrm{L}; \tag{13.3.16}$$

their hypercharge is Y_L; the neutrino has $I_3 = 1/2$, and the electron (or muon) has $I_3 = -1/2$. The right-handed leptons have no weak interactions described by charged currents; they are arranged into $\mathrm{SU}(2)_\mathrm{L}$ singlets ($I = 0$), and are assigned hypercharge Y_R. Note that $\chi_\mathrm{R}(e) = e_\mathrm{R}^-$, $\chi_\mathrm{R}(\mu) = \mu_\mathrm{R}^-$.

We shall denote the gauge fields of $\mathrm{SU}(2)_\mathrm{L}$ by W_a^μ ($a = 1, 2, 3$), and that of $U(1)_Y$ by B^μ. At the outset all these fields have zero mass. The same applies to the leptons, because in the absence of symmetry-breaking the masses of electron and neutrino must be the same. This is the case we start with, relegating the study of symmetry-breaking to a later stage.

Let us write down the Lagrangean $\mathscr{L}_l$ of the fermions; it is obtained from the Dirac Lagrangean through minimal coupling, equation (13.1.38). In the case of SU(2) it proves convenient to use a vector notation; W_a^μ can be considered as the a-component of a vector $\vec{W}^\mu$ in a three-dimensional space, because it transforms under the 3-dimensional representation of SU(2),

$$\mathscr{L}_l = \bar{\chi}_\mathrm{R} \mathrm{i}\gamma^\mu \left(\partial_\mu + \frac{\mathrm{i}}{2} g' Y_\mathrm{R} B_\mu \right) \chi_\mathrm{R}$$

$$+ \bar{\chi}_\mathrm{L} \mathrm{i}\gamma^\mu \left(\partial_\mu + \frac{\mathrm{i}}{2} g' Y_\mathrm{L} B_\mu + \frac{\mathrm{i}}{2} g \vec{\tau} \cdot \vec{W}_\mu \right) \chi_\mathrm{L}. \tag{13.3.17}$$

Next we couple the fermions to the neutral gauge fields B^μ and W_3^μ, dropping multiplicative factors irrelevant to the subsequent argument; thus the interaction reads

$$g' Y_\mathrm{R} j_\mu^\mathrm{R} B^\mu + g' Y_\mathrm{L} j_\mu^\mathrm{L} B^\mu + g j_\mu^\mathrm{L} \tau_3 W_3^\mu,$$

where j_μ^R and j_μ^L are proportional to $(1 - \gamma_5)$ and to $(1 + \gamma_5)$ respectively. For neutrinos ($Y_\mathrm{R} = 0$, $I_3 = +\frac{1}{2}$) and for electrons ($I_3 = -\frac{1}{2}$), this yields

$$\nu: g' Y_\mathrm{L} j_\mu^\mathrm{L} B^\mu + g j_\mu^\mathrm{L} W_3^\mu, \tag{13.3.18a}$$

$$e: g' [Y_\mathrm{R} j_\mu^\mathrm{R} + Y_\mathrm{L} j_\mu^\mathrm{L}] B^\mu - g j_\mu^\mathrm{L} W_3^\mu. \tag{13.3.18b}$$

Here one subjects the fields B^μ and W_3^μ to a canonical transformation, introducing two fields A^μ and Z^μ such that A^μ is not coupled to neutrinos, while its coupling to electrons conserves parity; therefore A^μ has all the attributes of the electromagnetic field. Accordingly we set

$$B^\mu = -\sin\theta\, Z^\mu + \cos\theta\, A^\mu, \tag{13.3.19a}$$

$$W_3^\mu = \cos\theta\, Z^\mu + \sin\theta\, A^\mu. \tag{13.3.19b}$$

The angle θ is called the *Weinberg angle*; it is often written θ_W. Substituting from (13.3.19) into (13.3.18) one obtains the couplings of the field A^μ,

$$\nu: [g' Y_L j_\mu^L \cos\theta + g j_\mu^L \sin\theta] A^\mu,$$

$$e: [g' Y_R j_\mu^R \cos\theta + (g' Y_L \cos\theta - g \sin\theta) j_\mu^L] A^\mu.$$

If A^μ is to have the properties appropriate to the electromagnetic field, then we must require that the neutrino charge be zero,

$$g' Y_L \cos\theta + g \sin\theta = 0,$$

and that the coupling of the photon be proportional to γ^μ, i.e. to $j_\mu^L + j_\mu^R$:

$$g' Y_R \cos\theta = g' Y_L \cos\theta - g \sin\theta.$$

For Y_L and Y_R these two equations yield

$$Y_L = -\frac{g}{g'} \tan\theta, \quad Y_R = 2Y_L.$$

The normalization of g' can always be chosen so that $Y_L = -1$ (whence $Y_R = -2$); then g and g' are related by

$$g' = g \tan\theta. \tag{13.3.20}$$

In order to relate g to the electric charge e, we write down the interaction of A_μ with electrons; restoring all multiplicative factors, we have

$$\bar{e}_L \gamma_\mu e_L = \frac{1}{2} \bar{\psi}(e) \gamma_\mu (1 - \gamma_5) \psi(e)$$

$$\bar{\psi}(e) \left\{ -\frac{1}{2} g' \gamma_\mu (1 + \gamma_5) \cos\theta - \frac{1}{4} (g' \cos\theta + g \sin\theta) \gamma_\mu \times (1 - \gamma_5) \right\} A^\mu \psi(e) = -g \sin\theta\, \bar{\psi}(e) \gamma_\mu \psi(e) A^\mu,$$

whence we identify

$$e = g \sin\theta. \tag{13.3.21}$$

It is now easy to complete the calculation and to determine the interactions of electrons and neutrinos with the Z^μ boson (Problem 13.9). For electrons for instance one finds

$$\mathscr{L} = \frac{-g}{\cos\theta} \bar{\psi}(e) \gamma_\mu (C_V - C_A \gamma_5) \psi(e) Z^\mu, \tag{13.3.22a}$$

where

$$C_V = -\frac{1}{4} + \sin^2\theta, \quad C_A = -\frac{1}{4}. \tag{13.3.22b}$$

At this stage our theory of weak interactions is seriously flawed, because one knows that these interactions, being short-ranged, must be mediated by massive vector bosons. Hence one must find some way of making the $W^{\pm}$ and Z_0 bosons massive, while preserving the zero mass of the photon. It will not surprise the reader that for generating these masses one appeals to the Higgs mechanism. In order to make the $W^{\pm}$ and the Z_0 massive, it is necessary, in view of the theorem from problem 13.6(b), that none of the generators of $SU(2)_L$ leave the vacuum invariant. By contrast, the linear combination

$$A^{\mu} = \sin\theta\, W_3^{\mu} + \cos\theta\, B^{\mu}$$

must preserve zero mass; since it is coupled to the electric charge Q, the vacuum must be invariant under Q. Notice that the values found above for Y_L and for Y_R entail

$$Q = I_3 + \tfrac{1}{2}Y. \tag{13.3.23}$$

The most economical solution is to introduce two doublets (h^+, h^0) and (h^0, h^-) of Higgs particles, described by complex fields $\varphi_c = (h^+, h^-)$ and $\varphi_0 = (h^0, \bar{h}^0)$ (four degrees of freedom in all),

$$\begin{pmatrix} h^+ \\ h^0 \end{pmatrix}: I = \frac{1}{2}, \quad Y = 1, \quad \begin{pmatrix} \bar{h}^0 \\ h^- \end{pmatrix}: \quad I = \frac{1}{2}, \quad Y = -1.$$

As regards the doublet (h^+, h^0), the operator Q is represented by the matrix

$$Q = e\begin{pmatrix} 1 & 0 \\ 0 & 0 \end{pmatrix}.$$

If one chooses to break the symmetry by assigning a nonzero vacuum expectation-value to φ_0 through

$$\varphi = \begin{pmatrix} \varphi_c \\ \varphi_0 \end{pmatrix}, \quad \langle\varphi\rangle = \frac{1}{\sqrt{2}}\begin{pmatrix} 0 \\ v \end{pmatrix}, \tag{13.3.24}$$

then one obtains a ground state that is invariant neither under $SU(2)_L$ nor under $U(1)_Y$, but is invariant under gauge transformations corresponding to the electric charge; Q is an appropriate linear combination of the infinitesimal generators of $SU(2)_L$ and of $U(1)_Y$. Let us explicate the interaction between Higgs bosons and the gauge fields as dictated by minimal coupling; it reads

$$\mathscr{L}_s = \left[\left(\partial_\mu + \frac{i}{2}g' B_\mu + \frac{i}{2}g\vec{\tau}\cdot\vec{W}_\mu\right)\varphi\right]^{\dagger} \times \left[\left(\partial^\mu + \frac{i}{2}g' B_\mu + \frac{i}{2}g\vec{\tau}\cdot\vec{W}^\mu\right)\varphi\right]. \tag{13.3.25}$$

Following the method of Section 13.3.1 we introduce, instead of the fields φ and $\varphi^\dagger$, three fields ξ_1, ξ_2, ξ_3, plus another field η,

$$\varphi = \mathrm{e}^{\mathrm{i}\tau\cdot\xi/2v}\begin{pmatrix} 0 \\ \dfrac{v+\eta}{\sqrt{2}} \end{pmatrix} = U^{-1}(\xi)\varphi,$$

and subject them to the $SU(2)_L$ gauge transformation

$$\varphi' = U(\xi)\varphi, \quad \chi'_L = U(\xi)\chi_L, \quad \chi'_R = \chi_R,$$

$$B'_\mu = B_\mu, \quad \vec{W}'_\mu = U(\xi)\vec{W}_\mu U^{-1}(\xi) + \frac{\mathrm{i}}{gv}(\partial_\mu U)(U^{-1}).$$

This transforms the fields ξ_1, ξ_2, ξ_3, which in the absence of gauge particles would be Goldstone bosons, into longitudinal components of vector bosons. There remains one scalar field, namely the Higgs boson field. The operation proceeds through exactly the same steps as it did for the Abelian case in Section 13.3.1, and turns the Lagrangean for the scalars into

$$\mathscr{L}_s = \frac{1}{2}(\partial_\mu\eta)(\partial^\mu\eta) + \frac{(v+\eta)^2}{8}\chi_-^\dagger(g'B + g\vec{\tau}\cdot\vec{W})^2\chi_-, \tag{13.3.26}$$

where χ_- is the (0, 1) element of the $SU(2)_L$ doublet. By expressing g' through (13.3.20) we evaluate the term proportional to v^2, which contains the information about the masses:

$$\mathscr{L}_s \to \frac{1}{8}v^2[g^2(W_1^2 + W_2^2) + g^2(W_3 - \tan\theta\, B)^2]. \tag{13.3.27}$$

One sees from this expression that W_1 and W_2 or their charged combinations

$$W^\pm = \frac{1}{\sqrt{2}}(W_2 \pm \mathrm{i}W_2)$$

have mass $m_W = gv/2$. To find the mass of the neutral boson one must remark that from (13.3.19) the contents of the second pair of brackets in (13.3.27) are proportional to Z^2: for the quadratic part of $\mathscr{L}_s$ this eventually yields

$$\mathscr{L}_s \to \frac{1}{2}(\partial_\mu\eta)(\partial^\mu\eta) + \frac{1}{4}g^2v^2W_\mu^\dagger W^\mu + \frac{g^2v^2}{8\cos^2\theta}Z_\mu Z^\mu. \tag{13.3.28}$$

As promised, the field A^μ remains massless. The Z_0 mass is identified as

$$m_Z = \frac{gv}{2\cos\theta}.$$

In brief, the Higgs mechanism has transferred three degrees of freedom from the Higgs fields to the longitudinal components of three massive gauge bosons;

the masses are given in terms of the vacuum expectation-value v of the Higgs field and of the Weinberg angle θ,

$$m_W = \frac{1}{2}gv; \quad m_Z = \frac{m_W}{\cos\theta} = \frac{1}{2}v(g^2 + g'^2)^{1/2}. \tag{13.3.29}$$

The total Lagrangean also contains couplings of the Higgs fields with each other, and between the Higgs fields and the fermions, namely

$$\mathscr{L}_{\text{int}} = \mu^2\varphi^\dagger\varphi - \lambda(\varphi^\dagger\varphi)^2 - g_l[\bar{\chi}_{\text{R}}(\varphi^\dagger\chi_{\text{L}}) + \text{h.c.}]. \tag{13.3.30}$$

The first two terms of $\mathscr{L}_{\text{int}}$ assign a mass $\sqrt{2}\mu$ to the neutral Higgs particle which survives the symmetry-breaking. The third term assigns masses to electrons and muons ($l = e$ or μ):

$$\mathscr{L}_{\text{int}} \to -\frac{g_l v}{\sqrt{2}}\bar{\psi}(l)\psi(l), \quad m_l = \frac{g_l v}{\sqrt{2}}. \tag{13.3.31}$$

To summarize (see (13.3.17), (13.3.25), (13.3.30)), the full Lagrangean of the GSW model reads

$$\mathscr{L} = -\frac{1}{4}\vec{F}_{\mu\nu}.\vec{F}^{\mu\nu} - \frac{1}{4}f_{\mu\nu}f^{\mu\nu} + \mathscr{L}_l + \mathscr{L}_{\text{s}} + \mathscr{L}_{\text{int}}, \tag{13.3.32}$$

where $\vec{F}_{\mu\nu}$ is the field strength associated with the gauge field $\vec{W}_\mu$, and $f_{\mu\nu}$ the field strength associated with the (Abelian) gauge field B_μ. In the limit of energies much smaller than the W mass one recovers the Fermi theory; for muon decay say (Fig. 13.6),

$$\frac{g^2}{(2\sqrt{2})^2}\bar{u}(\nu_\mu)\gamma_{\lambda'}(1-\gamma_5)u(\mu)\left[\frac{-g^{\lambda'\lambda} + q^{\lambda'}q^\lambda/m_W^2}{q^2 - m_W^2}\right]$$
$$\times\bar{u}(e)\gamma_\lambda(1-\gamma_5)v(\nu_e) \to \frac{g^2}{8m_W^2}(\bar{u}(\nu_\mu)\gamma_\lambda(1-\gamma_5)u(\mu))$$
$$\times(\bar{u}(e)\gamma^\lambda(1-\gamma_5)v(\nu_e)).$$

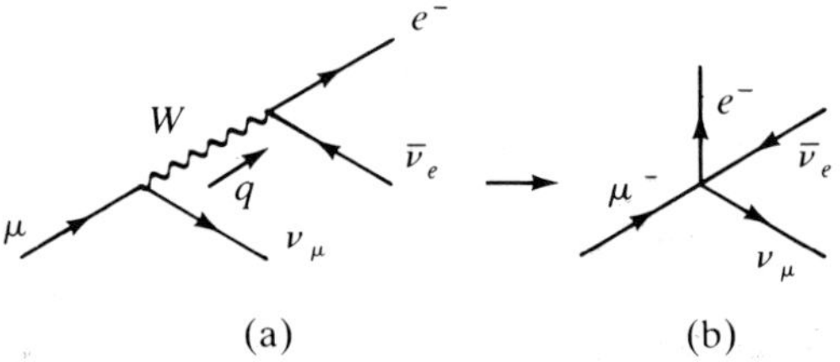

Fig. 13.6 Muon decay in the GSW model (a), and in Fermi theory (b)

Hence we can identify

$$\frac{G_F}{\sqrt{2}} = \frac{g^2}{8\,m_W^2} = \frac{1}{2v^2}. \tag{13.3.33}$$

This relation enables one to determine the vacuum expectation-value v,

$$v \approx 246\ \text{GeV},$$

while (13.3.29) yields the masses of the W and the Z in terms of the Weinberg angle, G_F, and e,

$$m_W^2 = \frac{e^2}{4\sqrt{2}G_F \sin^2\theta}, \quad m_Z^2 = \frac{m_W^2}{\cos^2\theta}. \tag{13.3.34}$$

Measurements give $m_W = 80.2 \pm 0.3\,\text{GeV}$ and $m_Z = 91.18 \pm 0.02\ \text{GeV}$.

The main test of the GSW model is agreement between the values of $\sin\theta$ measured in different experiments. One can summarize the situation by saying that $\sin\theta$ features in three different kinds of data:

(i) The W and the Z-masses

$$\sin^2\theta = \frac{e^2}{4\sqrt{2}G_F m_W^2} = \frac{\pi\alpha}{\sqrt{2}G_F m_W^2}.$$

(ii) The m_W/m_Z ratio

$$\sin^2\theta = 1 - m_W^2/m_Z^2.$$

(iii) The neutral current couplings: $\sin^2\theta$ may be, and has been, determined in a number of experiments:

- parity violation in atoms;
- scattering of polarized electrons by deuterium;
- elastic neutrino–electron scattering;
- elastic neutrino–proton scattering;
- deep inelastic scattering of neutrinos by nucleons;
- asymmetries at the Z^0 peak.

One can thus compare the values of $\sin^2\theta$ measured in a variety of experiments. However, before one can check the consistency of all determinations, radiative corrections must be taken into account. In order to display these corrections in a simple way, let us write the amplitude M_{AB} corresponding to Z^0-exchange between fermions of type (A) and (B). The neutral current couplings can be written, from the results of Problem 13.9, equations (13.3.33) and (13.3.34) as

$$M_{\mathrm{AB}} = \frac{\sqrt{2}G_{\mathrm{F}}m_Z^2}{D(s)}\,\rho_{\mathrm{AB}}(\gamma_\nu(1-\gamma_5)I_3^{\mathrm{A}} - 2k_{\mathrm{A}}\gamma_\nu Q^{\mathrm{A}}\sin^2\theta)$$
$$\times(\gamma^\nu(1-\gamma_5)I_3^B - 2k_{\mathrm{B}}\gamma^\nu Q^{\mathrm{B}}\sin^2\theta),$$

where $D(s)$ is the Z^0-propagator, I_3^{A} and $Q^{\mathrm{A}}(I_3^{\mathrm{B}}$ and $Q^{\mathrm{B}})$ are the third component of weak isospin and the charge of particle A(B). One has $I_3 = 1/2$ and $Q = 2/3$ for u, c and t quarks, $I_3 = -1/2$ and $Q = -1/3$ for d, s and b quarks. At the tree level, $\rho_{\mathrm{AB}} = 1$ and $k_{\mathrm{A}} = k_{\mathrm{B}} = 1$ in the 'minimal' GSW model with two Higgs doublets. In a nonminimal model (e.g. with Higgs triplets) one may have $\rho_{\mathrm{AB}} \neq 1$ even at the tree level. In what follows, we limit ourselves to the minimal model: in that case only radiative corrections may lead to $\rho_{\mathrm{AB}} \neq 1$, k_{A} and $k_{\mathrm{B}} \neq 1$. The above expression for M_{AB} is not exact; there are small corrections which do not factorize and cannot be written in a current–current form (Altarelli 1990).

When one wants to compute radiative corrections, it is necessary to fix a renormalization scheme. In the most popular (but not necessarily the most convenient) scheme, one *defines* $\sin^2\theta$ from the ratio of the experimentally measured W and Z masses, (Sirlin 1980; Marciano and Sirlin 1980)

$$\sin^2\theta \equiv 1 - m_W^2/m_Z^2.$$

By definition, there are thus no radiative corrections to (ii), but they are of course present in (i) and (iii). The most important contributions to radiative corrections come from large logarithms, of the kind $(\alpha/\pi\ln(m_Z/m_{\mathrm{f}}))^n$, where m_{f} is a fermion mass, and from quadratic terms in the top quark mass m_t. These contributions are universal, namely process independent (with one exception), and this allows us to write approximate forms of the radiative corrections in a simple way.

The large logarithms are well understood; one of the most spectacular consequences is the running of the fine structure constant α from the electron mass: $\alpha \simeq 1/137$ to the Z mass: $\alpha(m_Z) \simeq 1/128$, a 6% effect. The coupling constants of a renormalizable field theory really are running!

It may be surprising to discover that heavy quarks give rise to large corrections, since intuitively one might expect heavy quarks to decouple at energy scales much smaller than their masses. This is indeed what happens in a renormalizable field theory without spontaneous symmetry breaking. However, in the GSW model, one has to remember that the longitudinal modes of the W and Z bosons are generated by the Higgs mechanism, and their coupling increase with the masses: heavy quarks do not decouple in a spontaneously broken gauge theory. The main correction $\delta\rho_t$ to the ρ-parameter comes precisely from the top quark,

$$\delta\rho_t = \frac{3G_{\mathrm{F}}m_t^2}{8\pi^2\sqrt{2}} + \frac{19-2\pi^2}{3}\left(\frac{3G_{\mathrm{F}}m_t^2}{8\pi^2\sqrt{2}}\right)^2 + O(G_{\mathrm{F}}m_t^2)^3.$$

Radiative corrections are thus quite sensitive to the top quark mass, which is at present an unknown parameter of the model. There is a second very important unknown parameter, the Higgs boson mass m_H; fortunately this mass appears only logarithmically in the radiative corrections at one loop,

$$\delta\rho_H = -\frac{11G_F m_W^2}{24\pi^2\sqrt{2}}\tan^2\theta \ln\frac{m_H^2}{m_W^2}.$$

Power terms do appear at the two-loop level, but their effect is small for $m_H < 1\,\text{TeV}(= 10^3\,\text{GeV})$. One can now show that the 'universal' part of the radiative corrections may be written as follows:

$$\rho_{AB} = 1 + \delta\rho \simeq 1 + \delta\rho_t,$$

$$k_A = 1 + \text{ctn}^2\theta\,\delta\rho \simeq 1 + \text{ctn}^2\theta\,\delta\rho_t,$$

except when A (or B) is a b quark. The radiative corrections to neutral currents take a simple form, except when a b quark is involved (even in that case one can use a simple modification to the previous rules). It remains to write down the radiative corrections to the W mass,

$$m_W^2 = \frac{\pi\alpha}{\sqrt{2}G_F\sin^2\theta}\frac{1}{1-\Delta r},$$

with

$$\frac{1}{1-\Delta r} \simeq \frac{\alpha(m_Z)}{\alpha}\frac{1}{1+\text{ctn}^2\theta\,\delta\rho}.$$

At the time of writing, one can say that all experimental data are consistent with the minimal GSW model and the following values for $\sin^2\theta$ and m_t:

$$\sin^2\theta = 0.227 \pm 0.006,$$

$$m_t = 130 \pm 50\,(\text{GeV}/c^2).$$

It is important to note that radiative corrections put an upper limit on the top quark mass. Measurements of asymmetry parameters at LEP I will allow soon determinations of $\sin^2\theta$ with an accuracy $\delta\sin^2\theta = \pm\,0.001$. Another independent determination will come in a few years from an accurate measurement of m_W at LEP II. The GSW model will then be tested to an accuracy of order 10^{-3}.

To summarize, the GSW model is at present quite successful; it has allowed us to predict two phenomena which were far from evident in 1967, namely

(i) the existence of neutral currents;

(ii) the existence of two massive bosons, with masses larger by two orders of magnitude than any particle masses known in 1970.

To this one must add the prediction of the charmed quark, which we are not in a position to discuss here.

Admittedly, in spite of its successes the model has several weak points.

- The 'Higgs sector' is barely constrained: in particular the mass of the Higgs boson which survives the symmetry breaking and the masses of the leptons are completely arbitrary (since g_l in (13.3.31) is arbitrary).
- The model is not asymptotically free, on account of the gauge group $U(1)_Y$ and of the φ^4 coupling of the Higgs bosons.
- The gauge group is a direct product, thus admitting two independent coupling constants; further, the group $U(1)_Y$ is Abelian, and the quantization of electric charge remains unexplained.

13.4 Quantum chromodynamics

The theory which is presumed to explain the strong interactions is quantum chromodynamics (QCD). It is based on an internal symmetry called (for no good reason) *colour* symmetry. The symmetry group is SU(3), and the elementary constituents of hadrons, i.e. the quarks and antiquarks, are arranged in its fundamental representations 3 and $\bar{3}$ respectively. Recall that for $N > 2$ the groups SU(N) have two inequivalent N-dimensional representations each, called the fundamental representations N and $\bar{N}$. For $N = 2$ however, these two representations are equivalent. For SU(3), the basis vectors labelled by colour are, for instance.

$$|\text{blue}\rangle = \begin{pmatrix}1\\0\\0\end{pmatrix}, \quad |\text{red}\rangle = \begin{pmatrix}0\\1\\0\end{pmatrix}, \quad |\text{green}\rangle = \begin{pmatrix}0\\0\\1\end{pmatrix}.$$

Apart from their colour, the quarks are further differentiated by 'flavour'. We saw in the preceding chapter that there are five known quark flavours, up (u), down (d), strange (s), charmed (c), and beauty (b); probably there exists a sixth, the top (t) quark. The colour SU(3) symmetry must not be confused with the flavour SU(3) symmetry discovered by Gell-Mann and Neeman; the latter is a symmetry of the flavours (u, d, s), and commutes with the colour symmetry, so that every flavour exists in three colours. The flavour symmetry is approximate (e.g. the three quarks u, d, s have different masses), while the colour symmetry is exact. Moreover, and this is the most important point, the colour symmetry is *local*, and it underlies a non-Abelian gauge theory which is just the quantum chromodynamics we are discussing. Since SU(3) has 8 ($= 3^2 - 1$) generators, there are eight gauge bosons, called *gluons*.

To summarize, the strong interactions are described by a non-Abelian gauge theory whose gauge group is SU(3); the matter (spin 1/2) particles are the quarks and antiquarks, belonging to the representations 3 and $\bar{3}$ of SU(3). The gauge (spin 1) particles are the gluons, belonging to the 8-dimensional representation of SU(3). The hadrons observed in Nature are of two kinds; fermions, also called baryons (proton, neutron, . . .), and bosons, i.e. mesons (π-meson, K-meson, . . .). Baryons consist of three quarks whose wavefunction is an SU(3) singlet; mesons consist of a quark–antiquark pair whose wavefunction likewise belongs to a 1-dimensional representation of SU(3). In other words, *the hadrons are colour singlets.* The assumption that only colour singlets can be observed is called *confinement.*

Notice that if the confinement assumption is correct, then the asymptotic condition (10.3.24) cannot apply to the quark fields ψ and $\bar{\psi}$, because there are no free quarks. One assumes that this stems from the very severe infrared divergences in chromodynamics. The only combinations that can correspond to asymptotic states are colour singlets like $\bar{\psi}_i\psi_i$ or $\varepsilon_{ijk}\psi_i\psi_j\psi_k$.

13.4.1 Asymptotic freedom

The most important feature of QCD is its asymptotic freedom (which is a property of non-Abelian gauge theories generally). It is this that allows the perturbative calculation of certain high-energy reactions, like the e^+e^- annihilation we shall meet in Section 13.4.2. To demonstrate asymptotic freedom we must derive the relation between the bare and the renormalized coupling constants g_0 and g,

$$g_0 = Zg.$$

In view of the Ward identities (13.2.14) there are several ways of doing this calculation; the simplest is probably to exploit the quark–gluon coupling (see (13.2.13)),

$$g_0 = Z_{1\mathrm{F}} Z_3^{-1/2} Z_2^{-1/2} g, \quad Z = Z_{1\mathrm{F}} Z_3^{-1/2} Z_2^{-1}. \tag{13.4.1}$$

In order to avoid Wick rotations and other sources of factors i, we shall work with the Euclidean theory, which will also serve to prepare the way for Section 13.5. The Lagrangean of the Euclidean theory is found by performing the substitutions

$$A_\mu B^\mu \to -\boldsymbol{A}\cdot\boldsymbol{B} = -A_\mu B_\mu, \quad \partial_\mu A^\mu \to \nabla\cdot\boldsymbol{A} = \partial_\mu A_\mu,$$

$$\gamma_\mu A^\mu \to -\gamma\cdot\boldsymbol{A} = -\gamma_\mu A_\mu, \quad \gamma_\mu\partial^\mu \to \gamma\cdot\nabla = \gamma_\mu\partial_\mu,$$

and by then changing the sign of the result. ($\boldsymbol{A}$ here is a vector in D-dimensional Euclidean space, $\mu = 1, \ldots, D$.) For a scalar field say we have

$$\mathscr{L} = \frac{1}{2}(\partial_\mu\varphi)(\partial^\mu\varphi) - \frac{1}{2}m^2\varphi^2,$$

$$\mathscr{L}_\mathrm{E} = \frac{1}{2}(\nabla\varphi)^2 + \frac{1}{2}m^2\varphi^2 = \frac{1}{2}(\partial_\mu\varphi)(\partial_\mu\varphi) + \frac{1}{2}m^2\varphi^2.$$

The matrices γ_μ obey the anticommutation rules

$$\{\gamma_\mu, \gamma_\nu\} = -2\delta_{\mu\nu}, \tag{13.4.2}$$

and for momenta running into the vertex one must bear in mind that $\nabla \to i\mathbf{p}$ (instead of $\partial^\mu \to -\mathrm{i}p^\mu$). The Feynman rules follow from these remarks (we violate the vector notation, $\boldsymbol{k} \to k$):

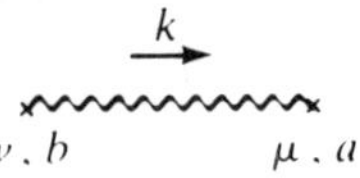

$$: \frac{\delta_{ab}}{k^2}\left(\delta_{\mu\nu} - (1-\alpha)\frac{k_\mu k_\nu}{k^2}\right), \tag{13.4.3a}$$

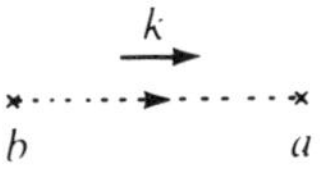

$$: \frac{\delta_{ab}}{k^2} \quad , \tag{13.4.3b}$$

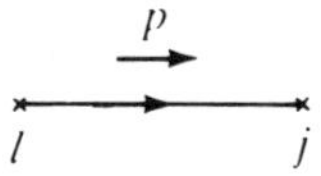

$$: \frac{\delta_{jl}}{\not{p} + m} = \frac{\delta_{jl}(m - \not{p})}{p^2 + m^2}. \tag{13.4.3c}$$

The vertices (13.2.7a, c, d) must be multiplied by i, and (13.2.7b) by $-$ i, and one must remember to replace $g_{\mu\nu} \to -\delta_{\mu\nu}$.

We proceed to evaluate the renormalization constants to one-loop order, starting with Z_3. The gluon self-energy draws contributions from a gluon loop, from a Faddeev–Popov loop, and from a quark loop. We shall work in the Feynman gauge ($\alpha = 1$) and use dimensional regularization, where tadpoles can be ignored (see Section 5.5.1); and we omit a multiplicative factor μ^ε which is irrelevant to what follows. The group-theory factors depend on C_F and C_A (see (13.2.11)). For SU(3) it will suffice to set

$$C_\mathrm{F} = 4/3, \quad C_\mathrm{A} = 3.$$

Start with the gluon loop (Fig. 13.7). The expression for this graph reads

$$\Pi^{(\mathrm{G})}_{\lambda\mu;ad}(k) = \frac{1}{2} g^2 \delta_{ad} C_\mathrm{A} \int \frac{N_{\lambda\mu}\,\mathrm{d}^D q}{(2\pi)^D q^2 (q+k)^2}, \tag{13.4.4}$$

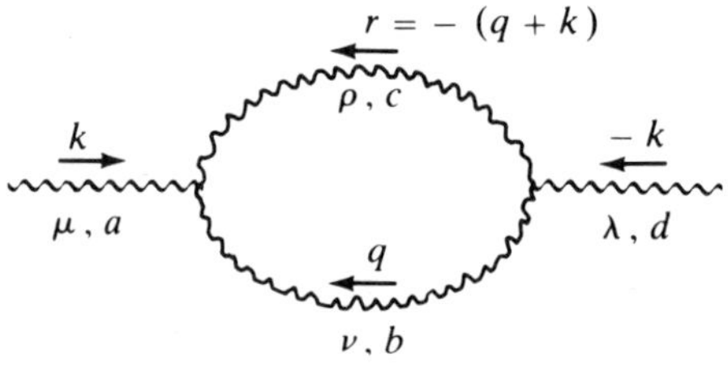

Fig. 13.7 Gluon contribution to the gluon self-energy

where 1/2 is the symmetry factor, C_A the group-theory factor (see (13.2.11a)), while $N_{\lambda\mu}$ is given by

$$N_{\lambda\mu} = \delta_{\lambda\mu}(5k^2 + 2q\,.\,k + 2q^2) + 2q_\lambda q_\mu(2D-3)$$
$$\qquad + k_\lambda k_\mu(D-6) + (q_\lambda k_\mu + k_\lambda q_\mu)(2D-3).$$

After using the Feynman identity to combine the denominators, one changes variables,

$q = q' - xk$, keeping only terms even in q; and one integrates over q by appeal to identities from Appendix B. The result reads

$$\Pi^{(\mathrm{G})}_{\lambda\mu;ad} = \frac{g^2 C_{\mathrm{A}} \delta_{ad} \Gamma(2 - D/2)}{4(4\pi)^{D/2}(k^2)^{2-D/2}} \frac{B\left(\frac{D}{2} - 1; \frac{D}{2} - 1\right)}{(D-1)} \times [(6D-5)k^2\delta_{\lambda\mu} - (7D-6)k_\lambda k_\mu]. \tag{13.4.5}$$

The calculation of the Fadeev–Popov loop (Fig. 13.8) is appreciably simpler:

$$\Pi^{(\mathrm{FP})}_{\lambda\mu;ad} = -g^2 C_{\mathrm{A}} \delta_{ad} \int \frac{\mathrm{d}^D q}{(2\pi)^D} \frac{q_\lambda (q+k)_\mu}{q^2 (q+k)^2}$$

$$= \frac{g^2 C_{\mathrm{A}} \delta_{ad} \Gamma(2 - D/2)}{(4\pi)^{D/2}(k^2)^{2-D/2}} \frac{B\left(\frac{D}{2} - 1; \frac{D}{2} - 1\right)}{4(D-1)} \times [k^2\delta_{\lambda\mu} + (D-2)k_\lambda k_\mu]. \tag{13.4.6}$$

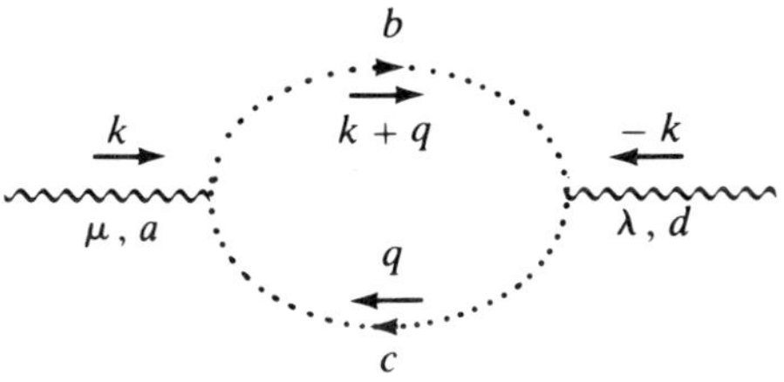

Fig. 13.8 The contribution of the Faddeev–Popov loop to the gluon self-energy

The sum of the two contributions (13.4.5) and (13.4.6) *is transverse* (conformably with a Ward identity, as can be seen from Problem 13.5(c)),

$$\Pi_{\lambda\mu;ad} = \frac{g^2 C_{\mathrm{A}} \delta_{ad} \Gamma(2 - D/2)}{(4\pi)^{D/2}(k^2)^{2-D/2}} \frac{B\left(\frac{D}{2} - 1; \frac{D}{2} - 1\right)}{2(D-1)} \times (3D-2)[k^2\delta_{\lambda\mu} - k_\lambda k_\mu], \tag{13.4.7}$$

whence one derives the divergent term

$$\Pi^{(\mathrm{div})}_{\lambda\mu;ad} = \frac{\alpha_s}{4\pi} C_{\mathrm{A}} \delta_{ad} \left(\frac{5}{3}\right)\left(\frac{2}{\varepsilon}\right)(k^2\delta_{\lambda\mu} - k_\lambda k_\mu). \tag{13.4.8}$$

Here, by analogy with electrodynamics, we have introduced a strong coupling constant $\alpha_s = g^2/4\pi$. It is unnecessary to (re)calculate the quark loop, which apart from a group-theory factor (13.2.11c) is identical to the vacuum-polarization loop in electrodynamics. With n_f the number of flavours, equation (12.3.23) yields for Z_3 in the minimal scheme MS the result

$$Z_3 = 1 + \frac{\alpha_s}{4\pi}\left(\frac{5}{3} C_{\mathrm{A}} - \frac{4}{3} T_{\mathrm{F}} n_{\mathrm{f}}\right)\frac{2}{\varepsilon}. \tag{13.4.9}$$

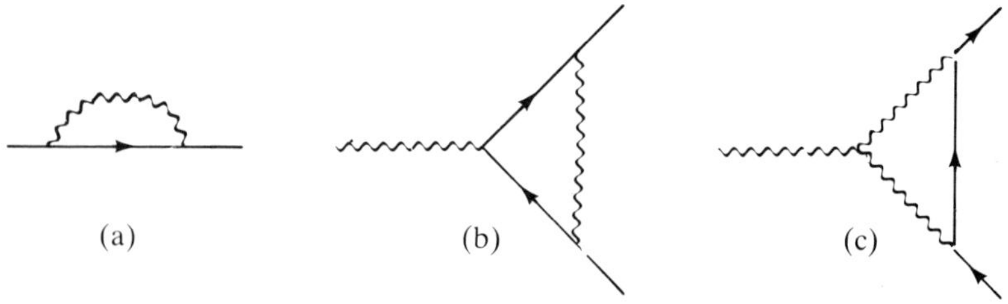

Fig. 13.9 Diagrams contributing to Z_2 and to $Z_{1\mathrm{F}}$

Next we must calculate the diagrams contributing to $Z_{1\mathrm{F}}$ and to Z_2 (Fig. 13.9). Apart from a group-theory factor, the diagrams (a) and (b) of Fig. 13.9 are identically the same as in electrodynamics. The factors in question are, respectively.

(a) C_F (see (13.2.11c)); (b) $C_\mathrm{F} - \frac{1}{2} C_\mathrm{A}$ (Problem 13.4(a)).

The factors C_F in (a) and (b) cancel from Z, leaving an effective contribution of

$$Z_2^{-1} Z_{1\mathrm{F}}^{(a+b)} = 1 - \frac{\alpha_\mathrm{s}}{4\pi}\left(-\frac{1}{2} C_\mathrm{A}\right)\frac{2}{\varepsilon}. \tag{13.4.10}$$

There remains diagram (c), which is rather cumbersome to evaluate. Luckily we need only its divergent part. The contribution of diagram (c) (Fig. 13.10) contains a factor

$$ig^3 f_{abc}\left(\frac{1}{2}\lambda_c\right)_{jl}\left(\frac{1}{2}\lambda_b\right)_{li},$$

where $\frac{1}{2}\lambda_a$ is an infinitesimal generator of the fundamental representation.

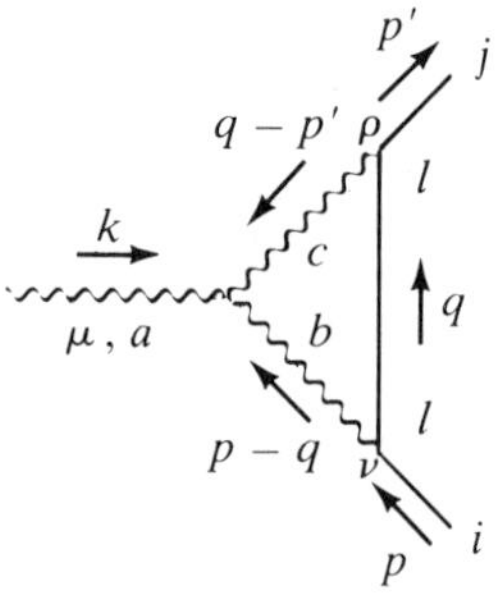

Fig. 13.10 The calculation of the diagram (c) of Fig. 13.9

According to Problem 13.4(b), this multiplicative factor is

$$\frac{1}{2} g^3 C_\mathrm{A}\left(\frac{1}{2}\lambda_a\right)_{ji}. \tag{13.4.11}$$

Let us now write down the numerator of the expression for the loop (Fig. 13.10): it reads

$$N_\mu = [\delta_{\mu\nu}(k - p + q)_\rho + \delta_{\nu\rho}(p + p' - 2q)_\mu + \delta_{\rho\mu}(q - p' - k)_\nu]\gamma_\rho(-\not{q})\gamma_\nu.$$

Since we require only the divergent term, we need keep only the part quadratic in q,

$$N_\mu \to 2q^2\gamma_\mu + 4q_\mu \not{q} \to 3q^2\gamma_\mu.$$

Then, for the divergent part of the diagram, a trivial integration yields

$$\Gamma^{(\text{div})}_{\mu,\, ji} = \left[\frac{1}{2}\, g(\lambda_a)_{ji}\gamma_\mu\right]\frac{\alpha_s}{4\pi}\left(\frac{1}{2}\, C_A\right)\left(3 \times \frac{2}{\varepsilon}\right);$$

to Z_{1F} this contributes an amount

$$Z^{(c)}_{1F} = 1 - \frac{\alpha_s}{4\pi}\left(\frac{3}{2}\, C_A\right)\frac{2}{\varepsilon}. \tag{13.4.12}$$

Collecting the results (13.4.9), (13.4.10), and (13.4.12), and substituting them into (13.4.1), we obtain our expression for Z,

$$Z = 1 - \frac{\alpha_s}{4\pi}\left[\frac{11}{6}\, C_A - \frac{2}{3}\, T_F n_f\right]\frac{2}{\varepsilon}. \tag{13.4.13}$$

(Z_3, Z_{1F}, and Z_2 depend on the gauge, i.e. on $\mathfrak{a}$, but, at least in the minimal scheme MS, Z is independent of the gauge parameter: see Problem 13.12.) The function $\beta(\alpha_s)$ is visible from (13.4.13) on inspection (with $C_A = N$, $T_F = 1/2$): using (7.5.10) and $\alpha^0_s = Z^2\alpha_s$, we find

$$\beta(\alpha_s) = \mu\,\frac{d\alpha_s}{d\mu} = -\frac{\alpha_s^2}{2\pi}\left(\frac{11}{3}\, N - \frac{2}{3} n_f\right) + O(\alpha_s^3). \tag{13.4.14}$$

The term featuring α_s^2 is negative if $n_f < 11N/2$, and in that case chromodynamics is asymptotically free.

There exists a (relatively) elementary explanation of (13.4.14). The fact that the coefficient of α_s^2 in $\beta(\alpha_s)$ is negative means that the dielectric constant of the vacuum, ε, is less than 1, which entails anti-screening of the colour charges. Since the vacuum is relativistically invariant, one must have $\varepsilon\mu = 1$, where μ is the magnetic permeability; therefore $\mu > 1$. One can calculate μ by evaluating the energy density in the presence of an external magnetic field. The result is

$$\mu = 1 + \chi, \quad \chi \propto (-1)^{2S} q^2 \sum_{S_3}\left(-\frac{1}{3} + \gamma^2 S_3^2\right),$$

where q, γ and S_3 are, respectively, the charge, the gyromagnetic ratio, and the spin projection along the external field. Up to a proportionality factor, we can recover (13.4.14) since we have, for scalar, spinor, and vector particles,

$$\text{scalar:}\quad S = 0, \quad \gamma = 0, \quad \chi \propto -\frac{1}{3};$$

$$\text{spinor:}\quad S = \frac{1}{2}, \quad \gamma = 2, \quad \chi \propto -\frac{4}{3};$$

$$\text{vector:}\quad S = 1, \quad \gamma = 2, \quad \chi \propto -\frac{22}{3};$$

the contribution of scalar particles to the β-function can be inferred from Problem 12.8. A little care is required with the group-theory factors in order to find the correct coefficients in (13.4.14): see, e.g., Cheng and Li (1984) chapter 10.

The literature often uses a coupling constant $\alpha_s(q^2)$, which depends on a mass q characteristic of the process under study; an appropriate choice of $\alpha_s(q^2)$ allows one to improve on perturbation theory through appeal to the renormalization group. Let us integrate the differential equation

$$\frac{\mathrm{d}\alpha_s}{\mathrm{d}\ln q^2} = -4\pi\beta_0\alpha_s^2,$$

where

$$\beta_0 = \frac{1}{16\pi^2}\left(\frac{11}{3}N - \frac{2}{3}n_f\right) \tag{13.4.15}$$

(the way we have chosen to normalize β_0 will be explained when we come to equation (13.4.17). The integration yields

$$\int_{\alpha_s(q^2)}^{\infty} \frac{\mathrm{d}\alpha_s}{4\pi\beta_0\alpha_s^2} = \ln\frac{q^2}{\Lambda^2},$$

where Λ is an integration constant, sometimes noted Λ_{QCD}, in order to avoid confusion with a cutoff. Thus the result reads

$$\alpha_s(q^2) = \frac{1}{4\pi\beta_0\ln(q^2/\Lambda^2)}. \tag{13.4.16}$$

The coupling constant decreases like $(\ln q^2)^{-1}$. One must be sure to understand the physical significance of (13.4.16): rather than choose a renormalization point μ^2 and a coupling constant $\alpha_s(\mu^2)$, we fix the strength of the coupling through a dimensional parameter Λ, which enters as an integration constant of the differential renormalization equation. This procedure is called 'dimensional transmutation'. Evidently it is not restricted to first order in α_s. Suppose we write

$$\beta(g) = -\beta_0 g^3 - \beta_1 g^5 + O(g^7), \tag{13.4.17}$$

where the coefficients β_0 and β_1 are universal (see Section 7.1.5), β_1 being given by (Caswell 1974)

$$\beta_1 = \left(\frac{1}{16\pi^2}\right)^2\left(\frac{34N^2}{3} - \frac{10Nn_f}{3} - \frac{n_f(N^2-1)}{N}\right); \tag{13.4.18}$$

then an appropriate choice of integration constant yields

$$\alpha_s(q^2) = \frac{1}{4\pi\beta_0\ln\dfrac{q^2}{\Lambda^2}}\left[1 - \frac{\beta_1\ln\ln q^2/\Lambda^2}{\beta_0^2\ln q^2/\Lambda^2} + O\left(\frac{\ln\ln q^2/\Lambda^2}{\ln^2 q^2/\Lambda^2}\right)\right] \tag{13.4.19}$$

(see Problem 13.11). The parameter Λ depends on the renormalization scheme. The value current in 1989 was, in the $\overline{\text{MS}}$ scheme (see the footnote on page 227 and Problem 13.14),

$$50 \text{ MeV} \lesssim \Lambda_{\overline{\text{MS}}} \lesssim 200 \text{ MeV}.$$

The importance of Λ lies in the fact that it determines the mass scale of the theory independently of the quark mass, in other words even if the quark mass is zero, indeed even if there are no quarks. We shall see this again from another point of view in Section 13.5.3.

13.4.2 e^+e^- Annihilation: kinematics

We set ourselves the task of improving the calculation of the ratio R in Section 12.2.3, by evaluating the correction of order α_s, and then discussing the validity of our approach. In the process we shall come across certain interesting phenomena, like infrared and collinear singularities: and shall have an opportunity to write down a special case of the very important Altarelli–Parisi equation.

Suppose that in e^+e^- annihilation into hadrons we observe a particular type of hadron H (e.g. a π^+ meson), with momentum p_H in the e^+e^- centre-of-mass frame. Then the energy p_{0H} of this hadron lies between m_H and $q/2$. Neglecting its mass ($m_H \ll q/2$, where q is the mass of the virtual photon), it is natural to introduce a scaling variable z_H lying between 0 and 1,

$$z_H = \frac{2p_{0H}}{q} = \frac{2p_H \cdot q}{q^2}, \quad 0 \leqslant z_H \leqslant 1. \tag{13.4.20}$$

One can measure the differential cross-section $\mathrm{d}\sigma/\mathrm{d}z_H$ for producing a hadron H with momentum $z_H\, q/2$, and define the *fragmentation function* $D_H(z_H)$,

$$D_H(z_H) = \frac{1}{\sigma_T} \frac{\mathrm{d}\sigma}{\mathrm{d}z_H}, \tag{13.4.21}$$

where σ_T is the total cross-section for $e^+e^- \to$ hadrons.

Start by describing the kinematics. The reaction

$$e^+e^- \to H(p_H) + X$$

is shown in Fig. 13.11; X represents jointly all the particles that are not observed.

Using the electromagnetic current j_μ we define a tensor $W^H_{\mu\nu}$,

$$W^H_{\mu\nu} = \frac{1}{4\pi} \int \mathrm{d}\Phi^{(X)} (2\pi)^4 \delta^{(4)}(q - p_H - p_X) \times \langle p_H, X | j_\mu(0) | 0 \rangle \langle 0 | j_\nu(0) | p_H, X \rangle, \tag{13.4.22}$$

where $\mathrm{d}\Phi^{(X)}$ is the element of phase space, and where the factor $1/4\pi$ is conventional. In terms of $W^H_{\mu\nu}$ and of the leptonic tensor

$$l_{\mu\nu} = 4(k_\mu k'_\nu + k_\nu k'_\mu) - 2q^2 g_{\mu\nu}$$

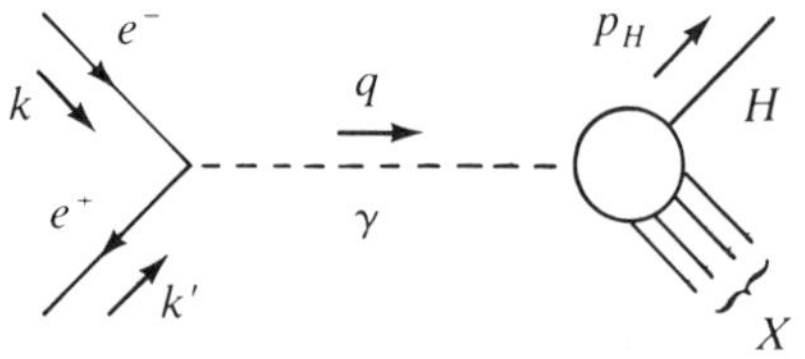

Fig. 13.11 The reaction $e^+ + e^- \to H(p_H) + X$

the cross-section $\mathrm{d}\sigma$ reads

$$\mathrm{d}\sigma = \frac{e^4}{8q^2}\frac{1}{q^4}\frac{\mathrm{d}^3p_H}{(2\pi)^3 2p_{0H}}\, 4\pi l^{\mu\nu} W^H_{\mu\nu}. \tag{13.4.23}$$

Because the current j_μ is conserved, we can express the tensor $W^H_{\mu\nu}$ as a function of two Lorentz scalars W^H_1 and W^H_2 called *structure functions*; they depend only on z_H and on q^2,

$$W^H_{\mu\nu} = -\left(g_{\mu\nu} - \frac{q_\mu q_\nu}{q^2}\right) W^H_1 + \left(p_{\mu H} - \frac{q_\mu(p_H\cdot q)}{q^2}\right)\left(p_{\nu H} - \frac{q_\nu(p_H\cdot q)}{q^2}\right) W^H_2, \tag{13.4.24}$$

and one finds

$$\frac{\mathrm{d}\sigma}{\mathrm{d}z_H\,\mathrm{d}\cos\theta_H} = \frac{\pi\alpha^2 z_H}{q^4}\, l^{\mu\nu} W^H_{\mu\nu} = \frac{\pi\alpha^2 z_H}{q^2}\left[W^H_1 + \frac{q^2 z_H^2}{8}\sin^2\theta_H W^H_2\right], \tag{13.4.25}$$

where θ_H is the angle between $\boldsymbol{k}$ and $\boldsymbol{p}_H$. If one is not interested in the angular distribution one can integrate over $\cos\theta_H$; this yields

$$\frac{\mathrm{d}\sigma}{\mathrm{d}z_H} = \frac{2\pi\alpha^2 z_H}{q^2}\left[W^H_1 + \frac{q^2 z_H^2}{12} W^H_2\right] = \frac{4\pi\alpha^2 z_H}{3q^2} W^H, \tag{13.4.26}$$

where we have defined

$$W^H = -\frac{1}{2} g^{\mu\nu} W^H_{\mu\nu}. \tag{13.4.27}$$

13.4.3 The Altarelli–Parisi equation

Equation (13.4.27) having completed our study of the kinematics, we proceed to the dynamics. Quantum chromodynamics tells us that the basic process is the production of quarks, antiquarks, and gluons; and that hadrons are formed only later, through a process of confinement which is not yet fully understood. Disregarding confinement for the moment, we study the production of a quark, for which we define the analogues z, $W_{\mu\nu}$, and W of z_H, $W^H_{\mu\nu}$, and W^H. The simplest process (of order $(\alpha_s)^0$ in QCD) is the reaction $e^+ + e^- \rightarrow q + \bar{q}$, which we considered in Chapter 12. The fragmentation function in this case is $\delta(1 - z)$, while the differential cross-section is

$$\frac{d\sigma}{dz} = \frac{4\pi\alpha^2}{3q^2} e_q^2 \delta(1 - z) \tag{13.4.28}$$

(see (12.2.12); e_q is the charge of the quark in units of the proton charge). Accordingly (see 13.4.26) ad (13.4.27)

$$W = \delta(1 - z). \tag{13.4.29}$$

The order-α_s correction stems from the process

$$e^+ + e^- \rightarrow q + \bar{q} + g,$$

where g represents a gluon. The two graphs contributing to this reaction are drawn in Fig. 13.12, which also defines the kinematics.

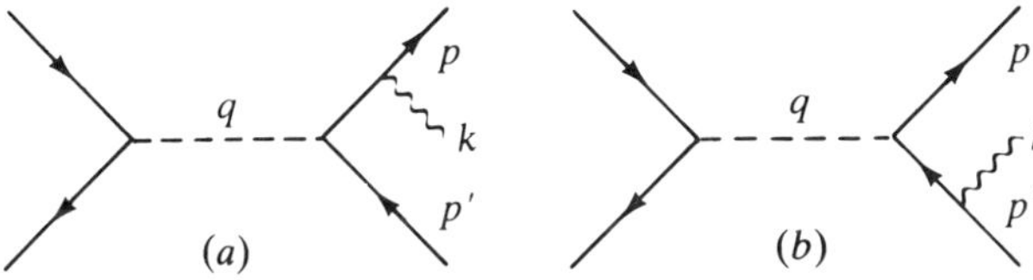

Fig. 13.12 The reaction $e^+e^- \rightarrow q\bar{q}g$ to order α_s

Since we are assuming that $q \gg 1$ GeV, we can neglect the quark masses. Under these conditions the matrix element for the reaction $\gamma^*(q) \rightarrow q(p) + q(p') + g(k)$ reads

$$\mathcal{M}_\mu = (-\mathrm{i}g)(-\mathrm{i}e_q)\bar{u}(p)\left[\not{\varepsilon}\frac{\lambda_a}{2}\frac{\mathrm{i}}{\not{p} + \not{k}}\gamma_\mu - \gamma_\mu\frac{\mathrm{i}}{\not{p}' + \not{k}}\not{\varepsilon}\frac{\lambda_a}{2}\right]v(p'), \tag{13.4.30}$$

where ε_μ is the gluon polarization. Squaring the matrix element, summing over final spins, contracting with $g^{\mu\nu}$, and calculating the requisite traces of γ matrices, we find straightforwardly (see Problem 13.13) that

$$|\mathcal{M}|^2 = -g^{\mu\nu}\sum_{\text{spins}}\mathcal{M}_\mu\mathcal{M}_\nu^* = 8C_F e_q^2 g^2 \frac{u^2 + s^2 + 2tq^2}{us}. \tag{13.4.31}$$

Here we have introduced the Mandelstam variables s, t, u,

$$s = (p' + k)^2, \quad t = (p + p')^2, \quad u = (p + k)^2, \tag{13.4.32}$$

and set $C_F = 4/3(= (N^2 - 1)/2N$ with $N = 3)$. The scaling variables z and z' are functions of s, u, and q^2,

$$z = \frac{2p \cdot q}{q^2} = 1 - \frac{s}{q^2}, \quad z' = \frac{2p' \cdot q}{q^2} = 1 - \frac{u}{q^2}; \tag{13.4.33}$$

they allow $|\mathcal{M}|^2$ to be expressed as

$$|\mathcal{M}|^2 = 8C_F e_q^2 g^2 X(z, z') = 8C_F e_q^2 g^2 \frac{z^2 + z'^2}{(1 - z)(1 - z')}. \tag{13.4.34}$$

Equation (13.4.34) deserves a pause for inspection. Suppose first that $z \neq 1$; then the quantity $X(z, z')$ is singular as $z' \to 1$. In that case the gluon and quark momenta are parallel and have the same sign; one says that they are collinear*. In order to deal with the consequent singularity one must regularize at some intermediate stage of the calculation. The most correct method uses a $(4 + \varepsilon)$-dimensional space (Problem 13.14). But to the order of perturbation theory to which we are working, one could alternatively assign a mass $\lambda \neq 0$ to the gluon, without contravening current conservation. We adopt this 'infrared regularization' because its physical interpretation is more transparent. We shall see presently that collinear configurations, namely $z \neq 1$, $z' \to 1$, and $z' \neq 1$, $z \to 1$, yield contributions proportional to $\ln q^2/\lambda^2$, i.e. singular in the limit $\lambda \to 0$; these are the *collinear singularities*, also called *mass singularities*. The region $z \to 1$, $z' \to 1$ corresponds to the emission of a gluon with very low momentum, appropriately called an *infrared* or a *soft gluon*. This region yields an *infrared singularity* proportional to $\ln^2(q^2/\lambda^2)$, which is characteristic of gauge theories. Indeed an ordinary renormalizable theory, like φ^3 theory in six dimensions, has only mass singularities.

Let us now evaluate the quantity $W = -\frac{1}{2} g_{\mu\nu} W^{\mu\nu}$ in full. According to (13.4.22) we must integrate over the momenta p' and k of antiquark and gluon. This is best done in their centre-of-mass frame, where $\boldsymbol{p}' + \boldsymbol{k} = 0$ (Fig. 13.13).

$$W = \frac{1}{64\pi^2} \frac{\| \boldsymbol{p}' \|}{\sqrt{s}} \int_{-1}^{+1} \mathrm{d}\cos\theta |\mathcal{M}|^2. \tag{13.4.35}$$

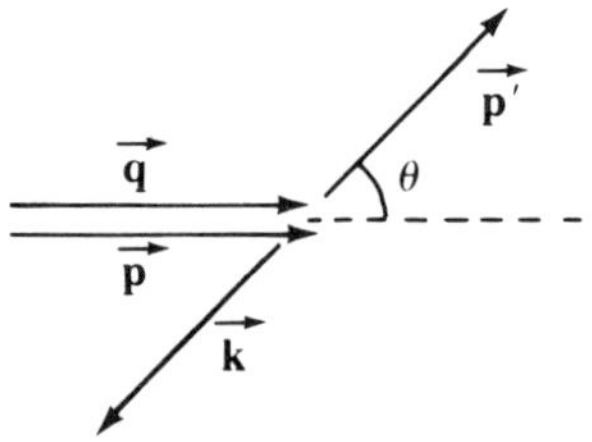

Fig. 13.13 Kinematics in the reference frame $\boldsymbol{p}' + \boldsymbol{k} = 0$

* It would be tempting to attribute the singularity to the graph (a) of Fig. 13.12. However, the contributions of individual graphs are not gauge-invariant. In fact it is easy to find gauges where the contribution of graph (a) vanishes (Problem 13.15).

A modicum of relativistic kinematics yields

$$q_0 = \frac{s+q^2}{2\sqrt{s}}, \quad p_0 = \|\boldsymbol{p}\| = \frac{q^2-s}{2\sqrt{s}}, \quad p'_0 = \|\boldsymbol{p}'\| = \frac{s-\lambda^2}{2\sqrt{s}}. \tag{13.4.36}$$

In (13.4.35), rather than integrate over $\cos\theta$, we integrate over a variable z' related to $\cos\theta$ by

$$z' = \frac{\|\boldsymbol{p}'\|}{q^2\sqrt{s}}[(q^2+s)-(q^2-s)\cos\theta] = \frac{s-\lambda^2}{2sq^2}[(q^2+s)-(q^2-s)\cos\theta]. \tag{13.4.37}$$

Hence

$$\left|\frac{\mathrm{d}z'}{\mathrm{d}\cos\theta}\right| = \frac{z\|\boldsymbol{p}'\|}{\sqrt{s}}$$

and

$$W = \frac{1}{64\pi^2 z}\int_{z'_m}^{z'_M} \mathrm{d}z' |\mathscr{M}|^2. \tag{13.4.38}$$

Equation (13.4.37) also yields the integration limits on z',

$$z'_m = 1 - z - \frac{\lambda^2}{q^2} \leqslant z' \leqslant 1 - \frac{\lambda^2}{q^2(1-z)} = z'_M. \tag{13.4.39}$$

On the other hand we must correct the squared matrix element $|\mathscr{M}|^2$ in (13.4.34) in order to take account of the gluon mass:

$$X(z, z') \to X(z, z') - \frac{\lambda^2}{q^2}\left(\frac{1}{(1-z)^2} + \frac{1}{(1-z')^2}\right). \tag{13.4.40}$$

Finally it proves convenient to define $1/(1-z)$ as a principal value; in view of $0 \leqslant z \leqslant 1 - \lambda^2/q^2$, we write

$$\int_0^{1-\lambda^2/q^2} \frac{\mathrm{d}z f(z)}{1-z} = \int_0^1 \mathrm{d}z \frac{f(z)-f(1)}{1-z} + f(1)\ln\frac{q^2}{\lambda^2} + O\left(\frac{\lambda^2}{q^2}\right). \tag{13.4.41}$$

In terms of the distribution $(1-z)_+^{-1}$ this reads

$$\frac{1}{1-z} = \frac{1}{(1-z)_+} + \delta(1-z)\ln\frac{q^2}{\lambda^2}. \tag{13.4.42}$$

Similarly one defines the distribution $[\ln(1-z)/(1-z)]_+$,

$$\left(\frac{\ln(1-z)}{1-z}\right) = \left(\frac{\ln(1-z)}{1-z}\right)_+ - \frac{1}{2}\delta(1-z)\ln^2\frac{q^2}{\lambda^2}. \tag{13.4.43}$$

One can observe the emergence of the anticipated singularities $\ln(q^2/\lambda^2)$ and $\ln^2(q^2/\lambda^2)$. With this groundwork done we are finally in a position to integrate over z'; terms that vanish in the limit $\lambda^2 \to 0$, like $(\lambda^2/q^2)\ln(q^2/\lambda^2)$, are simply ignored.

$$\int_{z'_m}^{z'_M} \mathrm{d}z' X(z, z') = \frac{1+z^2}{1-z}\int_{z'_m}^{z'_M} \frac{\mathrm{d}z'}{1-z'} + \frac{1}{1-z}\int_{z'_m}^{z'_M} \frac{(z'^2-1)\,\mathrm{d}z'}{1-z'} = I_1(z) + I_2(z).$$

Integration yields

$$I_1(z) = \frac{1+z^2}{(1-z)_+}\ln\frac{q^2}{\lambda^2} + (1+z^2)\left(\frac{\ln(1-z)}{1-z}\right)_+ + \frac{1+z^2}{1-z}\ln z + \delta(1-z)\ln^2(q^2/\lambda^2),$$

$$I_2(z) = -\frac{3}{2(1-z)_+} + \frac{3}{2} - \frac{1}{2}z - \delta(1-z)\left(\frac{3}{2}\ln\frac{q^2}{\lambda^2} - \frac{7}{4}\right).$$

From the second term of (13.4.40) we get the contributions

$$\frac{\lambda^2}{q^2(1-z)^2} \to \frac{1}{2}\delta(1-z), \quad \frac{\lambda^2}{q^2(1-z')^2} \to 1-z.$$

Collecting these results we obtain

$$W = \frac{2\alpha_s e_q^2}{3\pi z}\left[(1+z^2)\left[\frac{\ln(q^2/\lambda^2)}{(1-z)_+} + \left(\frac{\ln(1-z)}{1-z}\right)_+ + \frac{\ln z}{1-z}\right] - \frac{3}{2(1-z)_+} + \frac{1}{2}(1+z) + \delta(1-z)\left[\ln^2\frac{q^2}{\lambda^2} - \frac{3}{2}\ln\frac{q^2}{\lambda^2} + \frac{5}{4}\right]\right]. \quad (13.4.44)$$

This expression embodies the anticipated infrared and collinear singularities. However, our calculation of W is not yet complete. Still to be added is the diagram for the radiative correction to the photon–quark vertex (Fig. 13.14), which is of the same order in α_s; this contribution is called 'virtual' (since the gluon in Fig. 13.14 is virtual), by contrast to the contribution (13.4.44) which is called 'real'.

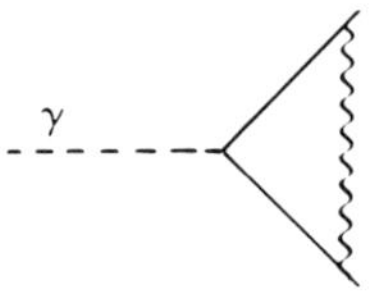

Fig. 13.14 Exchange of a virtual gluon in $\gamma^* \to q\bar{q}$

The contribution of Fig. 13.14 to W is simply $2\,\mathrm{Re}\,F_1(q^2)$ times the contribution from the diagram of order $(\alpha_s)^0$ (see (13.4.29)), where $F_1(q^2)$ is the form factor calculated in Chapter 12 (equation (12.3.50a)). From this expression one can show that the contribution to W from Fig. 13.14 is

$$-\frac{2\alpha_s}{3\pi}e_q^2\delta(1-z)\left[\ln^2\frac{q^2}{\lambda^2} - 3\ln\frac{q^2}{\lambda^2} + \frac{7}{2} - \frac{\pi^2}{3}\right] \quad (13.4.45)$$

(see Problem 12.7; and remember the analytic continuation from $q^2 < 0$ to $q^2 > 0$).

Since we have renormalized on the mass shell, there are no contributions from external lines.

Combining (13.4.29), (13.4.44), and (13.4.45) we obtain $d\sigma/dz$ in the form

$$\frac{d\sigma}{dz} = \frac{4\pi\alpha^2}{3q^2}e_q^2\left\{\delta(1-z)\left[1 + \frac{2\alpha_s}{3\pi}\left(\frac{\pi^2}{3} - \frac{9}{4}\right)\right] + \frac{2\alpha_s}{3\pi}\left[P_{qq}(z)\ln\frac{q^2}{\lambda^2} + (1+z^2)\left[\left(\frac{\ln(1-z)}{1-z}\right)_+ + \frac{\ln z}{1-z}\right] - \frac{3}{2(1-z)_+} + \frac{1}{2}(1+z)\right]\right\}, \quad (13.4.46)$$

where $P_{qq}(z)$ is the Altarelli–Parisi function

$$P_{qq}(z) = \frac{1+z^2}{(1-z)_+} + \frac{3}{2}\delta(1-z) = \left(\frac{1+z^2}{1-z}\right)_+. \qquad (13.4.47)$$

By combining the real and the virtual diagrams the infrared singularities have been made to cancel; the cancellation occurs through the so-called 'Bloch–Nordsieck mechanism'. Only mass singularities survive in (13.4.46), proportional to the function $P_{qq}(z)$. Let us try to interpret this equation. An interpretation directly in terms of cross-sections is evidently impossible, because the result depends on the gluon mass λ which is a mere intermediate artefact of the calculation. More generally, the result depends on what infrared regularization has been adopted. Nevertheless we can from (13.4.46) derive for the fragmentation function $D(z, q^2)$ the equation

$$\frac{\mathrm{d}D(z,q^2)}{\mathrm{d}\ln q^2} = \alpha_s \int_z^1 \frac{\mathrm{d}x}{x} P_{qq}\left(\frac{z}{x}\right) D(x,q^2) + O(\alpha_s^2), \qquad (13.4.48)$$

which holds to order α_s, *and which no longer depends on* λ^2. (On the right, $D(x, q^2) = \delta(1-x)$ is calculated to order $(\alpha_s)^0$.) In its turn, this equation suggests for the function $D_H(z_H, q^2)$ the evolution equation (called the Altarelli–Parisi equation)

$$\frac{\mathrm{d}D_H(z_H,q^2)}{\mathrm{d}\ln q^2} = \int_{z_H}^1 \frac{\mathrm{d}x}{x} \mathscr{P}\left(\frac{z_H}{x}\right) D_H(x,q^2), \qquad (13.4.49)$$

where

$$\mathscr{P}(z) = \frac{2\alpha_s(q^2)}{3\pi} P_{qq}(z) + O(\alpha_s^2). \qquad (13.4.50)$$

It is indeed possible to derive such an evolution equation, but the proof is beyond the scope of this book.

13.4.4 Corrections of order α_s to the ratio R

After this long detour (which did however acquaint us with several interesting features) we now revert to our initial aim of calculating the ratio R. Note that the integral of $\mathrm{d}\sigma/\mathrm{d}z$ over z is just the total cross-section σ_T, because only one quark is produced in the final state. By virtue of the identities

$$\int_0^1 \mathrm{d}z(1+z^2)\left(\frac{\ln(1-z)}{1-z}\right)_+ = \frac{7}{4},$$

$$\int_0^1 \mathrm{d}z \frac{1+z^2}{1-z} \ln z = -\frac{\pi^2}{3} + \frac{5}{4},$$

equation (13.4.46) yields

$$\sigma_T = \frac{4\pi\alpha^2}{3q^2} e_q^2 \left(1 + \frac{\alpha_s}{\pi}\right), \tag{13.4.51}$$

whence R is given by

$$R = 3\sum_i (e_q^{(i)})^2 \left(1 + \frac{\alpha_s(q^2)}{\pi} + O(\alpha_s^2)\right). \tag{13.4.52}$$

We have used the coupling constant $\alpha_s(q^2)$, because q is the only mass scale in the problem. There are three important points to note.

(i) At the highest energies accessible today one has $\alpha_s(q^2)/\pi \lesssim 0.1$; hence a perturbative calculation is legitimate, though its accuracy is much lower than in electrodynamics, where $\alpha/\pi \simeq 10^{-3}$.

(ii) The integration over all final states has freed (13.4.51) from all singularities, infrared and collinear. This amounts to verification, in our special case, of a theorem due to Kinoshita, Lee, and Nauenberg (called the KLN theorem) (Kinoshita 1962; Lee and Nauenberg 1964).

(iii) It can be shown that confinement leaves the result (13.4.52) unaffected in the limit $q^2 \to \infty$. In fact, by virtue of the energy scales characteristic of the two processes, the final quark–gluon state is formed in a time of order $1/q$, while confinement operates over times of order 1 $(\text{GeV})^{-1}$. One can imagine thought-experiments carried out over times t such that $1/q \ll t \ll 1$ $(\text{GeV})^{-1}$. Such an experiment would measure the cross-section (13.4.52); and, once this cross-section is determined, confinement cannot modify it subsequently.

The present Section 13.4 has given only very restricted glimpses of the extensive applications of perturbative chromodynamics, which include not only e^+e^- annihilation, but also deep-inelastic electron scattering, lepton pair production, W and Z production, jets, scattering with large momentum transfers, and so on. The interested reader is referred to the literature (Altarelli 1982; Yndurain 1983; Muta 1987).

13.5 Lattice gauge theories

Perturbative quantum chromodynamics simply ignores the problem of confinement. It can make predictions only for certain high-energy processes; more precisely, only for reactions with momentum transfers q such that $q^2 \gg \Lambda^2$, where Λ is the parameter characterizing the QCD coupling constant $\alpha_s(q^2)$ (see (13.4.16)). Λ is of the order of 0.1 GeV, which tallies with the energy scale typical of confinement. One cannot from perturbative chromodynamics calculate physical quantities related to confinement, like the fragmentation function

$D_H(z_H, q_0^2)$ (i.e. the initial condition for the evolution equation (13.4.49)), hadron masses, total hadronic cross-sections, etc. All such quantities must be calculated by nonperturbative techniques, i.e., at present, numerically.

The most fruitful idea is 'to put the theory on a lattice'; in other words, one writes down an action (called a Hamiltonian in Part I: see Section 10.2.3), which reduces to the action (13.1.48) of the continuum theory in the limit where the lattice parameter a tends to zero. In order to reduce the problem to a problem of classical statistical mechanics, we start with the four-dimensional Euclidean action, and then proceed to calculate free energies, correlation functions, and so on. Thus in a manner of speaking this final section closes a circle; having started with classical statistical mechanics, and having then abandoned it for a long excursion into quantum field theory, we now revert to it in our discussion of lattice gauge theories.

One preliminary comment is essential before tackling specific examples. In a lattice theory which does not consider massive quarks, the only dimensional parameter is the lattice spacing a; for instance, if we wish to calculate a mass m, then it will turn out proportional to $1/a$: $m = (\xi a)^{-1}$, where ξ is the correlation length in units of a. Since m must remain finite in the limit $a \to 0$, we must have $\xi \to \infty$; in other words we must be at a critical point of the theory under study. We have already seen in Chapters 6 and 7 that the construction of a renormalized theory presupposes the existence of such a point. In the case of chromodynamics, there is a critical point (hopefully the only one) at $g_0 = 0$, given that $g_0 = 0$ is a fixed point; the continuum limit will be such that $g_0(a) \to 0$. Note that the coupling constant here is the *bare* one; the lattice effectively serves to regularize the theory, and g_0 is the coupling constant of the theory so regularized.

13.5.1 The Wegner model

In order to introduce lattice gauge theories through an elementary example, we start by describing the two-dimensional Wegner model. Consider a square lattice with N sites ($N \gg 1$), and place Ising spins $S_i = \pm 1$ on the *bonds* (rather than on the sites) of the lattice (Fig. 13.15). The Hamiltonian (i.e. the action) is

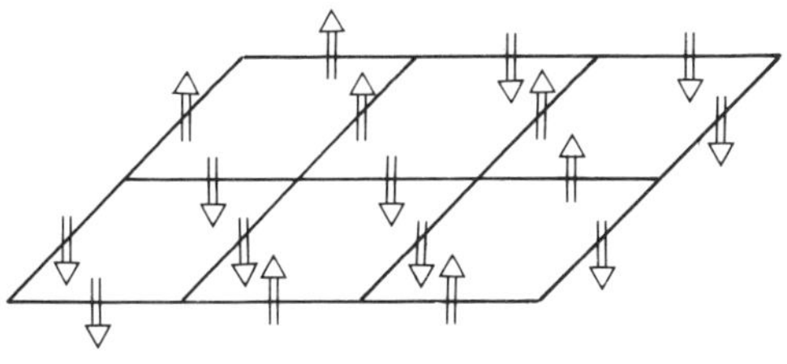

Fig. 13.15 The Wegner model in two dimensions

given by the sum of the products of the four spins along the edges of a square (called a plaquette (P); see Fig. 13.16),

$$H = -J\sum_{P} S_1(P)S_2(P)S_3(P)S_4(P); \tag{13.5.1}$$

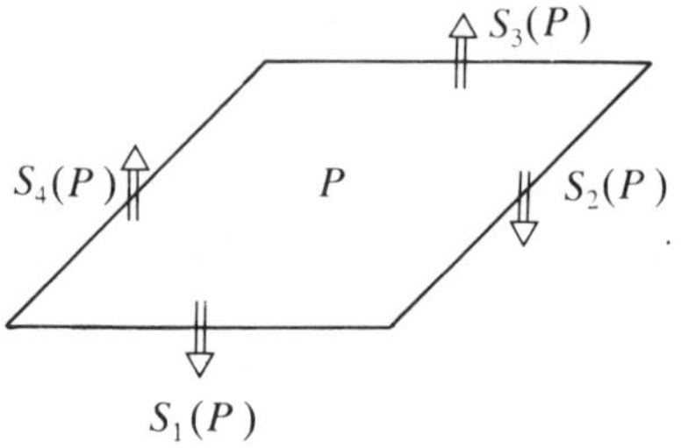

Fig. 13.16 A plaquette

the sum runs over the N plaquettes. The partition function reads

$$Z = \sum_{[S_i]} e^{-\beta H}. \tag{13.5.2}$$

Let $G(x)$ be the transformation that reverses the spins on all the bonds starting from the lattice point x (Fig. 13.17), but leaves other spins unchanged. The Hamiltonian (13.5.1) is invariant under $G(x)$, which can be interpreted as a local gauge transformation; with every point x one associates a reference frame allowing one to define the orientation of the four spins on the bonds starting from x. The fact that H is invariant shows that the relative orientation of two reference frames (x) and (x') is arbitrary. The gauge group here is the group whose elements allow two reference frames at a given point x to be compared with each other; it is the group with just the two elements $\{+1, -1\}$, called Z_2.

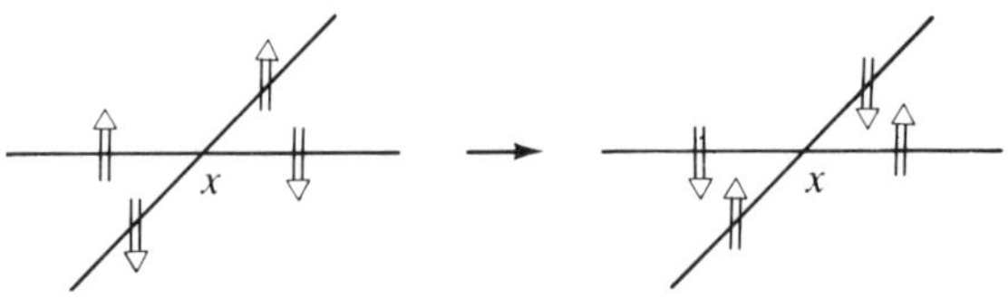

Fig. 13.17 The local gauge transformation $G(x)$

The Wegner model can be generalized to an arbitrary number D of dimensions, and it exhibits a phase transition when $D \geqslant 3$. Naturally one asks whether in the low-temperature phase the order parameter $\langle S_i \rangle$ differs from zero. We can easily see however that it cannot: subjecting the spin system to an infinitesimal magnetic field B, we observe that there is only an infinitesimal difference between the Hamiltonians of two configurations related through a local gauge transformation $G(x)$ (Problem 13.16). By contrast, in the Ising model the energy cost of reversing all the spins is NB. The argument can be made rigorous and generalized to any spin system having a local gauge symmetry (Elitzur's theorem). Hence we must find another kind of order parameter, which will in fact turn out to be nonlocal.

The partition function is easy to evaluate when $D = 2$; we need merely expand the exponential in (13.5.2) by using the identity

$$e^{\beta J S_1(P) \ldots S_4(P)} = \cosh(\beta J) + S_1(P) \ldots S_4(P) \sinh(\beta J).$$

For the sum over configurations not to vanish, it is necessary (by an argument similar to that in Section 1.2.2) either that there be no plaquette in the expansion, or that the plaquettes cover the entire lattice (subject to periodic boundary conditions). For $N \gg 1$ the end-result reads

$$Z = [4 \cosh(\beta J)]^N.$$

Consider next a rectangle $\mathscr{R}$ on the lattice, with sides of lengths T and R (Fig. 13.18), and the 'Wilson loop'

$$W(\mathscr{R}) = \frac{1}{Z} \sum_{[S_i]} \left(\prod_l S_l \right) e^{-\beta H}, \tag{13.5.3}$$

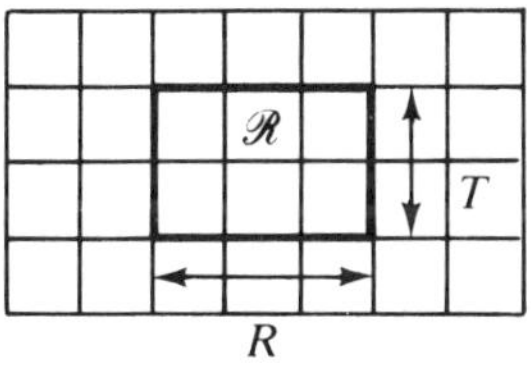

Fig. 13.18 The Wilson loop

where $\prod_l S_l$ stands for the product of all the spins on the bonds edging the rectangle. Two properties are easily derived.

(i) $W(\mathscr{R})$ is invariant under local gauge transformations.

(ii) $\ln W(\mathscr{R}) = (\ln \tanh \beta J)\, TR$; thus $\ln W(\mathscr{R})$ is proportional to the area of the rectangle $\mathscr{R}$. To see this one need merely observe that the plaquettes must cover all of $\mathscr{R}$.

The two-dimensional Wegner model is evidently reminiscent of the one-dimensional Ising model, and it is not difficult to show that the two are equivalent (Problem 13.17).

The quantity $W(\mathscr{R})$ is the desired order parameter. It is gauge invariant, and for $D \geqslant 3$ the phase transition manifests itself by the function $\ln W$ ceasing to be proportional to the area of the rectangle, and becoming proportional to the length of the periphery instead. The 'area law' $\ln W \sim TR$ follows from a high-temperature expansion, applicable when $\beta \to 0$; by contrast, the 'perimeter law' $\ln W \sim (T+R)$ follows from a low-temperature expansion, applicable when $\beta \to \infty$.

13.5.2 The Wilson action and strong coupling

We proceed to tackle more realistic theories. Let G be a compact Lie group, say $SU(N)$ to be definite, and let i and j be two neighbouring sites on a cubic lattice. The vector joining i and j, with direction specified, is denoted by μ,

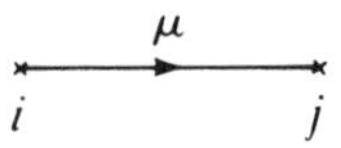

Now define an element $\mathscr{A}_\mu$ of the Lie algebra of G, as a function of the midpoint y of the bond (i, j), by

$$\mathscr{A}_\mu = \frac{1}{2}\lambda_b A_{\mu,b},$$

where the matrices $\frac{1}{2}\lambda_b$ are the infinitesimal generators of the fundamental representation of G; and let U_{ij} be the matrix

$$U_{ij} = e^{-ig_0 a \mathscr{A}_\mu}, \tag{13.5.4}$$

with a the lattice spacing. In this equation g_0 is the *bare* coupling constant, conformably with lattice regularization.

We have the matrix equation $U_{ij}^{-1} = U_{ji}$, and one often writes $U_{x,\mu}$ instead of U_{ij}:

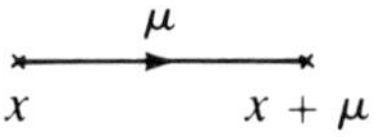

$$U_{x,\mu} = U_{x+\mu,\,-\mu}^{-1}.$$

A local gauge transformation g_i induces

$$U_{ij} \to g_i U_{ij} g_j^{-1}, \tag{13.5.5}$$

where g_i and g_j are matrices associated with the sites i and j respectively. It is clear (see (13.1.35)) that the matrix U_{ij} is the lattice analogue of the quantity $R(C; A)$ in Section 13.1.2. The method used in Section 13.1.3 suggests a gauge-invariant form for the action, namely

$$S = \sum_P S_P, \tag{13.5.6}$$

where

$$S_P = \beta\left(1 - \frac{1}{N}\operatorname{Re}\operatorname{Tr}(U_{ij}\,U_{jk}\,U_{kl}\,U_{li})\right); \tag{13.5.7}$$

the product of the U matrices is taken along the edges bounding a plaquette

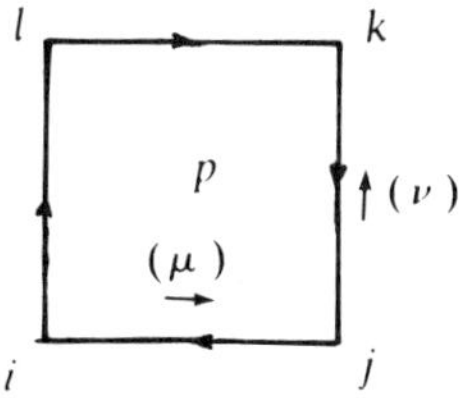

Fig. 13.19 Contour for (13.5.7)

(Fig. 13.19). On repeating for $a \to 0$ the calculation that led to (13.1.43) one finds

$$S_P = \beta\left(1 - \frac{1}{N} \operatorname{Re} \operatorname{Tr} \left(\exp(-\, \mathrm{i} g_0\, a^2 \mathscr{F}_{\mu\nu} + O(a^4))\right)\right).$$

Expanding the exponential, using $\operatorname{Tr} \lambda_a = 0$, summing over plaquettes, and taking the limit $a \to 0$, we obtain

$$S = \frac{\beta g_0^2}{2N} \frac{1}{2} \int \mathrm{d}^4 x [\operatorname{Tr} (\mathscr{F}_{\mu\nu} \mathscr{F}_{\mu\nu}) + O(a^2)];$$

comparison with (13.1.48) after a change of scale $A_\mu \to g A_\mu$ then leads one to identify

$$\beta = \frac{2N}{g_0^2}. \tag{13.5.8}$$

The action defined by (13.5.6) and (13.5.7) is called the *Wilson action*. It is not the only one to yield (13.1.48) in the continuum limit; one can modify it by adding terms in a^2, a^4, etc., which vanish in the limit (at least in the tree approximation). As regards the fixed point at $g_0 = 0$ these are irrelevant fields. They can however be used in practice to improve the rate of convergence towards the continuum limit. At this stage it would be proper to introduce fermions; but fermions on a lattice present several problems both theoretical and numerical, and we shall confine the present discussion to gauge fields.

As in the Wegner model we introduce a nonlocal order parameter, the Wilson loop for a closed contour C,

$$W(C) = \frac{1}{Z} \int \mathscr{D}g \left(\operatorname{Tr} \prod_C U_{kl}\right) \mathrm{e}^{-S}. \tag{13.5.9}$$

Here the product of U matrices is taken along a closed contour C formed by bonds of the lattice; thus $W(C)$ is manifestly gauge invariant. The integration measure $\mathscr{D}_g$ is a product of Haar measures, one for each bond. Because the integration volume is compact, there is no need to factor out an infinite volume (as in fixing a gauge).

Now consider the physical interpretation of the Wilson loop. In the continuum limit the product of the matrices U_{kl} in (13.5.9) becomes

$$\prod_C U_{kl} \to P\left(\mathrm{e}^{\mathrm{i}g_0 \oint_C \mathscr{A}_\mu \,\mathrm{d}x_\mu}\right)$$

(see (13.1.34)), representing the interaction of the field A_μ with a point source. For C we adopt a rectangular contour with one side (T) parallel to the (Euclidean) time axis, and the other (R) parallel to some coordinate axis; and we assume $T \gg R$. Then the point source is a static quark–antiquark pair created at a distance R at the time $T = 0$, and annihilated at time T. The quantity $W(C)$ may be interpreted as the ratio of two partition functions, one in the presence of a source J, and the other with $J = 0$:

$$W(C) = \frac{Z(J)}{Z(0)} = \mathrm{e}^{-[F(J)-F(0)]}.$$

When $T \gg R$, $[F(J) - F(0)]$ is proportional to T, since the free energy F is extensive; on the other hand we saw in Chapter 8 that the free-energy density may be identified with the energy of the ground state. Since the quark–antiquark pair is static, this energy is purely potential, and equals $V(R)T$. Accordingly

$$W(C) \sim \mathrm{e}^{-V(R)T}. \qquad (13.5.10)$$

Now assume (as will be shown presently for large g_0) that similarly to the Wegner model one finds an 'area law'

$$V(R)T = \sigma(TRa^2), \qquad (13.5.11)$$

where σ is a constant. Then the quark–antiquark potential would be linear, and we would indeed have confinement; an infinite amount of energy would be needed to achieve an arbitrarily large separation of quark from antiquark. The constant σ is called the 'string tension'.

To derive (13.5.11) in the high-temperature limit $g_0 \to \infty$ (see (13.5.8)) one uses a high-temperature expansion. It can be shown that the dominant term stems from paving the loop with plaquettes (Fig. 13.20). Indeed, the integrals over bonds satisfy

$$\int \mathrm{d}g = 1, \quad \int \mathrm{d}g\; U_{ij}\, U_{kl}^{-1} = \frac{1}{N}\delta_{il}\,\delta_{jk},$$

$$\int \mathrm{d}g\; U = \int \mathrm{d}g\; U^\dagger = \int \mathrm{d}g\; UU = \int \mathrm{d}g\; U^\dagger\, U^\dagger = 0. \qquad (13.5.12)$$

The proof hinges on the same principle as, for instance, the proof of the

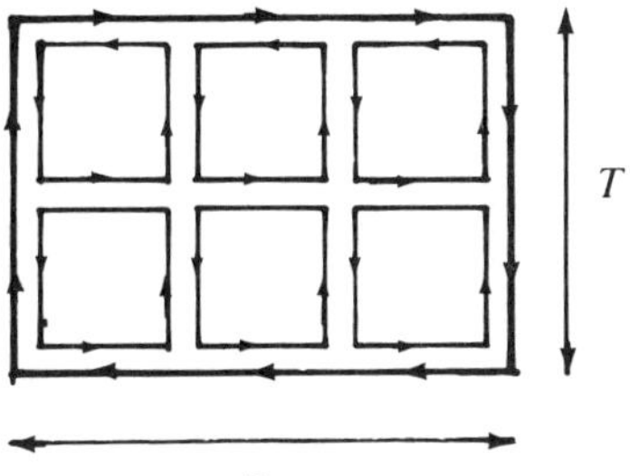

Fig. 13.20

high-temperature expansion of the XY model in Chapter 4. In the physically interesting case $N = 3$ one finds

$$W(T, R) \xrightarrow[g_0 \to \infty]{} \left(\frac{\beta}{18}\right)^{TR};$$

for the string tension this yields

$$\sigma = a^{-2} \ln (3 g_0^2). \tag{13.5.13}$$

It can be shown that the high-temperature expansion has a finite radius of convergence, whence the theory indeed exhibits confinement when g_0^2 is large enough. Unfortunately this theory with large g_0^2 is barely relevant to the continuum limit (Lorentz invariance for instance is badly broken), and we must establish some connection with small values of g_0, which as we have seen do relate to the continuum limit.

13.5.3 Weak coupling and asymptotic scale invariance

Suppose we wished to calculate some quantity having the dimension of mass, like the square root $\sigma^{1/2}$ of the string tension. By dimensional analysis $\sigma^{1/2}$ must take the form

$$\sigma^{1/2} = \frac{1}{a} f(g_0). \tag{13.5.14}$$

As $a \to 0$, which corresponds to a cutoff on k tending to infinity, $\sigma^{1/2}$ must become independent of a,

$$a \frac{\mathrm{d}}{\mathrm{d}a} \sigma^{1/2} = 0;$$

this yields a differential equation for $f(g_0)$,

$$f(g_0) + f'(g_0)\, \bar{\beta}(g_0) = 0, \tag{13.5.15}$$

where

$$\bar{\beta}(g_0) = -a\frac{\mathrm{d}}{\mathrm{d}a}g_0(a). \tag{13.5.16}$$

The differential equation (13.5.15) can be integrated straightforwardly by using the expansion (13.4.17) of $\bar{\beta}(g_0)$,

$$\bar{\beta}(g_0) = -\beta_0 g_0^3 - \beta_1 g_0^5 + O(g_0^7).$$

(We showed in Section 7.1.5 that the coefficients β_0 and β_1 are the same in $\beta(g)$ as in $\bar{\beta}(g_0)$.) The choice of some definite integration constant in (13.5.15) defines a dimensional parameter Λ_{L} (the suffix L alludes to the lattice),

$$\sigma^{1/2} = c_\sigma \Lambda_{\mathrm{L}}, \tag{13.5.17a}$$

$$\Lambda_{\mathrm{L}} = a^{-1}\mathrm{e}^{-1/(2\beta_0 g_0^2)}(\beta_0 g_0^2)^{-\beta_1/2\beta_0^2}(1 + O(g_0^2)). \tag{13.5.17b}$$

Note the non-analyticity in g_0 as $g_0 \to 0$, analogous to the behaviour found in Section 4.3. It shows that a perturbative calculation is a priori impossible. Λ_{L} is independent of the cutoff a^{-1}, and determines the mass scale of the theory; once again we observe the phenomenon of 'dimensional transmutation'. All variables having the dimensions of mass can be expressed in terms of Λ_{L},

$$m_i = c_i \Lambda_{\mathrm{L}}, \tag{13.5.18}$$

and the ratios m_i/m_j, $m_i/\sigma^{1/2}$, etc. are universal, depending only on the gauge group G. Of course (13.5.18) applies only when a is small enough for us to be in the continuum regime, where scale-invariance is satisfied. Asymptotic scale invariance imposes an even stronger constraint, requiring that the dependence on g_0 be given by (13.5.17b), or in other words that it be governed solely by the terms β_0 and β_1 of the function $\bar{\beta}(g_0)$.

Because a lattice regularizes, we can relate g_0 to the renormalized coupling constant g through $g_0 = \bar{Z}_1 \bar{Z}_3^{-3/2} g$, by calculating the renormalization constants $\bar{Z}_1$ and $\bar{Z}_3$. On the other hand we could equally well regularize dimensionally, and relate g_0 to g in say the $\overline{\mathrm{MS}}$ scheme. By comparing g_0(lattice) and $g_0(\overline{\mathrm{MS}})$ for a common value of g one can relate the (perturbatively defined) Λ to Λ_{L}; for SU(3) one finds

$$\Lambda_{\overline{\mathrm{MS}}} \simeq 29\,\Lambda_{\mathrm{L}}.$$

If $\Lambda_{\overline{\mathrm{MS}}}$ were well known experimentally (which it is not), then we could determine Λ_{L} unambiguously, and thus determine the mass scale on the lattice.

In the last ten years there have been many numerical calculations using the Monte-Carlo method. For detailed accounts we refer to Creutz, 1983; Kogut, 1979, Section VIII; Kogut, 1983. Since numerical results are apt to change rapidly (with the development of dedicated computers), we give just one illustrative result for the string tension (Schierholz 1985), Fig. 13.21.

Strong support for a uniquely defined theory of strong interactions (quarks confined at large distances but asymptotically free at short distances) would be

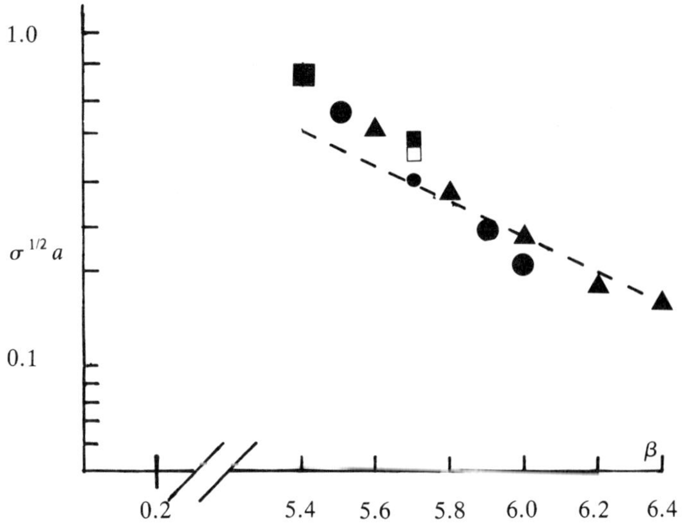

Fig. 13.21 A compilation of the string tension $\sigma^{1/2}$ in 1985. The numerical results suggest that asymptotic scale invariance (dashed line) applies when $\beta \geqslant 5.8$

furnished by a string tension different from zero all the way from the strong coupling region to the region where asymptotic scale invariance applies, and showing no indication of a phase transition in between. In that case (and for the moment there is no evidence whatever to the contrary) quantum chromodynamics would truly constitute *the* theory of strong interactions.

Our brief review of lattice gauge theories has described only a minute fraction of the work devoted to the problem. It was intended merely to show that such theories are based on a most remarkable synthesis of the corpus of ideas presented in this book.

Problems

13.1 (a) Prove the identities (13.1.50) and (13.1.51).

(b) Derive the equations of motion (13.1.52) and the continuity equation (13.1.53).

13.2 (a) Prove the identity

$$[D^\mu, \mathscr{F}^{\nu\rho}] = (D^\mu_{ab} F^{\nu\rho}_b) T_a.$$

(b) By using this identity and equation (13.1.50), prove the Bianchi identity (13.1.54).

13.3 Given that $\mathscr{F}_{\mu\nu} = 0$ in the vicinity of a point x, show that in this vicinity one has

$$\mathscr{A}_\mu(x) = \mathrm{i}[\partial_\mu U(g(x))]\, U^{-1}(g(x)),$$

and conversely.

13.4 (a) Derive equations (13.2.8).

(b) Determine the group-theory factors for the diagrams in Fig. 13.22. Solution:

$$\text{(a) } C_\mathrm{F} - \frac{1}{2} C_\mathrm{A}; \quad \text{(b) } -\frac{1}{2} C_\mathrm{A}; \quad \text{(c) } = C_\mathrm{F}^2; \quad \text{(d) } \frac{1}{2} C_\mathrm{A} C_\mathrm{F}.$$

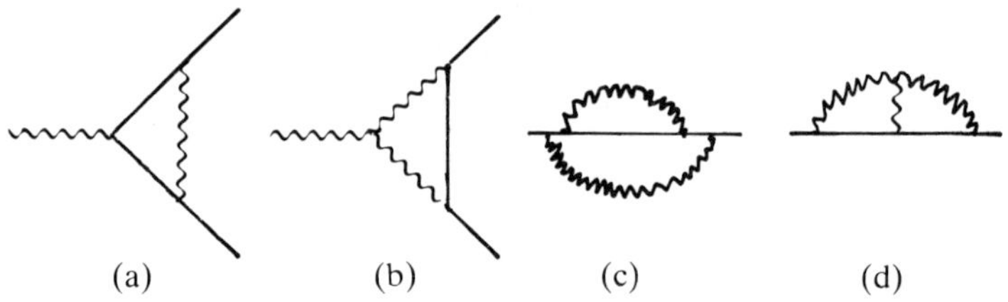

Fig. 13.22

13.5 The Becchi–Rouet–Stora (BRS) transformation

(a) Start by considering electrodynamics, with the gauge fixed quite generally by a condition $f(A) = 0$. (For definiteness one might envisage the 'pedagogic gauge' $\partial^\mu A_\mu + \frac{1}{2} g A_\mu A^\mu = 0$ from Problem 11.14.) To the Lagrangean one must add a term for the ghosts

$$\mathscr{L}_\mathrm{FP} = -\int \mathrm{d}^4 y\, \bar{\eta}(x)\, M_f(x, y) \eta(y),$$

where

$$M_f(x, y) = \left.\frac{\delta f(A^\Lambda(x))}{\delta \Lambda(y)}\right|_{\Lambda=0}.$$

We define the BRS transformation by

$$\delta A_\mu(x) = (\partial_\mu \eta(x))\,\delta\zeta, \quad \delta\psi(x) = -\, \mathrm{i}e\psi(x)\,\eta(x)\delta\zeta,$$

$$\delta\bar{\eta}(x) = \frac{1}{\mathrm{a}} f(A)\,\delta\zeta, \quad \delta\bar{\psi}(x) = \mathrm{i}e\bar{\psi}(x)\eta(x)\,\delta\zeta,$$

$$\delta\eta(x) = 0,$$

where $\delta\zeta$ is a Grassmann variable anticommuting with η and $\bar{\eta}$; $\delta X/\delta\zeta$ is defined by taking $\delta\zeta$ to the right and dividing δX by $\delta\zeta$: $\delta A_\mu/\delta\zeta = \partial_\mu\eta$. Note that the transformation law for A_μ is a gauge transformation with $\Lambda(x) = \eta(x)\delta\zeta$. Show that the Lagrangean

$$\mathscr{L} = -\frac{1}{4} F_{\mu\nu} F^{\mu\nu} - \frac{1}{2\mathrm{a}} (f(A))^2 + \mathscr{L}_{\mathrm{FP}}$$

is invariant under the BRS transformation, i.e. that $\delta\mathscr{L}/\delta\zeta = 0$. Show that the integration measure $\mathscr{D}(A, \eta, \bar{\eta})$ is likewise invariant. By considering the variation of the Green's function

$$\langle 0 | T(X(y)\,\bar{\eta}(x)) | 0 \rangle,$$

where $X(y)$ is a product of the fields A_μ, ψ and $\bar{\psi}$ (but not of η and $\bar{\eta}$), derive the Ward–Slavnov–Taylor identity

$$\langle 0 | T(X(y) f(A)) | 0 \rangle = \mathrm{a} \langle 0 | T\left(\frac{\delta X}{\delta\zeta}\,\bar{\eta}(x)\right) | 0 \rangle.$$

Recover (12.4.9) in the Lorentz gauge $\partial^\mu A_\mu = 0$.

(b) Now revert to non-Abelian theories, where the BRS transformation is defined by

$$\delta A_a^\mu(x) = D_{ab}^\mu \eta_b(x)\,\delta\zeta,$$

$$\delta\bar{\eta}_a(x) = \frac{1}{\mathrm{a}} f_a(A)\,\delta\zeta,$$

$$\delta\eta_b(x) = -\frac{g}{2} f_{abc}\,\eta_b(x)\,\eta_c(x).$$

Show that $\mathscr{L}_{\mathrm{G}} + \mathscr{L}_{\mathrm{FP}} + \mathscr{L}_{\mathrm{GF}}$ is BRS-invariant (fermions can be introduced without any difficulty). Show that the integration measure $\mathscr{D}(A, \eta, \bar{\eta})$ is invariant.

(c) Preliminary result: establish the equation of motion, which yields

$$\langle 0|\, T\left(\frac{\delta S}{\delta \bar{\eta}_a(x)}\, \bar{\eta}_b(x)\right)|0\rangle .$$

(Confine yourself to the Lorentz gauge $\partial_\mu A^\mu_a = 0$.) By writing

$$\frac{\delta}{\delta \zeta}\, \langle 0|\, T(\partial_\mu A^\mu_a(x)\, \bar{\eta}_b(x))|0\rangle = 0$$

show that the gauge-field propagator obeys equation (12.4.11): as in electrodynamics, the radiative corrections to the propagator are transverse.

13.6 Goldstone's theorem and the Higgs phenomenon in the general case (Abers and Lee 1973, Sections 2 and 3)

(a) The n *real* fields φ_i transform under a real representation of an N-dimensional group G, with generators T^a $(a = 1, \ldots, N)$. The effective action is

$$\Gamma = \int d^4x(\partial_\mu \varphi_i \partial_\mu \varphi_i + V(\varphi_i)).$$

Let v_i be some value of φ_i minimizing V,

$$(\partial V/\partial \varphi_i)|_{\varphi_i = v_i} = 0 .$$

Show that $V \simeq \frac{1}{2} M^2_{ij} (\varphi - v)_i\, (\varphi - v)_j$ in the vicinity of the minimum, and that $M^2_{ij}\, T^a_{jk}\, v_k = 0$.

Let H (M-dimensional) be that subgroup of G which leaves the vacuum invariant: i.e. $T^a\, v = 0$ if T^a is a generator of H. Thence show that the matrix M^2 has $(N - M)$ zero eigenvalues.

(b) Show that in a gauge theory $(N–M)$ gauge bosons become massive through the Higgs phenomenon.

13.7 Muon lifetime

Show that Fermi theory gives the mean life τ of the muon as

$$\tau \simeq \frac{192\, \pi^3}{G_F^2\, m_\mu^5} .$$

Show that the form of this result can be foreseen on dimensional grounds.

13.8 $\nu_e e$ and $\bar{\nu}_e e$ scattering according to Fermi theory

Using the Fermi theory, calculate the differential cross-sections for elastic $\nu_e e$ and $\bar{\nu}_e e$ scattering in the centre-of-mass frame. Show that the total cross-sections are given by

$$\sigma(\nu_e e) = \frac{4 G_F^2}{\pi} E_\nu^2,$$

$$\sigma(\bar{\nu}_e e) = \frac{4 G_F^2}{3\pi} \frac{3E_e^2 + E_\nu^2}{(E_e + E_\nu)^2} E_\nu^2,$$

where E_e and E_ν are the electron and the neutrino energies in the centre-of-mass frame.

13.9 Neutral-current interactions

(a) Show that the coupling between the Z^0 boson and leptons may be written as

$$-\frac{e}{\sin\theta\,\cos\theta}\,\bar{\psi}(l)\left[\frac{1}{2}\gamma_\mu(1-\gamma_5)I_3 - \gamma_\mu Q \sin^2\theta\right]\psi(l),$$

where I_3 is the third component of the weak isospin and Q the charge of the lepton in question.

(b) Recalculate the neutrino–electron and antineutrino–electron cross-sections of Problem 13.8, taking neutral currents into account. Confine yourself to the case where $E_e \gg m_e$, and use the Fierz identity

$$(\bar{\nu}\gamma_\mu(1-\gamma_5)e)(\bar{e}\gamma^\mu(1-\gamma_5)\nu) = (\bar{\nu}\gamma_\mu(1-\gamma_5)\nu)\,(\bar{e}\gamma^\mu(1-\gamma_5)e).$$

Show that the cross-sections now become

$$\sigma(\nu_e e) = \frac{G_F^2 s}{4\pi}\left[(1 + 2\sin^2\theta)^2 + \frac{4}{3}\sin^4\theta\right],$$

$$\sigma(\bar{\nu}_e e) = \frac{G_F^2 s}{4\pi}\left[4\sin^4\theta + \frac{1}{3}(1 + 2\sin^2\theta)^2\right],$$

where $s = (E_e + E_\nu)^2 \ll m_W^2$.

13.10 *W* and *Z* production

In the collider at CERN, *W* and *Z* bosons are produced by the 'fusion' of a quark–antiquark pair (Fig. 13.23).

One couples the bosons to leptons and to quarks through

$$\bar{v}(p')(C_V\gamma_\mu - C_A\gamma_\mu\gamma_5)u(p), \quad \text{(quarks)}$$

$$\bar{u}(k')(C'_V\gamma_\mu - C'_A\gamma_\mu\gamma_5)v(k), \quad \text{(leptons)}$$

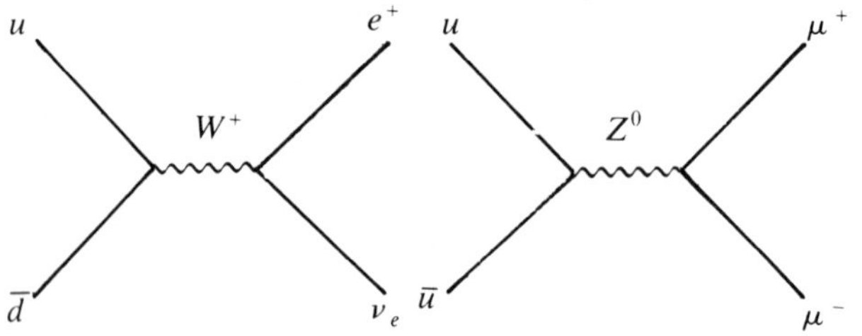

Fig. 13.23

defining

$$\alpha = \frac{2\,\mathrm{Re}\,C_V C_A^*}{|C_V|^2 + |C_A|^2}; \quad \alpha' = \frac{2\,\mathrm{Re}\,C'_V C'^*_A}{|C'_A|^2 + |C'_V|^2}.$$

Assuming quarks to be massless, show that the cross-section for observing a lepton at an angle φ to the direction of the u-quark in the boson rest frame, is proportional to

$$(1 + \cos^2\varphi) + 2\alpha\alpha'\cos\varphi.$$

In the W couplings one has $C_A = C_V$, and the angular distribution is proportional to $(1 + \cos\varphi)^2$.

13.11 The definition of Λ

The coupling constant α_s satisfies the evolution equation

$$\frac{d\alpha_s}{d\ln Q^2} = -\,4\,\pi\beta_0\,\alpha_s^2(Q^2) - 4\,\pi\beta_1\,\alpha_s^3(Q^2) + \cdots$$
$$= -\,\bar{\beta}_0\,\alpha_s^2(Q^2)(1 + \bar{\beta}_1\,\alpha_s(Q^2) + \cdots).$$

By integrating this equation in the form

$$\int_{\alpha_s(Q^2)} \frac{dx}{x^2(1 + \bar{\beta}_1 x)} = \bar{\beta}_0 \ln\frac{Q^2}{\Lambda^2} + \text{constant}$$

show that one may write

$$\alpha_s(Q^2) = \frac{1}{\bar{\beta}_0 y}\left(1 + \frac{c}{y} + \frac{d\ln y}{y} + \varepsilon\right),$$

where $y = \ln(Q^2/\Lambda^2)$. Determine c, d, and the order of magnitude of ε. Hence show that an appropriate choice of integration constant allows one to set $c = 0$. In this way recover (13.4.19).

13.12 The gauge independence of Z (Gross 1975, Section 5)

We aim to show that *in the minimal scheme* the renormalization constant Z (defined by $g_0 = \mu^{\varepsilon/2} Zg$) is independent of the gauge parameter $\mathfrak{a}$.

(a) Let $\hat{G}$ be an invariant charge (see Section 7.1.2), constructed from Green's functions of gauge invariant operators. Show that

$$\left.\frac{\mathrm{d}}{\mathrm{d}\mathfrak{a}}\right|_{g_0,\,\varepsilon} \hat{G}(g, \mathfrak{a}, \mu) = \left(\frac{\partial}{\partial \mathfrak{a}} + \rho(g, \mathfrak{a})\right) \hat{G}(g, \mathfrak{a}, \mu) = 0,$$

where $\rho(g, \mathfrak{a}) = \left.\dfrac{\partial g}{\partial \mathfrak{a}}\right|_{g_0,\,\varepsilon}$.

(b) Show that Z is independent of $\mathfrak{a}$, by using the fact that $\rho(g, \mathfrak{a})$ is independent of ε, together with the expansion (7.5.6) of Z in the minimal scheme. Why does this result not apply in general?

13.13 Derive equation (13.4.31).

13.14 Dimensional regularization of infrared singularities

(a) Calculation of the vertex.

Calculate the expression for the vertex in Fig. 13.14 (to one-loop order), by starting from (12.3.48) and integrating over the loop variable,

$$\Lambda_\mu = \frac{e_q^2\, g^2\, C_\mathrm{F}}{(4\pi)^{D/2}}\, \gamma_\mu ((4\,\pi)^{-1} \mu^2\, e^{\gamma})^{\varepsilon/2} \int \mathrm{d}x_1\, \mathrm{d}x_2\, \theta(1 - x_1 - x_2)$$
$$\times \left\{ \frac{(2 - D)^2 \Gamma(2 - D/2)}{2} (x_1 x_2\, Q^2)^{D/2 - 2} - Q^2\, \Gamma\left(3 - \frac{D}{2}\right) \right.$$
$$\left. \times \left[2(1 - x_1)(1 - x_2) + (D - 4) x_1\, x_2\right] (x_1 x_2\, Q^2)^{D/2 - 3} \right\},$$

where $\varepsilon = 4 - D$ and $Q^2 = -\,q^2 > 0$. The first term is UV divergent, and is to be renormalized in the $\overline{\mathrm{MS}}$ scheme (note the factor $(\mu^2\, e^{\gamma}/(4\pi)^{\varepsilon/2})$. After subtracting the UV pole proportional to $1/\varepsilon$, the result is continued analytically to values of $\varepsilon < 0$, $\varepsilon = -\,2\omega$. Show that $\Lambda_{\mathrm{R},\,\mu}$ reads

$$\Lambda_{\mathrm{R},\,\mu} = -\,\frac{\alpha_\mathrm{s}\, \gamma_\mu\, C_\mathrm{F}}{4\,\pi}\, e_q^2 \left[\left(\frac{Q^2}{\mu^2}\right)^{\omega} \left(\frac{2}{\omega^2} - \frac{4}{\omega} + 8 - \frac{\pi^2}{6}\right) + \ln \frac{Q^2}{\mu^2}\right]$$

(the formula (B.9) proves very useful here). Consider also the self-energy correction to external quark lines, and show, after subtraction of the

UV pole and analytic continuation to negative ε for $p^2 < 0$, that it satisfies

$$\left.\frac{\partial \Sigma_{\mathrm{R}}}{\partial \not p}\right|_{\not p = m} = -\frac{\alpha_s C_F}{4\pi}\frac{1}{\omega}.$$

(b) Calculation of gluon emission. Show that (13.4.31) must be replaced by

$$8\, C_F\, e_q^2 g^2 \left(\frac{\mu^2 e^{\gamma}}{4\pi}\right)^{-\omega} (1+\omega)\left[\frac{u^2 + s^2 + 2tq^2}{us} + \omega \frac{(u+s)^2}{us}\right],$$

and that (13.4.35) becomes

$$W = \frac{\| \boldsymbol{p}' \|^{2\omega}}{8(4\pi)^{D/2}\,\Gamma(1+\omega)} \int_{-1}^{1} \mathrm{d}(\cos\theta)\,(\sin\theta)^{2\omega}\, |\mathscr{M}|^2$$

(including a factor $z^{2\omega}$ due to (13.4.26)). Hence express W for real gluons as

$$W = \frac{2\alpha_s e_q^2}{3\pi z}\left(\frac{Q^2}{\mu^2}\right)^{\omega}\left\{\delta(1-z)\left(\frac{2}{\omega^2} - \frac{3}{2\omega} + \frac{7}{2} - \frac{\pi^2}{2}\right) + \frac{1}{\omega}\frac{1+z^2}{(1-z)_+}\right.$$
$$\left. + (1+z^2)\left(\frac{\ln(1-z)}{1-z}\right)_+ + 2\frac{1+z^2}{1-z}\ln z - \frac{3}{2(1-z)_+} + \frac{5}{2} - \frac{3}{2}z\right\}.$$

(c) By adjoining the virtual graph (a), and also the corrections due to external lines, derive the equation which, under dimensional regularization, is equivalent to (13.4.46). Check the cancellation of the IR divergences proportional to $1/\omega^2$ between the 'real' and the 'virtual' term, and identify the collinear singularities. Integrate over z and recover the value (13.4.52) of the ratio R.

13.15 The infrared approximation and the planar gauge

(a) Suppose that in calculating the matrix element (13.4.30), k_0, $\| \boldsymbol{k} \| \to 0$. Show that under these conditions

$$\mathscr{M}_\mu \simeq -\mathrm{i} e_q g \frac{\lambda_a}{2}\left[\frac{\varepsilon\cdot p}{p\cdot k} - \frac{\varepsilon\cdot p'}{p'\cdot k}\right]\bar{u}(p)\,\gamma_\mu v(p').$$

(b) Adopt an axial gauge $n^\mu A_\mu = 0$, where n_μ is a fixed vector. Show that one can sum over gluon polarizations by contracting with the tensor

$$d_{\mu\nu} = -g_{\mu\nu} + \frac{k_\mu n_\nu + k_\nu n_\mu}{k\cdot n} - \frac{n^2 k_\mu k_\nu}{(k\cdot n)^2}.$$

Calculate $|\mathscr{M}|^2$ by using $n_\mu = a p_\mu + b p'_\mu$, where a and b are constants.

Show that the interference term between the two diagrams of Fig. 13.12 vanishes, and that it is possible to choose a gauge where one of the two diagrams vanishes altogether.

13.16 Elitzur's theorem (Kogut 1979, Section V)

Consider the Wegner model in an arbitrary number D of dimensions. Let $S(n, \mu)$ be a spin S located on a bond μ starting from site n. Subject the system to an external magnetic field B, and calculate the mean value of $S(n, \mu)$,

$$\langle S(n, \mu)\rangle_B = \frac{\sum\limits_{[S(n,\mu)]} S(n, \mu) \exp\left[\beta J \sum\limits_{P} SSSS + B \sum\limits_{n,\mu} S(n, \mu)\right]}{\sum\limits_{[S(n,\mu)]} \exp\left[\beta J \sum\limits_{P} SSSS + B \sum\limits_{n,\mu} S(n, \mu)\right]}.$$

Let $\{l_n\}$ be the set of bonds starting from site n, and let $S'(l)$ arise from $S(l)$ by a gauge transformation which reverses the spins on the bonds l_n,

$$\delta S(l) = S'(l) - S(l) = -2S(l) \quad \text{if } l \in \{l_n\};$$

$$\delta S(l) = 0 \quad \text{if } l \notin \{l_n\}.$$

By changing variables $S \to S'$, prove the relation

$$\langle S(n, \mu)\rangle_B = \left\langle - S(n, \mu) \exp\left[- B \sum_{l \in \{l_n\}} \delta S(l)\right]\right\rangle ;$$

thence derive

$$2|\langle S(n, \mu)\rangle \leqslant |e^{4DB} - 1| \to 0 \quad \text{as } B \to 0.$$

13.17 Equivalence of the $D = 2$ Wegner model and the $D = 1$ Ising model (Kogut 1979, Section V)

For the two-dimensional lattice of Fig. 13.15 we choose two axes τ and x parallel to bond directions. Show that, by means of a sequence of local gauge transformations, and neglecting edge effects, one can reach a configuration where all the spins located on bonds parallel to the τ axis are equal to $+1$, i.e. $S(n, \tau) = 1$. Hence show that the partition function is a product of one-dimensional Ising partition functions. Using the result (1.2.9) for the $D = 1$ Ising correlation function, rederive the 'area law' for $W(\mathscr{R})$ (see (13.5.3)).

Further reading

The literature on non-Abelian gauge theories is enormous, and the references cited below are a very limited (and personal) selection. One classic reference for Lie groups is Pontryagin (1966). A discussion adapted to particle physics is given by Georgi (1982). A nice introduction to modern elementary-particle physics is Aitchison and Hey (1982). General references on non-Abelian gauge theories include Abers and Lee (1973); Itzykson and Zuber (1980, Chapter 12); Taylor (1976); Cheng and Li (1984); Zinn-Justin (1989, Chapter 18); Pokorski (1987); and Ramond (1980, Chapters VI, VII, VIII). For a full treatment of the renormalization of gauge theories, see Itzykson and Zuber (1980) or Collins (1983, Chapter 12). The GSW model is discussed by Abers and Lee (1973) and by Taylor (1976), and a good discussion of radiative corrections can be found in Altarelli (1990). The discussion of quantum chromodynamics by Altarelli (1982) is particularly accessible; see also the books by Yndurain (1983) and by Muta (1987). There are detailed calculations in de Rafael (1978) and in Le Bellac (1981). Finally, for the lattice gauge theories, see the book by Creutz (1983); also Kogut (1979, Section VIII, and 1983).

Appendices

A Fourier transforms and Gaussian integration

A.1 Fourier transforms

Proofs are given only in one dimension ($D = 1$); the generalization to an arbitrary number of dimensions is trivial.

A.1.1 Diagonalization of a matrix invariant under translations

Consider a (real) matrix A_{st}, $0 \leqslant s, t \leqslant N - 1$, such that A_{st} depends only on the difference $(s - t)$. Assume periodic boundary conditions: $p + N \equiv p$ for all integer p.

Theorem A.1: The matrix A_{st} is diagonalized by a Fourier transformation on a lattice, corresponding to the unitary transformation $U_{sq} = \dfrac{1}{\sqrt{N}} e^{iqx_s}$, with

$$x_s = sa; \quad q = 2\pi p/Na,$$

where s and p are integers between 0 and $N - 1$, and a is the lattice spacing.

This result is well known, for instance in the theory of vibrational normal modes. We rederive it briefly:

$$A_{q'q} = \frac{1}{N} \sum_{s,t} e^{iq'x_s} A_{st} e^{-iqx_t}$$

$$= \frac{1}{N} \sum_{s} e^{-i(q-q')x_s} \sum_{t} A(s-t) e^{-iq(x_t - x_s)}.$$

The sum over t is independent of s, by virtue of the condition $q = 2\pi p/Na$, and of the periodic boundary conditions. Call this sum $\tilde{A}(q)$:

$$A_{q'q} = \delta_{qq'} \tilde{A}(q). \tag{A.1.1}$$

In general $A(s - t)$ depends only on $|s - t|$, and $\tilde{A}(q)$ is real. Because the trace is invariant under a similarity transformation, we have

$$\operatorname{Tr} A_{st} = \sum_{q} \tilde{A}(q). \tag{A.1.2}$$

A.1.2 The continuum limit

To take the continuum limit, it proves convenient to define $\tilde{A}(q)$ by

$$\tilde{A}(q) = a\sum_t A(s-t)\mathrm{e}^{-\mathrm{i}q(x_t - x_s)} \to \int \mathrm{d}x\, A(x)\mathrm{e}^{\mathrm{i}qx}, \tag{A.1.3}$$

where we have gone from a Riemann sum to an integral. Similarly, the sum over q can be replaced by an integral on applying

$$\sum_q \to \frac{Na}{2\pi}\int \mathrm{d}q, \tag{A.1.4}$$

appropriate because successive values of q are spaced by $2\pi/Na$. On the other hand, instead of the limits $0 \leqslant q \leqslant 2\pi/a$, it is generally more convenient to choose

$$-\frac{\pi}{a} \leqslant q \leqslant \frac{\pi}{a}. \tag{A.1.5}$$

Under these conditions the inverse Fourier transform is given by

$$A(x) = \frac{1}{Na}\sum_q \mathrm{e}^{-\mathrm{i}qx}\tilde{A}(q) \to \int_{-\pi/a}^{\pi/a} \frac{\mathrm{d}q}{2\pi}\,\tilde{A}(q)\mathrm{e}^{-\mathrm{i}qx}, \tag{A.1.6}$$

and the trace relation becomes

$$\mathrm{Tr}\, A_{st} = N\int \frac{\mathrm{d}q}{2\pi}\,\tilde{A}(q). \tag{A.1.7}$$

In D dimensions these equations generalize to

$$\tilde{A}(\boldsymbol{q}) = \int \mathrm{d}^D x\, \mathrm{e}^{\mathrm{i}\boldsymbol{q}\cdot\boldsymbol{x}} A(\boldsymbol{x}), \tag{A.1.8a}$$

$$A(\boldsymbol{x}) = \int \frac{\mathrm{d}^D q}{(2\pi)^D}\,\mathrm{e}^{-\mathrm{i}\boldsymbol{q}\cdot\boldsymbol{x}}\tilde{A}(\boldsymbol{q}). \tag{A.1.8b}$$

A.1.3 The product of two matrices

A useful relation can be found by considering the product of two matrices. We look for the Fourier transform of $\sum_t A_{st}B_{tv} = \sum_{t,u} A_{st}\delta_{tu}B_{uv}$,

$$\frac{1}{N}\sum_{\substack{t,u\\s,v}} \mathrm{e}^{\mathrm{i}q'x_s}A_{st}\delta_{tu}B_{uv}\mathrm{e}^{-\mathrm{i}qx_v} = \frac{1}{N^2}\sum_{\substack{t,u\\s,v}}\sum_k \mathrm{e}^{\mathrm{i}q'x_s}A_{st}\mathrm{e}^{-\mathrm{i}kx_t}\mathrm{e}^{\mathrm{i}kx_u}B_{uv}\mathrm{e}^{-\mathrm{i}qx_v}$$

$$= \sum_k \delta_{q'k}\tilde{A}(q')\delta_{qk}\tilde{B}(q) = \delta_{qq'}\tilde{A}(q)\tilde{B}(q).$$

The Fourier transform of the matrix product is the product of the Fourier transforms. This is not surprising, given that the matrix product is just a convolution, by virtue of the translation invariance of the matrices A and B. It is interesting also to notice that the Fourier transform of A_{st}^{-1} is $1/\tilde{A}(q)$.

A.2 Gaussian integrals

A.2.1 One variable

Consider the quantity $Z(j)$:

$$Z(j) = \int \mathrm{d}x\, \mathrm{e}^{-(1/2)Ax^2 + jx},$$

$$Z(0) = \int \mathrm{d}x\, \mathrm{e}^{-(1/2)Ax^2} = \sqrt{(2\pi/A)}.$$

To calculate $Z(j)$ we change variables through $x = x' + j/A$:

$$\begin{aligned} -\frac{1}{2}xAx + jx &= -\frac{1}{2}x'Ax' + \frac{1}{2}j\frac{1}{A}j \\ Z(j) &= \mathrm{e}^{(1/2)j(1/A)j}Z(0). \end{aligned} \tag{A.2.1}$$

A.2.2 *N* variables

$$Z(j) = \int \prod_{i=1}^{N} \mathrm{d}x_i \exp\left(-\frac{1}{2}\sum_{i,j=1}^{N} x_i A_{ij} x_j + \sum_{i=1}^{N} j_i x_i \right),$$

where the matrix A_{ij} is *symmetric* and strictly positive*. To ease the notation we set (with T = transpose)

$$\sum_{i,j=1}^{N} x_i A_{ij} x_j = x^{\mathrm{T}}Ax, \quad \sum_{i=1}^{N} j_i x_i = j^{\mathrm{T}}x,$$

where x and j are column vectors $(x_1 \ldots x_N)$ and $(j_1 \ldots j_N)$, while x^{T} and j^{T} are row vectors. Change variables to x' given by

$$x = x' + A^{-1}j,$$

(the matrix A^{-1} exists because A is assumed positive); then

$$-\frac{1}{2}x^{\mathrm{T}}Ax + j^{\mathrm{T}}x = -\frac{1}{2}x'^{\mathrm{T}}Ax' + \frac{1}{2}j^{\mathrm{T}}A^{-1}j,$$

* Even if A_{ij} had an antisymmetric part, the contribution of this to $\sum_{i,j} x_i A_{ij} x_j$ would vanish.

whence

$$Z(j) = e^{(1/2) j^{\mathrm{T}} A^{-1} j} Z(0). \tag{A.2.2}$$

In many cases (A.2.2) is all one needs (as for instance in calculating correlation functions, from which $Z(0)$ cancels). It is not difficult to calculate $Z(0)$:

$$Z(0) = \int \prod_{i=1}^{N} \mathrm{d}x_i \, e^{-(1/2) x^{\mathrm{T}} A x}.$$

Let R be an orthogonal transformation ($RR^{\mathrm{T}} = \mathbb{1}$) diagonalizing A,

$$A = R^{\mathrm{T}} D R, \quad D = \begin{pmatrix} d_1 & & & \\ & d_2 & & 0 \\ & 0 & \ddots & \\ & & & d_N \end{pmatrix}, \quad d_i > 0 \quad \forall i.$$

Make the following change of variables with unit Jacobian:

$$x' = Rx \quad (\det R = 1),$$

$$\int \prod_{i=1}^{N} \mathrm{d}x_i \, e^{-(1/2) x^{\mathrm{T}} A x} = \int \prod_{i=1}^{N} \mathrm{d}x'_i \, e^{-(1/2) x'^{\mathrm{T}} D x'}.$$

The last integral is a product of N independent Gaussian integrals, and is given by

$$(2\pi)^{N/2} \prod_{i=1}^{N} (d_i)^{-1/2} = \frac{(2\pi)^{N/2}}{(\det A)^{1/2}},$$

$$Z(0) = \frac{(2\pi)^{N/2}}{(\det A)^{1/2}}. \tag{A.2.3}$$

Finally, let us derive a corollary of equations (A.2.2) and (A.2.3). Consider first a single complex variable $z = x + \mathrm{i}y$, and the integral

$$I = \int \mathrm{d}^2 z \, e^{-z^* A z + z^* j + z j^*} = \int \mathrm{d}x \, \mathrm{d}y \, e^{-A(x^2 + y^2) + 2 j_1 x + 2 j_2 y},$$

where $j = j_1 + \mathrm{i} j_2$ and $A = a_1 + \mathrm{i} a_2$, with $a_1 > 0$. Then I follows immediately:

$$I = \frac{\pi}{A} e^{j^* A^{-1} j}.$$

Next, consider the case of n complex variables z_i,

$$I = \int \prod_{i=1}^{N} \mathrm{d}^2 z_i \, e^{-z^{\dagger} A z + z^{\dagger} j + j^{\dagger} z},$$

where transposes have been replaced by the Hermitean conjugates. Assume that A can be diagonalized by a unitary transformation U,

$$A = U^{\dagger} D U,$$

where D is a diagonal matrix with elements d_i whose real parts are positive. Write

$$U = R + \mathrm{i}S,$$

where R and S are real matrices; the relation $U^{\dagger}U = \mathbb{1}$ entails

$$RR^{\mathrm{T}} + SS^{\mathrm{T}} = \mathbb{1}, \quad RS^{\mathrm{T}} - SR^{\mathrm{T}} = 0.$$

The transformation $z' = Uz$ amounts to

$$\begin{pmatrix} x' \\ y' \end{pmatrix} = \begin{pmatrix} R & -S \\ S & R \end{pmatrix} \begin{pmatrix} x \\ y \end{pmatrix},$$

and the matrix which transforms (x, y) into (x', y') is orthogonal, whence of determinant 1. The Jacobian of the transformation is 1, which leads to the end-result

$$\int \prod_{i=1}^{N} \mathrm{d}^2 z_i \mathrm{e}^{-z^{\dagger}Az + z^{\dagger}j + j^{\dagger}z} = \frac{\pi^N}{\det A} \mathrm{e}^{j^{\dagger}A^{-1}j}. \tag{A.2.4}$$

A.3 Integrals in *D* dimensions

In polar coordinates the D-dimensional volume element reads

$$\mathrm{d}^D x = r^{D-1} \mathrm{d}r \sin^{D-2}\theta_{D-1} \mathrm{d}\theta_{D-1} \sin^{D-3}\theta_{D-2} \mathrm{d}\theta_{D-2} \ldots \mathrm{d}\theta_1,$$

$$0 \leqslant \theta_1 \leqslant 2\pi, \quad 0 \leqslant \theta_k \leqslant \pi, \quad k \neq 1.$$

Very often the integrand is independent of the angles; then one needs only the surface area S_D of the D-dimensional sphere:

$$\mathrm{d}^D x \to S_D r^{D-1} \mathrm{d}r.$$

This is easily found by evaluating in two different ways the integral

$$J = \int \mathrm{d}^D x \mathrm{e}^{-(x_1^2 + \cdots + x_D^2)} = \pi^{D/2},$$

$$J = S_D \int_0^{\infty} r^{D-1} \mathrm{e}^{-r^2} \mathrm{d}r = \frac{1}{2} S_D \Gamma\left(\frac{D}{2}\right),$$

whence

$$S_D = \frac{2\pi^{D/2}}{\Gamma(D/2)}. \tag{A.3.1}$$

Examples: $D = 1$, $S_2 = 2\pi$; $D = 3$, $S_3 = 4\pi$; $D = 4$, $S_4 = 2\pi^2$.

As a rule the integration measure is $\mathrm{d}^D x/(2\pi)^D$, and it proves useful to define

$$K_D = \frac{S_D}{(2\pi)^D} = \frac{2}{(4\pi)^{D/2}\Gamma(D/2)}. \tag{A.3.2}$$

In particular,

$$K_4 = \frac{1}{8\pi^2}. \tag{A.3.3}$$

The expressions (A.3.1) and (A.3.2) can be continued analytically to nonintegral values of D.

B Feynman integrals with dimensional regularization (Euclidean case)

Amalgamation of denominators:

$$\frac{1}{A_1^{\alpha_1}A_2^{\alpha_2}\ldots A_n^{\alpha_n}}=\frac{\Gamma(\alpha_1+\alpha_2+\cdots+\alpha_n)}{\Gamma(\alpha_1)\Gamma(\alpha_2)\ldots\Gamma(\alpha_n)}\int\prod_{i=1}^{n}\mathrm{d}x_i\,\delta\left(1-\sum_{i=1}^{n}x_i\right)$$
$$\times x_1^{\alpha_1-1}\ldots x_n^{\alpha_n-1}[x_1A_1+x_2A_2+\cdots$$
$$+x_nA_n]^{-(\alpha_1+\cdots+\alpha_n)}. \tag{B.1}$$

Loop with two internal lines: $q'=q-xk$

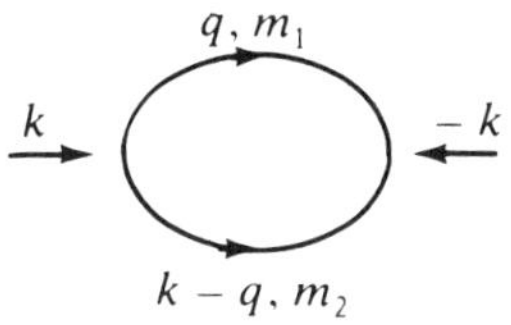

$$=\int_0^1\mathrm{d}x\int\frac{\mathrm{d}^Dq'}{(2\pi)^D}$$
$$\times[q'^2+x(1-x)k^2+xm_2^2+(1-x)m_1^2]^{-2}. \tag{B.2}$$

Loop with three internal lines: $q'=q-x_1p_3+x_3p_1$

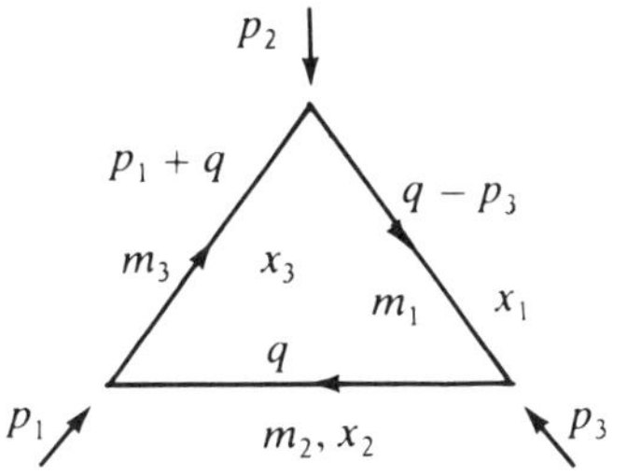

$$=2\int\prod_1^3\mathrm{d}x_i\,\delta\left(1-\sum_{i=1}^{3}x_i\right)\int\frac{\mathrm{d}^Dq'}{(2\pi)^D}\frac{1}{[q'^2+M^2]^3} \tag{B.3}$$

$$M^2=x_1m_1^2+x_2m_2^2+x_3m_3^2+x_2x_3p_1^2+x_1x_3p_2^2+x_1x_2p_3^2.$$

Integrals over q:

$$\int \frac{\mathrm{d}^D q}{(2\pi)^D} \frac{1}{(q^2 + m^2)^N} = \frac{\Gamma(N - D/2)}{(4\pi)^{D/2} \Gamma(N)} \frac{1}{(m^2)^{N - D/2}} \tag{B.4}$$

$$\int \frac{\mathrm{d}^D q}{(2\pi)^D} \frac{q^2}{(q^2 + m^2)^N} = \frac{\Gamma(N - 1 - D/2)}{2(4\pi)^{D/2} \Gamma(N)} \frac{D}{(m^2)^{N - 1 - D/2}} \tag{B.5}$$

$$\int \frac{\mathrm{d}^D q}{(2\pi)^D} \frac{q_i q_j}{(q^2 + m^2)^N} = \frac{\Gamma(N - 1 - D/2)}{2(4\pi)^{D/2} \Gamma(N)} \frac{\delta_{ij}}{(m^2)^{N - 1 - D/2}} \tag{B.6}$$

$$\int \frac{\mathrm{d}^D q}{(2\pi)^D} q_i q_j f(q^2) = \frac{\delta_{ij}}{D} \int \frac{\mathrm{d}^D q}{(2\pi)^D} q^2 f(q^2). \tag{B.7}$$

Expansion of the Γ function:

$$\Gamma\left(N - \frac{\varepsilon}{2}\right) = \Gamma(N)\left(1 - \frac{\varepsilon}{2}\psi(N) + O(\varepsilon^2)\right) \tag{B.8}$$

$$\psi(N) = \sum_{j=1}^{N-1} \frac{1}{j} - \gamma = S_{N-1} - \gamma; \quad S_0 = 0, \quad S_1 = 1, \quad S_2 = \frac{3}{2}, \quad S_3 = \frac{11}{6}$$

$$\Gamma\left(1 - \frac{\varepsilon}{2}\right) = 1 + \frac{\varepsilon}{2}\gamma + \frac{1}{8}\varepsilon^2\left(\gamma^2 + \frac{\pi^2}{6}\right) + O(\varepsilon^3)$$

$$= \exp\left(\frac{\varepsilon}{2}\gamma + \frac{\pi^2}{48}\varepsilon^2\right) + O(\varepsilon^3) \tag{B.9}$$

$$\Gamma\left(\frac{\varepsilon}{2}\right) = \frac{2}{\varepsilon} - \gamma + O(\varepsilon). \tag{B.10}$$

Expansion of the B function:

$$B(\alpha, \beta) = \frac{\Gamma(\alpha + \beta)}{\Gamma(\alpha)\Gamma(\beta)} = \int_0^1 \mathrm{d}x\, x^{\alpha - 1}(1 - x)^{\beta - 1} \tag{B.11}$$

$$B\left(N - \frac{\varepsilon}{2}, 1 - \frac{\varepsilon}{2}\right) = \frac{1}{N}\left(1 + \varepsilon S_N - \frac{\varepsilon}{2} S_{N-1} + O(\varepsilon^2)\right) \tag{B.12}$$

$$B\left(N - \frac{\varepsilon}{2}, 2 - \frac{\varepsilon}{2}\right) = \frac{1}{N(N + 1)}\left(1 - \frac{\varepsilon}{2} S_{N-1} - \frac{\varepsilon}{2} + \varepsilon S_{N+1} + O(\varepsilon^2)\right). \tag{B.13}$$

C Noether's theorem, conserved currents and Ward identities

In this Appendix, we examine the consequences of the invariance of the action under a continuous symmetry transformation. At the *classical* level, we derive Noether's theorem, namely the existence of a conserved current j_μ (or of several conserved currents j^a_μ if the symmetry has several independent generators). Current conservation

$$\partial^\mu j_\mu(x) = 0 \tag{C.0.1}$$

is valid when the fields obey the classical equations of motion. It is also valid (except in the case of anomalies) in operator form at the quantum level. However, if one chooses a path-integral approach, one has to integrate over field configurations which do not obey the equations of motion, and this will lead to Ward identities, which express current conservation at the quantum level (in the operator formalism, the Ward identities are obtained from the fact that the time derivative does not commute with the T-product).

C.1 Noether's theorem and conserved currents

In order to be specific, we consider in Minkowski space a Lagrangean $\mathscr{L}(x)$ invariant under an $O(n)$ symmetry ($l = 1, 2, \ldots, n$):

$$\mathscr{L} = \frac{1}{2}(\partial_\mu \varphi_l)(\partial^\mu \varphi_l) - V\left(\sum_{l=1}^{n} \varphi_l^2\right). \tag{C.1.1}$$

In Euclidean space, the Hamiltonian corresponding to (C.1.1) would describe a statistical system whose order parameter has dimension n. For simplicity we have assumed that $V(\varphi)$ does not depend on $\partial_\mu \varphi_l$, but all results in what follows are independent of this assumption. Let us consider an infinitesimal rotation around an order-parameter axis a ($a = 1, 2, \ldots, n(n-1)/2$); the rotation angle is denoted by ε. The transformation law of the fields reads

$$\delta^a_\varepsilon \varphi_l = -\mathrm{i}\varepsilon T^a_{lm}\varphi_m = \varepsilon\delta^a\varphi_l, \quad T^a_{lm} = T^{a*}_{ml}. \tag{C.1.2}$$

We have taken Hermitean generators T^a ($T^a = (T^a)^\dagger$). The angle ε is x-independent: we are dealing with a global symmetry, not a local one.

We have given in Section 11.3.3 a method for constructing the conserved current j^a_μ associated with the symmetry. Let us give here another derivation, which uses an x-dependent $\varepsilon(x)$, such that $\varepsilon(x) \to 0$ fast enough at infinity (or even an $\varepsilon(x)$ which is nonzero only in a finite region of space–time). In that case

$$\delta^a_\varepsilon(\partial_\mu \varphi_l) = -\mathrm{i}(\partial_\mu \varepsilon(x)) T^a_{lm} \varphi_m - \mathrm{i}\varepsilon(x) T^a_{lm}(\partial_\mu \varphi_m).$$

Since the action is invariant when ε is x-independent, we have

$$\delta S = \int \mathrm{d}^4 x (\partial^\mu \varepsilon(x)) B^a_\mu. \tag{C.1.3}$$

If the equations of motion are satisfied, we must have $\delta S = 0$ since $\delta^a \varphi_l$ is a variation around the classical field. We can integrate (C.1.3) by parts,

$$\delta S = -\int \mathrm{d}^4 x \varepsilon(x) \partial^\mu B^a_\mu,$$

and from this equation it follows that

$$\partial^\mu B^a_\mu = 0, \tag{C.1.4}$$

since $\varepsilon(x)$ is arbitrary. The conserved current j^a_μ is to be identified with B^a_μ.

In the present derivation, we have assumed only that $\delta S = 0$ under the symmetry operation, while in Chapter 11 we made the stronger assumption $\delta \mathscr{L} = 0$. In general $\delta S = 0$ implies only

$$\delta \mathscr{L} = \partial^\mu \Lambda^a_\mu,$$

and the conserved current is given by

$$j^a_\mu = \frac{\partial \mathscr{L}}{\partial(\partial_\mu \varphi_l)} \delta^a \varphi_l - \Lambda^a_\mu. \tag{C.1.5}$$

However, it may be simpler to identify the conserved current directly from (C.1.3): one has to integrate by parts until the coefficient of $\varepsilon(x)$ vanishes.

C.2 Energy–momentum tensor

As a straightforward application, let us derive the expression for the energy–momentum tensor. We want to find the conserved current associated with the invariance of the action under space–time translations; for simplicity we limit ourselves to the case of only one field. If we translate the origin of coordinates,

$$x^\mu \to x'^\mu = x^\mu + \varepsilon^\mu,$$

then we must have

$$\varphi'(x^\mu) = \varphi(x^\mu - \varepsilon^\mu) = \varphi(x) - \varepsilon^\alpha \partial_\alpha \varphi,$$

since $\varphi'(x') = \varphi(x)$. Thus

$$\delta_\varepsilon^\alpha \varphi = -\varepsilon^\alpha \partial_\alpha \varphi = -\varepsilon^\alpha \delta_\alpha \varphi. \tag{C.2.1}$$

We now compute δS for an x-dependent ε:

$$\delta S = \int \mathrm{d}^4 x\, \varepsilon^\alpha(x) \left[\frac{\partial \mathscr{L}}{\partial \varphi} \partial_\alpha \varphi + \frac{\partial \mathscr{L}}{\partial(\partial_\mu \varphi)} \partial_\alpha \partial_\mu \varphi \right] + \int \mathrm{d}^4 x (\partial_\mu \varepsilon^\alpha(x)) \frac{\partial \mathscr{L}}{\partial(\partial_\mu \varphi)} \partial_\alpha \varphi. \tag{C.2.2}$$

If S is to be translation invariant, $\mathscr{L}$ must depend only on φ and $\partial_\mu \varphi$: it must not depend *explicitly* on x. Then one recognizes in the first term of (C.2.2) the derivative $\partial_\alpha \mathscr{L}$ of $\mathscr{L}$. We can integrate by parts the last term in (C.2.2):

$$\delta S = \int \mathrm{d}^4 x\, \varepsilon^\alpha(x) \left(\partial_\alpha \mathscr{L} - \partial_\mu \left(\frac{\partial \mathscr{L}}{\partial(\partial_\mu \varphi)} \partial_\alpha \varphi \right) \right).$$

If the field $\varphi(x)$ obeys the classical equations of motion, then we have $\delta S = 0$, and, since $\varepsilon(x)$ is arbitrary, we obtain the conserved current $T^{\mu\alpha}$ (note that the Lorentz indices μ and α play quite different roles in the derivation):

$$T^{\mu\alpha} = \frac{\partial \mathscr{L}}{\partial(\partial_\mu \varphi)} \partial^\alpha \varphi - g^{\mu\alpha} \mathscr{L}, \tag{C.2.3}$$

$$\partial_\mu T^{\mu\alpha} = 0.$$

$T^{\mu\alpha}$ is the energy–momentum tensor of the field. The integrals of the time components $T^{0\alpha}$ give the energy–momentum 4-vector P^α,

$$P^\alpha = \int_{x^0 = t} \mathrm{d}^3 x\, T^{0\alpha}, \tag{C.2.4}$$

whose components are time-independent. Indeed

$$\frac{\mathrm{d}}{\mathrm{d}t} P^\alpha = \int \mathrm{d}^3 x\, \partial_0 T^{0\alpha} = -\int \mathrm{d}^3 x\, \partial_i T^{i\alpha},$$

and the last integral vanishes by virtue of the divergence theorem, assuming that the field decreases fast enough at infinity.

The tensor $T^{\mu\alpha}$ defined in (C.2.3) is not in general symmetric in the indices μ and α. However one can always modify $T^{\mu\alpha}$, without changing P^α, so as to make it symmetric. The interested reader is referred to the literature for further discussions of this property.

C.3 Ward identities

We start from the generating functional of Green's functions for the Lagrangean (C.1.1),

$$Z(j) = \int \mathscr{D}\varphi \exp\left(iS[\varphi] + i \int \mathrm{d}^4 x\, j_l(x) \varphi_l(x) \right), \tag{C.3.1}$$

and change variables

$$\varphi_l \to \varphi'_l = \varphi_l + \varepsilon(x)\delta^a\varphi_l, \tag{C.3.2}$$

with an x-dependent $\varepsilon(x)$. Using the invariance of the integration measure, namely

$$\mathscr{D}\varphi' = \mathscr{D}\varphi,$$

and (C.1.3), we get

$$Z(j) = \int \mathscr{D}\varphi \exp\left(iS[\varphi] + i \int d^4x\, j_l(x)\varphi_l(x) + (\partial^\mu \varepsilon(x)) j^a_\mu + \varepsilon(x) j_l(x)\delta^a\varphi_l \right) \tag{C.3.3}$$

(j_l should not be confused with j^a_μ!). We now expand the exponential in (C.3.3) to first order in $\varepsilon(x)$ and use the fact that $\varepsilon(x)$ is arbitrary:

$$\int \mathscr{D}\varphi \left[\exp\left(iS[\varphi] + i \int d^4x\, j_l(x)\varphi_l(x) \right) \right] (-\partial^\mu j^a_\mu(y) + j_l(y)\delta^a\varphi_l(y)) = 0. \tag{C.3.4}$$

This equation is a generating functional of Ward identities; when we take N derivatives with respect to $j_l(x)$ and set $j_l = 0$ we obtain the Ward identity:

$$\partial^\mu_y \langle 0 | T(j^a_\mu(y)\varphi_{i_1}(x_1) \ldots \varphi_{i_N}(x_N)) | 0 \rangle$$
$$= -i \sum_{p=1}^{N} \delta(x_p - y) \langle 0 | T(\varphi_{i_1}(x_1) \ldots \delta^a \varphi_{i_p}(x_p) \ldots \varphi_{i_N}(x_N)) | 0 \rangle. \tag{C.3.5}$$

One should note the close similarity between the derivation of (C.3.5) and the derivation of the equations of motion for Green's functions in Section 10.2.4. One can also see the quantum mechanical version of current conservation $\partial^\mu j^a_\mu = 0$: the divergence of the current, when inserted into a Green's function, vanishes, except at the points $x_1, \ldots, x_N$.

A deep and thorough account of the subject can be found in the two lectures by Jackiw (1985).

D Formulary

D.1 The Lorentz group

Four-vector:

$$V^\mu = (V^0, V^i) = (V^0, \vec{V}) = (V_t, V_x, V_y, V_z).$$

Metric tensor:

$$g^{\mu\nu} = \mathrm{diag}(1, -1, -1, -1).$$

Lorentz transformation:

$$x'^\mu = \Lambda^\mu_\nu x^\nu, \quad \Lambda^\mathrm{T} g \Lambda = g.$$

The proper Lorentz group:

$$\Lambda^0_0 \geqslant 1, \quad \det \Lambda = 1.$$

The orthochronous Lorentz group:

$$\Lambda^0_0 \geqslant 1, \quad \det \Lambda = \pm 1.$$

Gradient:

$$\partial_\mu = \frac{\partial}{\partial x^\mu} = \left(\frac{\partial}{\partial x^0}, \frac{\partial}{\partial x^i}\right) = (\partial_0, \nabla).$$

Scalar product:

$$x^\mu y_\mu = x^0 y^0 - \boldsymbol{x} \cdot \boldsymbol{y}$$

$$\partial^\mu V_\mu = \partial^0 V^0 + \nabla \cdot \boldsymbol{V}.$$

Completely antisymmetric tensor:

$$\varepsilon^{\mu\nu\sigma\rho} = \begin{cases} 1 & \text{if the permutation } 0123 \to \mu\nu\sigma\rho \text{ is even,} \\ -1 & \text{if the permutation } 0123 \to \mu\nu\sigma\rho \text{ is odd,} \\ 0 & \text{otherwise.} \end{cases} \qquad \text{(D.1.1)}$$

4-momentum operator:

$$p^\mu = \mathrm{i}\partial^\mu = (\mathrm{i}\partial^0, -\mathrm{i}\nabla).$$

D'Alembertian:

$$\square = \partial_\mu \partial^\mu = \frac{\partial^2}{\partial x_0^2} - \nabla^2.$$

D.2 Dirac matrices

Definitions:

$$\{\gamma_\mu, \gamma_\nu\} = \gamma_\mu \gamma_\nu + \gamma_\nu \gamma_\mu = 2g_{\mu\nu} \mathbb{1}$$

$$\sigma^{\mu\nu} = \frac{\mathrm{i}}{2}[\gamma^\mu, \gamma^\nu]; \quad \gamma_5 = \mathrm{i}\gamma^0\gamma^1\gamma^2\gamma^3 = \frac{-\mathrm{i}}{4!}\varepsilon_{\mu\nu\rho\sigma}\gamma^\mu\gamma^\nu\gamma^\rho\gamma^\sigma.$$

Dirac equation:

$$(\mathrm{i}\gamma^\mu\partial_\mu - m)\psi = (\mathrm{i}\vec{\not\partial} - m)\psi = 0.$$

Conjugate spinor:

$$\bar{\psi} = \psi^\dagger \gamma_0, \quad \bar{\psi}(\mathrm{i}\overleftarrow{\not\partial} + m) = 0.$$

Hermitean conjugation:

$$\gamma_\mu^\dagger = \gamma_0 \gamma_\mu \gamma_0, \quad \gamma_0 \gamma_5 \gamma_0 = -\gamma_5^\dagger = -\gamma_5.$$

If ψ_1 and ψ_2 are two Dirac spinors, and Γ is a 4×4 matrix, then

$$(\bar{\psi}_1 \Gamma \psi_2)^* = \bar{\psi}_2(\gamma_0 \Gamma^\dagger \gamma_0)\psi_1. \tag{D.2.1}$$

Charge-conjugation matrix:

$$C\gamma_\mu C^{-1} = -\gamma_\mu^{\mathrm{T}}, \quad C\gamma_5 C^{-1} = \gamma_5^{\mathrm{T}}. \tag{D.2.2}$$

In the usual representations

$$C^{-1} = C^{\mathrm{T}} = C^\dagger = -C. \tag{D.2.3}$$

Pauli matrices:

$$\sigma_1 = \begin{pmatrix} 0 & 1 \\ 1 & 0 \end{pmatrix}, \quad \sigma_2 = \begin{pmatrix} 0 & -\mathrm{i} \\ \mathrm{i} & 0 \end{pmatrix}, \quad \sigma_3 = \begin{pmatrix} 1 & 0 \\ 0 & -1 \end{pmatrix}. \tag{D.2.4}$$

The Dirac representation:

$$\gamma^0 = \begin{pmatrix} \mathbb{1} & 0 \\ 0 & -\mathbb{1} \end{pmatrix}, \quad \gamma^i = \begin{pmatrix} 0 & \sigma_i \\ -\sigma_i & 0 \end{pmatrix}, \quad \gamma_5 = \begin{pmatrix} 0 & \mathbb{1} \\ \mathbb{1} & 0 \end{pmatrix}. \tag{D.2.5}$$

The chiral representation:

$$\gamma^0 = \begin{pmatrix} 0 & -\mathbb{1} \\ -\mathbb{1} & 0 \end{pmatrix}, \quad \gamma^i = \begin{pmatrix} 0 & \sigma_i \\ -\sigma_i & 0 \end{pmatrix}, \quad \gamma_5 = \begin{pmatrix} \mathbb{1} & 0 \\ 0 & -\mathbb{1} \end{pmatrix}. \tag{D.2.6}$$

Useful identities in D dimensions ($\not{a} = \gamma_\mu a^\mu$):

$$\not{a}\not{b} + \not{b}\not{a} = 2a\cdot b$$

$$\gamma^\mu\gamma_\mu = D \tag{D.2.7}$$

$$\gamma^\mu\not{a}\gamma_\mu = -2\not{a} + (4 - D)\not{a} \tag{D.2.8}$$

$$\gamma^\mu\not{a}\not{b}\gamma_\mu = 4a\cdot b - (4 - D)\not{a}\not{b} \tag{D.2.9}$$

$$\gamma^\mu\not{a}\not{b}\not{c}\gamma_\mu = -2\not{c}\not{b}\not{a} + (4 - D)\not{a}\not{b}\not{c} \tag{D.2.10}$$

Trace identities:

$$\mathrm{Tr}\,\mathbb{1} = 4$$

$$\mathrm{Tr}\,\not{a}\not{b} = 4a\cdot b$$

$$\mathrm{Tr}\,\gamma_5\gamma_\mu = 0$$

$$\mathrm{Tr}(\not{a}\not{b}\not{c}\not{d}) = 4[(a\cdot b)(c\cdot d) - (a\cdot c)(b\cdot d) + (a\cdot d)(b\cdot c)] \tag{D.2.11}$$

$$\mathrm{Tr}(\gamma_5\not{a}\not{b}\not{c}\not{d}) = -4\mathrm{i}\varepsilon^{\mu\nu\rho\sigma}a_\mu b_\nu c_\rho d_\sigma \tag{D.2.12}$$

$$\mathrm{Tr}(\not{a}_1 \dots \not{a}_{2n-1}) = 0; \quad \mathrm{Tr}(\not{a}_1 \dots \not{a}_{2n}) = \mathrm{Tr}(\not{a}_{2n} \dots \not{a}_1) \tag{D.2.13}$$

$$\mathrm{Tr}(\not{a}_1 \dots \not{a}_{2n}) = (a_1\cdot a_2)\mathrm{Tr}(\not{a}_3 \dots \not{a}_{2n}) - (a_1\cdot a_3)\mathrm{Tr}(\not{a}_2\not{a}_4 \dots \not{a}_{2n}) + \cdots + (a_1\cdot a_{2n})\mathrm{Tr}(\not{a}_2\not{a}_3 \dots \not{a}_{2n-1}). \tag{D.2.14}$$

Solutions of the Dirac equation:

Positive energy: $(\not{p} - m)u(\boldsymbol{p}) = 0; \quad \bar{u}(\boldsymbol{p})(\not{p} - m) = 0.$

Negative energy: $(\not{p} + m)v(\boldsymbol{p}) = 0; \quad \bar{v}(\boldsymbol{p})(\not{p} + m) = 0.$

Normalization:

$$\bar{u}(\boldsymbol{p})u(\boldsymbol{p}) = -\bar{v}(\boldsymbol{p})v(\boldsymbol{p}) = 2m$$

$$\bar{u}(\boldsymbol{p})v(\boldsymbol{p}) = \bar{v}(\boldsymbol{p})u(\boldsymbol{p}) = 0.$$

Normalization of the densities:

$$\bar{u}(\boldsymbol{p})\gamma^0 u(\boldsymbol{p}) = \bar{v}(\boldsymbol{p})\gamma^0 v(\boldsymbol{p}) = 2E_p$$

$$\bar{u}(-\boldsymbol{p})\gamma^0 v(\boldsymbol{p}) = \bar{v}(-\boldsymbol{p})\gamma^0 u(\boldsymbol{p}) = 0.$$

Projection operators:

$$\sum_{r=1}^{2} u_\alpha^{(r)}(\boldsymbol{p})\bar{u}_\beta^{(r)}(\boldsymbol{p}) = (\Lambda_+)_{\alpha\beta} = (\not{p} + m)_{\alpha\beta}$$

$$-\sum_{r=1}^{2} v_\alpha^{(r)}(\boldsymbol{p})\bar{v}_\beta^{(r)}(\boldsymbol{p}) = (\Lambda_-)_{\alpha\beta} = (-\not{p} + m)_{\alpha\beta}.$$

Spinors with sharp helicity λ (Dirac representation):

$$u_\lambda(\boldsymbol{p}) = \frac{1}{\sqrt{E_p + m}} \begin{pmatrix} (E_p + m)R\chi_\lambda \\ 2p\lambda R\chi_\lambda \end{pmatrix},$$

$$v_\lambda(\boldsymbol{p}) = \frac{1}{\sqrt{E_p + m}} \begin{pmatrix} -2p\lambda R\mathrm{i}\sigma_2\chi_\lambda \\ (E_p + m)R\mathrm{i}\sigma_2\chi_\lambda \end{pmatrix}, \tag{D.2.15}$$

where

$$\boldsymbol{p} = (p\sin\theta,\, 0,\, p\cos\theta), \quad R = \exp\left(-\frac{\mathrm{i}}{2}\sigma_2\theta\right),$$

$$\chi_{1/2} = \begin{pmatrix} 1 \\ 0 \end{pmatrix}, \quad \chi_{-1/2} = \begin{pmatrix} 0 \\ 1 \end{pmatrix}.$$

Gordon identities:

$$\bar{u}(\boldsymbol{p})\gamma^\mu u(\boldsymbol{q}) = \frac{1}{2m}\bar{u}(\boldsymbol{p})[(p+q)^\mu + \mathrm{i}\sigma^{\mu\nu}(p-q)_\nu]u(\boldsymbol{q}) \tag{D.2.16}$$

$$\bar{u}(\boldsymbol{p})\gamma^\mu\gamma_5 u(\boldsymbol{q}) = \frac{1}{2m}\bar{u}(\boldsymbol{p})[(p-q)^\mu\gamma_5 + \mathrm{i}\sigma^{\mu\nu}(p+q)_\nu\gamma_5]u(\boldsymbol{q}) \tag{D.2.17}$$

$$\bar{v}(\boldsymbol{p})\gamma^\mu v(\boldsymbol{q}) = -\frac{1}{2m}\bar{v}(\boldsymbol{p})[(p+q)^\mu + \mathrm{i}\sigma^{\mu\nu}(p-q)_\nu]v(\boldsymbol{q}) \tag{D.2.18}$$

$$\bar{v}(\boldsymbol{p})\gamma^\mu\gamma_5 v(\boldsymbol{q}) = -\frac{1}{2m}\bar{v}(\boldsymbol{p})[(p-q)^\mu\gamma_5 + \mathrm{i}\sigma^{\mu\nu}(p+q)_\nu\gamma_5]v(\boldsymbol{q}). \tag{D.2.19}$$

D.3 Cross-sections

Normalization of the states ($\omega_p = \sqrt{\boldsymbol{p}^2 + m^2}$):

$$\langle \boldsymbol{p} | \boldsymbol{p}' \rangle = (2\pi)^3 2\omega_p \delta^{(3)}(\boldsymbol{p} - \boldsymbol{p}').$$

S-matrix:

$$S_{fi} = \delta_{fi} + i(2\pi)^4\delta^{(4)}(P_f - P_i)T_{fi}.$$

Cross-section for $1 + 2 \to 1' + 2' + \cdots + N'$:

$$\mathrm{d}\sigma = \frac{1}{4F}|T_{fi}|^2(2\pi)^4\delta^{(4)}(P_f - P_i)\frac{\mathrm{d}^3p_1'}{(2\pi)^3 2\omega_1'}\cdots\frac{\mathrm{d}^3p_N'}{(2\pi)^3 2\omega_N'}\mathscr{S}$$

$$F = [(p_1\cdot p_2)^2 - m_1^2 m_2^2]^{1/2};$$

$\mathscr{S}$ = factor due to the indistinguishability of particles (see (10.3.11)).

Decay rate for $1 \to 1' + 2' + \cdots + N'$:

$$\mathrm{d}\Gamma = \frac{1}{2m_1} |T_{fi}|^2 (2\pi)^4 \delta^{(4)}(P_f - P_1) \frac{\mathrm{d}^3 p_1'}{(2\pi)^3 2\omega_1'} \cdots \frac{\mathrm{d}^3 p_N'}{(2\pi)^3 2\omega_N'} \mathscr{S}.$$

Cross-section for $1 + 2 \to 1' + 2'$ in the centre-of-mass frame:

$$\frac{\mathrm{d}\sigma}{\mathrm{d}\Omega} = \frac{1}{64\pi^2 s} \frac{\|\boldsymbol{k}'\|}{\|\boldsymbol{k}\|} |T(s, \cos\theta)|^2, \quad s = (p_1 + p_2)^2$$
$$\|\boldsymbol{k}\| = \frac{1}{2\sqrt{s}} [(s - (m_1 + m_2)^2)(s - (m_1 - m_2)^2)]^{1/2}. \tag{D.3.1}$$

Optical theorem:

$$\sigma_{\mathrm{tot}} = \frac{1}{2\sqrt{s}\,\|\boldsymbol{k}\|} \operatorname{Im} T(s, \theta = 0). \tag{D.3.2}$$

D.4 Feynman rules

We give the Feynman rules for a connected Green's function $G_c^{(N)}(p_1, \ldots, p_N)$ without external lines, where the momenta p_i run inwards into the diagram, and where

$$(2\pi)^4 \delta^{(4)}\left(\sum_{i=1}^{N} p_i\right) G_c^{(N)}(p_i) = \int \left(\prod_{i=1}^{N} \mathrm{d}^4 x_i \mathrm{e}^{-\mathrm{i} p_i x_i}\right) G_c^{(N)}(x_i).$$

(i) Draw all topologically inequivalent diagrams.

(ii) To each line assign a propagator as follows:

spin 0: 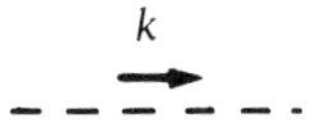

$$: \frac{\mathrm{i}}{k^2 - m^2 + \mathrm{i}\varepsilon};$$

spin 1/2: 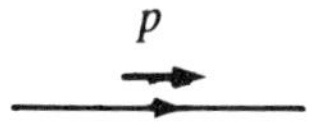

$$: \frac{\mathrm{i}}{\not{p} - m + \mathrm{i}\varepsilon} = \frac{\mathrm{i}(\not{p} + m)}{p^2 - m^2 + \mathrm{i}\varepsilon};$$

spin 1: 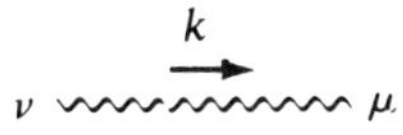

$$: \frac{\mathrm{i}}{k^2 - m^2 + \mathrm{i}\varepsilon}\left(-g_{\mu\nu} + \frac{k_\mu k_\nu}{m^2}\right);$$

gauge particle: 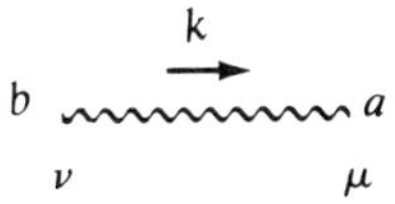

$$: \frac{\mathrm{i}\delta_{ab}}{k^2 + \mathrm{i}\varepsilon}\left(-g_{\mu\nu} + \frac{(1 - \alpha)k_\mu k_\nu}{k^2 + \mathrm{i}\varepsilon}\right);$$

Fadeev–Popov ghost: 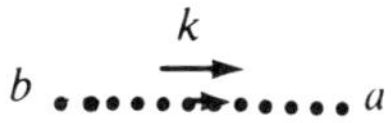

$$: \frac{\mathrm{i}\delta_{ab}}{k^2 + \mathrm{i}\varepsilon}.$$

(iii) To each vertex assign a factor determined by the interaction Lagrangean. Momentum is to be conserved at every vertex. To each ∂_μ assign a factor $-ik_\mu$, where k_μ is the momentum entering the vertex.

(iv) Integrate over all loops, with a factor $d^4q/(2\pi)^4$. Assign a factor -1 to every fermion loop.

(v) Multiply by a symmetry factor, and by an overall sign factor associated with the configuration of the external fermion lines.

The relation between Green's function and S-matrix: let T_{fi} be the connected T-matrix element for the reaction

$$p_1 + p_2 + \cdots + p_N \to p'_1 + p'_2 + \cdots + p'_M.$$

If all particles have spin zero and mass m, then

$$T_{fi} = -\mathrm{i}(z_3)^{(N+M)/2} \lim_{p_i^2 \to m^2} \lim_{p_i'^2 \to m^2} \prod_{i=1}^{N} \theta(p_{i0})$$
$$\times \prod_{j=1}^{M} \theta(p'_{j0}) \bar{G}^{(N+M)}_{\mathrm{c,R}}(-p'_1, \ldots, -p'_M; p_1, \ldots, p_M); \tag{D.4.1}$$

here $\bar{G}^{(N+M)}_{\mathrm{c,R}}$ is a renormalized connected Green's function shorn of its external propagators, and $\mathrm{i}z_3$ is the residue of the renormalized propagator at its pole at $k^2 = m^2$.

Factors assigned to external particles (spin 1/2 and spin 1).
Incoming fermion: $z_2^{1/2} u^{(r)}(\boldsymbol{p})$;
outgoing fermion: $z_2^{1/2} \bar{u}^{(r)}(\boldsymbol{p})$;
incoming antifermion: $z_2^{1/2} \bar{v}^{(r)}(\boldsymbol{p})$;
outgoing antifermion: $z_2^{1/2} v^{(r)}(\boldsymbol{p})$;
particle with spin 1: a factor $z_3^{1/2} \varepsilon_\mu^{(s)}$ $(z_3^{1/2} \varepsilon_\mu^{(s)*})$
for an incoming (outgoing) particle. If $\boldsymbol{k}$ is parallel to the z-axis, then

$$\text{helicity } +1\text{:} \quad \varepsilon_\mu^{(+)} = \left(0, \frac{-1}{\sqrt{2}}, \frac{-\mathrm{i}}{\sqrt{2}}, 0\right);$$

$$\text{helicity } 0\text{:} \quad \varepsilon_\mu^{(0)} = \left(\frac{\|\boldsymbol{k}\|}{m}, 0, 0, \frac{k_0}{m}\right);$$

$$\text{helicity } -1\text{:} \quad \varepsilon_\mu^{(-)} = \left(0, \frac{1}{\sqrt{2}}, \frac{-\mathrm{i}}{\sqrt{2}}, 0\right).$$

Rules for vertices.

(a) φ^4 theory:

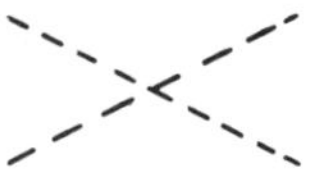

$: -\mathrm{i}g$

(b) Electrodynamics:

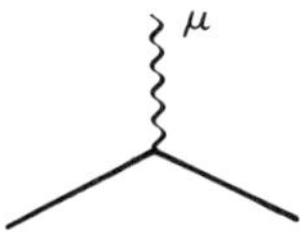

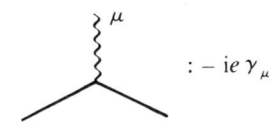

$: -\mathrm{i}e\gamma_\mu$

(c) Scalar electrodynamics:

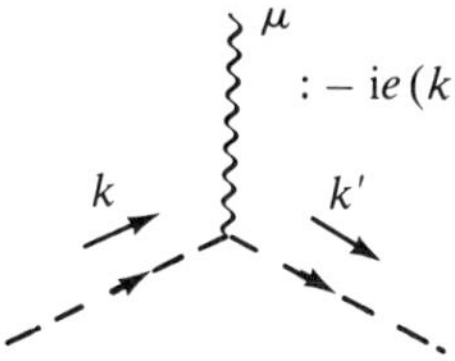

$: -\mathrm{i}e(k+k')_\mu$

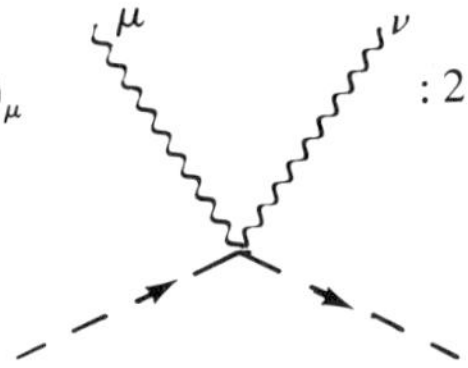

$: 2\,\mathrm{i}e^2 g_{\mu\nu}$

(d) Non-Abelian gauge theories: $[T_a, T_b] = \mathrm{i}f_{abc}T_c$

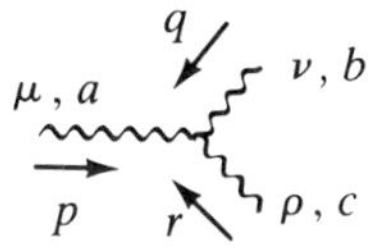

$$: -g f_{abc}[g_{\mu\nu}(p-q)_\rho + g_{\nu\rho}(q-r)_\mu + g_{\rho\mu}(r-p)_\nu]$$

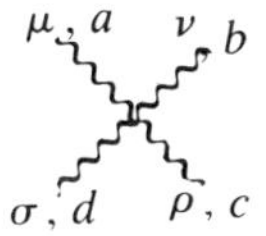

$$\begin{aligned} : -\mathrm{i}g^2[&f_{eab}f_{ecd}(g_{\mu\rho}g_{\nu\sigma} - g_{\mu\sigma}g_{\nu\rho}) \\ &+ f_{eac}f_{edb}(g_{\mu\sigma}g_{\rho\nu} - g_{\mu\nu}g_{\rho\sigma}) \\ &+ f_{ead}f_{ebc}(g_{\mu\nu}g_{\sigma\rho} - g_{\mu\rho}g_{\sigma\nu})] \end{aligned}$$

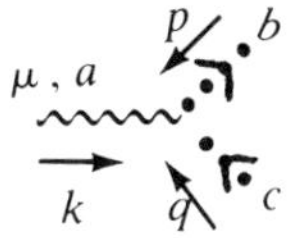

$: g f_{abc}p^\mu$

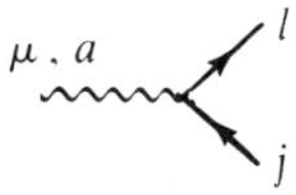

$: -\mathrm{i}g\gamma_\mu(T_a)_{lj}$

Coupling to spin-zero bosons:

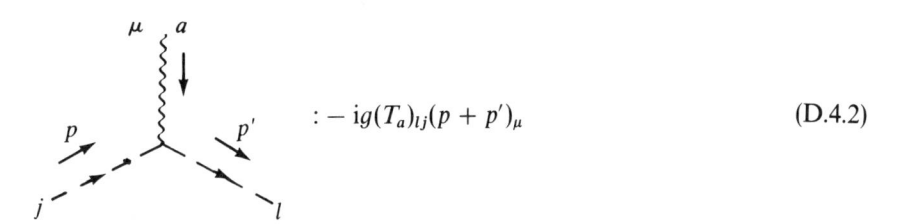

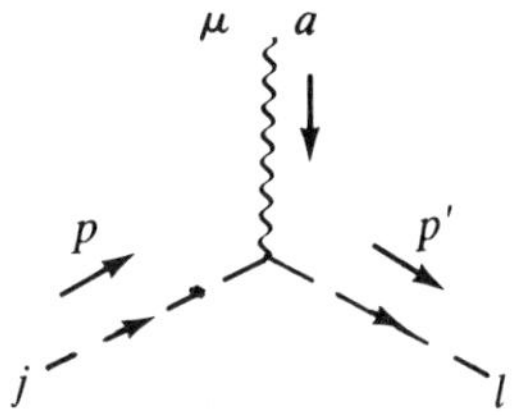

$$: -\mathrm{i}g(T_a)_{lj}(p+p')_\mu \tag{D.4.2}$$

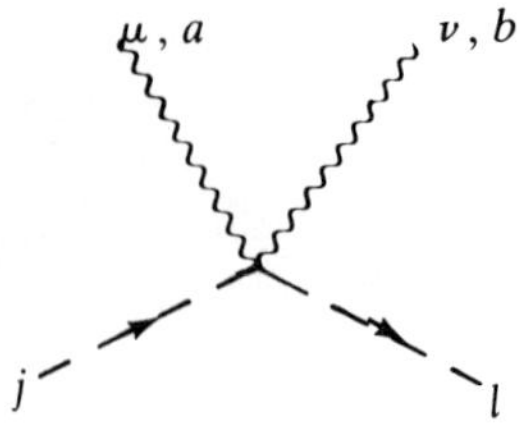

$$: ig^2 g_{\mu\nu}\{T_a, T_b\}_{lj} \tag{D.4.3}$$

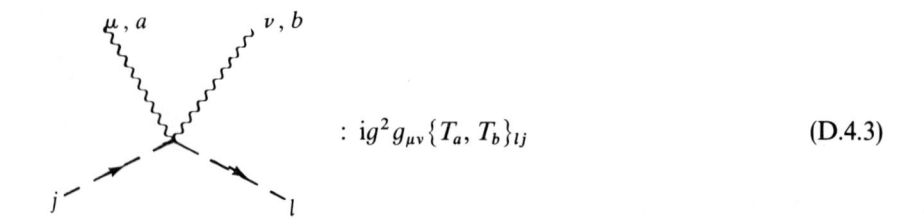

(e) Glashow–Salam–Weinberg model:

$$e = g\sin\theta; \quad \tan\theta = g'/g$$

$$B^\mu = \cos\theta A^\mu - \sin\theta Z^\mu$$

$$W^\mu = \sin\theta A^\mu + \cos\theta Z^\mu.$$

Coupling of Z^0 to fermions:

$$-\frac{e}{\sin\theta\cos\theta}\bar{\psi}(f)\left[\frac{1}{2}\gamma_\mu(1-\gamma_5)I_3 - \gamma_\mu Q\sin^2\theta\right]\psi(f) \tag{D.4.4}$$

I_3 = third component of weak isospin; Q = charge of fermion f.

References

Asterisks indicate the 'general references' mentioned in the Preface.

*Abers, E. and Lee, B. (1973). Gauge theories. *Physics Reports*, **9**, 141.

Aitchison, I. J. and Hey, A. J. (1982). *Gauge theories in particle physics.* Adam Hilger, Bristol.

Altarelli, G. (1982). Partons in QCD. *Physics Reports*, **81**, 1.

Altarelli, G. (1990). In *Proceedings of the XIVth International Symposium on Lepton and Photon Interactions, Stanford.* World Scientific, Singapore.

*Amit, D. (1984). *Field theory, the renormalization group and critical phenomena*, (2nd edn). World Scientific, Singapore.

Arnold, V. I. (1973). *Ordinary differential equations.* MIT Press, Cambridge, Massachusetts.

Arnold, V. I. (1983). *Geometrical methods in the theory of ordinary differential equations.* Springer-Verlag, Berlin.

Bell, T. and Wilson, K. (1975a). *Physical Review B*, **B10**, 3935.

Bell, T. and Wilson, K. (1975b). *Physical Review B*, **B11**, 3431.

Berestetskii, V., Lifschitz, E., and Pitaevskii, L. (1980). *Relativistic quantum theory.* Pergamon Press, Oxford.

Berezin, F. A. (1966). *The method of second quantization.* Academic Press, New York.

*Bjorken, J. D. and Drell, S. D. (1965). Vol. I, *Relativistic quantum mechanics*, Vol. II, *Relativistic quantum fields.* McGraw-Hill, New York.

*Bogoliubov, N. N. and Shirkov, D. V. (1959) *Introduction to the theory of quantized fields.* Interscience, New York.

Brézin, E. (1980). Critical behaviour from field theoretical renormalization group techniques. In *Phase transitions*, Cargèse Summer School. Plenum Press, New York.

*Brézin, E., Le Guillou, J. C., and Zinn-Justin, J. (1976). Field theoretical approach to critical phenomena. In *Phase transitions and critical phenomena*, Vol. VI. Academic Press, New York.

Bros, J. (1971). Lecture course, Ecole de Gif-sur-Yvette, pp. 147–254. IN2P3, Paris.

Brush, S. G. (1967). History of the Lenz–Ising model. *Reviews of Modern Physics*, **39**, 883.

Callaway, D. E. (1988), Triviality pursuit: can elementary scalars exist? *Physics Reports*, **167**, 241.

Caswell, W. (1974). *Physical Review Letters*, **33**, 244.

Cheng, T. P. and Li, L.-F. (1984). *Gauge theory of elementary particle physics.* Oxford University Press, Oxford.

Cohen-Tannoudji, C., Diu, B., and Laloë, F. (1977). *Quantum mechanics.* Wiley, New York.

Coniglio, A. (1989). *Physical Review Letters*, **62**, 3054.

Coleman, S. (1970). Renormalization and symmetry: a review for non-specialists. In *Proceedings of the Erice Summer School.* Editrice Compositori, Bologna.

Coleman, S. (1973). In *Properties of fundamental interactions*, Proceedings of the Erice Summer School (ed. A. Zichichi). Academic Press, New York.

Coleman, S. and Weinberg, E. (1973). *Physical Review D*, **D7**, 1888.
*Collins, J. C. (1983). *Renormalization.* Cambridge University Press, Cambridge.
Creutz, M. (1983). *Quarks, gluons and lattices.* Cambridge University Press, Cambridge.
Cvitanovic, P. (1976). *Physical Review D*, **D14**, 1536.
Dehmelt, H. (1988). *Physica Scripta*, **T22**, 102.
de Rafael, E. (1976). *Lectures on quantum electrodynamics.* University of Barcelona, Barcelona.
de Rafael, E. (1978). Cours de Gif-sur-Yvette, pp. 1–128. IN2P3, Paris.
De Witt, B. (1982). Path integrals. In *Fundamental interactions*, Cargèse lectures in theoretical physics. Plenum, New York.
Dolan, L. and Jackiw, R. (1974). *Physical Review D*, **D9**, 3320.
Duke, D. and Roberts, R. (1985) Determination of the QCD strong coupling α_s and the scale Λ_{QCD}. *Physics Reports*, **120**, 276.
Dyson, F. (1965). *Physics Today*, (June) 140.
Eden, R., Landshoff, P., Olive, D., and Polkinghorne, J. (1966). *The analytic S-matrix.* Cambridge University Press, Cambridge.
Epstein, H. and Glaser, V. (1973). *Annales de l' Institut Henri Poincaré*, **XIX**, 211.
Faddeev, L. D. (1975). Introduction to functional methods. In *Methods in field theory*, Les Houches Summer School, North Holland, Amsterdam.
Felsager, B. (1981). *Geometry, Particles and fields.* Odense University Press, Odense.
Fetter, A. and Walecka, J. D. (1971). *Quantum theory of many-particle systems.* McGraw-Hill, New York.
Feynman, R. P. (1964). *The Feynman lectures on physics*, Vol. II Addison-Wesley, New York.
Feynman, R. P. (1965). *The Feynman lectures on physics*, Vol. III. Addison-Wesley, New York.
Feynman, R. P. (1972). *Statistical mechanics.* Benjamin, New York.
Feynman, R. P. (1985). *Q.E.D.* Princeton University Press, Princeton, New Jersey.
Feynman, R. P. and Hibbs, R. (1965). *Quantum mechanics and path integrals.* McGraw-Hill, New York.
*Gasiorowicz, S. (1966). *Elementary particle physics.* Wiley, New York.
Gell-Mann, M. and Low, F. (1954). *Physical Review*, **95**, 1300.
Georgi, H. (1982). *Lie algebras in particle physics.* Benjamin, Reading, Massachusetts.
*Glimm, J. and Jaffe, A. (1987). *Quantum physics*, (2nd edn). Springer-Verlag, Berlin.
Goldstein, H. (1980). *Classical mechanics*, (2nd edn). Addison-Wesley, New York.
Gross, D. (1975). In *Methods in field theory*, Les Houches Summer School. North-Holland, Amsterdam.
Haber, H. and Kane, G. (1985). The search for supersymmetry: probing physics beyond the standard model. *Physics Reports*, **117**, 75.
Hahn, Y. and Zimmermann, W. (1968). Communications in Mathematical Physics, **10**, 330.
Hepp, K. (1965). Lecture notes, Brandeis Summer School, Gordon and Breach, New York.
't Hooft, G. (1980). Gauge theories of the forces between elementary particles. *Scientific American*, (June) 90.
*'t Hooft, G. and Veltman, M. (1974). Diagrammar. In *Particle interactions at very high energies.* Plenum Press, New York.
Iliopoulos, J. and Zumino, B. (1974). *Nuclear Physics B*, **B76**, 310.
*Itzykson, C. and Drouffe, J. M. (1989). *Statistical field theory.* Cambridge University Press, Cambridge.
*Itzykson, C. and Zuber, J. B. (1980). *Quantum field theory.* McGraw-Hill, New York.

Jackiw, R. (1985). In *Current algebra and anomalies.* World Scientific, Singapore.
Jost, R. (1965). *The general theory of quantized fields.* AMS, Providence, Rhode Island.
Kapusta, J. (1989). *Finite temperature field theory.* Cambridge University Press, Cambridge.
Kinoshita, T. (1962). *Journal of Mathematical Physics*, **3**, 650.
Kinoshita, T. (1983). *Physical Review D*, **D27**, 867.
Kinoshita, T. (1988). CERN preprint TH-5097 (unpublished).
Kittel, C. (1986). *Introduction to solid state physics*, (6th edn). Wiley, New York.
*Kogut, J. (1979). An introduction to lattice gauge theory and spin systems. *Reviews of Modern Physics*, **51**, 659.
Kogut, J. (1983). The lattice gauge theory approach to quantum chromodynamics. *Reviews of Modern Physics*, **55**, 775.
Kogut, J. and Wilson, K. (1973). The renormalization group and the ε-expansion. *Physics Reports*, **12C**, 76.
Kosterlitz, J. (1974). *Journal of Physics C*, **C7**, 1046.
Kosterlitz, J. and Thouless, D. (1973). *Journal of Physics C*, **C6**, 118.
Kupiainen, A. (1986). In *Proceedings of the VIIIth International Congress on Mathematical Physics.* World Scientific, Singapore.
*Landau, L. D. and Lifshitz, E. M. (1980). *Statistical physics, Part 1.* Pergamon Press, Oxford.
Landsman, N. and van Weert, Ch. (1987) Real- and imaginary-time field theory at finite temperature and density. *Physics Reports*, **145**, 142.
Lautrup, B. (1976). *Nuclear Physics B*, **B105**, 23.
Le Bellac, M. (1981), Cours de Gif-sur-Yvette, pp. 23–120. IN2P3, Paris.
Lee, B. W. (1975). Gauge theories. In *Methods in field theory*, Les Houches Summer School. North Holland, Amsterdam.
Lee, T. D. and Nauenberg, M. (1964). *Physical Review B*, **B133**, 1549.
Lévy-Leblond, J. M. (1967). *Communications in Mathematical Physics*, **6**, 286.
Ma, S. K. (1973). Introduction to the renormalization group. *Reviews of Modern Physics*, **45**, 589.
*Ma, S. K. (1976). *Modern theory of critical phenomena.* Benjamin, Philadelphia, Pennsylvania.
Ma, S. K. (1985). *Statistical mechanics.* World Scientific, Singapore.
Mandelbrot, B. (1982). *The fractal geometry of nature.* W. H. Freeman, New York.
Marciano, W. and Sirlin, A. (1980). *Physical Review D*, **D22**, 2695.
Mermin, N. (1979). The topological theory of defects in ordered media. *Reviews of Modern Physics*, **51**, 591.
Mermin, N. and Wagner, H. (1966). *Physical Review Letters*, **17**, 1133.
Messiah, A. (1961) *Quantum mechanics.* North-Holland, Amsterdam.
Muta, T. (1987). *Foundations of quantum chromodynamics.* World Scientific, Singapore.
Negele, J. W. and Orland, H. (1988). *Quantum many-particle systems.* Addison-Wesley, New York.
Nelson, D. R. (1983). Defect-mediated phase transitions. In *Phase transitions and critical phenomena*, Vol. VII. Academic Press, London.
Niemeijer, T. and Van Leuveen, J. (1975). In *Phase transitions and critical phenomena*, Vol. VI, Ch. VII. Academic Press, London.
*Parisi, G. (1988). *Statistical field theory.* Addison-Wesley, New York.
Pawley, G. S. *et al.* (1984). *Physical Review B*, **B29**, 4030.
*Pfeuty, P. and Toulouse, G. (1977). *Introduction to the renormalization group and to critical phenomena.* Wiley, New York.
Pokorski, S. (1987). *Gauge field theories*, Cambridge University Press, Cambridge.

Polchinski, J. (1984). *Nuclear Physics B*, **B231**, 269.
Pontryagin, L. (1966). *Topological groups*. Gordon and Breach, New York.
Popov, V. (1983). *Functional integration in quantum field theory and statistical mechanics*. Reidel, Dordrecht.
*Ramond, P. (1980). *Field theory: a modern primer*. Benjamin, Reading, Massachusetts.
Reif, F. (1965). *Fundamentals of statistical and thermal physics*. McGraw-Hill, New York.
Schierholz, G. (1985). CERN Preprint TH 4139, unpublished.
Schulman, L. (1981). *Techniques and applications of path integration*. Wiley, New York.
Schwinberg, P. *et al.* (1981). *Physical Review Letters*, **47**, 1679.
*Shenker, S. (1982). Field theories and phase transitions. In *Recent advances in field theory and statistical mechanics*, Les Houches Summer School. North Holland, Amsterdam.
Sirlin, A. (1980). *Physical Review D*, **D22**, 971.
*Streater, R. F. and Wightman, A. S. (1964). *PCT, Spin, statistics and all that*. Benjamin, New York.
Stueckelberg, E. and Petermann, A. (1953). *Helvetica Physica Acta*, **26**, 499.
Swendsen, R. (1980). In *Phase transitions*, Cargèse lectures in theoretical physics. Plenum Press, New York.
Taylor, J. C. (1976). *Gauge theories of weak interactions*. Cambridge University Press, Cambridge.
Villain, J. (1975). *Journal of Physics C*, **C36**, 581.
Weinberg, S. (1960). *Physical Review*, **118**, 838.
Wess, J. and Zumino, B. (1974). *Physics Letters B*, **131B**, 310.
Wightman, A. S. (1960). Invariance in relativistic quantum mechanics. In *Dispersion relations and elementary particles*, Les Houches Summer School, Hermann, Paris.
*Wilson, K. (1975). The renormalization group: critical phenomena and the Kondo problem. *Reviews of Modern Physics*. **47**, 774.
Wilson, K. (1979). *Scientific American*, (Aug.) 140.
Yang, C. N. and Mills, R. (1954). *Physical Review*, **96**, 191.
Yennie, D., Frautschi, S., and Suura, H. (1961). *Annals of Physics*, **13**, 379.
Yndurain, F. (1983). *Quantum chromodynamics*. Springer-Verlag, Berlin.
Young, A. (1979). *Physical Review B*, **B19**, 1855.
*Zinn-Justin, J. (1989). *Quantum field theory and critical phenomena*. Oxford University Press, Oxford.

Index

action 293, 315, 324
adjoint representation 498
Altarelli–Parisi equation 543
annihilation operator 311, 321, 403, 411
anomalous dimension 96, 248, 254
anomalous magnetic moment 436, 467
anomaly 206, 474
anticommutation relation 404
asymptotic condition 373
asymptotic freedom 250, 531, 535
axial gauge 426, 560

bare correlation function 210, 218
bare coupling constant 103, 203, 259, 545
bare field 376
bare mass 103, 203
baryon 531
Becchi–Rouet–Stora (BRS) transformation 523, 554
β-function 104, 111, 143, 241, 480, 535
Bloch–Nordsieck mechanism 543
blocks of spins 69, 87
Bogoliubov–Hepp–Parasiuk–Zimmermann (BPHZ) scheme 216
Bohm–Aharonov effect 424
boson 402
broken symmetry 13, 31, 40, 179, 515

Callan–Symanzik (CS) equations 241, 252, 257, 274
canonical commutation relations (CCR) 285, 317, 326
canonical dimension 95, 248
causality 371
charged scalar field 387
chirality 520
classical approximation 181
classical field 318, 323, 405
clustering property 21
coherent state 336, 343
Coleman–Weinberg potential 233, 278
colour 530
composite operator 203, 220
Compton effect 443, 487
confinement 531, 550
conjugate momentum 316, 319
connected correlation function 21, 171
connected diagram 171
connection 499
conserved current 422, 573
correlation function 10, 19, 50, 155, 165
correlation length 10, 24, 103, 211, 258, 545
Coulomb gauge 426
counterterm 212, 456, 461, 466
coupling constant 6, 71, 110, 151
covariant derivative 423, 502
creation operator 311, 321, 403, 411
critical exponents (or indices) 17, 25, 51
critical exponent η 24, 106, 190, 248, 272
critical exponent ν 24, 108, 253, 273
critical surface (or manifold) 74
critical temperature 30, 57
cross-section 364, 580
cubic anisotopy 122
cumulant 88, 171
curvature tensor 499
Cutkosky (or cutting) rules 386, 474
cutoff 46, 183, 206

decay rate 365, 581
derivative coupling (or interaction) 155, 191, 346, 392
differential renormalization equations 101
dilatation factor 29, 67
dimension of a Lie group 495
dimension of the order parameter 7, 108, 156
dimensional analysis 57, 95, 247
dimensional regularization 183, 206, 454, 532
dimensional transmutation 536, 552
Dirac equation 406
Dirac field 408, 410
Dirac matrices 406, 578
dispersion relation 380
Dyson's equation 332

effective action 175
effective potential 180
electromagnetic current 420
electromagnetic field 419
Elitzur's theorem 547, 561
energy-momentum tensor 574
ε-expansion 102, 273
equations of motion (Euler–Lagrange) 316, 324, 430

equations of motion for Green's functions 361
Euclidean
  action 285, 303
  continuation 361
  Lagrangean 303, 531
  postulate 361
  region 360
  theory 359, 531
evolution operator 331

Faddeev–Popov ghost 433, 508
Feynman contour 302, 329
Feynman diagram (or graph) 159
Feynman gauge 429
Feynman identity 183
Feynman rules 161, 170, 353, 444, 509, 532, 581
Fermi constant 519
Fermi theory 519
fermion 402
field strength tensor 503
fine structure constant 437
first-order phase transition 49
fixed point 75, 244, 260
  Gaussian 96
  infrared stable 246
  non-Gaussian 100
  ultraviolet stable 246
flavour 530
fluctuation-response theorem 22
flux factor 365
Fock space 323, 403
form factor 465
fragmentation function 537
free energy (Helmholtz) 9, 42, 84
free Lagrangean 346
functional differentiation (or derivative) 46, 65
functional integration (or integral) 47
fundamental representation 497

gauge
  field 401, 500
  group 500
  independence 430, 489
  invariance (global) 422
  invariance (local) 422
  transformation 419, 499, 501, 504
  transformation (global) 422, 500
  transformation (local) 422, 500
Gaussian model 56, 91, 93
Gaussian integration (or integral) 153, 567
Gell-Mann and Low's formula 351
generating function 21, 152
generating functional 154
  of connected correlation functions 173
  of correlation functions 154
  of proper vertices 175, 353
  of time-ordered products 298, 355, 418, 438, 508
Gibbs free energy (or thermodynamic potential) 26
Ginzburg–Landau Hamiltonian 43, 45, 108, 154, 202
Ginzburg's criterion 52, 56
Glashow–Salam–Weinberg model 521
gluon 402, 530
Goldstone boson 61, 515
Gordon identity 463, 580
Grassmann algebra 415
Grassmann integral 417
Grassmann variable (or number) 417
Green's function 300, 328

Hamiltonian 316, 318
Hamiltonian density 155, 212
Heisenberg model 7, 140
helicity 424, 520, 580, 582
Higgs boson 401, 525
Higgs mechanism (or phenomenon) 517, 524
high temperature expansion 8, 128, 550

in and out fields 336, 347, 370
in and out (or asymptotic) states 372
infinitesimal generator 496
infrared divergence 58, 183, 192, 238
infrared singularity 540, 559
interaction 155
interaction Lagrangean 346
interaction picture 332, 347
invariant charge 241
irrelevant (field or operator) 79, 255
Ising model 7, 11

Kinoshita–Lee–Nauenberg (KLN) theorem 544
Klein–Gordon equation 323
Klein–Gordon field 323

Lagrangean 293, 316, 323
Lagrangean density 318, 346, 407, 420, 508
Landau approximation 41, 44
Landau gauge 429
Landau theory of phase transitions 48
lattice gauge theory 544
lattice regularization 206
leading logarithm 265
Legendre transformation 25, 36, 175
Lie algebra 496
Lie group 495
light cone 330
locality 328, 371

loop 161, 163
loop expansion 55, 181
Lorentz covariance 324, 407
Lorentz gauge 421
low temperature expansion 129

magnetic susceptibility 16
marginal (field or operator) 79, 109
mass hyperboloid 324
mass insertion 222
mass (or collinear) singularity 540
massive vector field 390
mean field approximation (or theory) 13, 55
meson 531
metastability 49
minimal subtraction (MS) scheme 225, 267, 457, 533, 559
momentum 151, 327
muon 447

naïve scale invariance 82, 96, 459
Noether's theorem 421, 571
non-Abelian gauge field 495
non-exceptional configuration 193
non-linear σ-model 140
non-renormalizable theory 204
normal modes 96, 319
normal order (or form) 322
normal product 322
normalization conditions 211, 219, 240, 354

occupation number 321, 403
$O(n)$ symmetry 156
one-particle irreducible correlation function 174
order parameter 5
Osterwalder–Schrader axioms 361

parallel transport 499, 502
partition function 8, 40, 70
path integral 283, 293, 314, 355
Pauli matrices 496, 578
perturbation expansion 96, 157, 352
phonon 322
physical line 74
plaquette 70, 546
Poisson's summation formula 135
polarization 423
power counting 190, 451
probability amplitude 283
propagator 161, 164, 328, 414, 428, 438, 439, 581
proper vertex 174, 353
pure states 21

quantization
  of a classical field 313
  of the Dirac field 410
  of elastic vibrations 317
  of the electromagnetic field 427
  of the Klein–Gordon field 326
  of non-Abelian gauge fields 506
quantum chromodynamics (QCD) 402, 530
quantum electrodynamics (QED) 401, 436
quark 402, 449, 530

radiative corrections 436, 459
reduction formulae 373
relevant (field or operator) 79
regularization 205
renormalizable theories 204
renormalization 202
  constant 210
  of the field 209
  flow 74, 260, 265, 267
  of composite operators 220
  of the coupling constant 208
  group (RG) 67, 76, 256, 478
  group transformation (RGT) 67, 90
  of the mass 207
  on the mass shell 354
  of non-Abelian gauge theories 512
  of quantum electrodynamics 476
renormalized correlation functions 210, 218
renormalized coupling constant 103, 208, 259
renormalized mass 103, 207
renormalized trajectory 263
representation of a group 496

scale invariance 27, 81, 254
scaling field 79
scaling law 25, 86
Schwinger's regularization 186, 206
second-order phase transition 3, 48
self-energy 176, 353, 460
S-matrix 335, 349, 369, 377, 390, 442
source 155, 297, 323, 336, 418, 438
spin-statistics theorem 413
spin wave 132
spontaneous magnetization 4
spontaneous symmetry breaking 5
standard model 494
string tension 550
strong coupling expansion 548
structure constant 498
structure function 538
Stueckelberg Lagrangean 429
subtraction point 219, 354
superficial degree of divergence 190, 203, 451
super-renormalizable theories 204
symmetry factor 161, 164

tadpole 182, 353
time ordered (or T-) product 290, 296, 329, 413
T-matrix 363
tree approximation 181
tricritical point 80

ultraviolet divergence 183
unitarity 336, 377, 474
universality 28, 83

vacuum 296, 321, 349, 350
vacuum expectation-value 296
vacuum polarization 459, 478
vacuum to vacuum amplitude 297
vertex 158, 354, 440, 510, 582
Villain model 133
vortex 132

wave-packet 326
Ward identity 471, 555, 576
W-boson 401
weak coupling expansion 551
weak hypercharge 521
weak isospin 520
weak neutral current 520
Wegner model 545
Weinberg angle 523
Weinberg's theorem 195
Wick rotation 358
Wick's theorem 154, 337, 416
width of the critical region 81, 83
Wilson action 549
Wilson loop 547

XY-model 127

Z-boson 401, 523